AF358842

Applied Hedgehog Conservation Research

Applied Hedgehog Conservation Research

Editors

Anne Berger
Nigel Reeve
Sophie Lund Rasmussen

Basel • Beijing • Wuhan • Barcelona • Belgrade • Novi Sad • Cluj • Manchester

Editors

Anne Berger
Department Evolutionary
Ecology, Leibniz Institute for
Zoo and Wildlife Research
(IZW)
Berlin
Germany

Nigel Reeve
Independent Ecologist
London
UK

Sophie Lund Rasmussen
WildCRU, Department of
Biology
Abingdon
UK

Editorial Office
MDPI
St. Alban-Anlage 66
4052 Basel, Switzerland

This is a reprint of articles from the Special Issue published online in the open access journal *Animals* (ISSN 2076-2615) (available at: https://www.mdpi.com/journal/animals/special_issues/hedgehogs).

For citation purposes, cite each article independently as indicated on the article page online and as indicated below:

Lastname, A.A.; Lastname, B.B. Article Title. *Journal Name* **Year**, *Volume Number*, Page Range.

ISBN 978-3-7258-0935-6 (Hbk)
ISBN 978-3-7258-0936-3 (PDF)
doi.org/10.3390/books978-3-7258-0936-3

Cover image courtesy of Anne Berger

Contents

About the Editors

Anne Berger

Anne Berger is a behavioural biologist with a focus on chronobiological research on wild (mostly terrestrial) animals. As chronobiological studies require continuous and long-term measurements of behavioural or physiological data, part of her work also involves the development of small, wireless, and low-power devices that provide such data and can be used on wildlife. She is one of the pioneers in the analysis and application of 3D acceleration data in different wildlife species, with a focus on applied conservation research. Using machine learning algorithms, she automatically recognises different behaviours, and using chronobiological analysis, she detects stress and disturbances in the behaviour of wild animals who have been tagged. Since 2013, she has also been leading the research project "Applied Hedgehog Conservation Research" at the Leibniz Institute for Zoo and Wildlife Research and has since been increasingly focusing on the monitoring and protection of this unique wild animal species in Germany.

Nigel Reeve

Nigel Reeve (PhD, BSC (hons), and PGCE) studied for a BSc in Zoology at the Royal Holloway College, University of London, before researching hedgehog ecology and behaviour for his PhD. Having completed a PGCE at Garnett College (London), he taught and conducted research at Roehampton University (1982-2002), and in 1994, he published the book Hedgehogs in the Poyser Natural History series. From 2002 to 2013, he was the Head of Ecology for The Royal Parks in London. He is now retired but remains active as a Trustee of the British Hedgehog Preservation Society and is currently involved in a long-term project that began in 2014 to study and conserve hedgehogs in The Regent's Park, London.

Sophie Lund Rasmussen

Sophie Lund Rasmussen (PhD, MSc, BSc in Wildlife Ecology and Conservation, with a focus on European hedgehogs). Sophie, or Dr. Hedgehog, has engaged in multifaceted scientific research on hedgehogs since 2011, with the purpose of understanding why the European hedgehog is in decline and how the conservation initiatives directed at this species can be optimised accordingly. Sophie is leading "The Danish Hedgehog Project", a citizen science project where volunteers collected 697 dead hedgehogs to understand them. She has so far investigated the ecology of suburban juvenile hedgehogs; genetics; dental health; stress; personality; behaviour and survival of rehabilitated and wild juvenile hedgehogs; endoparasites; age distribution; eco-toxicology; MRSA; and ringworm prevalence in hedgehogs. Dr Hedgehog is currently leading research projects on the microbiome, diet choice, and coronavirus and cancer prevalence in hedgehogs, in addition to projects on the effects of robotic lawn mowers on hedgehogs. Moreover, she is engaged in several other research and conservation projects about hedgehogs.

Editorial

Applied Hedgehog Conservation Research

Nigel Reeve [1], Anne Berger [2] and Sophie Lund Rasmussen [3,4,5,*]

[1]   Independent Researcher, Woking GU21 5TR, UK
[2]   Department of Evolutionary Ecology, Leibniz Institute for Zoo and Wildlife Research,
      Alfred-Kowalke-Straße 17, 10315 Berlin, Germany
[3]   Wildlife Conservation Research Unit, Department of Biology, University of Oxford,
      The Recanati-Kaplan Centre, Tubney House, Abingdon Road, Abingdon OX13 5QL, UK
[4]   Department of Chemistry and Bioscience, Aalborg University, Fredrik Bajers Vej 7H,
      DK-9220 Aalborg, Denmark
[5]   Linacre College, St. Cross Road, Oxford OX1 3JA, UK
*   Correspondence: sophielundrasmussen@gmail.com

Citation: Reeve, N.; Berger, A.; Rasmussen, S.L. Applied Hedgehog Conservation Research. *Animals* **2024**, *14*, 976. https://doi.org/10.3390/ani14060976

Received: 1 March 2024
Accepted: 12 March 2024
Published: 21 March 2024

Hedgehogs (Order Eulipotyphla, Family Erinaceidae, Subfamily Erinaceinae) are familiar and popular spiny mammals, but they face many challenges in modern human-dominated environments. In this Special Issue, "Applied Hedgehog Conservation Research", we present an anthology of articles from the journal *Animals* which help to fill some of the many gaps in our knowledge of hedgehogs and describe new approaches to their conservation. Most articles in this collection focus on the West European hedgehog (*Erinaceus europaeus*), which remains by far the most researched species to date, but studies of other hedgehog species are also included when they are relevant and informative. The articles reflect a broad and diverse spectrum of current research that is relevant to the conservation of hedgehogs in the wild and can provide insights into their behaviour, genetics, disease, and mortality, including studies of hedgehogs in human care. We gratefully acknowledge the authors who have contributed articles, the peer reviewers, and the staff of the journal, who have facilitated the production of this Special Issue. All the articles are Open Access and free to download, ensuring that the studies are available to all and can contribute significantly to evidence-based hedgehog conservation.

There is growing evidence of a serious decline in the distribution and abundance of hedgehogs in Western European countries, such as the UK [1], the Netherlands [2], Belgium [3], and Switzerland [4]. This decline of the West European hedgehog within its native range seems to be caused by a variety of factors. Different influences may be at work in urban and rural environments, but generally, we can implicate the loss, degradation, and fragmentation of suitable habitats. The isolation of small hedgehog populations may also increase the likelihood of local extinctions from stochastic events and the effects of low genetic diversity. Other negative factors include the lethal or sub-lethal effects of environmental pollutants, hazards such as road traffic accidents, a decline in the abundance of invertebrate prey, high levels of disturbance, interspecific competition, and predation, e.g., by Eurasian badgers (*Meles meles*) [2,5,6]. Despite the many negative effects of anthropogenic hazards and environmental change, hedgehogs are often found living alongside humans, especially in low-density urban and suburban areas with gardens and plenty of greenspace, where they may benefit from supplementary food sources, a reduced risk of predation, additional shelter, and a warmer microclimate [6–8]. Hedgehogs are typically absent from highly urbanised environments such as town centres and cities with small, highly fragmented areas of greenspace [7,9], but the benefits of an urban lifestyle can be seen in northern Scandinavian countries, where hedgehogs are principally found in association with human settlements [10], provided that an appropriate number of small forest patches in urban areas ensure suitable hibernation habitats for hedgehogs [11]. Similarly, in Qatar, a population of Ethiopian hedgehogs (*Paraechinus aethiopicus*) benefit from visiting irrigated farms and a rubbish mound [12]. However, in a contrasting example, the formerly thriving urban hedgehog population in Zurich (Switzerland) has declined in the last 25 years in both distribution (down 17.6%) and abundance (down 40.6%), and further investigations are required to establish the reasons for this [4].

In this Anthropocene era, the existence of almost every wildlife species depends on how well they can cope with the conditions that are created by humans. Hedgehogs are generally popular animals, but just popularity is not enough when key groups in society are disconnected from and ignorant of the natural world [13]. Citizen Science and public engagement initiatives need to be sensitive to the social context to engage and motivate people to implement conservation actions, in order to make the changes that will allow wildlife to co-exist with us.

Urgent action is required to reverse the decline of the West European hedgehog, but there are still major gaps in our understanding of its population ecology, the underlying causes of decline, and the effectiveness of different conservation measures. Population modelling is an important strategic tool, but it requires data such as survival and reproductive rates, which are especially challenging to obtain from a secretive, nocturnal animal. If cause of death (or other additional) data can also be obtained, the models gain the power to examine further issues, such as the potential benefits of mitigations or the impact of certain causes of death (associated with certain environmental factors) on population trends [14,15]. Post-mortem age determinations of hedgehogs in Denmark [16] examined sex- and age-specific mortality rates in road-killed and other dead hedgehogs from urban and rural areas. Additional data on individual genetic heterozygosity revealed no significant association with either the age at death or the cause of death. In this case, it seems that low genetic diversity is not a principal risk factor. Moreover, hibernation is a key feature of West European hedgehogs' life history, and for many years, it has been assumed that hibernation was a high-risk period. However, recent studies [17] show that in fact, hibernation is a period of relatively low mortality. Death from collisions with road traffic is one of the most obvious mortality factors affecting hedgehogs, and the scale of the losses suggests population-level effects [1,9]; in addition, males are more likely to be traffic victims [14,16]. The population effects of road mortality and, where implemented, the benefits of road mitigation schemes such as crossing structures and fauna tunnels need to be assessed by detailed demographic research [14].

As hedgehogs are elusive animals, the use of environmental DNA (eDNA) to detect their presence has considerable potential. eDNA has now been used to detect the presence of many species, including a range of terrestrial mammal species in samples of river water [18], although no hedgehogs were detected in that study. Another approach that has been successfully trialled is to use DNA analysis to look for a hedgehog-specific parasitic nematode *Crenosoma striatum* (a lungworm) in slugs—the nematode's intermediate host [19]. Slugs are easy to collect, and this method has the potential to be used simply to confirm the presence of hedgehogs on a site, especially in habitats where fieldwork is difficult or where they are at low densities and hard to find.

Where populations of different hedgehog species meet, much can be learned about their ecological niche characteristics, the potential for interspecific competition, and the implications for conservation management. A detailed study of cranial and jaw morphology has revealed differences between allopatric and sympatric populations of the West European hedgehog and the Northern white-breasted hedgehog (*Erinaceus roumanicus*), with a convergence in jaw shape in the sympatric animals that may be a result of feeding niche overlap and competition [20]. In the contact zone between these two European hedgehogs, wildlife rescue centres may receive both species. If releasing individuals away from their point of origin, there is a potential to create artificial mixing between their populations, irrespective of natural landscape barriers [21]. Translocations by wildlife rescue centres may also explain the unexpected gene flow that has been observed between urban subpopulations of *E. europaeus* in Berlin, although natural movements through habitat patches in the city may also be possible [22].

Animal rescue and rehabilitation centres aim to contribute to hedgehog welfare and conservation by restoring ailing hedgehogs to health and returning them to the wild. Bearman-Brown and Baker estimated that in 2016, at least 40,000 hedgehogs were admitted to rescue centres in Britain and the Channel Islands, of which maybe 50% were released [23]. Such figures suggest potential population-level effects. Studies that test the efficacy of rehabilitation methods [24] and provide reliable reference data for physiological variables [25] play an important part in the development of good practice for hedgehog rehabilitation. The participation of rescue and rehabilitation centres in research can be extremely valuable and can reveal much about the problems facing hedgehogs in the wild [26,27], as can studies carried out in captivity examining behaviour and hibernation [28]. Rescue centre records can provide year-on-year data on the numbers and causes of admissions and can also contribute samples to studies of, for example, infectious and non-infectious diseases; the prevalence of environmental pollution by pesticides, heavy metals, organic pollutants, and other ecotoxins [29]; plastic waste [15]; and the occurrence and characteristics of cut injuries that are presumably caused by garden tools like robotic lawn mowers [30]. Studies by Rasmussen et al. of robotic lawn mowers have shown important differences between models in the risk of injury to hedgehogs, creating the basis for specific design improvements to mitigate such risks [31,32]. Moreover, investigations of the personality and reactions of live hedgehogs towards a disarmed, approaching robotic lawn mower are also applied in the design of a standardised hedgehog safety test, eventually serving to produce and approve hedgehog-friendly robotic lawn mowers [33].

A range of anthropogenic influences could potentially affect hedgehogs' behaviour. One of these is Artificial Light at Night (ALAN), which is known to disrupt the behaviour of a wide range of species. A study of hedgehogs using camera-trap videos at supplementary feeding stations in gardens found no consistent overall effect of ALAN on the feeding and general activity of hedgehogs [34]. However, a study in Berlin used GPS tracking and dataloggers and demonstrated a preference for movement in locations with lower light levels [35]. By attaching dataloggers to urban hedgehogs under different conditions, it was documented that the temporary disturbance of their habitat that was caused by a music festival had a more serious impact on hedgehog behaviour than a permanent disturbance caused by fragmentation [36]. Nevertheless, we stress the importance of further investigations on the effects of anthropogenic disturbances to the habitats before clear conclusions can be drawn.

The popularity of hedgehogs means that many people put out food for their much-loved garden visitors. Supplementary feeding could be an important factor for maintaining urban hedgehog populations, but feeding during the winter appears, at least in some cases, to be increasing the levels and duration of activity when hedgehogs would normally be hibernating [37]. Further research is needed to evaluate the risks and benefits of supplementary feeding, facilitating evidence-based advice to the public to benefit the hedgehogs. Furthermore, contrary to the heterogeneously dispersed invertebrate prey which hedgehogs naturally forage for, supplementary feeding is usually provided in food bowls, which may artificially attract a substantial number of hedgehogs to the feeding stations. This might trigger aggression between competing hedgehogs and increase the potential risks of injury and disease transmission from close, as well as aggressive, intraspecific encounters, or contact with other competing or predatory species that are attracted to the food [38]. However, ingestion of natural prey items may also cause risks to the hedgehogs, as Williams et al. demonstrated that molluscs, commonly eaten by hedgehogs, are potential vectors of rodenticide poisons [39], providing a potential explanation for the high prevalence of rodenticides that have been detected in hedgehogs in previous studies [29,40].

Clearly, there is a need for further, more detailed studies to answer the many remaining questions about hedgehogs' population ecology, habitat requirements, and behaviour, as well as the impacts of potential key causal factors in their decline. There is enough current knowledge to prescribe land management changes which are likely to benefit hedgehog populations in rural areas [41], but there is a serious lack of evidence underpinning this advice. Management interventions, or indeed any interventions aiming to benefit hedgehogs, need to be part of well-designed before/after studies to provide evidence of their effectiveness.

We hope that the studies published in this Special Issue, "Applied Hedgehog Conservation Research", will inspire forthcoming research and will contribute to an evidence-based optimisation of the conservation initiatives protecting this beloved species, thus preventing further population declines.

Conflicts of Interest: The authors declare no conflicts of interest.

References

1. Wembridge, D.; Johnson, G.; Al-Fulaij, N.; Langton, S. *The State of Britain's Hedgehogs 2022*; British Hedgehog Preservation Society and People's Trust for Endangered Species: London, UK, 2022; Available online: http://www.britishhedgehogs.org.uk/wp-content/uploads/2022/02/SoBH-2022-Final.pdf (accessed on 13 March 2024).
2. van de Poel, J.L.; Dekker, J.; van Langevelde, F. Dutch hedgehogs *Erinaceus europaeus* are nowadays mainly found in urban areas, possibly due to the negative effects of badgers *Meles meles*. *Wildl. Biol.* **2015**, *21*, 51–55. [CrossRef]
3. Holsbeek, L.; Rodts, J.; Muyldermans, S. Hedgehog and other animal traffic victims in Belgium: Results of a countrywide survey. *Lutra* **1999**, *42*, 111–119.
4. Taucher, A.L.; Gloor, S.; Dietrich, A.; Geiger, M.; Hegglin, D.; Bontadina, F. Decline in Distribution and Abundance: Urban Hedgehogs under Pressure. *Animals* **2020**, *10*, 1606. [CrossRef] [PubMed]
5. Hof, A.R.; Allen, A.M.; Bright, P.W. Investigating the Role of the Eurasian Badger (*Meles meles*) in the Nationwide Distribution of the Western European Hedgehog (*Erinaceus europaeus*) in England. *Animals* **2019**, *9*, 759. [CrossRef] [PubMed]
6. Williams, B.M.; Baker, P.J.; Thomas, E.; Wilson, G.J.; Judge, J.; Yarnell, R.W. Reduced occupancy of hedgehogs (*Erinaceus europaeus*) in rural England and Wales: The influence of habitat and an asymmetric intra-guild predator. *Sci. Rep.* **2018**, *8*, 12156. [CrossRef] [PubMed]
7. Hubert, P.; Julliard, R.; Biagianti, S.; Poulle, M.-L. Ecological factors driving the higher hedgehog (*Erinaceus europaeus*) density in an urban area compared to the adjacent rural area. *Landsc Urban Plan* **2011**, *103*, 34–43. [CrossRef]
8. Pettett, C.E.; Johnson, P.J.; Moorhouse, T.P.; Hambly, C.; Speakman, J.R.; Macdonald, D.W. Daily energy expenditure in the face of predation: Hedgehog energetics in rural landscapes. *J. Exp. Biol.* **2017**, *220*, 460–468. [CrossRef]
9. Huijser, M.P.; Bergers, P.J.M. The effect of roads and traffic on hedgehog (*Erinaceus europaeus*) populations. *Biol. Conserv.* **2000**, *95*, 111–116. [CrossRef]
10. Rautio, A.; Valtonen, A.; Auttila, M.; Kunnasranta, M. Nesting patterns of European hedgehogs (*Erinaceus europaeus*) under northern conditions. *Acta Theriol.* **2014**, *59*, 173–181. [CrossRef]
11. Korslund, L.M.; Floden, M.S.; Albertsen, M.M.S.; Landsverk, A.; Løkken, K.M.V.; Johansen, B.S. Home Range, Movement, and Nest Use of Hedgehogs (*Erinaceus europaeus*) in an Urban Environment Prior to Hibernation. *Animals* **2024**, *14*, 130. [CrossRef]
12. Pettett, C.; Macdonald, D.W.; Al-Hajiri, A.; Al-Jabiry, H.; Yamaguchi, N. Characteristics and Demography of a Free-Ranging Ethiopian Hedgehog, *Paraechinus aethiopicus*, Population in Qatar. *Animals* **2020**, *10*, 951. [CrossRef] [PubMed]
13. Ribeiro, Â.M.; Rodrigues, M.; Brito, N.V.; Mateus, T.L. Prickly Connections: Sociodemographic Factors Shaping Attitudes, Perception and Biological Knowledge about the European Hedgehog. *Animals* **2023**, *13*, 3610. [CrossRef] [PubMed]
14. Moore, L.J.; Petrovan, S.O.; Baker, P.J.; Bates, A.J.; Hicks, H.L.; Perkins, S.E.; Yarnell, R.W. Impacts and Potential Mitigation of Road Mortality for Hedgehogs in Europe. *Animals* **2020**, *10*, 1523. [CrossRef] [PubMed]
15. Thrift, E.; Nouvellet, P.; Mathews, F. Plastic Entanglement Poses a Potential Hazard to European Hedgehogs *Erinaceus europaeus* in Great Britain. *Animals* **2023**, *13*, 2448. [CrossRef] [PubMed]
16. Rasmussen, S.L.; Berg, T.B.; Martens, H.J.; Jones, O.R. Anyone Can Get Old—All You Have to Do Is Live Long Enough: Understanding Mortality and Life Expectancy in European Hedgehogs (*Erinaceus europaeus*). *Animals* **2023**, *13*, 626. [CrossRef] [PubMed]
17. Bearman-Brown, L.E.; Baker, P.J.; Scott, D.; Uzal, A.; Evans, L.; Yarnell, R.W. Over-Winter Survival and Nest Site Selection of the West-European Hedgehog (*Erinaceus europaeus*) in Arable Dominated Landscapes. *Animals* **2020**, *10*, 1449. [CrossRef]
18. Broadhurst, H.A.; Gregory, L.M.; Bleakley, E.K.; Perkins, J.C.; Lavin, J.V.; Bolton, P.; Browett, S.S.; Howe, C.V.; Singleton, N.; Tansley, D.; et al. Mapping differences in mammalian distributions and diversity using environmental DNA from rivers. *Sci Total Environ.* **2021**, *801*, 149724. [CrossRef]
19. Allen, S.; Greig, C.; Rowson, B.; Gasser, R.B.; Jabbar, A.; Morelli, S.; Morgan, E.R.; Wood, M.; Forman, D. DNA Footprints: Using Parasites to Detect Elusive Animals, Proof of Principle in Hedgehogs. *Animals* **2020**, *10*, 1420. [CrossRef]
20. Bolfíková, B.C.; Evin, A.; Knitlová, M.R.; Loudová, M.; Sztencel-Jabłonka, A.; Bogdanowicz, W.; Hulva, P. 3D Geometric Morphometrics Reveals Convergent Character Displacement in the Central European Contact Zone between Two Species of Hedgehogs (Genus Erinaceus). *Animals* **2020**, *10*, 1803. [CrossRef]

1. Ploi, K.; Curto, M.; Bolfíková, B.C.; Loudová, M.; Hulva, P.; Seiter, A.; Fuhrmann, M.; Winter, S.; Meimberg, H. Evaluating the Impact of Wildlife Shelter Management on the Genetic Diversity of *Erinaceus europaeus* and *E. roumanicus* in Their Contact Zone. *Animals* **2020**, *10*, 1452. [CrossRef]
2. Barthel, L.M.F.; Wehner, D.; Schmidt, A.; Berger, A.; Hofer, H.; Fickel, J. Unexpected Gene-Flow in Urban Environments: The Example of the European Hedgehog. *Animals* **2020**, *10*, 2315. [CrossRef] [PubMed]
3. Bearman-Brown, L.E.; Baker, P.J. An Estimate of the Scale and Composition of the Hedgehog (*Erinaceus europeaus*) Rehabilitation Community in Britain and the Channel Islands. *Animals* **2022**, *12*, 3139. [CrossRef] [PubMed]
4. Kadlecova, G.; Rasmussen, S.L.; Voslarova, E.; Vecerek, V. Differences in Mortality of Pre-Weaned and Post-Weaned Juvenile European Hedgehogs (*Erinaceus europaeus*) at Wildlife Rehabilitation Centres in the Czech Republic. *Animals* **2023**, *13*, 337. [CrossRef] [PubMed]
5. Rosa, S.; Silvestre-Ferreira, A.C.; Sargo, R.; Silva, F.; Queiroga, F.L. Hematology, Biochemistry, and Protein Electrophoresis Reference Intervals of Western European Hedgehog (*Erinaceus europaeus*) from a Rehabilitation Center in Northern Portugal. *Animals* **2023**, *13*, 1009. [CrossRef] [PubMed]
6. Garcês, A.; Soeiro, V.; Lóio, S.; Sargo, R.; Sousa, L.; Silva, F.; Pires, I. Outcomes, Mortality Causes, and Pathological Findings in European Hedgehogs (*Erinaceus europaeus*, Linnaeus 1758): A Seventeen Year Retrospective Analysis in the North of Portugal. *Animals* **2020**, *10*, 1305. [CrossRef] [PubMed]
7. Reeve, N.J.; Huijser, M.P. Mortality factors affecting wild hedgehogs: A study of records from wildlife rescue centres. *Lutra* **1999**, *42*, 7–24.
8. South, K.E.; Haynes, K.; Jackson, A.C. Hibernation Patterns of the European Hedgehog, *Erinaceus europaeus*, at a Cornish Rescue Centre. *Animals* **2020**, *10*, 1418. [CrossRef]
9. Rasmussen, S.L.; Pertoldi, C.; Roslev, P.; Vorkamp, K.; Nielsen, J.L. A Review of the Occurrence of Metals and Xenobiotics in European Hedgehogs (*Erinaceus europaeus*). *Animals* **2024**, *14*, 232. [CrossRef]
10. Berger, A. Occurrence and Characteristics of Cut Injuries in Hedgehogs in Germany: A Collection of Individual Cases. *Animals* **2024**, *14*, 57. [CrossRef]
11. Rasmussen, S.L.; Schrøder, A.E.; Mathiesen, R.; Nielsen, J.L.; Pertoldi, C.; Macdonald, D.W. Wildlife Conservation at a Garden Level: The Effect of Robotic Lawn Mowers on European Hedgehogs (*Erinaceus europaeus*). *Animals* **2021**, *11*, 1191. [CrossRef]
12. Rasmussen, S.L.; Schrøder, B.T.; Berger, A.; Sollmann, R.; Macdonald, D.W.; Pertoldi, C.; Alstrup, A.K.O. Testing the Impact of Robotic Lawn Mowers on European Hedgehogs (*Erinaceus europaeus*) and Designing a Safety Test. *Animals* **2024**, *14*, 122. [CrossRef]
13. Rasmussen, S.L.; Schrøder, B.T.; Berger, A.; Macdonald, D.W.; Pertoldi, C.; Briefer, E.F.; Alstrup, A.K.O. Facing Danger: Exploring Personality and Reactions of European Hedgehogs (*Erinaceus europaeus*) towards Robotic Lawn Mowers. *Animals* **2024**, *14*, 2. [CrossRef] [PubMed]
14. Finch, D.; Smith, B.R.; Marshall, C.; Coomber, F.G.; Kubasiewicz, L.M.; Anderson, M.; Wright, P.G.R.; Mathews, F. Effects of Artificial Light at Night (ALAN) on European Hedgehog Activity at Supplementary Feeding Stations. *Animals* **2020**, *10*, 768. [CrossRef]
15. Berger, A.; Lozano, B.; Barthel, L.M.F.; Schubert, N. Moving in the Dark—Evidence for an Influence of Artificial Light at Night on the Movement Behaviour of European Hedgehogs (*Erinaceus europaeus*). *Animals* **2020**, *10*, 1306. [CrossRef] [PubMed]
16. Berger, A.; Barthel, L.M.F.; Rast, W.; Hofer, H.; Gras, P. Urban Hedgehog Behavioural Responses to Temporary Habitat Disturbance versus Permanent Fragmentation. *Animals* **2020**, *10*, 2109. [CrossRef] [PubMed]
17. Gazzard, A.; Baker, P.J. Patterns of Feeding by Householders Affect Activity of Hedgehogs (*Erinaceus europaeus*) during the Hibernation Period. *Animals* **2020**, *10*, 1344. [CrossRef] [PubMed]
18. Scott, D.M.; Fowler, R.; Sanglas, A.; Tolhurst, B.A. Garden Scraps: Agonistic Interactions between Hedgehogs and Sympatric Mammals in Urban Gardens. *Animals* **2023**, *13*, 590. [CrossRef]
19. Williams, E.J.; Cotter, S.C.; Soulsbury, C.D. Consumption of Rodenticide Baits by Invertebrates as a Potential Route into the Diet of Insectivores. *Animals* **2023**, *13*, 3873. [CrossRef]
20. Dowding, C.V.; Harris, S.; Poulton, S.; Baker, P.J. Nocturnal ranging behaviour of urban hedgehogs, *Erinaceus europaeus*, in relation to risk and reward. *Anim. Behav.* **2010**, *80*, 13–21. [CrossRef]
21. Yarnell, R.W.; Pettett, C.E. Beneficial Land Management for Hedgehogs (*Erinaceus europaeus*) in the United Kingdom. *Animals* **2020**, *10*, 1566. [CrossRef]

Article

Decline in Distribution and Abundance: Urban Hedgehogs under Pressure

Anouk L. Taucher [1,*], **Sandra Gloor** [1], **Adrian Dietrich** [1], **Madeleine Geiger** [1,2], **Daniel Hegglin** [1,3,†] **and Fabio Bontadina** [1,4,†]

[1] SWILD—Urban Ecology & Wildlife Research, Wuhrstrasse 12, 8003 Zurich, Switzerland;
     sandra.gloor@swild.ch (S.G.); adrian.dietrich@swild.ch (A.D.); madeleine.geiger@swild.ch (M.G.);
     daniel.hegglin@swild.ch (D.H.); fabio.bontadina@swild.ch (F.B.)
[2] Palaeontological Institute and Museum, University of Zurich, Karl-Schmid-Strasse 4,
     8006 Zurich, Switzerland
[3] Institute of Parasitology, University of Zurich, Winterthurerstrasse 266a, 8057 Zurich, Switzerland
[4] Swiss Federal Research Institute WSL, Biodiversity and Conservation Biology, Zuercherstrasse 111,
     8903 Birmensdorf, Switzerland
*   Correspondence: anouk.taucher@swild.ch
†   Shared senior authorship.

Received: 13 August 2020; Accepted: 2 September 2020; Published: 9 September 2020

Simple Summary: Hedgehogs have been found in higher densities in urban compared to rural areas. Recent dramatic declines in rural hedgehog numbers lead us to pose the question: how are hedgehogs faring in urban areas? In this study, we examined how hedgehog numbers have changed in the city of Zurich, Switzerland, in the last 25 years. We compared data collected through citizen science projects conducted in 1992 and 2016–2018, including: observations of hedgehogs, data from footprint tunnels, and capture-mark recapture studies. We found that hedgehog numbers have declined by 41%, from the former average of more than 30 individuals per km^2, in the last 25 years. In the same time span, hedgehogs have lost 18% of their former urban distribution. The reasons for this decline are still unknown. Intensification of urban buildup, reduction of green space quality, the use of pesticides, parasites, or diseases, as well as increasing numbers of badgers, which are hedgehog predators, in urban areas are discussed as potential causes. Worryingly, these results suggest that hedgehogs are now under increasing pressure not only in rural but also in urban areas, their former refuges.

Abstract: Increasing urbanization and densification are two of the largest global threats to biodiversity. However, certain species thrive in urban spaces. Hedgehogs *Erinaceus europaeus* have been found in higher densities in green areas of settlements as compared to rural spaces. With recent studies pointing to dramatically declining hedgehog numbers in rural areas, we pose the question: how do hedgehogs fare in urban spaces, and do these spaces act as refuges? In this study, recent (2016–2018) and past (1992) hedgehog abundance and distribution were compared across the city of Zurich, Switzerland using citizen science methods, including: footprint tunnels, capture-mark recapture, and incidental sightings. Our analyses revealed consistent negative trends: Overall hedgehog distribution decreased by 17.6% ± 4.7%, whereas abundance declined by 40.6% (mean abundance 32 vs. 19 hedgehogs/km^2, in past and recent time, respectively), with one study plot even showing a 91% decline in this period (78 vs. 7 hedgehogs/km^2, respectively). We discuss possible causes of this rapid decline: increased urban densification, reduction of insect biomass, and pesticide use, as well as the role of increasing populations of badgers (a hedgehog predator) and parasites or diseases. Our results suggest that hedgehogs are now under increasing pressure not only in rural but also in urban areas, their former refuges.

Keywords: *Erinaceus europaeus*; population trend; population reduction; settlement area; citizen science; urban densification

---

1. Introduction

Currently, over half of the world's human population lives in cities and, by 2050, it is estimated that over 66% of people will do so [1]. With this current rise in population, the area covered by urban settlements is expected to triple by 2030 [2]. This massive increase in urbanized land cover has inevitably become one of the greatest concerns of modern conservation [3,4]. In addition, expanding urban spaces are also densifying, with the net result of smaller and more intensively used green spaces [5,6]. Understanding urban ecology is key to conservation efforts in these human-built and dominated landscapes.

Rather than being simply degraded landscapes, urban areas provide habitat for a wide array of wildlife. The urban environment's particular habitat characteristics render it a unique ecosystem. Although cities tend to be characterized by fewer natural resources, greater anthropogenic disturbance and higher levels of fragmentation than pristine systems, urban areas feature a greater diversity of habitats, more (often anthropogenic) resources, and fewer natural predators [7–10]. Anthropogenic disturbance creates an urban landscape that is highly variable in temperature, pollution level, habitat availability, and species composition across small spatial scales compared to the non-urban surroundings. These extreme pressures may result in profound behavioral adaptations in the animals inhabiting urban areas [11,12] and, in some cases, may promote rapid evolution [13–16]. Certain species benefit from the habitat mosaics that are urban spaces [17].

One such species is the European hedgehog *Erinaceus europaeus*, which is commonly associated with the agricultural landscape. This species, however, is also known to inhabit green areas of settlements. It can even reach higher densities in urban spaces than in rural ones [18–23]. Factors that provide more favorable conditions for hedgehogs in urban areas include: better habitat quality [19,24], higher anthropogenic food availability [25,26], higher availability of vegetation structures to build day nests [27], and more beneficial climatic conditions [28], coupled with lower risk of predation by badgers [20,29,30]. In multiple European countries, the distribution of hedgehogs has been declining over the past few decades [31–37]. While hedgehog populations seem to decline over a large range and in several countries, rural hedgehogs are affected particularly strongly by the decline [38]. In the UK, the hedgehog was recently classified as Vulnerable on the Red List [39]. Therefore, the question arises whether urban settlements act as shelters for hedgehog populations.

Although hedgehogs are regularly observed, systematic studies are not easy due to their nocturnal and secretive lifestyles. In addition, access to privately and semi-privately owned land is often limited for researchers. Citizen science is a particularly suitable approach to study urban wildlife because it enables access to privately and semi-privately owned lands, and it allows data collection in a scope otherwise not possible. In addition, the involvement of citizen scientists brings conservation aspects and local wildlife concerns closer to the people who live in cities and offers researchers access to local knowledge. Our recent survey was performed as part of the citizen science project StadtWildTiere ("urban wildlife") in Zurich, Switzerland, which was established in 2013. This project collects incidental observations of wild animals in various urban areas on an online platform with the aim to increase knowledge on occurrence and distribution, to raise awareness of wild animals in cities, and to promote conservation and mutually beneficial co-existence. Further, the project includes a well-established network of 60 volunteers, who have been trained in various wildlife research methods.

The aim of this study was to test how hedgehogs are doing in urban environments by evaluating the temporal changes in hedgehog distribution and abundance in the city of Zurich in the last 25 years in a case study. We were able to build upon a city-wide call for hedgehog sightings in 1992, combined with a capture-mark-recapture study, which resulted in the creation of a distribution map, as well as an estimate of abundance numbers for a period in the past. In the recent comparative study, we used footprint tunnels [31,40,41], collected hedgehog observations, and employed a capture-mark-recapture approach in four urban districts with the help of citizen scientists to estimate current distribution and abundance measures for urban hedgehogs. The resulting indicators of hedgehog density and distribution were compared over the 25-year period. Such long-term comparative datasets on population development are scarce, and the methodologies inevitably slightly differ between the study periods; thus (with certain caveats in mind), we performed a conservative comparison with the available data. In the wake of the modern biodiversity crisis, it is invaluable to have older data to compare with current estimates.

2. Materials and Methods

2.1. Study Species and Site

We studied temporal changes in the distribution and abundance of urban hedgehogs *Erinaceus europaeus*, in the city of Zurich (433,000 inhabitants, 92 km^2, 47.369, 8.539), Switzerland. We compared data from two studies, the first of which was carried out in 1992 [42] the second from 2016–2018. Both studies focused on the urbanized area of Zurich, excluding forests and agricultural lands. To allow complete comparability, we focused on a subset of the urbanized area of Zurich, the study area that was assessed in both time periods (46 km^2, hereafter referred to as the study area), for the abundance estimations (Figure 1).

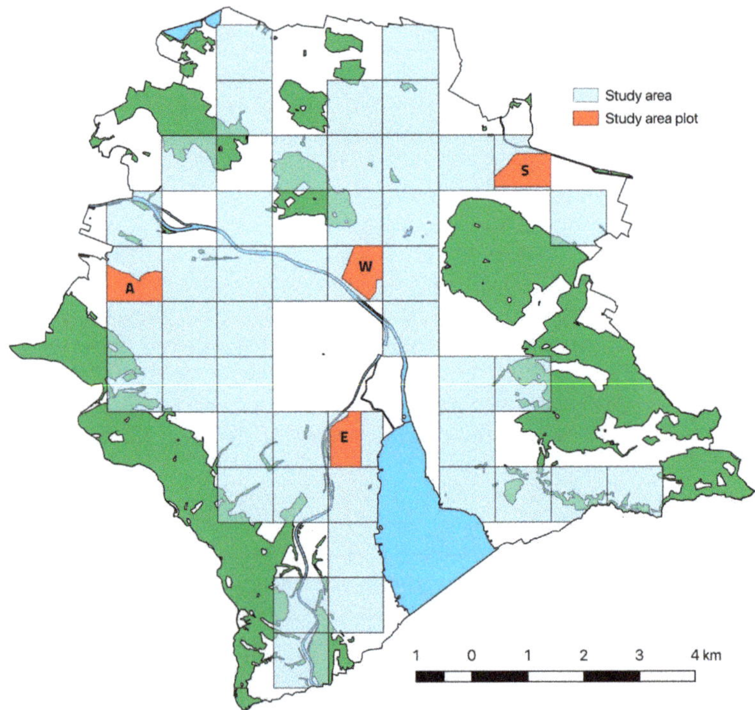

Figure 1. City of Zurich and study areas that were assessed in the course of 25 years (past (1992) vs. recent (2016/2017)). The outline delineates the municipal border of the city of Zurich with forest area (green) and water bodies (blue). Study areas are divided into 1 km^2 patches (light blue) and study plots (red: Altstetten (A), Enge (E), Schwamendingen (S), Wipkingen (W); in 1992, only Wipkingen was investigated; for details, see text).

The data for the abundance estimations were collected in the study area plots, which are sections of the study area (one in 1992 (size 0.23 km^2) and four in 2016–2018 (size 0.5 km^2)) and then extrapolated in accordance to the relative return rates of observations or footprint tunnel data to the study area (Figure 1). For the distribution maps, the entire urban space of the city of Zurich was considered (grid of 58 km^2 representing the urbanized area). To control for observation effort, the observation points from 1992 and 2016–2018 were classed as presence and absence points at a level of $\frac{1}{4}$ km^2. This level was chosen to allow for more in-depth comparisons than 1 km^2 would have done.

2.2. Past Study (1992)

In spring 1992, survey cards were sent by mail to approximately 18,000 households (mostly members of conservation and animal protection associations) all over the city of Zurich to ask for observations of hedgehogs. The obtained records of observations were then used to create a distribution map by classing them as presence and absence data. A $\frac{1}{4}$ km^2 grid size was chosen. As the cards were not evenly distributed over the city, the proportion of sent out and returned cards per $\frac{1}{4}$ km^2 grid cell was used to obtain a relative return rate. Cards were only returned when a hedgehog was seen. It was assumed that all people had the same intention in returning observations after seeing a hedgehog, as they all were members of nature conservation associations.

Intensive hedgehog searches were conducted in one specific plot within the above described study area (hereafter referred to as a study area plot, Wipkingen, 0.23 km^2) for abundance estimation (Figure 1). The whole area was scanned for hedgehogs once in the course of the night (n = 23), mainly by a single person walking in the dark but occasionally by searching larger areas with a torch. All hedgehogs encountered in this area during a radio telemetry study from beginning of June through end of July 1992 were uniquely marked by attaching six heat-shrink plastic tubes over individual spines with instant glue [19]. This marking lasted at least three months and made possible the unambiguous identification and count of all encountered hedgehog individuals. For estimating abundance, all encountered hedgehogs were summed up. The study area plot was searched intensely, so that it was assumed that the majority of hedgehogs had been encountered. The study area plot was surrounded by either train tracks or heavily trafficked on three sides, and only one hedgehog was observed crossing one of these streets. However, as the study area plot was not truly isolated, it was assumed that some hedgehogs were counted that did not really inhabit that area, while some living in the plot must have been missed. For example, some of the hedgehogs may not have regularly inhabited the study area but only appeared on an occasional exploratory trip or were young, dispersing individuals. To account for this uncertainty, a third of the total amount was used as error range of abundance.

2.3. Recent Study (2016–2018)

2.3.1. Surveys of Hedgehog Distribution and Abundance Today

To assess the distribution of hedgehogs, we divided the urbanized area of the city of Zurich into km-squares, 46 of which were surveyed with footprint tunnels made from corrugated plastic (1200 mm × 210 mm × 180 mm) between May and October 2016 [31,40,41]. These 46 km-squares were chosen as they contained the most suitable hedgehog habitat (e.g., no lake and forest) and only areas with the chance of regular sightings by citizens (no peripheral agricultural areas). The estimation of the abundance in both time periods was limited to the study area of the 46 km-squares (Figure 1). Thus, the study area of the early study period (1992) was pruned accordingly.

Within each study area square (1 km^2), we defined a 500 m × 500 m square (in cases where no settlement area was included in the square, we chose a 400 m × 600 m rectangle) near the center of that square. We placed ten tunnels within this square (>100 m apart from each other), and checked daily for five consecutive days for footprints and to top up bait and ink. A spoonful of bait (commercially available hedgehog food, Claus—Spezial-Igelfutter, Limburgerhof, Germany), two ink-pads (non-toxic ink: mix of carbon powder and vegetable oil), and two sheets of A4 paper were placed inside the tunnel, on a removable plate (Figure A1, Appendix A). It was expected that, if hedgehogs inhabit the study area, they would encounter these footprint tunnels during their nocturnal forays and enter them to reach the bait [40]. In doing so, their feet would touch the ink and leave species specific footprints on the paper. We placed the footprint tunnels along linear features (e.g., wall, hedgerow, fence, etc.) as these are structures that hedgehogs like to follow when foraging [43]. The proportion of these ten tunnels per km-square containing hedgehog footprints were used for further analyses of hedgehog abundance, 0 (no tunnel with hedgehog footprints) to 10 (all tunnel contain hedgehog footprints), hereafter referred to as hedgehog level; see below. The survey was conducted primarily by volunteers from the StadtWildTiere ("urban wildlife") project in Zurich, as well as interns, and the authors.

In addition to the footprint tunnels, we collected observations of hedgehogs by the general public on an observation platform, which is part of the citizen science project StadtWildTiere. Further, the project includes a well-established network of volunteers, who have been trained in various wildlife research methods. We asked people to send in observations of hedgehogs through the distribution of leaflets, hangouts, and articles in newspapers. From 2016 through 2018, 4125 observations from mammals, reptiles, and amphibians were collected by 449 users. For the recent study, we analyzed hedgehog observation data from 2016–2018 (n = 1096) to match the total number of observations in 1992 (n = 1011). Due to the differences in data collection between the first and the second study periods, we wanted to account for the possibility that there might be an underlying spatial clustering in the more recent data set resulting from a less systematic distribution of flyers, which may have caused us to underestimate the actual recent distribution. To account for such a bias, we examined the distribution of all observations other than hedgehogs (n = 3029). The mean of non-hedgehog observations per cell was 8.8 (= 3029/343 grid cells, median = 6). We removed all cells from both study periods with fewer than 4 observations of non-hedgehog species (in the recent time period) to ensure a minimum level of observation effort. We then examined distribution differences between the past and the recent study periods using these estimates and found that this method actually resulted in an even larger decline compared to the one described above, with all grid cells included. Therefore, we concluded that potential spatial clustering did not lead to an underestimation of the distribution in the recent study. Consequently, we chose to use our more conservative estimate with all grid cells included for the comparison of time periods.

2.3.2. Capture Mark Recapture Study

In 2017, we carried out a capture-mark-recapture (hereafter CMR) study to obtain estimates of abundance. We selected four study area plots, each 0.5 km^2 in size (districts Altstetten, Wipkingen, Enge, Schwamendingen, Figure 1). The one area where CMR was conducted in 1992 was 0.23 km^2 in size and delimited by large roads. In the recent study, the size of the study area plots were 0.5 km^2 to render them more representable for the area, while still being small enough to be searched within 4 h. These areas were chosen in 2017 for comparisons between study years because they all contained higher than average hedgehog density in 1992 but were found to show exceptionally low (Altstetten and Wipkingen) or high (Enge and Schwamendingen) relative density of hedgehogs, respectively, in the footprint tunnel study of 2016 (see Results). Therefore, we considered these areas appropriate for a comparison between years.

We surveyed each of the four study area plots eight times during 4 weeks in June 2017. Surveys were only conducted in good weather conditions, i.e., if there was no heavy rain, in order to not influence capture rates. While sometimes two areas were surveyed in the same night by different teams, the areas were generally surveyed in subsequent nights. For every survey, a researcher and a volunteer searched for hedgehogs with flashlights between 10 pm and 2 am, scanning the entire study area plot via public paths and green spaces. We captured all encountered hedgehogs and examined them to determine weight, sex, and health status (e.g., presence of injuries or an unusual amount of ectoparasites). We marked hedgehogs uniquely with shrink tubes (see methods from study in 1992).

Only adult hedgehogs were marked with shrink tubes, as juveniles could still be easily distinguished by their smaller size at this time of the year. The handling of each hedgehog lasted about 10 min, and the animals were subsequently released on site. The capturing protocols were in accordance with the regulations (1992) or approved by the Veterinary Office of the Canton Zurich through the animal experimentation authorization (ZH079/17).

2.4. Analyses

We analyzed the data from the CMR study from 2017 in the program MARK [44] to obtain estimates of hedgehog abundance. As all data was collected within four weeks, we assumed a closed population and used the closed capture model (Huggins' p and c). All four study area plots (Altstetten, Wipkingen, Enge, Schwamendingen) were analyzed together, nested according to study area plot. To test for overdispersion and the fit of the model to the data, we calculated median c-hat (c-hat; 0.987, sampling standard error = 0.046). Values over 1 indicate overdispersion or a bad fit of the model to the data. Models were ranked by Akaike information criterion (AIC). We averaged the three best models (AIC not more than 3 points apart) to get accurate abundance estimates for the high-density study area plots. The abundance in the two low-density areas was calculated assuming constant survival and recapture probability over time and among study sites. These two different approaches were necessary because the low recapture rates did not allow averaging the models.

To obtain population size estimates for the study areas where no CMR surveys were conducted (n = 42 km-squares) for the recent study period, we computed a linear regression between the abundance measures from the CMR study as the independent variable and the hedgehog levels from the footprint tunnel study in the respective areas as the dependent variable (Figure A2, Appendix A). Interpolating from the resulting regression equation, we subsequently assigned each hedgehog level across the study area to a corresponding estimate of the abundance. We then summed up all the estimates of the 46 km-squares in the study area, to get an estimate of the population size of hedgehogs in the city of Zurich for the recent study period. The confidence and prediction intervals were estimated using Monte-Carlo simulation in the software Stan [45].

Similarly, we calculated the hedgehog population size for the same study area (46 km-squares) in 1992 using the relative return rate (i.e., the proportion of survey cards returned per $\frac{1}{4}$ km^2). For this, we ran a linear regression using the relative return rate as the independent variable and the abundance estimate of the study area Wipkingen (where hedgehog abundance was estimated in 1992 by marking and summing up the encountered hedgehogs) as the dependent variable. Since abundance was only measured in one area, we improved the model by using the y-intercept from the 2016/2017 data (which is 3.4 hedgehogs/km^2; see results). We considered this a better fit for the model's intercept than setting it at zero, as even in areas where no hedgehog were detected by the footprint tunnels, there likely is a very low hedgehog density present. This was shown in the two study area plots, Altstetten and Wipkingen, where 1 and 0 out of 10 tunnels contained hedgehog tracks, respectively, yet some hedgehogs were found during the CMR study (4 and 3 individuals, respectively). The hedgehog abundance for each km-square (n = 46) was compared between time periods to get a relative percent change showing areas with relative declines or increases in hedgehog abundance compared to 1992. These estimates were obtained by using the relative return rate of the survey cards (for 1992), and the hedgehog level from the footprint tunnels (for 2016–2018) in combination with the CMR numbers from the respective year. In addition, the km-squares were divided in three abundance categories of low (0–19 hedgehogs per km^2), medium (20–39), and high (>40) and their frequency in each survey was plotted.

We then used the point observation data from 1992 (from survey cards) and 2016—2018 (from the StadtWildTiere platform) to construct hedgehog distribution maps for the two study periods based only on occurrence data (presence/absence of observations) on a $\frac{1}{4}$ km^2 level across the entire study area. We chose the level at $\frac{1}{4}$ km^2 in order to have enough resolution to compare the data and to have the same sized areas in both time periods. The presence data from the footprint tunnels are included in the data from 2016–2018. We modeled the distribution in both time periods with a subsampling of the data set to estimate the difference in the distribution between both time periods. Our subsampling consisted of randomly selecting 100 observations from the full sample (without resampling, using the sample-function from the r-package "base" [46] and seeing how many $\frac{1}{4}$ km^2 grid cells were occupied. We used this stepwise function to produce estimated asymptotic curve that then allowed us to calculate an estimate for the expected distribution (value estimating the horizontal asymptote for large x). We repeated the subsampling and asymptote modeling for subsample sizes of 200 to 900 at increments of 100 and repeated each subsampling 1000 times. With the resulting asymptote estimates (for the subsample sizes 500 to 900), we calculated differences in distribution estimates between the subsamples (Figure A3, Appendix A). This subsampling method was only used to validate the extent of the hedgehog decline in distribution between the two study periods. All analyses were conducted using the programs QGIS (Version 2.18) [47] and R (Version 3.6.2) [48].

3. Results

3.1. Change of Hedgehog Distribution in 25 Years

In the past study, 1011 observations of hedgehogs were received via returned survey cards (approximately 18,000 where sent in total). Hedgehogs were present in 87.1% of the 232 grid cells (circles in Figure 2). Only parts of the inner-city areas and the industrial areas along the Limmat river, which are densely built and contain little green space, were found to be unoccupied.

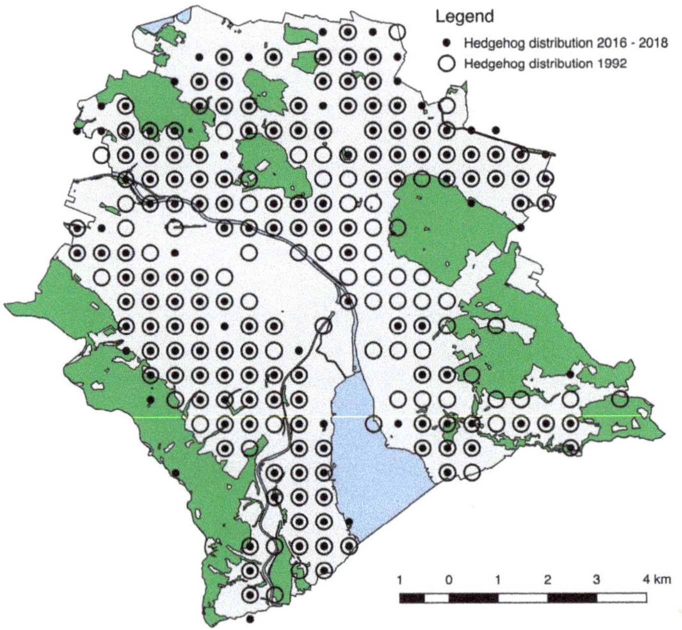

Figure 2. Distribution map of hedgehogs *Erinaceus europaeus* in the city of Zurich, Switzerland in the past study 25 years ago (1992, circles) and in the recent study (2016–2018, points). The outline delineates the municipal border of the city of Zurich with forest area (green) and water bodies (blue). Distribution circles represent the $\frac{1}{4}$ km^2 survey grid. The distribution of hedgehogs declined by −17.6% (± 4.7%) in the 25 years between the studies.

In the recent study from 2016–2018, we collected 1096 hedgehog observations on the citizen science platform stadtwildtiere.ch for the entire study area. In the footprint study from 2016, 121 of the 460 footprint tunnels (26.3%) contained hedgehog tracks. Taken together, these surveys indicate hedgehogs currently occupy 74.6% of the 232 grid cells (points in Figure 2).

A comparison of the hedgehog distributions between study periods showed that hedgehogs were found in 30 grid cells (12.9%) in the recent study, which were unoccupied in the past. However, in 25.4% of the 232 grid cells, hedgehogs were missing, despite having occurred there in 1992 (Figure 2). In summary, the presence of hedgehogs reduced by 12.5% of the 232 grid cells in the 25 years between the studies. A subsampling method revealed a reduction in median distribution estimates between the study periods of 17.6% ± 4.7% (mean ± confidence interval, Figure A3, Appendix A).

3.2. Change of Hedgehog Abundance and Density in 25 Years

In the past study, 18 individual hedgehogs were marked (June to August 1992) in the Wipkingen study area plot and the density was calculated to be 78.0 ± 26.0 hedgehogs/km^2 (mean ± confidence interval CI). To calculate population size across the total study area, the relative return rate of the survey cards, the estimate from Wipkingen, and the intercept estimate from the recent study period were used. This allowed us to estimate their abundance as 1477 ± 492 (total ± CI) hedgehogs for the study area in the city of Zurich in 1992.

In the recent study, we caught and marked a total of 57 hedgehogs in four study area plots (30 in Schwamendingen, 19 in Enge, 4 in Altstetten, and 3 in Wipkingen; S, E, A, and W, respectively, in Figure 1; Table A1, Appendix A). All hedgehogs were in relatively good condition. We recaptured individual hedgehogs on average 2.13 times (range: 1 to 6 times, Table A1, Appendix A). Using the estimates from the CMR models, we assigned each hedgehog level across the study area to a corresponding estimate of the abundance and added them to get an estimate for the entire study area. The hedgehog densities in the study area plots were 70.4 (62.6–78.1) hedgehogs/km^2 in Schwamendingen, 45.4 (38.1–52.7) hedgehogs/km^2 in Enge, 9.3 (7.5–11.2) hedgehogs/km^2 in Altstetten, and 7.0 (5.4–8.6) hedgehogs/km^2 in Wipkingen (estimate (confidence interval), Table A2, Appendix A). Both methods used in the recent study—footprint tunnel and mark-recapture—produced similar results for the different areas of Zurich. Areas with low hedgehog numbers in the footprint tunnels had also low numbers in the CMR study, and vice versa (Figure A2, Appendix A). The population size across the study area in 2017 was 878 (844–910) (total and prediction interval) hedgehogs. A comparison of the total estimated hedgehog abundance for the 46 km-squares studied in both study periods revealed a decline of 40.6% in the 25-year period.

The change in hedgehog abundance per km^2 varied between study plots and study periods. At the same time, some areas seem to have experienced a more pronounced decline in relative hedgehog abundance than other areas. When plotting the change in abundance for each km^2 grid, the changes between the recent and past become apparent (Figure 3). Very few areas have seen a relative increase in hedgehog abundance, while most study areas have fewer hedgehogs now than they used to. For the study area plot in Wipkingen, the hedgehog density in 1992 was calculated to be 78.0 ± 26.0 hedgehogs per km^2, while, in 2017, it was only 7.0 ± 1.6 hedgehogs, which corresponds to a ten-fold decline. In 1992, 76% of squares contained medium to high hedgehog abundances, while, in the recent period, 63% of squares contained low abundances, with only 10% containing high abundances (Figure 4).

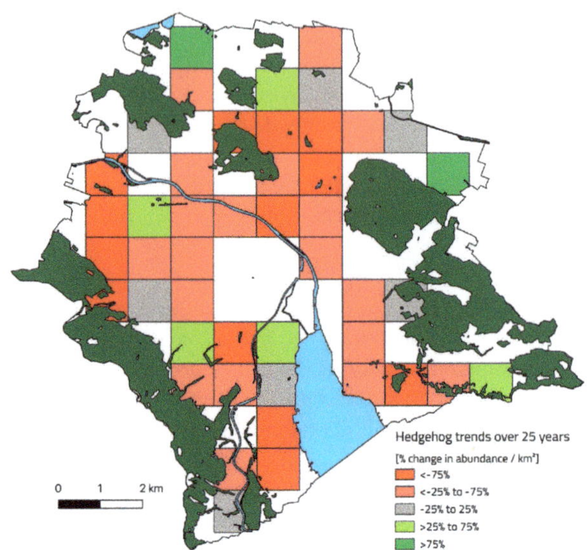

Figure 3. Changes in hedgehog abundance estimates between the past (1992) and recent (2016–2018) study (% change in abundance/km^2). The outline delineates the municipal border of the city of Zurich with forest area (green) and water bodies (blue). Squares signify the study area divided into km^2: declining hedgehog abundance (red), increasing hedgehog abundance (green).

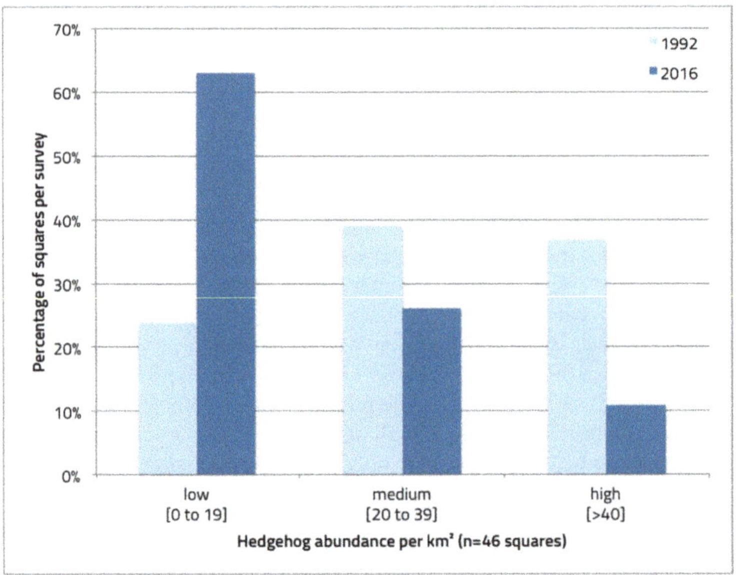

Figure 4. Proportion of km-squares per survey period (past in light blue; recent in dark blue) with low, medium, and high hedgehog abundance per km^2.

4. Discussion

Our results suggest that the distribution of hedgehogs in Zurich has reduced over the last 25 years by 17.6%, while hedgehog numbers have declined by 40.6%. This alarming result is in sharp contrast to the expectation that green urban areas provide an ideal habitat refuge for hedgehogs. In 1992, hedgehogs thrived in Swiss cities and occurred in all green areas of the city of Zurich [19,24,42]. The population size in the urban study area of Zurich (46 km^2) was estimated to be 1477 ± 492 hedgehogs in 1992 or on average 32.1 ± 4.8 hedgehogs per km^2, with varying hedgehog densities between areas across the city. In the study area plot for Wipkingen, where a CMR study was conducted, the hedgehog density was very high with 78 ± 26 hedgehogs per km^2.

Today, the situation in Zurich has dramatically changed. Hedgehog abundance has declined to an estimated 878 (844–910) hedgehogs for the same study area in Zurich, or 19.1 ± 5.0 hedgehogs per km^2. Hedgehogs can still be encountered in many parts of the city, and some areas harbor still high densities of hedgehogs (e.g., the highest density estimate was Schwamendingen: 70.4 hedgehogs per km^2). This density estimate is still high compared to the recent estimate from the urban areas of Sedan in France (36.5 ± 15.2 hedgehogs per km^2 [18]) or compared to amenity grassland in England (47.0 ± 9.0 [49]). However, overall abundance, densities and distribution in the city of Zurich are much lower now than they were 25 years ago (Figure 4). In certain study area plots with formerly high hedgehog densities, only very few hedgehogs were found today (e.g., Altstetten and Wipkingen). These data indicate that hedgehog densities have not declined equally across the city in this period. This spatially uneven decline suggests that factors which negatively affect hedgehog populations are not evenly distributed across the city (see below). Alternatively, the temporal onset of factors negatively affecting hedgehogs may be shifted between study area plots. However, this city-wide loss in hedgehog distribution does not seem to follow a pattern easily explainable by the data we currently have at hand.

Hedgehog populations have been declining across different habitats and geographical areas in the last few decades, and different rates of decline in distribution and abundance have been found. Hof and Bright (2016) calculated a hedgehog decline rate of 5.0 to 7.4% in occupied grid cells over a 40-year period in England (1960–1975 and 2000–2015 [32]). Davey and Aebischer (2006) report decline rates of 9.1% (in Scotland) to 37.3% (in Wales) and 30.0% (in England) in abundance over a period of nine years (1995–2004 [50]). Analyzing data sets from multiple studies in the UK, Roos et al. (2012) estimate the average decline in occupancy to be around 40% over ten years [37]. All these estimates come from studies in rural or larger geographical areas. The rate of decline in distribution found in the recent study (which corresponds to a decline of 7.0% over 10 years assuming a linear decline) is higher than the rates described by Hof and Bright (2016) in England, while certainly lower than Roos et al. 2012 rates for the UK. The rate of decline in abundance (which corresponds to a decline of 15% over 10 years assuming a linear decline) is slightly higher than the rates proposed by Davey and Aebischer (2006) for Scotland but lower than the rates of decline for Wales and England. Thus, these numbers from rural areas, when combined with the numbers of decline from our urban study, suggest that hedgehogs are declining in general. On the other hand, a recent study found high juvenile survival rates in a suburban area, pointing towards healthy populations in this habitat [23]. Therefore, our observed population decline might be caused by factors that are patchily distributed and which are not acting on all populations. Nevertheless, our study indicates that even cities, which are suggested to be refuges for hedgehogs, might be in danger of losing that status to some degree.

4.1. Potential Factors Negatively Affecting Hedgehog Populations

The reasons for the observed hedgehog decline are currently unknown and could even be multifactorial. Here, we will provide an overview of potential factors negatively affecting hedgehog populations dividing them into six topics: habitat, food, poison, predation, disease, and parasites. This list is not exhaustive; however, we discuss their relative importance, a possible contribution of an extinction debt, and suggest the major avenues of future research.

4.1.1. Habitat

Hedgehog habitat can become uninhabitable through either the loss of the habitat or its deterioration. Habitat may be lost by being rendered inaccessible via fragmentation or by the complete removal of the habitat itself, e.g., by sealing green spaces. Fragmentation can be caused by barriers [51,52], such as roads, train tracks, fences, or walls that cannot be overcome by hedgehogs [31,53–55]. The loss of dispersal structures (e.g., removal of hedges [56]) can contribute to the isolating effect of fragmentation on the populations. The major force bringing about such changes in urban areas is densification [50,57]. Densification of urban spaces is a process occurring around the globe to deal with growing urban populations while reducing urban sprawl. During the study period, the city of Zurich's population grew by 17.1% from 361,000 to 423,000 people, which is accompanied by an intensive densification process resulting in a loss of urban green space in Switzerland [58,59]. This might be one factor that contributed to the decrease of hedgehog populations in Zurich.

Besides the loss of habitat itself, the loss of habitat quality might be threatening hedgehog populations. The intense maintenance of "tidy" gardens and public green (e.g., dense hedges and lack of dead plant material, such as heaps of branches and leaves) leads to the loss of nesting opportunities [38], hiding spots, and shelter for hedgehogs in urban spaces. Additional habitat threats include garden hazards (e.g., pools without exit possibility, uncovered light shafts, electrical fencing, and automatic mowers), where hedgehogs can fall, get stuck, be hurt, or even be killed. Automobile traffic is also a known mortality factor for hedgehog populations, which might be of particular importance to urban and suburban hedgehog populations [54,60]. However, the moderate increase in vehicle numbers and the widespread introduction of zones with reduced driving speed in Zurich do not point towards a recently growing problem [61], but rather a constant risk and a source of background mortality for hedgehogs. Last but not least, changing summer and winter temperatures due to global climate change might limit the hedgehog's food and water availability in summer and disrupt its hibernating behavior in winter [23,38]. A detailed analysis of habitat characteristics and changes thereof through time are beyond the scope of this paper. However, analyses on temporal changes of the habitat quality in the study area are under way and will help to shed further light on the reasons behind the observed population decline of hedgehogs in the city of Zurich.

4.1.2. Food

The quality of the food and its availability are linked to the quality of the available habitat. Hedgehogs prey on a wide variety of arthropods and mollusks [62]. Therefore, the currently described global decline in insect biomass is likely to have an impact on insectivorous animals, such as the hedgehog [43,63,64].

In the urban habitat mosaic, we expect hedgehog food sources to be patchily distributed. Gardens and parks with local plant species provide valuable habitat elements for hedgehogs. Compost heaps, being humid and warm, contain a variety of insects and were often visited by hedgehogs in 1992. Although no systematic assessment has been conducted, the numbers of compost heaps in Zurich seem to have declined since the study in 1992 (pers. obs.). Since 2013, organic waste has been collected by the city government to produce biogas, and, since then, many people seem to have given up personal compost heaps in their gardens. This depletion of potential food sources might have contributed to the observed decline of hedgehog populations in Zurich and could also explain a spatially patchy pattern. Furthermore, some garden owners might provide artificial food sources for hedgehogs, which has been shown to increase activity levels during the winter when hedgehogs should be hibernating [65]. The importance of supplementary feeding, however, has not been investigated in detail.

4.1.3. Poison

As an insectivorous animal, the hedgehog undoubtedly suffers from the use of pesticides by a reduction in food availability. Further, as opportunistic feeders [62,66], hedgehogs might be exposed to pesticides either directly by ingesting poisoned bait or indirectly by ingesting poisoned prey. Even though the route is largely unknown, studies found residues of anticoagulant rodenticides (warfarin, coumetetralyl, difenacoum, bromadiolone, brodifacoum, flocoumafen) [67] and organochlorine compounds [68] in hedgehogs, even up to 9 months after the use of such in the study area [69]. Hedgehogs are believed to suffer the same exposure and potential effects of anticoagulant rodenticides as other non-target mammals and predatory birds [67]. A study looking at the casualties of the use of banned pesticides in the Canary Islands (Spain) found hedgehogs among the fatalities [70]. However, the illegal use of banned pesticides might just be a fraction of the problem regarding the potential effects of the use of legal pesticides. Previous studies on the effect of pesticides on hedgehogs were inconclusive [71]. Further research is needed to assess the effects of the ingestion (direct or indirect) of pesticides and environmental toxins by hedgehogs. Indirect effects of poisoning or heavy metal accumulation, such as reduced fecundity, reduced lifespan, impaired disease resistance, or poor growth, are likely hard to measure [72,73]. In Switzerland, the amount of herbicides sold decreased over the last ten years, while the amount of fungicide, bactericide, insecticide, akaricide, molluscicide, and growth regulators remained unchanged, even though many pesticides' effectiveness has increased in the same time period [74]. In Swiss urban areas, we would expect a generally smaller amount of pesticide used compared to rural areas, particularly less use of insecticide and fungicides. On the other hand, a higher rate of rodenticide use, probably the poison most likely to affect hedgehogs, is expected in more densely populated areas due to a higher abundance of rodents, such as rats (but this might not be a global pattern; also see Reference [75]).

4.1.4. Predation

With their spiny defenses, healthy adult hedgehogs are largely safe from most predators, although a few species can occasionally attack hedgehogs. Domestic dogs and cats, both of which might be encountered in higher densities in urban areas compared to rural ones, are known to be able to predate on young or injured hedgehogs [76]. The effect of such potential predation on hedgehog populations needs to be further explored. Predation of hedgehogs by badgers *Meles meles*, however, is often argued as being an important factor controlling hedgehog populations [21,43,77]. Indeed, badgers are not just predators but also food competitors of hedgehogs, and increasing badger numbers, as has been reported in the UK [78] and in Switzerland [61], could potentially cause hedgehog numbers to decline. These issues become even more relevant as badgers in Switzerland were found to increasingly venture into the urban landscape, areas that have so far been considered safe refuge for hedgehogs [20,61,79]. Hedgehogs were found to behave as in a landscape of fear [80]: they avoid areas with high badger densities [22,41,77,79,81,82], keep closer to hiding structures in areas with badgers [20,43], and increase in density after the removal of badgers [83]. On the other hand, there are areas where hedgehogs and badgers are sympatric and are both thriving, whereas there are suitable habitats with neither hedgehogs nor badgers recorded [22]. Therefore, increasing badger numbers and intraguild predation and competition by badgers is unlikely to be the sole explanation for the observed decline of hedgehog populations. However, increasing competition through increasing badger density in combination with a reduction in prey biomass might lead to an intensification of competition between hedgehogs and badgers [43]. In Zurich, badgers have been sighted across almost the entire city, although observations are less common in the city center for both badgers and hedgehogs (data from the StadtWildTiere platform). There are areas with both badger and hedgehog observations, but there are also areas where hedgehog numbers declined in this study without any badger sightings. We therefore conclude that, if badgers are contributing to hedgehog decline, it is not the single factor driving this pattern.

4.1.5. Diseases and Parasites

If diseases are on the rise or a new parasite is spreading this may negatively impact hedgehog populations, especially if animals are already weakened by high stress levels imposed by other factors [84]. With higher population densities in urban areas compared to rural areas [18–21], diseases and parasites are likely to spread more quickly within urban populations, especially at crowded feeding stations. Furthermore, high levels of parasitism in hedgehogs (e.g., *Ixodes hexagonus*, *Ixodes ricinus*, *Crenosoma striatum*, *Capillaria* spp.) in general [85], as well as documented increases in the abundance of some wildlife parasites relevant to hedgehogs, such as ticks [86,87] and gastropod-transmitted lungworms [88,89], support the idea of parasitism as contributing factor to population declines in this species. A recent study revealed high and fluctuating prevalence rates of *Capillaria aerophila* in the course of the last three decades ranging between 42.8% and 75% in foxes [88]. In the same period, foxes established in high densities in the middle of urban areas [90,91]. Therefore, an increased infection pressure with the infective stage of this parasite, which frequently also infests hedgehogs and can cause weight loss, bronchitis, and pulmonary damage, is likely [92]. At this point, however, no research has yet uncovered such a factor in the hedgehogs' distribution range, and the regional rescue center for hedgehogs has not recorded any apparent increase in the number or proportion of ill or heavily parasite affected individuals (Annekäthi Frei, pers. comm.). Furthermore, zoonoses originating from hedgehogs merit further research and monitoring, for example, surveying the prevalence of methicillin-resistent *Staphylococcus aureus*, due to its potential to cause severe infections in humans [93–95].

4.1.6. Extinction Debt

In addition to all the factors mentioned above, a potential extinction debt might further complicate the topic. An extinction debt describes emerging ecological cost from former habitat destruction [96]. Habitat fragmentation and isolation do not cause the extinction of species immediately, but produce smaller and potentially inbred subpopulations in smaller habitat islands, which may no longer be well adapted to the current conditions and suffer from inbreeding depression [97]. A study examining the spatial genetic structure of hedgehogs found three relative distinct sub-populations in the city of Zurich [52]. If these populations become increasingly isolated due to fragmentation and were to become inbred, they would be less able to adapt to (even small) environmental changes. A study on the extinction rate of urban plants showed that legacies of landscape transformations by agrarian and urban development can last for hundreds of years, and cities might carry a large extinction debt [98]. Therefore, the patterns of decline seen in current hedgehog populations might have been caused decades ago by habitat fragmentation. This could also explain why we see such patchy distribution and density patterns in cities. On the other hand, the relative high reproduction rate of hedgehogs coupled with a high potential in spatial exploration, might help hedgehogs to adjust their distribution to harmful factors more quickly compared to more stationary and slowly reproducing species. In general, it could prove valuable to take evolutionary principles into account when evaluating the causes of extinction [99].

Further research is necessary in order to study which habitat structures in urban areas support healthy hedgehog populations and enable co-occurrence of badgers and hedgehogs. It is crucial to know how hedgehogs navigate the patchy urban food mosaic and what influence individual gardens and supplemental food sources have. Monitoring efforts will have to be implemented to keep track of current and future levels of disease and parasites in urban hedgehog populations.

4.2. Comparability of the Studies

There are relatively few systematic studies of wild hedgehogs due to their nocturnal and secretive lifestyle. This is even more the case for systematic surveys on the distribution and density of hedgehogs. Therefore, there is very little data to investigate hedgehog population changes over time, and any such data is extremely valuable, especially regarding the current biodiversity crisis and already reported alarming hedgehog population declines in other parts of Europe. In Zurich, we were lucky to have such a longitudinal dataset. The data in both time periods were not collected with the exact same methodology. In 1992, observations of hedgehogs were elicited through systematic mailings and abundance was estimated through a CMR study in a small study area, delimited by large roads and measuring 0.24 km^2. In 2016–2018, when a postal questionnaire would no longer have worked, the people of Zurich were asked to send in observations in various ways (through flyers, media releases, and handouts). Additionally, we conducted a systematic study with footprint tunnels in an urban area of 46 km^2. Furthermore, CMR studies were conducted in four study plots (each 0.5 km^2). These study plots were purposefully chosen, as they were all areas that had high hedgehog abundances in 1992 (two with still high relative densities and two with low relative densities in the footprint tunnels in 2016). In both time periods, a high but similar motivation of people to report hedgehog sightings was assumed (similar target groups of nature lovers) and, to minimize the potential effect of differing observation effort, observations were analyzed as presence and absence data only. Therefore, we think that potential caveats by slightly different study designs are addressed sufficiently in order to render the different datasets comparable.

The abundance estimates for both years were extrapolated using relative return rates (in 1992) and footprint tunnel levels (in 2016–2018). This allowed us to get more accurate estimates rather than simply extrapolating the estimates from the CMR studies to all study areas. We used three years of data to construct the distribution map in the recent period, so that we could match the amount of observations in 1992 and make sure that with increasing numbers of observations the distribution was not increasing. In fact, the use of multiple years of observations should, if anything, lead to an overestimation of current hedgehog distribution. Even by using these conservative estimates, the pattern of decline in abundance and distribution was clearly confirmed.

5. Conclusions

This study is the first to quantify the decline rate of urban hedgehogs in a European city over time. In the light of continental-wide reports of declining hedgehog numbers in rural areas, urban areas have been seen to be the hedgehogs' refuge from habitat destruction, intensification of agriculture, and, in some places, the recent increase of badger populations. Our results, however, extend the alarming pattern of hedgehogs under pressure to urban areas.

After an evaluation of possible causes of the decline in urban habitat, the major reasons remain unclear. Further research is necessary to investigate the role of habitat deterioration, connectivity, and food supply, as well as the negative effects of predation, diseases and parasites, and pesticide use as potential causes. The patchy pattern of the decline suggests the influence of a single or combination of spatially unequally distributed factors. This evaluation of the causes of the decline is critical, given the alarming decline of this species in the urban area. Based on such further evaluations, conservation measures can be planned and implemented.

Citizen science proved to be a suitable method to investigate urban wildlife and is a promising tool to further investigate the causes of the decline, as well as an aid to implement measures to remedy this loss of urban wildlife. Particularly charismatic animals, such as hedgehogs, are well suited to such work, as they provide the perfect focus to engage a wide public and to raise awareness for conservation.

Author Contributions: Conceptualization, A.L.T., S.G., A.D., D.H., and F.B.; methodology, A.L.T., S.G., A.D., D.H., and F.B.; formal analysis, A.L.T.; investigation, A.L.T., S.G., M.G., and F.B.; data curation, A.L.T. and A.D.; writing—original draft preparation, A.L.T.; writing—review and editing, A.L.T., S.G., A.D., M.G., D.H., and F.B.; visualization, A.L.T.; project administration, A.L.T., S.G., A.D., D.H., and F.B.; funding acquisition, S.G. All authors have read and agreed to the published version of the manuscript.

Funding: The projects of the association StadtNatur are kindly supported, among others, by the Temperatio Foundation, the Vontobel Foundation, and the Ernst Göhner Foundation. The analyses and publication were financially supported by the Temperatio Foundation and Grün Stadt Zürich. This was an invited contribution by Animals.

Acknowledgments: This study was carried out in the framework of the citizen science projects stadtwildtiere.ch and wildenachbarn.ch of the association StadtNatur, c/o SWILD—Urban Ecology & Wildlife Research, Switzerland. The authors thank the volunteers involved in the projects in 1992, 2016, 2017, and 2018, as well as the students Chiara Baschung and Rolf Badat, and the team of Igelzentrum Zürich, particularly Simon Steinemann and Annekäthi Frei for their support in data collection, Max Ruckstuhl, Stefan Hose and Bettina Tschander of Grün Stadt Zürich for their support and access to data and GIS map layers, Fränzi Korner-Nievergelt of oikostat for her help with the Stan calculations, Mirco Lauper for his help with R and GIS, Kevin Vega for helpful comments and proofreading, four anonymous reviewers for helpful comments that served to improve the manuscript, and the assistant editors for their editorial contributions.

Conflicts of Interest: The authors declare no conflict of interest. The funders had no role in the design of the study; in the collection, analyses, or interpretation of data; in the writing of the manuscript, or in the decision to publish the results.

Appendix A

Figure A1. Footprint tunnel with removable plate in front, showing hedgehog footprints on them. Two paper sheets and two brushes of ink are applied to it. A spoon of bait is placed in the brown plate. © Thomas Massie/stadtwildtiere.ch.

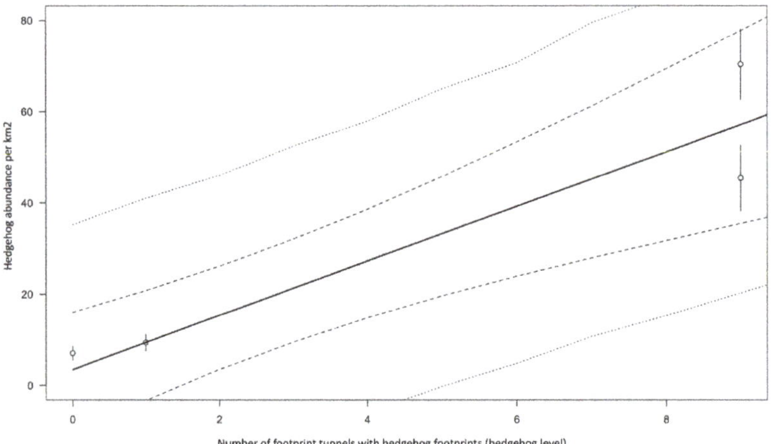

Figure A2. Linear regression between the abundance measures from the capture-mark-recapture (CMR) study (whiskers indicate confidence intervals of the CMR estimates) as the independent variable and the hedgehog levels from the footprint tunnel study (ranging from 1–10) in the respective area as the dependent variable. Regression line in black with confidence (dashed line) and prediction interval (dotted line).

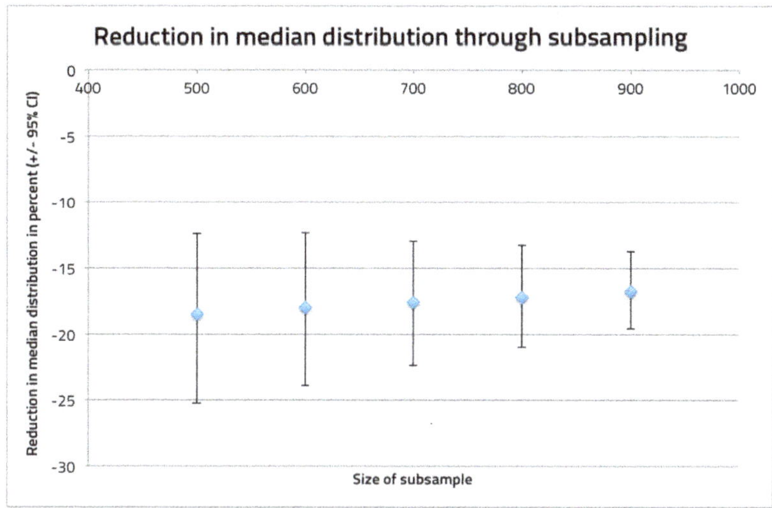

Figure A3. Estimation of the decrease in median distribution area (in percent (± 95% confidence interval), revealed through subsampling. With increasing size of the subsample, the estimate becomes more accurate, on the other side the degrees of freedom become more restricted (because overall sample size = 987 for 1992 and 1096 for 2016–2018). Our estimate resulted by averaging the values for subsample sizes 500 through 900.

Table A1. Data sheet of all hedgehog captured during the 2017 capture-mark-recapture (CMR) study.

Study Area Plot	Individual	Sex	Weight (Range in g)	Health Status	Number of Captures
Altstetten	A_A	female	1430	good	1
Altstetten	A_B	unknown	1460	good	1
Altstetten	A_C	unknown	1230	good	1
Altstetten	A_D	female	1000	good	1
Schwamendingen	S_A	female	580–1150	good	3
Schwamendingen	S_B	female	1030–1220	good	4
Schwamendingen	S_C	unknown	860–1010	good	2
Schwamendingen	S_D	unknown	1020	good	1
Schwamendingen	S_E	female	640	good	1
Schwamendingen	S_F	female	710–1120	good	4
Schwamendingen	S_G	female	820–1040	good	5
Schwamendingen	S_H	female	870–950	good	2
Schwamendingen	S_I	female	890–1050	good	3
Schwamendingen	S_J	male	840	good	1
Schwamendingen	S_K	male	790–820	good	2
Schwamendingen	S_L	female	1010–1040	good	2
Schwamendingen	S_M	female	910–980	good	4
Schwamendingen	S_N	unknown	1060	good	1
Schwamendingen	S_O	male	400–830	good	3
Schwamendingen	S_P	female	1130–1260	good	2
Schwamendingen	S_Q	female	1240–1340	good	4
Schwamendingen	S_R	unknown	700–900	good	4
Schwamendingen	S_S	female	1000	good	1

Table A1. *Cont.*

Study Area Plot	Individual	Sex	Weight (Range in g)	Health Status	Number of Captures
Schwamendingen	S_T	male	1050	good	2
Schwamendingen	S_U	male	920–1080	good	3
Schwamendingen	S_V	male	880–980	good	2
Schwamendingen	S_W	female	1040	good	1
Schwamendingen	S_X	female	470–500	good	2
Schwamendingen	S_Y	male	920	good	1
Schwamendingen	S_Z	female	840	good	1
Schwamendingen	S_AA	female	930	good	1
Schwamendingen	S_AB	female	1210	good	2
Schwamendingen	S_AC	male	1000–1120	good	2
Schwamendingen	S_AD	male	1070	good	1
Enge	E_1	female	1225–1125	good	5
Enge	E_2	male	1275–1375	good	6
Enge	E_3	female	825–875	good	4
Enge	E_4	male	875–1025	good	3
Enge	E_5	female	1225	good	1
Enge	E_6	female	875–925	good	3
Enge	E_7	female	1125	good	1
Enge	E_8	female	1275–1325	good	3
Enge	E_9	female	1275–1325	good	2
Enge	E_10	male	925–1025	good	3
Enge	E_11	unknown	825–875	good	2
Enge	E_12	unknown	975–1075	good	2
Enge	E_13	unknown	675	good	2
Enge	E_14	unknown	1025	good	1
Enge	E_15	female	925–1075	good, medium	2
Enge	E_16	unknown	1125	good	1
Enge	E_17	female	1175	good	1
Enge	E_18	male	1350	many flees	1
Enge	E_19	female	1125	good	1
Wipkingen	W_A	male	1275	good	1
Wipkingen	W_B	female	750	good	2
Wipkingen	W_C	male	1100	good	1

Table A2. Population size estimates for the study area plots resulting from the capture-mark-recapture (CMR) data (hedgehogs/$\frac{1}{2}$km^2).

Study Area	Estimate	Standard Error	Unconditional Standard Error
Enge	22.72	3.28	3.63
Schwamendingen	35.19	3.78	3.87
Wipkingen	3.52	0.79	NA
Altstetten	4.69	0.92	NA

References

1. *World Urbanization Prospects—The 2018 Revision*; United Nations: New York, NY, USA, 2018; Volume 12.
2. Seto, K.C.; Güneralp, B.; Hutyra, L.R. Global forecasts of urban expansion to 2030 and direct impacts on biodiversity and carbon pools. *Proc. Natl. Acad. Sci. USA* **2012**, *109*, 16083–16088. [CrossRef]
3. Shochat, E.; Warren, P.S.; Faeth, S.H.; McIntyre, N.E.; Hope, D. From patterns to emerging processes in mechanistic urban ecology. *Trends Ecol. Evol.* **2006**, *21*, 186–191. [CrossRef]
4. McKinney, M.L. Urbanization as a major cause of biotic homogenization. *Biol. Conserv.* **2006**, *127*, 247–260. [CrossRef]
5. Lin, B.; Meyers, J.; Barnett, G. Understanding the potential loss and inequities of green space distribution with urban densification. *Urban For. Urban Green.* **2015**, *14*, 952–958. [CrossRef]
6. Haaland, C.; van den Bosch, C.K. Challenges and strategies for urban green-space planning in cities undergoing densification: A review. *Urban For. Urban Green.* **2015**, *14*, 760–771. [CrossRef]
7. Fernandez-Juricic, E.; Telleria, J.L. Effects of human disturbance on spatial and temporal feeding patterns of Blackbird *Turdus merula* in urban parks in Madrid, Spain. *Bird Study* **2000**, *47*, 13–21. [CrossRef]
8. Reijnen, R.; Foppen, R.; Veenbaas, G. Disturbance by traffic of breeding birds: Evaluation of the effect and considerations in planning and managing road corridors. *Biodivers. Conserv.* **1997**, *581*, 567–581. [CrossRef]
9. Dowding, C.V.; Harris, S.; Poulton, S.; Baker, P.J. Nocturnal ranging behaviour of urban hedgehogs, *Erinaceus europaeus*, in relation to risk and reward. *Anim. Behav.* **2010**, *80*, 13–21. [CrossRef]
10. Werner, P. The ecology of urban areas and their functions for species diversity. *Landsc. Ecol. Eng.* **2011**, *7*, 231–240. [CrossRef]
11. Lowry, H.; Lill, A.; Wong, B.B.M. Behavioural responses of wildlife to urban environments. *Biol. Rev.* **2013**, *88*, 537–549. [CrossRef]
12. Evans, K.L.; Hatchwell, B.J.; Parnell, M.; Gaston, K.J. A conceptual framework for the colonisation of urban areas: The blackbird *Turdus merula* as a case study. *Biol. Rev.* **2010**, *85*, 643–667. [CrossRef]
13. Diamond, S.E.; Chick, L.; Perez, A.B.E.; Strickler, S.A.; Martin, R.A. Rapid evolution of ant thermal tolerance across an urban-rural temperature cline. *Biol. J. Linn. Soc.* **2017**, 1–10. [CrossRef]
14. Cheptou, P.-O.; Carrue, O.; Rouifed, S.; Cantarel, A. Rapid evolution of seed dispersal in an urban environment in the weed *Crepis sancta*. *Proc. Natl. Acad. Sci. USA* **2008**, *105*, 3796–3799. [CrossRef]
15. Reid, N.M.; Proestou, D.A.; Clark, B.W.; Warren, W.C.; Colbourne, J.K.; Shaw, J.R.; Karchner, S.I.; Hahn, M.E.; Nacci, D.; Oleksiak, M.F.; et al. The genomic landscape of rapid repeated evolutionary adaptation to toxic pollution in wild fish. *Science* **2016**, *354*, 1305–1308. [CrossRef] [PubMed]
16. Hendry, A.P.; Farrugia, T.J.; Kinnison, M.T. Human influences on rates of phenotypic change in wild animal populations. *Mol. Ecol.* **2008**, *17*, 20–29. [CrossRef] [PubMed]
17. Ineichen, S.; Ruckstuhl, M. *Stadtfauna—600 Tierarten der Stadt Zürich*; Haupt Verlag: Bern, Switzerland, 2010.
18. Hubert, P.; Julliard, R.; Biagianti, S.; Poulle, M.L. Ecological factors driving the higher hedgehog (*Erinaceus europeaus*) density in an urban area compared to the adjacent rural area. *Landsc. Urban Plan.* **2011**, *103*, 34–43. [CrossRef]
19. Zingg, R. *Aktivität Sowie Habitat-und Raumnutzung von Igeln (Erinaceus europaeus) in Einem Ländlichen Siedlungsgebiet*; Zürich Zentralstelle der Studentenschaft: Zurich, Switzerland, 1994.
20. Pettett, C.E.; Moorhouse, T.P.; Johnson, P.J.; Macdonald, D.W. Factors affecting hedgehog (*Erinaceus europaeus*) attraction to rural villages in arable landscapes. *Eur. J. Wildl. Res.* **2017**, *63*. [CrossRef]
21. Van de Poel, J.L.; Dekker, J.; van Langevelde, F. Dutch hedgehogs *Erinaceus europaeus* are nowadays mainly found in urban areas, possibly due to the negative Effects of badgers *Meles meles*. *Wildlife Biol.* **2015**, *21*, 51–55. [CrossRef]
22. Williams, B.M.; Baker, P.J.; Thomas, E.; Wilson, G.; Judge, J.; Yarnell, R.W. Reduced occupancy of hedgehogs (*Erinaceus europaeus*) in rural England and Wales: The influence of habitat and an asymmetric intra-guild predator. *Sci. Rep.* **2018**, *8*, 17–20. [CrossRef]
23. Rasmussen, S.L.; Berg, T.B.; Dabelsteen, T.; Jones, O.R. The ecology of suburban juvenile European hedgehogs (*Erinaceus europaeus*) in Denmark. *Ecol. Evol.* **2019**, 1–14. [CrossRef]
24. Egli, R.; Bontadina, F.; Deplazes, P.; Arlettaz, R. *Comparison of Physical Condition and Parasite Burdens in Rural, Suburban and Urban Hedgehogs Erinaceus europaeus: Implications for Conservation*; Universität Bern: Bern, Germany, 2004.

25. Doncaster, C.P.; Dickman, C.R.; Macdonald, D.W. Feeding ecology of red foxes (*Vulpes vulpes*) in the city of Oxford, England. *J. Mammal.* **1990**, *71*, 188–194. [CrossRef]

26. Contesse, P.; Hegglin, D.; Gloor, S.; Bontadina, F.; Deplazes, P. The diet of urban foxes (*Vulpes vulpes*) and the availability of anthropogenic food in the city of Zurich, Switzerland. *Mamm. Biol.* **2004**, *69*, 81–95. [CrossRef]

27. Baker, P.J.; Harris, S. Urban mammals: What does the future hold? An analysis of the factors affecting patterns of use of residential gardens in Great Britain. *Mamm. Rev.* **2007**, *37*, 297–315. [CrossRef]

28. Pickett, S.T.A.; Cadenasso, M.L.; Grove, J.M.; Nilon, C.H.; Pouyat, R.V.; Zipperer, W.C.; Costanza, R. Urban ecological systems: Linking terrestrial ecological, physical, and socioeconomic components of metropolitan areas. *Annu. Rev. Ecol. Syst.* **2001**, *32*, 127–157. [CrossRef]

29. Møller, A.P. Urban areas as refuges from predators and flight distance of prey. *Behav. Ecol.* **2012**, *23*, 1030–1035. [CrossRef]

30. Gering, J.C.; Blair, R.B. Predation on artificial bird nests along an urban gradient: Predatory risk or relaxation in urban environments? *Ecography* **1999**, *22*, 532–541. [CrossRef]

31. Huijser, M.P.; Bergers, P.J.M. The effect of roads and traffic on hedgehog (*Erinaceus europaeus*) populations. *Biol. Conserv.* **2000**, *95*, 6–9. [CrossRef]

32. Hof, A.R.; Bright, P.W. Quantifying the long-term decline of the West European hedgehog in England by subsampling citizen-science datasets. *Eur. J. Wildl. Res.* **2016**, *62*, 407–413. [CrossRef]

33. Krange, M. *Change in the occurrence of the West European Hedgehog (Erinaceus europaeus) in Western Sweden during 1950–2010*; Karlstad University: Karlstad, Sweden, 2015.

34. Harris, S.; Morris, P.; Wray, S.; Yalden, D. *A review of British mammals: Population estimates and conservation status of British mammals other than cetaceans*; Joint Nature Conservation Committee: Peterborough, UK, 1995; p. 216.

35. Battersby, J. *UK Mammals: Species Status and Population Trends*; JNCC: Peterborough, UK, 2005.

36. Wembridge, D. The State of Britain's hedgehogs 2011. Available online: https://ptes.org/wp-content/uploads/2015/11/SOBH2011.pdf (accessed on 23 March 2020).

37. Roos, S.; Johnston, A.; Noble, D. UK Hedgehog datasets and their potential for long-term monitoring. *British Trust for Ornithology Research Report 598*. 2012. Available online: https://www.bto.org/our-science/projects/gbw/publications/papers/monitoring/btorr598 (accessed on 15 May 2020).

38. Wilson, E. *Conservation Strategy for West-European Hedgehog* (Erinaceus europaeus*) in the United Kingdom (2015–2025)*; People's Trust for Endangered Species: London, UK, 2018.

39. Mathews, F.; Harrower, C. *Regional Red List of British Mammals*; The Mammal Society: Dorset, UK, 2020.

40. Yarnell, R.W.; Pacheco, M.; Williams, B.; Neumann, J.L.; Rymer, D.J.; Baker, P.J. Using occupancy analysis to validate the use of footprint tunnels as a method for monitoring the hedgehog *Erinaceus europaeus*. *Mamm. Rev.* **2014**, *44*, 234–238. [CrossRef]

41. Williams, B.; Mann, N.; Neumann, J.L.; Yarnell, R.W.; Baker, P.J. A prickly problem: Developing a volunteer-friendly tool for monitoring populations of a terrestrial urban mammal, the Western European hedgehog (*Erinaceus europaeus*). *Urban Ecosyst.* **2018**. [CrossRef]

42. Bontadina, F.; Gloor, S.; Hotz, T. *Igel—Wildtiere in der Stadt*; Gartenbau- und Landwirtschaftsamt: Zurich, Switzerland, 1993.

43. Hof, A.R.; Snellenberg, J.; Bright, P.W. Food or fear? Predation risk mediates edge refuging in an insectivorous mammal. *Anim. Behav.* **2012**, *83*, 1099–1106. [CrossRef]

44. White, G.C.; Burnham, K.P. Program MARK: Survival estimation from populations of marked animals. *Bird Study* **1999**, *46*, S120–S139. [CrossRef]

45. *RStan: The R Interface to Stan*, R package version 2.21.2; Stan Development Team, 2018. Available online: https://mc-stan.org/ (accessed on 13 March 2020).

46. Becker, R.A.; Chambers, J.M.; Wilks, A.R. *The New S Language: A Programming Environment for Data Analysis and Graphics*; Wadsworth & Brooks/Cole Computer Science Series; Chapman & Hall: London, UK, 1988.

47. *QGIS Geographic Information System*, Version 2.18, QGIS Development Team, Open Source Geospatial Foundation Project; 2009. Available online: https://qgis.org/en/site/ (accessed on 28 March 2020).

48. R: A Language and Environment for Statistical Computing. Version 3.6.2, R Core Team; R Foundation for Statistical Computing: Vienna, Austria, 2008.

49. Parrott, D.; Etherington, T.R.; Dendy, J. A geographically extensive survey of hedgehogs (*Erinaceus europaeus*) in England. *Eur. J. Wildl. Res.* **2014**, *60*, 399–403. [CrossRef]

50. Davey, P.A.; Aebischer, N.J. *Participation of the National Gamebag Census in the Mammal Surveillance Network*; JNCC: Peterborough, UK, 2006.

51. Hof, A.R.; Bright, P.W. The value of green-spaces in built-up areas for western hedgehogs. *Lutra* **2009**, *52*, 69–82.

52. Braaker, S.; Kormann, U.; Bontadina, F.; Obrist, M.K. Prediction of genetic connectivity in urban ecosystems by combining detailed movement data, genetic data and multi-path modelling. *Landsc. Urban Plan.* **2017**, *160*, 107–114. [CrossRef]

53. Pettett, C.E.; Johnson, P.J.; Moorhouse, T.P.; Macdonald, D.W. National predictors of hedgehog *Erinaceus europaeus* distribution and decline in Britain. *Mamm. Rev.* **2018**, *48*, 1–6. [CrossRef]

54. Rondinini, A.C.; Doncaster, C.P. Roads as barriers to movement for hedgehogs. *Funct. Ecol.* **2002**, *16*, 504–509. [CrossRef]

55. Braaker, S.; Moretti, M.; Boesch, R.; Ghazoul, J.; Obrist, M.K.; Bontadina, F. Assessing habitat connectivity for ground-dwelling animals in an urban environment. *Ecol. Appl.* **2014**, *24*, 1583–1595. [CrossRef]

56. Hof, A.R.; Bright, P.W. The value of agri-environment schemes for macro-invertebrate feeders: Hedgehogs on arable farms in Britain. *Anim. Conserv.* **2010**, *13*, 467–473. [CrossRef]

57. Hof, A.R. A Study of the current status of the hedgehog (*Erinaceus europaeus*), and its decline in Great Britain since 1960. Ph.D. Thesis, University of London, London, UK, 2009; pp. 15–50.

58. Zürich. *Statistisches Jahrbuch der Stadt Zürich 2017*; Stadt Zürich Präsidialdepartement: Zurich, Switzerland, 2017.

59. Felber, S. *Grünflächenveränderungen in Siedlungsgebieten der Schweiz: Mit Fernerkundungsdaten zum Vergleich der Grünflächenveränderungen auf Gemeindeebene*; Swiss Federal Institute of Technology: Zurich, Switzerland, 2020.

60. Wright, P.G.R.; Coomber, F.G.; Bellamy, C.C.; Perkins, S.E.; Mathews, F. Predicting hedgehog mortality risks on British roads using habitat suitability modelling. *Biodiver. Conserv.* **2019**, *2050*, 1–22. [CrossRef]

61. Geiger, M.; Taucher, A.L.; Gloor, S.; Hegglin, D.; Bontadina, F. In the footsteps of city foxes: Evidence for a rise of urban badger populations in Switzerland. *Hystrix* **2018**. [CrossRef]

62. Wroot, A.J. Feeding Ecology of the European Hedgehog. Ph.D. Thesis, University of London, London, UK, 1984.

63. Hallmann, C.A.; Sorg, M.; Jongejans, E.; Siepel, H.; Hofland, N.; Schwan, H.; Stenmans, W.; Müller, A.; Sumser, H.; Hörren, T.; et al. More than 75 percent decline over 27 years in total flying insect biomass in protected areas. *PLoS ONE* **2017**, *12*. [CrossRef] [PubMed]

64. Krebs, J.R.; Wilson, J.D.; Bradbury, R.B.; Gavin, M. The second silent spring? *Nature* **1999**, *400*. [CrossRef]

65. Gazzard, A.; Baker, P.J. Patterns of feeding by householders affect activity of hedgehogs (*Erinaceus europaeus*) during the hibernation period. *Animals* **2020**, *10*, 1344. [CrossRef] [PubMed]

66. Yalden, D.W. The food of the hedgehog in England. *Acta Theriol.* **1976**, *21*, 401–424. [CrossRef]

67. Dowding, C.V.; Shore, R.F.; Worgan, A.; Baker, P.J.; Harris, S. Accumulation of anticoagulant rodenticides in a non-target insectivore, the European hedgehog (*Erinaceus europaeus*). *Environ. Pollut.* **2010**, *158*, 161–166. [CrossRef]

68. D'Havé, H.; Mubiana, V.K.; Blust, R.; De Coen, W.; Scheirs, J.; Verhagen, R. Non-destructive pollution exposure assessment in the European hedgehog (*Erinaceus europaeus*): II. Hair and spines as indicators of endogenous metal and as concentrations. *Environ. Pollut.* **2006**, *142*, 438–448. [CrossRef]

69. Spurr, E.B.; Wright, G.R.G.; Radford, C.D.; Brown, L.E.; Maitland, M.J.; Taylor, G.E. Residues of brodifacoum and other anticoagulant pesticides in target and non-target species, Nelson Lakes National Park, New Zealand. *N. Z. J. Zool.* **2005**, *32*, 237–249. [CrossRef]

70. Ruiz-Suárez, N.; Boada, L.D.; Henríquez-Hernández, L.A.; González-Moreo, F.; Suárez-Pérez, A.; Camacho, M.; Zumbado, M.; Almeida-González, M.; del Mar Travieso-Aja, M.; Luzardo, O.P. Continued implication of the banned pesticides carbofuran and aldicarb in the poisoning of domestic and wild animals of the Canary Islands (Spain). *Sci. Total Environ.* **2015**, *505*, 1093–1099. [CrossRef]

71. Vermeulen, F.; Covaci, A.; D'Havé, H.; Van den Brink, N.W.; Blust, R.; De Coen, W.; Bervoets, L. Accumulation of background levels of persistent organochlorine and organobromine pollutants through the soil-earthworm-hedgehog food chain. *Environ. Int.* **2010**, *36*, 721–727. [CrossRef]

72. Reeve, N. *Hedgehogs (Poyser Natural History)*; Poyser: London, UK, 1994; ISBN 0-05661-081-X.

73. Rautio, A.; Kunnasranta, M.; Valtonen, A.; Ikonen, M.; Hyvärinen, H.; Holopainen, I.J.; Kukkonen, J.V.K. Sex, age, and tissue specific accumulation of eight metals, arsenic, and selenium in the European Hedgehog (*Erinaceus europaeus*). *Arch. Environ. Contam. Toxicol.* **2010**, *59*, 642–651. [CrossRef] [PubMed]

74. BLW (Bundesamt für Landwirtschaft). Verkaufsstatistik 2018 von Pflanzenschutzmitteln in der Schweiz. 2019. Available online: https://www.admin.ch/gov/de/start/dokumentation/medienmitteilungen.msg-id-77550.html (accessed on 20 April 2020).

75. Meftaul, I.M.; Venkateswarlu, K.; Dharmarajan, R.; Annamalai, P.; Megharaj, M. Pesticides in the urban environment: A potential threat that knocks at the door. *Sci. Total Environ.* **2020**, *711*. [CrossRef] [PubMed]

76. Reeve, N.J.; Huijser, M.P. Mortality factors affecting wild hedgehogs: A study of records from wildlife rescue centres. *Lutra* **1999**, *42*, 7–24.

77. Doncaster, C.P. Testing the role of intraguild predation in regulating hedgehog populations. *Proc. R. Soc. B* **1992**, *249*, 113–117. [PubMed]

78. Judge, J.; Wilson, G.J.; Macarthur, R.; Delahay, R.J.; McDonald, R.A. Density and abundance of badger social groups in England and Wales in 2011–2013. *Sci. Rep.* **2014**, *4*, 1–8. [CrossRef] [PubMed]

79. Young, R.P.; Davison, J.; Trewby, I.D.; Wilson, G.J.; Delahay, R.J.; Doncaster, C.P. Abundance of hedgehogs (*Erinaceus europaeus*) in relation to the density and distribution of badgers (*Meles meles*). *J. Zool.* **2006**, *269*, 349–356. [CrossRef]

80. Laundré, J.W.; Hernández, L.; Ripple, W.J. The Landscape of Fear: Ecological Implications of Being Afraid. *Open Ecol. J.* **2010**, *3*, 1–7. [CrossRef]

81. Micol, T.; Doncaster, C.; Mackinlay, L. Correlates of local variation in the abundance of hedgehogs *Erinaceus europaeus*. *J. Anim. Ecol.* **1994**, *63*, 851–860. [CrossRef]

82. Doncaster, C.P. Factors regulating local variations in abundance: Field tests on hedgehogs, *Erinaceus europaeus*. *Oikos* **1994**, *69*, 182–192. [CrossRef]

83. Trewby, I.D.; Young, R.; McDonald, R.A.; Wilson, G.J.; Davison, J.; Walker, N.; Robertson, A.; Patrick Doncaster, C.; Delahay, R.J. Impacts of removing badgers on localised counts of hedgehogs. *PLoS ONE* **2014**, *9*, 2–5. [CrossRef]

84. Gulland, F.M.D. The role of nematode parasites in Soay sheep (*Ovis aries* L.) mortality during a population crash. *Parasitology* **1992**, *105*, 493–503. [CrossRef] [PubMed]

85. Gaglio, G.; Allen, S.; Bowden, L.; Bryant, M.; Morgan, E.R. Parasites of European hedgehogs (*Erinaceus europaeus*) in Britain: Epidemiological study and coprological test evaluation. *Eur. J. Wildl. Res.* **2010**, *56*, 839–844. [CrossRef]

86. Daniel, M.; Danielová, V.; Kříž, B.; Jirsa, A.; Nožička, J. Shift of the tick *Ixodes ricinus* and tick-borne encephalitis to higher altitudes in Central Europe. *Eur. J. Clin. Microbiol. Infect. Dis.* **2003**, *22*, 327–328. [CrossRef] [PubMed]

87. Scharlemann, J.P.W.; Johnson, P.J.; Smith, A.A.; Macdonald, D.W.; Randolph, S.E. Trends in ixodid tick abundance and distribution in Great Britain. *Med. Vet. Entomol.* **2008**, *22*, 238–247. [CrossRef] [PubMed]

88. Gillis-Germitsch, N.; Tritten, L.; Hegglin, D.; Deplazes, P.; Schnyder, M. Conquering Switzerland: Emergence of *Angiostrongylus vasorum* over three decades and rapid regional increase in the fox population contrasts with the stable prevalence of lungworms. *Parasitology* **2020**, 1–28. [CrossRef]

89. Morgan, E.R.; Tomlinson, A.; Hunter, S.; Nichols, T.; Roberts, E.; Fox, M.T.; Taylor, M.A. *Angiostrongylus vasorum* and *Eucoleus aerophilus* in foxes (*Vulpes vulpes*) in Great Britain. *Vet. Parasitol.* **2008**, *154*, 48–57. [CrossRef]

90. Gloor, S.; Bontadina, F.; Hegglin, D.; Deplazes, P.; Breitenmoser, U. The rise of urban fox populations in Switzerland. *Mamm. Biol. Zeitschrift für Säugetierkd.* **2001**, *66*, 155–164.

91. Deplazes, P.; Hegglin, D.; Gloor, S.; Romig, T. Wilderness in the city: The urbanization of *Echinococcus multilocularis*. *Trends Parasitol.* **2004**, *20*, 77–84. [CrossRef]

92. Pfäffle, M.; Černábolfíková, B.; Hulva, P.; Petney, T. Different parasite faunas in sympatric populations of sister hedgehog species in a secondary contact zone. *PLoS ONE* **2014**, *9*, 1–14. [CrossRef]

93. Lund, S.; Id, R.; Id, J.L.; Wijk, R.E., Van; Jones, O.R.; Bj, T.; Angen, Ø.; Larsen, A.R. European hedgehogs (*Erinaceus europaeus*) as a natural reservoir of methicillin-resistant *Staphylococcus aureus* carrying mecC in Denmark. *PLoS ONE* **2019**, *114*, 1–13.

94. Monecke, S.; Gavier-Widen, D.; Mattsson, R.; Rangstrup-Christensen, L.; Lazaris, A.; Coleman, D.C.; Shore, A.C.; Ehricht, R. Detection of mecC-Positive *Staphylococcus aureus* (CC130-MRSA-XI) in diseased European hedgehogs (*Erinaceus europaeus*) in Sweden. *PLoS ONE* **2013**, *8*, 4–9. [CrossRef] [PubMed]

95. Bengtsson, B.; Persson, L.; Ekström, K.; Unnerstad, H.E.; Uhlhorn, H.; Börjesson, S. High occurrence of mecC-MRSA in wild hedgehogs (*Erinaceus europaeus*) in Sweden. *Vet. Microbiol.* **2017**, *207*, 103–107. [CrossRef] [PubMed]

96. Tilman, D.; Mayt, R.M.; Lehman, C.L.; Nowakt, M.A. Habitat destruction and the extinction debt. *Nature* **1994**, *371*, 65–66. [CrossRef]

97. Keller, L.F.; Waller, D.M. Inbreeding effects in the wild. *Trends Ecol. Evol.* **2002**, *17*, 230–241. [CrossRef]

98. Hahs, A.K.; McDonnell, M.J.; McCarthy, M.A.; Vesk, P.A.; Corlett, R.T.; Norton, B.A.; Clemants, S.E.; Duncan, R.P.; Thompson, K.; Schwartz, M.W.; et al. A global synthesis of plant extinction rates in urban areas. *Ecol. Lett.* **2009**, *12*, 1165–1173. [CrossRef]

99. Lambert, M.R.; Donihue, C.M. Urban biodiversity management using evolutionary tools. *Nat. Ecol. Evol.* **2020**. [CrossRef]

 *animals*

Article

Investigating the Role of the Eurasian Badger (*Meles meles*) in the Nationwide Distribution of the Western European Hedgehog (*Erinaceus europaeus*) in England

Anouschka R. Hof [1,2,*], **Andrew M. Allen** [3] **and Paul W. Bright** [4]

[1] Resource Ecology Group, Wageningen University, 6708 PB Wageningen, The Netherlands
[2] Department of Wildlife, Fish, and Environmental Studies, Swedish University of Agricultural Sciences, SE-907 36 Umeå, Sweden
[3] Department of Animal Ecology & Physiology, Radboud University, 6500 GL Nijmegen, The Netherlands; Andrew.Allen@science.ru.nl
[4] People's Trust for Endangered Species, 3 Cloisters House, 8 Battersea Park Road, London SW8 4BG, UK; drp.w.bright@gmail.com
* Correspondence: Anouschka.Hof@wur.nl

Received: 30 July 2019; Accepted: 24 September 2019; Published: 2 October 2019

Simple Summary: The hedgehog is a species known to many in society. What is perhaps less known, is that the hedgehog has been declining across large parts of Europe, including the United Kingdom. Effective hedgehog conservation requires a sound understanding of the causes of the decline. A potential cause is the badger, whose population has been recovering in recent years. The badger is an intraguild predator of the hedgehog, meaning that not only do the two species share the same food, like snails and earthworms, but badgers also predate on hedgehogs. Our study investigates how the presence of hedgehogs is related to the presence of badgers, along with other landscape features. Using information from two nationwide citizen science surveys, we first determine where both species can be found and then identify which factors best explain hedgehog presence. We found that the badger was indeed important, and hedgehogs were less likely to be found in areas where badgers were likely to be found. Interestingly, hedgehogs were also likely to be found in arable land, a habitat not directly thought to be favourable for hedgehogs. Badgers may, therefore, be an important consideration when designing hedgehog conservation plans, and further research of these impacts is needed.

Abstract: Biodiversity is declining globally, which calls for effective conservation measures. It is, therefore, important to investigate the drivers behind species presence at large spatial scales. The Western European hedgehog (*Erinaceus europaeus*) is one of the species facing declines in parts of its range. Yet, drivers of Western European hedgehog distribution at large spatial scales remain largely unknown. At local scales, the Eurasian badger (*Meles meles*), an intraguild predator of the Western European hedgehog, can affect both the abundance and the distribution of the latter. However, the Western European hedgehog and the Eurasian badger have shown to be able to co-exist at a landscape scale. We investigated whether the Eurasian badger may play a role in the likelihood of the presence of the Western European hedgehog throughout England by using two nationwide citizen science surveys. Although habitat-related factors explained more variation in the likelihood of Western European hedgehog presence, our results suggest that Eurasian badger presence negatively impacts the likelihood of Western European hedgehog presence. Intraguild predation may, therefore, be influencing the nationwide distribution of hedgehogs in England, and further research is needed about how changes in badger densities and intensifying agricultural practices that remove shelters like hedgerows may influence hedgehog presence.

Keywords: citizen science; conservation; displacement; predator-prey interaction; spatial use

1. Introduction

In a time of ongoing anthropogenic pressures on nature, a growing number of species throughout the world are facing population declines [1]. One of these species is the Western European hedgehog (*Erinaceus europaeus*). Albeit being classified as least concern on the International Union for Conservation of Nature (IUCN) red list of threatened species, hedgehog numbers appear to have fallen in several countries in Europe in the last couple of decades, such as in Belgium and the Netherlands [2,3], Sweden [4], and in the United Kingdom [5–10]. Reasons behind this decline are, however, currently unclear and several potential causes have been suggested. The network of roads throughout Europe has been increasing extensively over the past few decades [11], which may play a large role since hedgehogs often fall victim to traffic [2,3,12] and large roads may act as barriers [13]. Increasing demands for housing development may coincide with the loss of greenspaces in built-up areas that offer (sub)urban-dwelling species like hedgehog habitat and refuge [14,15]. Agricultural intensification may lead to decreasing habitat suitability and also reduce resource availability of macro-invertebrates [16–18]. Furthermore, agricultural intensification may lead to reduced landscape complexity, for example, the removal of hedgerows and coppices [19], which provide hedgehogs with shelter from predators and nesting habitats [14,20]. In addition, Eurasian badgers (*Meles meles*) may influence hedgehog populations at local scales [14,21], but largescale studies investigating the impact of such factors are currently lacking.

The role of the Eurasian badger as a potential driver for the nationwide decline of hedgehogs requires further research as it has been suggested as a potential driving factor of declining hedgehog populations following increased predation pressure [14]. Hedgehogs and badgers are in a guild of generalist predators of macro-invertebrate prey [20,22], but badgers may also be an intraguild predator of hedgehogs. Although many studies have not reported hedgehog remains in badger diet analyses [23–25], and incidences of badgers preying on hedgehogs are not thought to be common, there are several studies throughout Europe that do report hedgehog remains in the faeces or stomachs of badgers. Hedgehog occurrence in the diet of badgers varied from as much as four hedgehog remains in the stomach of one single adult badger found in England [26], to an 11.2% occurrence in badger scats in one of three study sites in Poland [27], 2.9% occurrence in badgers scats in Italy [28], and an unknown percentage of occurrence in an extensive review of dietary studies in the former Soviet Union [29].

Predation, both intraguild and interguild, is known to be important in shaping the local dynamics of predator–prey communities [30,31]. In fact, several fine-scale studies show that badgers can have large negative effects on local hedgehog populations [32–35]. Experimental evidence of the Randomised Badger Culling Trial in 100 km^2 large trial areas in England, set up to assess impacts of culling on the incidence of bovine tuberculosis, provided evidence that hedgehog numbers more than doubled over a five year period in areas with preferred habitat [36]. Furthermore, badgers appear to drive small scale movement patterns of hedgehogs [32,37,38]. Doncaster [21] showed, using an experimental setup in which a low-density hedgehog population was artificially increased, and a high-density population was artificially decreased, that predation by badgers can affect both hedgehog abundance and their distribution at a local scale. However, both species have shown to be able to coexist on a landscape scale in the recent past [34]. Yet, the number of badgers has been increasing in the past couple of decades; Judge et al. [39] estimated that there had been an 88% increase in badger numbers across England and Wales from the 1985–1988 to the 2011–2013 survey periods. The balance of co-existence between badgers and hedgehogs may, therefore, have been tipped. A study by Williams et al. [9], which investigated correlates of hedgehog presence in rural England and Wales using footprint tracking tunnels at 261 sites, found a strong negative relationship between hedgehog occupancy and badger sett density, but simultaneously that hedgehogs were absent from 71% of surveyed sites that had no badger setts. Consequently, factors driving the distribution of hedgehogs at large spatial scales, i.e., throughout the United Kingdom, remain uncertain. There is no indication that other potential predators of hedgehogs in the United Kingdom, such as the red fox (*Vulpes vulpes*), can regulate hedgehog presence [15,21,34].

As the abundance and distribution of species has a propensity to be linked, where widespread species tend to be more abundant, a thorough understanding of the drivers behind large scale, e.g., nationwide scale, distributions of species is highly valuable in complementing knowledge obtained from small scale studies and designing and implementing conservation measures at large scales [40]. The objective of our study was, therefore, to investigate the factors driving the distribution of hedgehogs on a large scale—the whole of England—and to investigate whether the presence of badgers and other landscape features explain some variation in the nationwide distribution of hedgehogs. We used two nationwide citizen science surveys and land cover data to investigate the impact of badgers and landscape features on the distribution of hedgehogs, which may provide valuable knowledge for the nationwide conservation of hedgehogs.

2. Materials and Methods

The current distribution of hedgehogs in England was obtained by a nationwide citizen science survey called 'HogWatch'. The survey was mainly designed by the British Hedgehog Preservation Society (BHPS, https://www.britishhedgehogs.org.uk/) and the People's Trust for Endangered Species (PTES, https://ptes.org/) in conjunction with Royal Holloway, University of London. It was both post and web based. Publicity was sought by means of (local) media, personal communication, and using existing member databases of the BHPS and the PTES. The survey, 'HogWatch' is no longer active but PTES collect annual hedgehog records through surveys like Mammals on Roads (https://ptes.org/get-involved/surveys/road/) and 'Living with Mammals' (https://ptes.org/get-involved/surveys/garden/living-with-mammals). PTES and BHPS also collect records of hedgehogs through their Big Hedgehog Map (https://bighedgehogmap.org/). A total of 19,184 people provided 25,911 presence and absence reports of hedgehogs that were distributed throughout England. The hedgehog distribution was based on 25,911 grid-referenced sightings and lack of sightings of living hedgehogs from 2005 (when the survey started) and 2006. Data on badger presence in England were derived from another survey called 'Living with Mammals' (https://ptes.org/get-involved/surveys/garden/living-with-mammals), which is open to every interested member of the public and was originally designed by the Mammals Trust UK in conjunction with Royal Holloway, University of London. Surveyors recorded badger presence throughout 13 consecutive weeks from the start of April, and they stated the approximate observation length during dawn, daytime, dusk, and night-time. The badger distribution was based on 2,703 sightings of living badgers recorded in 247 grid-referenced sites and the lack of sightings of living badgers recorded in 1464 sites throughout England in the years 2003–2006. We included badger data from 2003 (the first year the survey was held) and 2004 in our analyses as well since it allowed us to include a greater sample size and have a smaller discrepancy with the substantially larger 'HogWatch' dataset. In addition, we felt that the distribution of badgers would not rapidly change to a large extent from the years 2003–2004 to 2005–2006. Although respondents were asked to state the date, time, and approximate length of the observation, to get an estimate of effort, this effort was not considered, as such data were not collected for the 'HogWatch' dataset. We assumed that the relative density of hedgehogs/badgers is proportional to the actual density and that the rate of proportionality is constant [41].

As the surveys used to obtain data about hedgehog and badger presence/absence did not overlap with each other at a fine scale, since they were not especially designed for this study, we used ordinary kriging [42] to estimate the likelihood of the presence of hedgehogs and badgers throughout England at a 10 km^2 scale using ArcMap 10.5 (Environmental Systems Research Institute [ESRI], Redlands, CA, USA). We chose a 10 km^2 scale to obtain a reasonable number of species observations per cell, while still retaining some level of detail. Kriging is a geostatistical interpolation method that is based on linear regressions, and produces maps from irregular spatial data to visualize suggested trends and spatial differences in the likelihood of presence. The method is frequently used in spatial prediction applications in ecology [42–45] to, amongst others, predict species numbers in regions where data are not available [46–49]. Interpretations based on results derived from kriging must, however, be made with care. The proportion of respondents to the surveys that reported seeing hedgehogs/badgers per 10 km^2 were used as input for the kriging (Figures 1 and 2). To account for spatial autocorrelation, a semivariogram, which is a function describing the degree of spatial dependence of the data and characterizes the spatial continuity or roughness of data sets, was integrated in the kriging method, as is recommended [50,51]. An exponential model was used with 8 lags for the hedgehog (Figure 3) and a Gaussian model with 5 lags for the badger (Figure 4), and nuggets were enabled. The type of model and the amount of lags were chosen based on the smallest standard errors showing the uncertainty related to the predicted values. Maps, showing the likelihood of the presence of badgers and hedgehogs in each 10 km^2 grid-cell in England were produced in ArcMap10.5 (Environmental Systems Research Institute [ESRI], Redlands, CA, USA). The methodology allowed us to obtain maps depicting differences in the likelihood of the presence of the hedgehog and of the badger at a 10 km^2 scale for the entirety of England. Since the red fox does not appear to be able to regulate hedgehog presence [15,21,34], we did not include the likelihood of the presence of red foxes in our study.

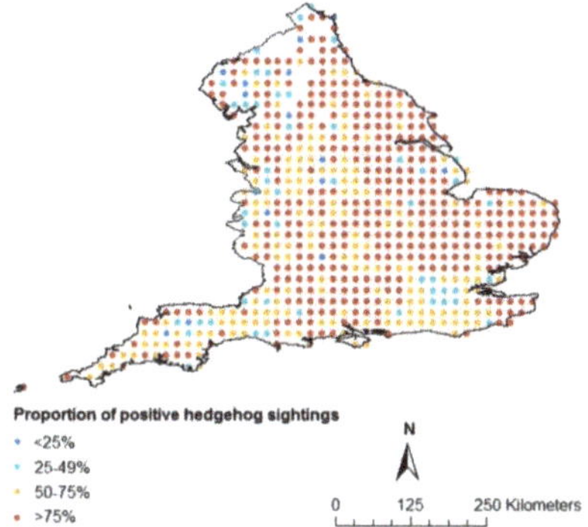

Figure 1. Proportion of positive hedgehog sightings according to the 'HogWatch' survey of 2005–2006.

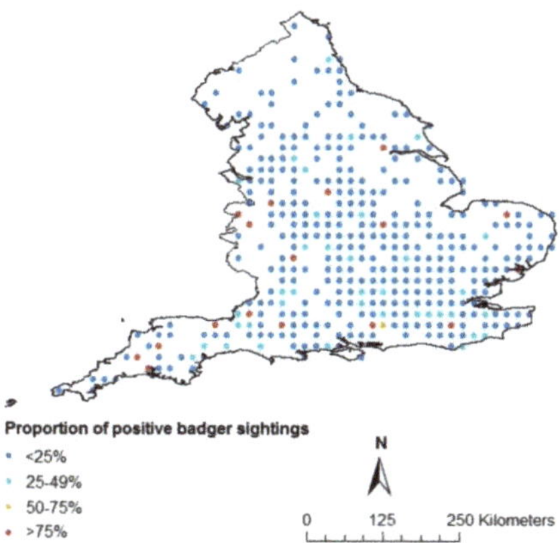

Figure 2. Proportion of positive badger sightings according to the 'Living with Mammals' survey of 2003–2006.

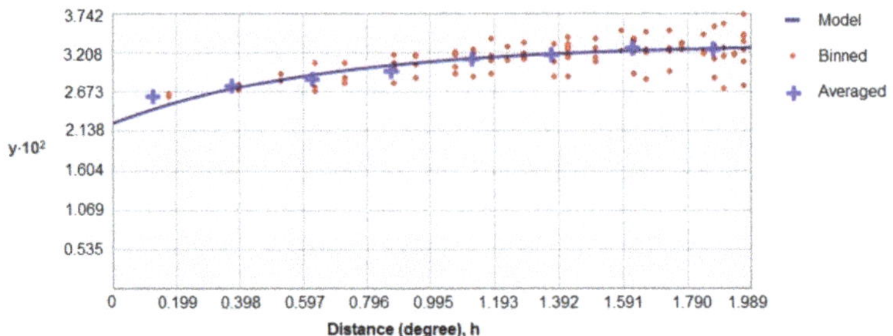

Figure 3. The exponential semivariogram model with 8 lags used for kriging the hedgehog data.

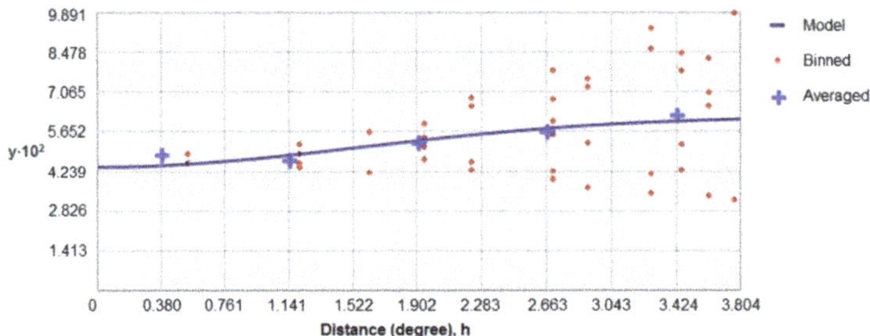

Figure 4. The Gaussian semivariogram model with 5 lags used for kriging the badger data.

Landscape features that may also be important in shaping the distribution of hedgehogs throughout England were obtained from various sources. In addition to the likelihood of the presence of badgers as described above, these variables included habitat-related variables and the density of built-up areas, which were obtained from the Land Cover Map 2000 (https://www.ceh.ac.uk/services/land-cover-map-2000) and the Land Cover Map 2007 (https://www.ceh.ac.uk/services/land-cover-map-2007) from the Centre for Ecology & Hydrology (https://www.ceh.ac.uk/), data on the human footprint obtained from the Socioeconomic Data and Applications Center (SEDAC, https://sedac.ciesin.columbia.edu/data/set/wildareas-v2-human-footprint-geographic) [52], and soil-related variables, which were obtained from the National Soil Resource Institute (NSRI, https://www.cranfield.ac.uk/centres/soil-and-agrifood-institute/research-groups/national-soil-resources-institute). The land cover data from the land cover maps are classifications of spectral data recorded by satellites and refined using external datasets. For more information about the methodology used to create the land cover maps, please refer to the final reports of the surveys [53,54]. Data were unfortunately not available for the exact timeframe when the hedgehog and badger data were collected (2003–2006), but only for 2000 and 2007. However, changes in landscape features between both timeframes were expected to be minor. We considered both the target (more detailed classification) and aggregate (less detailed classification) classes. The data were available at a scale of 1 km^2 and were converted to 10 km^2 by taking the mean and median of the values (if continuous) for each 10 km^2. In the case of soil data, presence (1) or absence (0) of a soil type was used. All variables used are shown in Table 1.

Table 1. Explanation of the response variables used (LCM: Land Cover Map 2000 and 2007, NSRI: National Soil Resource Institute, LWM: Living with Mammals survey, SEDAC: Socioeconomic Data and Applications Center [see text for explanation]) to study their impact on the relative hedgehog abundance throughout England in 2005–2006. Some land classes from the LCM were not included because they were extremely rare (absent from >90% of grid cells) or were not suitable, such as "Mountain, heath, bog", "Saltwater", "Freshwater", and "Coastal".

Variable	Explanation	Source
Arable land	Proportion of arable and horticultural area	LCM
Badger presence	Index of relative badger abundance 2003–2006	LWM
Broadleaf woodland	Proportion of broadleaved woodland	LCM
Built-up	Proportion of built-up area (includes target classes of urban and sub-urban)	LCM
Coniferous woodland	Proportion of coniferous woodland	LCM
Human footprint	Human Influence Index normalized by biome and realm	SEDAC
Improved grassland	Proportion of improved grassland	LCM
Semi-natural grassland	Proportion of semi-natural grassland (includes target classes of rough, neutral, calcareous, acid grassland, and fen, marsh and swamp)	LCM
Soil type	The soil types of England 1: soils with a clay texture, 2: soils with a peaty texture, 3: soils with a sandy texture, 4: soils with a loamy texture and rich in lime, 5: soils with a loamy texture and a low fertility, 6: soils with a loamy texture and a moderate to high fertility	NSRI

We used generalised linear modelling (GLM) to determine how the likelihood of hedgehog presence (non-kriged values, i.e., the proportion of respondents reporting a hedgehog sighting) was related to the (kriged) likelihood of badger presence and landscape features (Table 1). All explanatory variables were simultaneously tested for correlation and visualised using a correlation matrix. Any highly correlated ($r > 0.7$, [55]) variables were either excluded from the model or were not included simultaneously in the same model. We also checked the collinearity of explanatory variables in the final models using variation inflation factors (VIFs), where a VIF > 4 indicates collinearity and a VIF > 10 indicates severe collinearity [56]. An initial model was built that included all explanatory variables, and the most parsimonious model was obtained by stepwise backward selection, removing the least significant variable in each step. The Akaike Information Criterion (AIC) was compared to also determine whether the model-fit significantly changed (i.e., ΔAIC > 2; [57]). The final model would thus contain significant variables only.

We also considered a multi-model selection approach that compares all possible model combinations. Due to the large number of potential explanatory variables, we aimed to identify the most parsimonious model to avoid model overfitting. Models were ranked according to the Bayesian Information Criterion (BIC), which has a larger penalty for additional parameters compared with AIC and AICc [57]. Models with ΔBIC < 2 have substantial support, whilst models with a ΔBIC > 4 have considerably less support [57]. Therefore, we considered top models to be those within ΔBIC < 2. However, when considering variables that may be important for explaining the likelihood of hedgehog presence, we considered all variables included in models with ΔBIC < 4. The analysis was performed using the dredge function in the MuMIn package in R 3.5.1 [58,59]. We used hierarchical partitioning, which calculates the explained variation (R^2) for all combinations of the supplied variables in a regression hierarchy, to identify the individual contribution of each variable to the explained variation [60,61].

3. Results

We found that the likelihood of badger presence was highest in the southern and western parts of England (Figure 5). Simultaneously, there was a division in the likelihood of the presence of hedgehogs between the eastern and the southern and western parts of England, with a higher probability of hedgehog presence in eastern England.

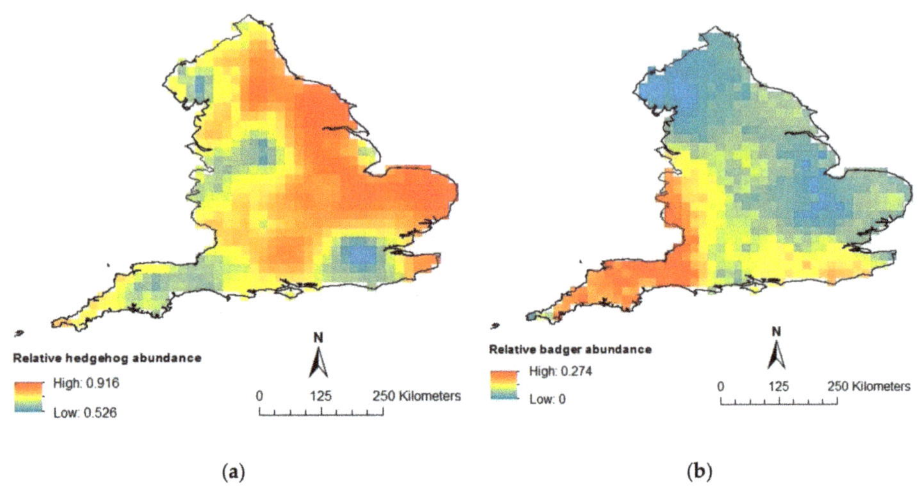

(a)　　　　　　　　　　　　　　(b)

Figure 5. Maps showing an index (low: 0, high: 1) of the likelihood of the presence of (**a**) hedgehogs and (**b**) badgers throughout England.

We present our results using median values for the land cover maps of 2007, but results were comparable when using either median or mean for both 2000 and 2007 land cover maps (Appendix A, Tables S1–S3). Both stepwise backwards selection and the multi-model selection approach included the same variables in the top model, meaning these variables explained most of the variation in hedgehog presence and that all variables were significant (Tables 2 and 3). Arable land had a positive relationship with hedgehog presence, whilst broadleaved woodland, improved grassland, built-up areas, and badger presence were negatively associated with hedgehog presence (Table 2). The multi-model selection confirmed the importance of badgers in explaining hedgehog presence, which was included in the top two models (Table 3). Human footprint and peaty soils were also amongst the top models with a ΔBIC < 4. Hierarchical partitioning of the seven variables included within models of ΔBIC < 4 (Table 3) indicated that arable land, built-up area, and improved grasslands explained the most variation, each with an independent contribution of more than 20% towards the explained variation (Total R^2 = 0.258; Figure 6). Amongst the other variables, broadleaved woodlands were the most important (13.4%) followed by badgers (7.8%), human footprint (4.9%), and peaty soils (1.6%; Figure 6).

Table 2. Model results of generalised linear modelling (GLM) explaining the likelihood of the presence of hedgehogs. Significant variables were determined using backwards stepwise selection. SE = standard error and p = p-value. The independent contribution of each variable towards the explained variation (R^2; total = 0.242) was measured using hierarchical partitioning and VIF is the variance inflation factor.

Variable	Coefficient	SE	p	R^2	VIF
Intercept	0.834	0.029	<0.001	-	-
Badger presence	−0.078	0.030	0.010	0.021	1.047
Arable land	0.090	0.036	0.012	0.073	1.671
Built-up	−0.238	0.056	<0.001	0.045	1.367
Improved grassland	−0.245	0.053	<0.001	0.068	1.422
Broadleaved woodland	−0.419	0.141	0.003	0.035	1.151

Table 3. Top-performing models with ΔBIC < 4 from a multi-model selection consisting of all possible explanatory variables. Shaded areas indicate that the variable was included in the model. All model variables had VIFs < 2. LogL is the log-likelihood, Arable is Arable Horticulture, Badgers is the likelihood of badger presence, BLwood is broadleaved woodland, ImprG is improved grasslands, HFI is human footprint index, and Peat is soils with a peaty texture.

ΔBIC	LogL	Arable	Badgers	BLwood	Built-Up	ImprGr	HFI	Peat
0.00 [1]	223.05	■	■	■	■	■		
0.46	219.85		■	■	■	■		
0.76	219.70	■		■	■	■		
0.87	216.68	■		■	■	■	■	
1.20	225.42	■	■	■	■	■	■	
1.81	222.15	■		■	■	■	■	
2.46	224.79	■	■		■	■		■
2.96	218.60	■		■	■	■		
2.99	221.56	■		■	■	■		■
3.22	221.44		■	■	■	■		■
3.40	218.39			■	■	■		■
3.84	221.13		■	■	■	■	■	■
3.98	227.00	■	■	■	■	■	■	■

[1] BIC = −404.54.

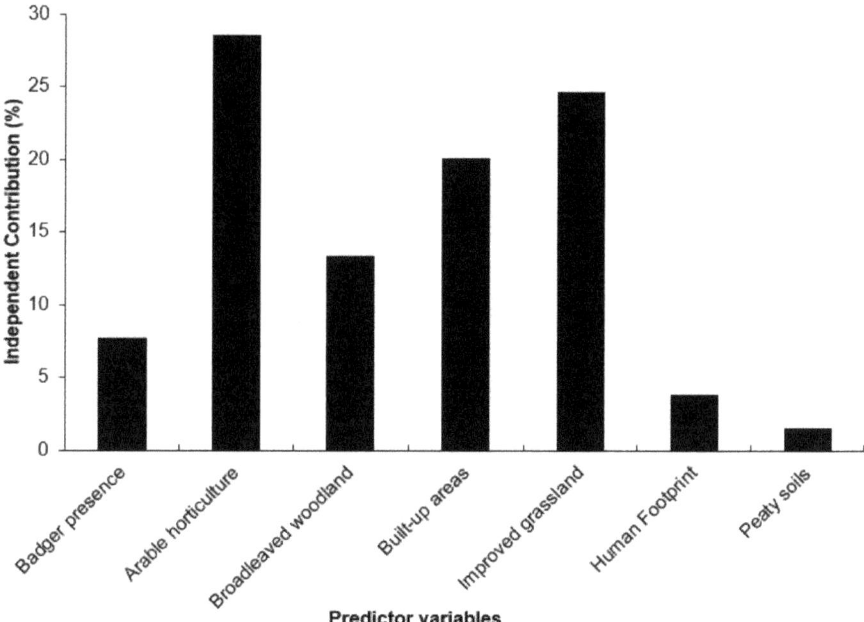

Figure 6. Results of hierarchical partitioning showing the individual contribution of each variable towards the total explained variation (R^2) of the model ($R^2 = 0.258$).

4. Discussion

Our results show that the nationwide distribution of hedgehogs was negatively related to that of its intraguild-predator, the badger. Arable land and the density of improved grassland were the strongest predictors of the likelihood of the presence of hedgehogs throughout England, followed by the amount of built-up area. The likelihood of hedgehog presence was negatively correlated with the density of improved grassland and the amount of built-up area but positively correlated with arable land. The likelihood of hedgehog presence was, thus, higher in the arable dominated landscapes of eastern England than in other parts of England, which agrees with previous findings [15,62]. Although badger presence was not the most important variable for explaining nationwide hedgehog presence, badger presence was, nonetheless, amongst the most important variables. When one considers that the likelihood of hedgehog presence was higher in areas that are supposedly less accommodating to their needs, i.e., areas dominated by arable land [20,63], it begs the question whether the lack of badger presence may lead to proportionally higher hedgehog numbers in less accommodating areas. Badgers were more likely to be present in the south-west, which is largely in agreement with previous findings [38,64] and correlates with the lower likelihood of the presence of hedgehogs in these regions. One of the reasons that badgers are likely to be present at higher densities in the south than in the east of England might be partly due to the higher density of broadleaved woodland and hillier and undulating habitat, environmental features favoured by badgers [65,66].

The negative relationship between hedgehog presence and the density of improved grassland and built-up areas was expected based on previous studies [20,63,67], but the positive relationship with arable land was unexpected given that radio-tracking studies indicate that hedgehogs often avoid these habitats [20,63]. The positive relationship with arable land was also contrary to general expectations based upon studies that investigated habitat preferences of hedgehogs at home range and landscape scales [20,68–70]. Which aspects of arable land were particularly related to hedgehog presence remains unclear due to the limited detail of the land cover data. Small scale, less intensively farmed areas with, for example, hedgerows, may be more attractive to hedgehogs than large scale intensively farmed areas with single crops. Unfortunately, detailed data about land-use intensity is lacking, and although we included human footprint, this also had a positive relationship with hedgehog presence, which is likely due to hedgehog's association with agricultural landscapes. Furthermore, detailed data about the presence of hedgerows throughout the country were not available and may further clarify the trends we observed. Further research at finer spatial studies may clarify which factors drive the positive relationship between arable land and hedgehog presence. A study by Hof et al. [37], however, indicates that the small-scale utilization of arable land can especially depend on the presence of badgers.

It is not known whether the geographical distribution of the hedgehog has changed in the recent past, or whether the likelihood of the presence of the hedgehog has been higher in the arable dominated areas in England for a long time already. However, whilst the numbers of hedgehogs seem to be falling in spite of protection [5–10], the number of badgers has been steadily increasing in recent decades following their protection [39,71]. This might have tipped the balance of co-existence between badgers and hedgehogs in favour of badgers and initiated a change in the likelihood of the presence of the hedgehog in England. That this may indeed be so is corroborated by research from Micol et al. [34], who predicted that apart from some isolated pockets, hedgehogs will be absent from most sites in the United Kingdom with badger sett densities above 0.23 per km^2. A recent study by Judge et al. [39] actually estimated that current badger sett densities were 0.49 per km^2 in England between 2011 and 2013, thus substantially larger than the 0.23 limit suggested by Micol et al. [34]. However, it cannot be ruled out that differences in survey protocols (partly) explain this difference in sett density as well. Furthermore, badger numbers are poorly predicted by sett characteristics [72].

Another line of explanation of the geographical distribution pattern of hedgehogs in England is offered by the hypothesis that the presence of an intraguild-prey species is partly dependent on the availability of alternative food sources for the intraguild-predator [73]. Generally, most studies have found that vertebrate prey, and especially mammalian prey, make up a small component of the badger's diet, which is generally dominated by cereals, fruits, and invertebrates [23,74,75]. Nonetheless, prey composition varies spatially, and our results may, therefore, suggest that western England currently offers a low abundance of alternative prey sources (e.g., macro-invertebrates). Low availability of prey sources would, thus, not only increase competition between badgers and hedgehogs but may also increase predation rates of the badger on hedgehogs. The low abundance of prey sources may possibly be an effect of intensified farming [76]. Detailed data on the England wide abundance of macro-invertebrates were, however, not available to include in this study. Yet, overwhelming evidence indicates that invertebrates are declining in agricultural landscapes across the globe [77,78], it can thus not be ruled out that declining invertebrate numbers did not only play a role in declining hedgehog numbers, but also in shaping their distribution. Other factors that may need investigation include possible impacts of disease, pesticides, and of potential predators other than the badger, such as feral/stray dogs and foxes. There are a large range of diseases and parasites that can negatively affect hedgehogs [20], but it thus far remains uncertain if they play a significant role in population declines [79]. We found no published evidence of disease transmission between badgers and hedgehogs. Disease transmission between the two species may be low, considering the fact that a prevalent disease in badgers, bovine tuberculosis, was not prevalent in wild hedgehogs in the United Kingdom [80]. It is known that pesticides can accumulate in hedgehogs [81]; the extent of negative effects and impacts on population declines, however, remain uncertain. There is currently no evidence that feral/stray dogs and foxes are able to regulate hedgehogs [15,21,34], they are, however, able to occasionally kill or inflict injury upon hedgehogs [20]. The potential impacts of feral/stray dogs and foxes are therefore expected to be more pronounced in small geographic scales rather than at large geographic scales.

Our study relies upon hedgehog sightings reported by citizen scientists through an initiative called 'HogWatch'. A benefit of using these is that we attained nationwide coverage with responses received from almost every 10 km^2 grid cell in England. Furthermore, nearly 26,000 responses were received from approximately 20,000 respondents. This attains a much larger spatial scale compared to field-based studies, such as the one recently conducted by Williams et al. [9]. However, citizen science survey data also have associated challenges that should be taken into account. For example, we made the assumption that a higher proportion of reports on hedgehog/badger presence per 10 km^2 grid-cell was positively related to the likelihood of presence. This assumption needs to be made with some care because of recordings of possible false absences and dissimilarity in visibility caused by differences in human habitation and environmental features. Furthermore, both species are nocturnal with higher chances of activity at night when people are less likely to see them. However, the 'HogWatch' survey was specifically designed for hedgehogs, and the respondents to the 'Living with Mammals' survey also surveyed during nightly hours. Considering the large amount of data collected and the spread of data throughout England in both surveys, it was assumed that the nocturnal activity pattern of both species had little effect on the spatial distribution patterns of sightings. Furthermore, both species are unmistakable with other mammals, so errors caused by misidentification are thought to be negligible. In addition, with regards to the survey to obtain data on hedgehogs, there is a possibility that possible trends may have emerged because of geographic differences in the eagerness of people to respond to surveys, although the participants of the survey were spread throughout England (Figure 1). Our analysis incorporates some of these challenges by, for example, only analysing the data at the 10 km^2 scale [82] and correcting for spatial-autocorrelation in the kriging process to improve confidence about the conclusions drawn from our study. All of these potential biases and challenges involved with citizen science data should not be underestimated and results coming from studies using citizen science data should be interpreted with caution. However, there are several benefits to such data as well, not least the potential to use data often collected at large geographic scales and on private lands, which can be difficult to obtain using more traditional survey approaches [83].

5. Conclusions

Whilst it was already known that badgers could regulate hedgehogs on smaller scales [33–37], the role badgers play at the nationwide scale was less clear. The negative relationship between the likelihood of hedgehog and badger presence observed in our study suggests that badgers may at least partly explain the variation in the presence of hedgehogs throughout England. These findings mirror findings by Williams et al. [9] who surveyed 261 sites in rural England and Wales using footprint tunnels to determine site occupancy by hedgehogs and badger sett presence as a proxy for the relative density of badgers. These combined results stress that it is imperative for the conservation of species to fully understand community interactions. Neglecting important community interactions, such as predation, may prevent us from recognising drivers of dynamics in species abundance and distribution, which may consequently lead to inappropriate conservation measures. In the case of the hedgehog, although nationwide hedgehog presence was negatively correlated with that of badgers, other, habitat-related, factors were stronger predictors. The potential role of declining macro-invertebrate abundance, therefore, needs further investigation.

Supplementary Materials: The following are available online at http://www.mdpi.com/2076-2615/9/10/759/s1, Table S1: results using mean values for the land cover maps of 2007, Table S2: results using median values for the land cover maps of 2000, Table S3: and results using mean values for the land cover maps of 2000.

Author Contributions: Conceptualization, A.R.H. and P.W.B.; methodology, A.R.H. and P.W.B.; formal analysis, A.R.H. and A.M.A.; writing—original draft preparation, A.R.H.; writing—review and editing, P.W.B. and A.M.A.; visualisation, A.R.H. and A.M.A.; funding acquisition, P.W.B.

Funding: This research was funded by the British Hedgehog Preservation Society and the People's Trust for Endangered Species.

Acknowledgments: We thank all participants of the surveys for reporting. We thank Dolly Jørgenson and Roland Jansson for much-appreciated comments to earlier versions of this manuscript. We thank three anonymous reviewers for valuable comments and suggestions.

Conflicts of Interest: The authors declare no conflict of interest.

Appendix A. Results Comparison Among Landcover Maps

The data collection of our study fell between two periods during which landcover maps were produced for the UK. Consequently, we compared our results from the landcover maps for both 2000 and 2007. Furthermore, landcover metrics can be calculated in different ways, for example taking the median percentage cover in a $10km^2$ grid cell or taking a mean. We present the results from median landcover in 2007 and here in this Appendix we include mean landcover for 2007, and also median and mean landcover for 2000.

The most important variables remained the same in all analyses, namely arable and horticulture, improved grassland, built-up areas, broad-leaved woodlands and badger presence (Tables S1–S3). Using a mean human footprint index explained more variation than a median human footprint index (Tables S1–S3). When using mean landcover data for 2007, semi-natural grassland and loamy soils were also included within models of $\Delta BIC < 4$, both of which had a positive relationship with hedgehog presence but had a low independent contribution towards the variation explained (Table S1).

The landcover map of 2000 included additional target classes for arable land, namely arable cereals, arable horticulture and arable non-rotational, which were not available in the 2007 data. Therefore, we determined whether these three separate variables performed better in explaining hedgehog presence than the aggregated landcover variables of arable and horticulture (which combined all three sub-classes). It was evident that arable non-rotational had a negative relationship with hedgehog presence, but this was not significant and was not included within any of the models with $\Delta BIC < 4$. The sub-class of arable cereal and arable horticulture were included, and both had a positive relationship with the likelihood of hedgehog presence, but only the sub-class of arable and horticulture was included in the top model. Despite the separate contributions of the arable sub-classes, the aggregated variable for arable and horticulture (includes all three arable types) provided a better model fit (BIC-400.93) than the target class arable and horticulture alone (BIC -401.96).

Using a median landcover value per grid cell, the most important variables remained the same as the main presented results (Table S2). An exception is that neutral grassland, coniferous forest and loamy soils were included in models with $\Delta BIC < 4$ and had a positive relationship with the likelihood of hedgehog presence, but these had a minor overall contribution towards the explained variation (Table S2).

The results of the models using a mean landcover value (Table S3), instead of a median (Table S2), for the landcover map 2000 were very similar, and also to the main results presented in the article. The only notable exception is that the variable improved grassland was excluded from the top models

(Table S3). The effect of improved grassland remained negative however the contribution towards model performance was reduced. This is somewhat evident in the results using a median value since improved grassland explained less variance and was not significant (Table S2) in comparison with the 2007 values (Table S1). This change may indicate that changes to grassland habitats, most notably intensifying use, may have increased in the period between 2000 and 2007. The inclusion of the variable neutral grasslands in top performing models using landcover maps from 2000 (Tables S2 and S3), and exclusion of neutral grasslands in top performing models using landcover maps from 2007 (Tables S1 and S3) may support this.

References

1. WWF. *Living Planet Report—2018: Aiming Higher*; Grooten, M., Almond, R.E.A., Eds.; WWF: Gland, Switzerland, 2018.
2. Holsbeek, L.; Rodts, J.; Muyldermans, S. Hedgehog and other animal traffic victims in Belgium: Results of a countrywide survey. *Lutra* **1999**, *42*, 111–119.
3. Huijser, M.P.; Bergers, P.J.M. The effect of roads and traffic on hedgehog (*Erinaceus europaeus*) populations. *Biol. Conserv.* **2000**, *95*, 111–116. [CrossRef]
4. Krange, M. *Change in the Occurrence of the West European Hedgehog (Erinaceus europaeus) in Western Sweden during 1950–2010*; Karlstad University: Karlstad, Sweden, 2015.
5. Battersby, J. *UK mammals: Species status and population trends: First report by the tracking mammals partnership*; JNCC, Tracking Mammals Partnership: Peterborough, UK, 2005.
6. Hof, A.R.; Bright, P.W. Quantifying the long-term decline of the West European hedgehog in England by subsampling citizen-science datasets. *Eur. J. Wildl. Res.* **2016**, *62*, 407–413. [CrossRef]
7. MTUK. *Mammals on Roads Survey—An Outline of 2004's Results*; Mammals Trust UK: London, UK, 2005.
8. Davey, P.A.; Aebischer, N.J. *Participation of the National Gamebag Census in the Mammal Surveillance Network*; JNCC: Peterborough, UK, 2006.
9. Williams, B.M.; Baker, P.J.; Thomas, E.; Wilson, G.; Judge, J.; Yarnell, R.W. Reduced occupancy of hedgehogs (*Erinaceus europaeus*) in rural England and Wales: The influence of habitat and an asymmetric intra-guild predator. *Sci. Rep.* **2018**, *8*, 17–20. [CrossRef] [PubMed]
10. Roos, S.; Johnston, A.; Noble, D. *UK Hedgehog Datasets and Their Potential for Long-Term Monitoring. BTO Research Report, 598*; The British Trust for Ornithology: Thetford, UK, 2012.
11. EU Open data Portal. Available online: https://data.europa.eu/euodp/en/data/publisher/estat (accessed on 27 September 2019).
12. Haigh, A.; O'Riordan, R.M.; Butler, F. Hedgehog *Erinaceus europaeus* mortality on Irish roads. *Wildl. Biol.* **2014**, *20*, 155–160. [CrossRef]
13. Rondinini, C.; Doncaster, C.P. Roads as barriers to movement for hedgehogs. *Funct. Ecol.* **2002**, *16*, 504–509. [CrossRef]
14. Hof, A.R. A Study of the Current Status of the Hedgehog (*Erinaceus europaeus*), and its Decline in Great Britain since 1960. Ph.D. Thesis, Roywal Holloway University of London, Egham, UK, 2009; pp. 1–192.
15. Hof, A.R.; Bright, P.W. The value of green-spaces in built-up areas for western hedgehogs. *Lutra* **2009**, *52*, 69–82.
16. Hof, A.R.; Bright, P.W. The impact of grassy field margins on macro-invertebrate abundance in adjacent arable fields. *Agric. Ecosyst. Environ.* **2010**, *139*, 280–283. [CrossRef]
17. Burel, F.; Butet, A.; Delettre, Y.R.; de La Peña, N.M. Differential response of selected taxa to landscape context and agricultural intensification. *Landsc. Urban Plan.* **2004**, *67*, 195–204. [CrossRef]
18. Fusser, M.S.; Pfister, S.C.; Entling, M.H.; Schirmel, J. Effects of landscape composition on carabids and slugs in herbaceous and woody field margins. *Agric. Ecosyst. Environ.* **2016**, *226*, 79–87. [CrossRef]
19. Burel, F. Hedgerows and their role in agricultural landscapes. *CRC. Crit. Rev. Plant Sci.* **1996**, *15*, 169–190. [CrossRef]
20. Reeve, N.J. *Hedgehogs*; T and A D Poyser Ltd.: London, UK, 1994.
21. Doncaster, C.P. Factors Regulating Local Variations in Abundance: Field Tests on Hedgehogs, Erinaceus europaeus. *Oikos* **1994**, *69*, 182–192. [CrossRef]
22. Neal, E.; Cheeseman, C.L. *Badgers*; T and A D Poyser Ltd.: London, UK, 1996.

23. Lüps, P.; Roper, T.J.; Stocker, G. Stomach contents of badgers (*Meles meles* L.) in central Switzerland. *Mammalia* **1987**, *51*, 559–570. [CrossRef]

24. Asprea, A.; de Marinis, A.M. The diet of the badger *Meles meles* (Mustelidae, Carnivora) on the Apennines (Central Italy). *Mammalia* **2005**, *69*, 89–95. [CrossRef]

25. Balestrieri, A.; Remonti, L.; Prigioni, C. Diet of the Eurasian badger (*Meles meles*) in an agricultural riverine habitat (NW Italy). *Hystrix Ital. J. Mammal.* **2005**, *15*, 3–12.

26. Middleton, A.D. The Food of a Badger (*Meles meles*). *J. Anim. Ecol.* **1935**, *4*, 291. [CrossRef]

27. Goszczyński, J.; Jędrzejewska, B.; Jędrzejewski, W. Diet composition of badgers (*Meles meles*) in a pristine forest and rural habitats of Poland compared to other European populations. *J. Zool.* **2000**, *250*, 495–505. [CrossRef]

28. Del Bove, E.; Isotti, R. The European badger (*Meles meles*) diet in a Mediterranean area. *Hystrix Ital. J. Mammal.* **2001**, *12*, 19–25.

29. Roper, T.J.; Mickevicius, E. Badger *Meles meles* diet: A review of literature from the former Soviet Union. *Mamm. Rev.* **1995**, *25*, 117–129. [CrossRef]

30. Polis, G.A.; Myers, C.A.; Holt, R.D. The ecology and evolution of intraguild predation: Potential competitors that eat each other. *Annu. Rev. Ecol. Syst.* **1989**, *20*, 297–330. [CrossRef]

31. Amarasekare, P. Trade-offs, temporal variation, and species coexistence in communities with intraguild predation. *Ecology* **2007**, *88*, 2720–2728. [CrossRef] [PubMed]

32. Young, R.P.; Davison, J.; Trewby, I.D.; Wilson, G.J.; Delahay, R.J.; Doncaster, C.P. Abundance of hedgehogs (*Erinaceus europaeus*) in relation to the density and distribution of badgers (*Meles meles*). *J. Zool.* **2006**, *269*, 349–356. [CrossRef]

33. Doncaster, C.P. Testing the role of intraguild predation in regulating hedgehog populations. *Proc. R. Soc. London. Ser. B Biol. Sci.* **1992**, *249*, 113–117.

34. Micol, T.; Doncaster, C.P.; Mackinlay, L.A. Correlates of local variation in the abundance of hedgehogs Erinaceus europaeus. *J. Anim. Ecol.* **1994**, *63*, 851–860. [CrossRef]

35. Parrott, D.; Etherington, T.R.; Dendy, J. A geographically extensive survey of hedgehogs (*Erinaceus europaeus*) in England. *Eur. J. Wildl. Res.* **2014**, *60*, 399–403. [CrossRef]

36. Trewby, I.D.; Young, R.; McDonald, R.A.; Wilson, G.J.; Davison, J.; Walker, N.; Robertson, A.; Patrick Doncaster, C.; Delahay, R.J. Impacts of removing badgers on localised counts of hedgehogs. *PLoS ONE* **2014**, *9*, 2–5. [CrossRef]

37. Hof, A.R.; Snellenberg, J.; Bright, P.W. Food or fear? Predation risk mediates edge refuging in an insectivorous mammal. *Anim. Behav.* **2012**, *83*, 1099–1106. [CrossRef]

38. Ward, J.F.; Macdonald, D.W.; Doncaster, C.P. Responses of foraging hedgehogs to badger odour. *Anim. Behav.* **1997**, *53*, 709–720. [CrossRef]

39. Judge, J.; Wilson, G.J.; Macarthur, R.; Delahay, R.J.; McDonald, R.A. Density and abundance of badger social groups in England and Wales in 2011–2013. *Sci. Rep.* **2014**, *4*, 3809. [CrossRef]

40. Gaston, K.J.; Blackburn, T.M.; Greenwood, J.D.; Gregory, R.D.; Quinn, M.; Lawton, J.H. Abundance—occupancy relationships. *J. Appl. Ecol.* **2000**, *37*, 39–59. [CrossRef]

41. Schwarz, C.J.; Seber, G.A.F. Estimating Animal Abundance: Review III. *Stat. Sci.* **1999**, *14*, 427–456.

42. Cressie, N.A.C. *Statistics for Spatial Data*; John Wiley and Sons Inc.: New York, NY, USA, 1991.

43. Fortin, M.J.; Dale, M.R. *Spatial Analysis: A Guide for Ecologists*; Cambridge University Press: Cambridge, UK, 2005.

44. Rossi, R.E.; Mulla, D.J.; Journel, A.G.; Franz, E.H. Geostatistical tools for modeling and interpreting ecological spatial dependence. *Ecol. Monogr.* **1992**, *62*, 277–314. [CrossRef]

45. Sutherland, W.J. *Ecological Census Techniques: A handbook*, 2nd ed.; Cambridge University Press: Cambridge, UK, 2006.

46. Certain, G.; Bellier, E.; Planque, B.; Bretagnolle, V. Characterising the temporal variability of the spatial distribution of animals: An application to seabirds at sea. *Ecography* **2007**, *30*, 695–708. [CrossRef]

47. Davis, S.E.; Newson, S.E.; Noble, D.G. The production of population trends for UK mammals using BBS mammal data: 1995–2005 update. *BTO Res. Rep.* **2007**, *462*, 54.

48. Carroll, S.S.; Pearson, D.L. Spatial modeling of butterfly species richness using tiger beetles (Cicindelidae) as a bioindicator taxon. *Ecol. Appl.* **1998**, *8*, 531–543. [CrossRef]

49. Villard, M.-A.; Maurer, B.A. Geostatistics as a Tool for Examining Hypothesized Declines in Migratory Songbirds. *Ecology* **1996**, *77*, 59–68. [CrossRef]

50. Isaaks, E.H.; Srivastava, R.M. *Applied Geostatistics*; Oxford University Press: New York, NY, USA, 1989.

51. Meyers, J.C. *Geostatistical Error Management: Quantifying Uncertainty for Environmental Sampling and Mapping*; Vam Nostrand Reinhold.: New York, NY, USA, 1997.

52. Sanderson, E.W.; Jaiteh, M.; Levy, M.A.; Redford, K.H.; Wannebo, A.V.; Woolmer, G. The human footprint and the last of the wild. *Bioscience* **2002**, *52*, 891–904. [CrossRef]

53. Fuller, R.M.; Smith, G.M.; Sanderson, J.M.; Hill, R.A.; Thomson, A.G.; Cox, R.; Brown, N.J.; Clarke, R.T.; Rothery, P.; Gerard, F.F. *Countryside Survey 2000 Module 7-LAND COVER MAP 2000—Final Report*; Centre for Ecology & Hydrology: Monks Wood, UK, 2002.

54. Morton, D.; Rowland, C.; Wood, C.; Meek, L.; Marston, C.; Smith, G.; Wadsworth, R.; Simpson, I.C. *Countryside Survey: Final Report for LCM2007—The new UK Land Cover Map*; Centre for Ecology & Hydrology: Monks Wood, UK, 2011.

55. Graham, M.H. Confronting Multicollinearity in Ecological. *Ecology* **2003**, *84*, 2809–2815. [CrossRef]

56. O'Brien, R.M. A caution regarding rules of thumb for variance inflation factors. *Qual. Quant.* **2007**, *41*, 673–690. [CrossRef]

57. Burnham, K.P.; Anderson, D.R. *Model Selection and Multiple Inference: A Practical Information-Theoretic Approach*, 2nd ed.; Springer: New York, NY, USA, 2002.

58. Barton, K. MuMIn: Multi-Model Inference. R package version 1.43.6. Available online: https://CRAN.R-project.org/package=MuMIn (accessed on 27 September 2019).

59. R Development Core Team. *R: A Language and Environment for Statistical Computing*; R Development Core Team: Vienna, Austria, 2018.

60. Chevan, A.; Sutherland, M. Hierarchical partitioning. *Am. Stat.* **1991**, *45*, 90–96.

61. Nally, R. MAC Hierarchical partitioning as an interpretative tool in multivariate inference. *Austral Ecol.* **1996**, *21*, 224–228. [CrossRef]

62. Hof, A.R.; Bright, P.W. Factors affecting hedgehog presence on farmland as assessed by a questionnaire survey. *Acta Theriol.* **2012**, *57*, 79–88. [CrossRef]

63. Zingg, R. *Aktivität Sowie Habitat-und Raumnutzung von Igeln (Erinaceus europaeus) in Einem Ländlichen Siedlungsgebiet*; University of Zürich: Zürich, Switzerland, 1994.

64. Etherington, T.R.; Ward, A.I.; Smith, G.C.; Pietravalle, S.; Wilson, G.J. Using the Mahalanobis distance statistic with unplanned presence-only survey data for biogeographical models of species distribution and abundance: A case study of badger setts. *J. Biogeogr.* **2009**, *36*, 845–853. [CrossRef]

65. Huck, M.; Davison, J.; Roper, T.J. Predicting European badger *Meles meles* sett distribution in urban environments. *Wildl. Biol.* **2008**, *14*, 188–198. [CrossRef]

66. Thornton, P.S. Density and distribution of Badgers in south-west England–a predictive model. *Mamm. Rev.* **1988**, *18*, 11–23. [CrossRef]

67. Huijser, M.P. *Life on the Edge. Hedgehog Traffic Victims and Mitigation Strategies in an Anthropogenic Landscape*; Wageningen University: Wageningen, The Netherlands, 2000.

68. Hof, A.R.; Bright, P.W. The value of agri-environment schemes for macro-invertebrate feeders: Hedgehogs on arable farms in Britain. *Anim. Conserv.* **2010**, *13*, 467–473. [CrossRef]

69. Riber, A.B. Habitat use and behaviour of European hedgehog *Erinaceus europaeus* in a Danish rural area. *Acta Theriol.* **2006**, *51*, 363–371. [CrossRef]

70. Pettett, C.E.; Moorhouse, T.P.; Johnson, P.J.; Macdonald, D.W. Factors affecting hedgehog (*Erinaceus europaeus*) attraction to rural villages in arable landscapes. *Eur. J. Wildl. Res.* **2017**, *63*, 54. [CrossRef]

71. Wilson, G.; Harris, S.; McLaren, G. *Changes in the British badger population, 1988 to 1997*; People's Trust for Endangered Species: London, UK, 1997.

72. Wilson, G.J.; Delahay, R.J.; De Leeuw, A.N.S.; Spyvee, P.D.; Handoll, D. Quantification of badger (*Meles meles*) sett activity as a method of predicting badger numbers. *J. Zool.* **2003**, *259*, 49–56. [CrossRef]

73. Heithaus, M.R. Habitat selection by predators and prey in communities with asymmetrical intraguild predation. *Oikos* **2001**, *92*, 542–554. [CrossRef]

74. Zabala, J.; Zuberogoitia, I. Badger, *Meles meles* (Mustelidae, Carnivora), diet assessed through scat-analysis: A comparison and critique of different methods. *Folia Zool.* **2003**, *52*, 23–30.

75. Cleary, G.P.; Corner, L.A.L.; O'Keeffe, J.; Marples, N.M. The diet of the badger *Meles meles* in the Republic of Ireland. *Mamm. Biol.* **2009**, *74*, 438–447. [CrossRef]

76. Robinson, R.A.; William, J. Sutherland Post-war changes in arable farming and biodiversity in Great Britain. *J. Appl. Ecol.* **2002**, *39*, 157–176. [CrossRef]

77. Sánchez-Bayo, F.; Wyckhuys, K.A.G. Worldwide decline of the entomofauna: A review of its drivers. *Biol. Conserv.* **2019**, *232*, 8–27. [CrossRef]

78. Dirzo, R.; Young, H.S.; Galetti, M.; Ceballos, G.; Isaac, N.J.B.; Collen, B. Defaunation in the Anthropocene. *Science* **2014**, *25*, 401–406. [CrossRef] [PubMed]

79. Gaglio, G.; Allen, S.; Bowden, L.; Bryant, M.; Morgan, E.R. Parasites of European hedgehogs (*Erinaceus europaeus*) in Britain: Epidemiological study and coprological test evaluation. *Eur. J. Wildl. Res.* **2010**, *56*, 839–844. [CrossRef]

80. Delahay, R.J.; De Leeuw, A.N.S.; Barlow, A.M.; Clifton-Hadley, R.S.; Cheeseman, C.L. The status of Mycobacterium bovis infection in UK wild mammals: A review. *Vet. J.* **2002**, *164*, 90–105. [CrossRef]

81. Dowding, C.V.; Shore, R.F.; Worgan, A.; Baker, P.J.; Harris, S. Accumulation of anticoagulant rodenticides in a non-target insectivore, the European hedgehog (*Erinaceus europaeus*). *Environ. Pollut.* **2010**, *158*, 161–166. [CrossRef]

82. Bird, T.J.; Bates, A.E.; Lefcheck, J.S.; Hill, N.A.; Thomson, R.J.; Edgar, G.J.; Stuart-Smith, R.D.; Wotherspoon, S.; Krkosek, M.; Stuart-Smith, J.F.; et al. Statistical solutions for error and bias in global citizen science datasets. *Biol. Conserv.* **2014**, *173*, 144–154. [CrossRef]

83. Dickinson, J.L.; Zuckerberg, B.; Bonter, D.N. Citizen Science as an Ecological Research Tool: Challenges and Benefits. *Annu. Rev. Ecol. Evol. Syst.* **2010**, *41*, 149–172. [CrossRef]

Article

Home Range, Movement, and Nest Use of Hedgehogs (*Erinaceus europaeus*) in an Urban Environment Prior to Hibernation

Lars Mørch Korslund [1,*], Marius Stener Floden [1], Milla Mona Sophie Albertsen [1], Amalie Landsverk [1], Karen Margrete Vestgård Løkken [1] and Beate Strøm Johansen [2,*]

[1]   Department of Natural Sciences, University of Agder, Universitetsveien 25, 4630 Kristiansand, Norway; marius.stener.floden@nmbu.no (M.S.F.); mmalbert@uio.no (M.M.S.A.); landsverk.amalie@gmail.com (A.L.)

[2]   Natural History Museum and Botanical Garden, Department of Natural Sciences, University of Agder, Gimleveien 27, Gimle Gård, 4630 Kristiansand, Norway

*   Correspondence: lars.korslund@uia.no (L.M.K.); beate.johansen@uia.no (B.S.J.); Tel.: +47-40215345 (L.M.K.); +47-93218653 (B.S.J.)

Simple Summary: Populations of West European hedgehogs (*Erinaceus europaeus*) are decreasing all over Europe, and we are urgently in need of more knowledge to understand the challenges they face. In the Nordic countries, the winter nest locations are of crucial importance for hedgehogs to survive the winter hibernation period. Using radio transmitters, we studied 9 adult hedgehogs during the pre-hibernation period from August–November in a typical residential area in the city of Kristiansand, Southern Norway. The hedgehogs had a highly variable home range size and displayed a large variation in distance moved per hour, with no clear difference between sexes. There were also large individual differences in the number of nest sites used and how often they changed nests. Although hedgehogs had nesting places in a variety of gardens and in hedgerows along roads, such places seemed to lack appropriate nesting materials, suggesting that this is not a habitat suitable for winter hibernation. In September, as they prepared for hibernation, hedgehogs rather chose permanent winter nests in natural forest patches within residential areas, often under tree roots. Our research highlights the importance of maintaining and increasing the number of smaller forested patches within urban regions to help conserve hedgehog populations.

Abstract: The West European hedgehog (*Erinaceus europaeus*) is in decline, and it is important to identify its challenges. We used VHF-telemetry to monitor pre-hibernation space use, nest use, and hibernation sites in a suburban area in Norway. Based on nine adult hedgehogs tracked between August and November 2002, we found that home range size was not dependent on individual sex or weight and that home ranges overlapped between individuals regardless of sex. The distance moved was not dependent on individual sex, but there was a tendency for increased movement before dawn. The number of nests used per individual (0–10) and the number of nest switches (0–14) varied greatly and did not differ significantly between sexes. Out of 28 nest sites, 16 were linked to buildings and 12 to vegetation, and nesting material was most often grass and leaves. Three hedgehogs monitored until hibernation established winter nests under tree roots in natural forest patches in September, and this suggests that establishing or maintaining forest patches in urban areas is important to ensure suitable hibernation habitat for hedgehogs. Our study was limited by a low sample size, and additional research is required to gain a deeper understanding of the challenges hedgehogs face in urban environments.

Keywords: hedgehog conservation; human–wildlife; urban wildlife; radio telemetry; home range; movements; nest use; hibernation nest

Citation: Korslund, L.M.; Floden, M.S.; Albertsen, M.M.S.; Landsverk, A.; Løkken, K.M.V.; Johansen, B.S. Home Range, Movement, and Nest Use of Hedgehogs (*Erinaceus europaeus*) in an Urban Environment Prior to Hibernation. *Animals* **2024**, *14*, 130. https://doi.org/10.3390/ani14010130

Academic Editors: Anne Berger, Nigel Reeve and Sophie Lund Rasmussen

Received: 17 November 2023
Revised: 20 December 2023
Accepted: 24 December 2023
Published: 29 December 2023

1. Introduction

Urbanization is often referred to as the process of the increasing concentration of people in cities and the transformation of natural environments into urban areas [1]. Urban environments have been rapidly expanding globally as a result of high population growth over the last decades [2]. The effect of urbanization can be complex and varied, but it generally tends to have a negative impact on biodiversity both locally and globally, mainly through habitat loss and fragmentation [3]. Several ecological studies on urban systems show that urban centers often have low biodiversity, with a few resilient species in high numbers [4]. Synanthropic species, species of wild animals that live in close proximity to humans and in environments that humans create [5], often reach higher densities in urban environments than in the wild [6]. A higher abundance of food, a lower abundance of predators, or a combination of these can result in increased population densities [7]. Small to medium-sized animals seem best suited for urban environments [8], and one such mammal that is common in urban environments in Europe is the West European hedgehog (*Erinaceus europaeus*, from here on called hedgehog) [9].

Results from multiple monitoring programs show that hedgehog populations are declining in many European countries [10–17]. According to the IUCN Red List assessment, the hedgehog's conservation status varies from "Near Threatened" in Sweden and Norway to "Vulnerable" in the UK and "Endangered" in the Netherlands [18–20]. It has previously been suggested that hedgehogs prefer to live in rural areas, but some studies report a substantial population decline in such areas and that hedgehogs prefer residential areas [7,15,16,21,22]. This decline in rural areas appears to be primarily caused by intensive agriculture and intraguild predators [16,23]. Hubert, Julliard, Biagianti, and Poulle [7] found that factors such as access to anthropogenic food sources and favorable micro-climatic conditions may be key indicators of the high hedgehog presence in urban areas, and garbage and food put out for pets or other animals are often available food sources for hedgehogs [9]. Hedgehogs living in urban environments tend to become active post-midnight and avoid foraging near roads as a response to human-related dangers such as pedestrians and vehicle traffic [24], and they are usually found in greenspaces such as parks, road verges, and gardens [25]. These habitats are well suited for hedgehogs, but the fragmentation between such habitats, caused by roads and fences, can pose a significant challenge to the survival of these populations [26]. The most important habitat in urban areas is private gardens [27], as these have a high structural complexity with different vegetation such as lawn, flowerbeds, hedges, and terraces, creating habitats for both nesting and foraging [9]. On and around gardens and lawns, hedgehogs can find valuable food such as insects, slugs, and snails [28]. However, gardens may pose a variety of threats to individual hedgehogs as well. Use of garden pesticides such as insecticides, molluscicides, and rodenticides will lead to a decrease in the availability of natural food sources and can also result in secondary poising [24]. Although hedgehogs are capable of swimming, garden ponds can pose a threat to them as they can drown if unable to find a way out onto solid ground [29,30], and uncovered window wells, basement stairs, tennis nets, and nets covering berry bushes can function as traps [14,31].

Hedgehogs are solitary mammals and are mostly active during the night, spending much of the day sleeping in a nest [32]. They are not territorial animals, and the home range of individuals of the same and opposite sexes can overlap (see, e.g., [32–39]). The size of the home range is dependent on food availability, season, and sex, and while some hedgehogs stay in the same area over several years, others may wander more erratically around [29]. Hedgehogs that live in less productive environments will typically have a larger home range in order to find enough food [38,40]. During spring, in the mating season, males can have considerably larger home ranges than females, while in autumn the differences get smaller, and females can even have larger home ranges than males [26,38]. This seasonal change in home range size is reflected in the travel distance, and while male hedgehogs are found to travel considerably longer distances than female hedgehogs during spring, distances are more similar in the post-mating season [36]. Doncaster, Rondinin and Johnson [21] found that some hedgehogs can move up to 9.9 km in total during the night, but during the summertime, hedgehogs often move only 2–3 km a night to forage and build up fat storage for their hibernation through the winter [29].

Nests are important for the nocturnal and solitary hedgehog, both for hibernation, protection, and breeding [41], and nests can be divided into three categories: breeding nests for the females and their litter, day or summer nests that are used as shelter during the days in the active season, and hibernation or winter nests where they spend up to several months undergoing hibernation [32]. Winter nests in mixed woodland habitat in England were often located at sites with structural support, such as under bramble bush or piles of logs, and the nesting material often consisted of grass or leaves packed together up to 20 cm in thickness [42,43]. Rautio et al. [41] found that urban hedgehogs in Finland moved to pine woods to establish winter nests and hibernate under the roots of large pine trees. The hedgehog hibernation period varies greatly with the local climate and, thus, geography. In Southern Europe, it only lasts for about two months, from January to February [33], while in Fennoscandia, at the northern boundary of the geographical range, it can last more than 200 days, starting as early as mid-September [41,44–46].

In coastal Southern Norway, the climate is mild relative to its latitude, but despite this, the hedgehog population seems to be low [47]. The Natural History Museum and Botanical Garden in Kristiansand is situated beside the campus of the University of Agder, and this area is one of the main areas for hedgehogs in the city [48]. The hedgehog became red-listed in Norway in 2021, and the aim of this study was to shed some light on the home range size, movements, and nest site selection of hedgehogs in a typical urban residential area in Kristiansand. By radio-tracking hedgehogs, we aimed to identify the challenges they face and the habitats that are especially important for their existence. We performed this study in late summer and autumn in order to identify where and when the hedgehogs chose to hibernate.

2. Methods

2.1. Study Area

This study took place in the Gimle/Lund area in the city of Kristiansand in Southern Norway (58.15° N, 8.00° E, Figure 1). Kristiansand has a population of 115,000 and covers an area of almost 644 km^2. There are approximately 53,000 residences, and the city has a population density of 186 citizens/km^2. Kristiansand is a coastal city that borders Skagerrak with relatively mild winters, given the latitude. The residential areas where this study took place are a mix of regular single-family homes with gardens, terraced houses, and apartment blocks. The terraced house gardens often connect to a larger communal lawn or park. Gardens are often divided by wooden fences, chain-link fences, and/or hedges, and most houses have open driveways without gates. The campus of the University of Agder Gimlemoen and the botanical garden, adjacent to the residential areas, have large, open lawns together with flower beds and bushes. Along roads and outside scattered office buildings, there are often beds of densely planted dense bush/hedge. In addition, hedgehogs, the most common mammals observed in the area are the Eurasian red squirrel (*Sciurus vulgaris*) and the brown rat (*Rattus norvegicus*). Badgers (*Meles meles*) are present but rarely reported or observed.

2.2. Citizen Science Initiative

The hedgehog population in the city of Kristiansand was mapped in 2019 using reports from the public [48]. This provided information that the Lund area, with the university campus and botanical garden, was one of the main areas where hedgehogs were observed. As preparations for the radio marking of hedgehogs from the 15 August 2022, we initiated a new citizen science campaign in June 2022 to obtain detailed information on where to find hedgehogs in this area, using radio stations, newspapers, Facebook groups, and posters along the roads. We received 45 observations of hedgehogs from the public. Every citizen hedgehog observation reported from this campaign was registered in the Norwegia[n] Species Observation System [47].

Figure 1. Map showing the location of Kristiansand in Norway (red dot), the rural areas (green) th[e] densely populated areas (light grey) and a blue circle surrounding the Lund area with the Universi[ty] of Agder and The Natural History Museum and Botanical Garden where the study was conducte[d] (Map sources: © Google, 2023, and © NordNordWest/Wikimedia Commons/CC-BY-SA-3.0).

2.3. Radio-Marked Hedgehogs

The field work lasted from 15 August to 5 November 2022 (Table 1). The registration[s] reported by citizens were used as a guideline when searching for hedgehogs with flashligh[t] in the evenings. When a hedgehog was spotted and hand-captured, we determined the se[x] and weighted it by placing it inside a plastic box on a scale (max = 5000 g, d = 1 g). We ha[d] in total six radio-transmitters (R1680 glue-on transmitter, Advanced Telemetry System[s] (ATS), Isanti, MN, USA) available, each weighing 3.6 g, including the 20 cm antenna, le[ss] than 0.4% of the body weight of any adult hedgehog. To fit the transmitter in a way th[at] would not inhibit a hedgehog's normal life, spikes in a concentrated area on the low[er] back were clipped 1–2 cm using scissors and clippers (as in [23,38,49]). The transmitt[er] was carefully glued onto the clipped spikes and to the spikes next to them using epo[xy] glue, making sure that no glue touched the skin of the animal, and was firmly held i[n] place for 10 min until the glue had cured properly. The transmitter was positioned wi[th] the antenna sticking out the back, so that it trailed behind the animal when it moved. Th[is] position enabled the animal to move freely without the transmitter effecting movement [in] any way. Of the originally six radio tagged hedgehogs, one was soon hit by a car, and tw[o] were killed by an unknown predator or died of other causes and scavenged after deat[h.] The three salvaged transmitters were therefore placed on three new hedgehogs as soo[n] as they were located. Three of the six transmitters lost the signal after a while, probab[ly] due to transmitter failure. Thus, in total, nine hedgehogs were tagged and tracked, but [at] the end of the tracking period, only three hedgehogs were still equipped with a workin[g] transmitter (Table 2).

Table 1. Overview of the six field periods, with start and end dates in 2022, type of field activity (M = marking individuals, T = radio tracking individuals), and the number of tracking sessions in each period. In period 2, the time alternated between early and late night. The shading of the cells with time indicates light conditions. Dark gray = nighttime, light gray = dusk, and white = daytime.

Period	Start Date	End Date	Time of Day	Field Activity	No. Days
1	15 August	30 August	22.30–07.00	M + T	16
2	31 August	13 September	22.00–03.30/02.30–07.00	M + T	14
3	24 September	10 October	10.00–14.30	T	17
4	11 October	20 October	17.00–20.00	T	9
5	23 October	29 October	00.00–02.00	T	4
6	5 November	5 November	11.00–14.00	T	1

Table 2. Detailed information on each individual hedgehog tracked, including sex, body mass, start and end of tracking, number of days tracked, and number of localizations (individual registrations) during the period between 15 August and 5 November 2022.

ID nr.	Sex	Body Mass (g)	Date Tagged (Start Tracking)	Last Tracking	Cause of End	Days Tracked	Localization
1	Female	958	15 August	30 August	Road killed	14	49
2	Female	1044	22 August	19 August	Predation	2	4
3	Male	1361	16 August	5 November	End of campaign	27	124
4	Female	928	17 August	3 September	Predation	17	62
5	Male	1244	17 August	9 September	Lost signal	22	88
6	Male	970	18 August	20 August	Lost signal	2	4
7	Male	1410	22 August	24 August	Lost signal	3	9
8	Female	1099	27 August	5 November	End of campaign	16	84
9	Male	1374	4 September	5 November	End of campaign	9	58

2.4. Radio Tracking

We used a scanning receiver (R410, ATS, USA) in conjunction with a smaller direction-based H-antenna (ATS, USA) and a much larger, five-element foldable Yagi Antenna (Model 17734, ATS, USA) to detect tagged hedgehogs. Tracking was carried out by car and by foot, depending on terrain, and the coordinates of every localization (or fix) were registered using a GPS. In the first period, from 15 to 30 August, the entirety of the night was spent looking for hedgehogs for tagging. A night typically started around 22:30–23:00 and lasted until 06:00–07:00. During the second period between 30 August and 13 September, a three-night rotation was performed, alternating between early half-night (22:00–23:00 to 03:00–03:30), late half-night (from 02:30–03:00 to 06:00–07:00), or full night (22.00–23.00 to 06.00–07.00, see Table 1). During all nighttime tracking, we rotated between all tagged individuals, and it usually took 1–2 h between every localization of an individual, depending on the number of hedgehogs marked at a given time.

After 13 September, a second phase of the project started to establish where and when the hedgehogs went hibernating. Between 24 September and 29 October (Table 1, periods 3 and 4), the individuals were tracked during the daytime. Since no nest site switches were observed after 29 September, we assumed hibernating was initiated, and to confirm this, we tracked once every night between 23 and 29 of October (Table 1, period 5). As there was no sign of hedgehog activity, neither day nor night, we ended the tracking on 5 November (Table 1, period 6). When hedgehog nest sites were investigated during the day, we noted the nesting type by category: Building (garage, porch, terrace, stairs, building materials) or Nature (vegetation, bush, forest). When the actual nest could be observed, we noted the nest materials as well.

Permissions to capture and tag hedgehogs were provided by the Norwegian Food Safety Authority (FOTS ID 27113) and the Norwegian Environment Agency (ref. 2022/7181) and permission to operate VHF tags was given by the Norwegian Communications Authority (PMR-no. 17808).

2.5. Data Analysis

All analyses were performed using R [50]. We estimated home-range size by both Minimum Convex Polygon (MCP) and Kernel Utilization Distribution (KUD) estimators following Riber [51] using the sp [52] and adehabitatHR packages [53]. To examine to what extent home range size estimates were sensitive to localization outliers, we performed an incremental analysis by calculating MCP sizes based on 50 to 100 percent of localizations (with 5 percent increments) from each individual and plotted the relationship between the number of localizations and the estimated MCP area. We also calculated 95% and 50% of the kernel utilization distribution (KUD) area, where the smoothing parameter was set to 46.2, based on the reference bandwidth method. The 95% KUD excludes 5% of the most extreme localization outliers in order to better represent the "true home range", while the 50% KUD is expected to represent the core of the home range [51]. As suggested by Seaman et al. [54], we only included individuals with more than 30 localizations when calculating home range sizes. We used a simple two-sided *t*-test to investigate if kernel home range sizes were different between sexes and linear regression to see if kernel home range sizes were affected by initial body size. We included (MCP) as this is used in many other relevant studies and therefore is good for comparison [22,24,33,35,36,38,39,51], but also KUD since there is a debate regarding which home range estimator is least biased [55,56].

All localizations of the six individuals between 22.00 and 07.00 were used to investigate the distance moved at night as a function of time of night and sex. The time in hours and the distance in meters between two successive localizations of the same individual on the same night were calculated. The distance was calculated as the shortest distance between the two localizations, ignoring physical structures like buildings, fences, etc., by using the distHaversine function in the geosphere R-package [57]. The distance in meters divided by the time in hours, was used as the response variable. This variable was highly right-skewed, with many values being equal to or close to zero but with some extreme values, and therefore we applied a negative binomial mixed model, with individual ID as a random intercept (to account for multiple observations of the same individual), using the glmer.nb-function in the lme4-package [58]. We investigated the effect of the explanatory variables sex, as a binomial variable, and time in hours, relative to 22.00 (the earliest time of radio tracking at night), as a continuous variable. We started with the full model, including sex, time, and their interaction, and fitted all nested simpler models, including the null model. We used Akaike's Information Criterion, corrected for small sample size (AICc) [59] to compare the fit of the different models. We also calculated the minimum distance moved per night per individual by adding all consecutive distances between pairs of localizations. This variable was highly right-skewed, with many values being close to zero, and therefore we applied a logistic regression with individual ID as a random intercept (to account for multiple observations of the same individual), investigating the effect of sex as a binomial variable on total meters moved per night. The model took into account that the number of pairs of localizations varied between nights and individuals.

3. Results

3.1. Radio Tracking

Nine hedgehogs (4 females ♀, 5 males ♂) were tracked during the field campaign, for a total of 482 localizations (see map in Figure 2 and Table 2). During this period, individuals 1, 2, and 4 died (see methods), and individuals 6 and 7 lost their radio transmitter after a few days. The transmitters were all recovered and then reused to mark other individuals. This allowed us to track nine different hedgehogs in total, despite only having six radio transmitters. However, due to being tracked only for a few days and thus not being localized many enough times, individuals 2, 6, and 7 were removed from the spatial analysis. At the time of tagging, the weights of the hedgehogs ranged from 928 g to 1410 g (Tabel 2), and the average weight of the females and males was 1007 g and 1271 g, respectively.

Figure 2. Map of Lund and Gimle in Kristiansand (Southern Norway) with 100% minimum convex polygon (MCP) home ranges of six hedgehogs, displayed with color-codes for each individual (1, 4, 8 = females, and 3, 5, 9 = males). The home ranges overlap both within and between sexes.

3.2. Home Range

Pairs of hedgehogs had overlapping 100% MCP home ranges with either the same or the opposite sex (Figure 2). In the middle of the area, two male home ranges overlapped (Male 5 and Male 9). Male 7 was also found in this area, as were males 5 and 9, but its transmitter fell off after only three days. In the south-eastern part of the area, two females had overlapping home ranges (Female 1 and Female 8). The size difference in home range between sexes was neither significant for 95% (t = 0.08, df = 4, $p > 0.05$) nor 50% (t = 0.5, df = 4, $p > 0.05$) kernel sizes (Figure 3, Table 3). There was no significant relationship between weight at capture and home range size at neither 95% (F (1,4) = 0.05, $p > 0.05$) nor 50% (F (1,4) = 0.9, $p > 0.05$).

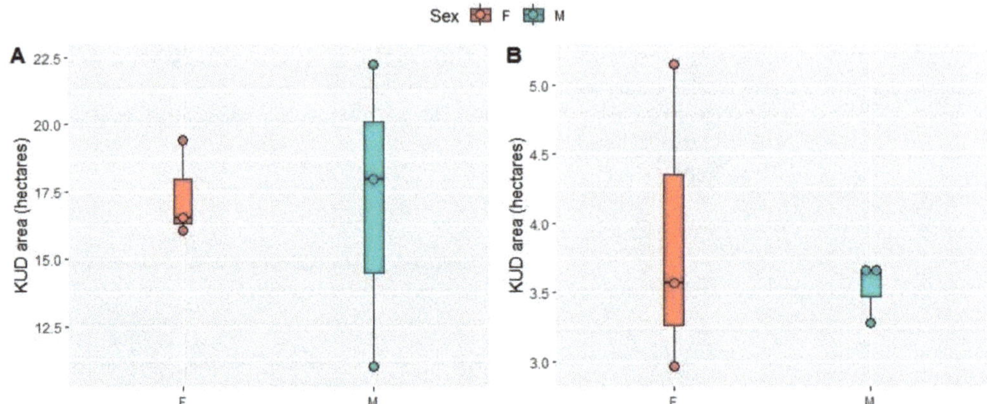

Figure 3. Approximately 95% (**A**) and 50% (**B**) Kernel Utilization Distribution (KUD) estimates, i hectares, of three male (blue) and three female (red) adult hedgehogs in the pre-hibernation peric Points represent individual estimates, and the horizontal line in the box represents the median.

Table 3. Approximately 50% and 95% kernel utilization density (KUD) home range estimates, é well as 100% maximum convex polygon (MCP) home range estimate (in hectares), for each of the s individuals, based on all localizations.

ID	95 KUD	50 KUD	100% MCP	Sex
1	16.0	3.6	8.8	F
3	22.2	3.7	15.6	M
4	19.4	5.1	9.0	F
5	11.0	3.3	6.3	M
8	16.5	3.0	10.7	F
9	18.0	3.7	8.7	M

3.3. Movement

The distance moved was analyzed based on 196 pairs of localizations from th six individuals. The number of pairs per individual per night varied between 1 and (median = 3). The time between two localizations of the same individual varied betwee 0.29 and 3.54 h (median = 0.87) and the distance moved per hour varied between 0 ar 420.5 m/h (median = 42.0). Even with such large overall variation in distance moved, th variation was high within most individuals (Figure 4A), and only 21% and 7% of the vari tion were explained by individual differences in the full model and null model, respective There were some differences in median distance moved between the six individuals, wit individual numbers 3 (a male) and 4 (a female) standing out as individuals showing low and high median distance moved, respectively (Figure 4A). However, there was n clear tendency between the other four individuals for a systematic sex effect, and the AIC value of the sex model (Table 4, model 4) was lower than that of the null model (model : indicating low support for a systematic difference between sexes in distance moved p hour. There were clearly more longer distances moved per hour during the last part of th night, especially between 04:00 and 06:00, and these longer bursts were more common i some individuals than others, but there were also many observations with no moveme between two successive localizations during the whole night (Figure 4B). The model wi only the time effect was the one with the lowest AICc value (Table 4, model 1), but th model was not substantially better than the null model (deltaAICc = 1). All in all, the mod selection shows that there is low support for any sex difference and some support for an increase in movement per hour at the end of the night.

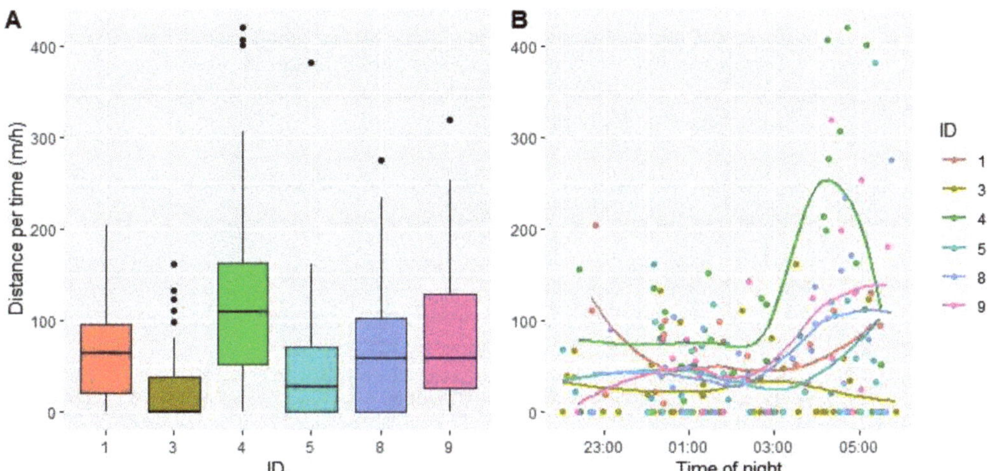

Figure 4. Distance moved (m/h) between successive relocations the same night, as a box plot with each individual (**A**) and as a function of time at night for each individual (**B**). Time in (**B**) represents the first of the two successive localizations. Lines are created by loess-smoothing.

Table 4. A list of negative binomial mixed effects models was used to investigate the effect of sex and time of night and the two-way interaction (Time:Sex) on the distance moved by hedgehogs between successive localizations. The inclusion of a term or model is indicated by an "x". AICc represents Akaike's Information Criterion (a lower value indicates better model fit), and deltaAICc represents the difference in AICc-value between the current model and the model with the lowest AICc-value (model 1).

Model	Time	Sex	Time:Sex	Df	AICc	deltaAICc
1	x			4	1785.9	0
2	x	x		5	1786.4	0.5
3				3	1786.9	1
4		x		4	1787.5	1.6
5	x	x	x	6	1787.8	1.9

The total distance moved during a single night for one individual varied between 0 and 667 m. Despite both females and males being localized on average an equal number of times per night (mean pairs of localizations per night = 2.48 and 2.53 for females and males, respectively), and the fact that the maximum distance moved was more or less similar for both sexes (Female 4: 605 m and male 9: 667 m), female total movement per night was overall higher than that of males (X^2 = 4.65. df = 1, p = 0.03). The median number of meters moved per night for females was 227, and the median for males was only 70, and this difference was largely due to the fact that males often did not move, or moved very little, between pairs of localizations. However, the variation within sexes was large, and p-values from mixed logistic regression models must be interpreted with some caution [60].

3.4. Nest Use

The six hedgehogs used a total of 28 different nest sites (see Table 5); 18 of these nests were used by males and 10 by females. The number of nest sites used per individual varied from one to 10, and the maximum of 10 nest sites was that of a male, and the second of seven nest sites was that of a female. The individual hedgehogs switched nests from zero to 14 times during the tracking periods; the maximum of 14 switches was both by a male and a female. Three females did a total of 15 switches, and four males did 21 switches, so the total number of nest switches was 36. No hedgehog used the nest of another hedgehog. Approximately 16 of the 28 nests were under/inside a building (garage, porch, veranda) and 12 were in natural habitat (hedge, bush, forest). Nest sites of six hedgehogs are shown in Supplementary Figure S1.

Table 5. Nest use by the seven hedgehogs of which this was documented. The table shows the total number of nests (Nests) and the number of nest switches (Nest switches), as well as the nest location and nest material when this was possible to observe. Numbers in parenthesis in nest location and nest material represent the number of nests observed per individual. A date of hibernation is provided for those three individuals that were monitored until this happened. Explanation of abbreviations. Nest location: B = Building, V = Vegetation. Nest material: P = Pine needles, T = Trash, L = Leaves, G = Grass, M = Moss, Y = Yew needles, and N = No material.

Individual	Nests	Nest Switches	Nest Location	Nest Material	Date Hibernation
F1	2	1	B (2)	-	-
M3	10	14	B (5), V (5)	P (1), T (1), L (1), G/L/M (2)	12 September
F4	1	0	V (1)	L (1)	-
M5	3	2	B (3)	G/L (2)	-
M7	1	0	B (1)	G/L (1)	-
F8	7	14	B (4), V (3)	G (1), G/L (2)	29 September
M9	4	5	B (1), V (3)	N (1), Y/T (1)	16 September
TOTAL	28	36			

Only three individuals remained tagged with a radio transmitter until hibernation: female nr 8; male nr 3; and male nr 9, and these individuals did not have overlapping home ranges. Male nr 3 stayed in the area around the campus and botanical garden at the University of Agder. The first observations were made in a residential area south of the university campus, but later they moved to campus and went hibernating under the root of a pine tree (*Pinus sylvestris*) in a small patch of natural wood with pine trees, scrub, heather and tall grass. The 15th of September was the last time the hedgehog switched nests, and we suppose that he went hibernating. Female nr 8 moved back and forth between nest sites under buildings by crossing a road that has high traffic during the daytime. The 29th of September was the last time this hedgehog switched nests, and it established itself in a small patch of natural habitat with blackberries (*Rubus* sp.), under the root of a big deciduous tree, next to a rock, and we assume that she went hibernating from this date. Male nr 9 was moving between a nest in a private garage and a nest in a field of planted bushes along a road. There was no nesting material under these bushes, apart from a few small, dry leaves and a little trash. For hibernation, he found a patch of planted yew (*Taxus baccata*) along a road where he made a burrow under the roots. The 16th of September was the last time this hedgehog switched nests. All three hedgehogs chose hibernation sites under tree roots in as many natural environments as they could possibly find.

3.5. Nesting Material

We observed the nesting material in 14 of the 28 nests, belonging to six of the individuals (Table 5). The nesting material was primarily sourced from locally available vegetation, but two nests consisted completely or partially of plastic and paper trash, and at one nest site, the hedgehog slept openly in a flowerbed. One nest consisted only of grass, two consisted of pine/yew needles mixed with trash, two consisted of leaves, and eight consisted of a mix of grass, leaves, and/or moss. Two of the three hibernating hedgehogs found sufficient nesting material in the surroundings in the chosen small patches of woodland, but the last hedgehog (in a planted yew-bed) only found yew needles and some trash as nesting material. We were worried that this would not be sufficient protection during the winter, but fortunately this individual managed to dig down under the roots after a few days, and we hope that this was enough for it to survive the winter.

4. Discussion

4.1. Home Range

To investigate hibernation nest sites, our data were collected in late summer and autumn. We did not find a statistical difference in the home range size between the two sexes in our study at this time of year. This lack of difference could be due to the low sample size in our study, as the variation in estimates was great, especially among males. In Italy, Bottani and Reggiani [33] did not find any significant sex difference in home range size, but that home range sizes varied greatly (more than 10-fold), with the largest being more than 100 ha. Although we did not detect home range sizes near this size, we also found substantial variation in home range sizes for both sexes (Figure 3, Table 3). Reeve and Morris [61] found that male hedgehogs on a golf course in England had larger home ranges and faster and longer movements than females, and both Reeve and Morris [61] and Bottani and Reggiani [33] report a mating season from May until September, which is different from the seasonality in home range size between the genders that we find in the Nordic countries.

Male and female home ranges of similar sizes late in the season, as we found in our study, are also described elsewhere in Fennoscandia. Close to the northern boundary of the species' distribution, hedgehogs seem to terminate the mating season earlier, and there is a seasonal change in home range sizes and movements, with males having the larger home ranges and traveling the longest in spring/early summer and with less or no sex difference in autumn [26,36,38]. The study area of our choice in Kristiansand, Norway (58° N) is comparable to that of Rautio, Valtonen, Auttila, and Kunnasranta [38] and Rautio, Valtonen, Auttila, and Kunnasranta [41] in Joensuu, Finland (63° N), though Joensuu is situated even further north and east in Fennoscandia and thus probably has a somewhat colder climate. Rautio, Valtonen, and Kunnasranta [38] found that, in the pre-hibernation period, adult males and females had a home range of 17 and 29 ha, respectively, and our home range estimates are of similar size (95% KUD ranges from 11 to 22.2 ha). Also, the body weight of the hedgehogs in the pre-hibernation period was very similar in Finland and Norway, with males being heavier than females. In Kristiansand, the male average weight was 1272 g ($n = 5$) and female 1007 g ($n = 4$), and in Finland, the male average weight was 1286 g ($n = 5$), females 958 g ($n = 4$) [38]. Food availability within the home range might affect home range size, and the availability of food in urban and wild environments might be quite different [28]. Uneven food availability within the study area might be the reason for the individual differences in home range size that we have observed, but unfortunately, we do not have data to test this.

The precision of home range size estimates is sensitive to the sample size [54]. In Finland, Rautio, Valtonen, and Kunnasranta [38] showed that less than 30 localizations from individual hedgehogs should be enough to provide unbiased home range sizes, and Pettett, Moorhouse, Johnson, and Macdonald [22] found that 20 localizations were sufficient. However, our incremental analysis indicates that we, with as many as 50+ localizations, do not have a sufficient sample size to ensure unbiased estimates (Supplementary Figure S2). Indeed, for some of our individuals, the localizations are quite uniformly distributed within the home range, with no clear core areas (see individuals 4 and 5 in Figure 2), while others have two core areas and/or MCP home ranges that are clearly affected by one or a few outliers (see individuals 2 and 9 in Figure 2). We therefore expect our estimates to be somewhat negatively biased.

We found that our hedgehogs in Kristiansand of both the same and opposite sex ha overlapping home ranges, which support the findings in a large number of studies (se e.g., Campbell [34]; Parkes [37]; Reeve [39]; Bottani & Reggiani [33]; Kristiansson [36 Riber [51]; Rautio et al. [38]). Studies in our neighboring countries, Finland and Denmar have shown that females tend to have fewer overlapping home ranges and core areas tha males during late summer and autumn, as mutual avoidance can ensure enough foo availability to increase fat deposits before winter [38,51]. In Ireland, Haigh et al. [62] foun that each hedgehog occupied a distinct area of the arable field and rarely crossed the pat of another. Cassini and Föger [63] found that hedgehogs showed mutual avoidance an suggested that this imposes a limit on the number of animals in an area. We did not observ any hedgehogs at the same spot as another hedgehog in our study, which might indicat that they are solitary animals practicing mutual avoidance in time, despite having, at leas partially, overlapping home ranges.

Two of the marked hedgehogs were found dead and scavenged 3 and 16 days afte marking, respectively, and although we could not establish the cause of death, it is possib that these two individuals were killed by a predator. Predators are generally very rare i the study area. The presence of badgers, however, has been shown to restrict hedgeho movement and foraging and lead to smaller home ranges [22,64]. Badgers have bee observed within our study area, but rare reports and the fact that we did not observ any during this study suggest that direct or indirect interactions between badgers an hedgehogs were uncommon. None the less, it is possible that some of the variation in spac use among the hedgehogs included in our study can be explained by the local presence o one or more badgers.

4.2. Movement

We did not find any significant difference in the distance moved per hour betwee the sexes. Both males and females moved as much as 300–400 m/h, but this was n common, and we also observed individuals (mostly males) not moving at all betwee successive localizations at any given time of night. We believe that the timing of our stud the pre-hibernation period, August to September, is the explanation for this observe pattern. Adult females may use longer time to feed and build up a fat layer before wint after reproduction in summer and therefore need to move more actively around in th pre-hibernation period in order to locate enough food, while males have had longer time feed since the mating period in the spring. This is partially supported by Kristiansson [36 who found that males in Sweden traveled longer than females in the mating season bu that males and females traveled similar distances in the post-mating season.

Based on this, we would expect similar distances between males and females, bu in our study, both sexes moved considerably shorter than what has been reported fro Sweden [36]. This difference in distance moved between two otherwise quite simila climatic regions may depend on both the nature of the habitat, food availability, and th time interval and length of time of the registration points. The frequency of registratio during the night was higher in the study of Kristiansson [36] with locations every 15 mi while we usually had more than one hour between our registrations and sometimes sever hours. We expect that these longer intervals will lead to a higher degree of underestimatic since this method only registers the shortest distance between two localizations. In additic Kristiansson [36], did his study in a small town that could be considered rural. Further, has been suggested that there is higher food availability in urban environments compare to rural environments [9], and this could lead to hedgehogs in our study not having to mo as much to find sufficient food. Hedgehogs spend most of the night in the post-matir season to forage [29,35,36], and we therefore suppose that the shorter distance travele in our study is related to easier access to food. On the other hand, a study in an urba environment in the UK [24] found much higher travel distances than in our study, bu gardens in our study were mostly open and easy to access, while urban gardens in the U are usually smaller and harder to access [65], and this could result in the hedgehogs having to travel longer distances to find sufficient food.

We detected no active hedgehogs before 22:00, and most individuals left the nest around 23:00, when the neighborhood traffic, by car or foot, was strongly reduced. Almost all anthropogenic disturbances in this area had ceased by midnight. Dowding, Harris, Poulton, and Baker [24] found that hedgehogs were more active after midnight as a risk-reducing behavior. Based on this, we suspect that the hedgehogs in our study may have started their nightly movements as late as 23:00 to minimize exposure to human activity. Most of our hedgehogs increased their average movements after 03:00, and this was the portion of the nights with minimum disturbances. We did observe that the hedgehogs consistently began moving towards their nests from around 03:00, and thus it seems that this increased movement in the late part of the night is linked to the need to find shelter before dawn when human activity increased.

4.3. Nests

We did not find any evidence of nest sharing in our study, and this might be the normal pattern of nest use by hedgehogs. Nest-sharing simultaneously in wild-living hedgehogs is rare, though non-simultaneous nest-sharing has been documented [32,41,51,61,62,66]. Nest sharing can increase transmission of ectoparasites [41,61], and this could be the reason why nest sharing is rarely observed among hedgehogs. Ectoparasite exposure might also explain why hedgehogs alternate between several nests, as this can reduce parasite exposure (see, e.g., Stanback and Dervan [67]; Bize et al. [68]). However, we never observed ectoparasites in our study, suggesting that ectoparasite burdens overall were low.

In total, we found 28 different nest sites used by seven hedgehogs (the remaining two individuals were never observed in a nest during the short period they were tracked), and 18 and 10 nests were used by males and females, respectively. Three males and three females switched nests a total of 21 and 14 times, respectively. Bottani and Reggiani [33] in Italy and Reeve and Morris [61] in England found that females used the same nest repeatedly for periods significantly longer than the males, and that this was due to reproduction or because the males, with larger home ranges, used several nests rather than having to move long distances to reach the same one. In our study, there were no clear differences between sexes, neither in the number of switches nor the number of nests used, but large differences between individuals overall (Table 5), and this is in accordance with several other studies [35,41,61,69].

The nest is a very significant feature in a hedgehog's life, particularly during hibernation [42], determining both its habitat choice and distribution. The nest location habitat has been studied in several countries: the UK ([22,24,42,61,62], New Zealand [69], Denmark [26,44,51], and Finland [41]. The most common nesting habitat is hedgerow or forest [22,51,62,70,71] and forest is especially preferred for hibernation [44,70,72]. In Finland, Rautio et al. [72] observed that the hedgehogs in urban environments still preferred to hibernate under tree roots in forest patches, exactly as we observed for our three hedgehogs in Kristiansand. Such habitats are usually limited in urban environments, and our result indicates that maintenance of natural forest patches in urban areas where hedgehogs exist may improve habitat suitability, and thus survival, and help mitigate the present decline in hedgehog numbers. Hedges and hedgerows, often common in urban environments, are known to be of special importance for hedgehogs during the active season since they offer shelter, nest locations, and food [22,25,51,62,70,71,73]. We also observed that hedgehogs were often found in hedges, both during their nightly movements and when sleeping during the day. Manmade constructions, such as playhouses, sheds, porches, and terraces, were also often used as nest locations.

Leaves or grass are known to be the two most important nesting materials for hedgehogs [22,26,41,42,44,51,61,71,73], and we also found that leaves and grass were most often used, but that hedgehogs used other material available close to the nest site, regardless of source. Nest sites under dense hedgerows, which do provide protection from predators, did not necessarily offer sufficient nesting material for hibernation, and the hedgehogs abandoned such habitats and rather utilized the more limited forest patches when preparing for hibernation. We worry that lack of proper nesting material and limited availability of forest patches may increase winter mortality in urban environments, especially in more northern regions where winter temperatures can be more challenging.

The hedgehogs in our study went hibernating in September, comparable to what is described in Finland [41], but much earlier than what is common in other places in Europe [33,62]. We do not know what initiated the start of hibernation in our study, but we assume that a combination of sufficient energy storage and external zeitgebers triggered this. The fact that hedgehogs in Finland, with comparable body weights, also initiated hibernating in September suggests that this is the normal behavior of hedgehogs in the extreme north of the species distribution in Europe.

5. Conclusions

This study offers new knowledge of hedgehog home range sizes, movements, and nesting behavior in suburban areas close to this species' northern boundary. Home range size, distance moved, and the number of nests varied greatly between individuals, and no sex effect was observed. Our results indicate that hedgehogs in urban environments prefer forest patches and use natural nesting material when preparing for hibernation, and this supports the findings from other studies in Fennoscandia. We therefore conclude that it is important to maintain existing and establish new forest patches in urban environments to ensure hedgehog winter survival in the north. It would also be beneficial to encourage garden owners to contribute by providing shelters with ample natural nesting material in their gardens.

Supplementary Materials: The following supporting information can be downloaded at: https://www.mdpi.com/article/10.3390/ani14010130/s1, Figure S1: 100% MCP home ranges of all six individuals. Open symbols represent localization where the animal was active (during night) and closed symbols represent nest sites. Due to GPS-inaccuracy multiple nest sites positioned close together represent a nest with several localization, but at exactly the same place. Figure S2: Home range asymptote figure with maximum convex polygon (MCP) area as a function of percentage of data used for each individual hedgehog. The shape of the curves suggest that our home range estimates are underestimated, as the asymptotes still rise at 100% MCPs.

Author Contributions: Conceptualization, L.M.K. and B.S.J.; Methodology, L.M.K. and B.S.J.; Software, L.M.K., M.S.F., M.M.S.A., A.L. and K.M.V.L.; Validation, L.M.K. and B.S.J.; Formal Analysis, L.M.K.; Investigation, M.S.F., M.M.S.A., A.L., K.M.V.L., L.M.K. and B.S.J.; Data curation and statistics, L.M.K.; Writing—original draft, M.S.F., M.M.S.A., A.L. and K.M.V.L.; Writing—Review and Editing, B.S.J. and L.M.K.; Visualization, L.M.K. and B.S.J.; Supervision, L.M.K. and B.S.J.; Project Administration, B.S.J. and L.M.K.; Funding Acquisition, B.S.J. All authors have read and agreed to the published version of the manuscript.

Funding: This research was funded by the Natural History Museum and Botanical Garden and Department of Natural Sciences, UiA.

Institutional Review Board Statement: The animal study protocol, including capture and marking, was approved by the Norwegian Food Safety Authority (FOTS ID 27113, 24 March 2021) and the Norwegian Environment Agency (ref. 2022/7181, 7 July 2022).

Informed Consent Statement: Not applicable.

Data Availability Statement: The data presented in this study are available on request from the corresponding author.

Acknowledgments: The authors would like to thank the reviewers for their valuable feedback during the submission process. Thanks to Raul Ramirez, museum director, who supported the project by providing project funding and positivity. We also thank the citizens and house owners in Kristiansand for providing information about hedgehogs and letting us enter their gardens.

Conflicts of Interest: The authors declare no conflicts of interest. The funders had no role in the design of this study, in the collections, analyses, or interpretation of data, in the writing of this manuscript, or in the decision to publish the results.

References

Antrop, M. Landscape change and the urbanization process in Europe. *Landsc. Urban Plan.* **2004**, *67*, 9–26. [CrossRef]

Sun, L.; Chen, J.; Li, Q.; Huang, D. Dramatic uneven urbanization of large cities throughout the world in recent decades. *Nat. Commun.* **2020**, *11*, 5366. [CrossRef] [PubMed]

Elmqvist, T.; Goodness, J.; Marcotullio, P.J.; Parnell, S.; Sendstad, M.; Wilkinson, C.; Fragkias, M.; Güneralp, B.; McDonald, R.I.; Schewenius, M.; et al. *Urbanization, Biodiversity and Ecosystem Services: Challenges and Opportunities: A Global Assessment*; Springer Nature: New York, NY, USA, 2013; p. 755.

Shochat, E.; Lerman, S.B.; Anderies, J.M.; Warren, P.S.; Faeth, S.H.; Nilon, C.H. Invasion, Competition, and Biodiversity Loss in Urban Ecosystems. *BioScience* **2010**, *60*, 199–208. [CrossRef]

Rodewald, A.D.; Shustack, D.P. Consumer resource matching in urbanizing landscapes: Are synanthropic species over-matching. *Ecology* **2008**, *89*, 515–521. [CrossRef] [PubMed]

Chace, J.F.; Walsh, J.J. Urban effects on native avifauna: A review. *Landsc. Urban Plan.* **2006**, *74*, 46–69. [CrossRef]

Hubert, P.; Julliard, R.; Biagianti, S.; Poulle, M.-L. Ecological factors driving the higher hedgehog (*Erinaceus europeaus*) density in an urban area compared to the adjacent rural area. *Landsc. Urban Plan.* **2011**, *103*, 34–43. [CrossRef]

McCleery, R. Urban Mammals. In *Urban Ecosystem Ecology*; Aitkenhead-Peterson, J., Volder, A., Eds.; Agronomy Monographs; Wiley: Hoboken, NJ, USA, 2010; pp. 87–102.

Baker, P.J.; Harris, S. Urban mammals: What does the future hold? An analysis of the factors affecting patterns of use of residential gardens in Great Britain. *Mammal Rev.* **2007**, *37*, 297–315. [CrossRef]

Hof, A.R.; Bright, P.W. Quantifying the long-term decline of the West European hedgehog in England by subsampling citizen-science datasets. *Eur. J. Wildl. Res.* **2016**, *62*, 407–413. [CrossRef]

Huijser, M.P.; Bergers, P.J.M. The effect of roads and traffic on hedgehog (*Erinaceus europaeus*) populations. *Biol. Conserv.* **2000**, *95*, 111–116. [CrossRef]

Krange, M. *Change in the Occurrence of the West European Hedgehog (Erinaceus europaeus) in Western Sweden during 1950–2010*; Karlstad University: Karlstad, Sweden, 2015.

Müller, F. Langzeit-Monitoring der Strassenverkehrsopfer beim Igel (*Erinaceus europaeus* L.) zur Indikation von Populations-dichteveränderungen entlang zweier Teststrecken im Landkreis Fulda. *Beiträge Naturkunde Osthess.* **2018**, *54*, 21–26.

Taucher, A.L.; Gloor, S.; Dietrich, A.; Geiger, M.; Hegglin, D.; Bontadina, F. Decline in Distribution and Abundance: Urban Hedgehogs under Pressure. *Animals* **2020**, *10*, 1606. [CrossRef]

van de Poel, J.L.; Dekker, J.; van Langevelde, F. Dutch hedgehogs *Erinaceus europaeus* are nowadays mainly found in urban areas, possibly due to the negative effects of badgers *Meles meles*. *Wildl. Biol.* **2015**, *21*, 51–55. [CrossRef]

Williams, B.M.; Baker, P.J.; Thomas, E.; Wilson, G.; Judge, J.; Yarnell, R.W. Reduced occupancy of hedgehogs (*Erinaceus europaeus*) in rural England and Wales: The influence of habitat and an asymmetric intra-guild predator. *Sci. Rep.* **2018**, *8*, 12156. [CrossRef] [PubMed]

Wilson, E.; Wembridge, D. *The state of Britain's Hedgehogs 2018*; People's Trust for Endangered Species: London, UK, 2018.

Eldegard, K.; Syvertsen, P.O.; Bjørge, A.; Kovacs, K.; Støen, O.-G.; van der Kooij, J. Pattedyr: Vurdering av Piggsvin *Erinaceus europaeus* for Norge. Rødlista for Arter 2021. Artsdatabanken. Available online: https://www.artsdatabanken.no/lister/rodlisteforarter/2021/6278 (accessed on 23 March 2023).

Mathews, F.; Harrower, C. *IUCN—Compliant Red List for Britain's Terrestrial Mammals. Assessment by the Mammal Society under Contract to Natural England, Natural Resources Wales and Scottish Natural Heritage*; Natural England: Worcester, UK, 2020.

Meinig, H.; Boye, P.; Dähne, M.; Hutterer, R.; Lang, J. *Rote Liste und Gesamtartenliste der Säugetiere (Mammalia) Deutschlands*; BfN-Schriften Vertrieb im Landwirtschaftsverlag: Bonn, Germany, 2020.

Doncaster, C.P.; Rondinini, C.; Johnson, P.C.D. Field test for environmental correlates of dispersal in hedgehogs *Erinaceus europaeus*. *J. Anim. Ecol.* **2001**, *70*, 33–46. [CrossRef]

Pettett, C.E.; Moorhouse, T.P.; Johnson, P.J.; Macdonald, D.W. Factors affecting hedgehog (*Erinaceus europaeus*) attraction to rural villages in arable landscapes. *Eur. J. Wildl. Res.* **2017**, *63*, 1–12. [CrossRef]

Young, R.P.; Davison, J.; Trewby, I.D.; Wilson, G.J.; Delahay, R.J.; Doncaster, C.P. Abundance of hedgehogs (*Erinaceus europaeus*) in relation to the density and distribution of badgers (*Meles meles*). *J. Zool.* **2006**, *269*, 349–356. [CrossRef]

Dowding, C.V.; Harris, S.; Poulton, S.; Baker, P.J. Nocturnal ranging behaviour of urban hedgehogs, *Erinaceus europaeus*, in relation to risk and reward. *Anim. Behav.* **2010**, *80*, 13–21. [CrossRef]

Hof, A.R.; Bright, P.W. The value of green-spaces in built-up areas for western hedgehogs. *Lutra* **2009**, *52*, 69–82.

Rasmussen, S.L.; Berg, T.B.; Dabelsteen, T.; Jones, O.R. The ecology of suburban juvenile European hedgehogs (*Erinaceus europaeus*) in Denmark. *Ecol. Evol.* **2019**, *9*, 13174–13187. [CrossRef]

Turner, J.; Freeman, R.; Carbone, C. Using citizen science to understand and map habitat suitability for a synurbic mammal in an urban landscape: The hedgehog *Erinaceus europaeus*. *Mammal Rev.* **2022**, *52*, 291–303. [CrossRef]

Yalden, D.W. The food of the hedgehog in England. *Acta Theriol.* **1976**, *21*, 401–424. [CrossRef]

29. Morris, P. *Piggsvinboka*; Aschehoug: Oslo, Norway, 1987.
30. Reeve, N.J.; Huijser, M.P. Mortality factors affecting wild hedgehogs: A study of records from wildlife rescue centres. *Lutra* 199 42, 7–24.
31. Johansen, B.S. *Barnas Bok om Piggsvin*; Damm: Oslo, Norway, 2000.
32. Reeve, N. *Hedgehogs*; T & D Poyser: London, UK, 1994.
33. Bottani, L.; Reggiani, G. Movements and activity patterns of hedgehogs (*Erinaceus europaeus*) in Mediterranean coastal habitats. *Säugetierkunde* **1984**, *49*, 193–206.
34. Campbell, P.A. *Feeding Behaviour of Erinaceus europaeus in New Zealand Pasture*; University of Canterbury: Christchurch, Ne Zealand, 1973.
35. Haigh, A.J. *The Ecology of the European hedgehog (Erinaceus europaeus) in Rural Ireland*; University College: Cork, UK, 2011.
36. Kristiansson, H. *Ecology of a Hedgehog Erinaceus europaeus Population in Southern Sweden*; University of Lund: Lund, Sweden, 198
37. Parkes, J. Some aspects of the biology of the hedgehog (*Erinaceus europaeus* L.) in the Manawatu, New Zealand. *N. Z. J. Zool.* **197** 2, 463–472. [CrossRef]
38. Rautio, A.; Valtonen, A.; Kunnasranta, M. The Effects of Sex and Season on Home Range in European Hedgehogs at the Norther Edge of the Species Range. *Ann. Zool. Fenn.* **2013**, *50*, 107–123. [CrossRef]
39. Reeve, N.J. The home range of the hedgehog as revealed by a radio tracking study. In Proceedings of the Symposia of th Zoological Society of London, London, UK, 11–12 November 1982; pp. 207–230.
40. Harestad, A.S.; Bunnel, F.L. Home Range and Body Weight—A Reevaluation. *Ecology* **1979**, *60*, 389–402. [CrossRef]
41. Rautio, A.; Valtonen, A.; Auttila, M.; Kunnasranta, M. Nesting patterns of European hedgehogs (*Erinaceus europaeus*) und northern conditions. *Acta Theriol.* **2014**, *59*, 173–181. [CrossRef]
42. Morris, P. Winter nests of the hedgehog (*Erinaceus europaeus* L.). *Oecologia* **1973**, *11*, 299–313. [CrossRef] [PubMed]
43. Morris, P. *Hedgehogs*; Whittet Books Ltd: Stansted Mountfitchet, UK, 2014; Volume 4.
44. Jensen, A.B. Overwintering of European hedgehogs *Erinaceus europaeus* in a Danish rural area. *Acta Theriol.* **2004**, *49*, 145–15 [CrossRef]
45. Kristiansson, H. Distribution of the European hedgehog (*Erinaceus europaeus* L.) in Sweden and Finland. *Ann. Zool. Fenn.* **1981**, *1* 115–119.
46. Walhovd, H. Partial arousals from hibernation in hedgehogs in outdoor hibernacula. *Oecologia* **1979**, *40*, 141–153. [CrossRef]
47. Artsobservasjoner. Reported Findings of Erinaceus Europaeus (in the Period: 01.05.2022–10.10.2022). Available online: http //www.artsobservasjoner.no/ (accessed on 24 March 2023).
48. Johansen, B.S. Piggsvin i Kristiansand. 2019. Available online: https://naturblogg.uia.no/piggsvin-i-kristiansand/ (accessed c 15 September 2023).
49. Morris, P.A. A study of home range and movements in the hedgehog (*Erinaceus europaeus*). *J. Zool.* **1988**, *214*, 433–449. [CrossRe
50. *R Core Team R: A Language and Environment for Statistical Computing*; R Foundation for Statistical Computing: Vienna, Austr 2022.
51. Riber, A.B. Habitat use and behaviour of European hedgehog *Erinaceus europaeus* in a Danish rural area. *Acta Theriol.* **2006**, *5* 363–371. [CrossRef]
52. Bivand, R.S.; Pebesma, E.; Gómez-Rubio, V. *Applied Spatial Data Analysis with R*, 2nd ed.; Springer: New York, NY, USA, 2013.
53. Calenge, C. The package "adehabitat" for the R software: A tool for the analysis of space and habitat use by animals. *Ecol. Mod* **2006**, *197*, 516–519. [CrossRef]
54. Seaman, D.E.; Millspaugh, J.J.; Kernohan, B.J.; Brundige, G.C.; Raedeke, K.J.; Gitzen, R.A. Effects of Sample Size on Kernel Hom Range Estimates. *J. Wildl. Manag.* **1999**, *63*, 739–747. [CrossRef]
55. Row, J.R.; Blouin-Demers, G. Kernels Are Not Accurate Estimators of Home-range Size for Herpetofauna. *Copeia* **2006**, *200* 797–802. [CrossRef]
56. Nilsen, E.B.; Pedersen, S.; Linnell, J.D.C. Can minimum convex polygon home ranges be used to draw biologically meaningf conclusions? *Ecol. Res.* **2008**, *23*, 635–639. [CrossRef]
57. Hijmans, R.J. Geosphere: Spherical Trigonometry, R Package Version 1.5–18. 2022. Available online: https://CRAN.R-proje org/package=geosphere (accessed on 3 October 2023).
58. Bates, D.; Mächler, M.; Bolker, B.M.; Walker, S.C. Fitting Linear Mixed-Effects Models Using lme4. *J. Stat. Softw.* **2015**, *67*, 1–4 [CrossRef]
59. Burnham, K.P.; Anderson, D.R. *Model Selection and Multimodel Inference—A Practical Information-Theoretic Approach*, 2nd e Springer: New York, NY, USA, 2002; p. 488.
60. Zuur, A.F.; Ieno, E.N.; Walker, N.J.; Saveliev, A.A.; Smith, G.M. *Mixed Effects Models and Extensions in Ecology with R*; Spring New York, NY, USA, 2009.
61. Reeve, N.J.; Morris, P.A. Construction and use of summer nests by the hedgehog (*Erinaceus europaeus*). *Mammalia* **1985**, *49*, 187–1 [CrossRef]
62. Haigh, A.; O'Riordan, R.M.; Butler, F. Nesting behaviour and seasonal body mass changes in a rural Irish population of th Western hedgehog (*Erinaceus europaeus*). *Acta Theriol.* **2012**, *57*, 321–331. [CrossRef]
63. Cassini, M.H.; Föger, B. The effect of food distribution on habitat use of foraging hedgehogs and the ideal non-territorial despo distribution. *Acta Oecologica* **1995**, *16*, 657–669.

4. Pettett, C.E.; Johnson, P.J.; Moorhouse, T.P.; Hambly, C.; Speakman, J.R.; Macdonald, D.W. Daily energy expenditure in the face of predation: Hedgehog energetics in rural landscapes. *J. Exp. Biol.* **2017**, *220*, 460–468. [CrossRef] [PubMed]
5. Gazzard, A.; Boushall, A.; Brand, E.; Baker, P.J. An assessment of a conservation strategy to increase garden connectivity for hedgehogs that requires cooperation between immediate neighbours: A barrier too far? *PLoS ONE* **2021**, *16*, e0259537. [CrossRef]
6. Recio, M.R. Crowded house: Nest sharing among solitary European hedgehogs in New Zealand. *Front. Ecol. Environ.* **2016**, *14*, 225–226. [CrossRef]
7. Stanback, M.T.; Dervan, A.A. Within-season nest-site fidelity in Eastern Bluebirds: Disentangling effects of nest success and parasite avoidance. *Auk* **2001**, *118*, 743–745. [CrossRef]
8. Bize, P.; Roulin, A.; Richner, H. Adoption as an offspring strategy to reduce ectoparasite exposure. *Proc. R. Soc. Lond. Ser. B Biol. Sci.* **2003**, *270*, S114–S116. [CrossRef] [PubMed]
9. Moors, P.J. Observations on the nesting habits of the European hedgehog in the Manawatu sand country, New Zealand. *N. Z. J. Zool.* **1979**, *6*, 489–492. [CrossRef]
10. Huijser, M.P. *Life on the Edge—Hedgehog Traffic Victims and Mitigation Strategies in an Anthropogenic Landscape*; Wageningen University & Research: Wageningen, The Netherlands, 2000.
11. Yarnell, R.W.; Pettett, C.E. Beneficial land management for hedgehogs (*Erinaceus europaeus*) in the United Kingdom. *Animals* **2020**, *10*, 1566. [CrossRef]
12. Rautio, A. *On the Northern Edge: Ecology of Urban Hedgehogs in Eastern Finland*; University of Eastern Finland: Kuopio, Finland, 2014.
13. Bearman-Brown, L.E.; Baker, P.J.; Scott, D.; Uzal, A.; Evans, L.; Yarnell, R.W. Over-Winter Survival and Nest Site Selection of the West-European Hedgehog (*Erinaceus europaeus*) in Arable Dominated Landscapes. *Animals* **2020**, *10*, 1449. [CrossRef]

 *animals*

Article

Characteristics and Demography of a Free-Ranging Ethiopian Hedgehog, *Paraechinus aethiopicus,* Population in Qatar

Carly Pettett [1], David W. Macdonald [1], Afra Al-Hajiri [2], Hayat Al-Jabiry [2] and Nobuyuki Yamaguchi [2,3,*]

[1] Wildlife Conservation Research Unit, Department of Zoology, Recanati-Kaplan Centre, University of Oxford, Tubney House, Abingdon Road, Tubney, Oxfordshire OX13 5QL, UK; carly.pettett@gmail.com (C.P.); david.macdonald@zoo.ox.ac.uk (D.W.M.)

[2] Department of Biological and Environmental Sciences, Qatar University, P.O. Box 2713 Doha, Qatar; xxafra@yahoo.com (A.A.-H.); haljabiry@qu.edu.qa (H.A.-J.)

[3] Institute of Tropical Biodiversity and Sustainable Development, Universiti Malaysia Terengganu, Kuala Nerus 21030, Terengganu, Malaysia

* Correspondence: nobuyuki.yamaguchi@umt.edu.my; Tel.: +60-9-668-3629

Received: 3 May 2020; Accepted: 28 May 2020; Published: 30 May 2020

Simple Summary: Information on population characteristics of *Paraechinus* is valuable for ensuring long term survival of populations, however, studies are currently lacking. Here we investigate the population dynamics of Ethiopian hedgehogs based on a capture study in Qatar by fitting several statistical models. Over the 19 months of the study, we estimate a mean population of 60 hedgehogs, giving a density of 7 hedgehogs per km^2 in our 8.5 km^2 search area. The monthly abundance of hedgehogs decreased over the study and although survival was constant over the study period, with a mean monthly rate of 75%, there was a decline in the number of new entrants over time. We also studied these parameters over one year, excluding winter, and found that monthly estimates of juvenile and subadult survival decreased over time. We surmise that survival of juveniles may be a factor in the decrease in abundance and there may be implications for the persistence of this population in the future, with human influenced resources playing an important role. We caught between 91.3% and 100% of the estimated population at this site, indicating that our capture methodology was efficient. We conclude that the methodology used here is transferrable to other hedgehog species.

Abstract: Information on population characteristics of *Paraechinus* is valuable for ensuring long term survival of populations, however, studies are currently lacking. Here we investigate the population dynamics of Ethiopian hedgehogs based on a capture-mark-recapture study in Qatar by fitting Jolly-Seber and Cormack-Jolly-Seber models. Over the 19 months of the study, we estimate a mean population of 60 hedgehogs, giving a density of 7 hedgehogs per km^2 in our 8.5 km^2 search area. The monthly abundance of hedgehogs decreased over the study and although survival was constant over the study period, with a mean monthly rate of 75%, there was a decline in the number of new entrants over time. We also studied these parameters over one year, excluding winter, and found that monthly estimates of juvenile and subadult survival decreased over time. We surmise that survival of juveniles may be a factor in the decrease in abundance and there may be implications for the persistence of this population, with anthropogenic influenced resources playing an important role. We caught between 91.3% and 100% of the estimated population at this site, indicating that our capture methodology was efficient. We conclude that the methodology used here is transferrable to other hedgehog species.

Keywords: Arabia; arid environment; desert; Middle East; density; survival; capture; abundance; population dynamics; small mammal

---

1. Introduction

Hedgehogs are small terrestrial mammals with a spiny integument in the subfamily Erinaceinae, of which 16 species in five genera are currently recognized [1,2]. Extensive research on the European hedgehog (*Erinaceus europaeus*) has led to the common notion that hedgehogs are characteristic of the moist temperate environments of the world. However, many hedgehog species occur in arid and semi-arid environments, such as the "desert hedgehogs" of the genus *Paraechinus*, and yet, little is known about their ecology and behavior in these arid environments [1–10]. More than 25 years ago, in his monograph of hedgehogs, Reeve (1994) [1] expressed his frustration by stating "There is a frustrating lack of further studies ... in non-European hedgehogs ... There is a clear need for much more fundamental work on all these and other, as yet unstudied, hedgehog species". Sadly, although there has been some work on ecology, behavior, and physiology of non-European hedgehog species in the past 25 years [3–13], basic information about their population characteristics is still largely lacking.

The Ethiopian Hedgehog (*Paraechinus aethiopicus*, Ehrenberg, 1832), which is well-adapted to arid environments, has a wide distribution across North Africa and the Middle East, including the Arabian Peninsula [1,2], and is the only native hedgehog species in Qatar [14]. There has been some recent study on the habitat use and home range of the species [6,13], as well as on the timing of breeding [5,9], hibernation [3,8,9], and behavior in winter [11]. However, there are no previous studies on the population density and dynamics of *Paraechinus* hedgehogs.

There are several sampling methods that are applicable to hedgehogs that have previously been used to investigate local and national population density, mainly for *Erinaceus* species. These include spotlight surveys [15–18], footprint tunnel surveys [19–22], citizen science surveys [23–25], game bag surveys [26] and roadkill surveys [27]. These surveys are often used to assess occupancy rather than population density and demography. There has been a sparsity of long term demographic studies in all hedgehog species. There are some valuable capture-mark-recapture studies [28–30] that have investigated population dynamics and density of *E. europaeus*, but there are no such studies for other hedgehog species, including *Paraechinus*. Capture-mark-recapture methodology entails capturing and marking individuals then releasing them to re-mix with the local population. Individuals are then recaptured regularly over the study period, giving each individual a capture history. Two types of models can be fitted to these capture histories in order to estimate population size. The first are those for closed populations, where population size is assumed to be constant throughout the study period and there is no emigration or immigration [31,32]. The second is open population models such as Jolly-Seber and Cormack-Jolly-Seber models that can be used to estimate population size and parameters for survival and capture probability in an open population [33–36].

In this paper, we report, for the first time, the population dynamics of a free-ranging Ethiopian hedgehog population based on a capture-mark-recapture study in Qatar. We present data from a two year study to estimate hedgehog population size, growth rate, capture rate and survivability, in a discrete study area. The study of this population has not only resulted in population census methodology that is transferable to other hedgehog species but also allows for comparison of population density and dynamics with that of the better studied European hedgehog.

2. Materials and Methods

2.1. Study Area and Animal Capture

The study area consisted of ~15 km^2 of arid land around the Qatar University Farm (25°48′ N, 51°20′ E) in northern Qatar. The area included 11 active farms that were irrigated daily using underground water extracted through deep wells. Except for those farms, the area was an arid plain with a total annual precipitation of less than 100 mm, and the surface was predominantly covered by desert pavement with exposed loose gravels. The ambient air temperature ranges between ~5 °C in the early morning in winter and ~50 °C in the early afternoon in summer. There was little vegetation except for isolated short acacia tress and ephemeral grass patches emerging after rains in cold months. Various structures created by human activities, such as rubbish dumps, piles of abandoned building materials, and soil mounds, were ubiquitously found across the study area. Fieldwork was carried out between April 2010 and April 2012.

A consecutive four-night hedgehog capture survey was conducted, from dusk until dawn, once a month in an area of ~8.5 km^2 (regular survey area) by a field team of 1–3 individuals. Hedgehogs were captured by hand, usually curling into a ball, and were processed at the capture sites without anesthesia or sedation. Hedgehogs were individually marked by painting the spines with unique combinations of nail polishes of different colors, and sexed before they were released. A hedgehog was classified as a juvenile if an animal was less than six months old or if it was a new individual and weighed less than 200 g. Each hedgehog was only processed once during the four night survey. A substantial amount of capturing efforts was made (1) around the "Rubbish Mound" (Figure 1, location ①) where a higher concentration of hedgehogs was found throughout the year probably due to year-round availability of food resources (although the rubbish mound was partially cleared in March 2011, and a further major cleaning operation started in March 2012); (2) "Municipal Farm" (Figure 1, location ②) where permanent grass fields seemed to produce rich invertebrate communities seasonally; (3) "Rawdat Al Faras Farm" (Figure 1, location ③); and (4) Qatar University Farm where the field station was located (Figure 1, location ④). In addition to captures at those sites, hedgehogs were captured wherever and whenever they were found in the regular survey area. Searching was carried out as follows: firstly each farm and the rubbish mound were searched because hedgehogs tended to nest in these areas and could be captured as they emerged from their nests. Each farm was searched once and the rubbish mound twice during this time, finishing around 22:00. The rest of the study area was then searched from the north to the south, with the aim of randomly encountering hedgehogs. This area was completely searched over the four days but the whole area was not covered every evening. We divided the year into four hedgehog seasons as follows; Early Breeding Season; February–April, Late Breeding Season; May–July, Autumn Season; August–October and Winter Season; November–January [5].

2.2. Data Analysis

The statistical analysis of the capture-recapture data was carried out in the R package "RCapture" (R Core Team 2014), following the paper 'Rcapture: Loglinear Models for Capture-Recapture in R' by Baillargeon and Rivest [37]. This package cannot handle irregular capture intervals, and because some months of captures were missed in early 2010, we decided to subset the data into two blocks; all continuous months in the study (19 months from October 2010 to April 2012) and excluding the Winter Season (nine months from February 2011 to October 2011). Analyzing the data excluding the Winter Season allowed us to look at population demographics over one year of hedgehog activity, where the capture rate is not affected by the change of behavior that this hedgehog species (and other hedgehog species) exhibit over winter [3,6,11]. The analysis was also run separately for males and females. We also ran separate analyses for hedgehogs that were juveniles at first capture versus those deemed to be adults at first capture. The analyses on these different age cohorts were only performed on one year of data excluding winter months (2011) because over more than one year those hedgehogs deemed juveniles or at first capture would have become adults during the course of the study. However, the juvenile category does include those deemed a juvenile at first capture in 2010, i.e., by summer 2011 they would be subadults that have overwintered once. Therefore, in these analyses juveniles and these subadults were combined into one group and are hereafter called juveniles.

The RCapture package fits both open (Jolly-Seber and Cormack-Jolly-Seber) and closed population models to estimate N (the population size) along with parameters for capture probability at each sampling occasion, and survival and the number of new entrants between sampling occasions [37]. The study area was not a closed population and therefore the open models are most likely suitable for this data. However, we did some exploratory analysis using both closed and open models to confirm which fit the data best. Model fit was judged on Akaike Information Criterion (AIC) values, the lowest AIC being deemed as the best fitting model. Following Baillargeon and Rivest 2007 [37], we also examined heterogeneity plots of the capture histories, plotted the Pearson residuals from each model, and performed tests of model fit. If Pearson residuals were high or there appeared to be heterogeneity in the data then models were adjusted in a number of ways, for example, the model was re-run with capture histories with high residuals removed [37]. We also checked whether individuals captured at all sampling occasions or at only one may have had a big influence on the model fit. The output from the open population models in the Rcapture package includes a test for trap effect. The AIC value for the model including the trap effect was compared with that for a homogenous trap effect to investigate whether there was a difference in capture probability through time because of a behavioral response to capture. Finally, we ran the same models with equal capture probabilities defined and compared their fit to all of the models.

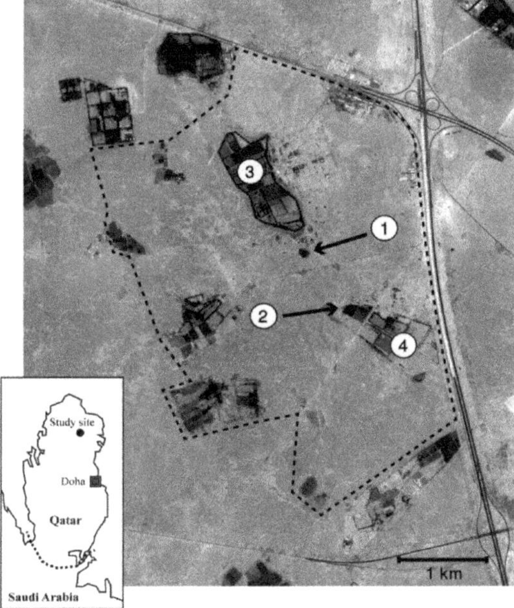

Figure 1. Map of the 15 km^2 study site in Qatar where Ethiopian hedgehogs were captured by hand as part of a monthly capture-recapture experiment from April 2010 to April 2012 (GoogleEarth Image Copyright 2018 DigitalGlobe). The dashed line indicates an 8.5 km^2 focal search area. Numbers indicate: ①: The "Rubbish Mound" where a higher concentration of hedgehogs was found throughout the study probably due to year-round availability of food resources; ②: "Municipal Farm" where permanent grass fields attracted hedgehogs; ③: Rawdat Al-Faras Research Station where street lights across the farm increased the chance of locating hedgehogs; ④: Qatar University Farm where the field station was located.

After selecting the best fitting models for each subset of our data, we obtained the total population estimate along with monthly population size, survival and capture rate for each block of data analyzed. We then tested potential differences in these estimates between males and females and adults and juveniles by constructing a series of linear models in R (R Core Team, Vienna, Austria, 2014). We examined the residuals of these linear models for normality and transformed the dependent where appropriate. Density was calculated based on the mean monthly N over the 8.5 km^2 regular search area. All means are presented as mean ± the standard error of the mean.

3. Results

We recorded 1190 captures between April 2010 and April 2012. Males were highly statistically significantly more likely to be captured (744 times) in comparison to females (427 times) (binominal test: $p < 0.001$), although the overall sex ratio of captured animals did not statistically significantly deviate from 1:1 (87 males and 74 females, binominal test: $p = 0.34$). Based on monthly data larger numbers of males were recorded than those of females throughout the study period. This bias towards male capture was statistically significantly more obvious during the Winter Season (ANOVA, $F_3 = 6.6$, $p = 0.003$) whilst there was no significant difference amongst the other hedgehog seasons (ANOVA, $F_2 = 0.52$, $p = 0.61$). This seasonal difference might be related to a statistically significantly smaller average number of female hedgehogs caught during the Winter Season (ANOVA, $F_3 = 5.3$, $p = 0.008$), whilst there was no such seasonality detected in males (ANOVA, $F_3 = 2.2$, $p = 0.12$). The foregoing results may suggest that the catchability of males was higher than females, and also female catchability decreased in winter. Therefore, this finding supports our decision to include some population estimates of males and females separately, in order to distinguish different patterns in male and female demography, and also to perform a separate analysis with winter excluded.

3.1. Model Fit

As predicted, open population models fit the data better than closed population models. In some cases exploratory plots of heterogeneity were u-shaped rather than linear; therefore showing that heterogeneity in the capture probabilities may be an issue (e.g., Figure 2). We were able to improve model fit by adjusting the models, for example removing those animals captured on every occasion to reduce heterogeneity in the data. We also removed capture histories where plots of residuals revealed very large residuals. The best-fitting models and the adjustments made to them are presented in Table 1, alongside the models with no modifications. These adjustments did not dramatically alter population estimates but did improve the standard error of the population estimates and the model fit (Table 1). In all cases, the p-value of goodness of fit testing, based on the deviance of the models, was >0.05, indicating that our models adequately fit our data.

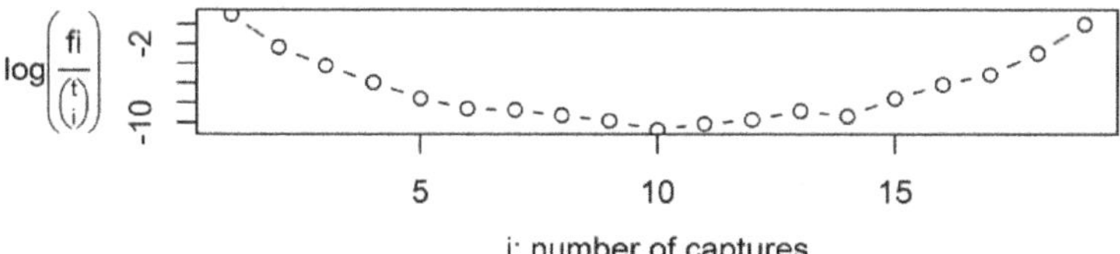

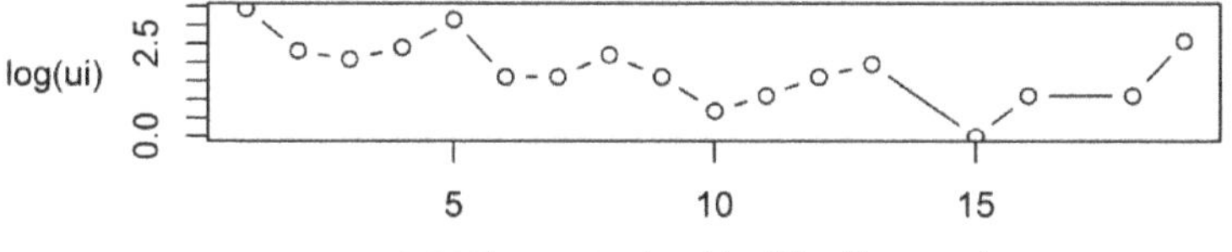

Figure 2. Exploratory heterogeneity graph showing descriptive data from the capture histories of Ethiopian hedgehogs caught as part of a capture-mark-recapture study in Qatar over 19 months (October 2010 to April 2012).

Table 1. Estimated population size of Ethiopian hedgehogs at a study site in Qatar from a series of open population models constructed in the R package RCapture. Results are presented from the models with no modifications and the best fitting models, determined by AIC values.

Subset	Block	No. in Model [1]	Adjustments to Improve Model Fit	AIC	N_{tot} [2]	SE	N [3]	SE	Density (km⁻²) [4]
All	All 19 months	144	None	1180	149	2.7	55	2.4	6.5
All	All 19 months	144	Excluding those captured all 19 times and including residuals <10	439	151	0	60	2.9	7.0
Males	All 19 months	75	None	787	77	1.6	32	1.1	3.8
Males	All 19 months	75	Excluding those captured all 19 times and including residuals <50	489	77	0	33	1.0	3.9
Females	All 19 months	62	None	540	65	2.3	22	1.5	2.6
Females	All 19 months	62	Capture constant on model and including residuals <800 [5]	511	65	1.9	21	1.5	2.5
All	Excluding winter	112	None	346	115	2.3	52	3.8	6.1
All	Excluding winter	112	Excluding those captured all 9 times	327	117	1.9	54	3.7	6.3
Males	Excluding winter	59	None	244	60	1.2	30	1.9	3.5
Males	Excluding winter	59	Excluding those captured 8 or 9 times and including residuals <6	177	61	0	30	1.7	3.5
Females	Excluding winter	47	None	180	47	1.1	19	1.9	2.2
Females	Excluding winter	47	Excluding those captured 8 or 9 times	142	47	0	19	1.9	2.2
Adults	Excluding winter	92	None	304	93	1.5	45	3.9	5.3
Adults	Excluding winter	92	Excluding those caught all 9 times and including residuals <10	266	94	0	46	3.9	5.4
Juveniles	Excluding winter	21	None	100	26	5.1	7	1.8	0.8 [6]

[1] The number of hedgehogs captured in this time period and included in the model—note, not all hedgehogs could be aged and/or sexed hence males and females do not add up to the total number of hedgehogs captured. [2] The total number of hedgehogs estimated to be on site during the period analyzed. [3] The mean monthly number of hedgehogs estimated to be on the site during the period analyzed. [4] Density was calculated based on the mean monthly N over the 8.5 km² regular search area. [5] Note this model had some extremely high residuals but the lowest number at which the model would converge was <800. [6] The null model was deemed the best fitting model for juveniles.

3.2. Population Size

Our models resulted in a range of population estimates for our study site (Table 1). If we are to include all 19 continuous months and hedgehogs caught during this period (144 hedgehogs) then we estimate a mean monthly population of 60 ± 2.9 hedgehogs on our study site. If we exclude the Winter Season (112 hedgehogs included), the number of hedgehogs is slightly smaller at 54 ± 3.7 hedgehogs. There was no statistically significant difference between the monthly abundance estimates from these two time periods (ANOVA, $F_{1,22}$ = 1.67, p = 0.21). These estimates give a density of 7 hedgehogs per km^2 in our 8.5 km^2 focal search area, or a density of 6.3 hedgehogs per km^2 excluding winter. As expected from observations on the ground, the estimated population size was larger for males than females in all our models (Table 1).

Twenty-two percent of hedgehogs were caught on the very first sampling occasion. This figure was higher for males (24%) than it was for females (21%). This pattern was followed when excluding the Winter Season, with 63% of male hedgehogs captured in the study caught on the first occasion compared with 53% of females. Only 4.8% of hedgehogs caught at the very first sampling occasion were not captured again throughout the study. The figure was 12.8% for the analysis performed excluding the Winter Season.

The monthly estimation of N at each sampling occasion decreased throughout both time periods analyzed (all 19 months: ANOVA, $F_{1,10}$ = 17.78, p = 0.0002, excluding the Winter Season: ANOVA, $F_{1,10}$ = 33.68, p = 0.0002), suggesting that the population size was decreasing at the study site (Figure 3). Over the full 19 months, the monthly population estimate fell by 20%. The average monthly growth rate was −0.86%. The abundance of males at each sampling occasion was significantly higher over both analyzed time periods (all 19 months: ANOVA, $F_{1,10}$ = 70.40, p < 0.0001, excluding the Winter Season: ANOVA, $F_{1,10}$ = 75.64, p < 0.0001). When all 19 months were included, there was a significant interaction between sex and sampling period (ANOVA, $F_{1,30}$ = 8.08, p = 0.008), with a steeper decline in the estimated number of females each month over the course of the study (Figure 4). On average there were estimated to be over seven times more adults (46.36 ± 3.87) than juveniles (7.33 ± 1.75) when excluding the Winter Season, which was highly statistically significant (ANOVA, $F_{1,10}$ = 179.67, p < 0.0001, Figure 5). There was a statistically significant interaction between age and sampling occasion (ANOVA, $F_{1,10}$ = 6.60, p = 0.03). The estimated number of juveniles and subadults increased during the first four sampling periods then leveled off, whereas the estimated number of adults decreased (Figure 5).

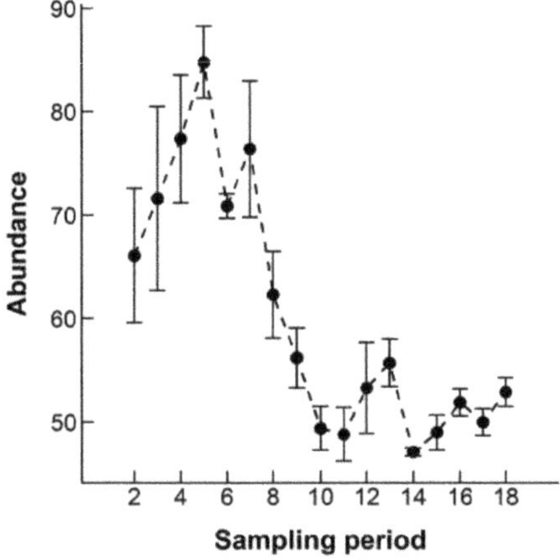

Figure 3. The estimated abundance of Ethiopian hedgehogs in an 8.5 km^2 search area in Qatar at each monthly sampling occasion over a 19 month period (October 2010 to April 2012). Bars indicate the standard error of each abundance estimate.

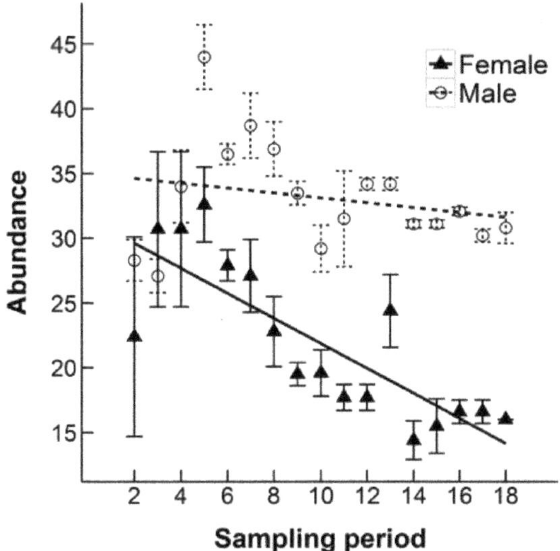

Figure 4. The estimated abundance of male and female Ethiopian hedgehogs in an 8.5 km^2 search area in Qatar at each monthly sampling occasion over a 19 month period (October 2010 to April 2012). Error bars shown are the standard error of the estimate. Linear regression lines are also displayed for each sex.

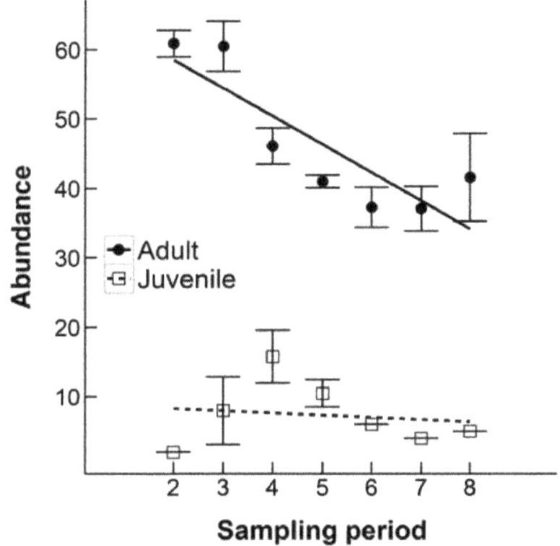

Figure 5. The estimated abundance of adult and juvenile (including subadult) Ethiopian hedgehogs in an 8.5 km^2 search area in Qatar at each monthly sampling occasion over a nine-month period (February 2011-October 2011). Linear regression lines are displayed for each group, error bars are the standard error of the monthly abundance estimate.

3.3. Capture Probability

Mean capture probability was higher for males (0.70 ± 0.04) than females (0.60 ± 0.06) but this difference was not statistically significant (ANOVA, $F_{1,30}$ = 2.06, p = 0.16). Surprisingly, when excluding the Winter Season, the mean capture probability from the best fitting models was higher for females (0.81 ± 0.07) than males (0.63 ± 0.11), however, the difference was also not statistically significant (ANOVA, $F_{1,10}$ = 2.40, p = 0.15). Looking at the whole study, there was no effect of sampling period on the capture rate (ANOVA, $F_{1,30} = 1.03$, $p = 0.32$). However, it may be more appropriate to look at this over the awake period for hedgehogs, indeed there was a near significant effect of sampling period on capture probability when excluding the Winter Season (ANOVA, $F_{1,10} = 4.02$, $p = 0.07$). Capture probability decreased through time (Figure 6). There was a higher probability of catching juveniles (0.80 ± 0.097) than adults (0.70 ± 0.07), but this was not statistically significant (ANOVA, $F_{1,10} = 0.85$, $p = 0.38$). Furthermore, there was a near statistically significant interaction between age (adults versus juveniles) and sampling period over this awake period (ANOVA, $F_{1,10} = 7.46$, $p = 0.066$). Capture probability of adults declined over the awake period, whereas juveniles increased. We also tested to see if there was a trap effect over the course of the study. The AIC value for the models including trap effect was higher than that for a homogenous trap effect in all our models, indicating that any differences in capture probability through time were not down to a behavioral response to capture.

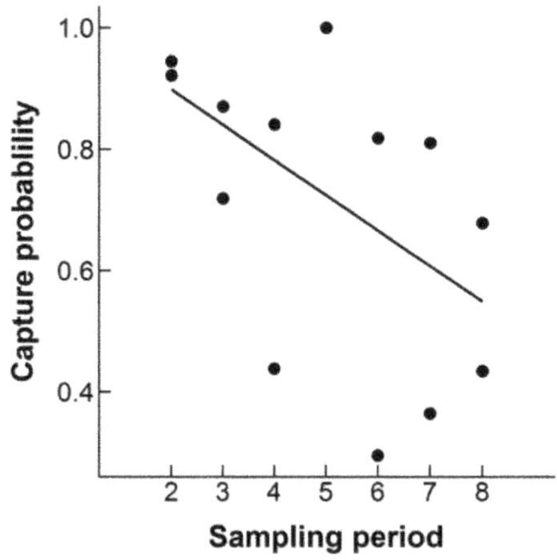

Figure 6. The estimated capture probability of Ethiopian hedgehogs sampled monthly from February 2011 to October 2011 in an 8.5 km^2 search area in Qatar. A linear regression line is displayed.

3.4. Survival

Mean survival between sampling occasions was similar for males (0.89 ± 0.02) and females (0.88 ± 0.04). When the Winter Season was excluded, the mean survival between sampling periods was higher for males (0.81 ± 0.05) than females (0.74 ± 0.06). However, this was not statistically significant (ANOVA, $F_{1,10} = 0.60$, $p = 0.46$). There was also no difference in survival between sampling periods both including and excluding the Winter Season (all 19 months: ANOVA, $F_{1,30} = 2.14$, $p = 0.15$, excluding winter: ANOVA, $F_{1,10} = 0.31$, $p = 0.59$). Mean adult survival between monthly captures (0.85 ± 0.04) was greater than mean juvenile survival (0.68 ± 0.10) but this was not statistically significant (ANOVA, $F_{1,10} = 0.77$, $p = 0.40$). However, there was again a statistically significant interaction between age and sampling period (ANOVA, $F_{1,10} = 9.51$, $p = 0.01$). Juvenile survival decreased over time whereas adult survival increased (Figure 7), although note there is an outlier in the juvenile estimates and the standard error bars are very large. There were only a few cases where we were able to confidently identify causes of mortality during the study. The main causes were traffic accidents and starvation/exhaustion.

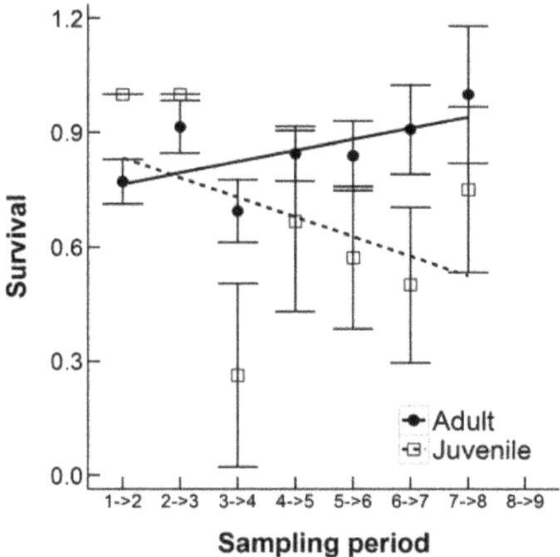

Sampling period

Figure 7. The estimated survival rate of Ethiopian hedgehogs between monthly samples from February 2011 to October 2011 in an 8.5 km^2 search area in Qatar, showing the survival of adults versus juveniles (including subadults). Linear regression lines are shown for each group, error bars are the standard error of the monthly survival estimate.

3.5. New Entrants

The number of new entrants decreased over time and this was statistically significant when including all 19 sampling periods (ANOVA, $F_{1,28}$ = 7.22, p = 0.01, Figure 8). This finding again supports that the population size on the site was decreasing. The mean rate of new arrivals to the study site between sampling occasions was higher for males (all 19 months: 3.92 ± 1.03, excluding the Winter Season: 4.83 ± 0.99) than females (all 19 months: 2.35 ± 0.66, excluding the Winter Season: 3.38 ± 1.1) over both sampling periods, but again this was not statistically significant (all 19 months: ANOVA, $F_{1,28}$ = 2.06, p = 0.16, excluding the Winter Season: ANOVA, $F_{1,8}$ = 0.97, p = 0.35, Figure 8). The mean number of new entrants between monthly sampling occasions was similar for adults (3.63 ± 0.43) and juveniles (3.47± 1.98). However, note the high standard error for the mean rate of juvenile new entrants.

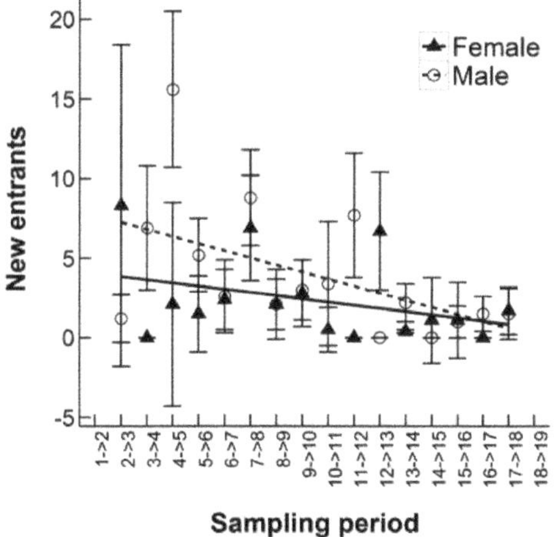

Sampling period

Figure 8. The estimated number of new entrants to a population of Ethiopian hedgehogs between monthly samples from October 2010 to April 2012 in an 8.5 km^2 search area in Qatar, showing the estimates for females versus males with a linear regression line for each group.

3.6. Evaluation of Methodology

When marking hedgehogs, we found that black, yellow, and non-metallic green nail polish tended to disappear soon, whilst white, red, metallic green, and blue lasted for longer (some even lasted for more than six months). We found that applying the nail polish along the entire length of the spines was more successful because the colors tended to wear off towards the tip of the spines.

According to our population estimates, we captured between 91.3% and 100% of the local hedgehog population at our study site (Table 2).

Table 2. Percentage of a hedgehog population estimated from Jolly-Seber modeling in the R package Rcapture that was captured in the field.

Subset	Block	No. Hedgehogs [1]	N_{tot}	% of Population Captured
All	All 19 months	144	151	95.36
Males	All 19 months	75	77	97.40
Females	All 19 months	62	65	95.38
All	Excluding winter	112	117	95.73
Males	Excluding winter	59	61	96.72
Females	Excluding winter	47	47	100.00
Adults	Excluding winter	92	94	97.87
Juveniles	Excluding winter	21	23	91.30

[1] The number of hedgehogs captured in this time period and included in the model—note, not all hedgehogs could be aged and/or sexed hence males and females do not add up to the total number of hedgehogs captured.

4. Discussion

Our results indicate a mean population of 60 hedgehogs at the study site at one time over a 19 month period. The outputs from our models suggest that the monthly population estimate declined over the study period. Survival appeared to be stable throughout the study, potentially a result of a lack of predators of hedgehogs in Qatar [11] and plentiful food resources provided by the rubbish mound and irrigated farms at the site. The rubbish mound was partially cleared in March 2011 (month six of the study), and a further major cleaning operation started in March 2012, potentially resulting in a reduction of resources. Indeed the large drop in abundance was during months seven to ten of the study (Figure 3) and coincides with this change at the study site. However, it was still observed to be the activity center of quite a few animals until it was totally cleared. There were many more adults in the population than juveniles and juvenile survival slightly decreased over time, which could indicate that juvenile survival could be a potential reason for the decrease in monthly abundance estimates. However, there were large standard errors for the estimates of juvenile survival so this result must be interpreted with caution. This decline in juvenile survival could also potentially be down to individuals emigrating from the site as resources at the rubbish mound were reduced. The finding of a population decline is also supported by a statistically significant decline in new entrants (births and/or immigrants) to the population over the study period. The mean number of new entrants between sampling occasions was similar for adults and juveniles. Over one year numbers of juveniles increased initially and then levelled off, as would be expected after breeding activity first peaks in March [5]. However, Ethiopian hedgehogs are thought to have two litters [5] so we would expect to see a second peak in the abundance of juveniles later on in the season and we did not, suggesting the success of the second breeding attempt could be an issue for population recruitment. Conversely, the expected second peak is substantially smaller than the first one [5], and potentially this was not picked up in our models.

Survival in *E. europaeus* has been shown to be impacted by a high predation rate [38], road casualties [29,39], and poor survival over winter, particularly for juveniles [29,40]. There were only a few cases where we were able to identify mortality and its cause during the study and these included traffic accidents and starvation/exhaustion, which we suggest is a result of searching for mates in males and rearing young in females. In all 19 months of our study (including two winters), the estimated survival between monthly sampling occasions ranged from 38.6% to 100% (mean 75%). When excluding the Winter Season, the mean survival between sampling periods was higher ranging from 66% to 100% (mean 84%). These results indicate mortality over winter, may be an important factor in survival in the population studied here.

The reduction in new entrants over the study period could also be a result of a lack of immigration to the population. Only 4.8% of hedgehogs caught at the very first sampling occasion were not captured again throughout the study, indicating that there was not a high number of transients at the site and therefore not much immigration and emigration. The area surrounding the study site consisted of similar habitat with irrigated farms surrounded by arid desert. Released Ethiopian hedgehogs have been shown to travel 131.4–426.7 m per evening and utilize more than one irrigated farm [41], so dispersal between farms is likely and we would expect some immigration from outside of the study area. However, we can assume that distances of greater than the 8.5 km^2 search area would be much less likely and although wild individuals of this species may have home ranges of up to 230 ha [6,13], their home ranges have been shown to be smaller in resource-rich habitats, such as around irrigated farms [13]. It may be that the species is reasonably site faithful where resources are plentiful and radio-tracking studies at the same study site found that home ranges centered around the rubbish dump and irrigated farms, and hedgehogs did not appear to leave the 15 km^2 study site [6].

There was a higher abundance of males at the study site, which has been found for populations of *Erinaceus* species e.g., References [1,38,42]. This finding could be down to the higher capture probability of males because, like Erinaceous hedgehogs, they tend to range further distances than females [6]. This idea is supported by the sex ratio of 1:1 in raw captures. However, it could also be due to the decline in estimated monthly abundance being steeper for females than males. There were not enough juveniles in the dataset to investigate whether survival was lower for female juveniles than male juveniles at the site.

Another potential factor in the population decline is that, when the Winter Season was excluded, capture probability decreased throughout the year. This finding could be down to the reduction of the population size but we must also consider that it somehow became harder to capture hedgehogs; perhaps they began to avoid the site to evade being captured, or became 'trap happy'. The AIC values for the models including trap effect was higher than the AIC values for the same models with a homogenous trap effect included, indicating that there was not a difference in capture probability through time because of a behavioral response to capture. Another potential reason for this is that hedgehog activity peaked in the early breeding season, resulting in less hedgehog movement later in the season and thus fewer captures [5,6]. Capture probability of juveniles increased over one season, likely because as the hedgehogs begin to breed after emergence from torpor, juveniles enter the population and more will be captured as they come out of the nest(s) and start to move around the site.

Studies of hedgehog demography over longer periods have shown that big fluctuations in population size are common. Kristiansson studied a population of *E. europaeus* in Sweden over eight years and found the population was in decline for the first three years, increased for the next three years and then declined again [29,43,44]. Akin to our study, a decreasing population was linked to low numbers of juveniles. Again, one potential factor in the survival of juveniles indicated in the study was survival over winter, especially with respect to colder winters. Although winters are not particularly cold at our study site and some hedgehogs, particularly males, remain active to some degree over winter [6], hedgehogs do enter torpor for short periods [3]. To ascertain whether the population on our study site is in long term decline or this is merely a fluctuation in population size, a much longer-term study is needed, including observations on causes of mortality in different age classes, principally over winter.

4.1. Comparison with Other Hedgehog Species

Using the population estimate from all months of the study, the density of hedgehogs in the regular search area was 7 hedgehogs per km^2. Comparative estimates of European hedgehog density vary greatly with habitat type and methodology. For example, Hubert et al. 2011 [45] found that mean hedgehog density, was 4.4 individuals per km^2 in rural areas of France and 36.5 individuals per km^2 in urban areas. Young et al. 2006 [46] found a mean density of nine individuals per km^2 in a survey of pasture fields in England, the same survey in amenity grassland found up to 154 hedgehogs per km^2. In their capture-mark-recapture study, Reeve 1981 found a population size of 82.5 hedgehogs per km^2 on a suburban golf course in England. Jackson et al. 2007 [18] studied the abundance of introduced hedgehogs on the Scottish island of South Uist (where hedgehogs are thriving) and found 31.8 hedgehogs per km^2 in sandy-soiled flat dune grassland habitat and 15.4 peaty-soiled pastureland. It seems the density found in this study is fairly low compared to that of Erinaceus species in a reasonably productive habitat, which may be expected given the nature of a less productive hyper-arid environment at the study site. This idea is supported by the finding that Ethiopian hedgehogs in Qatar have been shown to have a larger home range than their European counterparts, in spite of the latter being substantially heavier, likely because of these dispersed resources [6]. The density estimate presented here may also be inflated because of the artificial food sources at the site and it is likely in areas of 'natural desert' in Qatar the hedgehog density is lower.

Survival rates also vary between studies of *E. europaeus* and we must be careful about drawing conclusions from these comparisons as figures are presented from a range of habitats and methodologies. Translocated and released hedgehogs have a survival rate of between 40% and 77% after several weeks in the wild [47–50]. Survival of individuals in extant populations over short periods is higher, for example nearly 95% during an eight week study in urban habitats in the UK [49]. Reeve 1981 [28] found a survival rate of 62%, over one year including winter, however over two winters this was reduced to 37%. Over the whole of our study, including two winters, the mean monthly survival rate between sampling occasions was higher at 75%. Kristiansson 1990 [29] found an average annual survival of 66% in juveniles and 55% in adults. Whereas we found a similar mean monthly survival rate of 64% for juveniles and subadults over nine months, our figure for adults was much higher at 85%. Rasmussen et al. 2019 [51] found a slightly higher juvenile survival rate of 70% over one year, with an over winter survival rate of 89%, they attribute this high winter survival to the suburban habitat type.

4.2. Evaluation of Methodology

Our results suggest that we were able to capture most of the local population and therefore we conclude that sampling at key habitats with spotlights for hedgehogs seems to be sufficient in capturing most of the hedgehog population at a given site. Other capture-mark-recapture experiments on *E. europaeus* have also found that they were also efficiently able to catch a high proportion of the estimated population size [28,29]. Like the hedgehogs in this study that were attracted to the 'rubbish mound', European hedgehogs may also be attracted to areas of abundant food e.g., urban habitats with plentiful pet food [42,45,52], and we conclude that sampling in these areas may be sufficient to gain knowledge of the local hedgehog population. However, we must take note that our density estimate cannot be extrapolated over the whole 15 km^2 study site because we cannot assume the rest of the habitat at the study area to be of the same quality as our focal search area, and likewise we cannot assume all hedgehogs in the 15 km^2 area were attracted to our key survey areas. Further study could be conducted by carrying out a mark-recapture-study on the less utilized arid areas of the study site. Additionally, it would be interesting for further study to compare the density found here with areas of 'natural' desert in Qatar where anthropogenic intervention to the habitat is minimal.

Jolly-Seber and Cormack-Jolly-Seber modelling may underestimate the population size at a study site, especially in short-lived species [53]. However, hedgehogs were observed to live multiple years during the study with some animals first encountered as adults in April 2010 still alive when the project finished in April 2012, which may reduce the risk of us underestimating the population. However, a future study could include performing a "robust design" analysis, whereby hedgehogs would be processed multiple times during the four-day survey and two levels of modeling are carried out; closed population modeling between each consecutive night of the survey and open models between each monthly visit [37,53,54]. Robust design methodology is less likely to underestimate the population size [53] and would make an interesting comparison with the modeling presented here.

We were able to sufficiently identify the marked hedgehogs in the study using colored nail polish, but this method requires re-application. We found that Since our study improved ways of marking hedgehogs, such as using numbered plastic tubing, have been successfully tested and could be used to increase confidence in animal identification in future studies [55].

5. Conclusions

We successfully used capture-mark-recapture methodology to come up with a range of population estimates for the Ethiopian hedgehog. We found that the estimated monthly population at our study site had decreased over the 19 months of our study and potential causes of this include poor juvenile survival and a lack of immigration in to the study area. However, a longer term study is needed to ascertain if this is a sustained population decline and to confirm the causes. As well as obtaining a range of population estimates, our methodology allowed us to report a range of valuable demographic parameters that give the first insight into the population dynamics of Ethiopian hedgehogs in Qatar. The methods presented here are transferrable to other hedgehog species in a range of habitats.

Author Contributions: Conceived and designed the experiment: N.Y. Data collection: N.Y., A.A.-H. and H.A.-J. Data analysis: C.P. Writing of the manuscript: C.P., N.Y. and D.W.M. Creation of figures for the paper: C.P., N.Y. Review of the paper: N.Y. and D.W.M. All authors have read and agreed to the published version of the manuscript.

Funding: This report was made possible by an Undergraduate Research Experience Programme (UREP) grant (UREP07-071-1-016) from the Qatar National Research Fund (a member of The Qatar Foundation), and Qatar University Student Grants (QUST-CAS-BES-10/11-3 and QUST-CAS-BES-10/11-2). Qatar Natural History Group kindly provided financial support for renting a 4 × 4 vehicle for the fieldwork.

Acknowledgments: We thank Hamda Al Naemi and Abubaker el Taeb for their constant and vital logistic support for the project including allowing us to use the field station, Qatar University Farm. We thank the staff of the farms in the study area to allow us to access their farms. We also thank Lissa Barrows, April Conkey, Rita Chan, and John Tribuna for organizing teams of volunteers for helping our fieldwork.

Conflicts of Interest: The authors declare no conflict of interest.

Correction Statement: This article has been republished with a minor change. The change does not affect the scientific content of the article and further details are available within the backmatter of the website version of this article.

References

1. Reeve, N. *Hedgehogs*; Poyser: London, UK, 1994; ISBN 085661081X.
2. He, K.; Chen, J.-H.; Gould, G.C.; Yamaguchi, N.; Ai, H.-S.; Wang, Y.-X.; Zhang, Y.-P.; Jiang, X.-L. An Estimation of Erinaceidae Phylogeny: A Combined Analysis Approach. *PLoS ONE* **2012**, *7*, e39304. [CrossRef] [PubMed]
3. Al-Musfir, H.M.; Yamaguchi, N. Timings of hibernation and breeding of Ethiopian Hedgehogs, *Paraechinus aethiopicus,* in Qatar. *Zool. Middle East* **2008**, *45*, 3–10. [CrossRef]
4. Zapletal, M.; Sodnompil, B.; Atwood, J.; Murdoch, J.D.; Reading, R.P. Home range characteristics and habitat selection by Daurian hedgehogs (*Mesechinus dauuricus*) in Ikh Nart Nature Reserve, Mongolia. *Sciences* **2012**, *10*, 41–50.
5. Yamaguchi, N.; Al-Hajri, A.; Al-Jabiri, H. Timing of breeding in free-ranging Ethiopian hedgehogs, *Paraechinus aethiopicus,* from Qatar. *J. Arid Environ.* **2013**, *99*, 1–4. [CrossRef]
6. Pettett, C.E.; Al-Hajri, A.; Al-Jabiry, H.; Macdonald, D.W.; Yamaguchi, N. A comparison of the Ranging behaviour and habitat use of the Ethiopian hedgehog (*Paraechinus aethiopicus*) in Qatar with hedgehog taxa from temperate environments. *Sci.Rep.* **2018**, *8*, 1–10. [CrossRef]
7. Murdoch, J.D.; Buyandelger, S.; Kenny, D.; Reading, R.P. Ecology of the Daurian hedgehog (*Hemiechinus dauuricus*) in Ikh Nart Nature Reserve, Mongolia: Preliminary findings. *Mong. J. Biol. Sci.* **2006**, *4*, 25–32.

8. Boyles, J.G.; Bennett, N.C.; Mohammed, O.B.; Alagaili, A.N. Torpor patterns in Desert Hedgehogs (*Paraechinus aethiopicus*) represent another new point along a thermoregulatory continuum. *Physiol. Biochem. Zool.* **2017**, *90*, 445–452. [CrossRef]

9. Alagaili, A.N.; Bennett, N.C.; Mohammed, O.B.; Hart, D.W. The reproductive biology of the Ethiopian hedgehog, *Paraechinus aethiopicus*, from central Saudi Arabia: The role of rainfall and temperature. *J. Arid Environ.* **2017**, *145*, 1–9. [CrossRef]

10. Alagaili, A.N.; Bennett, N.C.; Amor, N.M.; Hart, D.W. The locomotory activity patterns of the arid-dwelling desert hedgehog, *Paraechinus aethiopicus*, from Saudi Arabia. *J. Arid Environ.* **2020**, *177*, 104141. [CrossRef]

11. Abu Baker, M.A.; Reeve, N.; Mohedano, I.; Conkey, A.A.T.; Macdonald, D.W.; Yamaguchi, N. Caught basking in the winter sun: Preliminary data on winter thermoregulation in the Ethiopian hedgehog, Paraechinus aethiopicus, in Qatar. *J. Arid Environ.* **2016**, *125*, 52–55. [CrossRef]

12. Zapletal, M.; Sodnompil, B.; Atwood, J.L.; Murdoch, J.D.; Reading, R.P. Fine-scale habitat use by Daurian hedgehogs in the Gobi-steppe of Mongolia. *J. Arid Environ.* **2015**, *114*, 100–103. [CrossRef]

13. Abu Baker, M.A.; Reeve, N.; Conkey, A.A.T.; Macdonald, D.W.; Yamaguchi, N. Hedgehogs on the move: Testing the effects of land use change on home range size and movement patterns of free-ranging Ethiopian hedgehogs. *PLoS ONE* **2017**, *12*, e0180826. [CrossRef] [PubMed]

14. Harrison, D.L.; Bates, P.J.J. *The mammals of Arabia*; Harrison Zoological Museum: Sevenoaks, UK, 1991; Volume 354.

15. Micol, T.; Doncaster, C.P.; Mackinlay, L.A. Correlates of local variation in the abundance of hedgehogs *Erinaceus europaeus*. *J. Anim. Ecol.* **1994**, *63*, 851–860. [CrossRef]

16. Haigh, A.; Butler, F.; Riordan, R.M.O. An investigation into the techniques for detecting hedgehogs in a rural landscape. *J. Negat. Results Ecol. Evol. Biol.* **2012**, *9*, 15–26.

17. Doncaster, C.P. Factors regulating local variations in abundance: Field tests on hedgehogs, *Erinaceus europaeus*. *Oikos* **1994**, *69*, 182–192. [CrossRef]

18. Jackson, D.B. Factors affecting the abundance of introduced hedgehogs (*Erinaceus europaeus*) to the Hebridean island of South Uist in the absence of natural predators and implications for nesting birds. *J. Zool.* **2007**, *271*, 210–217. [CrossRef]

19. Yarnell, R.W. Using occupancy analysis to validate the use of footprint tunnels as a method for monitoring the hedgehog *Erinaceus europaeus*. *Mamm. Rev.* **2014**, *44*, 234–238. [CrossRef]

20. Huijser, M.P.; Bergers, P.J.M. The effect of roads and traffic on hedgehog (*Erinaceus europaeus*) populations. *Biol. Conserv.* **2000**, *95*, 6–9. [CrossRef]

21. Williams, B.; Mann, N.; Neumann, J.L.; Yarnell, R.W.; Baker, P.J. A prickly problem: Developing a volunteer-friendly tool for monitoring populations of a terrestrial urban mammal, the West European hedgehog (*Erinaceus europaeus*). *Urban Ecosyst.* **2018**, *21*, 1075–1086. [CrossRef]

22. Williams, B.M.; Baker, P.J.; Thomas, E.; Wilson, G.; Judge, J.; Yarnell, R.W. Reduced occupancy of hedgehogs (*Erinaceus europaeus*) in rural England and Wales: The influence of habitat and an asymmetric intra-guild predator. *Sci. Rep.* **2018**, *8*, 1–10. [CrossRef]

23. Hof, A.R.; Bright, P.W. The value of green-spaces in built-up areas for western hedgehogs. *Lutra* **2009**, *52*, 69–82.

24. Van De Poel, J.L.; Dekker, J.; Van Langevelde, F. Dutch hedgehogs *Erinaceus europaeus* are nowadays mainly found in urban areas, possibly due to the negative effects of badgers *Meles meles*. *Wildl. Biol.* **2015**, *21*, 51–55. [CrossRef]

25. Hof, A.R.; Bright, P.W. Quantifying the long-term decline of the West European hedgehog in England by subsampling citizen-science datasets. *Eur. J. Wildl. Res.* **2016**, *62*, 407–413. [CrossRef]

26. Aebischer, N.J.; Davey, P.D.; Kingdon, N.G. National Gamebag Census: Mammal Trends to 2009. Available online: http://www.gwct.org.uk/ngcmammals (accessed on 23 April 2020).

27. Pettett, C.E.; Johnson, P.J.; Moorhouse, T.P.; Macdonald, D.W. National predictors of hedgehog *Erinaceus europaeus* distribution and decline in Britain. *Mamm. Rev.* **2018**, *48*, 1–6. [CrossRef]

28. Reeve, N.J. A Field Study of the Hedgehog (*Erinaceus europaeus*) with Particular Referance to Movements and Behaviour. Ph.D. Thesis, University of London, London, UK, 1981.

29. Kristiansson, H. Population variables and causes of mortality in a hedgehog (*Erinaceous europaeus*) population in southern Sweden. *J. Zool.* **1990**, *220*, 391–404. [CrossRef]

30. Jackson, D.B.; Green, R.E. The importance of the introduced hedgehog (*Erinaceus europaeus*) as a predator of the eggs of waders (Charadrii) on machair in South Uist, Scotland. *Biol. Conserv.* **2000**, *93*, 333–348. [CrossRef]

31. White, G.C. *Capture-Recapture and Removal Methods for Sampling Closed Populations*; Los Alamos National Laboratory: Los Alamos, NM, USA, 1982.

32. Otis, D.L.; Burnham, K.P.; White, G.C.; Anderson, D.R. Statistical inference from capture data on closed animal populations. *Wildl. Monogr.* **1978**, *62*, 3–135.

33. Cormack, R.M. Examples of the use of GLIM to analyse capture-recapture studies. In *Statistics in Ornithology*; Springer: Berlin/Heidelberg, Germany, 1985; pp. 243–273.

34. Cormack, R.M. Log-linear models for capture-recapture. *Biometrics* **1989**, 395–413. [CrossRef]

35. Rivest, L.; Daigle, G. Loglinear models for the robust design in mark–recapture experiments. *Biometrics* **2004**, *60*, 100–107. [CrossRef]

36. Seber, G.A.F. *The Estimation of Animal Abundance and Related Parameters*; Blackburn Press: Caldwell, NJ, USA, 1982; Volume 8.

37. Baillargeon, S.; Rivest, L.-P. Rcapture: Loglinear models for capture-recapture in R. *J. Stat. Softw.* **2007**, *19*, 1–31. [CrossRef]

38. Hof, A.R.; Bright, P.W. The value of agri-environment schemes for macro-invertebrate feeders: Hedgehogs on arable farms in Britain. *Anim. Conserv.* **2010**, *13*, 467–473. [CrossRef]

39. Wembridge, D.E.; Newman, M.R.; Bright, P.W.; Morris, P.A. An estimate of the annual number of hedgehog (*Erinaceus europaeus*) road casualties in Great Britain. *Mammal Commun.* **2016**, *2*, 8–14.

40. Walhovd, H. Records of young hedgehogs (*Erinaceus europaeus* L.) in a private garden. *Z. Säugetierkd.* **1990**, *55*, 289–297.

41. Baker, M.A.A.; Villamil, L.A.; Reeve, N.; Karanassos, C.; Mahtab, H.; Yamaguchi, N. Into the wild: Survival, movement patterns, and weight changes in captive Ethiopian hedgehogs, Paraechinus aethiopicus following their release. *J. Arid Environ.* **2018**, *158*, 43–46. [CrossRef]

42. Pettett, C.E.; Moorhouse, T.P.; Johnson, P.J.; Macdonald, D.W. Factors affecting hedgehog (*Erinaceus europaeus*) attraction to rural villages in arable landscapes. *Eur. J. Wildl. Res.* **2017**, *63*, 54. [CrossRef]

43. Kristiansson, H. Ecology of a Hedgehog (*Erinaceus europaeus*) Population in Southern Sweden. Ph.D. Thesis, Department of Animal Ecology, University of Lund, Lund, Sweden, 1984.

44. Kristiansson, H. Young Production of European Hedgehog in Sweden and Britain. *Acta Theriol.* **1981**, *26*, 504–507. [CrossRef]

45. Hubert, P.; Julliard, R.; Biagianti, S.; Poulle, M.L. Ecological factors driving the higher hedgehog (*Erinaceus europeaus*) density in an urban area compared to the adjacent rural area. *Landsc. Urban Plan.* **2011**, *103*, 34–43. [CrossRef]

46. Young, R.P.; Davison, J.; Trewby, I.D.; Wilson, G.J.; Delahay, R.J.; Doncaster, C.P. Abundance of hedgehogs (*Erinaceus europaeus*) in relation to the density and distribution of badgers (*Meles meles*). *J. Zool.* **2006**, *269*, 349–356. [CrossRef]

47. Warwick, H.; Morris, P.; Walker, D. Survival and weight changes of hedgehogs (*Erinaceus europaeus*) translocated from the Hebrides to Mainland Scotland. *Lutra* **2006**, *49*, 89–102.

48. Morris, P.A.; Warwick, H. A study of rehabilitated juvenile hedgehogs after release into the wild. *Anim. Welf.* **1994**, *3*, 163–177.

49. Molony, S.E.; Dowding, C.V.; Baker, P.J.; Cuthill, I.C.; Harris, S. The effect of translocation and temporary captivity on wildlife rehabilitation success: An experimental study using European hedgehogs (*Erinaceus europaeus*). *Biol. Conserv.* **2006**, *130*, 530–537. [CrossRef]

50. Morris, P.A. Released, rehabilitated hedgehogs: A follow-up study in Jersey. *Anim. Welf.* **1997**, *6*, 317–327.

51. Rasmussen, S.L.; Berg, T.B.; Dabelsteen, T.; Jones, O.R. The ecology of suburban juvenile European hedgehogs (*Erinaceus europaeus*) in Denmark. *Ecol. Evol.* **2019**, *9*, 13174–13187. [CrossRef] [PubMed]

52. Pettett, C.E. *Factors Affecting Hedgehog Distribution and Habitat Selection in Rural Landscapes*; University of Oxford: Oxford, UK, 2016.

53. Amori, G.; De Silvestro, V.; Ciucci, P.; Luiselli, L. Quantifying whether different demographic models produce incongruent results on population dynamics of two long-term studied rodent species. *Eur. J. Ecol.* **2017**, *3*, 18–26. [CrossRef]

54. Casula, P.; Luiselli, L.; Amori, G. Which population density affects home ranges of co-occurring rodents? *Basic Appl. Ecol.* **2019**, *34*, 46–54. [CrossRef]
55. Reeve, N.; Bowen, C.; Gurnell, J. *An Improved Identification Marking Method for Hedgehogs*; The Mammal Socity: Southampton, UK, 2019.

Article

Prickly Connections: Sociodemographic Factors Shaping Attitudes, Perception and Biological Knowledge about the European Hedgehog

Ângela M. Ribeiro [1,*], Micaela Rodrigues [2], Nuno V. Brito [1,3] and Teresa Letra Mateus [1,4,5,*]

1   CISAS—Center for Research and Development in Agrifood Systems and Sustainability, Polytechnic Institute of Viana do Castelo, NUTRIR (Technological Center for AgriFood Sustainability), Monte de Prado, 4960-320 Melgaço, Portugal; nunobrito@esa.ipvc.pt
2   Escola Superior Agrária, Polytechnic Institute of Viana do Castelo, Rua Escola Industrial e Comercial de Nun'Àlvares, 4900-347 Viana do Castelo, Portugal; mrmicarodrigues7@gmail.com
3   1H-TOXRUN—One Health Toxicology Research Unit, University Institute of Health Sciences, Cooperativa de Ensino Superior Politécnico e Universitário, 4585-116 Gandra, Portugal
4   CECAV-Animal and Veterinary Research Centre, Associate Laboratory for Animal and Veterinary Sciences (AL4AnimalS), University of Trás-os-Montes and Alto Douro, Quinta de Prados, 5000-801 Vila Real, Portugal
5   EpiUnit-Instituto de Saúde Pública da Universidade do Porto, Laboratório Para a Investigação Integrativa e Translacional em Saúde Populacional (ITR), Universidade do Porto, Rua das Taipas, nº 135, 4050-091 Porto, Portugal
*   Correspondence: angelaribeiro@ipvc.pt (Â.M.R.); tlmateus@esa.ipvc.pt (T.L.M.)

Citation: Ribeiro, Â.M.; Rodrigues, ; Brito, N.V.; Mateus, T.L. Prickly Connections: Sociodemographic Factors Shaping Attitudes, Perception and Biological Knowledge about the European Hedgehog. *Animals* 2023, 13, 3610. https://doi.org/10.3390/ani13233610

Academic Editors: Anne Berger, Nigel Reeve and Sophie Lund Rasmussen

Received: 11 October 2023
Revised: 10 November 2023
Accepted: 17 November 2023
Published: 22 November 2023

Simple Summary: The modern lifestyle, including indoor-centric living, urbanization and limited exposure to nature, contributes to the estrangement of humans from nature and a rapid decline in people's natural history knowledge. Meanwhile, several wild species are adapting and thriving in urban environments alongside humans. How should we see the rising human disconnect with nature, even while urban wildlife increases and environmental education programmes are deployed? How does this lack of connection affect perception and attitude towards wildlife? What is the role of the sociodemographic context? To address these questions, we used a keystone species as a study model—the European hedgehog. We collected data via online questionnaires that comprised four main sections: (i) socio-demographic features; (ii) feelings, attitude and perception; (iii) natural history knowledge about species; and (iv) self-evaluation about the extent of knowledge and past experience. The data indicate generally positive feelings and attitudes towards hedgehogs. We found that academic qualifications and past experience with the species shaped people's attitudes and natural history knowledge; however, the extent of knowledge, overall, was low and the study population was self-aware of this. We discuss the relevance of citizen profiling and possible avenues to enhance nature experience, improve knowledge, and increase public support for conservation measures.

Abstract: The modern lifestyle of humans is leading to a limited exposure to nature. While several wild species are adapting and thriving in anthropic environments, natural history knowledge is declining, and positive attitudes and behaviours towards nature are facing challenges. Because anticipating attitudes and engendering broad-based support for nature-related measures requires a good grasp of social contexts, we set out to evaluate the sociodemographic factors driving the perception, attitudes towards, and natural history knowledge of a keystone species—the European hedgehog. In 2022, we conducted a questionnaire answered by 324 Portuguese adults. We found generally positive feelings and attitudes towards this species. A higher degree of academic qualifications and previous personal experience with the species seem to play a role in (i) people's perception about human impacts on hedgehogs and (ii) positive attitudes, especially during encounters where the animals were in difficulty. Despite this, the extent of natural history knowledge was low overall, and the study population was self-aware of this. Our insights underline the need to tailor educational programmes if we are to encourage people to re-establish meaningful connections with nature, to foster social support for biodiversity stewardship, and to implement the One Health approach in way that resonates with distinct social groups.

Keywords: awareness; biodiversity stewardship; common knowledge; *Erinaceus europaeus*; human nature interactions; perception; Portugal; questionnaire

1. Introduction

"We are human in good part because of the particular way we affiliate with other organisms". E. O. Wilson. 1984. Biophilia—the human bond with other species, page 139.

Our experience of nature is declining. Regular interactions with nature have bee progressively diminishing due to growing urbanization, indoor-centric living, sedentar lifestyles, and technological distractions. The consequence of this is an extinction experience [1–3]. Despite the hardwired biophilic responses of humans to nature [4,5 the recognition of the positive effects of human–nature interactions on human health an wellbeing [6,7], and the fundamental role of these interactions for the future of ecosystem and biodiversity [8,9], there is mounting evidence pointing to the diminishing connectio between people and nature [9].

As people become disconnected from the natural world, their familiarity with th natural environment, as assessed through their level of natural history knowledge (fe tures of wildlife/ecosystems), declines. Therefore, there is a sense of detachment, littl appreciation of the natural world and, ultimately, a lessening of positive attitudes an behaviours towards nature [3]. The consequences of experience extinction and the rapi decline in people's natural history knowledge might be particularly challenging for wildli conservation, as well as for the One Health framework. Traditionally, the approach negative attitudes and behaviours is reinforcing conservation and health strategies wi ecological knowledge. However, support has increased for the explicit incorporation social dimensions, namely factors such as age, gender, level of education, urban vs. rur area of residence, and people's perceptions [10,11]. Understanding stakeholders' soci contexts and perceptions makes it possible to anticipate the attitudes and behaviour [12–1 necessary for implementing design strategies that generate engagement and consequent ensure conservation and the success of public health programmes [15–17].

While the gap between people and nature is widening mostly due to the urbanizatic of human life, many wildlife species are colonizing and thriving in urban areas [18–2(This is an unprecedented paradox: urban wildlife is increasing, efforts to raise awarene: through public outreach campaigns are on the rise, and yet humans have less direct conta with nature than ever before. To explore this paradigm and its underlying drivers, we u: the European hedgehog (*Erinaceus europaeus*, hereafter hedgehog) as a model species. Th hedgehog, a ground-dwelling nocturnal mammal that is widespread across Europe in rur and urban habitats [21], was selected due to conservation concerns and its synanthropic b haviour (propensity to live in anthropic environments). Although the International Unic for the Conservation of Nature (IUCN; [21]) and Red Book of Vertebrates of Portugal [2 list the species status as "least concern", recent trends indicate that populations are plur meting in rural areas, while high densities near urbanized/humanized areas are increasir (e.g., [23–25]). Several factors contribute to these observations: exposure to pesticides ar rodenticides in agricultural areas [26]; traffic collisions and mortality affecting dispersic and population dynamics [27]; and the decreased risk of predation in villages [28]. Regar ing synanthropic behaviour, compelling studies demonstrate that hedgehogs are ecosyste sentinels for heavy metal(loid) pollution [29] and human health threats associated wi zoonotic diseases [30,31]. Both hedgehogs' ecology and synanthropic behaviour make th mammal a keystone species for agroecosystems and a sentinel for ecosystems and huma health (One Health framework).

The concomitant evidence that estrangement from nature is increasing, interactions with natural world are enhancing emotional ties and positive attitudes, and European agroecosystems deserve conservation efforts [32] makes it timely and necessary to evaluate perception and natural history knowledge of a sentinel and keystone species. For this study, we set out to (i) assess citizens' natural history knowledge about the hedgehog and (ii) evaluate the sociodemographic factors driving citizens' perception and attitudes towards a common species whose populational trends have been affected, in some instances severely, by anthropogenic disturbances.

2. Materials and Methods

2.1. Data Collection: Questionnaire Design and Survey

The questionnaire design was based on a two-step approach: first, based on a bibliographic review of the topic, we designed a draft questionnaire that was tested with 15 respondents. After tailoring the questions, the final survey occurred in summer 2022 (July–August) using Google Forms and was rolled out following a "snowball" approach [33]. The "virtual snowball" sampling survey was disseminated via email through the mailing lists of researchers; these participants (primary respondents) were asked to disseminate the questionnaire with at least one of their personal contacts to proceed with the snowball and reach secondary respondents. All respondents were age > 18 and residents in Portugal. A small portion of respondents (seniors) requested a verbal survey. The final questionnaire included 35 questions organized into four main sections: the first block of questions collected information about respondents' sociodemographic features; the second section gathered information on the respondents' feelings, attitudes and perceptions towards the study species; the third part targeted respondents' natural history knowledge of the hedgehog; and finally, in the fourth block, we assessed respondents' self-evaluations about the species' biology and their past experience with the study mammal (see Table S1). We adhered to Likert-type/scale questions, with a 6-point scale in which 1 is very negative/unimportant and 6 is very positive/important. The full questionnaire is provided as Supplementary Materials.

While conducting the questionnaires, we adhered to ethical principles as follows: (i) full disclosure—the respondents were fully informed about the scope and goal of the research; (ii) prior informed voluntary consent—consent was verbally/tacitly obtained from each respondent before conducting the questionnaire; and (iii) confidentiality—we ensured anonymity and privacy of the respondents.

2.2. Feelings, Attitudes Towards, and Perceptions about Hedgehogs

First, participants were questioned about their feelings towards the species (negative to positive on a scale from 1 to 6; Table S1). Next, we asked about their (i) attitudes towards a potential encounter with a distressed hedgehog, (ii) perceptions relating to the human impacts on the species mortality, (iii) perceptions of the need for management/conservation measures, and (iv) perceptions regarding the species' impact on agriculture.

2.3. Biological Knowledge about Hedgehogs

To summarize the correctness of citizen natural history knowledge about hedgehog we used the information gathered in 19 questions to derive the *Erinaceus* Biological Knowledge Index (EBKI). The index was estimated as EBKI (1-i) = n° of correct answers/total n° of questions, where i is the total number of respondents; it is a continuous variable ranging from 0 to 1, where 1 indicates that all questions had a correct answer and 0 reflects completely incorrect answers.

2.4. Self-Evaluation and Past Experience with the Species

We asked the respondents to self-evaluate their knowledge about the species' natural history (from "very poor" to "very good"), and subsequently inquired about whether they have ever seen a hedgehog alive in the wild, in a zoo/wildlife rescue centre, or in the media (Table S1).

2.5. Predictors of Attitudes, Perceptions, and Natural History Knowledge about Hedgehog

Sociodemographic features. We gathered sociodemographic information about the participants, such as age, gender, academic qualification, and profession/occupation. Afterwards, based on the professional activity/occupation reported (following the formal Portuguese classification of professions), we defined three social groups according to the potential to encounter/interact with/require information about hedgehogs in their professional/daily activity: 1—farmers ($n = 18$), 2—veterinary assistants, nurses and doctors and biologists ($n = 31$), and 3—others ($n = 240$).

Urban–rural classification. To characterize the level of urbanization of the participant's area of residence, we used the Portuguese classification of urban areas; each parish of residence was assigned to one of three possible categories (Table S1).

2.6. Data Analysis

Descriptive statistics were used to summarize information about the study population. We used non-parametric tests as the data did not comply with normality as assessed by quantile–quantile plots and Tukey's test. Kruskal–Wallis rank tests (χ^2 value reported) were used to evaluate whether sociodemographic features, the level of urbanization of the residency area, and past experience impacted attitudes and perceptions. We tested (Student's t-test) the hypothesis that the observed mean EBKI is the same as expected when assuming a normal distribution centred on 0.5 (sufficient knowledge). Next, because we were interested in understanding the effect of participants' occupation on the level of EBKI, we performed a Kruskal–Wallis test followed by Dunn's test for multiple pairwise comparisons. Finally, to predict whether the level of natural history knowledge about hedgehogs (EBKI) was explained by sociodemographic features and/or past experience of respondents, we used a partitioning approach through a regression tree, as implemented in rpart R package [34] and rpart.plot [35]. In brief, the tree was built by recursively identifying variables that cluster the dataset into two groups ("branches"), while minimizing the dissimilarity at the terminal nodes, according to the Gini criterion [36]. The partition ceases when no additional variables achieve further reductions in node impurity, as per the Gini criterion. To optimize the predictive performance, the trees were pruned to achieve minimal expected error and a 10-fold cross validation was implemented. All statistical analyses were performed using R (version 4.2.3) and R Studio (version 2022.02.3+492) with the packages gghalves, ggplot2, and ggstatsplot; p-values < 0.05 were considered statistically significant.

3. Results

3.1. Socio-Demographic Features of Respondents

We had 324 participants in our survey aged between 18 and 93 years old (237 females, 86 males, and 1 non-binary participant) and living in 82 Portuguese municipalities (from north to south). Most of the sample corresponded to young adults (66.7%; 18–44 years old) and 14.5% were seniors (>65 years old). While 76.5% of the respondents lived in urban areas, 3.4% resided in rural parishes. More than half of the sample (60.5%) reported having higher academic qualifications (honours/licentiate, master's or doctoral degrees) and a reduced number (2.5%) were illiterate. The respondents more likely to deal with hedgehogs due to their profession/occupation (i.e., farmers ($n = 18$), biologists ($n = 5$), veterinary assistants/nurses/doctors ($n = 26$)) represented 23.8% of the sample. This differential participation is noteworthy and thus discussed later.

3.2. Feeling, Attitude and Perception Regarding Hedgehogs

Many respondents revealed mostly positive feelings about the species (83.3%, scores 4, 5 and 6); only 16.7% reported mostly negative feelings (scores 1, 2 and 3). Attitude towards hedgehogs was, in general, positive. The most popular attitude towards a likely encounter with a hedgehog in difficulty was to "seek help" (74.4%), either by contacting the authorities or a wildlife rescue centre; on the other side of the scale, 1.4% reported an attitude of killing the animal. Ignoring the situation was the attitude supported by 12.8% of the respondents.

Less than half of the participants (42.9%) perceive humans as a factor of high impact on hedgehog mortality, whereas 6.6% agreed that human impacts are "very unimportant" (Table S1). There was a general consensus for the need to protect the species (90.4%), but some participants thought that it needed to be controlled (7.7%) or eliminated (1.9%). The impact of the species on farming was unevenly appreciated: 36.1% of the participants were worried about the negative impacts (scores 1, 2, and 3; Table S1), whereas 63.9% recognized the benefits of hedgehogs in agriculture (scores 4, 5, and 6; Table S1).

3.3. Common Knowledge about Hedgehog Biology and Conservation

Only 289 respondents fully answered the 19 questions regarding the natural history of the species, and 64.1% showed a low level of knowledge (EBKI < 0.5; Figure 1a). The knowledge about hedgehogs' natural history (mean EBKI = 0.43) was significantly lower than expected if assuming a population with normal distribution cantered on 0.5 ($t(288) = -7.46$, $p < 0.001$). A close inspection of the results revealed the following: most respondents (91.7%) recognized hedgehogs as a rural dweller species; in several questions, participants reported "do not know", with the highest frequency for two questions inquiring about breeding biology (for litter number per year and number of offspring, 68.8% and 46.6% reported a lack of knowledge, respectively).

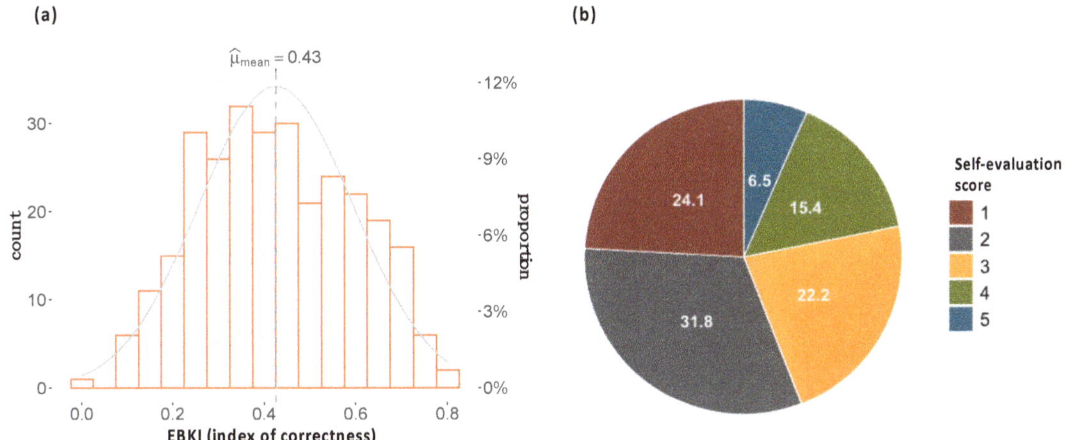

Figure 1. Assessment of common knowledge about hedgehog's biology. (**a**) Distribution of citizen knowledge correctness estimated as EBKI (ranges: 0–1). μ = observed mean value for the index estimated based on 289 respondents. Student's t-test revealed that EBKI was significantly lower than the value expected, assuming a normal distribution centred around 0.5 (sufficient knowledge). (**b**) Respondents' knowledge self-evaluation. The scores provided in the questionnaire ranged from 1 to 6, yet category 6 "very good" had zero observations.

3.4. Self-Knowledge and Past Experience

In general, the respondents self-evaluated their knowledge about hedgehogs' biology as quite poor (scores 1 and 2 = 55.9%; Figure 1b). For 9.0% of the respondents, the hedgehog has become a species primarily accessed through the filters of media (i.e., books, magazines, TV, or online) and 7.6% had never had an encounter with the species; however, most participants (83.4%) reported observations in the wild or in captive conditions (zoos or rescue centres).

3.5. Citizen Profile as a Driver of Attitudes, Perceptions and Common Knowledge

Our results show that attitude or perception is neither affected by gender nor the level of urbanization in the parish of residency of participants (Table 1). In contrast, the academic qualification of the respondents is related to the attitude and the three assessments of perception (all χ^2 tests of independence are significant; Table 1). Attitude and perception are associated with occupation/profession, except for the perceived impact that hedgehogs have on farming (Table 1). Upon closer examination, the results reveal that, generally, the proportions inside each explanatory category of perception were not equal and thus significant (see portions test in Figures S1–S5). For instance, the perceptions that humans have no impact on hedgehog mortality, that the species needs culling, and that hedgehogs have a negative impact on farming were mostly conveyed by participants with basic levels of education (academic qualification: first cycle). Regarding past experiences and encounters, i.e., type of previous observation of the species, it was also related to attitude towards a hedgehog in difficulty, the perception of hedgehogs' impact on farming, and the perception of the measures of conservation/management, but did not relate to the perception that humans influence species mortality (Table 1).

Table 1. Summary of the Chi-square tests of independence for attitudes and perceptions from sociodemographic factors. Dependent and explanatory variables are all categorial (details in Table S1).

Explanatory	Attitude		Perception	
	Towards a Hedgehog in Difficulties	Management/Conservation Measures	Human Effect on Mortality	Hedgehog Impact on Farming
Gender	$\chi^2 = 10.19$ $p = 0.25$ VCramer = 0.06	$\chi^2 = 0.49$ $p = 0.97$ VCramer = 0.00	$\chi^2 = 4.38$ $p = 0.93$ VCramer = 0.00	$\chi^2 = 6.85$ $p = 0.74$ VCramer = 0.00
Level of urbanization	$\chi^2 = 5.77$ $p = 0.67$ VCramer = 0.00	$\chi^2 = 2.26$ $p = 0.69$ VCramer = 0.00	$\chi^2 = 10.94$ $p = 0.36$ VCramer = 0.04	$\chi^2 = 6.51$ $p = 0.85$ VCramer = 0.00
Academic qualification	$\chi^2 = 171.56$ $p = 1.43 \times 10^{-22}$ VCramer = 0.35	$\chi^2 = 78.19$ $p = 6.12 \times 10^{-11}$ VCramer = 0.33	$\chi^2 = 153.08$ $p = 1.03 \times 10^{-18}$ VCramer = 0.29	$\chi^2 = 72.11$ $p = 2.23 \times 10^{-4}$ VCramer = 0.16
Occupation	$\chi^2 = 77.82$ $p = 1.34 \times 10^{-13}$ VCramer = 0.35	$\chi^2 = 41.95$ $p = 1.71 \times 10^{-8}$ VCramer = 0.26	$\chi^2 = 70.39$ $p = 3.72 \times 10^{-11}$ VCramer = 0.32	$\chi^2 = 14.47$ $p = 0.15$ VCramer = 0.09
Past experience	$\chi^2 = 43.20$ $p = 2.61 \times 10^{-4}$ VCramer = 0.15	$\chi^2 = 16.71$ $p = 0.03$ VCramer = 0.12	$\chi^2 = 25.40$ $p = 0.19$ VCramer = 0.07	$\chi^2 = 39.32$ $p = 6.08 \times 10^{-3}$ VCramer = 0.13

VCramer [0.10–0.20] indicates a weak association; VCramer [0.20–0.40] indicates a moderate association.

Most participants knew that hedgehogs are solitary mammals (51.6%; 26.6% reported no knowledge) that hibernate (72.3%; 21.1% indicated no knowledge) and were confident that it is mostly a rural-dwelling species (91.7%). Some participants (5.0%) expressed that keeping a hedgehog as a pet animal is a legal practice (21.5% reported an absence of knowledge). The regression tree analysis for predicting the factors shaping EBKI included five splits with six leaf nodes and three variables (Figure 2). The provided model includes the five explanatory variables (Table 1), but only three are needed to explain the variation in the dataset. The first split in the decision tree is associated with academic qualification and explains 21.5% of the variance in the data; the second split relates to past experience and helps to explain another 4.8% of variance; and the fourth split accounts for the effect of occupation. Higher levels of academic qualification, seeing the species live (either in the wild or captive) and belonging to an occupation group other than "farmers" are the predictors for largest EBKI (0.55). Further exploring the occupation/profession effect, we found significant differences in EBKI among the three functional groups (Kruskal–Wallis $\chi^2 = 28.89$, $p < 0.001$), with "farmers" presenting the lowest index of correctness (Figure 3).

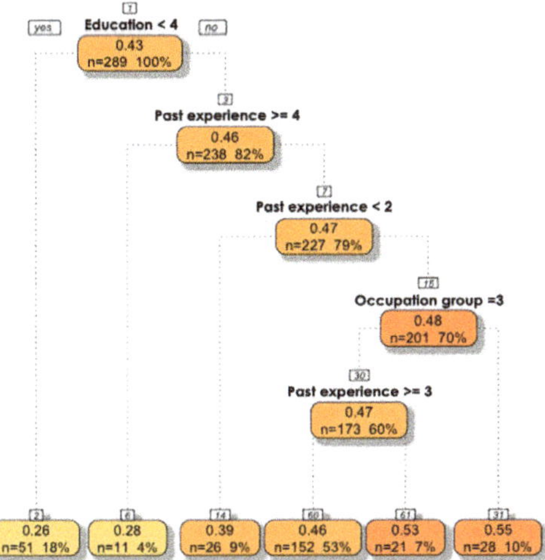

Figure 2. Decision tree to predict EBKI. *Academic qualification* (Ac. Qualif.) is self-explanatory (see Table S1 for levels). *Occupation/profession* pertains the social functional group used as a proxy for participants' likelihood of encountering/interacting with hedgehogs in their professional or daily activities: 1: farmers, 2: biologist and veterinary doctors/nurses/assistant, 3: others. *Past experience* refers to the participants past observation of a hedgehog either in real life or through the media, or never seen it. Reading the tree: the first split, at the root node, asked "at mean EBKI (0.43), is the academic qualification of the participant in basic levels (illiterate + 1st cycle + 2nd cycle)"? Negative or "no" responses branched to the right. If "no" (negative responses branch to the right), the second question inquired whether "the respondent had never seen a hedgehog" or if "yes" (left branch) only via media. In participants with higher levels of academic qualification who have seen the species before, a third question followed: "does the respondent belong to the occupation group less likely to encounter a hedgehog in their professional or occupational activities?", if "no" participants were classified as the ones with the largest EBKI (0.55; 10% of the sampled population).

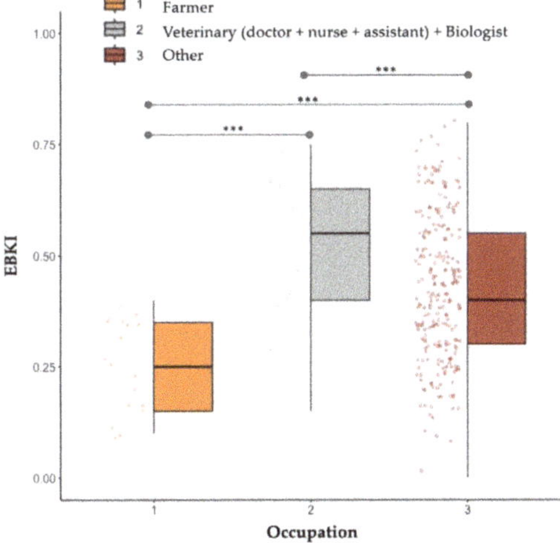

Figure 3. Effect of occupation on EBKI. Testing the role of participant's occupation on the correctness of knowledge about species biology; the three groups were defined according to their likelihood of encountering/interacting with a hedgehog during their professional/daily activities. Horizontal lines connecting boxplots indicate pairwise comparisons (1–2; 1–3, 2–3) and *** indicates $p < 0.001$.

4. Discussion

The Western European hedgehog is a cosmopolitan mammal found in the countryside as well as being common in suburban and urban areas. In Portugal, it is a popular animal which (i) evokes positive feelings, (ii) generally receives positive attitudes, (iii) prompts positive perception regarding the needs for conservation, and (iv) leads to high levels of awareness regarding the negative impact that humans have on the species. The general positive attitudes and feelings towards hedgehogs did not equate to natural history knowledge; the lack of knowledge was evident by the mean EBKI = 0.43 and by the high frequency of replies of "do not know". The participants were self-aware about their limited knowledge and surprisingly honest about it.

4.1. Citizen Profile

The social perception and natural history knowledge about hedgehogs are basic issues related to education and occupation/profession. There is a discrepancy in both perception and EBKI between participants with lower degrees of qualification (i.e., illiterate, first, second) vs. people with high school or university qualifications. Respondents with lower academic qualifications (all age > 65) perceive hedgehogs as detrimental for agriculture and in need of culling, and they believe that their decreasing populational trend is not affected by humans. Contrastingly, participants with higher academic qualifications mostly reported positive attitudes and perceptions. This pattern may reflect (i) farmers' general intolerance and negative attitudes towards wildlife due to potential agricultural damages or simply due to social/cultural norms, as noted by Jordan et al. [37], as well as (ii) the awareness of wildlife conservation and environmental issues taught in high schools and universities. The apparent generational shift may result from wider societal experiences or shifting cultural norms. It is widely known that the way individuals observe, understand, interpret and evaluate a given object/experience/outcome (i.e., perception) and the culmination of feelings or opinions regarding that same issue (i.e., attitude) is shaped by several personal factors, as well as cultural norms and beliefs [10]. Considering the social context in a given ecological system can provide insights to create opportunities to reconnect with nature, design awareness-raising programmes, and avoid polarized opinions that perpetuate human–wildlife conflicts.

4.2. Opportunities for Reconnecting with Nature—Education

Resolving the lack of experience of nature requires opportunities for meaningful interactions with the natural world [2]. Although 76% of the citizens reported an encounter with a wild hedgehog and nearly half of the participants reported positive feelings about the species, the low level of natural history knowledge is striking and indicates the pressing need to implement education programmes. For instance, in Portugal, a country where hedgehogs are identified as reservoirs of zoonotic diseases [38,39], half of the respondents (53%) reported no knowledge about this issue, and a quarter stated that hedgehogs do not transmit any diseases, either to humans or other animals. Furthermore, the lack of knowledge of simple biological facts, such as habitat preferences (cosmopolitan), social behaviour (solitary mammal), seasonal dormancy (hibernation), and the illegality of keeping a captive animal, call for urgent action.

The general public still perceives hedgehogs as a rural mammal, when several studies have highlighted their increasing presence, sometimes reaching high densities, in human environments [20,39]. Their synanthropy increases the likelihood and frequency of contact with pathogens from domestic animals and humans, increasing the potential for zoonotic transmission [25,31]. Given the above, we not only need to revert the alienation from nature by providing opportunities to experience nature [14], but also need to deploy bold educational policy changes that shift from "one fits all" paradigm. We require macro (European) and meso (national) One Health and biodiversity conservation educational and outreach programmes to be tailored to the micro scale, i.e., to meet the regional/local community features (perception, attitudes, knowledge).

4.3. Implications for Conservation Strategies and Eco-Schemes

The hedgehog is a key indicator of a healthy and sustainable farmland (arable land and pastures), so its absence must be a serious concern for agriculture. In fact, along with urban expansion and traffic accidents, farming intensification is pointed out as one of the main factors threating hedgehog populations [24,26,39]. Notwithstanding, the EU Common Agricultural Policy eco-schemes (designation under CAP 2023-2027; previously agri-environment schemes) have no direct measures for targeting this keystone species. Our results, which indicate a public perception that hedgehogs have positive impacts on farming despite an evident lack of natural history knowledge, highlight the need for national and regional authorities to implement CAP to incorporate farmland management measures that benefit hedgehogs [39]. The negative perceptions and smaller EBKI of Portuguese farmers should grant further research to assess farmers willingness to adopt eco-schemes dedicated to hedgehogs, so the national authorities can tailor effective conservation strategies [40,41]. It is evident that only a good grasp of the local social context of human–nature interactions will allow the implementation of widely accepted biodiversity conservation plans [9,10,42].

4.4. Study Caveats

The survey was distributed in two formats: online and face-to-face (for elderly and mostly illiterate). We were aware of the bias it could introduce, hence we read the questions to the participants very carefully, did not express any opinions ourselves, and ensured that we had their verbal informed consent. Moreover, we contend that our survey strategy may have caused some misrepresentation of groups, which might have partially affected the conclusions. For instance, males were under-represented (females: 73.5% vs. males: 22.8%) when considering the Portuguese population as reference (females: 52.4% vs. males: 47.6%); however, this gender difference in response rates is not exclusive to our study; in fact, it is well-known and widely discussed [43]. Additionally, there was a lower representation of seniors (14.5%, age > 65) when the recent reference demographic parameter is 23.4% [44]. Therefore, we refrained from further considerations regarding gender and age. Despite an apparent small sample size for farmers in the profession/occupational groups (6.2%), it fits well the Portuguese population with full-time employment in agriculture (<5%) [45]. Notably, the farmer participants were mostly seniors and had the lowest academic qualifications in the sample (illiterate or first cycle). Hence, once again, we are cautious with the interpretation of our results.

5. Conclusions

It is important to recognize that attitudes towards nature are not static and can be influenced and shaped over time through targeted efforts. Although direct experiences with nature foster a sense of appreciation and empathy and deem the natural world fundamental to people's lives, they may not be sufficient. In light of our findings, we advocate for the creation of strategies to identify and engage with local and relevant stakeholders. For instance, it is crucial that the sector with the largest power helps to reverse the current declining trends of rural hedgehog populations, i.e., encourage the farming sector to get involved with the development of integrated and sustainable management of the rural landscape. We suggest addressing the gap between the growth of scientific knowledge production and farming practices via the new eco-scheme policy instruments. Likewise, we call for the inclusion of social groups' perceptions when tailoring educational programmes pertaining One Health topics, particularly with regard to wild species that are becoming urban dwellers. Although synanthropy can bring benefits, such as increasing biodiversity in urban areas and improving human physical and mental health, there is also a risk of exposure to pathogens due to extensive lack of knowledge. We strongly believe that studying citizen profiles and social groups is crucial for the development of strategies that account for the diverse ways in which humans, animals, and the environment interact, ultimately leading to improved health outcomes for all, as desired in a One Health approach.

Raising awareness towards conservation needs and deploying interventions targetin
specific players are crucial for hedgehog conservation.

Supplementary Materials: The following supporting information can be downloaded at: http
//www.mdpi.com/article/10.3390/ani13233610/s1. Supplementary Text (informed consent an
questionnaire), Table S1: Details about the data collected. Variables used for the statistical analyse
and their attributes. Figure S1: Proportions test for gender in each category (defined in Table S
for three perception variables assessed; Figure S2: Proportions test for Academic Qualification i
each category (defined in Table S1) for three perception variables assessed; Figure S3: Proportio
test for Occupation in each category (defined in Table S1) for three perception variables assesse
Figure S4: Proportions test for Occupation in each category (defined in Table S1) for three perceptic
variables assessed; Figure S5: Proportions test for the level of urbanization in the participan
residency area—residency typology—for each category (defined in Table S1) for three perceptic
variables assessed.

Author Contributions: Conceptualization: Â.M.R. and T.L.M.; data collection: M.R. and T.L.N
methodology: Â.M.R., M.R. and T.L.M.; writing—original draft preparation: Â.M.R.; writing—revie
and editing: all authors; funding acquisition: N.V.B. and T.L.M. All authors have read and agreed
the published version of the manuscript.

Funding: Norte Portugal Regional Operational Programme (NORTE 2020), under the PORTUGA
2020 Partnership Agreement, through the European Regional Development Fund (ERDF), wit
project number NORTE-06-3559-FSE-000204, which includes the contract of Ângela M. Ribeiro.

Institutional Review Board Statement: The study was conducted in accordance with the Declaratic
of Helsinki and approved by the Ethics Committee the Social, Life and Health Sciences of Polytechn
Institute of Viana do Castelo (protocol code 15/A/2022).

Informed Consent Statement: Informed consent was obtained from all subjects involved in the stuc

Data Availability Statement: Anonymized datasets are available from the corresponding autho
upon reasonable request.

Acknowledgments: We thank Clarisse Rodrigues from Centro de Recuperação e Interpretação d
Ouriço (CRIDO) and all the staff at Centro de Recuperação de Animais Selvagens da Universidac
de Trás-os-Montes e Alto Douro (CRAS-HVUTAD). We extend our gratitude to Associação Soci
e Cultural de Louredo, Associação de Solidariedade Social de Nespereira, Associação de Apo
à 3.ª Idade S. Miguel de Beire and Centro Social e Paroquial de Sousela for their willingness
help us with the elderly participants. Micaela Rodrigues participated in this study in the conte
of her final internship in veterinary nursing licentiate degree. We acknowledge the Portugue
Foundation for Science and Technology (FCT) for the financial support to CISAS (UIDB/05937/20
and UIDP/05937/2020). Three anonymous reviewers are thanked for their helpful insights.

Conflicts of Interest: The authors declare no conflict of interest. The funders had no role in the desig
of the study; in the collection, analyses, or interpretation of data; in the writing of the manuscript;
in the decision to publish the results.

References

1. Pyle, R.M. *The Thunder Tree*; Houghton Mifflin: Boston, MA, USA, 1993.
2. Miller, J.R. Biodiversity Conservation and the Extinction of Experience. *Trends Ecol. Evol.* **2005**, *20*, 430–434. [CrossRef]
3. Soga, M.; Gaston, K.J. Extinction of Experience: The Loss of Human–Nature Interactions. *Front. Ecol. Environ.* **2016**, *14*, 94–1(
[CrossRef]
4. Kellert, S.R.; Wilson, E.O. *The Biophilia Hypothesis*; Island Press: Washington, DC, USA, 1993; ISBN 1559631473.
5. Wilson, E.O. *Biophilia*; Harvard University Press: Cambridge, MA, USA, 1986; ISBN 0674268393.
6. Hartig, T.; Mitchell, R.; de Vries, S.; Frumkin, H. Nature and Health. *Annu. Rev. Public Health* **2014**, *35*, 207–228. [CrossRe
[PubMed]
7. White, M.P.; Elliott, L.R.; Grellier, J.; Economou, T.; Bell, S.; Bratman, G.N.; Cirach, M.; Gascon, M.; Lima, M.L.; Lõhmus, M.; et
Associations between Green/Blue Spaces and Mental Health across 18 Countries. *Sci. Rep.* **2021**, *11*, 8903. [CrossRef] [PubMe
8. Otto, S.; Pensini, P. Nature-Based Environmental Education of Children: Environmental Knowledge and Connectedness
Nature, Together, Are Related to Ecological Behaviour. *Glob. Environ. Chang.* **2017**, *47*, 88–94. [CrossRef]

29. Zaradic, P.A.; Pergams, O.R.W.; Kareiva, P. The Impact of Nature Experience on Willingness to Support Conservation. *PLoS ONE* **2009**, *4*, e7367. [CrossRef]
30. Dickman, A.J. Complexities of Conflict: The Importance of Considering Social Factors for Effectively Resolving Human–Wildlife Conflict. *Anim. Conserv.* **2010**, *13*, 458–466. [CrossRef]
31. Šūmane, S.; Kunda, I.; Knickel, K.; Strauss, A.; Tisenkopfs, T.; Rios, I.d.I.; Rivera, M.; Chebach, T.; Ashkenazy, A. Local and Farmers' Knowledge Matters! How Integrating Informal and Formal Knowledge Enhances Sustainable and Resilient Agriculture. *J. Rural Stud.* **2018**, *59*, 232–241. [CrossRef]
32. Castillo-Huitrón, N.M.; Naranjo, E.J.; Santos-Fita, D.; Estrada-Lugo, E. The Importance of Human Emotions for Wildlife Conservation. *Front. Psychol.* **2020**, *11*, 1277. [CrossRef] [PubMed]
33. Rosa, C.D.; Collado, S. Experiences in Nature and Environmental Attitudes and Behaviors: Setting the Ground for Future Research. *Front. Psychol.* **2019**, *10*, 763. [CrossRef]
34. Soga, M.; Gaston, K.J. The Ecology of Human–Nature Interactions. *Proc. R. Soc. B* **2020**, *287*, 20191882. [CrossRef] [PubMed]
35. Straka, T.M.; Miller, K.K.; Jacobs, M.H. Understanding the Acceptability of Wolf Management Actions: Roles of Cognition and Emotion. *Hum. Dimens. Wildl.* **2020**, *25*, 33–46. [CrossRef]
36. Ceríaco, L.M.; Marques, M.P.; Madeira, N.C.; Vila-Viçosa, C.M.; Mendes, P. Folklore and Traditional Ecological Knowledge of Geckos in Southern Portugal: Implications for Conservation and Science. *J. Ethnobiol. Ethnomedicine* **2011**, *7*, 26. [CrossRef]
37. Manfredo, M.J. *Who Cares About Wildlife?* Springer: Berlin/Heidelberg, Germany, 2008; pp. 1–27. [CrossRef]
38. Ives, C.D.; Lentini, P.E.; Threlfall, C.G.; Ikin, K.; Shanahan, D.F.; Garrard, G.E.; Bekessy, S.A.; Fuller, R.A.; Mumaw, L.; Rayner, L.; et al. Cities Are Hotspots for Threatened Species: The Importance of Cities for Threatened Species. *Glob. Ecol. Biogeogr.* **2015**, *25*, 117–126. [CrossRef]
39. Alagona, P.S. *The Accidental Ecosystem: People and Wildlife in American Cities*; University of California Press: Berkeley, CA, USA, 2022.
40. Hubert, P.; Julliard, R.; Biagianti, S.; Poulle, M.-L. Ecological Factors Driving the Higher Hedgehog (Erinaceus Europeaus) Density in an Urban Area Compared to the Adjacent Rural Area. *Landsc. Urban Plan.* **2011**, *103*, 34–43. [CrossRef]
41. Amori, G. *Erinaceus Europaeus. The IUCN Red List of Threatened Species*; IUCN: Gland, Switzerland, 2016; e.T29650A2791303.
42. Mathias, M.d.L.; Fonseca, C.; Rodrigues, L.; Grilo, C.; Lopes-Fernandes, M.; Palmeirim, J.M.; Santos-Reis, M.; Alves, P.C.; Cabral, J.A.; Ferreira, M. *Livro Vermelho Dos Mamíferos de Portugal Continental*; Fciências, I.D., Ed.; ICNF: Lisboa, Portugal, 2023; ISBN 9895372426.
43. Hof, A.R.; Bright, P.W. Quantifying the Long-Term Decline of the West European Hedgehog in England by Subsampling Citizen-Science Datasets. *Eur. J. Wildl. Res.* **2016**, *62*, 407–413. [CrossRef]
44. Pettett, C.E.; Johnson, P.J.; Moorhouse, T.P.; Macdonald, D.W. National Predictors of Hedgehog Erinaceus Europaeus Distribution and Decline in Britain. *Mammal Rev.* **2018**, *48*, 1–6. [CrossRef]
45. Taucher, A.L.; Gloor, S.; Dietrich, A.; Geiger, M.; Hegglin, D.; Bontadina, F. Decline in Distribution and Abundance: Urban Hedgehogs under Pressure. *Animals* **2020**, *10*, 1606. [CrossRef] [PubMed]
46. Dowding, C.V.; Shore, R.F.; Worgan, A.; Baker, P.J.; Harris, S. Accumulation of Anticoagulant Rodenticides in a Non-Target Insectivore, the European Hedgehog (Erinaceus Europaeus). *Environ. Pollut.* **2010**, *158*, 161–166. [CrossRef]
47. Moore, L.J.; Petrovan, S.O.; Baker, P.J.; Bates, A.J.; Hicks, H.L.; Perkins, S.E.; Yarnell, R.W. Impacts and Potential Mitigation of Road Mortality for Hedgehogs in Europe. *Animals* **2020**, *10*, 1523. [CrossRef] [PubMed]
48. Pettett, C.E.; Moorhouse, T.P.; Johnson, P.J.; Macdonald, D.W. Factors Affecting Hedgehog (Erinaceus Europaeus) Attraction to Rural Villages in Arable Landscapes. *Eur. J. Wildl. Res.* **2017**, *63*, 54. [CrossRef]
49. Jota Baptista, C.; Seixas, F.; Gonzalo-Orden, J.M.; Patinha, C.; Pato, P.; Ferreira da Silva, E.; Casero, M.; Brazio, E.; Brandão, R.; Costa, D.; et al. High Levels of Heavy Metal(loid)s Related to Biliary Hyperplasia in Hedgehogs (Erinaceus europaeus). *Animals* **2023**, *13*, 1359. [CrossRef]
50. Ruszkowski, J.J.; Hetman, M.; Turlewicz-Podbielska, H.; Pomorska-Mól, M. Hedgehogs as a Potential Source of Zoonotic Pathogens-A Review and an Update of Knowledge. *Animals* **2021**, *11*, 1754. [CrossRef]
51. Jota Baptista, C.; Oliveira, P.A.; Gonzalo-Orden, J.M.; Seixas, F. Do Urban Hedgehogs (Erinaceus europaeus) Represent a Relevant Source of Zoonotic Diseases? *Pathogens* **2023**, *12*, 268. [CrossRef] [PubMed]
52. Pe'er, G.; Finn, J.A.; Díaz, M.; Birkenstock, M.; Lakner, S.; Röder, N.; Kazakova, Y.; Šumrada, T.; Bezák, P.; Concepción, E.D.; et al. How Can the European Common Agricultural Policy Help Halt Biodiversity Loss? Recommendations by over 300 Experts. *Conserv. Lett.* **2022**, *15*, e12901. [CrossRef]
53. Aguiar, A.; Pinto, M.; Duarte, R. Psychological Impact of the COVID-19 Pandemic and Social Determinants on the Portuguese Population: Protocol for a Web-Based Cross-Sectional Study. *JMIR Res. Protoc.* **2021**, *10*, e28071. [CrossRef]
54. Therneau, T.; Atkinson, B. _rpart: Recursive Partitioning and Regression Trees_. Rpackage version 4.1.19. 2022. Available online: https://CRAN.R-project.org/package=rpart (accessed on 15 September 2023).
55. Milborrow, S. _rpart.plot: Plot 'rpart' Models: An Enhanced Version of 'plot.rpart'_. R package version 3.1.1. 2022. Available online: https://CRAN.R-project.org/package=rpart.plot (accessed on 15 September 2023).
56. Breiman, L. *Classification And Regression Trees*; Routledge: New York, NY, USA, 2022. [CrossRef]
57. Jordan, N.R.; Smith, B.P.; Appleby, R.G.; Eeden, L.M.; Webster, H.S. Addressing Inequality and Intolerance in Human–Wildlife Coexistence. *Conserv. Biol.* **2020**, *34*, 803–810. [CrossRef] [PubMed]

38. Barradas, P.F.; Mesquita, J.R.; Mateus, T.L.; Ferreira, P.; Amorim, I.; Gärtner, F.; Sousa, R. de Molecular Detection of Rickettsia Spp in Ticks and Fleas Collected from Rescued Hedgehogs (Erinaceus Europaeus) in Portugal. *Exp. Appl. Acarol.* **2021**, *83*, 449–46? [CrossRef] [PubMed]
39. Yarnell, R.W.; Pettett, C.E. Beneficial Land Management for Hedgehogs (Erinaceus Europaeus) in the United Kingdom. *Animal* **2020**, *10*, 1566. [CrossRef] [PubMed]
40. Kusmanoff, A.M.; Fidler, F.; Gordon, A.; Garrard, G.E.; Bekessy, S.A. Five Lessons to Guide More Effective Biodiversity Conservation Message Framing. *Conserv. Biol.* **2020**, *34*, 1131–1141. [CrossRef]
41. Massfeller, A.; Meraner, M.; Hüttel, S.; Uehleke, R. Farmers' Acceptance of Results-Based Agri-Environmental Schemes: A German Perspective. *Land Polic.* **2022**, *120*, 106281. [CrossRef]
42. Bennett, N.J.; Roth, R.; Klain, S.C.; Chan, K.; Christie, P.; Clark, D.A.; Cullman, G.; Curran, D.; Durbin, T.J.; Epstein, G.; et al. Conservation Social Science: Understanding and Integrating Human Dimensions to Improve Conservation. *Biol. Conserv.* **2017**, *205*, 93–108. [CrossRef]
43. Tourangeau, R.; Rips, L.J.; Rasinski, K. *The Psychology of Survey Response*; Cambridge University Press: Cambridge, UK, 2000. [CrossRef]
44. INE—National Statistics Institute. Available online: https://censos.ine.pt (accessed on 15 September 2023).
45. FFMS—Fundação Francisco Manuel dos Santos. Available online: https://www.pordata.pt/en (accessed on 15 September 2023).

Review

Impacts and Potential Mitigation of Road Mortality for Hedgehogs in Europe

Lauren J. Moore [1,*], **Silviu O. Petrovan** [2], **Philip J. Baker** [3], **Adam J. Bates** [1], **Helen L. Hicks** [1], **Sarah E. Perkins** [4] and **Richard W. Yarnell** [1]

[1] School of Animal, Rural and Environmental Sciences, Nottingham Trent University, Southwell NG25 0QF, Nottinghamshire, UK; adam.bates@ntu.ac.uk (A.J.B.); helen.hicks@ntu.ac.uk (H.L.H.); richard.yarnell@ntu.ac.uk (R.W.Y.)

[2] Department of Zoology, University of Cambridge, The David Attenborough Building, Cambridge CB2 3QZ, Cambridgeshire, UK; sop21@cam.ac.uk

[3] School of Biological Sciences, University of Reading, Reading RG6 6AH, Berkshire, UK; p.j.baker@reading.ac.uk

[4] School of Biosciences, Cardiff University, Sir Martin Evans Building, Cardiff CF10 3AX, UK; perkinss@cardiff.ac.uk

[*] Correspondence: lauren.moore@ntu.ac.uk

Received: 29 June 2020; Accepted: 26 August 2020; Published: 28 August 2020

Simple Summary: The environmental impacts of transport infrastructure are attracting substantial research focus and road-induced mortality of wildlife is perhaps the most conspicuous impact of roads. Hedgehogs are a common victim of traffic collisions in Europe and several hedgehog species are showing marked population declines across their range. This review aims to consolidate current knowledge on the impacts of road mortality on the viability of populations of the five hedgehog species in Europe and identify research gaps. Previous studies have shown that roads are a major source of mortality for hedgehogs and that individuals with greater net movement, generally males, have the greatest likelihood of mortality. Road mortality also contributes to population isolation. More research is needed into how different individuals perceive, use and cross roads, as well as the efficacy of different mitigation measures (e.g., wildlife crossing structures) designed to reduce road mortality and population isolation. Assessing whether local hedgehog populations are at risk of extirpation or further declines due to road mortality is a prerequisite for effective conservation in environments affected by continuously developing road networks.

Abstract: Transport infrastructure is a pervasive element in modern landscapes and continues to expand to meet the demands of a growing human population and its associated resource consumption. Road-induced mortality is often thought to be a major contributor to the marked declines of European hedgehog populations. This review synthesizes available evidence on the population-level impacts of road mortality and the threat to population viability for the five hedgehog species in Europe. Local and national studies suggest that road mortality can cause significant depletions in population sizes, predominantly removing adult males. Traffic collisions are a probable cause of fragmentation effects, subsequently undermining ecological processes such as dispersal, as well as the genetic variance and fitness of isolated populations. Further studies are necessary to improve population estimates and explicitly examine the consequences of sex- and age-specific mortality rates. Hedgehogs have been reported to use crossing structures, such as road tunnels, yet evaluations of mitigation measures for population survival probability are largely absent. This highlights the need for robust studies that consider population dynamics and genetics in response to mitigation. In light of ongoing declines of hedgehog populations, it is paramount that applied research is prioritised and integrated into a holistic spatial planning process.

Keywords: road mortality; collision; fragmentation; movement; demography; population viability; mitigation; road ecology; hedgehogs

---

1. Introduction

The last century has been characterised by intense modification of the natural landscape, and road networks are now pervasive in most landscapes on Earth [1,2]. Interest in the ecological impacts of roads has grown since the mid-20th century, with formal recognition of a new field, road ecology, by Forman and Alexander in 1998 [3]. This branch of ecological research has revealed the extensive role that roads play in direct and indirect habitat loss and alteration. Traffic noise, light pollution and chemical pollution (salt, heavy metals, herbicides) are all identified as important correlates of habitat modification, fragmentation and changes in animal movement in road-dominated environments [4].

Perhaps the most conspicuous impact of roads are wildlife–vehicle collisions (WVCs) that result in the death of billions of animals worldwide every year [5]. Biological characteristics of the animals themselves (e.g., age, sex and movement), biotic factors (e.g., time of day, season), traffic and road characteristics (e.g., traffic volume, road width, tortuosity) and environmental characteristics (e.g., topography, neighbouring habitat structure) all interact to form a species-specific spatiotemporal distribution of WVCs [6]. The consequences of road mortality are typically two-fold: (1) direct depletion of individuals from a population and/or (2) fragmentation of populations and reduced gene flow [7–9]. Importantly, these consequences can alter meta-population structure and population fitness, in turn increasing the risk of local extinction [10]. Roads are therefore considered responsible for the nationwide decline of certain species and a limiting factor in the recovery of others [11,12]. The growing literature on road ecology has been largely motivated by WVCs that are of legislative or conservation concern and/or which give rise to economic or human safety issues, such as collisions with deer [6]. In comparison, fewer studies have examined smaller mammal species, such as hedgehogs.

There are five species of hedgehog with all or part of their range in Europe, although hedgehog taxonomy has been debated due to contradictions between molecular and morphological phylogenies [13]. The West European hedgehog (*Erinaceus europaeus*) is distributed over Ireland, Britain and western mainland Europe. The Algerian hedgehog (*Atelerix algirus*) is present in North Africa and was introduced to Spain and several Mediterranean islands [14]. Moreover, the northern white-breasted hedgehog (*Erinaceus roumanicus*) is distributed throughout Central and Eastern Europe, whilst the southern white-breasted hedgehog (*Erinaceus concolor*) is present in Eastern Europe and Southwestern Asia [15]. The long-eared hedgehog (*Helmiechinus auritus*) maintains part of its predominately Middle Eastern range in Cyprus and Ukraine [13].

Although hedgehog density has been reported to be up to 35% lower near roads [16], road-killed hedgehogs are a very familiar sight across Europe and are frequently the main mammal roadkill recorded in citizen science projects and expert multispecies roadkill surveys [17,18]. For example, an estimated 113,000–340,000 *E. europaeus* individuals are killed on roads every year in the UK [19] and The Netherlands [20], and 230,000–350,000 individuals every year in Belgium [21]. Comparisons between short-term studies are difficult as roadkill rates can fluctuate with changes in hedgehog density, road conditions and traffic volume [22]. Alternatively, long-term roadkill data are valuable to observe changes in temporal behaviour or monitor population trends [23]. For example, Recihholf [24] and Müller [25] found hedgehog road mortality to have steadily decreased since the 1970s in Germany, and Wilson and Wembridge [26] found similar patterns in the UK since 2001. It is claimed that these changes reflect the marked declines over the past two decades of *E. europaeus* in several countries across Europe [25,27]. *A. algirus* has also shown reduced abundance and local extinctions in its introduced range in Europe [14]. However, sufficient hedgehog population data to identify declines are currently limited to the UK [28,29], the Netherlands [30] and, to a lesser extent, Denmark [31] and Germany [25]. Traffic-related mortality has been implicated as a significant component of hedgehog population declines and also constitutes a welfare concern [8,19,24,32]. In recent times, the field has used nationwide monitoring schemes such as "Project Splatter", a citizen science study in the UK that collates such data [33]. Studies using nationwide data have demonstrated broad spatiotemporal patterns; hedgehog roadkill hotspots are associated with suburban areas and grassland, as well as the breeding season in late spring and early summer [22,32]. Records of hedgehog road mortality have also been used to estimate annual road mortality [19–21], track epizootics [22] and have the potential to estimate population abundance [34,35]. Substantial gaps in knowledge remain, however, about whether roads affect long-term population persistence. Likewise, the use of appropriate techniques to evaluate the complexity of the impact (e.g., population modelling using collected demographic data) have received little attention [36].

Investigating the population-level impacts of road mortality is of both theoretical and applied importance. It is likely that Europe is already the most fragmented continent due to transport infrastructure [2,37] and road networks continue to expand rapidly. In the UK alone, an average of 70,000 km of new roads are built every year [38] and many existing roads are modernised or widened throughout Europe [39]. Road development, however, is not consistent across European countries [37]. Coupled with the assertion that road mortality is the leading cause of human-induced vertebrate mortality on land [3], road ecology is a critical frontier of applied scientific research. As several European hedgehog species are declining and disproportionately represented in roadkill records [21], understanding how important road mortality is for population trends is a necessary step for hedgehog conservation. This review aims to consolidate the current knowledge on the consequences of road mortality for the viability of hedgehog populations in Europe. We used online databases to search for and appraise published, peer-reviewed articles on hedgehog road ecology, complemented by government reports on road statistics. This review synthesises information on the possible direct role that road mortality plays in population declines. It then discusses the individual-level risk of road mortality and the contribution of hedgehog–vehicle collisions to much-discussed fragmentation effects and associated genetic heterozygosity. Finally, this review identifies opportunities for road mitigation for hedgehogs, current knowledge gaps and priorities for future research.

2. Does Road Mortality Really Reflect Population Persistence?

It is difficult to confirm or refute the impact of road mortality on population trends because survival probabilities depend on a complex set of inter-related factors [40]. Several criteria exist to evaluate the ecological effects of road mortality. For example, the total number of road-killed animals must be considered in the context of population size [19], reproductive output, immigration and emigration rates [41] and whether WVCs are compensatory or additive to other forms of mortality [5]. To date, research has only partially met these criteria. Recent year-round studies have evidenced an average 0.001–3.65 hedgehog casualties/km/year for all European species across several European countries (Table 1).

Table 1. Estimates of the mean number of hedgehogs killed on roads per kilometre per year in different European countries. Data are derived from systematic, year-round studies.

Species	Country	Year(s) of Study	Mean Number of Road Casualties/km/Year	Surveyed Road Types	Habitats along Survey Route	Reference
E. europaeus	Ireland	2008–2010	0.001	National and regional roads	Residential, rural, woodland and built-up areas	[42]
	Finland	2004–2005	0.007	National, regional and local roads	Residential, forests and built-up areas	[43]
	Spain	2001–2003	0.76–1.42	National roads	Forests	[44]
	Slovakia	2000–2002	1.6	National, regional and local roads	Agricultural land, forests and nature preserves	[45]
E. europaeus and *E. concolor*	Poland	2001–2003	0.07	National, regional and local roads	Arable land and built-up areas	[46]
A. algirus	Lanzarote	2010–2011	0.65	National, regional and local roads	Urban areas and Badlands	[18]
E. roumanicus	Bulgaria	2017	0.06–0.08	National and regional roads	Rural areas	[47]
	Bulgaria	2015–2017	0.23	National roads	Arable land, riparian vegetation and forests	[17]
	Ukraine	2000–2009	0.73	National, regional and local roads	Urban areas and forests	[48]
	Slovakia	2008–2012	1.52–2.68	National roads	Agricultural land, grasslands, forest, riparian and built-up areas	[49]
	Ukraine	2002–2005	3.65	National roads	Marshes, forests and urban areas	[50]

Counts of the hedgehogs killed on roads indicate the extent of (lethal) collisions with vehicles and can be used to quantify differences between species, countries and road types if the survey methodology is clearly described. However, they do not indicate the relative importance of traffic-related deaths in the context of populations as a whole [5], and there are issues of standardisation between studies due to differences in study design, effort, frequency and duration. Notably, these issues include accurately accounting for variable carcass persistence [51,52]. Examining the proportion of a population killed on roads every year is more informative. Previous studies of *E. europaeus* have calculated that traffic casualties amount to 9–26% of the total (nationwide) population size in The Netherlands [53] and 10–30% in the UK [19], assuming the population estimates are accurate [28]. At the local scale, previous studies have used capture–mark–recapture methods to identify an annual loss of 3–22% of local *E. europaeus* populations on roads in Sweden [54] and 24% in Poland [46]. Examining the proportional loss at the local scale is instrumental for targeted conservation action. This is because the impact of roads may be different between local populations due to regional variation in habitat type, quality, population densities and road networks [55].

Another promising indication of the population-level effects of road mortality is to compare it with mortality from other sources and identify its contribution to cumulative annual mortality. This can be used to assess the impact of traffic collisions on the mortality–recruitment ratio [5]. *E. europaeus* is the most studied hedgehog species worldwide and mortality of the species has been investigated using radio-tracking methods [43], capture–mark–recapture methods [54] and data from rescue centres [56] (Table 2). It should be noted that the small sample sizes in the reported studies in Table 2 and their study design can skew the relative importance of a cause of death, and that there will naturally be local variation in the occurrence of each mortality factor. Although the studies are an important first step in refining an understanding on mortality, the results should be interpreted with caution. The studies suggest that road traffic is consistently in the top three most common causes of death for hedgehogs, alongside illness and natural predation, supporting the narrative that traffic mortality potentially places substantial pressure on population dynamics. The magnitude of this effect will depend on the ability of populations to compensate for additional mortality by increased survival and/or reproduction, for example, with second litters [57]. Determining how plausible compensation is for hedgehogs is hampered by a lack of data on female hedgehog fecundity, such as the proportion of females that breed successfully, the mean number of litters per female annually and mean litter size, as well as juvenile survival rates. However, the evidence for ongoing hedgehog declines suggests that compensation might not be occurring [27]. It is likely that the declines of hedgehog populations across Europe are a result of a combination of factors. For example, intensified agricultural practices, molluscicide and rodenticide poisoning, badger predation and loss of habitat have also been raised as important correlates of reduced population density and local extinction risk [24,28,58–60]. Disentangling the relative impact of factors to population demography, which is likely to be area-specific, remains a principal goal to improve hedgehog conservation.

Table 2. Percentage of deaths by different sources for studied *E. europaeus* individuals in Europe.

Location	Year of Study	Number of Individuals	Roads	Natural Predation (Badger and Fox)	Unnatural Predation (Dog and Cat)	Illness [a]	Poison	Other [b]	Unknown	Reference
UK and The Netherlands	1999	~83580	8.8	Not included	2.3	58.5	4.3	26.0	Not included	[56]
UK	1988	109	78.0	Not included	Not included	Not included	2.8	1.8	17.4	[61]
UK	1981	22	18.2	31.8	13.6	Not included	4.5	31.8	Not included	[62]
UK	1992	8	25.0	75.0	Not included	Not included	Not included	Not included	Not included	[63]
UK	2019	7	43.0	43.0	Not included	14.0	Not included	Not included	Not included	[64]
UK	1998	7	57.0	14.0	Not included	14.0	Not included	14.0	Not included	[65]
Finland	2016	106	72.6	Not included	0.9	20.7	Not included	1.8	4.0	[43]
Denmark	2019	9	Not included	Not included	22.0	22.0	11.0 [c]	44.0	Not included	[66]

[a] includes parasitological, pathological (e.g., starvation and gangrenous limb) and bacteriological findings (e.g., *Salmonella*); [b] includes drowning, injury from agricultural or garden tools and fire; [c] speculated but not confirmed.

3. The Risk of Road Mortality Is Not Equal. Which Are the Risk-Prone Individuals?

The risk of road mortality over time varies spatially and between individuals in a population [67]. Differential risk is a function of risk per crossing, which largely depends on animal crossing speed, traffic volume and road width, multiplied by the frequency of crossing. This is associated with individual responses to roads and biological characteristics, such as reproductive strategy and pre-hibernation foraging [68]. Individual-based movement patterns cause different exposure to traffic in the environment [41], which has important repercussions for reproductive output [10]. For example, for species such as hedgehogs that have a promiscuous mating system and maternal natal care, adult females have a more important role in population growth than males [69]. Moreover, the frequency distribution of age-at-death in a population is central to life history evolution and population dynamics [10].

No studies have empirically examined the individual-based risk of road mortality over time for hedgehogs, nor the potential variation in carcass detectability or persistence between different age groups. Current knowledge relies on limited data on the sex ratio and age structure of casualties. During a study of *E. europaeus* over 259.5 km of road in Ireland, Haigh et al. [42] revealed that 65% (67 out of 103) of those killed on roads were male. Moreover, Haigh et al. [70] tested several techniques to age hedgehogs, such as dentary bone analysis, jaw and hind foot length. These produced accurate age assessments and identified that the mean age of road-killed hedgehogs was 1.94 years. These findings were similar to those of Goransson et al. [71], who found that 80% of *E. europaeus* traffic casualties in Sweden were males who had survived one winter. To understand the significance of sex- and age-specific road mortality to population dynamics, these figures should be considered in the context of the number of individuals in that sex/age class in the wider population. Moreover, it is possible that, due to their small size, juvenile hedgehogs are readily scavenged or not detected during driving surveys.

The majority of hedgehogs are reproductively active in their second year (after one successful hibernation) [54,70]. Although research into the road mortality of different sexes and age groups is sparse, the majority of studies indicate that reproductively active males are most commonly killed on roads. Male hedgehogs have larger home ranges and nightly movements than females [14,72], particularly during the breeding season [73]. This would, all other conditions being equal, increase the number of roads that males must cross each night. Conversely, females are most likely to be involved in traffic collisions in autumn after intensive natal care as their net-movement increases to build fat reserves for hibernation [42]. The removal of reproductively active individuals carries a greater threat to hedgehog population viability because it can skew the age ratio and cause a decline in recruitment [74]. On the one hand, the disproportionate loss of adult males may not be as consequential for population growth as adult female deaths [5]. On the other hand, males are more commonly killed before or during breeding season, unlike females [42]. There is a possibility that fewer males successfully contribute to the gene pool and the relatedness in a population increases over time. If severe enough, this may cause a decrease in population fitness associated with inbreeding depression [75] (see Section 4), although research on the topic remains limited.

4. The Role of Road Mortality in Fragmentation Effects

Habitat fragmentation by transport infrastructure and the associated development has become one of the greatest threats to biodiversity [39]. The consequences of road-induced fragmentation for the integrity of natural environments are well-researched [76,77]. Several different, yet not mutually exclusive, mechanisms restrict animal dispersal across roads—lethal road collisions, the avoidance of the road or roadside habitat and the inability to traverse the road or nearby area, such as due to a central median or parallel drainage ditch [78]. Road mortality is likely to act as a filter to movement for many species, rather than an absolute barrier, as animals may be able to make successful journeys across the road, even across large roads and bridges [79]. For hedgehogs, road mortality is considered a more severe restriction to dispersal on smaller roads. For example, *E. roumanicus* in Bulgaria [33] and

E. concolor in Turkey [80] were shown to have greater casualty rates on quieter, regional roads than highways. This may result from quieter roads allowing more crossing attempts [58], having fewer physical barriers than major roads and/or their placement in areas with higher hedgehog densities. In severe cases, increased road mortality could lead to death rates exceeding birth rates, which may change a local population to a sink [81].

Road mortality has been shown to be the largest contributor to population fragmentation [81,82], albeit not always [3]. It is possible that physical barriers such as roadside fencing and road avoidance behaviour cause fragmentation via more stringent restrictions to movement. Both barriers and road avoidance behaviour are particularly common on roads with higher traffic volumes and speeds [78]. Dowding et al. [58] reported avoidance of foraging near roads, but not of crossing quieter roads, by *E. europaeus*. Moreover, Rondinini and Doncaster [83] compared observed *E. europaeus* movements in Southampton, UK, with "random walks" and identified clear road avoidance behaviour that increased with road width (and associated higher traffic). In corroboration with Rondinini and Doncaster [83], a traffic volume of 3000 vehicles/day (common for busy urban roads) in New Zealand led to the isolation of *E. europaeus* populations [22].

This combined effect of road mortality and avoidance for fragmentation is readily explained by traffic flow theory, which postulates a positive and asymptotic relationship between traffic volume and roadkill counts. Road mortality will increase with rising traffic volume until reaching an asymptote, when the busy roads (with greater noise levels) form complete barriers and are avoided, or the roads suppress population size and reduce the number of individuals crossing roads [84]. It is clear that roads constitute semi-permeable barriers for hedgehogs and that the extent of fragmentation is context-specific.

Biomolecular Insights into Fragmentation

Recent advances in genetic approaches have bridged the gap between molecular and road ecology to address the chronic impacts of fragmentation [85]. Insights into the genetic effects of hedgehog population fragmentation have grown since the development of eleven nuclear microsatellite primers (genetic markers) for *E. europaeus* by Becher and Griffiths [86] and Henderson et al. [87]. The markers have been used to genotype several closely related hedgehog species and can identify genetic similarities between individuals and, therefore, the level of inbreeding [15]. The variability of genetic markers is particularly important for small mammals such as hedgehogs, where fragmentation is likely to act at microspatial scales [88]. Braaker et al. [89] reported that two main rivers and major transport infrastructure (a four-lane highway and railroads) separate three genetic clusters of the *E. europaeus* population in Zurich. Moreover, combined movement models and microsatellite data indicated that fragmentation and high resistance in the urban matrix of Zurich, predominately from highways, footpaths, buildings and water bodies, contribute to the genetic structure of the hedgehog population at the local level, i.e., within clusters [89]. The weak correlation between genetic structure and geographical distance in several additional hedgehog studies indicates that linear infrastructure restricts gene flow enough to affect genetic heterozygosity [88,90,91]. However, the hedgehog's promiscuous mating system and ability for heteropaternal superfecundity (a litter fertilised by different males) may partly counteract the genetic effects of isolation [92]. Inbreeding coefficients would be reduced as a litter can consist of several half-siblings. The reality of this, however, remains untested, and Barthel [93] reported potentially early signs of inbreeding in *E. europaeus* subpopulations in Berlin. A promising, relatively unused strategy for examining population isolation is genetic pedigree analysis, which uses microsatellites to detect migration rates (e.g., across roads) and local geographies of closely related individuals. This forms a quantitative tool to identify the likelihood of inbreeding and whether the population is acting as a sink population [76,94].

5. Potential Road Mitigation Measures for Hedgehog Populations

As road construction and traffic volumes continue to grow, accommodating the increase in human activity without jeopardising the viability of wild populations remains a major challenge. Approaches for sustainable infrastructure development should tackle both the local (mortality and habitat degradation) and landscape (fragmentation and population viability) impacts of roads, yet there is no simple solution or decision-making framework [11]. A growing number of legal imperatives, such as Article 10 of the European Union's Habitat Directive (92/43/EEC) and the National Environmental Policy Act (1969), as well as international guidelines, such as the United Nations' Sustainable Development Goals, motivate transport planners to safeguard habitat connectivity and ecosystem functioning. This means newer major roads, in particular those built in Central and Eastern Europe, often have integrated wildlife crossings, such as underpasses or overpasses [39]. Minor roads, however, receive less attention despite the majority of road networks consisting of these low-traffic roads [82]. The range of mitigation measures can be classified using four main criteria; road crossing structures, traffic calming measures, habitat management and configuration of the road network [95].

5.1. Road Crossing Structures

Exclusionary fencing is a dominant strategy to impede an animal's attempt to cross a road. However, fencing was shown to cause a 30% reduction in *E. europaeus* population viability in The Netherlands by intensifying population isolation [96]. Instead, combining fencing with road tunnels or green bridges such as overpasses is widely advocated for many species [36,97,98]. This method strives to reduce barrier effects by providing both a reduction in road mortality and conserving or increasing landscape permeability [5]. Several studies have documented varied levels of crossing structure use by *E. concolor* in Greece [99], *E. europaeus in* Spain [44,100], Portugal [101], the UK [102,103] and Poland [104,105], and *Erinaceinae* sp. in Spain [106] (see review by De Vries [107]). This variation of use is likely due, in part, to differences in tunnel design, location and surrounding habitat, suggesting that the uptake of mitigation depends on optimality of species-specific features. For example, hedgehogs have been shown to frequent tunnels with a greater openness ratio (short in length, high and wide) nearer urban areas [101]. Moreover, previous studies demonstrate that hedgehogs avoid areas with predator (badger *Meles meles*) odour, although the avoidance did not always persist [108,109]. Badgers are known to utilise road tunnels [103], sometimes very regularly, and whether this negatively influences hedgehog use of road mitigation structures remains unknown.

5.2. Traffic Calming Measures

Crossing structures are often concentrated at clusters of roadkill [110]. However, this hotspot approach is contentious; several authors propose that a lack of road mortality may signal a previously declined population or a population that exhibits high road avoidance behaviour [111,112]. If so, the necessity for mitigation to assist in population recovery or protection is overlooked. Similarly, the fencing associated with crossing structures could block locations of frequent successful crossings if inappropriately placed. Instead, smaller-scale traffic calming measures that increase driver awareness may be equally effective and substantially cheaper. These aim to enhance preferred crossing sites, which do not necessarily correspond with roadkill hotspots, in order to dissuade the use of riskier crossing locations [7,81]. Traffic calming measures adopted in the past include speed bumps, speed restrictions and warning signs [95]. These initiatives may be particularly effective for hedgehogs given that they frequently attempt to cross quieter roads [58]. Whilst a reduction in speed would be expected to result in a substantial reduction in roadkill [113], the realised effect depends on whether drivers adhere to the speed regulations, which can be difficult to govern [114], and whether, even at a slower speed, a driver can see and avoid a small animal at night.

5.3. Habitat Management

Additional mitigation possibilities include managing roadside habitats by increasing habitat quality, local connectivity [95] and changing road verge management [115]. These improve the core habitat and allow individuals to locate sufficient resources whilst crossing fewer roads. Several authors recommend removing or reducing shrubbery in central medians to reduce road mortality [116,117] (but see [118]). The use of central medians by fauna has not been well-studied and, if they are in fact beneficial to animal movement across a road, their removal may exacerbate barrier effects [78]. Modifying hedgerows, which act as conduits of hedgehog movements, near roads is also likely to be an important action. For example, Huijser [53] identified that, out of 942 traffic victims, 20–27% and 140% more *E. europaeus* road casualties were found in areas where hedgerows and railroads, respectively, were perpendicular to roads rather than parallel. Therefore, how roads and local landscape features are orientated in relation to one another warrants consideration.

5.4. Road Configuration

In Western Europe, many major roads were built more than 40 years ago with little consideration for wildlife [37]. Retrofitting crossing structures can be an expensive undertaking, and their construction is often logistically challenging [119]. It is therefore essential to consider how landscape configuration can be designed to meet the needs of human settlements, associated road systems and habitat networks simultaneously [120]. Previous multispecies simulation studies have reported that road mortality rates and population persistence were improved when traffic volume was concentrated on fewer roads [121,122]. Surprisingly, van Strien and Grêt-Regamey [119] reported opposite results for hedgehogs. These studies reinforce the significance of whole landscape planning; the high rates of new road development in Central and Eastern Europe provide the opportunity to consider road configuration and maintain suitable habitat matrices for *E. roumanicus* and *E. concolor* [17,49].

6. Current Knowledge Gaps and Future Directions

Major impediments to furthering knowledge on hedgehog road ecology are the high labour and monetary costs linked to collecting relevant data for at least one population—that is, road casualty rates, movement and population structure data (and optionally genetic information). Moreover, although GPS devices are increasingly utilised for movement studies [10,123], including for hedgehogs [89,93], the high initial costs often reduce sample sizes and lead to results with poor statistical inference (see [124] for full review). Understanding the ramifications of hedgehog road mortality is further hindered by the lack of basic biological and ecological knowledge on some species such as *E. concolor*, as well as uncertain rigor of population and road casualty estimations for other species. Current population estimates are from citizen science surveys and extrapolations of presence-only density estimates in different habitat types [125,126]. The assumptions associated with these methods make estimates of population size equivocal [19]. Improved population estimates are critical to validate existing findings and could be achieved by large-scale collaborations or more standardised citizen science, such as using camera traps and random encounter methods [127,128]. Moreover, roadkill estimates of many species are likely to be underestimated due to scavengers removing carcasses and varying carcass detectability due to factors such as carcass decay, the driver's speed and the animal's body size [81,129]. As a result, raw carcass data must be corrected for carcass persistence and detection probability to obtain accurate estimates of the number of animals killed on roads, as demonstrated by Péron et al. [130] and Santos et al. [52]. Similarly, it is likely that a small proportion of hedgehog–vehicle collisions do not result in instantaneous death and that a hedgehog's delayed traffic-induced death off the road is not counted. The possible role of wildlife hospitals in affecting estimates of mortality rates and genetic fragmentation is also important to consider. Particularly common for *E. europaeus* in Western Europe, wildlife rehabilitators care for and release injured hedgehogs that would otherwise die [64]. While this is undoubtedly valuable for the species' conservation, future road ecology analyses must consider confounding factors such as these.

Of particular significance is that studies seldom examine road mortality in the context of a population's intrinsic growth rate. Considering growth rates reveals less of a "snapshot" of mortality and determines whether populations can sustain current and future road casualty rates. Future research should explicitly model the sensitivity of population growth curves to sex- and age-specific road mortality, using methods such as population viability models and elasticity analysis [41]. Population modelling could be further used on existing data sets, such as from nationwide citizen science projects, to accurately estimate yearly road mortality or, for populations with both road mortality and density estimates, an estimate of local demographic compensation. Another informed approach could incorporate population density, the sex and age of casualties and other sources of mortality into the framework of compensation-additive mortality [131]. This explores whether road mortality is compensatory and removes the already "doomed surplus" in a population or is additive by increasing total mortality [55]. For example, if road-killed individuals have a poor body condition (e.g., they are affected by parasites or other diseases), the severity of road mortality is reduced as their likelihood of long-term survival is low regardless of traffic [43].

The efficacy of road mitigation measures for wildlife is rarely tested; this poses significant constraints on justifying mitigation efforts and adapting strategies for maximum benefit. Many studies are either too short or adopt study designs that cannot demonstrate causality to population viability, such as gene flow or lasting reductions in road mortality [132]. In the future, studies should employ long-term monitoring of mitigation measures and before-and-after-control-impact (BACI) or control-impact experimental designs, where possible. These studies allow for changes in the investigated population parameters, such as density, sex ratio or genetic diversity, to be soundly attributed to the mitigation measures [133]. Future research should also present more holistic mitigation recommendations by examining socioeconomic factors such as vehicle and pedestrian travel efficiency [119] and the cost-effectiveness of strategies [98] (Table 3). The challenge of accommodating both hedgehog and anthropogenic demands on the landscape highlights the crucial role of interdisciplinary and collective thinking in road ecology [11].

Table 3. Summary of published findings, as well as gaps in the literature and recommendations for future research as discussed in this review.

Published Findings	Gaps in Understanding as Revealed by This Review	Directions for Future Research as Recommended by This Review
• Traffic collisions may cause an annual loss of 3–24% of a local hedgehog population, and 9–30% of a nationwide population [19,46,53,54]. • Road mortality is consistently in the top three contributors to total mortality [40,60,61].	• The accuracy of current local and total population estimates. • Whether populations can compensate for road mortality with increased survival and/or fecundity.	• Establishing standardised surveys for improved population estimates. • Long-term population studies to evaluate road mortality in the context of population growth.
• Hedgehog roadkill is disproportionately clustered in suburban areas and consists predominately of males and adults [28,30,39,68].	• Whether carcass detectability and persistence vary between age groups. • How road and habitat characteristics influence road mortality risks between demographic groups over time.	• Studies into the road crossing behaviour of different demographic groups. • Evaluating the consequences of sex- and age-specific road mortality on hedgehog population trends.
• Hedgehog populations appear particularly vulnerable to fragmentation effects [30,81]. • Hedgehog populations exhibit distinct genetic substructure, often in relation to linear infrastructure [86,90,91].	• Whether the hedgehog's promiscuity and heteropaternal superfecundity can lessen the impacts of isolation on genetic structure.	• Establishing isolation effects from roads, such as using inbreeding coefficients or genetic pedigree analysis.
• Exclusionary fences alone are not an appropriate mitigation measure for hedgehog road mortality [96]. • Hedgehogs infrequently use crossing structures [96,98,101].	• The population-level responses to mitigation measures. • Whether the use of crossing structures by badgers impacts their efficacy for hedgehogs. • Whether traffic-calming methods are an effective and cheaper option for road mitigation.	• Quantification of population viability in relation to mitigation using BACI or control/impact studies, such as using roadkill counts, population density and gene flow. • Integration of ecological and socioeconomic perspectives on road mitigation and construction.

7. Conclusions

As hedgehogs remain a prominent victim of WVCs and road infrastructure continues to expand in Europe [38,39], evaluating whether hedgehog populations are vulnerable to the long-term negative impacts of roads is urgently needed. The literature presents several evaluative criteria for this purpose, including proportional loss, differential risk between demographic groups and the fecundity of the remaining population. Previous studies are in general agreement that adult males are more prone to road mortality and that hedgehog–vehicle collisions can disrupt population dynamics, for example, by fragmentation. However, barriers exist to understanding whether this translates to population decline and to disentangling the relative impact of road mortality on population viability compared to other factors. These difficulties remain the primary challenges for hedgehog conservation throughout Europe. Future research should prioritise the inclusion of sex- and age-specific fecundity and survival rates in population models and analyses. This review highlights the importance of long-term monitoring and robust experimental design such as BACI for effective decision-making by conservation practitioners and policy makers. Moreover, considerations of wildlife must be integrated into the early planning stages of road construction to meet the goals of sustainable development. Collaboration between ecologists, engineers and spatial planners is not only good practice, but likely to be indispensable in achieving a reduction in the conflict for space that characterises the 21st century.

Author Contributions: Conceptualisation, L.J.M. and R.W.Y.; writing—original draft preparation, L.J.M.; writing—review and editing, L.J.M., S.O.P., P.J.B., A.J.B., H.L.H., S.E.P. and R.W.Y.; supervision, S.O.P., P.J.B., A.J.B., H.L.H., S.E.P. and R.W.Y.; funding acquisition, R.W.Y. All authors have read and agreed to the published version of the manuscript.

Funding: This research was funded by the People's Trust for Endangered Species and Nottingham Trent University.

Acknowledgments: We would like to thank the two anonymous reviewers for their insightful comments that greatly improved this manuscript.

Conflicts of Interest: The authors declare no conflict of interest. The funders had no role in the conceptualisation or writing of the manuscript.

References

1. Ibisch, P.L.; Hoffmann, M.T.; Kreft, S.; Pe'Er, G.; Kati, V.; Biber-Freudenberger, L.; DellaSala, D.A.; Vale, M.M.; Hobson, P.R.; Selva, N. A global map of roadless areas and their conservation status. *Science* **2020**, *80*, 1423–1427. [CrossRef] [PubMed]

2. Meijer, J.R.; Huijbregts, M.A.J.; Schotten, K.C.G.J.; Schipper, A.M. Global patterns of current and future road infrastructure. *Environ. Res. Lett.* **2018**, *13*, 6. [CrossRef]

3. Forman, R.T.T.; Alexander, L.E. Roads and their major ecological effects. *Ann. Rev. Ecol. Syst.* **1998**, *29*, 207–231. [CrossRef]

4. Coffin, A.W. From roadkill to road ecology: A review of the ecological effects of roads. *J. Transp. Geogr.* **2007**, *15*, 396–406. [CrossRef]

5. Seiler, A.; Helldin, J.O. Mortality in wildlife due to transportation. In *The Ecology of Transportation: Managing Mobility for the Environment*; Davenport, J., Davenport, J.L., Eds.; Springer: New York, NY, USA, 2006; pp. 165–189.

6. Morelle, K.; Lehaire, F.; Lejeune, P. Spatio-temporal patterns of wildlife-vehicle collisions in a region with a high-density road network. *Nat. Conserv.* **2013**, *5*, 53–73. [CrossRef]

7. Meek, R. Patterns of reptile road-kills in the Vendée region of western France. *Herpetol. J.* **2009**, *19*, 135–142.

8. Huijser, M.P.; Bergers, P.J.M. The effect of roads and traffic on hedgehog (*Erinaceus europaeus*) populations. *Biol. Conserv.* **2000**, *95*, 111–116. [CrossRef]

9. Fahrig, L.; Rytwinski, T. Effects of roads on animal abundance: An empirical review and synthesis. *Ecol. Soc.* **2009**, *14*, 1. [CrossRef]

10. Dexter, C.E.; Appleby, R.G.; Scott, J.; Edgar, J.P.; Jones, D.N. Individuals matter: Predicting koala road crossing behaviour in south-east Queensland. *Aust. Mammal.* **2018**, *40*, 67–75. [CrossRef]

11. Trombulak, S.C.; Frissell, C.A. Review of ecological effects of roads on terrestrial and aquatic communities. *Conserv. Biol.* **2000**, *14*, 18–30. [CrossRef]

12. Carvalho, F.; Mira, A. Comparing annual vertebrate road kills over two time periods, 9 years apart: A case study in Mediterranean farmland. *Eur. J. Wildl. Res.* **2011**, *57*, 157–174. [CrossRef]

13. Bannikova, A.; Lebedev, V.; Abramov, A.; Rozhnov, V. Contrasting evolutionary history of hedgehogs and gymnures (Mammalia: Erinaceomorpha) as inferred from a multigene study. *Biol. J. Linn. Soc.* **2014**, *112*, 499–519. [CrossRef]

14. García-rodríguez, S.; Puig-montserrat, X. Algerian hedgehog (*Atelerix algirus* Lereboullet, 1842) habitat selection at the northern limit of its range. *Galemys* **2014**, *26*, 49–56. [CrossRef]

15. Bolfíková, B.; Hulva, P. Microevolution of sympatry: Landscape genetics of hedgehogs *Erinaceus europaeus* and *E. roumanicus* in Central Europe. *Heredity* **2012**, *108*, 248–255. [CrossRef]

16. Huijser, M.P.; Bergers, P.J.M.; De Vries, J.G. Hedgehog traffic victims: How to quantify effects on the population level and the prospects for mitigation. In Proceedings of the International Conference on Wildlife Ecology and Transportation, Fort Myers, FL, USA, 9–12 February 1998; Evink, G., Garrett, P., Zeigler, D., Berry, J., Eds.; Florida Department of Transportation: Tallahassee, FL, USA, 1998.

17. Gruychev, G.V. Animal road mortality (Aves & Mammalia) from the new section of the Maritsa Highway (South Bulgaria). *Ecol. Balk.* **2018**, *10*, 11–18.

18. Tejera, G.; Rodríguez, B.; Armas, C.; Rodríguez, A. Wildlife-vehicle collisions in Lanzarote Biosphere Reserve, Canary Islands. *PLoS ONE* **2018**, *13*, e0192731. [CrossRef] [PubMed]

19. Wembridge, D.E.; Newman, M.R.; Bright, P.W.; Morris, P.A. An estimate of the annual number of hedgehog (*Erinaceus europaeus*) road casualties in Great Britain. *Mammal. Commun.* **2016**, *2*, 8–14.

20. Huijser, M.P.; Bergers, P.J.M. The mortality rate in a hedgehog (Erinaceus europaeus) population: The relative importance of road kills. In *Habitat Fragmentation & Infrastructure, Proceedings of the International Conference on Habitat Fragmentation, Infrastructure and the Role of Ecological Engineering, Maastricht, The Netherlands, 18–21 September 1995*; Canters, K., Ed.; Ministry of Transport, Public Works and Water Management: The Hague, The Netherlands, 1997; pp. 98–104.

21. Holsbeek, L.; Rodts, J.; Muyldermans, S. Hedgehog and other animal traffic victims in Belgium: Results of a countryside survey. *Lutra* **1999**, *42*, 111–119.

22. Brockie, R.E.; Sadleir, R.M.F.S.; Linklater, W.L. Long-term wildlife road-kill counts in New Zealand. *N. Z. J. Zool.* **2009**, *36*, 123–134. [CrossRef]

23. Schwartz, A.L.W.; Shilling, F.M.; Perkins, S.E. The value of monitoring wildlife roadkill. *Eur. J. Wildl. Res.* **2020**, *66*, 18. [CrossRef]

24. Reichholf, J.H. Starker Rückgang der Häufigkeit überfahrener Igel Erinaceus europaeus in Südostbayern und seine Ursachen. *Mitteilungen der Zool Gesellschaft Braunau* **2015**, *113*, 309–314.

25. Müller, F. Langzeit-Monitoring der Strassenverkehrsopfer beim Igel (*Erinaceus europaeus* L.) zur Indikation von Populationsdichteveränderungen entlang zweier Teststrecken im Landkreis Fulda. *Beiträge zur Naturkd Osthessen* **2018**, *54*, 21–26.

26. Wilson, E.; Wembridge, D.E. *The State of Britain's Hedgehogs 2018*; The People's Trust for Endangered Species: London, UK, 2018.

27. Hof, A.R.; Bright, P.W. Quantifying the long-term decline of the West European hedgehog in England by subsampling citizen-science datasets. *Eur. J. Wildl. Res.* **2016**, *62*, 407–413. [CrossRef]

28. Pettett, C.E.; Johnson, P.J.; Moorhouse, T.P.; Macdonald, D.W. National predictors of hedgehog *Erinaceus europaeus* distribution and decline in Britain. *Mamm. Rev.* **2018**, *48*, 1–6. [CrossRef]

29. Roos, S.; Johnston, A.; Noble, D. *UK Hedgehog Datasets and Their Potential for Long-Term Monitoring*; The British Trust of Ornithology: Norfolk, UK, 2012.

30. Van de Poel, J.L.; Dekker, J.; van Langevelde, F. Dutch hedgehogs *Erinaceus europaeus* are nowadays mainly found in urban areas, possibly due to the negative effects of badgers *Meles meles*. *Wildl. Biol.* **2015**, *21*, 51–55. [CrossRef]

31. Krange, M. Change in the Occurrence of the West European Hedgehog (*Erinaceus europaeus*) in Western Sweden during 1950–2010. Master's Thesis, University of Karlstad, Karlstad, Sweden, 2015.

32. Wright, P.G.R.; Coomber, F.G.; Bellamy, C.C.; Perkins, S.E.; Mathews, F. Predicting hedgehog mortality risks on British roads using habitat suitability modelling. *PeerJ* **2020**, *7*, e8154. [CrossRef]

33. Bil, M.; Heigl, F.; Janoska, Z.; Vercayie, D.; Perkins, S.E. Benefits and challenges of collaborating with volunteers: Examples from National wildlife roadkill reporting systems in Europe. *J. Nat. Conserv.* **2020**, *54*, 125798. [CrossRef]

34. George, L.; Macpherson, J.L.; Balmforth, Z.; Bright, P.W. Using the dead to monitor the living: Can road kill counts detect trends in mammal abundance? *Appl. Ecol. Environ. Res.* **2011**, *9*, 27–41. [CrossRef]

35. Baker, P.J.; Harris, S.; Robertson, C.P.J.; Saunders, G.; White, P.C.L. Is it possible to monitor mammal population changes from counts of road traffic casualties? An analysis using Bristol's red foxes Vulpes vulpes as an example. *Mamm. Rev.* **2004**, *34*, 115–130. [CrossRef]

36. Van der Ree, R.; Heinze, D.; Mccarthy, M.; Mansergh, I. Wildlife tunnel enhances population viability. *Ecol. Soc.* **2009**, *14*, 7–15. [CrossRef]

37. Selva, N.; Kreft, S.; Kati, V.; Schluck, M.; Jonsson, B.-G.; Mihok, B.; Okarma, H.; Ibisch, P.L. Roadless and low-traffic areas as conservation targets in Europe. *Environ. Manag.* **2011**, *48*, 865–877. [CrossRef] [PubMed]

38. Department of Transport UK. Road Traffic Forecasts 2018: Moving Britain Ahead. 2018. Available online: https://assets.publishing.service.gov.uk/government/uploads/system/uploads/attachment_data/file/740399/road-traffic-forecasts-2018.pdf (accessed on 29 March 2020).

39. Dom, A.; De Ridder, W. *Paving the Way for EU Enlargement*; European Environment Agency: Copenhagen, Denmark, 2002.

40. Bonnet, X.; Naulleau, G.; Shine, R. The dangers of leaving home: Dispersal and mortality in snakes. *Biol. Conserv.* **1999**, *89*, 39–50. [CrossRef]

41. Row, J.R.; Blouin-Demers, G.; Weatherhead, P.J. Demographic effects of road mortality in black ratsnakes (*Elaphe obsoleta*). *Biol. Conserv.* **2007**, *137*, 117–124. [CrossRef]

42. Haigh, A.; O'Riordan, R.M.; Butler, F. Hedgehog *Erinaceus europaeus* mortality on Irish roads. *Wildl. Biol.* **2014**, *20*, 155–160. [CrossRef]

43. Rautio, A.; Isomursu, M.; Valtonen, A.; Hirvelä-Koski, V.; Kunnasranta, M. Mortality, diseases and diet of European hedgehogs (*Erinaceus europaeus*) in an urban environment in Finland. *Mammal. Res.* **2016**, *61*, 161–169. [CrossRef]

44. Puig, J.; Sanz, L. Wildlife roadkills and underpass use in Northern Spain. *Environ. Eng. Manag. J.* **2012**, *11*, 1141–1147. [CrossRef]

45. Hell, P.; Plavý, R.; Slamečka, J.; Gašparík, J. Losses of mammals (Mammalia) and birds (Aves) on roads in the Slovak part of the Danube Basin. *Eur. J. Wildl. Res.* **2005**, *51*, 35–40. [CrossRef]

46. Orłowski, G.; Nowak, L. Road mortality of hedgehogs *Erinaceus* spp. in farmland in lower Silesia (south-western Poland). *Pol. J. Ecol.* **2004**, *52*, 377–382.

47. Mikov, A.M.; Georgiev, D.G. On the road mortality of the northern white-breasted hedgehog (*Erinaceus roumanicus* Barrett-Hamilton, 1900) in Bulgaria. *Ecol. Balk.* **2018**, *10*, 19–23.

48. Пархоменко, В. Загибель ссавців на автошляхах північно-східної України (In English: Death of mammals on the roads of northeastern Ukraine). *Proc. Theriol. Sch.* **2017**, *15*, 139–149.

49. Bitušík, P.; Kocianová-adamcová, M.; Brabec, J.; Malina, R.; Tesák, J.; Urban, P. The effects of landscape structure and road topography on mortality of mammals: A case study of two different road types in Central Slovakia. *Lynx* **2017**, *48*, 39–51. [CrossRef]

50. Загороднюк, І. Загибель тварин на дорогах: Оцінка впливуавтотранспорту на популяції диких і свійських тварин (In English: Mortality of animals on roads: Assessment of vehicle traffic's influence at populations of wild and domestic animals.). *Proc. Theriol. Sch.* **2006**, *8*, 120–125.

51. Santos, S.M.; Carvalho, F.; Mira, A. How long do the dead survive on the road? Carcass persistence probability and implications for road-kill monitoring surveys. *PLoS ONE* **2011**, *6*, e25383. [CrossRef] [PubMed]

52. Santos, R.A.L.; Santos, S.M.; Santos-Reis, M.; Picanço de Figueiredo, A.; Bager, A.; Aguiar, L.M.S.; Ascensão, F. Carcass persistence and detectability: Reducing the uncertainty surrounding wildlife-vehicle collision surveys. *PLoS ONE* **2016**, *11*, e0165608. [CrossRef] [PubMed]

53. Huijser, M.P. Life on the Edge. Hedgehog Traffic Victims and Mitigation Strategies in an Anthropogenic Landscape. Ph.D. Thesis, Wageningen University, Wageningen, The Netherlands, 2000.

54. Kristiansson, H. Population variables and causes of mortality in a hedgehog (*Erinaceus europaeus*) population in southern Sweden. *J. Zool.* **1990**, *220*, 391–404. [CrossRef]

55. Seiler, A. The Toll of the Automobile: Wildlife and Roads in Sweden. Ph.D. Thesis, Swedish University of Agricultural Sciences, Uppsala, Sweden, 2003.

56. Reeve, N.J.; Huijser, M.P. Wildlife rescue centre records as a means of monitoring relative change in mortality factors affecting hedgehogs (*Erinaceus europaeus*). *Lutra* **1999**, *42*, 7–24.

57. Sutherland, W.J. Sustainable exploitation: A review of principles and methods. *Wildl. Biol.* **2001**, *7*, 131–140. [CrossRef]

58. Dowding, C.V.; Harris, S.; Poulton, S.; Baker, P.J. Nocturnal ranging behaviour of urban hedgehogs, *Erinaceus europaeus*, in relation to risk and reward. *Anim. Behav.* **2010**, *80*, 13–21. [CrossRef]

59. Hof, A.R.; Allen, A.M.; Bright, P.W. Investigating the role of the eurasian badger (*Meles meles*) in the nationwide distribution of the western european hedgehog (*Erinaceus europaeus*) in England. *Animals* **2019**, *9*, 759. [CrossRef]

60. Hubert, P.; Julliard, R.; Biagianti, S.; Poulle, M.L. Ecological factors driving the higher hedgehog (*Erinaceus europeaus*) density in an urban area compared to the adjacent rural area. *Landsc. Urban. Plan.* **2011**, *103*, 34–43. [CrossRef]

61. Dickman, C.R. Age-related dietary change in the European hedgehog, *Erinaceus europaeus*. *J. Zool.* **1988**, *215*, 1–14. [CrossRef]

62. Reeve, N.J. A Field Study of the Hedgehog (*Erinaceus europaeus*) with Particular Reference to Movements and Behaviour. Ph.D. Thesis, Royal Holloway University of London, London, UK, 1981.

63. Doncaster, C.P. Testing the role of intraguild predation in regulating hedgehog populations. *Proc. R. Soc. B Biol. Sci.* **1992**, *249*, 1324.

64. Yarnell, R.W.; Surgey, J.; Grogan, A.; Thompson, R.; Davies, K.; Kimbrough, C.; Scott, D.M. Should rehabilitated hedgehogs be released in winter? A comparison of survival, nest use and weight change in wild and rescued animals. *Eur. J. Wildl. Res.* **2019**, *65*, 6. [CrossRef]

65. Reeve, N.J. The survival and welfare of hedgehogs (*Erinaceus europaeus*) after release back into the wildlife. *Anim. Welf.* **1998**, *7*, 189–202.

66. Rasmussen, S.L.; Berg, T.B.; Dabelsteen, T.; Jones, O.R. The ecology of suburban juvenile European hedgehogs (*Erinaceus europaeus*) in Denmark. *Ecol. Evol.* **2019**, *9*, 13174–13187. [CrossRef]

67. Aresco, M.J. The effect of sex-specific terrestrial movements and roads on the sex ratio of freshwater turtles. *Biol. Conserv.* **2005**, *123*, 37–44. [CrossRef]

68. Hels, T.; Buchwald, E. The effect of road kills on amphibian populations. *Biol. Conserv.* **2001**, *99*, 331–340. [CrossRef]

69. Rautio, A.; Valtonen, A.; Auttila, M.; Kunnasranta, M. Nesting patterns of European hedgehogs (*Erinaceus europaeus*) under northern conditions. *Acta Theriol.* **2014**, *59*, 173–181. [CrossRef]

70. Haigh, A.; Kelly, M.; Butler, F.; O'Riordan, R.M. Non-invasive methods of separating hedgehog (*Erinaceus europaeus*) age classes and an investigation into the age structure of road kill. *Acta Theriol.* **2014**, *59*, 165–171. [CrossRef]

71. Goransson, G.; Karlsson, J.L. Road mortality of the hedgehog *Erinaceus europaeus* in southern Sweden. *Fauna Flora* **1976**, *71*, 1–6.

72. Glasby, L.; Yarnell, R.W. Evaluation of the performance and accuracy of Global Positioning System bug transmitters deployed on a small mammal. *Eur. J. Wildl. Res.* **2013**, *59*, 915–919. [CrossRef]

73. Rautio, A.; Valtonen, A.; Kunnasranta, M. The effects of sex and season on home range in European hedgehogs at the northern edge of the species range. *Ann. Zool. Fenn.* **2013**, *50*, 107–123. [CrossRef]

74. Marchand, M.N.; Litvaitis, J.A. Effects of habitat features and landscape composition on the population structure of a common aquatic turtle in a legion undergoing rapid development. *Conserv. Biol.* **2004**, *18*, 758–767. [CrossRef]

75. Jaeger, J.A.G.; Bowman, J.; Brennan, J.; Fahrig, L.; Bert, D.; Bouchard, J.; Charbonneau, N.; Frank, K.; Gruber, B.; Tluk von Toschanowitz, K. Predicting when animal populations are at risk from roads: An interactive model of road avoidance behavior. *Ecol. Model.* **2005**, *185*, 329–348. [CrossRef]

76. Kormann, U.; Gugerli, F.; Ray, N.; Excoffier, L.; Bollmann, K. Parsimony-based pedigree analysis and individual-based landscape genetics suggest topography to restrict dispersal and connectivity in the endangered capercaillie. *Biol. Conserv.* **2012**, *152*, 241–252. [CrossRef]

77. Cullen, L.; Stanton, J.C.; Lima, F.; Uezu, A.; Perilli, M.L.L.; Akcakaya, H.R. Implications of fine-grained habitat fragmentation and road mortality for jaguar conservation in the Atlantic forest, Brazil. *PLoS ONE* **2016**, *11*, e0167372. [CrossRef]

78. Plante, J.; Jaeger, J.A.G.; Desrochers, A. How do landscape context and fences influence roadkill locations of small and medium-sized mammals? *J. Environ. Manag.* **2019**, *235*, 511–520. [CrossRef] [PubMed]

79. Bontadina, F. Straßenüberquerungen von Igeln (*Erinaceus europaeus*). Ph.D. Thesis, University of Zurich, Zurich, Switzerland, 1991.

80. Özcan, A.U.; Özkazanç, N.K. Identifying the hotspots of wildlife–vehicle collision on the Çankırı–kırıkkale highway during summer. *Turk. J. Zool.* **2017**, *41*, 722–730. [CrossRef]

81. Grilo, C.; Sousa, J.; Ascensão, F.; Matos, H.; Leitão, I.; Pinheiro, P.; Costa, M.; Bernardo, J.; Reto, D.; Lourenço, R.; et al. Individual spatial responses towards roads: Implications for mortality risk. *PLoS ONE* **2012**, *7*, e43811. [CrossRef]

82. Jackson, N.D.; Fahrig, L. Relative effects of road mortality and decreased connectivity on population genetic diversity. *Biol. Conserv.* **2011**, *144*, 3143–3148. [CrossRef]

83. Rondinini, C.; Doncaster, C.P. Roads as barriers to movement for hedgehogs. *Funct. Ecol.* **2002**, *16*, 504–509. [CrossRef]

84. Van Langevelde, F.; Jaarsma, C.F. Using traffic flow theory to model traffic mortality in mammals. *Landsc. Ecol.* **2005**, *19*, 895–907. [CrossRef]

85. Balkenhol, N.; Waits, L.P. Molecular road ecology: Exploring the potential of genetics for investigating transportation impacts on wildlife. *Mol. Ecol.* **2009**, *18*, 4151–4164. [CrossRef] [PubMed]

86. Becher, S.A.; Griffiths, R. Isolation and characterization of six polymorphic microsatellite loci in the European hedgehog *Erinaceus europaeus*. *Mol. Ecol.* **1997**, *6*, 89–90. [CrossRef] [PubMed]

87. Henderson, M.; Becher, S.A.; Doncaster, C.P.; Maclean, N. Five new polymorphic microsatellite loci in the European hedgehog *Erinaceus europaeus*. *Mol. Ecol.* **2000**, *9*, 1932–1934. [CrossRef] [PubMed]

88. Becher, S.A.; Griffiths, R. Genetic differentiation among local populations of the European hedgehog (*Erinaceus europaeus*) in mosaic habitats. *Mol. Ecol.* **1998**, *7*, 1599–1604. [CrossRef] [PubMed]

89. Braaker, S.; Kormann, U.; Bontadina, F.; Obrist, M.K. Prediction of genetic connectivity in urban ecosystems by combining detailed movement data, genetic data and multi-path modelling. *Landsc. Urban. Plan.* **2017**, *160*, 107–114. [CrossRef]

90. Rasmussen, S.; Nielsen, J.; Jones, O.R.; Berg, T.B.; Pertoldi, C. Genetic structure of the European hedgehog (*Erinaceus europaeus*) in Denmark. *PLoS ONE* **2020**, *15*, e0227205. [CrossRef]

91. Curto, M.; Winter, S.; Seiter, A.; Schmid, L.; Scheicher, K.; Barthel, L.M.F.; Plass, J.; Meimberg, H. Application of a SSR-GBS marker system on investigation of European hedgehog species and their hybrid zone dynamics. *Ecol. Evol.* **2019**, *9*, 2814–2832. [CrossRef]

92. Moran, S.; Turner, P.D.; O'Reilly, C. Multiple paternity in the European hedgehog. *J. Zool.* **2009**, *278*, 349–353. [CrossRef]

93. Barthel, L.M.F. Population Structure and Behaviour of the European Hedgehog in an Urbanized World. Ph.D. Thesis, Freie Universität, Berlin, Germany, 2019.

94. Proctor, M.F.; Kasworm, W.F.; Teisberg, J.E.; Servhen, J.E.; Radandt, T.G.; Lamb, C.T.; Kendall, K.C.; Mace, R.D.; Paetkau, D.; Boyce, M.S. American black bear population fragmentation detected with pedigrees in the transborder Canada–United States region. *Ursus* **2020**, *31*, 1–15. [CrossRef]

95. Lin, S.C. Landscape and traffic factors affecting animal road mortality. *J. Environ. Eng. Landsc. Manag.* **2016**, *24*, 10–20. [CrossRef]

96. Bergers, P.J.M.; Nieuwenhuizen, W. Viability of hedgehog populations in central Netherlands. *Lutra* **1999**, *42*, 65–75.

97. Clevenger, A.P.; Waltho, N. Factors influencing the effectiveness of wildlife underpasses in Banff National Park. *Cons. Biol.* **2010**, *14*, 47–56. [CrossRef]

98. Helldin, J.O.; Petrovan, S.O. Effectiveness of small road tunnels and fences in reducing amphibian roadkill and barrier effects at retrofitted roads in Sweden. *PeerJ* **2019**, *7*, e7518. [CrossRef] [PubMed]

99. Tritsis, K.M. Assessment of the Use of Crossing Structures by Mammals along Egnatia Motorway in Northern Greece. Master's Thesis, University of Kiel, Kiel, Greece, 2011.

100. Mata, C.; Hervás, I.; Herranz, J.; Suárez, F.; Malo, J. Are motorway wildlife passages worth building? Vertebrate use of road-crossing structures on a Spanish motorway. *J. Environ. Manag.* **2008**, *88*, 407–415. [CrossRef]

101. Ascensão, F.; Mira, A. Factors affecting culvert use by vertebrates along two stretches of road in southern Portugal. *Ecol. Res.* **2007**, *22*, 57–66. [CrossRef]

102. Jarvis, L.E.; Hartup, M.; Petrovan, S.O. Road mitigation using tunnels and fences promotes site connectivity and population expansion for a protected amphibian. *Eur. J. Wildl. Res.* **2019**, *65*, 27. [CrossRef]

103. Eldridge, B.; Wynn, J. Use of badger tunnels by mammals on highways agency schemes in England. *Conserv. Evid.* **2011**, *8*, 53–57.

104. Ważna, A.; Kaźmierczak, A.; Cichocki, J.; Bojarski, J.; Gabryś, G. Use of underpasses by animals on a fenced expressway in a suburban area in western Poland. *Nat. Conserv.* **2020**, *39*, 1–18. [CrossRef]

105. Myslajek, R.W.; Nowak, S.; Kurek, K.; Tolkacz, K.; Gewartowska, O. Utilisation of a wide underpass by mammals on an expressway in the Western Carpathians, S Poland. *Folia Zool.* **2016**, *65*, 225–232. [CrossRef]

106. Yanes, M.; Velasco, J.M.; Suárez, F. Permeability of roads and railways to vertebrates: The importance of culverts. *Biol. Conserv.* **1995**, *71*, 217–222. [CrossRef]

107. De Vries, J.G. Hedgehogs on the road: From research to practice. *Lutra* **1999**, *42*, 99–110.

108. Ward, J.F.; Macdonald, D.W.; Doncaster, C.P. Responses of foraging hedgehogs to badger odour. *Anim. Behav.* **1997**, *53*, 709–720. [CrossRef]

109. Doncaster, C.P. Can badgers affect the use of tunnels by hedgehogs? A review of the literature. *Lutra* **1999**, *42*, 59–64.

110. Clevenger, A.; Waltho, N. Performance indices to identify attributes of highway crossing structures facilitating movement of large mammals. *Biol. Conserv.* **2005**, *121*, 453–464. [CrossRef]

111. Eberhardt, E.; Mitchell, S.; Fahrig, L. Road kill hotspots do not effectively indicate mitigation locations when past road kill has depressed populations. *J. Wildl. Manag.* **2013**, *77*, 1353–1359. [CrossRef]

112. Ascensão, F.; Kindel, A.; Teixeira, F.Z.; Barrientos, R.; D'Amico, M.; Borda-de-Água, L.; Pereira, H.M. Beware that the lack of wildlife mortality records can mask a serious impact of linear infrastructures. *Glob. Ecol. Conserv* **2019**, *19*, e00661.

113. Hobday, A.J.; Minstrell, M.L. Distribution and abundance of roadkill on Tasmanian highways: Human management options. *Wildl. Res.* **2008**, *35*, 712–726. [CrossRef]

114. Dique, D.S.; Thompson, J.; Preece, H.J.; Penfold, G.C.; de Villiers, D.L.; Leslie, R.S. Koala mortality on roads in south-east Queensland: The koala speed-zone trial. *Wildl. Res.* **2003**, *30*, 419–426. [CrossRef]

115. Clevenger, A.P.; Chruszcz, B.; Gunson, K.E. Spatial patterns and factors influencing small vertebrate fauna road-kill aggregations. *Biol. Conserv.* **2002**, *109*, 15–26. [CrossRef]

116. Boves, T.J.; Belthoff, J.R. Roadway mortality of barn owls in Idaho, USA. *J. Wildl. Manag.* **2012**, *76*, 1381–1392. [CrossRef]

117. Clevenger, A.P.; Kociolek, A.V. Potential impacts of highway median barriers on wildlife: State of the practice and gap analysis. *Environ. Manag.* **2013**, *52*, 1299–1312. [CrossRef] [PubMed]

118. Canal, D.; Camacho, C.; Martín, B.; de Lucas, M.; Ferrer, M. Fine-scale determinants of vertebrate roadkills across a biodiversity hotspot in Southern Spain. *Biodivers. Conserv.* **2019**, *28*, 3239–3256. [CrossRef]

119. Van Strien, M.J.; Grêt-Regamey, A. How is habitat connectivity affected by settlement and road network configurations? Results from simulating coupled habitat and human networks. *Ecol. Modell.* **2016**, *342*, 186–198. [CrossRef]

120. Van der Ree, R.; Jaeger, J.A.G.; van der Grift, E.A.; Clevenger, A.P. Effects of roads and traffic on wildlife populations and landscape function: Road ecology is moving toward larger scales. *Ecol. Soc.* **2011**, *16*, 1. [CrossRef]

121. Jaeger, J.A.G.; Fahrig, L.; Ewald, K.C. Does the configuration of road networks influence the degree to which roads affect wildlife populations? In Proceedings of the 2005 International Conference on Ecology and Transportation, San Diego, CA, USA, 29 August 2005; Irwin, C.L., Garrett, P., McDermott, K.P., Eds.; Center for Transportation and the Environment, North Carolina State University: Raleigh, NC, USA,, 2005.

122. Rhodes, J.R.; Lunney, D.; Callaghan, J.; McAlpine, C.A. A few large roads or many small ones? How to accommodate growth in vehicle numbers to minimise impacts on wildlife. *PLoS ONE* **2014**, *9*, 3. [CrossRef]

123. Bencin, H.L.; Prange, S.; Rose, C.; Popescu, V.D. Roadkill and space use data predict vehicle-strike hotspots and mortality rates in a recovering bobcat (*Lynx rufus*) population. *Sci. Rep.* **2019**, *9*, 1. [CrossRef]

124. Hebblewhite, M.; Haydon, D.T. Distinguishing technology from biology: A critical review of the use of GPS telemetry data in ecology. *Philos. Trans. R. Soc. B Biol. Sci.* **2010**, *365*, 2303–2312. [CrossRef]

125. Harris, S.; Morris, P.; Wray, S.; Yalden, D. *A Review of British Mammals Population Estimates and Conservation Status of British Mammals Other Than Cetaceans*; Joint Nature Conservation Committee: Peterborough, UK, 1995.

126. Mathews, F.; Kubasiewicz, L.; Gurnell, J.; Harrower, C.; McDonald, R.; Shore, R.A. *Review of the Population and Conservation Status of British Mammals*; A report by the Mammal Society under contract to Natural England, Natural Resources Wales and Scottish Natural Heritage; Natural England: Peterborough, UK, 2018.

127. Schaus, J.; Uzal, A.; Gentle, L.K.; Baker, P.J.; Bearman-Brown, L.; Bullion, S.; Gazzard, A.; Lockwood, H.; North, A.; Reader, T.; et al. Application of the random encounter model in citizen science projects to monitor animal densities. *Remotes Sens. Ecol. Conserv.* **2020**. [CrossRef]

128. Petrovan, S.O.; Vale, C.G.; Sillero, N. Using citizen science in road surveys for large-scale amphibian monitoring: Are biased data representative for species distribution? *Biodivers. Conserv.* **2020**, *29*, 1767–1781. [CrossRef]

129. Schwartz, A.L.W.; Williams, H.F.; Chadwick, E.; Thomas, R.J.; Perkins, S.E. Roadkill scavenging behaviour in an urban environment. *J. Urb. Ecol.* **2018**, *4*, 1–7. [CrossRef]

130. Péron, G.; Hines, J.E.; Nichols, J.D.; Kendall, W.L.; Peters, K.A.; Mizrahi, D.S. Estimation of bird and bat mortality at wind-power farms with superpopulation models. *J. Appl. Ecol.* **2013**, *50*, 902–911. [CrossRef]

131. Péron, G. Compensation and additivity of anthropogenic mortality: Life-history effects and review of methods. *J. Anim. Ecol.* **2013**, *82*, 408–417. [CrossRef]

132. Van der Grift, E.A.; van der Ree, R.; Fahrig, L.; Findlay, S.; Houlaham, J.; Jaeger, J.A.G.; Klar, N.; Madriñan, L.F.; Olson, L. Evaluating the effectiveness of road mitigation measures. *Biodivers. Conserv.* **2013**, *22*, 425–448. [CrossRef]

133. Glista, D.J.; DeVault, T.L.; DeWoody, J.A. A review of mitigation measures for reducing wildlife mortality on roadways. *Landsc. Urban. Plan.* **2009**, *91*, 1–7. [CrossRef]

Communication

Plastic Entanglement Poses a Potential Hazard to European Hedgehogs *Erinaceus europaeus* in Great Britain

Emily Thrift, Pierre Nouvellet and Fiona Mathews *

School of Life Sciences, University of Sussex, Brighton BN1 9RH, UK; ed343@sussex.ac.uk (E.T.); pierre.nouvellet@sussex.ac.uk (P.N.)
* Correspondence: f.mathews@sussex.ac.uk

Simple Summary: Plastic entanglement is well known for causing both conservation and welfare issues for marine mammals, but little is known about the impacts on terrestrial species. Following anecdotal reports in the media, we assessed the prevalence and consequences of plastic entanglement for the European hedgehog (*Erinaceus europaeus*) in Great Britain. Based on data provided by rescue centres and population modelling, we estimate that 4000–7000 hedgehog deaths occur annually occur as a consequence of plastic entanglement, representing a significant welfare issue and placing additional pressure on a declining species.

Abstract: A questionnaire to gather evidence on the plastic entanglement of the European hedgehog (*Erinaceus europaeus*) was sent to 160 wildlife rehabilitation centres in Great Britain. Fifty-four responses were received, and 184 individual admissions owing to plastic entanglement were reported. Death was the outcome for 46% (*n* = 86) of these cases. A high proportion of Britain's hedgehogs enter rehabilitation centres annually (approximately 5% of the national population and potentially 10% of the urban population), providing a robust basis for assessing the minimum impacts at a national level. We estimate that 4000–7000 hedgehog deaths per year are attributable to plastic, with the true rate likely being higher, since many entangled hedgehogs—in contrast to those involved in road traffic accidents—will not be found. Population modelling indicates that this excess mortality is sufficient to cause population declines. Although the scale of the impact is much lower than that attributable to traffic, it is nevertheless an additional pressure on a species that is already in decline and presents a significant welfare issue to a large number of individuals.

Keywords: wildlife; rehabilitation centres; plastic waste; population modelling

Citation: Thrift, E.; Nouvellet, P.; Mathews, F. Plastic Entanglement Poses a Potential Hazard to European Hedgehogs *Erinaceus europaeus* in Great Britain. *Animals* **2023**, *13*, 2448. https://doi.org/10.3390/ani13152448

Academic Editors: Anne Berger, Nigel Reeve and Sophie Lund Rasmussen

Received: 29 June 2023
Revised: 24 July 2023
Accepted: 28 July 2023
Published: 28 July 2023

1. Introduction

Plastic production has increased significantly since the 1950s, resulting in a global accumulation of plastic waste, which totalled 379.3 megatons (MT) in 2021 alone [1]. There is now concern about the ecological impacts of this waste [2]. The size of the plastic waste is important when considering what risks it poses; for example, macroplastics (defined as pieces of plastic >10 mm) [3] pose entanglement and gut blockage risks, whereas other risks may be presented by mesoplastics (size range $1 \leq 10$ mm) and microplastics (MP; <1 mm) [3]. There are also concerns about the leaching of plastic additives and plasticizers such as bisphenol A and phthalates, which are potential endocrine-disrupting chemicals (EDCs), refs. [4–7] and the adsorption and accumulation of toxins, including EDCs and heavy metals [8–11] on plastic particles. Recently, there has been substantial research on the impacts of all these types of plastic waste within marine habitats [12–27]. A recent review indicated, for example, that almost 40% of the 123 known marine mammals have been reported as becoming entangled in mesoplastics [28]. A further study by Butterworth and colleagues reports that entanglement has caused widespread suffering and death among marine mammals and birds [17]. Our understanding of the scale and impact of plastic entanglement on terrestrial species is, by comparison, very limited. Most studies on this topic have focused on birds [29–32], though entanglement is also reported in mammals such as the black howler monkey (*Alouatta caraya*), white-tailed opossum (*Didelphis albiventris*), fat mice (*Thylamys* sp.), polar bear (*Ursus maritimus*), and artic fox (*Alopex lagopus*) [19,33,34]. A study of a range of mammals and birds in Argentina found that 60% of the reported entanglement cases resulted in death due to asphyxiation or starvation [33]. A further study of both agricultural and urban crows indicated that 100% of nestlings that become entangled (*n* = 11) were unable to fledge and, in most cases, had long-term injuries, mainly to their toes [29]. The injuries and deaths reported in these terrestrial studies indicate that plastic entanglement potentially poses a risk to terrestrial wildlife. However, the scale of these risks relative to other factors, and the types of plastic involved, are poorly understood.

In Great Britain, there is growing concern about the impacts of plastic entanglement on the European hedgehog (*Erinaceus europaeus*). Plastic poses risks to both populations and individuals. Potentially even small increases in fatality rates could be important at a national level in a species classed as vulnerable to extinction [35], and entanglement also compromises individual welfare. This study, therefore, seeks to quantify the rate and severity of plastic entanglement of the European hedgehog and to estimate the likely impacts at the population level.

2. Materials and Methods

2.1. Questionnaire

An eight-question survey was created using Qualtrics software v. 2021 (Qualtrics, Provo, UT, USA) to gather data on the number of annual admissions, plastic-related admissions, and outcomes at British wildlife rehabilitation centres (Table A1). The survey was sent via email to 160 centres identified from a British database (directory.helpwildlife. co.uk accessed on 15 July 2019) and Google searches using the terms 'hedgehog' AND 'hospital' OR 'rescue'. The survey covered the period from August 2019 to August 2020 and from October 2021 to October 2022, with the latter period targeting centres that had not previously completed the questionnaire to increase the response rate. The centres were asked to report on any cases they had seen in the previous 12 months. Incidents were recorded as entanglement regardless of whether it was the primary or secondary reason for admission. Furthermore, if an individual was known to have been released from plastic prior to admission, this was also recorded as an entanglement case.

2.2. Statistical Analyses

Analyses were carried out in R Studio base package [36]. Chi-square tests were used to assess the relationships between survival outcome and the predictor variables habitat type and plastic type. Wilson's 95% confidence intervals were calculated using the Wilson.ci function.

2.3. Habitat Type

The locations at which hedgehogs were reported were mapped using the Geographical Information System Package QGIS (QGIS 3.28.3, 2019. The rescue centres (*n* = 52) were asked to provide the location of the site at which the hedgehog was reported as entangled; however, in some cases, this was not possible, and the site of the rescue centre was used (*n* = 5 centres). These habitats were then classified into three types (urban, peri-urban, and rural) using Google Earth (Google LLC, Google Earth version 7.3 2023). The peri-urban locations were classified as areas with less than 30% built cover.

2.4. Mortality Model

A population dynamics model was developed to assess the likely impacts of plastic entanglement-associated fatalities on population stability. We describe the dynamics of the hedgehog population as a (Leslie) matrix population:

$$N_t = \begin{pmatrix} n_{1,t} \\ n_{2,t} \\ n_{3,t} \end{pmatrix}$$

with $n_{i,t}$ representing the number of hedgehogs in age class 'i', i.e., 1 for individuals aged between 1 day to 15 weeks, 1 and 2 for those less than 2 years old, and 3 for adult and sexually active individuals. We implicitly assume an age class 0 for individuals 0–15 weeks old, which is not observed. The population number refers to the sizes 15 weeks after the birth pulse.

We define the mortality for each age class 'i' as μ_i; female fecundity of sexually mature females as f, and assume a 1:1 sex ratio. Therefore, we have:

$$N_{t+1} = M \, N_t$$

$$\text{with } M = \begin{bmatrix} 0 & \frac{f(1-\mu_0)}{2}(1-\mu_2) & \frac{f(1-\mu_0)}{2}(1-\mu_3) \\ (1-\mu_1) & 0 & 0 \\ 0 & (1-\mu_2) & (1-\mu_3) \end{bmatrix}.$$

The dominant eigenvalue, λ, of M gives us the population growth rate (i.e., $\lambda = N_{t+1}/N_t$) while the associated eigenvector, $v = \begin{pmatrix} v_1 \\ v_2 \\ v_3 \end{pmatrix}$, gives us the stable age distribution (e.g., the proportion of individuals in each age class).

Hedgehog population dynamics in Great Britain are uncertain, but some key estimated parameters relevant to us are highlighted in Table 1.

Table 1. Parameters used for the model.

Parameter		Reference
Sexual maturity	From 2 years old	[37]
Average litter size—f	4–5	[38,39]
Mortality to 15 weeks μ_0	0.5	[38,39]
Mortality to 1 year old—μ_1	0.5	[39,40]
Exponential growth rate,	−8% to 0%	[41]
Mortality for individuals >1 year old, $\mu_2 = \mu_3$	To be estimated	
Total population size, N	879,000	[42]
Deaths linked to Road Traffic Accidents D_R	1.1% Low 8.8% Medium 55% high	[37,43,44]

Using the population dynamics model, the combinations of estimated parameters from Table 1, and the 2 estimated growth rates, we estimate the overall mortality of adult hedgehogs (i.e., $\mu_2 = \mu_3$), and the stable age distribution (v) by numerically solving the equation $M \, v = \lambda \, v$.

Then, we aim to attribute all deaths (D_T) to 3 main causes: road traffic accidents (RTA) (D_R), plastic entanglement (D_P), and other causes (D_O).

It has recently been reported that 50% of hedgehogs in rescue centres are released [28], implying that 50% die, so the number of deaths in our sample (D_t^{rescue}) was 6273. The all-cause fatality rate is similar to that reported by our respondents (53.2% total mortality). Rescue centres provided estimates of plastic entanglement and other causes of death for 12,546 individual admissions. This revealed that plastic entanglement occurred in 184 hogs from which $D_P^{rescue} = 85$ died. For RTAs, we took estimated figures from 3 studies, owing to the absence of information from our dataset. This provides a low, medium and high range.

Assuming these figures reflect deaths in a natural environment, we estimate that among hedgehogs not dying from RTA, a proportion

$$p_O^{rescue} = \frac{D_O^{rescue}}{D_O^{rescue} + D_P^{rescue}}$$ die from other causes.

Taken altogether, we have:

$$D_P = D_T - D_R - D_O$$

with

- The total number of deaths: $D_T = N \sum \left(\begin{bmatrix} 0 & 0 & 0 \\ (1-\mu_1) & 0 & 0 \\ 0 & (1-\mu_2) & (1-\mu_3) \end{bmatrix} v \right)$, with
  N The total population size and μ_1 from Table 2, and μ_2 and v (age class stable distribution, i.e., eigenvector) estimated above;
- The total number of RTA (D_R) informed by the literature (Table 2);
- The total number of deaths linked to other causes: $D_O = (D_T - D_R)\, p_O^{rescue}$.

Table 2. Estimated adult mortality in Great Britain and cause of deaths in hedgehogs based on assumptions. f = average litter size, μ_0 = mortality to 15 weeks, μ_1 = mortality to 1 year old, $\log(\lambda)$ = Exponential growth, N = Total population size, D_R = Deaths linked to road traffic accidents, $\mu_2 = \mu_3$ = mortality for individuals >1 year old, D_T = Deaths associated with all causes, D_p = deaths associated with plastic and D_o deaths associated with other causes.

		Assumed					Estimated		
f	μ_0	μ_1	$\log(\lambda)$	N	D_R	$\mu_2 = \mu_3$	D_T	D_O	D_P
4	0.5	0.5	−0.08	879,000	9.67×10^3	0.401221	599,448.5	581,787.9	7991.592
					7.73×10^4	0.401221	599,448.5	515,022	7074.478
					4.38×10^5	0.401221	599,448.5	114,426.7	1571.795
			0	879,000	9.67×10^3	0.333322	597,719.7	580,082.6	7968.167
					7.73×10^4	0.333322	597,719.7	513,316.7	7051.053
					4.38×10^5	0.333322	597,719.7	112,721.4	1548.37
4.5	0.5	0.5	−0.08	879,000	9.67×10^3	0.426385	594,704.7	577,108.4	7927.313
					7.73×10^4	0.426385	594,704.7	510,342.5	7010.199
					4.38×10^5	0.426385	594,704.7	109,747.2	1507.516
			0	879,000	9.67×10^3	0.359991	592,813.7	575,243	7901.69
					7.73×10^4	0.359991	592,813.7	508,477.1	6984.576
					4.38×10^5	0.359991	592,813.7	107,881.8	1481.893
5	0.5	0.5	−0.08	879,000	9.67×10^3	0.449562	590,586.8	573,046.3	7871.515
					7.73×10^4	0.449562	590,586.8	506,280.4	6954.402
					4.38×10^5	0.449562	590,586.8	105,685.1	1451.719
			0	879,000	9.67×10^3	0.384617	588,547.9	571,035	7843.887
					7.73×10^4	0.384617	588,547.9	504,269.1	6926.773
					4.38×10^5	0.384617	588,547.9	103,673.8	1424.09

3. Results

Of the 160 rescue centres contacted, responses were received from 52 (Figure 1). These centres provided data on 12,546 hedgehog admissions. Ten centres reported zero admissions of plastic-entangled hedgehogs, and all of these reported annual admissions of fewer than 200 individuals; however, there were also centres with similar admission rates which did reported entanglement incidents. Data from the 44 centres with entanglement cases showed 184 admissions (1.4% of their total admissions) were a consequence of plastic entanglement.

More cases were recorded in urban than in other habitat types (Figure 2A). However, habitat was not associated with survivorship ($\chi^2 = 1.27$, df = 2, $p = 0.52$). Figure 2A indicates that although fewer cases were reported in rural locations, these tended to have a higher mortality rate.

The main sources of plastics reported were netting, bags, fencing, and rings from bottles. Survivorship varied between plastic types ($\chi^2 = 14.47$, df = 5, $p = 0.01$). Although plastic netting was the most frequently recorded type of plastic ($n = 114/184$) (Figure 2B), most individuals entangled in this way recovered (69/114). The highest mortality rates were associated with bands (hair bands and elastic bands) and yoghurt pots. The category 'Other' included a variety of sources that were generally associated with high fatality rates such as plastic fencing (2/5), bailer twine (2/2), and cables ties (2/5).

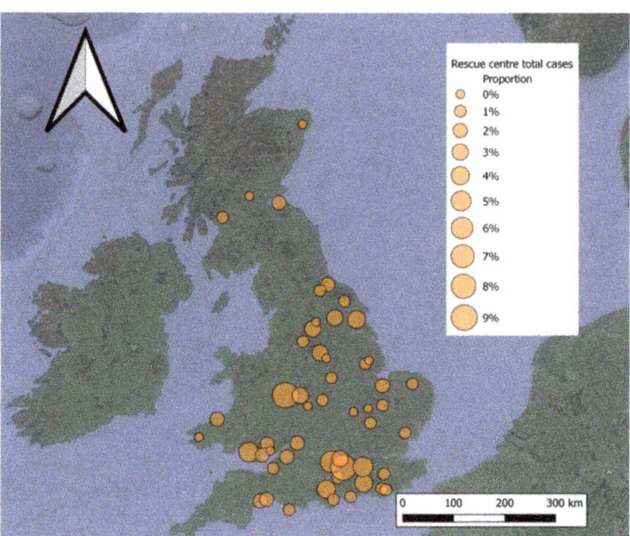

Figure 1. Map showing the proportion of admissions recording entanglement for each of the 52 rescue centres.

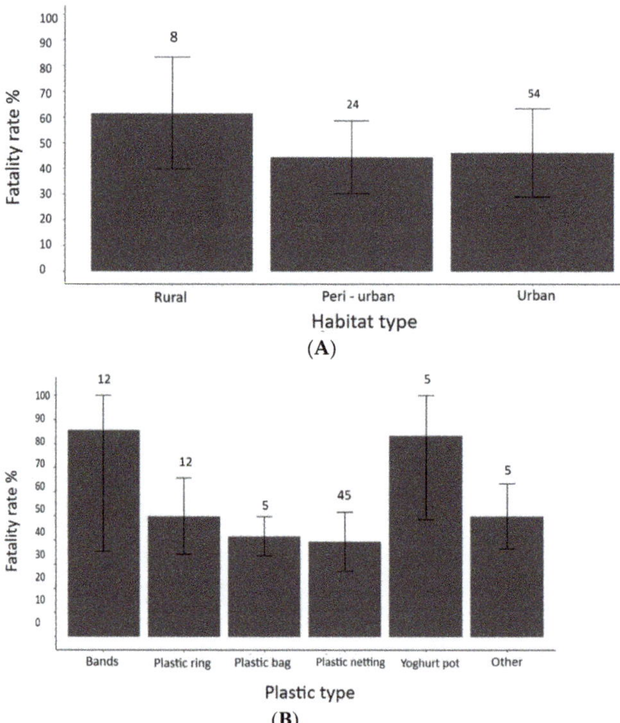

Figure 2. Plot showing the fatality rate from plastic entanglement, (**A**) by habitat type, (**B**) by plastic type. Error bars show 95% Wilson Confidence Intervals.

Mortality Model

Using the population dynamics model, key parameter estimates from the literature, and the survey of 52 rescue centres, we estimate that plastic entanglement in hedgehogs is responsible for between 1400 and 7999 hedgehog deaths annually in Great Britain (Table 2). This is an additional 1.4% compared with baseline estimated mortality.

The mortality can be attributed to three causes in proportions: D_i/D_T for cause 'i'. allowing us to predict the population dynamics (e.g., growth rate) if one cause was removed.

If the population suffers all causes of deaths, then $\lambda_{all\ causes}$ is found as the dominant eigenvector when μ_2 takes the values found in Table 2. As expected, this leads us to retrieve an exponential growth rate from -8% to 0%. If the population does not suffer from RTAs, then λ_{noRTA} is found as the dominant eigenvector when μ_s takes the values $\mu_s^{noRTA} = \mu_s \frac{D_R + D_O}{D_T}$. Similarly, we can obtain λ_{noP} and λ_{noRTA_P}, or the growth rates when no plastic entanglement occurs, or no RTAs nor plastic entanglement occur. Assuming the baseline mid-points for f and D_R (i.e., $f = 4.5$ and $D_R = 74,000$), and two scenarios of high and low growth rate (solid vs. dashed lines in the figure for low and high growth rate), we can predict the growth rate, population structure, and population dynamics for each of the four mortality scenarios (baseline with all causes of mortality, and three counterfactual scenarios).

The predicted dynamics (Figure 3) indicate that if plastic entanglement deaths are removed, the wild population of hedgehogs is likely to increase slightly, although this increase is much smaller than would be observed with the cessation of road traffic accidents.

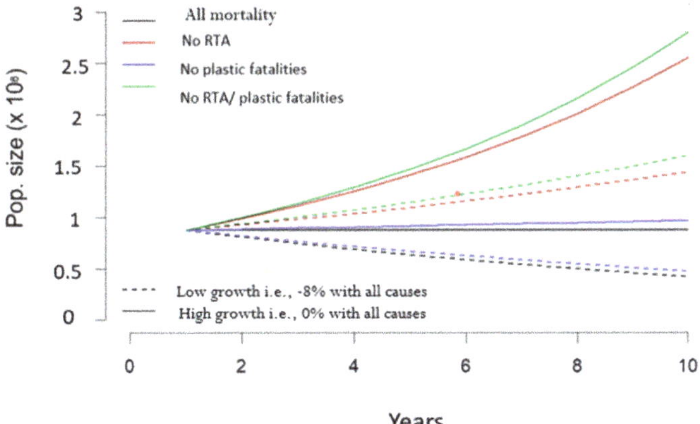

Figure 3. Predicted dynamics (population size over the 10 years) when growth rate is low or high., i.e., leading to a [0; −8%] growth rate when all mortality causes are accounted for (solid vs. dashed lines). The predicted dynamics are shown when all mortality is present (black), as well as 3 counterfactual scenarios, e.g., (1) if no RTAs occur (red), (2) if no plastic entanglement occurs (blue), (3) if no RTAs nor plastic entanglement fatalities occur (green).

4. Discussion

This study demonstrates that plastic entanglement accounts for 1.4% of hedgehog admissions to British wildlife rehabilitation centres annually, and 46% of these animals die. Responses were received from one-third of the centres contacted—a good response rate for a questionnaire survey—and even had the non-responding centres recorded zero cases, appreciable numbers of hedgehogs would still be affected annually. The true rate of entanglement will be higher than that reported here, since members of the public may assist without contacting rescue centres, and entangled hedgehogs are less likely to be found than road traffic casualties, which have been widely studied [45–47]. An estimated 879,000 hedgehogs are present in Britain [42], and our models indicate that plastic entanglement is likely to have a negative overall impact on the population. While the scale of the impact is much lower than

that attributable to road traffic accidents, it is nevertheless an additional pressure on a speci
that is considered vulnerable to extinction on the GB Mammal Red List [35].

Plastic netting often used in gardens, agriculture, and allotments was the most cor
mon cause of plastic entanglement, which is unsurprising as hedgehogs are likely to spen
significant time in these locations, and the netting is deployed low to the ground. As th
type of plastic is so often used by gardeners and farmers, this could suggest hedgehog
entangled in this plastic are more frequently found, and therefore have a better surviv
rate compared to those entangled in other plastics. Studies of marine mammals have als
identified netting as one of the most common causes of entanglement and one that is linke
with higher mortality rates [12,22,27,48]. Plastic bags and plastic rings from can holde
were found to cause most of the remaining entanglement cases, which is also comparab
with evidence of high rates of marine mammal entanglement in single-use plastics [17,4
The plastics with the highest mortality rates, however, were bands, including elastic ban
and hair bands. These cases had an 85% fatality rate. This is comparable with the serio
and often life-threatening injuries reported in sea lions and seals because of entanglem
with bands [49,50]. It is possible that elastic bands are particularly problematic for tw
reasons: first, it is difficult for wild animals to extricate themselves once the band is
place; and second because the bands cause damage to tissues and nerves and can als
constrict the blood supply. Recent studies of pinnipeds have indicated that bands are ofte
the most common type of plastic for them to become entangled in [12,48]. These studi
have also found that juveniles are five times more likely to become entangled, wherea
pups and adults are the least likely [12,48]. Other studies investigating species includir
crows, turtles, blue sharks and Antarctic fur seals also reported that juveniles are the mo
common age class to be entangled in plastic [29,51–53].

Entangled hedgehogs were rescued less frequently in rural than urban areas, pote
tially reflecting lower hedgehog population densities in rural areas and also the low
probability of being found owing to the lower human population. This is similar to mar
other studies that also report higher admission rates from urban than rural locations [54–5
The prognosis was slightly poorer for animals in rural areas, possibly because there is like
to be, on average, a longer interval between entanglement and being found. The preci
geographical location of the casualty was unknown for five individuals, so the habit
type with a 15 km radius of the centre (the maximum distance from which casualties we
accepted) was used as a proxy. Given the small numbers of individuals involved, it
unlikely that this would materially affect the results.

This study is the first to indicate that plastic entanglement is causing serious welfa
issues for the European hedgehog and results in high mortality rates. Therefore, we sugge
that a national database is established to enable rescue centres and members of the public
record all incidents of plastic entanglement, allowing for future assessments to be made c
a wider geographical scale. The database could also collate information on the sex and ag
profiles of casualties, together with more detailed information on the type of plastic involve
to enable more comprehensive assessments to be made of the risks of plastic entanglemen

5. Conclusions

This study indicates that, although understudied within terrestrial environmen
plastic entanglement poses a welfare issue for an estimated 1400–7999 hedgehogs in Gre
Britain alone, and poses a conservation threat to populations already at risk. The dev
opment and use of the national database would facilitate better understanding of the tr
rates of plastic entanglement in wild populations.

Author Contributions: Conceptualization, E.T. and F.M.; methodology, E.T.; software, E.T., F.M. a
P.N.; validation, E.T., F.M. and P.N.; formal analysis, E.T., F.M. and P.N.; investigation, E.T.; resources, E.
data curation, E.T., F.M. and P.N.; writing—original draft preparation, E.T.; writing—review and editi
E.T., F.M. and P.N.; visualization, E.T. and P.N.; supervision, F.M.; project administration, E.T.; fundi
acquisition, E.T. and F.M. All authors have read and agreed to the published version of the manuscript

Funding: This research was funded by UFAW.

Institutional Review Board Statement: The study was conducted in accordance with the Declaration of Helsinki, and approved by the Ethics Committee of the University of Sussex (protocol code ER/ED343/2 and date of approval 1 May 2020).

Informed Consent Statement: Informed consent was obtained from all subjects involved in the study. Written informed consent has been obtained from the participants to publish this paper.

Data Availability Statement: Available online: https://doi.org/10.6084/m9.figshare.23600328 (accessed on 19 June 2023).

Acknowledgments: Thank you to the UFAW who funded the research and all the rescue centres that completed the questionnaire.

Conflicts of Interest: The authors declare no conflict of interest.

Appendix A

Table A1. Questionnaire provided to the rescue centres to record entanglement cases.

Number	Question
1	Name of centre
2	Approximate number of admissions per year
3	Have you had animals admitted in the last 12 months with an injury or illness linked with plastics?
4	If yes, tell us abut the cases in the table below. An example would be species, hedgehog, entanglement, head trapped in plastic can holder which resulted in neck lacerations, outcome recovered and released
5	If you deal with large numbers of plastic related casualties and do not have time to tell us about each case, please summarise the data for each species e.g., hedgehogs 4 gut blockages, 2 died, 2 released, 2 foxes, 2 entanglement, lacerations, 2 released
6	If you have photographs of the incidents you are happy to share, we would be very grateful to receive them.
7	If you are happy to do so, please give a contact email address
8	Please tick the box to confirm that you are willing for anonymised information provided in this questionnaire to be analysed and published

Table A2. Responses from the 52 rescue centres that completed the questionnaire.

Rescue Centre	Number of Hedgehog Admissions	No. of Admissions of Plastic-Entangled Hedgehogs	Number of Survivors (and Rereleased Animals)
1	100	0	0
2	50	0	0
3	150	0	0
4	70	0	0
5	50	0	0
6	200	0	0
7	150	0	0
8	100	0	0
9	150	0	0
10	100	0	0
11	92	1	1
12	500	1	1
13	300	18	14
14	150	1	0
15	100	4	1
16	200	11	7
17	150	7	3
18	300	1	0
19	700	11	4

Table A2. *Cont.*

Rescue Centre	Number of Hedgehog Admissions	No. of Admissions of Plastic-Entangled Hedgehogs	Number of Survivors (and Rereleased Animals)
20	140	16	11
21	160	1	0
22	225	1	1
23	100	3	2
24	300	4	1
25	500	3	2
26	100	2	2
27	300	5	4
28	300	2	1
29	100	1	1
30	180	1	1
31	100	3	1
32	100	2	1
33	120	2	1
34	300	7	6
35	30	1	1
36	250	2	1
37	350	1	0
38	1000	4	1
39	100	2	0
40	100	7	6
41	250	3	2
42	200	2	1
43	300	4	3
44	125	3	1
45	150	5	2
46	250	10	4
47	250	5	2
48	850	5	3
49	259	1	0
50	375	3	1
51	250	2	1
52	520	8	3

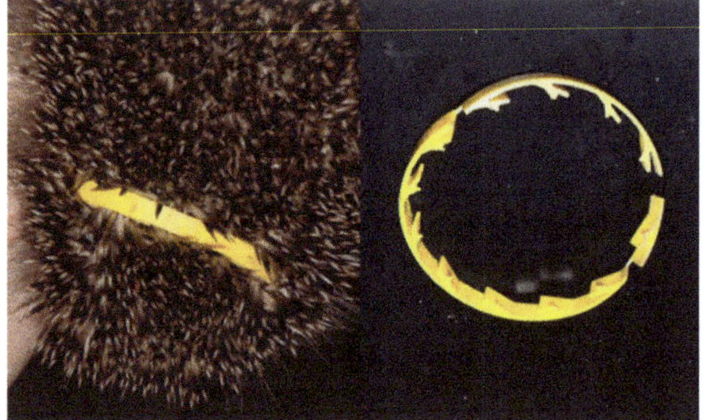

Figure A1. *Cont.*

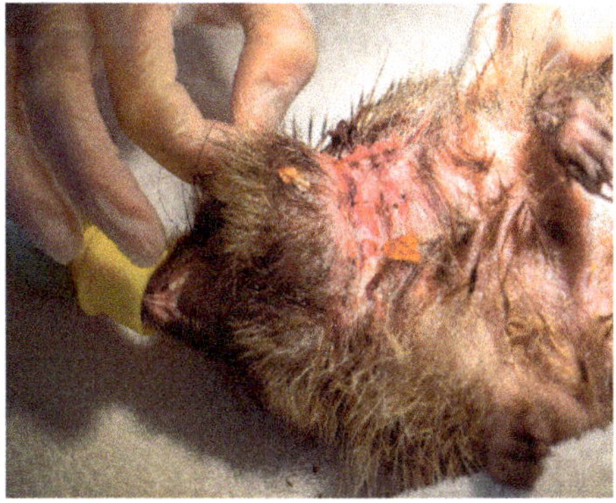 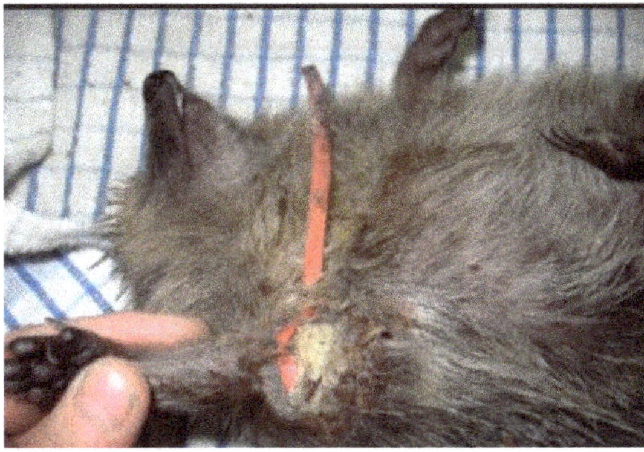

Figure A1. Images of some of the entanglement cases with credit to Hedgehog Bottom, Micklefield Hedgehog Rescue and Wildlife Aid.

References

Wilson, D.; Rodic, L.; Modak, P.; Soss, R.; Rogero, A.; Velis, C.; Iyer, M.; Simonett, O. *Global Waste Management Outlook*; UNEP: Nairobi, Kenya, 2022.

de Souza Machado, A.A.; Kloas, W.; Zarfl, C.; Hempel, S.; Rillig, M.C. Microplastics as an Emerging Threat to Terrestrial Ecosystems. *Glob. Change Biol.* **2018**, *24*, 1405–1416. [CrossRef] [PubMed]

Hartmann, N.B.; Hüffer, T.; Thompson, R.C.; Hassellöv, M.; Verschoor, A.; Daugaard, A.E.; Rist, S.; Karlsson, T.; Brennholt, N.; Cole, M.; et al. Are We Speaking the Same Language? Recommendations for a Definition and Categorization Framework for Plastic Debris. *Environ. Sci. Technol.* **2019**, *53*, 1039–1047. [CrossRef] [PubMed]

Rubin, B.S. Bisphenol A: An endocrine disruptor with widespread exposure and multiple effects. *Steroid Biochem.* **2011**, *127*, 27–34. [CrossRef] [PubMed]

Avio, C.G.; Gorbi, S.; Regoli, F. Plastics and Microplastics in the Oceans: From Emerging Pollutants to Emerged Threat. *Mar. Environ. Res.* **2017**, *128*, 2–11. [CrossRef]

Avio, C.G.; Gorbi, S.; Milan, M.; Benedetti, M.; Fattorini, D.; D'Errico, G.; Pauletto, M.; Bargelloni, L.; Regoli, F. Pollutants Bioavailability and Toxicological Risk from Microplastics to Marine Mussels. *Environ. Pollut.* **2015**, *198*, 211–222. [CrossRef]

Rochman, C.M.; Hoh, E.; Kurobe, T.; Teh, S.J. Ingested Plastic Transfers Hazardous Chemicals to Fish and Induces Hepatic Stress. *Sci. Rep.* **2013**, *3*, 3263. [CrossRef]

Xia, Y.; Zhou, J.J.; Gong, Y.Y.; Li, Z.J.; Zeng, E.Y. Strong Influence of Surfactants on Virgin Hydrophobic Microplastics Adsorbing Ionic Organic Pollutants. *Environ. Pollut.* **2020**, *265*, 115061. [CrossRef]

Liu, G.; Zhu, Z.; Yang, Y.; Sun, Y.; Yu, F.; Ma, J. Sorption Behavior and Mechanism of Hydrophilic Organic Chemicals to Virgin and Aged Microplastics in Freshwater and Seawater. *Environ. Pollut.* **2019**, *246*, 26–33. [CrossRef]

Cole, M.; Lindeque, P.; Halsband, C.; Galloway, T.S. Microplastics as Contaminants in the Marine Environment: A Review. *Mar. Pollut. Bull.* **2011**, *62*, 2588–2597. [CrossRef]

Gunaalan, K.; Fabbri, E.; Capolupo, M. The Hidden Threat of Plastic Leachates: A Critical Review on Their Impacts on Aquatic Organisms. *Water Res.* **2020**, *184*, 116170. [CrossRef]

Butterworth, A. A Review of the Welfare Impact on Pinnipeds of Plastic Marine Debris A Review of the Welfare Impact on Pinnipeds of Plastic Marine Debris. *Front. Mar. Sci.* **2016**, *3*, 93–102. [CrossRef]

Good, T.P.; Samhouri, J.F.; Feist, B.E.; Wilcox, C.; Jahncke, J. Plastics in the Pacific: Assessing Risk from Ocean Debris for Marine Birds in the California Current Large Marine Ecosystem. *Biol. Conserv.* **2020**, *250*, 108743. [CrossRef]

Casabianca, S.; Capellacci, S.; Grazia, M.; Dell, C.; Tartaglione, L.; Varriale, F.; Narizzano, R.; Risso, F.; Moretto, P.; Dagnino, A.; et al. Plastic-Associated Harmful Microalgal Assemblages in Marine. *Environ. Pollut.* **2019**, *244*, 617–626. [CrossRef]

Bellingeri, A.; Casabianca, S.; Capellacci, S.; Faleri, C.; Paccagnini, E.; Lupetti, P.; Koelmans, A.A.; Penna, A. Impact of Polystyrene Nanoparticles on Marine Diatom Skeletonema Marinoi Chain Assemblages and Consequences on Their Ecological Role in Marine Ecosystems. *Environ. Pollut.* **2020**, *262*, 114268. [CrossRef]

Mishra, S.; charan Rath, C.; Das, A.P. Marine Microfiber Pollution: A Review on Present Status and Future Challenges. *Mar. Pollut. Bull.* **2019**, *140*, 188–197. [CrossRef]

Butterworth, A.; Clegg, I. *Marine Debris: A Global Picture of the Impact*; WSPA International: London, UK, 2015.

Nelms, S.E.; Galloway, T.S.; Godley, B.J.; Jarvis, D.S.; Lindeque, P.K. Investigating Microplastic Trophic Transfer in Marine Top Predators. *Environ. Pollut.* **2018**, *238*, 999–1007. [CrossRef]

Bergmann, M.; Lutz, B.; Tekman, M.B.; Gutow, L. Citizen Scientists Reveal: Marine Litter Pollutes Arctic Beaches and Affects Wild Life. *Mar. Pollut. Bull.* **2017**, *125*, 535–540. [CrossRef]

20. López-Martínez, S.; Morales-Caselles, C.; Kadar, J.; Rivas, M.L. Overview of Global Status of Plastic Presence in Marine Vertebrate. In *Global Change Biology*; Blackwell Publishing Ltd.: Hoboken, NJ, USA, 2021; pp. 728–737. [CrossRef]
21. Magnusson, K.; Eliasson, K.; Fråne, A.; Haikonen, K.; Hultén, J.; Olshammar, M.; Stadmark, J.; Voisin, A. *Swedish Sources an Pathways for Microplastics to the Marine Environment*; IVL Swedish Environmental Research Institute: Stockholm, Sweden, 2017
22. Donnelly-Greenan, E.L.; Nevins, H.M.; Harvey, J.T. Entangled Seabird and Marine Mammal Reports from Citizen Science Surve from Coastal California (1997–2017). *Mar. Pollut. Bull.* **2019**, *149*, 110557. [CrossRef]
23. Parton, K.J.; Galloway, T.S.; Godley, B.J. Global Review of Shark and Ray Entanglement in Anthropogenic Marine Debris. *Endang Species Res.* **2019**, *39*, 173–190. [CrossRef]
24. Provencher, J.F.; Bond, A.L.; Avery-Gomm, S.; Borrelle, S.B.; Bravo Rebolledo, E.L.; Hammer, S.; Kühn, S.K.; Lavers, J.I Mallory, M.L.; Trevail, A.; et al. Quantifying Ingested Debris in Marine Megafauna: A Review and Recommendations f Standardization. *Anal. Methods* **2017**, *9*, 1454–1469. [CrossRef]
25. Costa, R.A.; Sá, S.; Pereira, A.T.; Ângelo, A.R.; Vaqueiro, J.; Ferreira, M.; Eira, C. Prevalence of Entanglements of Seabirds Marine Debris in the Central Portuguese Coast. *Mar. Pollut. Bull.* **2020**, *161*, 111746. [CrossRef] [PubMed]
26. Brentano, R.; Petry, M.V. Marine Debris Ingestion and Human Impacts on the Pygmy Sperm Whale (Kogia Breviceps) in Southe Brazil. *Mar. Pollut. Bull.* **2020**, *150*, 110595. [CrossRef] [PubMed]
27. Salazar-Casals, A.; de Reus, K.; Greskewitz, N.; Havermans, J.; Geut, M.; Villanueva, S.; Rubio-Garcia, A. Increased Incidence Entanglements and Ingested Marine Debris in Dutch Seals from 2010 to 2020. *Oceans* **2022**, *3*, 389–400. [CrossRef]
28. Kühn, S.; van Franeker, J.A. Quantitative Overview of Marine Debris Ingested by Marine Megafauna. *Mar. Pollut. Bull.* **2020**, *151*, 1108. [CrossRef]
29. Townsend, A.K.; Barker, C.M. Plastic and the Nest Entanglement of Urban and Agricultural Crows. *PLoS ONE* **2014**, *9*, e8800 [CrossRef]
30. Mallet, J.; Liébana, M.S.; Santillán, M.Á.; Grande, J.M. Raptor Entanglement with Human Debris at Nests: A Patchy and Specie Specific Problem. *J. Raptor Res.* **2020**, *54*, 316–318. [CrossRef]
31. Ayala, F.; Vizcarra, J.K.; Castillo-Morales, K.; Torres-Zevallos, U.; Cordero-Maldonado, C.; Ampuero-Merino, L.; Herrera-Peralta, I De-la-Torre, G.E.; Angulo, F.; Cárdenas-Alayza, S. From Social Networks to Bird Enthusiasts: Reporting Interactions between Plas Waste and Birds in Peru. *Environ. Conserv.* **2023**, *50*, 136–141. [CrossRef]
32. Thomas, L.; Dobbs, P.; Ashfield, S. Digit Entrapment Due to Plastic Waste in a Verreaux's Eagle Owl (*Bubo lacteus*). *J. Zool. B Gard.* **2022**, *3*, 442–447. [CrossRef]
33. Blettler, M.C.M.; Mitchell, C. Dangerous Traps: Macroplastic Encounters Affecting Freshwater and Terrestrial Wildlife. *Sci. Tot Environ.* **2021**, *798*, 149317. [CrossRef]
34. Bergmann, M.; Collard, F.; Fabres, J.; Gabrielsen, G.W.; Provencher, J.F.; Rochman, C.M.; van Sebille, E.; Tekman, M.B. Plast Pollution in the Arctic. *Nat. Rev. Earth Environ.* **2022**, *3*, 323–337. [CrossRef]
35. Mathews, F.; Harrower, C. An IUCN-Compliant Red List for British Mammals. 2020. Available online: https://nora.nerc.ac.u id/eprint/530791/3/N530791CR.pdf (accessed on 1 June 2018).
36. R Studio. RStudio: Integrated Development Environment for R. RStudio: Boston. Available online: http://www.rstudio.con (accessed on 26 June 2023).
37. Rasmussen, S.L.; Berg, T.B.; Martens, H.J.; Jones, O.R. Anyone Can Get Old—All You Have to Do Is Live Long Enoug Understanding Mortality and Life Expectancy in European Hedgehogs (*Erinaceus europaeus*). *Animals* **2023**, *13*, 626. [CrossRef]
38. Mammal Society. *Hedgehog Fact Sheet*; Mammal Society: Dorset, UK, 2018.
39. British Hedgehog Preservation Society. Caring for Hoglets. 2016. Available online: https://www.britishhedgehogs.org.u caring-for-hoglets/ (accessed on 1 June 2018).
40. Bexton, S.; Couper, D. Veterinary Care of Free-Living Hedgehogs. *Practice* **2019**, *41*, 420–432. [CrossRef]
41. Wembridge, D.; Johnson, G.; Al-Fulajim, N.; Langton, S. *The State of Britain's Hedgehogs 2022*; British Hedgehog Preservatic Society and People's Trust for Endangered Species: London, UK, 2022.
42. Mathews, F.; Kubasiewicz, L.; Gurnell, J. A Review of the Population and Conservation Status of British Mammals. 2018. Availab online: http://publications.naturalengland.org.uk/publication/5636785878597632?category=47019 (accessed on 1 June 2018)
43. Emma Bearman-Brown, L. Anthropogenic Factors Associated with West-European Hedgehog (*Erinaceus europaeus*) Surviv Ph.D. Thesis, University of Reading, Reading, UK, 2020.
44. Reeve, N.J. The Survival and Welfare of Hedgehogs (*Erinaceus europaeus*) After Release Back Into the Wild. *Anim. Welf.* **1998**, 189–202. [CrossRef]
45. Wright, P.G.R.; Coomber, F.G.; Mathews, F. Predicting Hedgehog Mortality Risks on British Roads Using Habitat Suitabili Modelling. *PeerJ* **2020**, *2020*, e8154. [CrossRef]
46. Wembridge, D.; Newman, M.; Bright, P.; Morris, P. Hedgehog Road Casualties in Great Britain. *Mammal Commun.* **2016**, *2*, 8–1
47. Moore, L.J.; Petrovan, S.O.; Baker, P.J.; Bates, A.J.; Hicks, H.L.; Perkins, S.E.; Yarnell, R.W. Impacts and Potential Mitigation Road Mortality for Hedgehogs in Europe. *Animals* **2020**, *10*, 1523. [CrossRef]
48. Jepsen, E.M.; de Bruyn, P.J.N. Pinniped Entanglement in Oceanic Plastic Pollution: A Global Review. *Mar Pollut Bull* **2019**, *1* 295–305. [CrossRef]
49. Senko, J.F.; Nelms, S.E.; Reavis, J.L.; Witherington, B.; Godley, B.J.; Wallace, B.P. Understanding Individual and Population-Lev Effects of Plastic Pollution on Marine Megafauna. *Endanger. Species Res.* **2020**, *43*, 234–252. [CrossRef]

20. Franco-Trecu, V.; Drago, M.; Katz, H.; Machín, E.; Marín, Y. With the Noose around the Neck: Marine Debris Entangling Otariid Species. *Environ. Pollut.* **2017**, *220*, 985–989. [CrossRef]
21. Colmenero, A.I.; Barría, C.; Broglio, E.; García-Barcelona, S. Plastic Debris Straps on Threatened Blue Shark Prionace Glauca. *Mar. Pollut. Bull.* **2017**, *115*, 436–438. [CrossRef]
22. Duncan, E.M.; Botterell, Z.L.R.; Broderick, A.C.; Galloway, T.S.; Lindeque, P.K.; Nuno, A.; Godley, B.J. A Global Review of Marine Turtle Entanglement in Anthropogenic Debris: A Baseline for Further Action. *Endanger. Species Res.* **2017**, *34*, 431–448. [CrossRef]
23. Waluda, C.M.; Staniland, I.J. Entanglement of Antarctic Fur Seals at Bird Island, South Georgia. *Mar. Pollut. Bull.* **2013**, *74*, 244–252. [CrossRef] [PubMed]
24. Panter, C.T.; Allen, S.; Backhouse, N.; Mullineaux, E.; Rose, C.A.; Amar, A. Causes, Temporal Trends, and the Effects of Urbanization on Admissions of Wild Raptors to Rehabilitation Centers in England and Wales. *Ecol. Evol.* **2022**, *12*, e8856. [CrossRef] [PubMed]
25. Cope, H.R.; McArthur, C.; Dickman, C.R.; Newsome, T.M.; Gray, R.; Herbert, C.A. A Systematic Review of Factors Affecting Wildlife Survival during Rehabilitation and Release. *PLoS ONE* **2022**, *17*, e0265514. [CrossRef] [PubMed]
26. Mullineaux, E. Veterinary Treatment and Rehabilitation of Indigenous Wildlife. *J. Small Anim. Pract.* **2014**, *55*, 293–300. [CrossRef] [PubMed]

Article

Anyone Can Get Old—All You Have to Do Is Live Long Enough: Understanding Mortality and Life Expectancy in European Hedgehogs (*Erinaceus europaeus*)

Sophie Lund Rasmussen [1,2,*], Thomas B. Berg [3,4], Helle Jakobe Martens [5] and Owen R. Jones [4,6]

[1] Wildlife Conservation Research Unit, Department of Biology, University of Oxford, The Recanati-Kaplan Centre, Tubney House, Abingdon Road, Tubney, Abingdon OX13 5QL, UK
[2] Department of Chemistry and Bioscience, Aalborg University, Fredrik Bajers Vej 7H, DK-9220 Aalborg, Denmark
[3] Naturama, 30 Dronningemaen, DK-5700 Svendborg, Denmark
[4] Department of Biology, University of Southern Denmark, 55 Campusvej, DK-5230 Odense M, Denmark
[5] Department of Geosciences and Natural Resource Management, Section for Forest, Nature and Biomass, Copenhagen University, 23 Rolighedsvej, DK-1958 Frederiksberg C, Denmark
[6] Interdisciplinary Centre on Population Dynamics (CPop), University of Southern Denmark, 55 Campusvej, DK-5230 Odense M, Denmark
* Correspondence: sophielundrasmussen@gmail.com

Citation: Rasmussen, S.L.; Berg, T.B.; Martens, H.J.; Jones, O.R. Anyone Can Get Old—All You Have to Do Is Live Long Enough: Understanding Mortality and Life Expectancy in European Hedgehogs (*Erinaceus europaeus*). *Animals* **2023**, *13*, 626. https://doi.org/10.3390/ani13040626

Academic Editor: Luciano Bani

Received: 18 December 2022
Revised: 4 February 2023
Accepted: 6 February 2023
Published: 10 February 2023

Simple Summary: As populations of European hedgehogs are declining, it is vital that we monitor and understand the population dynamics of this species to optimise conservation initiatives to protect the hedgehogs in the wild. We determined the age of 388 dead European hedgehogs, collected by volunteers from all over Denmark, by counting periosteal growth lines, a method similar to counting year rings in trees. The overall mean age was 1.8 years (1.6 years for females and 2.1 years for males), and the range was between 0 and 16 years. We found the oldest scientifically confirmed hedgehogs in Europe among our samples (11, 13, and 16 years), with previous research recording a maximum age of 9 years. We constructed life tables showing life expectancies at 2.1 years for females and 2.6 years for males. We found that male hedgehogs were more likely to have died in traffic than females and that traffic-related deaths peaked in July for both sexes. For non-traffic deaths, most males died in July, and most females died in September. Most of the road-killed individuals in the study died in rural habitats. The degree of inbreeding did not influence longevity. These new insights may be used to improve future conservation strategies protecting the European hedgehog.

Abstract: The European hedgehog is in decline, triggering a need to monitor population dynamics to optimise conservation initiatives directed at this species. By counting periosteal growth lines, we determined the age of 388 dead European hedgehogs collected through citizen science in Denmark. The overall mean age was 1.8 years (1.6 years for females and 2.1 years for males), ranging between 0 and 16 years. We constructed life tables showing life expectancies at 2.1 years for females and 2.6 years for males. We discovered that male hedgehogs were more likely to have died in traffic than females, but traffic-related deaths peaked in July for both sexes. A sex difference was detected for non-traffic deaths, as most males died in July, and most females died in September. We created empirical survivorship curves and hazard curves showing that the risk of death for male hedgehogs remains approximately constant with age. In contrast, the risk of death for females increases with age. Most of the collected road-killed individuals died in rural habitats. The degree of inbreeding did not influence longevity. These new insights are important for preparing conservation strategies for the European hedgehog.

Keywords: sex-biased longevity; age structure; wildlife conservation; age; matrix models; life table; sex-biased mortality; European hedgehogs; periosteal growth lines; urban and rural

1. Introduction

The life history strategy of a species—the age- or stage-specific patterns of events in a life cycle—is shaped by evolution to maximise fitness [1]. Variation across species and environments and the trade-off between reproduction and survival have led to a striking diversity in life history strategy (and its component traits) across the tree of life. For example, short-lived species such as small rodents tend to be precocious and prodigious breeders, while large-bodied mammals such as elephants and whales grow and reproduce slowly [2]. Similar patterns are also apparent within species: variation in the risk of predation, which alters the survival–reproduction trade-off, may result in shifts towards a "fast" strategy with a shorter generation time, higher reproduction, and shorter life span [3]. Understanding the life history strategy of a species is beneficial for developing robust species-specific conservation practices.

Population models are important tools in species conservation that are particularly relevant when a species appears to be in decline, as in the example of the European hedgehog (*Erinaceus europaeus*). Ecological population modelling can be used to quantify the relative importance of different parts of the life cycle for population growth to predict changes in parameters such as population size and age distribution and understand the impact of exogenous drivers [4]. Population modelling approaches have been widely applied in tackling the conservation of countless species, such as the Pacific fisher (*Pekania pennanti*) [5] and California spotted owl (*Strix occidentalis occidentalis*) [6], leading to increased political priority and conservation efforts [7,8]. These modelling efforts require parameter estimates of key factors such as survival probability and reproductive output at different life cycle stages or ages. For the European hedgehog, a species of growing conservation concern, there is still considerable uncertainty about these key traits. This lack of knowledge has so far hampered the construction of robust population models to explore the dynamics and conservation status of this species.

The European hedgehog is a widely distributed species that can survive across diverse habitat types in rural as well as urban landscapes [9,10]. However, recent research on national and local scales has documented a decline or indicated concerns for a decline in their populations in several western European countries [11–21]. The suspected reasons for the decline include the following: habitat loss; habitat fragmentation; inbreeding; intensified agricultural practices; road traffic accidents; a reduction in suitable nest sites in residential gardens, as well as biodiversity, and hence food items; accidents caused by garden tools, netting, and other anthropogenic sources in residential gardens; molluscicide and rodenticide poisoning; and, in some areas, badger predation [12,22–35].

Hedgehogs hibernate to conserve energy during colder periods during which food availability is low [36]. In Denmark, they usually hibernate from late September (male adults), late October (female adults), or mid-November (juveniles) to around mid-April or mid-May [33,37,38]. However, juveniles may extend their activity period until mid-December if the weather conditions are mild, leaving food items such as slugs, snails, and insects available [33]. The period of winter inactivity is reduced for European hedgehogs residing in milder climates, such as Southern Europe and New Zealand. In New Zealand, hedgehogs may even stay active throughout the year [36]. Hibernating hedgehogs experience periodic arousals every 7–11 days on average, and during these periodic arousals they may remain active for a few days, leaving the nest to forage, or simply change nests [10]. Up to eight nest changes during a Danish winter have been recorded [33].

During hibernation, when the calcium metabolism in the hedgehog is modified, bone growth is reduced markedly or even stopped. This causes densification of the bone resulting in periosteal growth lines, or lines of arrested growth (LAGs), which are formed in the periosteum of the bones. The periosteal growth lines appear to be caused by the arrest of cartilage growth, leading to the infusion of the cartilage plate with apatite (calcium phosphate) [39]. In general, growth lines are developed in vertebrates as the metabolism and growth are inhibited by seasonal cycles in the environment [40]. They are comparable to tree annual growth rings and become visible in stained sections of bones such as the mandible, the lower jaw [41,42]. This phenomenon allows researchers to count the numbe of hibernations a hedgehog has survived. However, bone growth or bone deposition i the periosteum may also become reduced during periods of stress, potentially formir accessory lines in the bone, which may complicate the interpretation of growth lines fc age determination [41]. Previous studies on growth lines in mammals, or LAGs, fro known-age individuals indicate that the number of LAGs in bones positively correlate with age [43–45], though the evidence base is larger for herpetofauna [46].

Data on age-specific survival and reproduction aid the parameterisation of matr models providing insight into the population dynamics of hedgehogs. For hedgehog one approach to accommodate the demand for such age-related data is to conduct ag determination studies on deceased individuals. Previous research into the age structure European hedgehogs is based on four different methods: capture–mark–recapture, countir of periosteal growth lines in the mandible of the hedgehog, measuring the extent of too wear (tooth abrasions), and estimating the number of growth lines in the teeth, typically the molar cement of the M^1 [41,42,47–64]. As an alternative method for determining th life expectancy of hedgehogs, Parkes [65] used a formula by Petrides [66] to calculate a average life expectancy of 1.97 years based on a sample of 144 individuals divided in adults and juveniles (N = 73). Morris [67] also developed a method for age determinatic in hedgehogs by using X-rays to measure epiphyseal fusion in the forefoot, as the presen of epiphyseal cartilage in the metacarpal bones is a juvenile characteristic. This meth classifies juvenile hedgehogs into four age categories and distinguishes juvenile hedgehog from adults. Table 1 provides an overview of the data from the present and previous ag determination studies on hedgehogs [36,41,42,47–56,58–65,68].

Our aim with this study was to add to this information by collecting and analysing substantial dataset, including data on inbreeding, which have hitherto not been include in previous age structure studies on European hedgehogs, providing the opportuni to study a combination of longevity, inbreeding, and cause and timing of death in th Danish hedgehog population. We obtained our data from a large sample of decease Danish hedgehogs collected via a citizen science study. Specifically, we investigated th ages of deceased individuals collected from the Danish hedgehog population by countir periosteal growth lines in transverse sections of the mandibles and creating life tabl based on these data. We also explored the timing and cause of death in our sample understand whether sex, season, or habitat type (urban/rural) influenced the cause ar frequency of death. Finally, we aspired to understand whether the degree of inbreedir influenced longevity in European hedgehogs. The research was performed to achieve deeper understanding of the population structure and specific factors influencing hedgeh ecology and to provide data that improve the conservation initiatives directed at th declining species.

Table 1. Overview of age determination studies on hedgehogs, including the present study. * signifies that individuals of the age group 0–1 year have been omitted from the calculation of mean age. A blank field indicates missing data. Measures of age are presented in years. Brockie [47]: Detailed information was limited. One hundred forty-three individuals, 11% (n = 16) over 18 months (counted as 1-year-olds) and 7 over 2.5 years counted as 2-year-olds. The mean value is based on 120 individuals of 0–1 years, 16 individuals of 1–2 years, and 7 individuals of 2–3 years. Brockie [47] estimated that average life expectancy from birth was around 18 months. Brockie [64] estimated that the life expectancy was less than 2.7 years and probably closer to 2 years. Calculating the life expectancy excluding the individuals below 1 year of age, the mean is 2.7 years. When including individuals younger than 1 year of age, the mean is 1.8 years.

Author	Country	Technique	Period	Sample Size			Mean Age			Age Span
				Female	Male	Total	Female	Male	Total	
Brockie [47]	New Zealand	Tooth abrasions				143			0.4 (see table text)	0–2
Brockie [64]	New Zealand	Mandible (periosteal growth lines)				83			<2.7* or 1.8 (see table text)	0–7
Dickman [48]	United Kingdom	Mandible (periosteal growth lines)	1982–84			87			3.3*	1–8
Döpke [49]	Germany	Mandible (periosteal growth lines)	1980–2001	32	34	66	0.8	1.7	1.2	0–6
Haigh, Kelly, Butler and O'Riordan [50]	Ireland	Mandible (periosteal growth lines)	2008–2011	31	46	83 (6 of unknown sex)	2.1	1.9	1.9	0–9
Heddergott [51]	Thüringen, Germany	Tooth growth lines	1993–2002	81	99	185 (5 of unknown sex)	2.5	2.1	2.3*	1–7
Heddergott [52]	Harz, Germany	Tooth growth lines	1997–2002	49	57	106	1.6	1.5	1.5*	1–4
Heddergott [53]	Germany, Saalfeld-Rudolstadt (Thüringen)	Tooth growth lines	1992–2003	44	37	81	1.9	2.5	2.2*	1–8
Heddergott and Müller [54]	Germany, Osthessen	Tooth growth lines	1980–2006	62	61	124	1.9	1.7	1.8*	1–6
Heddergott, Steinbach and Heddergott [55]	Germany, Heilbad Heiligenstadt (Thüringen)	Tooth growth lines	1998–2004	62	51	113	2.2	2.5	2.3*	1–7
Heyne [56]	Germany	Tooth abrasions+ Tooth growth lines	1981–89			62			2*	1–5

Table 1. *Cont.*

Author	Country	Technique	Period	Sample Size			Mean Age			Age Span
				Female	Male	Total	Female	Male	Total	
Kristoffersson [58]	Finland	Mandible (periosteal growth lines)	1960–61			67			2.7* (individuals < 500 g excluded)	1–7
Scott-Hayward [68]	Uists, Outer Hebrides, UK	Mandible (periosteal growth lines)				66			2.6*	1–6
Morris [36], Morris [41], Morris [60]	Southern England, United Kingdom	Mandible (periosteal growth lines)	1965–68	102	127	244 (15 of unknown sex)			1.7	0–6
Neuschulz and Schubert [60]	Dresden-Bühlau, Germany	Capture-mark-recapture	1984–89			125 (60 adults and 65 juveniles marked, 53 recaptured over 5 years)			2.1*	1–6
Parkes [65]	Manawatu, New Zealand	Capture-mark-recapture + calculation using a formula [66]	1970–1971			144 (73 juveniles)			2.0	
Rasmussen et al. 2023 (present study)	Denmark	Mandible (periosteal growth lines)	2016	109	177	388 (101 of unknown sex)	1.6	2.2 (1.6 for unknown sex)	1.8	0–16
Rautio, Kunnasranta, Valtonen, Ikonen, Hyvarinen, Holopainen and Kukkonen [61]	Finland	Mandible (periosteal growth lines)	2004–2005	31	34	65	1.5	1.7	1.6	0–7
Reeve, Love and Shore [42]	United Kingdom	Mandible (periosteal growth lines)	1988–89			68			1.6	0–7
Skoudlin [62]	Czech republic	Tooth abrasions		43	63	106			2.5*	1–5
Skoudlin [63]	Czech republic, Bohemina and Moravia	Tooth abrasions				215			2.1	0–6

2. Materials and Methods

We obtained samples of dead hedgehogs from the Danish public as part of the citizen science project "The Danish Hedgehog Project", which aimed to use dead hedgehogs to assess the general health of the Danish hedgehog population. The public outreach was primarily based on >200 features in the national and local media from April to December 2016. With approximately 400 volunteers from throughout Denmark collecting 697 dead hedgehogs, during May to December 2016, there was an excellent geographical representation of the entire country's population. The hedgehogs were primarily road kills, but the sample set also included individuals dying in care and individuals dying from other causes in the wild.

For each specimen, we collected geolocation (latitude and longitude), sex, cause of death, and date of death. We also took tissue samples (skin, fat, and muscle) for genetic analysis to quantify inbreeding. The individuals were in different stages of preservation, and many of the road-killed individuals were not intact. However, heads, front legs, and any organs present were extracted for future research. Out of the 697 dead individuals, 388 were adequately intact to have mandibles of a sufficient quality to use for age-at-death determination.

To determine the age-at-death of the hedgehogs, we counted periosteal growth lines in transverse sections of the mandibles, with each growth line indicating one hibernation season, as described by Morris [41]. To do this, we first macerated the bones to obtain clean skeletal remains. This was achieved by placing the hedgehog heads and legs in gauze with knots between the body parts and on the ends and placing the gauze packs in a bucket of water with a constant temperature of 40 °C for 10 days in a heating cabinet. After 10 days, the water was changed three times at 2–3 h intervals. Then the gauze packs were boiled for 10 min at 90–95 °C in clean water, followed by a drying period of 5 days. After drying, the remaining fat on the jawbones was removed with a cotton bud. Some samples were placed in clean water supplemented with 1:100 H_2O_2 for 30 min after boiling, after which the water was changed again. After 24 h in the water, the samples were then removed and left to dry before the bones were ready for decalcification.

To decalcify the mandibles, the samples were first placed in 2 cl glasses and covered in 10% formalin for 24 h. Afterward, the jaws were rinsed in running water and put in 2 cl glasses with water for 24 h. Finally, the mandibles were rinsed in running water and stored in 2 cl glasses with Rapid Decalcifier (diluted 2:1, with two parts distilled water and one part decalcifier) added. The mandibles were removed from the decalcifier whenever they were bendable and cuttable with scalpel knives. The jawbones were checked regularly with 30 min intervals. Smaller mandibles from juveniles were only left in the decalcifier for 2 h; the larger ones were left for up to 9 h. On average, 4–5 h was sufficient, with a room temperature of 15 °C, even though previous studies describe a mean time of 6 h using the same product [50]. When the bones were sufficiently decalcified, they were rinsed in water, stored in plastic cups with screw lids, and covered in formalin ready for slide preparation.

We prepared the slides by taking an approximately 10 mm transverse section of the jawbone near the last molar with a scalpel knife and mounting it on a cutting plate with a drop of Optimal Cutting Temperature (O.C.T.) Compound (polyvinyl alcohol <11%, Carbovax <5%, nonreactive agents >85%), with the weakest part of the bone pointing upwards. The section was placed in the cryostat microtome (Leica CM3050 S) to be sufficiently frozen before cutting, making it as hard as possible, lasting approximately 3–3.5 min. The jawbone sections were then sectioned in the cryostat microtome at a temperature of −20 °C with a thickness of 50 μm.

The 50 μm jawbone sections were placed on SuperFrost Plus slides, and distilled water (Milli-Q) drops were added on top of the sections to keep them from drying out.

The water was then removed with an automatic pipette, and the sections were stained with crystal violet (0.005%) for 5 min, after which excess dye was removed with an automatic pipette. Distilled water (Milli-Q) drops were then added to rinse away the dye; afterward, the water was removed with an automatic pipette. Subsequently, the slide was coated with the mounting agent Aquatex to preserve the sample, and a cover slip was added on top of the slide. Compared to previous studies using the mounting agent Eupara our samples did not require suction, but could be placed under the microscope immediate and were completely dry and ready for age determination within 24 h.

To estimate the age-at-death, we analysed the prepared and mounted sections under microscope at 20× objective giving a 200 times enlargement (Leica DM 5000B Fluorescen Microscope). A Canon EOS 1200 D Camera was connected to the microscope and was use for image recordings. Scientific Focus Acquisition Software 1 was used to view and ed the pictures. Contrast adjustments were carried out to improve the clarity of images. Fin image processing and cropping and mounting of the images were performed with Adok Photoshop CS5 and Illustrator CS5. We were then able to estimate age by counting tl periosteal growth lines in the prepared transverse section of the lower jaw. At least tw biologists evaluated each sample, counting the periosteal growth lines, and the results we compared. In the few cases of disagreement, the slides were reviewed and discussed, and third observer was involved when needed. Discontinued and imperceptible lines were n included in the counts. After applying these methods, we had a full representation of tl age structure in our sample of the Danish hedgehog population.

We analysed the age-at-death records, alongside covariates of habitat type, timir and cause of death, and sex. Our analyses of the timing of death and cause of death we restricted to individuals of known sex. We divided cause of death into two categories: traff and non-traffic-related deaths. Our analyses and figure preparations were performed i R [69].

To allow us to explore the influence of habitat on mortality, we classified each hedgeh sample into "urban" or "rural" habitat types. We did this for each geolocated sample base on land use types within a 500 m radius around the spot where the hedgehog was foun This area is roughly equivalent to a large hedgehog home range [10]. We obtained the lar use data from the EU CORINE land cover dataset, which has a 100×100 m resolutic (CLC 2012, Version 18.5.1), which resulted in 81 squares for each hedgehog. The CORIN data use satellite imagery and post-processing to assign habitat types as artificial surfac industry, agricultural areas, forest and semi-natural areas, wetlands, and water bodi We then analysed the data using the raster package [70] in R [69] and reclassified tl habitat types as "urban", "rural", or "other", following the method of Rasmussen, et al. [7 (S1). We categorised individual hedgehogs as "urban" or "rural" based on the percenta; representation of the two categories among 81 squares per individual. Combined with tl information on the sex of the individuals killed in traffic, we investigated whether habit type (urban or rural) or sex influenced the likelihood of being killed in traffic.

In a previous publication, we estimated inbreeding as the degree of individual he erozygosity (iH_O) for a subsample of 151 aged individuals (78 males, 50 females, 23 unknown sex) [31]. We tested whether inbreeding was associated with age-at-death usir GLMs with quasi-Poisson error structure. We included the integer estimated age-at-dea as the response variable and the inbreeding coefficient as the explanatory variable. We d this for both sexes combined, and for each sex independently.

We used quasi-Poisson GLMs to explore the association between age-at-death ar inbreeding (degree of individual heterozygosity (iH_O)), cause of death (traffic/non-traffi and the interaction between them.

Lastly, to assess how the mortality risk (probability of death) changed with age, v constructed empirical actuarial life tables using the age-at-death information obtained fro the dental analysis described above. We did this for males and females separately ar excluded data from individuals of unknown sex. We used standard life table approach and nomenclature [72,73]. Using these methods, we calculated empirical survivorship ar hazard trajectories for both sexes.

3. Results

We determined the age-at-death of 388 hedgehogs by counting the periosteal growth lines in prepared transverse sections of the lower jaws (Table 2). The sample set consisted of 109 females, 177 males, and 102 of unknown sex. The mean age-at-death was 1.8 years (22 months) $\pm$ 95% CI [1.62, 2.04], distributed between the ages of 0 and 16 years (Table 2). The mean age-at-death was 1.6 years $\pm$ 95% CI [1.32, 1.93] for females and 2.1 years $\pm$ 95% CI [1.73, 2.47] for males. Dividing the individuals into categories based on the cause of death (traffic, dying from other causes in the wild, in-care), the mean age-at-death was 2.1 years $\pm$ 95% CI [1.80, 2.32], 2.0 years $\pm$ 95% CI [1.51, 2.54], and 1.3 years $\pm$ 95% CI [0.75, 1.83], respectively.

Although the samples collected had a wide geographic coverage across Denmark, they were not evenly distributed (Figure 1).

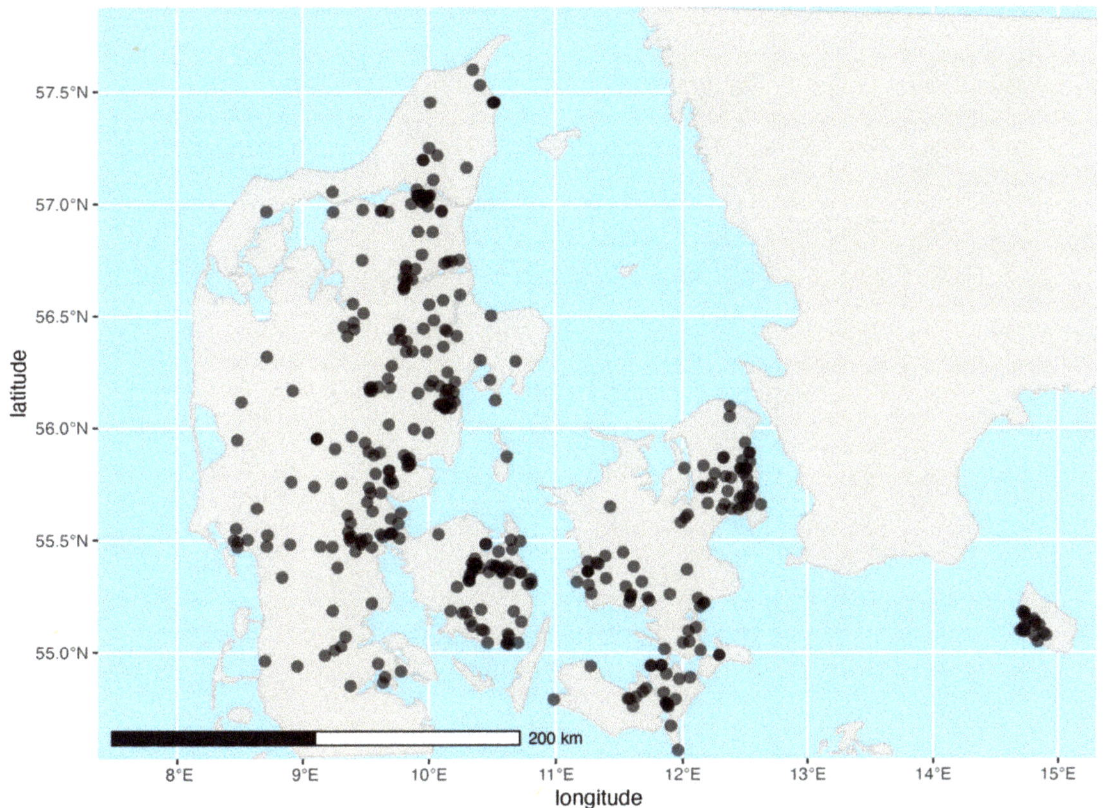

Figure 1. The distribution of the 369 geolocated samples collected during the study. Each point represents a single sample. Points are opaque and thus multiple samples from the same or nearby locations result in darker points.

Figure 2 presents illustrated examples of the twelve age classes determined in the present study.

Table 2. Overview of age estimation data showing the number of individuals at each age, in years, out of a total sample of 388 hedgehogs. Age was estimated by counting the periosteal growth lines in transverse sections of the lower jaws. Rows are omitted where no individuals of that age were found.

Age in Years	Number of Individuals	Sex	Distribution by Sex
0	109		
		Male	43
		Female	39
		Unknown	27
1	115		
		Male	54
		Female	21
		Unknown	40
2	54		
		Male	23
		Female	15
		Unknown	16
3	53		
		Male	29
		Female	18
		Unknown	6
4	22		
		Male	6
		Female	9
		Unknown	7
5	15		
		Male	7
		Female	6
		Unknown	2
6	11		
		Male	8
		Female	1
		Unknown	2
9	2		
		Male	2
		Female	0
		Unknown	0
10	4		
		Male	2
		Female	0
		Unknown	2
11	1		
		Male	1
		Female	0
		Unknown	0
13	1		
		Male	1
		Female	0
		Unknown	0
16	1		
		Male	1
		Female	0
		Unknown	0
Total individuals	**388**		

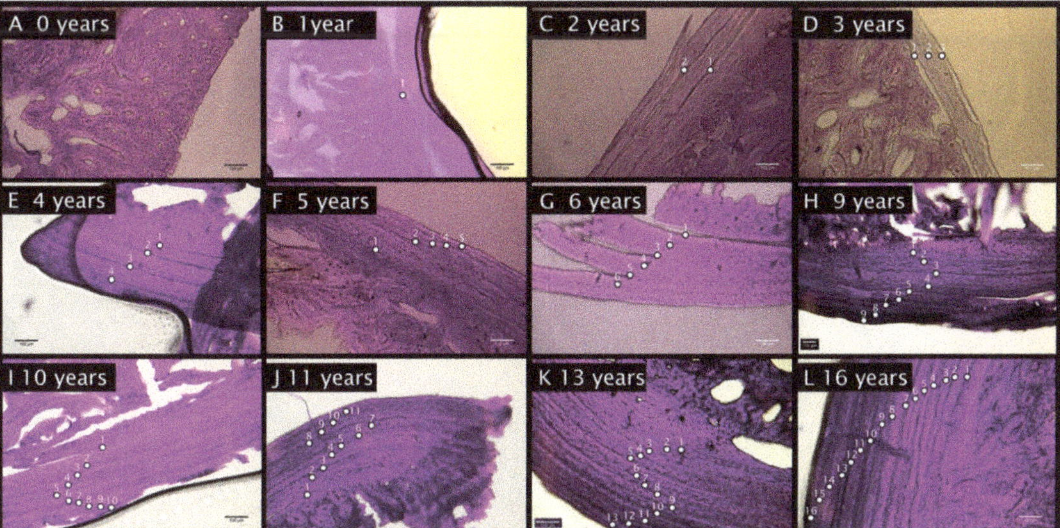

Figure 2. Illustrated examples of the 12 age categories observed in prepared transverse sections of the lower jaws of European hedgehogs from Denmark. Each white dot shows a growth line, and the small numbers next to each point indicate the numbers of growth lines, and their order, counted in each section. The age categories ranged from 0 years (dying before first hibernation), labeled A, to 16 years, labeled L.

3.1. Timing and Cause of Death

Males accounted for 45.6% (177) of the samples, and females accounted for 28.1% (109). The remaining 26.3% (102) of the samples could not be sexed. Out of the 388 individuals, 55.7% (216) were road kills, 22.2% (86) died in care, 21.6% (84) died of natural causes in the wild, and 0.5% (2) could not be categorised due to unknown causes of death. The time distribution of samples through the year showed a clear modal distribution. The timing of the peak in deaths in general varied between the sexes, with the deaths of males being most prevalent in July and the deaths of females being most prevalent in September (Figure 3A). We divided causes of death into two categories: road traffic death and non-traffic-related death. There was a clear sex difference in this cause of death, with male samples being more likely to have died in traffic than females (Figure 3B). A sex difference was detected in the timing of the "peak death" month for non-traffic-related deaths (Figure 3C). For non-traffic-related deaths, males peaked earlier (July) than females (September), but for traffic deaths, the peaks were synchronous (July).

3.2. The Influence of Habitat Type (Urban/Rural) and Sex on the Amount of Road-Killed Individuals

Of the 369 dead individuals with precise geolocation information, 49.6% (183) were found in urban habitats and 50.4% (186) were found in in rural habitats, with a further relatively even distribution of individuals dying in urban or rural habitats when categorised into males and females (Figure 4). Of the 206 road-killed individuals with known geolocations, 37.5% (78) died in urban habitats and 62.5% (130) died in rural habitats, with approximately the same pattern showing when looking at the sexes separately.

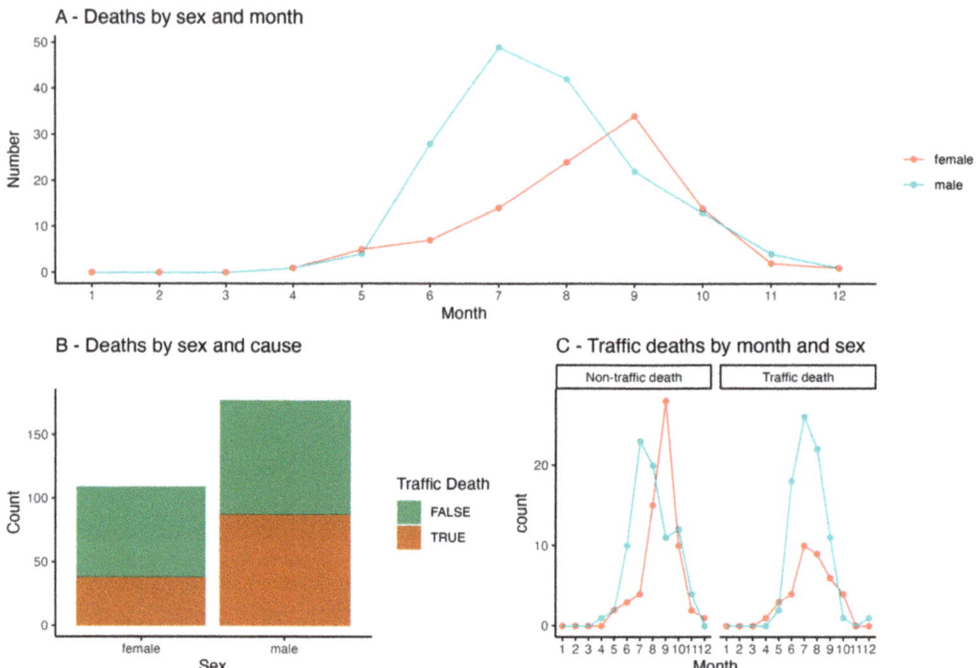

Figure 3. Proportion of deaths across the months of the year, disaggregated by sex (**A**): red colour represents females and blue colour represents males. Cause of death disaggregated by sex (**B**): orange colour is for traffic deaths and green colour is for non-traffic deaths (in-care, dying of other causes the wild). Deaths disaggregated by both month and sex (**C**): red colour represents females and blue colour represents males.

All causes of death						
	🦔	%	♀	%	♂	%
🏢	183	49.6	62	58.5	84	49.7
🏠	186	50.4	44	41.5	85	50.3
Total	**369**	100	**106**	100	**169**	100

Traffic deaths						
	🦔	%	♀	%	♂	%
🏢	78	37.5	16	43.2	32	37.6
🏠	130	62.5	21	56.8	53	62.4
Total	**208**	100	**37**	100	**85**	100

Figure 4. An overview of the dataset showing whether the 369 hedgehogs with geolocations die in rural (barn symbol) or urban (high-rise buildings symbol) habitats, divided into categories all sexes, including the individuals of unknown sexes (hedgehog symbol), and each separate se respectively, including percentage representation of the 208 geolocated traffic-related deaths for ea sex and habitat type (urban/rural).

3.3. The Effects of Genetic Heterozygosity on Age-at-Death

The mean inbreeding score, observed individual heterozygosity (iH$_O$), was 0.2 (SD = 0.074), which indicates a moderate degree of inbreeding. There was no significant a sociation between inbreeding and age-at-death (all animals: quasi-Poisson GLM, t = −0.3 residual d.f. = 149, p = 0.798; males only: quasi-Poisson GLM, t = 0.272, residual d.f. = 7

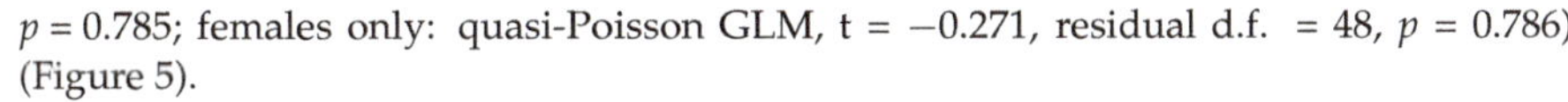

$p = 0.785$; females only: quasi-Poisson GLM, t = −0.271, residual d.f. = 48, $p = 0.786$) (Figure 5).

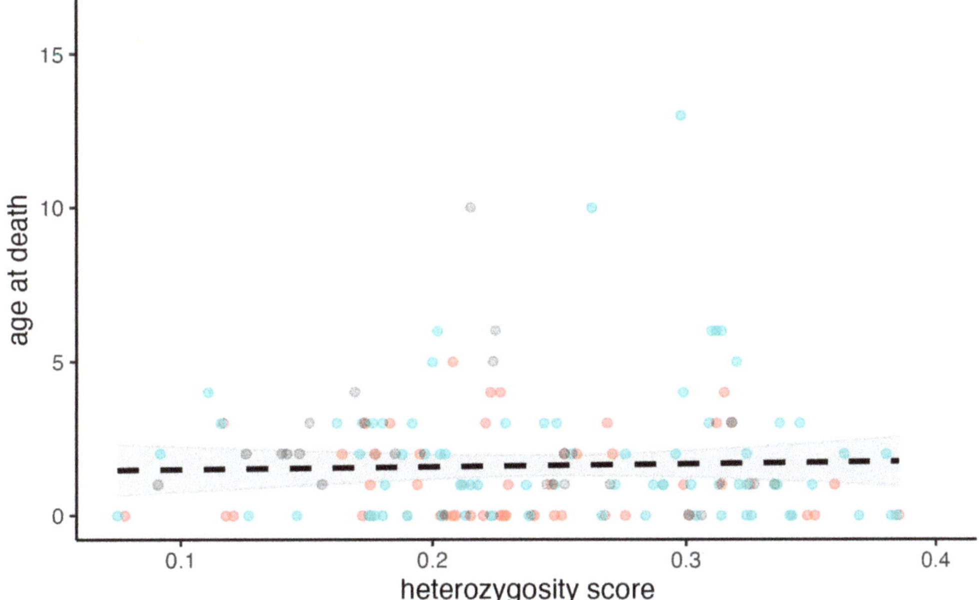

Figure 5. The relationship between individual heterozygosity (iH$_O$) and age-at-death of the hedgehogs studied (N = 151) is not statistically significant. The red points represent females, the blue points males, and the grey points unknown sex. The line and ribbon represent the fitted model without sex included (sex does not influence the fit).

3.4. Exploring the Association between Age-at-Death and Degree of Inbreeding, Cause of Death (Traffic/Non-Traffic), and the Interaction between Them

The results showed no significant interaction between the cause of death and the degree of inbreeding (quasi-Poisson GLM: t = −0.321, d.f. = 147, $p = 0.749$), indicating that the effect of inbreeding was independent of the cause of death. Although the age-at-death tends to be a little higher in hedgehogs killed in traffic (1.972 years; 95% CI [1.667–2.277]) compared to those dying by other causes (1.645 years; 95% CI [1.333–1.957]), it is not significantly so (quasi-Poisson GLM: t = 1.481, d.f. = 386, $p = 0.139$).

3.5. Life Table Analyses

The life table calculations for males (Table 3) and females (Table 4) showed that life expectancy at age 0 (e_x) was 2.6 years for males and 2.1 years for females, indicating that newborn male hedgehogs can expect a 24% longer life than females.

The empirical survivorship and hazard trajectories (also known as mortality rate or death rate curve) showed a classic Type II survivorship for males and a Type I survivorship curve for females (Figure 6A). Type II survivorship is indicative of a constant risk of death with age while Type I is indicative of an increasing risk of death with age (Figure 6B), indicating that the risk of death in males is approximately constant despite increasing age and the risk of death in females increases with age.

Table 3. An empirical life table for male hedgehogs. Nomenclature follows Jones [73]: x: the exact age at the start of the interval; n: the length of the interval in person-years (the difference between the values of x in consecutive rows); l_x: the number of individuals entering the interval at age x, with the first entry being the number of individuals in the entire cohort and subsequent entries being the number surviving to each age (x); $_nd_x$: the number of individuals dying between ages x and $x + n$; $_nq_x$: the probability of dying, calculated as $_nd_x/l_x$; $_np_x$: the probability of surviving, calculated as $- (_nd_x/lx)$; $_nL_x$: person-years lived between ages x and $x + n$; T_x: person-years lived above age x; e_x: life expectancy from age x; $_nm_x$: death rate in the cohort between ages x and $x + n$; $_na_x$: the average number of years lived in the time interval by those dying in the time interval, was assumed to be 0.5 throughout.

x	l_x	$_nd_x$	$_nq_x$	$_np_x$	$_nL_x$	T_x	e_x	$_nm_x$	$_na_x$
0	177	43	0.243	0.757	155.5	460.5	2.602	0.277	0.5
1	134	54	0.403	0.597	107.0	305.0	2.276	0.505	0.5
2	80	23	0.287	0.713	68.5	198.0	2.475	0.336	0.5
3	57	29	0.509	0.491	42.5	129.5	2.272	0.682	0.5
4	28	6	0.214	0.786	25.0	87.0	3.107	0.240	0.5
5	22	7	0.318	0.682	18.5	62.0	2.818	0.378	0.5
6	15	8	0.533	0.467	11.0	43.5	2.900	0.727	0.5
7	7	0	0.000	1.000	7.0	32.5	4.643	0.000	0.5
8	7	0	0.000	1.000	7.0	25.5	3.643	0.000	0.5
9	7	2	0.286	0.714	6.0	18.5	2.643	0.333	0.5
10	5	2	0.400	0.600	4.0	12.5	2.500	0.500	0.5
11	3	1	0.333	0.667	2.5	8.5	2.833	0.400	0.5
12	2	0	0.000	1.000	2.0	6.0	3.000	0.000	0.5
13	2	1	0.500	0.500	1.5	4.0	2.000	0.667	0.5
14	1	0	0.000	1.000	1.0	2.5	2.500	0.000	0.5
15	1	0	0.000	1.000	1.0	1.5	1.500	0.000	0.5
16	1	1	1.000	0.000	0.5	0.5	0.500	2.000	0.5
17	0	0	-	-	-	0.0	-	-	0.5

Table 4. An empirical life table for female hedgehogs. Nomenclature follows Jones [73]: x: the exact age at the start of the interval; n: the length of the interval in person-years (the difference between the values of x in consecutive rows); l_x: the number of individuals entering the interval at age x, with the first entry being the number of individuals in the entire cohort and subsequent entries being the number surviving to each age (x); $_nd_x$: the number of individuals dying between ages x and $x + n$; $_nq_x$: the probability of dying, calculated as $_nd_x/l_x$; $_np_x$: the probability of surviving, calculated as $- (_nd_x/lx)$; $_nL_x$: person-years lived between ages x and $x + n$; T_x: person-years lived above age x; e_x: life expectancy from age x; $_nm_x$: death rate in the cohort between ages x and $x + n$; $_na_x$: the average number of years lived in the time interval by those dying in the time interval, was assumed to be 0.5 throughout.

x	l_x	$_nd_x$	$_nq_x$	$_np_x$	$_nL_x$	T_x	e_x	$_nm_x$	$_na_x$
0	109	39	0.358	0.642	89.5	231.5	2.124	0.436	0.5
1	70	21	0.300	0.700	59.5	142.0	2.029	0.353	0.5
2	49	15	0.306	0.694	41.5	82.5	1.684	0.361	0.5
3	34	18	0.529	0.471	25.0	41.0	1.206	0.720	0.5
4	16	9	0.562	0.438	11.5	16.0	1.000	0.783	0.5
5	7	6	0.857	0.143	4.0	4.5	0.643	1.500	0.5
6	1	1	1.000	0.000	0.5	0.5	0.500	2.000	0.5
7	0	0	-	-	-	0.0	-	-	0.5

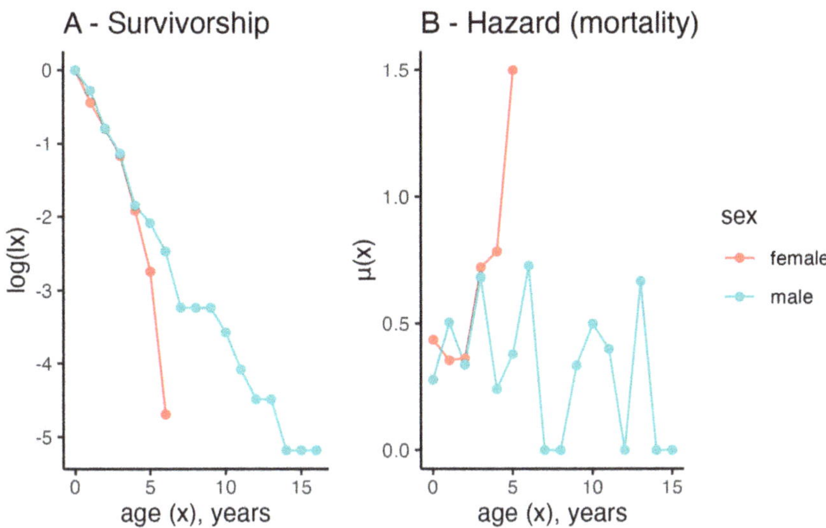

Figure 6. Empirical survivorship curves (**A**) and mortality rates (**B**) for male and female hedgehogs, calculated using standard life table methods.

4. Discussion

This research elucidated several interesting results describing the population dynamics of European hedgehogs, which will be applicable for the preparation of future conservation strategies for the declining population of this species. By counting the periosteal growth lines and creating life tables, we found a mean age for all individuals of 1.8 years, 1.6 years for females and 2.1 years for males, as well as 2.1 years for females and 2.6 years for males, respectively, the latter measured as the life expectancy at age 0, indicating that male hedgehogs live longer than females. We found that male hedgehogs were more likely to have died in traffic than females but that traffic-related deaths peaked in July for both sexes. A sex difference was detected in the timing of the "peak death" month for non-traffic deaths, as males peaked earlier (July) than females (September). Based on the life tables, we created empirical survivorship curves and hazard curves indicating that the risk of death in males is more or less constant despite increasing age and the risk of death in females increases with increasing age. We found a 50/50 distribution of hedgehogs dying in urban and rural habitats, with a further relatively even distribution of individuals dying in urban or rural habitats when categorised into males and females. Focusing on road-killed individuals, we found that approximately one-third of the individuals (37.5%; 78) died in urban habitats and two-thirds (62.5%; 130) died in rural habitats, with roughly the same pattern showing when looking at the sexes separately. Finally, we detected no significant association between the degree of inbreeding and age-at-death or between the degree of inbreeding and cause of death.

4.1. Periosteal Growth Lines

Periosteal growth lines are excellent instruments for age determination, but the method comes with certain caveats which should be considered when interpreting results [45]. Firstly, it is important to be aware that some growth lines may not represent hibernation but perhaps a period of stress for the animal, where the bone growth may be reduced or stopped altogether [74]. These so-called accessory lines tend to be less visible than hibernation lines after the staining of the jawbone sections. Secondly, the growth lines may be absent in cases where the individuals do not hibernate. This statement is supported by Morris [41], who noted that a young hedgehog kept active and well-fed throughout winter, and which did not hibernate, failed to form any growth line in the bone. This means that the assessment of periosteal growth lines is not appropriate for age determination in hedgehogs that do not hibernate, such as many individuals in New Zealand [75]. This problem may be increasingly relevant in other populations too. For example, in recent years during exceptionally mild winters, hedgehog rehabilitators in the UK and Italy have recorded much more activity among hedgehogs during periods where they were supposed to be hibernating.

It is currently unknown how much activity it takes to prevent the growth lines from forming and whether additional growth lines could form when individuals experience interrupted hibernation (e.g., if an individual's winter hibernation is split into two hibernation periods by an active period caused by disturbance). Because these factors may compromise the validity of the age determination method, it is important to keep them in mind when interpreting the results. In the current Danish climate, it would be highly unlikely that hedgehogs would remain active throughout winter, and to our knowledge non-hibernation in these animals has never been recorded in Denmark, despite ad libitum winter food availability at local feeding stations in residential gardens. Studies of hibernation in Danish hedgehogs have failed to provide evidence of extended periods of activity between multiple distinct periods of dormancy [33,38,76,77]. Nevertheless, it is possible that during recent years with the extremely mild winter conditions, and therefore periods of potential food availability, the hedgehogs may have become active for longer periods in Denmark as well. The current study benefitted from having a large sample size consisting mainly of individuals aged 0 and 1 years (n = 109 and n = 115, respectively) with known dates of death. In addition to the jaw bones being markedly smaller than those of adult hedgehogs, we could simply cross-check the growth line count with the age-at-death determined by other means. For example, we know that juveniles are almost all born in late July or during August in Denmark and that females may produce second litters which are born from September to late October [33]. This clearly indicated that small individuals with no visible growth lines, killed in the autumn of 2016, were certainly less than one year old. In the cases where a single growth line was detected, we studied the death date of the individual to determine whether it had reached the age of one year (dead before or after June 2016). In none of the cases where we counted growth lines in small jaw bones, which were easily recognisable as being from juveniles or subadults, did we detect more than one clear growth line. Therefore, based on this knowledge, we have concluded that hedgehogs in our study were unlikely to have formed accessory growth lines during the hibernation period. However, what we cannot conclude is whether the reason was a lack of activity during winter or that they simply did not form extra growth lines in cases where the hibernation had been interrupted for a longer period of time. Further research on the dormancy period required for bone deposition to become slow enough to allow the formation of growth lines is required.

Our estimate of a mean age-at-death of 1.8 years is roughly equivalent to the findings of previous studies (presented in Table 1), ranging from 0.4 years to 3.3 years with a median of 2. However, several of these studies excluded individuals aged <1 year. If the individuals <1 year (109) were excluded from the present dataset, the mean age-at-death would be 2.5 years (279). Additionally, it should be noted that the previous studies used four different age determination methods, which could lead to a variation among results. Furthermore, the method of age determination would suggest that we have recorded the oldest confirmed European hedgehogs among our samples (11, 13, and 16 years), with previous research discovering a maximum age of 9 years [50]. This could be explained by our larger sample size (388) compared to the range of 62–244 individuals in previous studies (see Table 1) with a median of 97 hedgehogs and a mean of 112 hedgehogs ± 95% CI (87,136)

4.2. Timing and Cause of Death

Although our study used a large sample size of dead hedgehogs collected by volunteers throughout the whole of Denmark from May to December 2016, we must be aware of the potential bias in the distribution of roadkill per calendar month, because the public awareness of the project gradually built up as more news features were launched during the year. There is also potential bias in the cause of death among the hedgehogs because the road-killed individuals are easier to find than those with natural death and did represent a majority of the sample size.

Our results indicate that male hedgehogs are more prone to death in traffic than females, especially during the mating season, which is usually in June and July, but sometimes extends into August, in Denmark [33]. This is explained by the fact that males tend to have larger home ranges than females [10,33,36] and likely move over larger areas bringing them into contact more frequently with roads and road traffic. Previous research has also demonstrated that male hedgehogs tend to cross roads more often than females [29]. It is notable that the home ranges of males, but not females, increase during the mating season [33], which would exacerbate the male–female difference. However, traffic deaths were also higher for females during July and August, which includes the mating season, probably because they also tend to move over relatively larger areas in the search for mates, although not to the same degree as males [10,33,36]. Females may increase movements, and thus road-crossing behavior, during these periods to either search for mates or forage more widely during lactation. Previous research on traffic deaths of Danish wildlife in 1957–1958 and 1964–1965 [78], showed similar patterns: of 178 road-killed hedgehogs, most deaths occurred in October (66, i.e., 37%), and counts from July to September showed approximately 30 individuals per month (17% per month). This means that collectively July–October was the most dangerous time of the year for hedgehogs. These earlier results contrast with the present study where we found that road-kill deaths were concentrated in June–September, with a clear peak in July. A possible explanation for this difference between the studies could be that only a few road-killed hedgehogs were counted in May during the years of 1957–1958 and 1964–1965 (3), indicating that the hedgehogs may have become active after hibernation considerably later during those years compared to 2016, when the samples from the present study were collected, and that this delay could have influenced the results. Furthermore, Hansen [78] was limited to a local scale compared to the present study which represented road kills from all over Denmark. In comparison, Holsbeek, et al. [79] also found a peak in the number of road-killed hedgehogs during July, in Belgium, with a large sample size of 1281 road-killed hedgehogs in 1995–1996. Another possible influence potentially adding to the peak in road-killed individuals collected during the summer could be the longer daytime hours, making the hedgehog cadavers more visible to the volunteers collecting them.

Of the non-traffic-related deaths, incidences peaked in July for males and August and September for females, which is likely explained by the physical exhaustion caused by the mating season leaving males more vulnerable to infections, starvation, and dehydration at this time. This is evidenced by the fact that male hedgehogs are commonly brought into care during this period. There was a clear increase in non-traffic-related deaths among females in August and September. This is likely associated with greatly elevated energetic costs during lactation and compounded by parental care responsibilities and, consequently, the associated reduced time allocated to foraging. In September, the higher death rates could also be explained by a reduced body, and general health, condition after raising the young.

4.3. Testing the Influence of Habitat Type (Urban/Rural) and Sex on the Number of Road-Killed Individuals and Using Data on Road-Killed Individuals for Population-Level Research

Monitoring patterns and trends in road-kill data can be useful for many ecological purposes including tracking population trends, mapping native and invasive species distributions, studying animal behaviour, and monitoring pollutants and wildlife disease [80]. In European hedgehogs, longitudinal studies have highlighted a drastic population decline in the UK [20,81]. Estimating and understanding the risk of death experienced by individual animals requires a carefully planned and well-controlled study. It is well known that road-kill data include potential sampling biases as simple counts of dead individuals do not provide valid information about the probability of hedgehogs being killed in traffic. Risk calculations require both the numerator (number killed) and denominator (number exposed to risk). Nevertheless, the spatial pattern of road-kill risk is certainly closely related to hedgehogs' live distribution, indicated by significant live distribution–road-kill overlap in a UK study [82].

Another potential bias in our dataset is the relatively large proportion of males in our sample of road-killed individuals. Unfortunately, we are not aware of the spatial distribution of males versus females, so this pattern could be caused by there being more males in the sampled populations. Alternatively, it is possible that males are more prone to taking risks such as crossing roads [83]. Male hedgehogs have larger home ranges than females [84–86], and their home ranges appear to be larger in rural areas compared to urban [33], so by traveling further the males may be more exposed to being killed on roads, especially in landscapes that are heavily fragmented by roads. This line of thinking supports the results from previous studies on hedgehog road casualties, where a higher proportion of males were recorded [87–89], and is consistent with our own dataset.

The tendency in our dataset for rural areas to have a higher proportion of road-kill deaths is interesting. There are numerous potential explanations for this observation. It could for example be due to a sampling bias whereby rural cadavers may have remained intact on rural roads that have lighter traffic [90], leaving them easier to collect. Alternatively, our result could be explained by differences in road crossing behaviour between urban and rural areas. This idea is supported by previous studies, which have shown that hedgehogs tend to cross smaller and less busy roads more frequently, leading to a higher proportion of hedgehogs being killed in traffic on the smaller rural roads [29,88,89,91–93]. Kent, Schwartz and Perkins [82] also found that hedgehog road-kill risk was reduced at high urbanity levels. Other factors, such as speed of traffic, type and width of the road, and density of scavenger species that could move the carcass before collection, may all have influenced the results in terms of the distribution of road-killed individuals between urban and rural habitats. Slater [94] documented that simple counts of wildlife corpses found on roads are severe underestimates of the actual road casualty rate, with death rates up to 12–16 times that observed by simply counting corpses, due to the removal of carcasses by scavengers. The removal rates depend on different factors such as species of scavengers, road structure, road traffic, season, time of day, and weather conditions. In a study on the influence of scavengers on road-kill data, Ratton, et al. [95] found that 89% of carcasses were removed by scavengers in the first 24 h and 66% were removed within 12 h. This could potentially have caused a sampling bias in our dataset, as scavengers increasingly tend to inhabit urban areas [96]. Other factors which may have influenced the sampling of the present dataset through citizen science could be the number of people that would see the hedgehog carcass and the probability that they would collect the carcass, which is contingent upon the ease of collection related to road type, safety, and the state of the carcass after being run over. Lastly, several factors intrinsic to the hedgehog population could also have influenced the sampling, such as population density, population structure (age/sex) and behaviour including avoidance and acclimatisation to traffic. When considering all these different factors, it would be possible to invent many scenarios that would produce the observed results, even if there were no differences in individual risk between rural and urban settings.

It is a general challenge to properly interpret data from sample sets of wildlife casualties, as there will always be biases connected to the collection of samples, not only for road-kill data. Another example could be bycatch and stranding data of marine mammals because one must ask whether the sample is representative of the wild population in general and whether there is a tendency for individuals of certain categories of, e.g., age, sex or even degree of boldness to become stranded. Regardless of the limitations mentioned, the conclusion must be that it is still preferable to work with the available data compared to having no research at all, and to interpret these datasets with caution, being aware of and articulating the potential caveats when presenting the results.

4.4. The Effects of Genetic Heterozygosity on Longevity

From a conservation perspective, it was a positive discovery that the degree of inbreeding did not seem to influence the longevity of European hedgehogs, because previous research has found that Danish hedgehogs have low genetic heterozygosity [31,32], which is indicative of a high degree of inbreeding. A high degree of inbreeding could lead to inbreeding depression including hereditary, and potentially lethal, health conditions in the hedgehogs [97].

4.5. Exploring the Association between Age-at-Death and Degree of Inbreeding, Cause of Death (Traffic/Non-Traffic), and the Interaction between Them

We found a tendency for age-at-death to be a little higher in hedgehogs killed in traffic compared to hedgehogs dying from other causes (in-care, natural causes in the wild), but the results were not significant. This difference could be influenced by the large number of orphaned juvenile hedgehogs dying in care in the sample set, constituting 59% (51/86) of the individuals in the category "dying in care". Furthermore, males in the study were more likely to have died in traffic and showed a higher age-at-death than females, which could also have influenced the results.

4.6. Life Table Analyses and the Life Expectancy of Females and Males

Our life table analyses showed that life expectancy at age 0 (e_x) was 2.6 years for males and 2.1 years for females. During the manual calculation of periosteal growth lines, we also found that all individuals >6 years were males. In the age determination study, the mean age was 1.8 years, which is lower than the results of the life table analyses. This is likely caused by the inclusion of a large number of individuals of unknown sex in the age determination study (27/109 individuals aged 0 years and 40/115 individuals aged 1 year out of a total sample size of 388 individuals) compared to the life table analyses only including individuals of known sex.

Our study revealed that male hedgehogs are more prone to death in traffic than females, especially during the mating season, but they also appear to live longer than females. This is surprising given that road traffic is thought to be a major cause of hedgehog deaths [25,87,98]. The mean age-at-death of road-killed individuals was higher (2.1 years) than that for the categories of non-traffic deaths (2.0) and individuals dying in care (1.3). Therefore, it is not surprising that males (87, 69.6%), which were more likely to be road-killed than females (38, 30.4%), also had a higher age-at-death than females. A large proportion of the hedgehogs dying in care are orphaned hoglets, which likely explains the lower age-at-death for this category of individuals.

Previous research on the life span of European hedgehogs has had mixed findings, with some studies of relatively balanced, or equal, male–female ratios showing that females have a longer life span [50–52,54] and others concluding that males live longer [49,53,55,61]. One possible explanation for the difference in life expectancy between male and female hedgehogs in the present study could be the larger sample size of males compared to females, which may have introduced a bias by increasing the likelihood of longer-lived individuals in the larger sample of males.

However, with these potential biases set aside, the finding that males live longer than females is surprising, because females of other vertebrate species, including humans, tend to live longer than males [99,100]. Clutton-Brock and Isvaran [100] found that reduced longevity in adult males compared to females is common in polygynous vertebrate species due to an earlier onset and more rapid progression of senescence in males, with a possible explanation for this being the intense intra-sexual competition for breeding opportunities between males in polygynous species. This is supported by Trivers [101], suggesting that the higher mortality of males in polygynous mammals is driven by the larger potential reproductive benefits of winning competitive encounters for males than females, and that the behavioural traits which are thought to enhance competitive success usually trade off against survival. According to Clutton-Brock and Isvaran [100], the magnitude of sex differences in life expectancy is determined by the degree of sex differences in the duration of effective breeding (DEB), defined as the period over which individuals are likely to breed successfully, causing the sex differences in aging and adult longevity to be more pronounced in polygynous species. However, DEB does not seem to be lower in male European hedgehogs compared to females. Furthermore, European hedgehogs are polygynandrous as both males and females mate with several partners [10,36], which means that the males are not territorial and have no harems to protect, decreasing the intra-sexual competition compared to polygynous species. The male hedgehogs do occasionally fight each other for the right to mate with a female, which can lead to injuries and other fitness costs. However, at the same time, the hedgehogs are solitary, and the males do not take part in the rearing of offspring [10,36], which could potentially give the males an important fitness advantage compared to the females. In badgers, *Meles meles*, another polygynandrous mammal species, which are intraguild predators of, and share habitats with, European hedgehogs [26], males show higher body mass senescence rates compared to females [102]. The authors found that this sex difference was largely caused by the negative fitness effect of the intensity of intra-sexual competition in males, especially that experienced during early adulthood. Male European badgers explore mating opportunities both within and among social groups, which leads to a high rate of extra-group paternity [103] with a higher incidence of bite-wounding and mortality among males, suggesting that the intra-sexual reproductive competition is more intense among males than females [104]. Macdonald et al. [105] detected no significant sex-bias in badger mortality rate, and Sugianto, et al. [106] found that badgers showed similar senescence schedules (somatic and hormonal) between the sexes, in the same study population. This difference in longevity among sexes compared to hedgehogs could therefore be explained by the fact that male badgers experience higher levels of intra-sexual competition, often live in complex social groups, and appear to take part in the rearing of the young, with some female individuals even showing alloparental behaviour [107]. These contrasting factors compared to hedgehogs highlight our previously mentioned possible explanations for the male-biased difference in longevity in hedgehogs.

4.7. Life Expectancy, Risks, and the Effects on Breeding

Our counting of periosteal growth lines and the life table analyses showed slightly different results. Our analysis using periosteal growth lines showed a mean life expectancy of 1.8 years overall (including individuals of unknown sex), with 1.6 years for females and 2.1 years for males. Discovering that the hedgehogs, on average, reach an approximate age of two years in Denmark is positive in the sense that these individuals will have the opportunity to take part in two mating and breeding seasons during their life time, since most individuals will become sexually mature during their second summer (aged one year) [10,36]. This is an important insight into the population dynamics of hedgehogs, as this will in theory positively influence the population growth rate and hence the survival of the population.

We saw a high proportion of individuals dying at the age of one year (n=115/383, 30%). Fortunately, our data also showed that if the individuals survived this life stage, they could potentially live to become 16 years old and contribute to the production of offspring for several years, compared to individuals dying at an earlier age. This hurdle of surviving over the age of two could be tied with risks that are particularly high for young hedgehogs. Recently published research based on the same sample set showed that the endoparasite occurrence in hedgehogs aged one year was significantly higher (n = 52/58, 90%) compared to that of hedgehogs of two years of age (n = 20/31, 65%) [108]. It may also be the case that individuals gradually gain more experience as they grow older. If the individuals have managed to survive to reach the age of two years or more, they would have likely learned to avoid dangers such as cars and predators and have established good nests and foraging routines throughout their home range [109].

5. Conclusions

Analysing this comprehensive dataset of the Danish population of European hedgehogs, we found that the mean age-at-death of all individuals was 1.8 years, 1.6 years for females and 2.1 years for males, through the counting of periosteal growth lines. The age-at-death ranged from 0 to 16 years, and the hedgehog reaching 16 years is currently the oldest scientifically documented individual of the species. Our actuarial life tables showed that life expectancy at age 0 (e_x) was 2.6 years for males and 2.1 years for females. The results of both methods indicate that male hedgehogs live longer than females in terms of maximum recorded life span, which stands in contrast to most other mammal species. The fact that an average hedgehog reaches an approximate age of two years in Denmark means that most individuals will have the opportunity to take part in at least two breeding seasons during their lifetime.

Future work could use our results, alongside additional data on age-specific reproductive output data, to construct matrix population models that could be used for more robust modelling of European hedgehog populations. These would indicate whether or not the populations of hedgehogs are in decline, enabling an evaluation of the conservation status of the European hedgehog. However, in spite of the great concern for the populations in more recent times, we do not yet have sufficient demographic data to adequately determine the demographic health of the population through a population matrix model. We would therefore like to address a need for representative demographic data on age/stage-specific survival and fecundity to be able to assess the current demographic trends of European hedgehogs.

In conclusion, the various findings of this study have improved our understanding of the basic life history of hedgehogs in Denmark. This information, and in particular the mortality trajectories of males and females, will eventually generate improved modelling of population dynamics to inform the important conservation management for this declining species.

Author Contributions: S.L.R. conceived and led the project under the supervision of O.R.J. and T.B.B.; S.L.R. conducted the age determination lab work under the supervision of H.J.M.; H.J.M. and S.L.R. provided high-resolution images of the analysed jaw bones; O.R.J. and T.B.B. analysed the data with input from S.L.R.; S.L.R. wrote the manuscript with input from all other authors. All authors have read and agreed to the published version of the manuscript.

Funding: This research was funded by Beckett-Fonden, grant number 44200; Ingeniør K. A. Rohde og hustrus Legat; Fonden til Værn for Værgeløse Dyr; Iwan Kliem-Larsens Mindelegat; Svalens Fond and Carlsberg Foundation, grant number CF20_0443.

Institutional Review Board Statement: Ethical review and approval were waived for this study as the hedgehogs used for the research were already dead when included in the study (dying in the wild or in care at a hedgehog rehabilitation centre).

Informed Consent Statement: Not applicable.

Data Availability Statement: The dataset is available from Zenodo (https://doi.org/10.5281/zenodo.7615782).

Acknowledgments: The authors thank Rien van Wijk and Camilla Birch for performing the necropsies; volunteers Ida Brund, Terese Egebjerg, Louise Marie Jønsson, Emilie Sandberg Lund, Mathilde Skovborg, and Kamma Westenholz for performing bone macerations; Rien van Wijk, Terese Egebjerg, and Louise Marie Jønsson for assisting with the age determinations; Natural History Museum of Denmark for providing facilities for the necropsies and bone macerations; Daniel Klingenberg Johansen and Mikkel Høegh Post for sharing their expert knowledge on bone maceration and their assistance with practicalities during the work; and lab technician Catherine Nielsen and Center for Advanced Bioimaging, University of Copenhagen, for help with microscopy work.

Conflicts of Interest: The authors declare no conflict of interest.

References

1. Roff, D.A. *Life History Evolution*; Sinauer Associates: Sunderland, MA, USA, 2002.
2. Roff, D. *Evolution of Life Histories: Theory and Analysis*; Chapman and Hall: New York, NY, USA, 1992.
3. Reznick, D.A.; Bryga, H.; Endler, J.A. Experimentally induced life-history evolution in a natural population. *Nature* **1990**, *34*, 357–359. [CrossRef]
4. Neal, D. *Introduction to Population Biology*; Cambridge University Press: Cambridge, UK, 2003.
5. Spencer, W.; Rustigian-Romsos, H.; Strittholt, J.; Scheller, R.; Zielinski, W.; Truex, R. Using occupancy and population models t assess habitat conservation opportunities for an isolated carnivore population. *Biol. Conserv.* **2011**, *144*, 788–803. [CrossRef]
6. Tempel, D.J.; Peery, M.; Gutierrez, R.J. Using integrated population models to improve conservation monitoring: Californi spotted owls as a case study. *Ecol. Model.* **2014**, *289*, 86–95. [CrossRef]
7. Sierra Forest Legacy. Pacific fisher (Pekania pennanti). Available online: https://www.sierraforestlegacy.org/FC SierraNevadaWildlifeRisk/PacificFisher.php (accessed on 10 January 2023).
8. Sierra Forest Legacy. California Spotted Owl (Strix occidentalis occidentalis). Available online: https://www.sierraforestlegac org/FC_SierraNevadaWildlifeRisk/CaliforniaSpottedOwl.php (accessed on 10 January 2023).
9. Morris, P. *Hedgehogs*; Whittet Books Ltd.: Stansted, Essex, UK, 2014.
10. Reeve, N. *Hedgehogs*; Poyser: London, UK, 1994.
11. Müller, F. Langzeit-Monitoring der Strassenverkehrsopfer beim Igel (*Erinaceus europaeus* L.) zur Indikation von Population dichteveränderungen entlang zweier Teststrecken im Landkreis Fulda. *Beiträge Nat. Osthess.* **2018**, *54*, 21–26.
12. SoBH. *The state of Britain's Hedgehogs 2011*; British Trust for Ornithology (BTO) commissioned by People's Trust for Endangere Species (PTES) and the British Hedgehog Preservation Society (BHPS): London, UK, 2011.
13. SoBH. *The State of Britain's Hedgehogs 2015*; British Hedgehog Preservation Society and People's Trust for Endangered Specie London, UK, 2015.
14. SoBH. *The State of Britain's Hedgehogs 2018*; British Hedgehog Preservation Society and People's Trust for Endangered Specie London, UK, 2018.
15. Hof, A.R.; Bright, P.W. Quantifying the long-term decline of the West European hedgehog in England by subsampling citizer science datasets. *Eur. J. Wildl. Res.* **2016**, *62*, 407–413. [CrossRef]
16. Krange, M. Change in the occurrence of the West European Hedgehog (Erinaceus europaeus) in western Sweden during 1950–201 Ph.D. Thesis, Karlstad University, Karlstad, Sweden, 2015.
17. van de Poel, J.L.; Dekker, J.; van Langevelde, F. Dutch hedgehogs *Erinaceus europaeus* are nowadays mainly found in urban area possibly due to the negative effects of badgers Meles meles. *Wildl. Biol.* **2015**, *21*, 51–55. [CrossRef]
18. Williams, B.M.; Baker, P.J.; Thomas, E.; Wilson, G.; Judge, J.; Yarnell, R.W. Reduced occupancy of hedgehogs (*Erinaceus europaeu* in rural England and Wales: The influence of habitat and an asymmetric intra-guild predator. *Sci. Rep.* **2018**, *8*, 12156. [CrossRe
19. Taucher, A.L.; Gloor, S.; Dietrich, A.; Geiger, M.; Hegglin, D.; Bontadina, F. Decline in Distribution and Abundance: Urba Hedgehogs under Pressure. *Animals* **2020**, *10*, 1606. [CrossRef]
20. Wembridge, D.; Johnson, G.; Al-Fulaij, N.; Langton, S. *The State of Britain's Hedgehogs 2022*; British Hedgehog Preservation Societ and People's Trust for Endangered Species: London, UK, 2022.
21. Mathews, F.; Harrower, C. *IUCN—Compliant Red List for Britain's Terrestrial Mammals. Assessment by the Mammal Society unde contract to Natural England, Natural Resources Wales and Scottish Natural Heritage*; Natural England: Peterborough, UK.
22. Brakes, C.R.; Smith, R.H. Exposure of non-target small mammals to rodenticides: Short-term effects, recovery and implication for secondary poisoning. *J. Appl. Ecol.* **2005**, *42*, 118–128. [CrossRef]
23. Haigh, A.; O'Riordan, R.M.; Butler, F. Nesting behaviour and seasonal body mass changes in a rural Irish population of th Western hedgehog (*Erinaceus europaeus*). *Acta Theriol.* **2012**, *57*, 321–331. [CrossRef]
24. Hof, A.R.; Bright, P.W. The value of agri-environment schemes for macro-invertebrate feeders: Hedgehogs on arable farms i Britain. *Anim. Conserv.* **2010**, *13*, 467–473. [CrossRef]
25. Huijser, M.P.; Bergers, P.J.M. The effect of roads and traffic on hedgehog (*Erinaceus europaeus*) populations. *Biol. Conserv.* **2000**, *9* 111–116. [CrossRef]
26. Young, R.P.; Davison, J.; Trewby, I.D.; Wilson, G.J.; Delahay, R.J.; Doncaster, C.P. Abundance of hedgehogs (*Erinaceus europaeus*) i relation to the density and distribution of badgers (*Meles meles*). *J. Zool.* **2006**, *269*, 349–356. [CrossRef]
27. Hubert, P.; Julliard, R.; Biagianti, S.; Poulle, M.-L. Ecological factors driving the higher hedgehog (*Erinaceus europeaus*) density i an urban area compared to the adjacent rural area. *Landsc. Urban Plan.* **2011**, *103*, 34–43. [CrossRef]
28. Pettett, C.E.; Moorhouse, T.P.; Johnson, P.J.; Macdonald, D.W. Factors affecting hedgehog (Erinaceus europaeus) attraction to rur villages in arable landscapes. *Eur. J. Wildl. Res.* **2017**, *63*, 54. [CrossRef]
29. Dowding, C.V.; Harris, S.; Poulton, S.; Baker, P.J. Nocturnal ranging behaviour of urban hedgehogs, *Erinaceus europaeus*, in relatio to risk and reward. *Anim. Behav.* **2010**, *80*, 13–21. [CrossRef]
30. Dowding, C.V.; Shore, R.F.; Worgan, A.; Baker, P.J.; Harris, S. Accumulation of anticoagulant rodenticides in a non-targe insectivore, the European hedgehog (*Erinaceus europaeus*). *Environ. Pollut.* **2010**, *158*, 161–166. [CrossRef]
31. Rasmussen, S.L.; Nielsen, J.L.; Jones, O.R.; Berg, T.B.; Pertoldi, C. Genetic structure of the European hedgehog (*Erinaceus europaeu* in Denmark. *PLoS ONE* **2020**, *15*, e0227205. [CrossRef]

Rasmussen, S.L.; Yashiro, E.; Sverrisdóttir, E.; Nielsen, K.L.; Lukassen, M.B.; Nielsen, J.L.; Asp, T.; Pertoldi, C. Applying the GBS Technique for the Genomic Characterization of a Danish Population of European Hedgehogs (*Erinaceus europaeus*). *Genet. Biodivers. (GABJ)* **2019**, *3*, 78–86. [CrossRef]

Rasmussen, S.L.; Berg, T.B.; Dabelsteen, T.; Jones, O.R. The ecology of suburban juvenile European hedgehogs (*Erinaceus europaeus*) in Denmark. *Ecol. Evol.* **2019**, *9*, 13174–13187. [CrossRef] [PubMed]

Burroughes, N.D.; Dowler, J.; Burroughes, G. Admission and Survival Trends in Hedgehogs Admitted to RSPCA Wildlife Rehabilitation Centres. *Proc. Zool. Soc.* **2021**, *74*, 198–204. [CrossRef]

Lukešová, G.; Voslarova, E.; Vecerek, V.; Vucinic, M. Trends in intake and outcomes for European hedgehog (*Erinaceus europaeus*) in the Czech rescue centers. *PLoS ONE* **2021**, *16*, e0248422. [CrossRef] [PubMed]

Morris, P. *Hedgehog*; Collins: London, UK, 2018; pp. 1–404.

Rasmussen, S.L.; Kalliokoski, O.; Dabelsteen, T.; Abelson, K. An exploratory investigation of glucocorticoids, personality and survival rates in wild and rehabilitated hedgehogs (*Erinaceus europaeus*) in Denmark. *BMC Ecol. Evol.* **2021**, *21*, 96. [CrossRef]

Jensen, A.B. Overwintering of European hedgehogs *Erinaceus europaeus* in a Danish rural area. *Acta Theriol.* **2004**, *49*, 145–155. [CrossRef]

Park, E.A. The imprinting of nutritional disturbances on the growing bone. *Pediatrics* **1964**, *33*, 815–862. [CrossRef]

Peabody, F.E. Annual growth zones in living and fossil vertebrates. *J. Morphol.* **1961**, *108*, 11. [CrossRef]

Morris, P.A. A method for determining absolute age in the hedgehog. *J. Zool.* **1970**, *161*, 277–281. [CrossRef]

Reeve, N.J.; Love, B.; Shore, R. X-ray measures of long bones as an aid to age-determination in hedgehogs (*Erinaceus europaeus*). In Proceedings of the 1st European Hedgehog Research Group Meeting, Arendal, Norway; 1996.

Ohtaishi, N.; Hachiya, N.; Shibata, Y. Age determination of the hare from annual layers in the mandibular bone. *Acta Theriol.* **1976**, *21*, 168–171. [CrossRef]

Henderson, B.A.; Bowen, H.M. Estimating the age of the European rabbit, *Oryctolagus Cuniculus*, by counting the adhesion lines in the periosteal zone of the lower mandible. *J. Appl. Ecol.* **1979**, *16*, 393–396. [CrossRef]

Castanet, J.; Croci, S.; Aujard, F.; Perret, M.; Cubo, J.; de Margerie, E. Lines of arrested growth in bone and age estimation in a small primate: *Microcebus murinus*. *J. Zool.* **2004**, *263*, 31–39. [CrossRef]

Sander, P.M.; Andrássy, P. Lines of arrested growth and long bone histology in Pleistocene large mammals from Germany: What do they tell us about dinosaur physiology? *Palaeontogr. Abt. A* **2006**, *277*, 143–159. [CrossRef]

Brockie, R.E. Ecology of the hedgehog *Erinaceus europaeus* in New Zealand. Master's Thesis, Victoria University of Wellington, Wellington, New Zealand, 1958.

Dickman, C.R. Age-related dietary change in the European hedgehog, *Erinaceus europaeus*. *J. Zool.* **1988**, *215*, 1–14. [CrossRef]

Döpke, C. Kasuistische Auswertung der Untersuchungen von Igeln (*Erinaceus europaeus*) im Einsendungsmaterial des Instituts für Pathologie von 1980 bis 2001. Ph.D. Thesis, Tierärztliche Hochschule Hannover, Hannover, Germany, 2002.

Haigh, A.; Kelly, M.; Butler, F.; O'Riordan, R.M. Non-invasive methods of separating hedgehog (*Erinaceus europaeus*) age classes and an investigation into the age structure of road kill. *Acta Theriol.* **2014**, *59*, 165–171. [CrossRef]

Heddergott, M. Zur Altersstruktur des Igels *Erinaceus europaeus* (L., 1758) in Thüringen anhand von Schädeln (Mammalia: Insectivora, Erinaceidae). *Veröffentlichungen Naturhist. Mus. Schleus.* **2003**, *18*, 79–82.

Heddergott, M. Age-structure of hedgehog *Erinaceus europaeus* (L., 1758) in the Harz Mountains. *Abh. Und Ber. Aus Dem Mus. Heine.* **2004**, *6*, 125–130.

Heddergott, M. Age structure of the hedgehog *Erinaceus europaeus* L., 1758, in the district Saalfeld-Rudolstadt (Thuringia). *Hercynia* **2005**, *38*, 113–118.

Heddergott, M.; Müller, F. Zur Altersstruktur zweier Populationen des Braunbrustigels *Erinaceus europaeus* L., 1758 (Mammalia: Insectivora) in Osthessen. *Beiträge Zur Nat. Osthess.* **2008**, *45*, 63–71.

Heddergott, M.; Steinbach, O.; Heddergott, C. Zur Altersstruktur des Braunbrustigels *Erinaceus europaeus* (Linnaeus, 1758) im Stadtgebiet Heilbad Heiligenstadt (Thüringen) (Mammalia: Insectivora, Erinaceidae)*. *Mauritiana* **2010**, *21*, 231–239.

Heyne, P. Beitrag zur Altersstruktur von Erinaceus europaeus L., 1758. Populationsökologie von Kleinsäugerarten, Martin-Luther-Universitaet Halle-Wittenberg Wissenschaftliche Beitrage. *SLUB* **1990**, *1990*, 49–54.

Kratochvil, J. Hedgehogs of the genus *Erinaceus* in Czechoslovakia insectivora mammalia. *Zool. Listy* **1975**, *24*, 297–312.

Kristoffersson, R. A note on the age distribution of hedgehogs *Erinaceus europaeus* in Finland. *Ann. Zool. Fenn.* **1971**, *8*, 554–557.

Morris, P.A. Some Aspects of the Ecology of the Hedgehog (*Erinaceus europaeus*). University of London: London, UK, 1969.

Neuschulz, N.; Schubert, M. Altersermittlung bei *Erinaceus europaeus* L., 1758 an einer Igelfütterung. *Popul. Von Kleinsäugerarten Wiss. Beitr. Univ. Halle* **1990**, *34*, 42.

Rautio, A.; Kunnasranta, M.; Valtonen, A.; Ikonen, M.; Hyvarinen, H.; Holopainen, I.J.; Kukkonen, J.V.K. Sex, Age, and Tissue Specific Accumulation of Eight Metals, Arsenic, and Selenium in the European Hedgehog (*Erinaceus europaeus*). *Arch. Environ. Contam. Toxicol.* **2010**, *59*, 642–651. [CrossRef]

Skoudlin, J. On determining the age in erinaceus-europaeus and erinaceus-concolor insectivora erinaceidae. *Vestn. Ceskoslovenske Spol. Zool.* **1976**, *40*, 300–306.

Skoudlin, J. Age structure of czechoslovak populations of Erinaceus-Europaeus and Erinaceus-Concolor Insectivora Erinaceidae. *Vestn. Ceskoslovenske Spol. Zool.* **1981**, *45*, 307–313.

64. Brockie, R.E. *Studies of the Hedgehog Erinaceus europaeus L. in New Zealand*; Victoria University of Wellington: Wellington, New Zealand, 1974.
65. Parkes, J. Some aspects of the biology of the hedgehog erinaceus-europaeus in the manawatu new-zealand. *New Zealand J. Zo.* **1975**, *2*, 463–472. [CrossRef]
66. Petrides, G.A. The determination of sex and age ratios in the cottontail rabbit. *Am. Midl. Nat.* **1951**, *46*, 312–336. [CrossRef]
67. Morris, P.A. Epiphyseal fusion in the forefoot as a means of age determination in the hedgehog (*Erinaceus europaeus*) *J. Zool.* **197** *164*, 254–259. [CrossRef]
68. Scott-Hayward, L. Data on age determination (pers. comm. p. 224). In *Hedgehogs*; Morris, P.A., Ed.; Collins: London, UK, 201
69. R Core Team. *A Language and Environment for Statistical Computing*; R Foundation for Statistical Computing: Vienna, Austria, 20.
70. Hijmans, R.J.; raster: Geographic Data Analysis and Modeling. R package. Available online: https://cran.r-project.org/wel packages/raster/index.html (accessed on 2 September 2022).
71. Rasmussen, S.L.; Larsen, J.; van Wijk, R.E.; Jones, O.R.; Berg, T.B.; Angen, O.; Larsen, A.R. European hedgehogs (*Erinace. europaeus*) as a natural reservoir of methicillin-resistant Staphylococcus aureus carrying mecC in Denmark. *PLoS ONE* **2019**, *1* e0222031. [CrossRef]
72. Samuel Preston, P.H.; Guillot, M. *Demography: Measuring and Modeling Population Processes*; Wiley-Blackwell: Hoboken, NJ, USA, 202
73. Jones, O.R. *Life tables: Construction and interpretation In Demographic Methods Across the Tree of Life*; Oxford University Press: Oxfo. UK, 2021.
74. Kulus, M.J.; Dąbrowski, P. How to calculate the age at formation of Harris lines? A step-by-step review of current methods and proposal for modifications to Byers' formulas. *Archaeol. Anthropol. Sci.* **2019**, *11*, 1169–1185. [CrossRef]
75. Foster, N.J.; Maloney, R.F.; Recio, M.R.; Seddon, P.J.; van Heezik, Y. European hedgehogs rear young and enter hibernation New Zealand's alpine zones. *New Zealand J. Ecol.* **2021**, *45*, 1–6. [CrossRef]
76. Walhovd, H. Winter activity of Danish hedgehogs in 1973-74 with information on the size of the animals observed and location the recordings. *Flora Og Fauna* **1976**, *82*, 35–42.
77. Walhovd, H. The overwintering pattern of Danish hedgehogs in outdoor confinement, during three successive winters. *Nat. J.* **1978**, *20*, 273–284.
78. Hansen, L. Trafikdræbte dyr i Danmark. *Dan. Ornitol. Foren. Tidsskr.* **1982**, *76*, 96–110.
79. Holsbeek, L.; Rodts, J.; Muyldermans, S. Hedgehog and other animal traffic victims in Belgium: Results of a countrywide surv. *Lutra* **1999**, *42*, 111–119.
80. Schwartz, A.L.; Shilling, F.M.; Perkins, S.E. The value of monitoring wildlife roadkill. *Eur. J. Wildl. Res.* **2020**, *66*, 1–12. [CrossRe
81. Wembridge, D.; Newman, M.R.; Bright, P.; Morris, P. *An estimate of the annual number of hedgehog (Erinaceus europaeus) road casualt. in Great Britain*; Mammal Society: London, UK, 2016.
82. Kent, E.; Schwartz, A.L.; Perkins, S.E. Life in the fast lane: Roadkill risk along an urban–rural gradient. *J. Urban Ecol.* **2021**, juaa039. [CrossRef]
83. Raynaud, J.; Schradin, C. Experimental increase of testosterone increases boldness and decreases anxiety in male African strip. mouse helpers. *Physiol. Behav.* **2014**, *129*, 57–63. [CrossRef]
84. Morris, P.A. A study of home range and movements in the hedgehog (*Erinaceus europaeus*). *J. Zool.* **1988**, *214*, 433–449. [CrossRe
85. Reeve, N.J. The home range of the hedgehog as revealed by a radio tracking study. *Symp. Zool. Soc. Lond.* **1982**, *49*, 207–230.
86. Kristiansson, H. Ecology of a hedgehog (*Erinaceus europaeus*) population in Southern Sweden. PhD Thesis, Department of Anim Ecology, University of Lund, Lund, Sweden, 1984.
87. Haigh, A.; O'Riordan, R.M.; Butler, F. Hedgehog *Erinaceus europaeus* mortality on Irish roads. *Wildl. Biol.* **2014**, *20*, 155–16 [CrossRef]
88. Göransson, G.; Karlsson, J.; Lindgren, A. Road mortality of the hedgehog *Erinaceus europaeus* in southern Sweden (Igelkotten o. biltrafiken). *Fauna Och Flora* **1976**, *71*, 1–6.
89. Huijser, M.P. *Life on the Edge: Hedgehog Traffic Victims and Mitigation Strategies in an Anthropogenic Landscape*; Wageningen Universi and Research: Wageningen, The Netherlands, 2000.
90. Vejdirektoratet. Trafik på strækninger 2021. Available online: http://vej08.vd.dk/stroemkort/nytui/kort/Stroemkort.html?ic 201 (accessed on 10 January 2023).
91. Rondinini, C.; Doncaster, C.P. Roads as barriers to movement for hedgehogs. *Funct. Ecol.* **2002**, *16*, 504–509. [CrossRef]
92. Brockie, R.E.; Sadleir, R.M.; Linklater, W.L. Long-term wildlife road-kill counts in New Zealand. *New Zealand J. Zool.* **2009**, *.* 123–134. [CrossRef]
93. Moore, L.J.; Petrovan, S.O.; Baker, P.J.; Bates, A.J.; Hicks, H.L.; Perkins, S.E.; Yarnell, R.W. Impacts and potential mitigation of roa mortality for hedgehogs in Europe. *Animals* **2020**, *10*, 1523. [CrossRef]
94. Slater, F.M. An assessment of wildlife road casulties—the potential discrepancy between numbers counted and numbers kille. *Web Ecol.* **2002**, *3*, 33–42. [CrossRef]
95. Ratton, P.; Secco, H.; Da Rosa, C.A. Carcass permanency time and its implications to the roadkill data. *Eur. J. Wildl. Res.* **2014**, *.* 543–546. [CrossRef]
96. Schwartz, A.L.W.; Williams, H.F.; Chadwick, E.; Thomas, R.J.; Perkins, S.E. Roadkill scavenging behaviour in an urban envirc ment. *J. Urban Ecol.* **2018**, *4*. [CrossRef]
97. Keller, L.F.; Waller, D.M. Inbreeding effects in wild populations. *Trends Ecol. Evol.* **2002**, *17*, 230–241. [CrossRef]

8.   Wright, P.G. , Coomber, F. G., Bellamy, C. C., Perkins, S. E., Mathews, F. Predicting hedgehog mortality risks on British roads using habitat suitability modelling. *PeerJ* **2020**, *7*, e8154. [CrossRef]

9.   Austad, S.N. Why women live longer than men: Sex differences in longevity. *Gend. Med.* **2006**, *3*, 79–92. [CrossRef] [PubMed]

30.   Clutton-Brock, T.H.; Isvaran, K. Sex Differences in Ageing in Natural Populations of Vertebrates. *Proc. Biol. Sci.* **2007**, *274*, 3097–3104. [CrossRef] [PubMed]

31.   Trivers, R.L. Parental investment and sexual selection. In *Sexual Selection and the Descent of Man*; Campbell, B., Ed.; Aldine-Atherton: Chicago, IL, USA, 1972; pp. 136–179.

32.   Beirne, C.; Delahay, R.; Young, A. Sex differences in senescence: The role of intra-sexual competition in early adulthood. *Proc. R. Soc. B Biol. Sci.* **2015**, *282*, 20151086. [CrossRef]

33.   Dugdale, H.L.; Macdonald, D.W.; Pope, L.C.; Burke, T. Polygynandry, extra-group paternity and multiple-paternity litters in European badger (*Meles meles*) social groups. *Mol. Ecol.* **2007**, *16*, 5294–5306. [CrossRef]

34.   Delahay, R.; Walker, N.; Forrester, G.; Harmsen, B.; Riordan, P.; Macdonald, D.; Newman, C.; Cheeseman, C. Demographic correlates of bite wounding in Eurasian badgers, *Meles meles* L., in stable and perturbed populations. *Anim. Behav.* **2006**, *71*, 1047–1055. [CrossRef]

35.   Macdonald, D.W.; Newman, C.; Nouvellet, P.M.; Buesching, C.D. An Analysis of Eurasian Badger (*Meles meles*) Population Dynamics: Implications for Regulatory Mechanisms. *J. Mammal.* **2009**, *90*, 1392–1403. [CrossRef]

36.   Sugianto, N.A.; Newman, C.; Macdonald, D.W.; Buesching, C.D. Reproductive and somatic senescence in the European badger (*Meles meles*): Evidence from lifetime sex-steroid profiles. *Zoology* **2020**, *141*, 125803. [CrossRef] [PubMed]

37.   Woodroffe, R. Alloparental behaviour in the European badger. *Anim. Behav.* **1993**, *46*, 413–415. [CrossRef]

38.   Rasmussen, S.L.; Hallig, J.; van Wijk, R.E.; Petersen, H.H. An investigation of endoparasites and the determinants of parasite infection in European hedgehogs (*Erinaceus europaeus*) from Denmark. *Int. J. Parasitol. Parasites Wildl.* **2021**, *16*, 217–227. [CrossRef]

39.   Marshall, H.H.; Vitikainen, E.I.; Mwanguhya, F.; Businge, R.; Kyabulima, S.; Hares, M.C.; Inzani, E.; Kalema-Zikusoka, G.; Mwesige, K.; Nichols, H.J. Lifetime fitness consequences of early-life ecological hardship in a wild mammal population. *Ecol. Evol.* **2017**, *7*, 1712–1724. [CrossRef]

 *animals*

Article

Over-Winter Survival and Nest Site Selection of the West-European Hedgehog (*Erinaceus europaeus*) in Arable Dominated Landscapes

Lucy E. Bearman-Brown [1,*], **Philip J. Baker** [2], **Dawn Scott** [3], **Antonio Uzal** [4], **Luke Evans** [2] **and Richard W. Yarnell** [4]

[1] Department of Animal & Agriculture, Hartpury University, Gloucestershire GL19 3BE, UK
[2] School of Biological Sciences, University of Reading, Reading RG6 6AH, UK; p.j.baker@reading.ac.uk (P.J.B.); L.C.Evans@pgr.reading.ac.uk (L.E.)
[3] School of Life Sciences, Keele University, Staffordshire ST5 5BG, UK; d.scott@keele.ac.uk
[4] School of Animal, Rural & Environmental Sciences, Nottingham Trent University, Southwell, Nottinghamshire NG25 0QF, UK; antonio.uzal@ntu.ac.uk (A.U.); richard.yarnell@ntu.ac.uk (R.W.Y.)
* Correspondence: lucy.bearman-brown@hartpury.ac.uk; Tel.: +44-1452-702465

Received: 22 July 2020; Accepted: 13 August 2020; Published: 19 August 2020

Simple Summary: Hedgehogs (*Erinaceus europaeus*) have declined markedly in the UK in recent decades. One key stage that could affect their population dynamics is the annual winter hibernation period. Therefore, we studied two contrasting populations in England to examine patterns of winter nest use, body mass changes and survival during hibernation. On average, animals at both sites weighed the same prior to, and used the same number of nests, during hibernation. There was a marked difference in survival rates between the two sites, but no animals died during hibernation; all deaths occurred prior to or after the hibernation period, mainly from predation or vehicle collisions. Hedgehogs consistently nested in proximity to some habitats (hedgerows, roads, woodlands) but avoided others (pasture fields); the use of other habitats (arable fields, amenity grassland, buildings) varied between the two sites. These data suggest: (i) that hibernation was not a period of significant mortality at either site for individuals that had attained a sufficient weight (>600 g) in autumn; but that (ii) habitat composition did significantly affect the positioning of winter nests, such that different land management practices (historic and current) could influence hibernation success.

Abstract: The West-European hedgehog (*Erinaceus europaeus*) has declined markedly in the UK. The winter hibernation period may make hedgehogs vulnerable to anthropogenic habitat and climate changes. Therefore, we studied two contrasting populations in England to examine patterns of winter nest use, body mass changes and survival during hibernation. No between-site differences were evident in body mass prior to hibernation nor the number of winter nests used, but significant differences in overwinter mass change and survival were observed. Mass change did not, however, affect survival rates; all deaths occurred prior to or after the hibernation period, mainly from predation or vehicle collisions. Hedgehogs consistently nested in proximity to hedgerows, roads and woodlands, but avoided pasture fields; differences between sites were evident for the selection for or avoidance of arable fields, amenity grassland and buildings. Collectively, these data indicate that hibernation was not a period of significant mortality for individuals that had attained sufficient weight (>600 g) pre-hibernation. Conversely, habitat composition did significantly affect the positioning of winter nests, such that different land management practices (historic and current) might potentially influence hibernation success. The limitations of this study and suggestions for future research are discussed.

Keywords: *Erinaceus europaeus*; farmland; habitat fragmentation; hedgerow; hibernacula; hibernation; mammal; nest

---

144

1. Introduction

Agricultural intensification and climate alteration are two anthropogenic processes that have profound impacts on natural ecological systems [1–7]. The effects arise from a wide range of underlying causal factors including: habitat destruction, fragmentation and degradation [8,9]; the introduction of livestock, diseases and non-native biological control agents [10–14]; the management of wildlife where they conflict with human interests [15–18]; the application of chemical biocides [19]; and changes in the phenology of key biological events [20,21]. Collectively, these factors have led to the decline, extirpation and extinction of large numbers of species [22–27], but also increases in the abundance and geographic range of others (e.g., [28,29]).

One group of species that might be expected to be particularly affected by agricultural practices and changing climatic conditions are hibernators [30–33]. Hibernation has typically evolved to enable species to survive periods of prolonged food shortages by dramatically reducing levels of energy expenditure [34,35]. One consequence of this is that hibernating species tend to have slower reproductive rates [36], potentially increasing their long-term vulnerability to human activities.

The West-European hedgehog (*Erinaceus europaeus*, hereafter 'hedgehog') is a medium-sized (<1.2 kg) insectivorous mammal found from the Iberian Peninsula and Italy northwards into Scandinavia [37]. In Britain, hedgehogs were historically found throughout a broad range of agricultural landscapes [38–41], but rural populations have declined markedly in recent decades [42–44]. Consequently, hedgehogs are now increasingly found within areas of human habitation in this country [45–47] and elsewhere mboxciteB48-animals-891721,B49-animals-891721. Associated with this decline has been a substantial reduction in the availability [50] and quality [51–53] of hedgerows, an important habitat for foraging [54], dispersal [55] and refuge [56], and a substantive increase in the numbers of badgers (*Meles meles*) [57,58], an intra-guild predator [59].

During hibernation, hedgehogs face specific challenges. First, they need to accumulate sufficient fat reserves to survive for a period of many months; in Britain, hedgehogs typically hibernate from October/November to March/April [37], although the exact timing is dependent upon a combination of both temperature and food availability [60]. Second, they need to find enough appropriate building material(s) to construct a hibernaculum that will maintain the environment within the nest at an appropriate temperature; nests are preferentially constructed from the leaves of broadleaved trees [61]. Third, the habitat must be sufficiently diverse that it offers a range of nesting locations in close proximity to one another so that an individual can relocate safely if necessary. In addition, by nesting at ground level, hedgehogs are susceptible to a range of other factors such as flooding, trampling by livestock, and disturbance by e.g., land managers, walkers and domestic dogs (*Canis familiaris*). Finally, changes in temperature patterns throughout winter may cause hedgehogs to rouse from hibernation when natural food availability is limited.

Hibernation success is, therefore, dependent on several factors, all of which may be negatively affected by agricultural intensification and/or climate change. For example: hot dry summers, soil compaction from heavy machinery and the application of pesticides and molluscicides may all reduce food availability prior to hibernation and, therefore, limit the ability of animals to acquire sufficient fat reserves to successfully complete hibernation; habitat loss and degradation may limit the number of suitable sites for hibernacula, meaning that hedgehogs may be forced to use alternative locations/habitats where preferred nesting materials are not available or where the risk of disturbance is greater; and warmer, wetter and/or more variable winters may cause animals to rouse more often and move between nests more frequently thereby depleting fat reserves and increasing susceptibility to some forms of mortality. Ultimately, such effects would be evident as: reductions in body mass before, and increased mass loss during, hibernation; an increase in the number of winter nests used and their placement in the environment; and an increase in over-winter mortality rates. These parameters would be expected to vary between areas undergoing different types of land management practice, and potentially between sexes (e.g., females may enter hibernation in poorer condition because of the energetic burden of rearing offspring, whilst males may finish hibernating earlier so that they can put on weight before the mating season).

Given the wide range of ways in which human activities could affect this phase, hibernation could represent a key critical period in the dynamics of hedgehog populations [62,63]. Despite its potential importance, little research has been conducted on the hibernation behaviour of hedgehogs in Britain in the last 40 years [37,64]. Therefore, in this study, we radio-tracked hedgehogs at one arable-dominated and one pasture-dominated site in England over the hibernation period to quantify differences in: (i) the number of winter nest sites used; (ii) patterns of habitat selection for nests; (iii) over-winter survival rates; and (iv) over-winter changes in body mass.

2. Materials and Methods

Data were collected from: (1) the Brackenhurst Campus (332 ha) of Nottingham Trent University, Nottinghamshire, UK (National Grid reference: SK695523); and (2) Hartpury University and College campus (339 ha), Gloucestershire, UK (National Grid reference: SO785237). Both sites were mixed commercial farms alongside a university campus, managed under the Entry level Environmental Stewardship Scheme [65]. Brackenhurst is dominated by arable fields (68.7%), with pasture fields, amenity grassland and woodland covering 24.4%, 1.9% and 2.7% of total land area, respectively. In contrast, Hartpury is dominated by pasture (34.8%) and amenity grassland (16.8%), with higher woodland (8.0%) and lower arable (30.8%) coverage than at Brackenhurst. Hedgerow length at each site is 27.1 km (Brackenhurst) and 16.9 km (Hartpury). Badgers were present at both locations: based on the numbers of setts at each site, and the frequency with which they have been photographed on motion activated trail cameras, badger density was considered comparable between the two locations. Hedgehog densities estimated in 2017 using two different methods (random encounter model based on data from trail cameras; spatial capture-recapture based on the capture history of animals along standardized transect routes) were 5.6–9.4 km^{-2} at Brackenhurst and 4.3–12.5 km^{-2} at Hartpury [66].

Fieldwork was conducted from August 2015–May 2016 and August 2016–May 2017, inclusive. Hedgehogs were captured by hand at night under licence from Natural England (ref: 20130866-0-0-0-3) using a 1-million candlepower spotlight to systematically search arable fields, pasture fields and areas of amenity grassland. Sites were surveyed at least twice per week during August and September. Once captured, animals were sexed, given a visual health check and weighed using digital scales (Salter 1035 platform scales, Salter, UK). Healthy animals weighing ≥600 g were fitted with a VHF radio transmitter (10 g: <2% of body mass; Biotrack Ltd., Wareham, UK) glued to a region of clipped dorsal spines. All animals, regardless of body mass, were marked with coloured heat shrink tubing attached to 10 dorsal spines in a unique location; tubing was attached using a portable soldering iron. The capture location was recorded with a handheld GPS unit (Garmin GPS 60, Garmin, UK). Animals were released at the point of capture, typically within 15 min.

2.1. Nesting Behaviour

Determining the onset of hibernation for each individual using radio-tracking is difficult. Previous authors have tended to use either a criterion based on the number of successive days a single nest was used, although these have been variable (e.g., seven days [67], one month [68]), or based upon a defined time period [64]. In this study, the latter approach was used as it was not possible to definitively identify the onset of hibernation based upon patterns of nest use alone (see Results) and because it was plausible that hibernating animals may have moved nests following e.g., disturbance by human activities.

Consequently, radio-tracking data were divided into three phases in line with the time periods defined by Yarnell et al. [64]: August–October (pre-hibernation); November–March (hibernation period); and April (post-hibernation). In the pre-hibernation phase, animals were located one night each week to record body mass and check transmitter attachment, and once per week during the day to determine the position of nests. In the hibernation phase, animals were located two-three times each week to determine the position of nests: searches were a minimum of two days apart. Radio-tracking was conducted using a Sika radio-tracking receiver and handheld, three element Yagi antenna (Biotrack).

The location of nests was recorded with a GPS unit and marked with a cane close to the nest for future identification. The position of nests was considered in the context of its specific location (e.g., in an animal burrow, hedgerow, next to or underneath a building) and the surrounding habitats (e.g., gardens, pasture, woodland). Where possible, nests were examined once they had been vacated to identify the dominant and secondary nesting materials. After examination, all nest material was left in position for future use, as hedgehogs have been found to return to nests or to occupy those of other individuals [69].

The number of nests used by each hedgehog was calculated for the time period 1 November–31 March inclusive. Where an individual had not been tracked before 1 November (n = 3) or up to 31 March (n = 3), one extra nest was added to the actual number recorded in line with the pattern of nest use observed for other animals. Differences in the number of nests used by males and females within and between the two sites were analysed using a Kruskal–Wallis test as the data were not normally distributed.

Patterns of habitat selection for winter nests were quantified by comparing the characteristics of observed (used) nest locations with those of randomly selected locations within the area available to hedgehogs. Data for each site were analysed separately. The available area was defined as the minimum convex polygon (MCP) encompassing all the diurnal and nocturnal locations from all hedgehogs radio-tracked during the study period at that site; this was used to incorporate areas outside each individual's home range [70], and is a more objective reflection of the area used by each hedgehog population collectively than an arbitrarily predefined study area [71]. Available nest locations were randomly sampled (10 times the number of used locations) within the MCP for each study area to create an available versus used dataset. The habitat characteristics of used and available nest locations were obtained by calculating the minimum Euclidian distances to each of the seven main land cover types (amenity grassland, arable fields, buildings and associated hard-standing (hereafter 'buildings'), hedgerows, pasture fields, roads and road verges (hereafter 'roads'), woodland) found in both areas. All GIS analyses were carried out using ArcMap 10.3.1 software [72].

Resource Selection Functions (RSFs, [73]) based on generalised linear models for each site were used to quantify habitat selection. A logistic regression for each site was fitted, with the response variable being the used (1: GPS nesting locations) and available locations (0: random location within the MCP area defined above). Collinearity among explanatory variables was assessed using the Pearson correlation coefficient. At Brackenhurst, but not Hartpury, the minimum distances to amenity grassland and buildings were highly correlated (r = 0.7). Therefore, two different RSFs were built: Model A included amenity grassland but not buildings; Model B included buildings but not amenity grassland. Both amenity grassland and buildings were included in the Hartpury model.

Akaike's Information Criterion (AIC) [74,75] was used for model selection. Parameter values were averaged across models within two AIC units of the best fitting model [74].

2.2. Patterns of Survival

Survival rates were compared between sites using Kaplan–Meier analysis [76]. Sexes and years were combined because of relatively small sample sizes (Brackenhurst n = 10; Hartpury: n = 21), and because there was no apparent difference in the number of males and females that died at each site (see Results). Because animals were captured at different times, a staggered entry [77] design was used: the first animal was captured (Day 1) on 1 August. To avoid potential biases associated with the ad hoc recovery of untagged individuals, only radio-tagged individuals were included in this analysis. Differences in survival between the two sites were quantified using a log-rank test.

2.3. Body Mass Changes

Differences in overwinter changes in mass were compared between sites and sexes using a series of general linear models. Mass loss was calculated using each individual's mass at capture as close to the start and end of the hibernation period as possible; on average, animals were captured 15.5 days before 1 November and 2.6 days after 31 March. Statistical models compared differences in body mass at the start of hibernation, and mass change and percentage mass change during hibernation. All models included SITE and SEX as fixed factors and included a SEX*SITE interaction term. Linear correlation was also used to compare the number of nest sites used during hibernation with mass change over the hibernation period.

2.4. Data Analysis

General linear modelling and Kruskal–Wallis analyses were conducted using MINITAB version 19.1.1 and SPSS version 25, respectively. Survival analysis and RSF analyses were undertaken in R 3.3.3 [78] using lme4 and MuMIn packages [79,80]. All data were checked to ensure they conformed to the underlying assumptions of the tests used. All results are presented as mean ($\pm$SD) unless otherwise specified. As it was not possible to e.g., re-capture all tagged animals or access all nest sites, and because some animals perished during the course of the study, sample sizes vary between analyses.

3. Results

Forty hedgehogs were found during nocturnal surveys: 33 were fitted with radio transmitters (Table 1). Data on nesting behaviour during the hibernation period were collected from 21 hedgehogs. In total, 448 nocturnal locations, 138 nests, and 1028 diurnal locations were recorded.

Table 1. Number of hedgehogs captured and radio-tagged at each site, the total number of nocturnal and diurnal locations recorded, and the number of nest sites identified.

	Brackenhurst		Hartpury		Total
	2015–2016	2016–2017	2015–2016	2016–2017	
No. captured & marked	7 (4♀:3♂)	3 (2♀:1♂)	22 (12♀:10♂)	8 (3♀: 5♂)	40 (21♀:19♂)
No. radio-tagged	7 (4♀:3♂)	3 (2♀:1♂)	18 (9♀:9♂)	5 (3♀:2♂)	33 (18♀:15♂)
No. tracked during hibernation	7 (4♀:3♂)	3 (2♀:1♂)	7 (4♀:3♂)	4 (2♀:2♂)	21 (12♀:9♂)
Total no. of nests recorded (% accessible for recording composition)	54 (59%)	12 (100%)	50 (66%)	16 (75%)	138 (65%)
No. of nocturnal locations recorded	103	74	210	61	448
No. of diurnal locations recorded	408	114	360	146	1028

3.1. Nesting Behaviour

The pattern of nest use was highly variable, with several animals using the same nest site for extended periods before and/or during the hibernation period (Figure 1). There was no significant difference in the number of nests used by males and females within and between the two sites (Kruskal–Wallis test: H = 0.60, DF = 3, p = 0.896). Combining the data, hedgehogs used a median of five nests (mean $\pm$ SD = 5.5 $\pm$ 2.3) across the 151-day hibernation period. Thirteen animals (62%) used at least one site for $\geq$89 days.

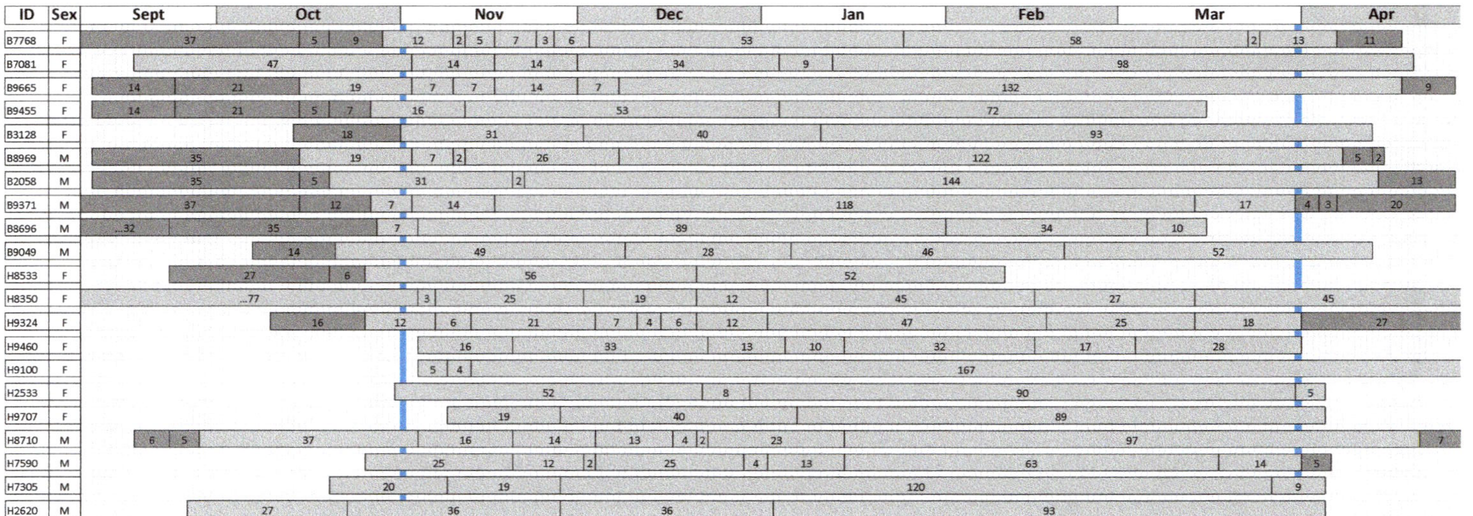

Figure 1. Pattern of occupation of winter nests by hedgehogs at Brackenhurst (ID numbers prefixed by "B") and Hartpury (ID numbers prefixed by "H"). Figures in horizontal bars indicate the number of days that each nest was estimated to be occupied based upon the sampling regime (see text for details). Vertical blue columns indicate the start (1 November) and end (31 March) of the hibernation period: dark and light shaded bars indicate nests excluded from and included in the analysis of the number of nests used over the hibernation period, respectively.

RSF analyses indicated that woodland, roads, pasture and, to a lesser extent, hedgerows, were consistently included in the top (ΔAIC < 2) ranked models at both sites (Figure 2; Table 2). At both sites, hedgehogs selected nest locations closer to hedgerows, in vegetation alongside roads and in woodlands, but avoided pasture fields (Table 3). Between-site differences were evident for arable fields (neither selected nor avoided at Brackenhurst; avoided at Hartpury) and both amenity grassland and buildings (both selected for at Brackenhurst in each model where these habitats were included; neither selected nor avoided at Hartpury, or not retained in top-ranked models).

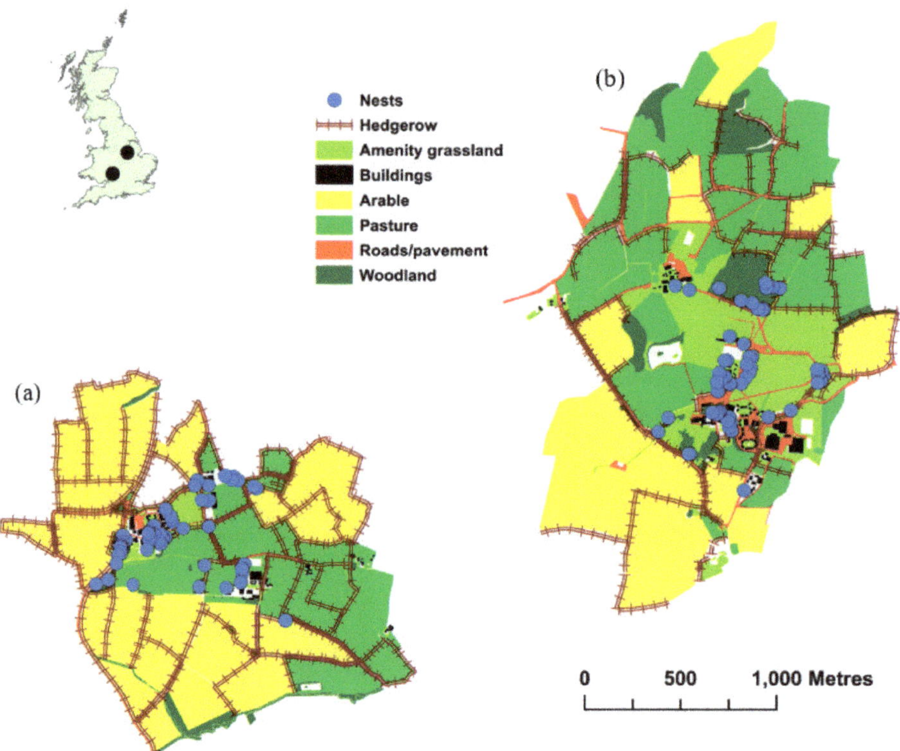

Figure 2. Position of hedgehog winter nest sites (blue dots) at (**a**) Brackenhurst and (**b**) Hartpury in relation to habitat composition.

Table 2. Results of the top five *a-priori* models for predictors of habitat selection of hedgehog winter nests. Models are ranked based on their AIC values. Null model is also provided for comparison. Models indicated in bold were selected to build average models. Brackenhurst had two alternative maximal models, one including distance to amenity grassland (Brackenhurst Model A) and another including distance to buildings (Brackenhurst Model B). Habitats included in each of the top-ranking models are indicated by the "✓" symbol. Bold indicates top ranked models at each site (ΔAIC < 2).

Brackenhurst Model A									
Models (N = 64)									
Amenity grassland	Buildings	Hedgerows	Pastures	Roads	Woodland	Arable	AIC	ΔAIC	AIC$_w$
✓	Not included	✓	✓	✓	✓		**357.5**	**0.00**	**0.38**
✓	Not included	✓	✓	✓	✓	✓	**358.2**	**0.75**	**0.26**
✓	Not included		✓	✓	✓	✓	**359.4**	**1.94**	**0.14**
✓	Not included		✓	✓	✓	✓	360.8	3.33	0.07
✓	Not included	✓	✓		✓	✓	362.2	4.67	0.04
NULL							491.2	134.00	<0.01
Brackenhurst Model B									
Models (N = 64)									
Amenity grassland	Buildings	Hedgerows	Pastures	Roads	Woodland	Arable	AIC	ΔAIC	AIC$_w$
Not included	✓	✓	✓	✓	✓		**350.2**	**0.00**	**0.41**
Not included	✓	✓	✓	✓	✓	✓	**351.1**	**0.90**	**0.26**
Not included	✓		✓	✓	✓		**352.1**	**1.89**	**0.16**
Not included	✓		✓	✓	✓	✓	352.6	3.44	0.07
Not included	✓	✓	✓		✓	✓	354.3	4.09	0.05
NULL							491.2	141.00	<0.01
Hartpury									
Models (N = 128)									
Amenity grassland	Buildings	Hedgerows	Pastures	Roads	Woodland	Arable	AIC	ΔAIC	AIC$_w$
		✓	✓	✓	✓	✓	**395.6**	**0.00**	**0.49**
	✓	✓	✓	✓	✓	✓	**397.4**	**1.80**	**0.20**
✓		✓	✓	✓	✓	✓	397.6	2.04	0.18
✓	✓	✓	✓	✓	✓	✓	399.4	3.84	0.07
		✓	✓		✓	✓	401.2	5.61	0.03
NULL							464.4	68.8	<0.01

Table 3. Model averaged values of the best *a-priori* models (ΔAIC < 2) investigating habitat selection for winter nest sites. SE = standard error. Brackenhurst had two alternative models, one including distance to amenity grassland but excluding buildings (Brackenhurst Model A) and another including distance to buildings but excluding amenity grassland (Brackenhurst Model B). Negative values indicate a higher probability of nesting closer to that specific habitat.

Variable	Brackenhurst Model A (3 Best *a-priori* Models)				Brackenhurst Model B (3 Best *a-priori* Models)				Hartpury (2 Best *a-priori* Models)			
	Estimate	SE	z	*p*-Value	Estimate	SE	z	*p*-Value	Estimate	SE	z	*p*-Value
(Intercept)	−0.281	0.439	0.640	0.522	−0.113	0.432	0.261	0.794	−2.514	0.515	4.879	<0.001
Hedgerows	−0.013	0.006	2.000	<0.05	−0.013	0.006	2.000	<0.05	−0.008	0.003	3.204	<0.01
Pasture	0.017	0.006	2.942	<0.01	0.017	0.006	2.877	<0.01	0.010	0.003	3.748	<0.001
Roads	−0.012	0.005	2.544	<0.05	−0.010	0.004	2.443	<0.05	−0.016	0.006	2.590	<0.01
Woodland	−0.020	0.003	5.919	<0.001	−0.020	0.003	5.607	<0.001	−0.013	0.003	3.774	<0.001
Arable	0.002	0.002	1.127	0.260	0.002	0.002	1.062	0.288	0.005	0.001	3.436	<0.001
Buildings	Not included				−0.01	0.003	3.412	<0.001	0.001	0.001	0.488	0.626
Amenity grassland	−0.008	0.003	2.527	<0.05	Not included				Not included			

At both sites, winter nests were primarily constructed from broad leaves (major component in 45% and 51% of nests, respectively: Supplementary Table S1). Major differences in the relative proportion of nests containing different materials were, however, evident. For example, litter and/or plastic waste was present in 20 nests (24%) at Hartpury, although never as the dominant material, but was never recorded at Brackenhurst.

3.2. Patterns of Survival

Nine animals died during the study, with no apparent sex difference in mortality risk (Brackenhurst: $1\male$; Hartpury: $4\female{:}4\male$). The overall survival rate was significantly lower at Hartpury (Log-rank test: $X^2_1 = 9.46$, $p = 0.002$). All deaths occurred before or after the hibernation period (Figure 3). The most common single known cause of death was predation by badgers (3 of 9 deaths; see Supplementary Table S2).

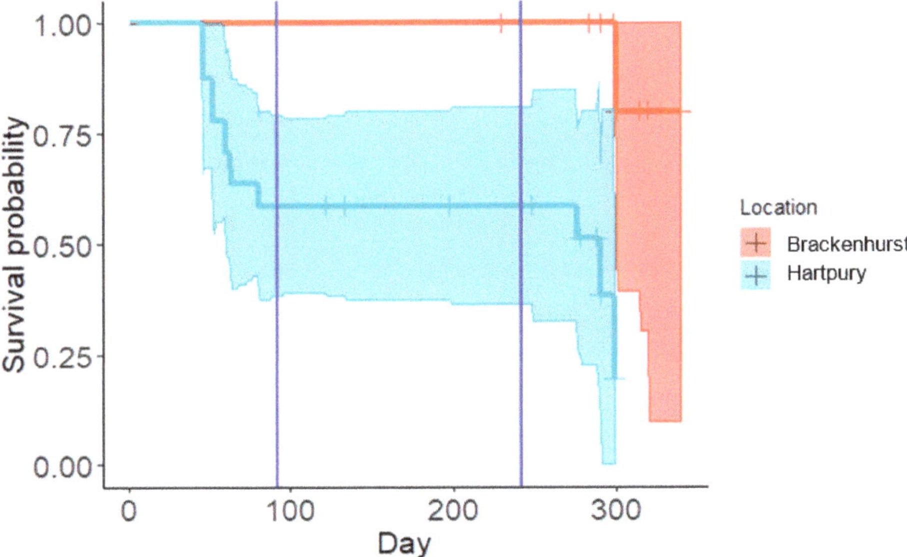

Figure 3. Kaplan–Meir survival functions for hedgehogs at Brackenhurst (n = 10) versus Hartpury (n = 21). Data from sexes and years (2015–2016 and 2016–2017) combined. Vertical blue lines indicate the start (1 November) and end (31 March) of the hibernation period.

3.3. Body Mass Changes

Data on body mass changes across the study were available for 21 individuals. There was no significant SITE ($F_{1,17} = 3.75$, $p = 0.069$), SEX ($F_{1,17} = 0.78$, $p = 0.389$) or SITE*SEX ($F_{1,17} = 3.75$, $p = 0.943$) differences in mean body mass at the start of the hibernation period (Supplementary Table S3); collectively, hedgehogs weighed 869 ± 133 g (females: 843 ± 144 g; males: 898 ± 120 g). During hibernation, 16 individuals lost mass (Brackenhurst—$5\female{:}3\male$; Hartpury—$5\female{:}3\male$), whilst five (Brackenhurst—$2\male$; Hartpury—$1\female{:}2\male$) gained mass. Mass change ($F_{1,17} = 4.65$, $p = 0.046$) but not percentage mass change ($F_{1,17} = 4.22$, $p = 0.056$) differed significantly between the sexes at each site, although the latter was close to significance. At Brackenhurst, females lost 242 ± 150g on average whilst males gained a small amount of weight (4 ± 89 g; Figure 4); male and female hedgehogs at Hartpury lost 117 ± 121 g and 110 ± 141 g, respectively. These figures are equivalent to average percentage mass changes of −25%, +1%, −14% and −15%, respectively (Supplementary Figure S1).

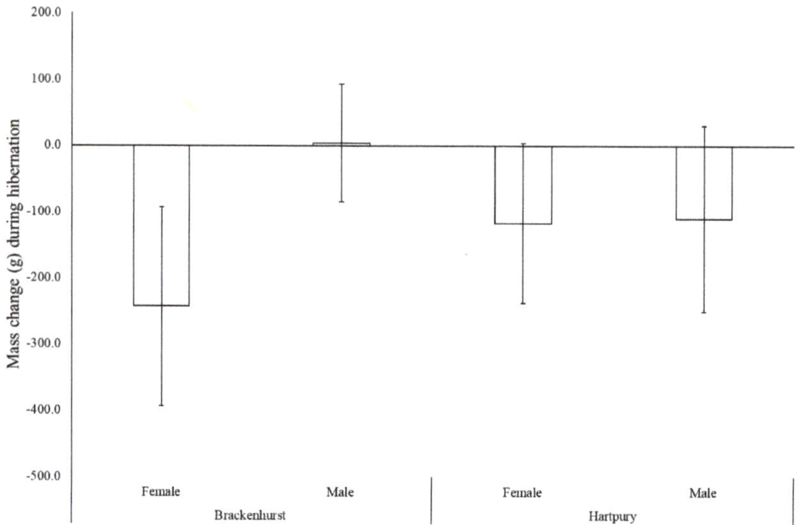

Figure 4. Mean (±SD) mass change during the hibernation period (1 November–31 March) in relation to site and sex (Brackenhurst: n = 5♀:5♂; Hartpury: n = 6♀:5♂).

There was a negative correlation between the number of nest sites used and the loss in body mass, although this was not significant ($r = -0.409$, n = 21, $p = 0.066$; Figure 5). However, this was dependent on the extreme loss exhibited by a single female at Brackenhurst (432 g); excluding this female, the relationship is significant ($r = -0.561$, n = 20, $p = 0.010$).

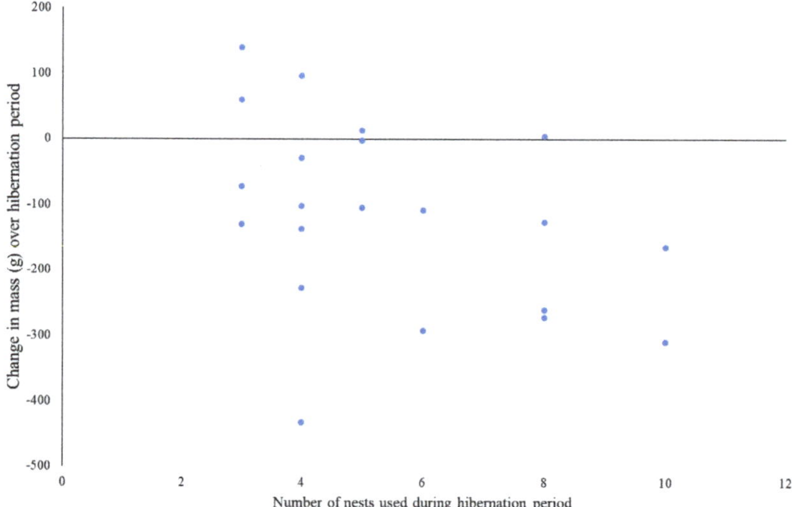

Figure 5. Relationship between number of nests used during the hibernation period (1 November –31 March) and the corresponding change in mass (g) over the hibernation period (n = 21).

4. Discussion

In this study, we investigated four factors associated with the winter hibernation period of hedgehogs that could potentially be affected by agricultural land-use and climate change: (i) patterns of body mass change; (ii) frequency of winter nest use; (iii) habitat selection for winter nest sites; and (iv) over-winter survival. Between the two sites studied, one dominated by arable crop production and the other by pasture and amenity grasslands, there were no apparent differences in body mass at the start of hibernation, the number of nest sites used during winter, and the selection for and avoidance of many, but not all, major habitats as nesting locations. In contrast, there were significant differences between the study sites with respect to sex-specific changes in body mass, the use of hedgerows and buildings for nesting, and patterns of survival.

4.1. Change in Body Mass

Estimated body mass of radio-tagged animals at the outset of the hibernation period was not significantly different between Brackenhurst and Hartpury, with animals weighing, on average 869 ± 133 g. This is likely due, in part, to the fact that we only radio-tagged individuals ≥600 g in accordance with guidance relating to the release of rehabilitated hedgehogs by the major wildlife welfare organisation in the UK [81]. This reliance on radio-tagged individuals to ensure that individuals captured before hibernation could be re-captured afterwards does, however, preclude obtaining data on animals below this threshold weight.

Acknowledging this caveat, the general pattern of mass loss observed (mean of 100–240 g within most site-sex divisions, equivalent to a mean of 14–25% of pre-hibernation mass) is within the range recorded in previous studies (Table 4). However, there was a substantial difference in sex-specific patterns of mass change at the two sites. At Hartpury, both males and females lost approximately the same amount of weight (Figure 4). Conversely, females at Brackenhurst lost markedly more weight than any other division, whereas males, on average, gained a small amount of weight. In fact, five (23.8%) animals across both sites gained weight across the hibernation period. This could indicate that individuals may have been able to access sufficient food resources during the winter period to offset the fat reserves used during hibernation, or that some animals may have already stopped hibernating and resumed typical foraging activity before they were recaptured in March/April. Although we are not able to discriminate between these possibilities, it is clear that the magnitude of these average changes are within the survivable range documented for this species.

Table 4. Summary of body mass changes recorded in previous studies of the West-European hedgehog over the winter hibernation period.

Country	Habitat	Years Studied	Sample Size& Composition	Mass Loss Recordedover Winter	Minimum Weightto SurviveHibernation	Reference
England	Urban parkland	1963–1968	105	25%	Recommends 450 g (550 g in more northern areas)	[63]
Denmark	Rural	2001–2002	10 (5♀:5♂); (3A:7J)	30.2 ± 7.1% (A) 22.1 ± 10.1% (J)	513 g	[82]
Ireland	Rural	2008–2009	8 (7A:1J)	301 ± 3.9g (♀) (range: 15–38%) 108 ± 2.6g (♂) (range: 3–6%)	475 g in Nov	[67]
Denmark	Suburban	2014–2015	8 (8J)	16 ± 2.9% (J)	-	[83]
England	Various	2010–2014	55 (19♀:30♂:16?); (20A:35J)	98.6 ± 35.6 g (♀) 160.8 ± 40.5 g (♂) 111.4 ± 33.0 g (A) 162.2 ± 43.3 g (J) 14.1 ± 3.1% (All animals)	Recommends >600 g for release, but one individual weighing 391 g survived release and hibernation	[64]
England	Various	2015–2017	21 (11♀:10♂)	Site 1: 240 ± 150 g (25 ± 13%) (♀) Site 1: −4 ± 89 g (1 ± 9%) (♂) Site 2: 117 ± 121 g (14 ± 16%) (♀) Site 2: 110 ± 141 g (15± 19%) (♂)	-	Present study

Mass loss was also negatively correlated with the number of nests used in the winter period (Figure 5), although not significantly ($p = 0.066$). The lack of significance may, in part, be attributable to the relatively small sample size (n = 21), the highly variable changes in mass recorded, and the presence of one female that lost >400 g (40% of her body mass). Although this is among one of the largest percentage mass losses ever recorded (Table 4) and was >100 g more than any other individual in this study, this individual survived to spring. As rousing from hibernation is energetically expensive [84], hedgehogs would be expected to avoid doing so unnecessarily to avoid depleting their fat reserves. Rousing is likely to occur in response to environmental fluctuations, including both rises or falls in temperature [60], but in anthropogenic landscapes, it may also occur in response to human disturbance. To date, however, there are very few data on the extent to which disturbances affect hedgehog hibernation, either by causing them to move nests or rouse but remain in the same nest [85], and what impacts these may have on energy consumption and mortality risk.

4.2. Nesting Behaviour

Hedgehogs used a median of 5 (mean: 5.5) nests during the 151-day hibernation period. This is markedly higher than that observed in other studies (Table 5). Drawing direct comparisons between the number of nests used in such studies is, however, problematic because of the methodological differences used to define the onset and duration of hibernation, coupled with latitudinal differences in weather and/or temperature which extend or shorten the overall length of the hibernation period. It is worth noting, however, that the mean number of nests used by the animals in this study was more than twice that (1.74 nests per 100 days = 2.6 nests over 151 days) recorded in the most recent study of hedgehogs in England and which utilized the same dates for defining the hibernation period [64].

Table 5. Summary of over-winter nesting behaviour in previous studies of the West-European hedgehog. Studies are listed in chronological order.

Country	Habitat	Years Studied	Sample Size& Composition [1]	Duration of Hibernation(Days)	Number of Nests Used	Reference
England	Urban park	1963–1967	167 nests	Not recorded	Mean occupation time = 1.4 months (range 0–6 months)	[61]
Denmark	Rural	2001–2002	10 (3A:7J)	197.7 ± 2.2 (A) 178.8 ± 13.1 (J)	2.2 (range: 1–4)	[82]
Ireland	Rural	2008–2010	8 (7A:1J)	167.3 ± 10.5 (♀) 148.6 ± 10.2 (♂) 155.4 ± 9.0 (A) 157 (J)	2.0 ± 0.6 (♀) 3.2 ± 0.6 (♂) 2.4 ± 0.7 (A) 5.0 (J)	[67]
Finland	Urban	2004–2006	11 (11A) (5♀:6♂)	223 ± 2.5 (♀) 224 ± 4.8 (♂)	1.0 (♀) 1.0 (♂)	[68]
Denmark	Urban	2014–2015	8 (8J)	138.0 ± 5.6 (J)	1.8 ± 0.14 (J)	[83]
England	Various	2010–2014	55 (20A:35J); (19♀:30♂:16?)	Not recorded	2.2 ± 0.5 (♀) [2] 1.7 ± 0.4 (♂) [2] 1.8 ± 0.4 (A) [2] 2.6 ± 0.6 (J) [2]	[64]
England	Arable	2015–2017	21A (12♀:9♂)	Not recorded	5.8 ± 2.6 (♀) 5.0 ± 1.9 (♂)	Present study

[1] Data were recorded by the authors either in terms of the number of nests studied or the number of individuals studied: A = adult; J = juvenile; ? = unknown sex. [2] The number of nests used per 100 days.

The increased number of nests used in our study was associated with periods during November, December and/or January where several individuals used a series of nests in quick succession (Figure 1). Although some of these periods of frequent movements between nests could be interpreted as indicating that an individual had not yet started hibernating, the patterns of nest retention exhibited throughout the study as a whole were extremely variable such that it is difficult to identify clear general trends. The possible exception to this is that the majority (62%) of animals used a single nest location for >89 days, with many of these used for the first time in November or December; this is markedly higher than the 21% of nests (n = 167) occupied for ≥3 months reported by Morris [61] in west London.

Clear patterns in nest location were evident for most, but not all, habitats. Hedgehogs consistently avoided nesting near pasture fields, whilst favouring hedgerows, woodlands and roads. In contrast, differing patterns of selection were evident for arable fields, buildings and amenity grassland. At Brackenhurst, nests were preferentially located near to amenity grassland and near buildings, although these habitats were strongly correlated with one another, whereas arable fields were neither selected nor avoided. Conversely, at Hartpury, arable fields were avoided, buildings were neither selected nor avoided and amenity grassland was not retained in the top-ranked models. These data imply that agricultural habitats were generally unsuitable for hibernation, a finding consistent with behaviour outside the hibernation period that has been attributed to a combination of reduced food availability [86] and increased risk of predation from and competition with badgers [44–49,59,87].

Hedgerows and woodland were an important habitat for nesting, a pattern that is evident in both summer and winter seasons in other studies [47,68,82,88]. Similarly, the selection for roads in this study is also most probably associated with the presence of hedgerows as borders along roads at both sites. In addition to acting as nesting sites, hedgerows are also recognised as an important refuge habitat whilst foraging where badgers are present [47,56] and for orientation through fragmented landscapes [55]. As such, the general loss and degradation of hedgerows in the UK [50,89,90] is likely to have negatively affected hedgehog populations due to impacts at multiple stages in their annual cycle, although the exact mechanisms are unknown because of the relative paucity of data on rural hedgehog populations and behaviour since the 1950s [91].

Similarly, there are few data on the importance of woodlands for hedgehogs. For example, woodlands were not identified as a factor affecting patterns of occupancy in a national survey of England and Wales [44], they were the least selected habitat in a radio-tracking study in arable landscapes [47], and no hedgehogs were detected in woodland in a pilot project on the Hartpury campus investigating the efficacy of three different methods for surveying hedgehogs [92]: all these studies were, however, conducted in the summer. The preference for woodlands as sites for hibernation observed in this study, and the reliance on broad leaves as nesting material, may suggest that hedgehogs tend to avoid woodlands during the summer months but use them as sites for hibernating during the winter months. As outlined above, one possible reason for these seasonal differences is the presence of badgers, which favour woodlands and plantations as sites for their setts [93] but undergo a period of torpor in winter [94]. Consequently, hedgehogs could be avoiding woodlands during the summer when badgers are active but using them as hibernation sites in the winter when the risk from badgers is markedly lower. As such, woodlands may represent a key resource for hedgehogs but only during one phase of their annual cycle. The impact of historical changes in the coverage of different types of woodland [95,96], their management and their interaction with an increasing badger population [57,58] on hedgehog populations are unknown but require investigation. For example, in their recent report, Mathews et al. [43] estimated that 37% of the British hedgehog population was supported by broadleaved woodland.

The affinity for amenity grassland as a foraging habitat has been well documented in Britain, most notably in the context of responses to the culling of badgers as a means for managing bovine tuberculosis in cattle [45,46,59]. During winter these areas are likely to be associated with low levels of badger activity (due to torpor) but also possibly marginally higher average temperatures than surrounding areas due to their proximity to buildings, and provision of food either accidentally (discarded refuse) or deliberately (although we were not aware of anyone deliberately feeding hedgehogs on either campus). However, amenity areas on university campuses are likely to experience high levels of pedestrian activity except in particularly poor weather and over the Christmas holiday period. The presence of buildings on these two sites also enabled hedgehogs to use some unusual nest locations, including piles of building materials and underground heating tunnels.

4.3. Over-Winter Survival

Survival across the study period as a whole (August–April) was significantly lower at Hartpury versus Brackenhurst. However, this was not associated with differences in mortality during the hibernation period itself, but rather mortality prior to the onset of hibernation and in the period after animals had resumed foraging in spring: in fact, none of the tagged animals in this study (n = 31) died during the hibernation period itself (Figure 3). Consequently, mortalities were not related to body mass *per se* but stochastic events such as predation by badgers and road traffic accidents (although it could be argued that animals which have not yet accumulated sufficient fat reserves and/or those that leave hibernation having lost a large amount of might be expected to take greater risks when foraging). However, it must be emphasised that these survival data are based on animals that were in good physical condition (visually health-checked and ≥600 g) prior to hibernation in accordance with welfare guidelines; this is substantially higher than the minimum threshold of 450–513 g outlined in Table 4, and which would tend to elevate survival rates.

The survival rate observed at Hartpury, when measured from August to April (approximately 65%), was lower than that recorded in Sweden (57–96%, mean = 71%) over seven years in the 1970s [62], whereas the survival rates at both sites when measured from October to April were comparable to studies from England (83%) Ireland (100%), Denmark (89–90%) and Finland (100%) conducted between 2001 and 2017 [64,67,82,97]. Overall, this body of evidence suggests that, in general terms, the survival rate of animals that have accumulated sufficient fat reserves prior to hibernation is likely to be high, but that site-specific pressures associated with movements in autumn and spring can substantially increase mortality rates [64].

5. Conclusions

This study has identified key similarities and differences in four key parameters associated with the winter hibernation of hedgehogs across two sites associated with different patterns of land management. Most notably, the period of hibernation itself, when hedgehogs are generally inactive within hibernacula, is not associated with high levels of mortality. Conversely, it is the periods before and after entering hibernation that pose significant risks, predominantly from stochastic factors such as badger predation and vehicle collisions. In addition, hedgehogs at both sites consistently avoided nesting in proximity to pastoral fields during winter, but favoured locations near to hedgerows, woodlands and roads. Selection for or avoidance of arable fields, buildings, and amenity grasslands varied between the two sites.

However, this study was associated with several practical limitations. Data could only be reliably collected from radio-tagged individuals and radio-tags can only be fitted to animals weighing ≥600 g for welfare reasons. Radio-tracking is also limited in the extent to which the start and end of the hibernation period (for each individual) can be identified reliably, and the ease with which data on short-term patterns of movement between nests can be collected given that animals are inactive for many successive days. Future studies, therefore, need to consider the use of other technologies, such as GPS tracking devices [98] and animal-mounted bio-loggers [99], to overcome these constraints. In particular, such studies need to focus on: (i) quantifying patterns of survival of animals weighing <600 g; (ii) identifying factors associated with nest movements and whether this affects mass change during hibernation; and (iii) the role of woodlands in the annual cycle of hedgehogs in both arable and pastoral dominated landscapes.

Supplementary Materials: The following are available online at http://www.mdpi.com/2076-2615/10/9/1449/s1, Figure S1: Mean (±SD) (a) body mass (g) at the start of the hibernation season, and (b) percentage mass change during the hibernation period in relation to site and sex (Brackenhurst: n = 5♀:5♂; Hartpury: n = 6♀:5♂), Table S1: Summary of the dominant materials used in winter nest construction at Brackenhurst and Hartpury. Data for 2015–2016 and 2016–2017 combined. Figures in parentheses are the number of nests where the material was recorded as a secondary material. Sample sizes are less than the total number of nests used by study animals as not all nests were accessible, Table S2: Cause of death (n = 9) from a sample of 31 individuals followed over two winter hibernation periods (2015–2016 or 2016–2017), Table S3: General linear models comparing site and sex differences in (a) body mass (g) at the start of the hibernation season, and (b) mass change and (c) percentage mass change during the hibernation period (n = 21).

Author Contributions: Conceptualization, R.W.Y.; methodology, R.W.Y., D.S., A.U., L.E.B.-B., L.E. and P.J.B.; validation, R.W.Y., L.E.B.-B. and P.J.B.; formal analysis, R.W.Y., L.E.B.-B., A.U., L.E. and P.J.B.; investigation, R.W.Y., D.S., A.U. and L.E.B.-B.; resources, R.W.Y., and L.B.B.; data curation, R.W.Y., L.E.B.-B., and P.J.B.; writing—original draft preparation, L.E.B.-B., P.J.B.; writing—review and editing, all authors; visualization, R.W.Y., A.U., L.E., L.E.B.-B., P.B.; supervision, R.W.Y., P.J.B.; project administration, R.WY.; funding acquisition, R.W.Y., D.S., A.U., L.E.B.-B., P.J.B. All authors have read and agreed to the published version of the manuscript.

Funding: This research was funded by the Peoples' Trust for Endangered Species and British Hedgehog Preservation Society.

Acknowledgments: The authors would like to thank the volunteers who assisted with data collection, the land owners where data collection was undertaken and Vicky Boult who assisted with some GIS mapping. Particular thanks go to our funders; People's Trust for Endangered Species and British Hedgehog Preservation Society.

Conflicts of Interest: The authors declare no conflict of interest. The funders had no role in the design of the study; in the collection, analyses, or interpretation of data; in the writing of the manuscript, or in the decision to publish the results.

References

1. Parmesan, C. Ecological and evolutionary responses to recent climate change. *Annu. Rev. Ecol. Evol. Syst.* **2006**, *37*, 637–669. [CrossRef]

2. Firbank, L.G.; Petit, S.; Smart, S.; Blain, A.; Fuller, R.J. Assessing the impacts of agricultural intensification on biodiversity: A British perspective. *Philos. Trans. R. Soc. B Biol. Sci.* **2008**, *363*, 777–787. [CrossRef]

3. Tscharntke, T.; Clough, Y.; Wanger, T.C.; Jackson, L.; Motzke, I.; Perfecto, I.; Vandermeer, J.; Whitbread, A. Global food security, biodiversity conservation and the future of agricultural intensification. *Biol. Conserv.* **2012**, *151*, 53–59. [CrossRef]

4. Tuck, S.L.; Winqvist, C.; Mota, F.; Ahnström, J.; Turnbull, L.A.; Bengtsson, J. Land-use intensity and the effects of organic farming on biodiversity: A hierarchical meta-analysis. *J. Appl. Ecol.* **2014**, *51*, 746–755. [CrossRef]

5. Veach, V.; Moilanen, A.; Di Minin, E. Threats from urban expansion, agricultural transformation and forest loss on global conservation priority areas. *PLoS ONE* **2017**, *12*, 1–14. [CrossRef] [PubMed]

6. Zabel, F.; Delzeit, R.; Schneider, J.M.; Seppelt, R.; Mauser, W.; Václavík, T. Global impacts of future cropland expansion and intensification on agricultural markets and biodiversity. *Nat. Commun.* **2019**, *10*, 1–10. [CrossRef] [PubMed]

7. Firbank, L.G.; Attwood, S.; Eory, V.; Gadanaski, Y.; Lynch, J.M.; Sonnino, R.; Takahashi, T. Grand challenges in sustainable intensification and ecosystem services. *Front. Sustain. Food Syst.* **2018**, *2*, 7. [CrossRef]

8. Ellis, E.C.; Goldewijk, K.K.; Siebert, S.; Lightman, D.; Ramankutty, N. Anthropogenic transformation of the biomes, 1700 to 2000. *Glob. Ecol. Biogeogr.* **2010**, *19*, 589–606. [CrossRef]

9. Crooks, K.R.; Burdett, C.L.; Theobald, D.M.; Rondinini, C.; Boitani, L. Global patterns of fragmentation and connectivity of mammalian carnivore habitat. *Philos. Trans. R. Soc. B Biol. Sci.* **2011**, *366*, 2642–2651. [CrossRef]

10. Robinson, T.P.; Wint, G.R.; Conchedda, G.; Van Boeckel, T.P.; Ercoli, V.; Palamara, E.; Cinardi, G.; D'Aietti, L.; Hay, S.I.; Gilbert, M. Mapping the global distribution of livestock. *PLoS ONE* **2014**, *9*. [CrossRef]

11. Wiethoelter, A.K.; Beltrán-Alcrudo, D.; Kock, R.; Mor, S.M. Global trends in infectious diseases at the wildlife-livestock interface. *Proc. Natl. Acad. Sci. USA* **2015**, *112*, 9662–9667. [CrossRef] [PubMed]

12. Gordon, I.J. Review: Livestock production increasingly influences wildlife across the globe. *Animal* **2018**, *12*, S372–S382. [CrossRef] [PubMed]

13. Howell, H.J.; Mothes, C.C.; Clements, S.L.; Catania, S.V.; Rothermel, B.B.; Searcy, C.A. Amphibian responses to livestock use of wetlands: New empirical data and a global review. *Ecol. Appl.* **2019**, *29*, 1–15. [CrossRef] [PubMed]

14. Öllerer, K.; Varga, A.; Kirby, K.; Demeter, L.; Biró, M.; Bölöni, J.; Molnár, Z. Beyond the obvious impact of domestic livestock grazing on temperate forest vegetation—A global review. *Biol. Conserv.* **2019**, *237*, 209–219. [CrossRef]

15. Baker, P.J.; Boitani, L.; Harris, S.; Saunders, G.; White, P.C.L. Terrestrial carnivores and human food production: Impact and management. *Mamm. Rev.* **2008**, *38*, 123–166. [CrossRef]

16. Eklund, A.; López-Bao, J.V.; Tourani, M.; Chapron, G.; Frank, J. Limited evidence on the effectiveness of interventions to reduce livestock predation by large carnivores. *Sci. Rep.* **2017**, *7*, 1–9. [CrossRef] [PubMed]

17. van Eeden, L.M.; Eklund, A.; Miller, J.R.B.; López-Bao, J.V.; Chapron, G.; Cejtin, M.R.; Crowther, M.S.; Dickman, C.R.; Frank, J.; Krofel, M.; et al. Carnivore conservation needs evidence-based livestock protection. *PLoS Biol.* **2018**, *16*, 1–8. [CrossRef]

18. van Eeden, L.M.; Crowther, M.S.; Dickman, C.R.; Macdonald, D.W.; Ripple, W.J.; Ritchie, E.G.; Newsome, T.M. Managing conflict between large carnivores and livestock. *Conserv. Biol.* **2018**, *32*, 26–34. [CrossRef]

19. Garcês, A.; Pires, I.; Rodrigues, P. Teratological effects of pesticides in vertebrates: A review. *J. Environ. Sci. Heal.Part. B Pestic. Food Contam. Agric. Wastes* **2020**, *55*, 75–89. [CrossRef]

20. Brown, D.D.; Kays, R.; Wikelski, M.; Wilson, R.; Klimley, A. Observing the unwatchable through acceleration logging of animal behavior. *Anim. Biotelemetry* **2013**, *1*. [CrossRef]

21. Kharouba, H.M.; Ehrlén, J.; Gelman, A.; Bolmgren, K.; Allen, J.M.; Travers, S.E.; Wolkovich, E.M. Global shifts in the phenological synchrony of species interactions over recent decades. *Proc. Natl. Acad. Sci. USA* **2018**, *115*, 5211–5216. [CrossRef] [PubMed]

22. Butchart, S.H.M.; Walpole, M.; Collen, B.; van Strien, A.; Scharlemann, J.P.W.; Almond, R.E.A.; Baillie, J.E.M.; Bomhard, B.; Brown, C.; Bruno, J.; et al. Global Biodiversity: Indicators of recent declines. *Science* **2010**, *1164*, 1164–1169. [CrossRef] [PubMed]

23. Hoffmann, M.; Hilton-Taylor, C.; Angulo, A.; Böhm, M.; Brooks, T.M.; Butchart, S.H.M.; Carpenter, K.E.; Chanson, J.; Collen, B.; Cox, N.A.; et al. The impact of conservation on the status of the world's vertebrates. *Science* **2010**, *330*, 1503–1509. [CrossRef] [PubMed]

24. Di Marco, M.; Boitani, L.; Mallon, D.; Hoffmann, M.; Iacucci, A.; Meijaard, E.; Visconti, P.; Schipper, J.; Rondinini, C. A Retrospective evaluation of the global decline of carnivores and ungulates. *Conserv. Biol.* **2014**, *28*, 1109–1118. [CrossRef] [PubMed]

25. Maxwell, S.L.; Fuller, R.A.; Brooks, T.M.; Watson, J.E.M. Biodiversity: The ravages of guns, nets and bulldozers. *Nature* **2016**, *536*, 143–145. [CrossRef]

26. Dudley, N.; Alexander, S. Agriculture and biodiversity: A review. *Biodiversity* **2017**, *18*, 45–49. [CrossRef]

27. Stanton, R.L.; Morrissey, C.A.; Clark, R.G. Analysis of trends and agricultural drivers of farmland bird declines in North America: A review. *Agric. Ecosyst. Environ.* **2018**, *254*, 244–254. [CrossRef]

28. Clout, M.N.; Russell, J.C. The invasion ecology of mammals: A global perspective. *Wildl. Res.* **2007**, *35*, 180–184. [CrossRef]

29. Long, J.L. *Introduced Mammals of the World: Their History, Distribution and Influence*; CSIRO Publishing: Victoria, Australia, 2003.

30. Inouye, D.W.; Barr, B.; Armitage, K.B.; Inouye, B.D. Climate change is affecting altitudinal migrants and hibernating species. *Proc. Natl. Acad. Sci. USA* **2000**, *97*, 1630–1633. [CrossRef]

31. Lane, J.E. Evolutionary Ecology of Mammalian Hibernation Phenology. In *Living in a Seasonal World*; Ruf, T., Bieber, C., Arnold, W., Millesi, E., Eds.; Springer: Berlin, Germany, 2012; pp. 51–61. ISBN 9783642286780.

32. Lane, J.E.; Kruuk, L.E.B.; Charmantier, A.; Murie, J.O.; Dobson, F.S. Delayed phenology and reduced fitness associated with climate change in a wild hibernator. *Nature* **2012**, *489*, 554–557. [CrossRef]

33. Geiser, F. Hibernation. *Curr. Biol.* **2013**, *23*, 188–193. [CrossRef] [PubMed]

34. Geiser, F. Hibernation: Endotherms. *eLS* **2011**, *1*. [CrossRef]

35. Staples, J.F. Metabolic suppression in mammalian hibernation: The role of mitochondria. *J. Exp. Biol.* **2014**, *217*, 2032–2036. [CrossRef] [PubMed]

36. Turbill, C.; Bieber, C.; Ruf, T. Hibernation is associated with increased survival and the evolution of slow life histories among mammals. *Proc. R. Soc. B Biol. Sci.* **2011**, *278*, 3355–3363. [CrossRef] [PubMed]

37. Morris, P.A.; Reeve, N.J. Hedgehog Erinaceus europaeus. In *Mammals of the British Isles: Handbook*; Harris, S.J., Yalden, D.W., Eds.; The Mammal Society: Southampton, UK, 2008; pp. 241–249.

38. Burton, M. *The Hedgehog*; Andre Deutsch: London, UK, 1969.

39. Tapper, S.S. *Game Heritage: An Ecological Review from Shooting and Gamekeeping Records*; Game Conservancy Ltd.: Fordingbridge, Hampshire, UK, 1992.

40. Arnold, H.R. *Atlas of Mammals of Britain*; HMSO: London, UK, 1993.

41. Lovegrove, R. *Silent Fields: The Long Decline of a Nation's Wildlife*; Oxford University Press: Oxford, UK, 2007.

42. Hof, A.R.; Bright, P.W. Quantifying the long-term decline of the West European hedgehog in England by subsampling citizen-science datasets. *Eur. J. Wildl. Res.* **2016**, *62*, 407–413. [CrossRef]

43. Mathews, F.; Kubasiewicz, L.M.; Gurnell, J.; Harrower, C.A.; McDonald, R.A.; Shore, R.F. *A Review of the Population and Conservation Status of British Mammals. A report by the Mammal. Society under contract to Natural England, Natural Resources Wales and Scottish Natural Heritage*; Natural England: Peterborough, UK, 2018; ISBN 9781783544943.

44. Williams, B.M.; Baker, P.J.; Thomas, E.; Wilson, G.J.; Judge, J.; Yarnell, R.W. Reduced occupancy of hedgehogs (Erinaceus europaeus) in rural England and Wales: The influence of habitat and an asymmetric intra-guild predator. *Sci. Rep.* **2018**, *8*, 12156. [CrossRef]

45. Young, R.P.; Davison, J.; Trewby, I.D.; Wilson, G.J.; Delahay, R.J.; Doncaster, C.P. Abundance of hedgehogs (Erinaceus europaeus) in relation to the density and distribution of badgers (Meles meles). *J. Zool.* **2006**, *269*, 349–356. [CrossRef]

46. Parrott, D.; Etherington, T.R.; Dendy, J. A geographically extensive survey of hedgehogs (Erinaceus europaeus) in England. *Eur. J. Wildl. Res.* **2014**, *60*, 399–403. [CrossRef]

47. Pettett, C.E.; Moorhouse, T.P.; Johnson, P.J.; Macdonald, D.W. Factors affecting hedgehog (Erinaceus europaeus) attraction to rural villages in arable landscapes. *Eur. J. Wildl. Res.* **2017**, *63*, 54. [CrossRef]

48. Hubert, P.; Julliard, R.; Biagianti, S.; Poulle, M.L. Ecological factors driving the higher hedgehog (Erinaceus europeaus) density in an urban area compared to the adjacent rural area. *Landsc. Urban. Plan.* **2011**, *103*, 34–43. [CrossRef]

49. Van de Poel, J.L.; Dekker, J.; Langevelde, F.V. Dutch hedgehogs Erinaceus europaeus are nowadays mainly found in urban areas, possibly due to the negative effects of badgers Meles meles. *Wildl. Biol.* **2015**, *21*, 51–55. [CrossRef]

50. Robinson, R.A.; Sutherland, W.J. Post-war changes in arable farming and biodiversity in Great Britain. *J. Appl. Ecol.* **2002**, *39*, 157–176. [CrossRef]

51. Carey, P.D.; Wallis, S.M.; Emmett, B.; Maskell, L.C.; Murphy, J.; Norton, L.R.; Simpson, I.C.; Smart, S.M. *Countryside Survey: UK Headline Messages from 2007*; Centre for Ecology & Hydrology: Wallingford, UK, 2008.

52. Wright, J. *A Natural History of the Hedgerow and Ditches, Dykes and Dry Stone Walls*; Profile Books Ltd.: London, UK, 2016.

53. Dover, J.W. *The Ecology of Hedgerows and Field Margins*; Routledge: Abingdon, UK, 2019.

54. Hof, A.R.; Bright, P.W. The value of agri-environment schemes for macro-invertebrate feeders: Hedgehogs on arable farms in Britain. *Anim. Conserv.* **2010**, *13*, 467–473. [CrossRef]

55. Moorhouse, T.P.; Palmer, S.C.F.; Travis, J.M.J.; Macdonald, D.W. Hugging the hedges: Might agri-environment manipulations affect landscape permeability for hedgehogs? *Biol. Conserv.* **2014**, *176*, 109–116. [CrossRef]

56. Hof, A.R.; Snellenberg, J.; Bright, P.W. Food or fear? Predation risk mediates edge refuging in an insectivorous mammal. *Anim. Behav.* **2012**, *83*, 1099–1106. [CrossRef]

57. Judge, J.; Wilson, G.J.; Macarthur, R.; Delahay, R.J.; McDonald, R.A. Density and abundance of badger social groups in England and Wales in 2011-2013. *Sci. Rep.* **2014**, *4*, 3809. [CrossRef]

58. Judge, J.; Wilson, G.J.; Macarthur, R.; McDonald, R.A.; Delahay, R.J. Abundance of badgers (Meles meles) in England and Wales. *Sci. Rep.* **2017**, *7*, 276. [CrossRef]

59. Trewby, I.D.; Young, R.P.; McDonald, R.A.; Wilson, G.J.; Davison, J.; Walker, N.; Robertson, P.A.; Doncaster, C.P.; Delahay, R.J. Impacts of removing badgers on localised counts of hedgehogs. *PLoS ONE* **2014**, *9*, 2–5. [CrossRef]

60. Morris, P.A. *Hedgehog*; HarperCollins: London, UK, 2018.

61. Morris, P.A. Winter nests of the hedgehog. *Oecologia* **1973**, *11*, 299–313. [CrossRef]

62. Kristiansson, H. Population variables and causes of mortality in a hedgehog (Erinaceus europaeus) population in Southern Sweden. *J. Zool.* **1990**, *220*, 391–404. [CrossRef]

63. Morris, P.A. An estimate of the minimum body weight necessary for hedgehogs (Erinaceus europaeus) to survive hibernation. *J. Zool.* **1984**, *203*, 291–294.
64. Yarnell, R.W.; Surgey, J.; Grogan, A.; Thompson, R.; Davies, K.; Kimbrough, C.; Scott, D.M. Should rehabilitated hedgehogs be released in winter? A comparison of survival, nest use and weight change in wild and rescued animals. *Eur. J. Wildl. Res.* **2019**, *65*. [CrossRef]
65. *Natural England Entry Level Stewardship*; Natural England: Sheffield, UK, 2013; ISBN 978-1-84754-239-7.
66. Schaus, J.; Uzal, A.; Gentle, L.K.; Baker, P.J.; Bearman-Brown, L.; Bullion, S.; Gazzard, A.; Lockwood, H.; North, A.; Reader, T.; et al. Application of the Random Encounter Model in citizen science projects to monitor animal densities. *Remote Sens. Ecol. Conserv.* **2020**. [CrossRef]
67. Haigh, A.; O'Riordan, R.M.; Butler, F. Nesting behaviour and seasonal body mass changes in a rural Irish population of the Western hedgehog (Erinaceus europaeus). *Acta Theriol.* **2012**, *57*, 321–331. [CrossRef]
68. Rautio, A.; Valtonen, A.; Auttila, M.; Kunnasranta, M. Nesting patterns of European hedgehogs (Erinaceus europaeus) under northern conditions. *Acta Theriol.* **2014**, *59*, 173–181. [CrossRef]
69. Morris, P.A. *Hedgehogs*, 4th ed.; Whittet Books: Stansted, UK, 2014.
70. McClean, S.A.; Rumble, M.A.; King, R.M.; Baker, W.L. Evaluation of Resource Selection Methods with Different Definitions of Availability. *J. Wildl. Manag.* **1998**, *62*, 793. [CrossRef]
71. Uzal, A.; Walls, S.; Stillman, R.A.; Diaz, A. Sika deer distribution and habitat selection: The influence of the availability and distribution of food, cover, and threats. *Eur. J. Wildl. Res.* **2013**, *59*, 563–572. [CrossRef]
72. Environmental Systems Resource Institute. *ArcGIS 2015*; ESRI: Redlands, CA, USA, 2015.
73. Manly, B.F.L.; McDonald, L.; Thomas, D.L.; McDonald, T.L.; Erickson, W.P. *Resource Selection by Animals: Statistical Design and Analysis for Field Studies*; Springer: London, UK, 2007.
74. Burnham, K.P.; Anderson, D.R. *Model. Selection and Multi-Model Inference: A Practical Information-Theoretic Approach*; Springer: New York City, NY, USA, 2002.
75. Zuur, A.F.; Ieno, E.N.; Walker, N.; Saveliev, A.A.; Smith, G.M. *Mixed Effects Models and Extensions in Ecology with R*; Springer Science & Business Media: London, UK, 2009.
76. Kaplan, E.L.; Meier, P. Nonparametric estimation from incomplete observations. *J. Am. Stat. Assoc.* **1958**, *53*, 457. [CrossRef]
77. Pollock, K.H.; Winterstein, S.R.; Bunck, C.M.; Curtis, P.D. Survival analysis in telemetry studies: The staggered entry design. *J. Wildl. Manage.* **1989**, *53*, 7–15. [CrossRef]
78. R Core Team. *R: A Language and Environment for Statistical Computing*; R Core Team R: Vienna, Austria, 2016.
79. Bates, D.; Machler, M.; Bolker, B.; Walker, S. Fitting Linear Mixed-Effects Models Using lme4. *J. Stat. Softw.* **2015**, *67*, 1–48. [CrossRef]
80. Barton, K. *Package "MuMIn"*; R Package Version 1(6); CRAN: 2019. Available online: https://cran.r-project.org/web/packages/MuMIn/MuMIn.pdf (accessed on 19 August 2020)
81. RSPCA. *RSPCA Wildlife Rehabilitation Protocol: Hedgehogs*; RSPCA: Southwater, UK, 2013.
82. Jensen, A.B. Overwintering of European hedgehogs Erinaceus europaeus in a Danish rural area. *Acta Theriol.* **2004**, *49*, 145–155. [CrossRef]
83. Rasmussen, S.L.; Berg, T.B.; Dabelsteen, T.; Jones, O.R. The ecology of suburban juvenile European hedgehogs (Erinaceus europaeus) in Denmark. *Ecol. Evol.* **2019**, *9*, 13174–13187. [CrossRef]
84. Tähti, H.; Soivio, A. Respiratory and circulatory differences between induced and spontaneous arousals in hibernating hedgehogs (Erinaceus europaeus L.). *Ann. Zool. Fenn.* **1977**, *14*, 198–203.
85. Walhovd, H. Partial arousals from hibernation in hedgehogs in outdoor hibernacula. *Oecologia* **1979**, *153*, 141–153. [CrossRef]
86. Hof, A.R.; Bright, P.W. The impact of grassy field margins on macro-invertebrate abundance in adjacent arable fields. *Agric. Ecosyst. Environ.* **2010**, *139*, 280–283. [CrossRef]
87. Pettett, C.E.; Johnson, P.J.; Moorhouse, T.P.; Hambly, C.; Speakman, J.R.; Macdonald, D.W. Daily energy expenditure in the face of predation: Hedgehog energetics in rural landscapes. *J. Exp. Biol.* **2017**, *220*, 460–468. [CrossRef]
88. Riber, A. Habitat use and behaviour of European hedgehog Erinaceus europaeus in a Danish rural area. *Acta Theriol.* **2006**, *51*, 363–371. [CrossRef]
89. Sutherland, W.J.; Armstrong-Brown, S.; Armsworth, P.R.; Brereton, T.; Brickland, J.; Campbell, C.D.; Chamberlain, D.E.; Cooke, A.I.; Dulvy, N.K.; Dusic, N.R.; et al. The identification of 100 ecological questions of high policy relevance in the UK. *J. Appl. Ecol.* **2006**, *43*, 617–627. [CrossRef]

90. Cornulier, T.; Robinson, R.A.; Elston, D.; Lambin, X.; Sutherland, W.J.; Benton, T.G. Bayesian reconstitution of environmental change from disparate historical records: Hedgerow loss and farmland bird declines. *Methods Ecol. Evol.* **2011**, *2*, 86–94. [CrossRef]

91. Harris, S.J.; Morris, P.A.; Wray, S.; Yalden, D.W. *A Review of British Mammals: Population Estimates and Conservation Status of British Mammals Other than Cetaceans*; JNCC: Peterborough, UK, 1995; ISBN 1-873701-68-3.

92. Bearman-Brown, L.E.; Wilson, L.E.; Evans, L.; Baker, P.J. Comparing non-invasive surveying techniques for elusive, nocturnal mammals: A case study of the West European hedgehog (Erinaceus europaeus). *J. Vertebr. Biol.* in press.

93. Wilson, G.; Harris, S.J.; McLaren, G. *Changes in the British Badger Population, 1988 to 1997*; Kerenza J Ltd.: London, UK, 1997.

94. Roper, T.J. *Badger*; HarperCollins: London, UK, 2010.

95. Hopkins, J.J.; Kirby, K.J. Ecological change in British broadleaved woodland since 1947. *Ibis* **2007**, *149*, 29–40. [CrossRef]

96. Amar, A.; Smith, K.W.; Butler, S.; Lindsell, J.A.; Hewson, C.M.; Fuller, R.J.; Charman, E.C. Recent patterns of change in vegetation structure and tree composition of British broadleaved woodland: Evidence from large-scale surveys. *Forestry* **2010**, *83*, 345–356. [CrossRef]

97. Rautio, A.; Valtonen, A.; Kunnasranta, M. The effects of sex and season on home range in European hedgehogs at the northern edge of the species range. *Ann. Zool. Fenn.* **2013**, *50*, 107–123. [CrossRef]

98. Glasby, L.; Yarnell, R.W. Evaluation of the performance and accuracy of Global Positioning System bug transmitters deployed on a small mammal. *Eur. J. Wildl. Res.* **2013**, *59*, 915–919. [CrossRef]

99. Chmura, H.E.; Glass, T.W.; Williams, C.T. Biologging physiological and ecological responses to climatic variation: New tools for the climate change era. *Front. Ecol. Evol.* **2018**, *6*. [CrossRef]

Article

DNA Footprints: Using Parasites to Detect Elusive Animals, Proof of Principle in Hedgehogs

Simon Allen [1,2,*], Carolyn Greig [3], Ben Rowson [4], Robin B. Gasser [5], Abdul Jabbar [5], Simone Morelli [6], Eric R. Morgan [2,7], Martyn Wood [1] and Dan Forman [3]

[1] Gower Bird Hospital, Sandy Lane, Parkmill, Gower, Swansea SA3 2EW, UK; martyn@gowerbirdhospital.org.uk

[2] School of Biological Sciences, University of Bristol, Bristol Life Sciences Building, 24, Tyndall Avenue, Bristol BS8 1TQ, UK; eric.morgan@qub.ac.uk

[3] College of Science, Swansea University, Singleton Park, Swansea, Wales SA2 8PP, UK; c.greig@swansea.ac.uk (C.G.); d.w.forman@swansea.ac.uk (D.F.)

[4] Department of Natural Sciences, National Museum of Wales, Cardiff, Wales CF10 3NP, UK; ben.rowson@museumwales.ac.uk

[5] Faculty of Veterinary and Agricultural Sciences, The University of Melbourne, Parkville 3010, Australia; robinbg@unimelb.edu.au (R.B.G.); jabbara@unimelb.edu.au (A.J.)

[6] Faculty of Veterinary Medicine, Teaching Veterinary Hospital, University of Teramo, 64100 Teramo, Italy; smorelli@unite.it

[7] School of Biological Sciences, Queen's University Belfast, 19 Chlorine Gardens, Belfast BT9 5DL, Northern Ireland, UK

* Correspondence: simon@gowerbirdhospital.org.uk; Tel.: +44-179-237-1630

Received: 30 April 2020; Accepted: 12 August 2020; Published: 14 August 2020

Simple Summary: Nocturnal and elusive animals are notoriously difficult to count—hedgehogs being a prime example. Therefore, any reliable way to demonstrate the presence of a particular animal, within a given area, would be a valuable addition to many ecologists' tool kits. The proposed method is based upon the idea that you can find a parasite, specific to a vertebrate animal of interest that has a life stage within an invertebrate host. Molecular detection of these parasites is then carried out in the more abundant and easily collected invertebrate intermediate host. The key to this proposed method is the specificity of the parasite to the vertebrate animal and its detection in the invertebrate intermediate hosts. *Crenosoma striatum* is specific to hedgehogs and was chosen as the parasite to develop the molecular survey tool for hedgehogs, an elusive nocturnal species of considerable interest at present. Results revealed the presence of the nematode only at a site known to be inhabited by hedgehogs confirming the potential of this method to improve the accuracy of recording hedgehog populations.

Abstract: The Western European Hedgehog (*Erinaceous europaeus*) is a nocturnal animal that is in decline in much of Europe, but the monitoring of this species is subjective, prone to error, and an inadequate basis for estimating population trends. Here, we report the use of *Crenosoma striatum*, a parasitic nematode specific to hedgehogs as definitive hosts, to detect hedgehog presence in the natural environment. This is achieved through collecting and sampling the parasites within their intermediate hosts, gastropoda, a group much simpler to locate and sample in both urban and rural habitats. *C. striatum* and *Crenosoma vulpis* were collected post-mortem from the lungs of hedgehogs and foxes, respectively. Slugs were collected in two sessions, during spring and autumn, from Skomer Island ($n = 21$), which is known to be free of hedgehogs (and foxes); and Pennard, Swansea ($n = 42$), known to have a healthy hedgehog population. The second internal transcribed spacer of parasite ribosomal DNA was used to develop a highly specific, novel, PCR based multiplex assay. *Crenosoma striatum* was found only at the site known to be inhabited by hedgehogs, at an average prevalence in gastropods of 10% in spring and autumn. The molecular test was highly specific: One mollusc was positive for both *C. striatum* and *C. vulpis*, and differentiation between the two nematode species was clear. This study demonstrates proof of principle for using detection of specific parasite DNA in easily sampled intermediate hosts to confirm the presence of an elusive nocturnal definitive host species. The approach has great potential as an adaptable, objective tool to supplement and support existing ecological survey methods.

Keywords: Hedgehog; PCR; *Crenosoma striatum*; rDNA; Gastropod; Nematode; Biological tag

1. Introduction

Objective methods for monitoring wild animals are needed to support management efforts, but are rarely straightforward, especially for elusive and nocturnal species. A complete census is usually impossible, and surveys more often rely on observations of individuals and indirect evidence of their presence, such as faecal counts or tracks [1]. With regards to elusive nocturnal animals specifically, even detection can be difficult, as exemplified by carnivore species that are widely dispersed, solitary and nocturnal [1–3]. Locating even the largest of terrestrial mammals, for example, the African forest elephant, can be a difficult task fraught with contestable results [4].

Western European Hedgehogs (*Erinaceus europaeus* Linnaeus, 1758) are classified as a species of least concern [5]; however, there is strong evidence of a recent decline in numbers across mainland Europe and in the UK [6–10]. Estimates suggest a reduction in UK populations within the range of 5–7% in the last 50 years [11], with one study suggesting a potential 25% reduction over the last decade [12]. Current survey methods rely on physical sightings and subjective evidence, such as scats (faecal deposits), tracks and carcases from road deaths, to determine the presence of hedgehogs [13–16]. Given the difficulties in sighting and correctly monitoring nocturnal animals, such as hedgehogs, there is a need to develop a wider panel of objective, evidence-based survey methods to supplement and confirm the findings of those currently used [17].

The use of parasites to monitor host populations has long been employed in the aquatic environment for fish populations [18–21], and more recently to quantify the presence of the elusive diamondback terrapin [22]. The use of parasites and their DNA as biological markers, however, remains underdeveloped in terrestrial environments. The parasitic nematode *Crenosoma striatum* is a lungworm highly specific to hedgehogs [23–28], and common in most populations. In a study of 74 dissected hedgehogs in the UK, 71% were found to be infected with *C. striatum* [29]. While hedgehogs are the sole definitive hosts for *C. striatum*, the available intermediate host range is much wider. Experimental infections comprising species from several gastropod (slug and snail) families of the orders Stylommatophora and Hygrophila [26,30] suggest a large number of potential hosts in hedgehog environments. Terrestrial molluscs are an integral part of many ecosystems and can be found across a diverse range of habitats throughout the British Isles [31–33].

It is here proposed that a polymerase chain reaction (PCR) based test could be used to rapidly and effectively determine the presence of *C. striatum* in local slug and snail populations, thereby indicating the presence or absence of hedgehogs within a given geographical area. If effective, this test would greatly facilitate monitoring of hedgehog distribution, and could potentially be adapted and developed for use in the monitoring of other species of interest. In the present study, this approach is evaluated by first devising a PCR assay specific for *C. striatum*, and then comparing results from areas of known hedgehog presence.

2. Materials and Methods

2.1. Isolation of DNA from Nematodes for Molecular Test Development

Adult worms of *C. striatum* were collected from the lungs of hedgehogs *post mortem*, and identified morphologically [29]. *Crenosoma vulpis*, a closely related species, collected from the lungs of red foxes

(*Vulpes vulpes*) *post mortem*, was also used. DNA was extracted using the DNeasy Blood and Tissue Kit (Qiagen, Hilden, Germany); according to the manufacturer's instructions, except that adult worms were initially ground in ATL buffer using a microfuge pestle. DNA was eluted in 100 µL and stored at −20 °C prior to analysis.

2.2. Primer Design and Multiplex Assay Development

The second internal transcribed spacer (ITS-2) of ribosomal DNA (rDNA) was chosen as the primary region of interest for primer design, due to its successful use in distinguishing between closely related nematodes in numerous previous studies [34–40]. To obtain sequence information for primer design, primer sequences NC1 and NC2 (Table 1, from Gasser et al. [41] 1993) were used to amplify the ITS-2 region of selected parasite DNA for sequencing. PCR conditions were optimised to achieve a single band of the expected size on an agarose gel. Specific products were purified by mini-column (Qiagen) and sequenced in both directions (Eurofins). Sequences obtained were aligned using the ClustalW function in BioEdit software [42], and a consensus sequence established for each species. Sequences from *C. striatum* (*n* = 2) and *C. vulpis* (*n* = 2) were compared with each other and with sequences from *Angiostrongylus vasorum* (a metastrongylid nematode using gastropod intermediate hosts and common in the study area [43], and *Aelustrongylus abstrusus* (a metastrongylid feline lungworm also using gastropod intermediate hosts). This was done to find suitable regions for the design of primers that would allow species differentiation by sequence and PCR product size (as illustrated in Supplementary Figure S1). The ITS2 sequences of *Crenosoma* spp. Were submitted to GenBank with accession numbers MT808322 to MT808325. Primers were designed using Oligo6 (Molecular Biology Insights, Colorado Springs, CO, USA) to uniquely amplify a 157 bp region of *C. striatum* ITS-2 (C.St), and a 207 bp region of *C. vulpis* ITS-2 (C.Vu) (Table 1). Primers were checked with NCBI basic local alignment search tool (BLAST) for species specificity. An independent pair of primers for the amplification of a 710-bp fragment of the invertebrate mitochondrial cytochrome c oxidase subunit I gene (COX1) was selected [44] (henceforth termed COI) as a control to verify that DNA could be amplified from each sample. PCR conditions were optimised for both individual and multiplexed PCRs.

Table 1. The primers and their loci used in PCR tests to identify the presence or absence of *Crenosoma striatum* in slugs.

Primer	Sequence (5′-3′)	Locus	References
NC1	ACGTCTGGTTCAGGGTTGTT	5.8S nuclear rRNA gene	[41]
NC2	TTAGTTTCTTTTCCTCCGCT	28S nuclear rRNA gene	[41]
C.St-ITS2F	CGATTCCCGTTCTAGTTGAGAC	ITS-2 nuclear rDNA	this study
C.St-ITS2R	AAAACCACCTCGACGACATC	ITS-2 nuclear rDNA	this study
C.Vu-ITS2F	CGATTCCCGTTTTAGTTAAGGA	ITS-2 nuclear rDNA	this study
C.Vu-ITS2R	GCTTATCAATCGTCGAATATCATGC	ITS-2 nuclear rDNA	this study
LCO1490	GGTCAACAAATCATAAAGATATTGG	Mitochondrial *cox1* gene	[44]
HC02198	TAAACTTCAGGGTGACCAAAAAATCA	Mitochondrial *cox1* gene	[44]

C.St-ITS2F/R and C.Vu-ITS2F/R = forward and reverse primers for *Crenosoma striatum* and *Crenosoma vulpis*.

PCRs were performed in a volume of 15 µL including 2 µL of template DNA, 2.5 mM MgCl$_2$, 0.2 mM dNTPs (Thermo Fisher, Loughborough, UK) 0.025 µ/µL GoTaq® Flexi polymerase and 1× buffer (Promega, Southampton, UK) and 1× primer mix. 10× primer mixes were COI: 10 mM each primer, optimised multiplex 5 mM each C.St primer and 3 mM each C.Vu primer. The PCRs were carried out on a Biorad T100 Thermal Cycler using a touchdown profile, consisting of an initial denaturation at 95 °C for 3 min followed by nine cycles of 94 °C for 30 s, 65 °C (1 °C decrease per cycle) for 20 s, 72 °C extension then 33 cycles of 94 °C for 30 s 55 °C for 20 s and 72 °C extension. Extension at 72 °C was for 30 s for the multiplex PCR and 1 min for the COI PCR. The final extension was 10 min at 72 °C. PCR products were examined on 1% agarose gels stained with GelRed™ (Biotium Inc., Fremont, CA, USA).

The multiplex PCR was initially checked for analytical specificity by testing against a species panel of DNA isolated from morphologically identified adult lungworms, and confirmed to be diagnostic for C.St and C.Vu (see Supplementary Figure S2).

For PCR testing of slug DNA, an initial control COI PCR was performed prior to the test C.St-C.Vu multiplex, and was negative for some samples, mostly from *Arion ater* slugs, and some appeared tinged with a dark colour. For these, 2 µL of genomic DNA was examined on an agarose gel, and the presence of high molecular weight DNA in the extraction was confirmed. Attempts were made to re-purify to negate the effects of inhibitors. For most samples, PCR was successful with the addition of PCRboost® (Biomatrica, San Diego, CA, USA). Multiplex PCRs were carried out under the same conditions for these samples.

PCRs were repeated twice to verify an amplification (test positivity). Test-negative PCRs were scored only if samples with a positive PCR for the control invertebrate COI PCR. The results of the C.St-C.Vu multiplex on positive COI PCR's were analysed using an exact binomial test.

2.3. Slug Samples

In order to demonstrate the correlation between *C. striatum* incidence and the presence of hedgehogs, slugs were collected in autumn from Skomer Island, covering an area of approximately 160 ha, and in both spring and autumn in Pennard, covering an area of 0.36 ha: Both areas are in south-west Wales, UK. There are no known reports of hedgehogs (or Foxes) on Skomer Island (personal communication with Mark Hodgson, Wildlife Trust South West Wales), whereas Pennard is an area with an abundant local hedgehog population; more than 180 individuals from this particular region were admitted to Gower Bird Hospital wildlife rehabilitation centre between 2001 and 2017. The slugs collected were identified morphologically by BR[4] and then stored at −20 °C before processing. The posterior foot section of each slug was removed and macerated prior to tissue lysis.

2.4. Gastropod DNA Extraction

Genomic DNAs from 80 slugs were extracted from slug tissue using Dneasy Blood and Tissue Kits (Qiagen, Hilden, Germany) employing a Maxwell® 16 MDx Research System (Promega, Maddison, WI, USA) as recommended by the manufacturers. Any undigested tissue and pigment from the larger *Arion ater* specimens were removed by centrifugation before spin column purification. DNA was eluted in 100 µL and stored at −20 °C prior to further analysis.

2.5. Sample Size Calculator

The number of slugs required to be sampled to provide a reliable indicator of the absence of *C. striatum* infection, and hence, the absence of hedgehogs, was simulated using the binomial distribution. Thus, the required sample size was defined as that yielding a <0.05 probability of zero successes (=detected infections), at a given above-zero true prevalence (p. 64, [45]). This is the sample size needed to avoid a type II error, i.e., falsely declaring the absence of *C. striatum* when actually present, at $p = 0.05$.

3. Results

Out of the 80 slugs 17 were excluded (Table 2), due to negative COI result. Slug samples from Pennard collected in spring ($n = 20$) and autumn ($n = 22$) represented nine species. Overall, the prevalence of *C. striatum* in this sample set was 10% (95% exact binomial confidence bounds 3–23%). Species infected with *C. striatum* were *Arion subfuscus* (spring; $n = 1$), *Arion ater* agg. (autumn; $n = 1$) and *Tandonia sowerbyi* (autumn, $n = 2$). Additionally, the *A. ater* agg. Individual was concurrently infected with *C. vulpis*, confirming the sensitivity of the assay without cross-species amplification. The Skomer slug samples collected in autumn ($n = 21$) comprised two species: *A. ater* and *Lehmannia marginata*. Neither *C. striatum* nor *C. vulpis* was detected in any of these samples. Results of the sample size simulation are presented in Figure 1. At the 10% prevalence observed in this study, a sample of

29 slugs would be needed to reasonably (at $p = 0.05$) avoid a false negative, i.e., erroneously conclude that infection is absent. The number of slugs needed would rise at lower prevalence, and fall at higher prevalence.

Table 2. All slug samples included in the proof of principle with their nucleic purity, tissue weight and COI result.

Sample ID	Slug Species	DNA µg/mL	260/280 Ratio	Weight mg	COI	Excluded
	Pennard Spring					
1	*Tandonia sowerbyi*	59.12	1.54	50	+ve	
2	*Tandonia sowerbyi*	86.36	1.60	34	+ve	
3	*Tandonia sowerbyi*	45.64	1.60	40	+ve	
4	*Tandonia sowerbyi*	102.74	1.83	47	−ve	x
5	*Deroceras panormitanum*	151.01	1.91	42	−ve	x
6	*Deroceras panormitanum*	179.24	2.00	63	+ve	
7	*Deroceras panormitanum*	180.09	1.97	55	−ve	x
8	*Deroceras panormitanum*	94.04	1.79	48	−ve	x
9	*Lehmannia marginata*	7.19	1.37	50	+ve	
10	*Lehmannia marginata*	13.98	1.42	56	+ve	
11	*Lehmannia marginata*	108.34	1.77	42	−ve	x
12	*Lehmannia marginata*	54.50	1.89	48	+ve	
13	*Arion hortensis agg.*	30.28	1.55	45	+ve	
14	*Arion hortensis agg.*	37.04	1.70	41	+ve	
15	*Arion hortensis agg.*	25.47	1.63	38	+ve	
16	*Arion hortensis agg.*	26.16	1.49	50	−ve	x
SL-1	*Tandonia sowerbyi*	32.00	2.05	200	+ve	
SL-2	*Tandonia sowerbyi*	325.69	2.03	141	+ve	
SL-3	*Tandonia sowerbyi*	310.11	2.04		+ve	
SL-4	*Arion subfuscus*	464.97	2.00		+ve	
SL-5	*Arion subfuscus*	300.12	2.04		−ve	x
SL-6	*Arion subfuscus*	73.10	1.84		−ve	x
SL-7	*Arion subfuscus*	173.21	1.98		−ve	x
SL-8	*Arion subfuscus*	128.40	1.94		−ve	x
SL-9	*Arion flagellus*	45.35	1.89		+ve	
SL-10	*Arion flagellus*	226.78	2.02		+ve	
SL-11	*Arion flagellus*	153.25	1.90		+ve	
SL-12	*Arion flagellus*	107.87	1.89		+ve	
SL-13	*Arion flagellus*	107.57	1.89		+ve	
SL-14	*Arion flagellus*	200.06	1.99		+ve	

Table 2. Cont.

Sample ID	Slug Species	DNA μg/mL	260/280 Ratio	Weight mg	COI	Excluded
	Pennard Autumn					
11.1	*Arion ater agg.*	27.62	1.46	60	+ve	
11.2	*Arion ater agg.*	46.00	1.81	50	+ve	
11.3	*Arion ater agg.*	23.00	1.61	50	+ve	
11.4	*Arion ater agg.*	57.00	1.71	50	+ve	
28.1	*Arion ater agg.*	107.77	1.70	50	−ve	x
28.2	*Arion ater agg.*	44.54	1.68	50	+ve	
28.3	*Arion ater agg.*	31.04	1.41	50	−ve	x
28.4	*Arion ater agg.*	44.49	1.61	50	+ve	
15.1	*Arion ater agg.*	32.01	1.49	50	+ve	
15.2	*Arion ater agg.*	62.77	1.67	50	+ve	
15.3	*Arion ater agg.*	10.13	1.12	50	+ve	
15.4	*Arion ater agg.*	61.56	1.80	80	+ve	
2.1	*Limax flavus*	0.00	−0.69	50	−ve	x
2.2	*Limax flavus*	17.91	1.26	50	+ve	
2.3	*Limax flavus*	6.27	1.12	60	+ve	
2.4	*Limax flavus*	93.23	1.77	60	+ve	
1.1	*Tandonia sowerbyi*	76.78	0.92	58	+ve	
1.2	*Tandonia sowerbyi*	43.57	1.57	54	+ve	
1.3	*Tandonia sowerbyi*	56.92	1.70	44	+ve	
1.4	*Tandonia sowerbyi*	40.62	1.49	47	+ve	
22.1	*Arion rufus*	2.55	0.72	58	+ve	
22.2	*Arion rufus*	0.59	0.61	48	−ve	x
22.3	*Arion rufus*	19.40	1.21	41	+ve	
22.4	*Arion rufus*	11.28	1.06	52	+ve	
23.1	*Arion rufus*	7.08	1.08	49	+ve	
23.2	*Arion rufus*	25.60	1.34	57	+ve	
	Skomer Autumn					
SK1	*Arion ater*	179.7	1.8	30	+ve	
SK2	*Arion ater*	114	1.85		+ve	
SK3	*Arion ater*	53.2	1.87		+ve	
SK4	*Arion ater*	105.1	1.83		+ve	
SK5	*Arion ater*	129.1	1.85	60	+ve	
SK6	*Arion ater*	172.3	1.75		+ve	
SK7	*Arion ater*	160.3	1.64	100	−ve	x
SK8	*Arion ater*	271.4	1.97	20	−ve	x
SK9	*Arion ater*	266.8	1.93	41	−ve	x
SK10	*Arion ater*	118.8	1.86		+ve	
SK11	*Lehmannia marginata*	107.4	1.86	48	+ve	
SK12	*Lehmannia marginata*	150.8	1.95		+ve	
SK13	*Lehmannia marginata*	174.7	1.93	19	+ve	
SK14	*Lehmannia marginata*	166.4	2.02		+ve	
SK15	*Lehmannia marginata*	161.3	1.93		+ve	
SK16	*Lehmannia marginata*	279.7	1.95		+ve	
SK17	*Lehmannia marginata*	355.2	1.95		+ve	
SK18	*Arion ater*	100.1	1.91		+ve	
SK19	*Arion ater*	73.1	1.82		+ve	
SK20	*Arion ater*	97.2	1.87	83	+ve	
SK21	*Arion ater*	108.7	2.1		+ve	
SK22	*Arion ater*	233.4	1.92		+ve	
SK23	*Arion ater*	130.9	1.99		+ve	
SK24	*Arion ater*	130.6	1.9		+ve	

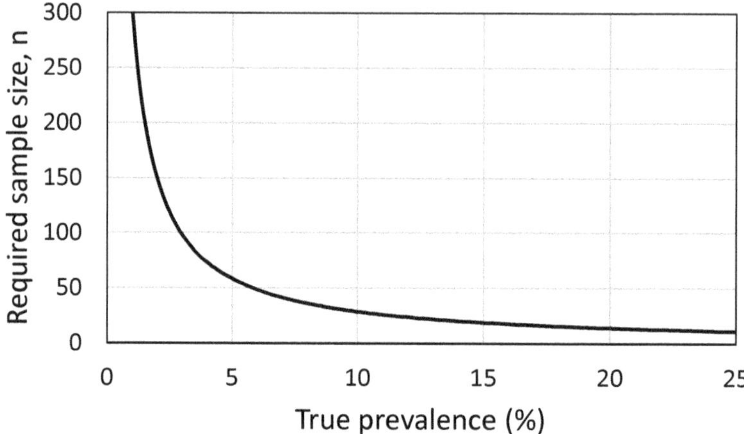

Figure 1. The sample size (=number of slugs) required to detect at least one infected slug, given true prevalence from 1% (*n* = 299) to 25% (*n* = 11). Higher prevalence omitted for clarity: *n* declines further to 5 (at 50% prevalence) and 3 (75%).

4. Discussion

This study demonstrates the use of a multiplex test for *Crenosoma* species, which can accurately identify and discriminate between closely related species *C. striatum* and *C. vulpis* from slug tissues. The fact that no *C. striatum* was detected in the Skomer sample set indicates the potential of *C. striatum* as an indicator species for the presence of elusive hedgehogs in any given locale. Furthermore, the sensitivity of the assay suggests that other parasites highly specific to host species of interest could be used in this way for monitoring and surveillance, for instance as part of management programmes for endangered or invasive species [4,46]. Direct detection of environmental DNA also has potential for monitoring of elusive species [47,48]. Detection of host-specific parasites within intermediate hosts, as proposed here, has the advantages of focusing sampling and potentially longer persistence of DNA in the form of living immature parasite stages.

The methodology described here may need refinement in terms of sample preparation. Some parasites have a preferred site within their host; for instance, *Angiostrongylus vasorum* occupies the right ventricle and pulmonary arteries in its vertebrate hosts [49], whilst *C. striatum* prefers the bronchioles and bronchi of the lungs [26]. The affinity of these parasites to particular sites within the host may extend to the intermediate host, such that sub-sampling of tissue could bias results and affect method sensitivity. Further research needs to be carried out to determine if *C. striatum* has a predilection site in slugs, to increase the efficacy of detection in slug tissue.

To increase the chances of detecting a parasitised slug, species that have been active the longest, and therefore, had the greatest opportunity to acquire parasite infections should, in principle, be targeted for sampling. For example, *A. subfuscus* activity has been seen to peak between May and June with little between-year deviation [50], making it an ideal candidate for spring and summer sampling. The present study found *A. subfuscus* to be the only species with a positive *C. striatum* result in spring sampling. Similarly, *A. ater* and *T. sowerbyi* would be of major interest in autumn and winter sampling, with their peak activity being in January or between August and October, respectively [50]. *Arion ater* may be of particular interest in future research, as it was the only species that presented simultaneous infection with both *C. striatum* and *C. vulpis*. Additionally, the detection of *C. striatum* in *A. ater*, *A. subfuscus* and *T. sowerbyi* appears to be the first confirmed report of infection in these species [30]. This suggests that the potential intermediate host range of *C. striatum* could be much greater than previously thought. Extensions to the present study could further develop the test for hedgehog monitoring through targeting particular slug species and anatomical sites, and by matching the target sample size to the expected prevalence and required precision. The number and cost of PCR assays performed per geographical site could also be reduced by pooling samples from different slugs. These refinements require validation and could establish whether parasite abundance in slugs is related to hedgehog population density, which if it were found to be the case, would enhance its utility as a monitoring tool. Regardless of this relationship, however, results here suggest that presence or absence of *C. striatum* correlates, as expected, with that of its hedgehog definitive host, and can, therefore, be used as a robust indirect indicator of hedgehog presence. The required number of slugs to be sampled in order to reasonably exclude the possibility of *C. striatum* depends on the underlying prevalence, which is unlikely to be known in a newly surveyed site. Further information on the range of prevalence of *C. striatum* infection in gastropods in areas inhabited by hedgehogs would, therefore, be useful to evaluate the feasibility and efficiency of the present approach across the species range.

The approach presented here could be extended to other systems, where highly host-specific parasites are present at reasonably high prevalence, distinguishable from closely related species, and accessible, for example in easily sampled intermediate hosts. The most fundamental of these factors is the host specificity. Host specificity is often under or over-estimated for parasitic species [51], and parasite-host interactions are rarely well-understood in wild animals [52]. Most parasites can infect multiple host species [53–55], albeit to a highly varied extent [56], rendering most as unsuitable for host population studies. Helminths, however, often demonstrate high host-specificity, with nearly 50% of those reported in one study of primates inhabiting a single host species [54]. The sensitivity of the assay presented herein demonstrates that quick and accurate delineation between closely related parasite species can be achieved. It is entirely possible that this methodology could be adapted to other vertebrate species of conservation concern, wherever a suitable parasite species can be identified. To date, only a small number of parasites with singular definitive hosts have been described; Table 3 provides examples of such species. It may be the case that host-specific helminths occur commonly; however, further research is needed in order to clarify this. Furthermore, taxonomic revision frequently leads to a reassessment of host specificity: For example, many nematodes found in amphibia had been previously identified as *Rhabdias rana*, molecular analysis later demonstrated historical misidentification [57], and new species were described as a result. Therefore, it is quite possible that many parasitic species identified before the modern molecular biology era, may have been incorrectly described, increasing the possibility of detecting species-specific and molecularly distinct parasites with potential as indicators of host presence. In addition to taxonomy, ecological factors determine the realisation of potential host range, and are changing in many systems [58]. Shifts in prevalence and host range might have to be taken into account during parasite-based monitoring programmes, and at the same time can provide additional information on host ecology and infection patterns.

Further improvements could be made through development as a loop-mediated isothermal amplification (LAMP-PCR), using similar methodology to that previously described [59,60]. This has potential for a test which could be used in a field setting: Feng et al. [59] found the LAMP-PCR method had lower, but adequate sensitivity for the specific detection of cestode DNA as compared to multiplex PCR, while Abbasi et al. [60] demonstrated 10-fold increased sensitivity over PCR for the detection of *Schistosoma* spp. in infected snails.

Table 3. Examples of helminth species that to date, have only been identified in a singular definitive host.

Parasite Group	Helminth Species	Definitive Host	Common Name	Host Class	Reference
Nematode					
	Abbreviata perenticola	*Varanus giganteus*	Perentie	Reptilia	[61]
	Abbreviata physignathi	*Physignathus lesueurii*	Australian Water Dragon	Reptilia	[61]
	Abbreviata glebopalmae	*Varanus glebopalma*	Black Palmed Rock Monitor	Reptilia	[61]
	Abbreviata barrowi	*Pseudechis australis*	Mulga Snake	Reptilia	[61]
	Nematodirus davtiani alpinus *	*Capra ibex*	Alpine Ibex	Mammalia	[62]
	Filaroides martes	*Martes americana*	American Pine Marten	Mammalia	[63]
	Perostrongylus falciformis	*Meles meles*	European badger	Mammalia	[64]
	Crenosoma striatum *	*Erinaceus europaeus*	European Hedgehog	Mammalia	
	Rhabdias bakeri	*Rana sylvatica*	Wood Frog	Amphibia	[57]
	Rhabdias ambystomae	*Ambystoma maculatum*	Spotted Salamander	Amphibia	[65]
	Parachordatortilis mathevossianae	*Falco tinnunculus*	Common Kestrel	Aves	[66]
	Physaloptera apivori	*Pernis apivorus*	European Honey Buzzard	Aves	[66]
	Physaloptera Mexicana *	*Buteo buteo*	Common Buzzard	Aves	[66]
	Serratospiculum tendo	*Falco peregrinus*	Peregrine falcon	Aves	[66]
Trematode					
	Urotrema scabridum *	*Anolis sagrei*	Brown Anole	Reptilia	[67]
	Pleurogonius malaclemys	*Malaclemys terrapin*	Diamondback Terrapin	Reptilia	[22]
	Mesocoelium lanfrediae	*Rhinella marina*	Cane Toad	Amphibia	[68]
	Parastrigea intermedia	*Circus aeruginosus*	Western Marsh Harrier	Aves	[66]
Cestode					
	Schistotaenia tenuicirrus	*Podiceps grisegena*	Red Necked Grebe	Aves	[69]
	Cladotaenia foxi	*Falco Peregrinus*	Peregrine falcon	Aves	[66]

* denotes parasite species for which the definitive host is geographically isolated from other host species.

5. Conclusions

We conclude that proof of principle has been demonstrated in using terrestrial parasite DNA to confirm the presence of hedgehogs in a given locale. PCR tests can be used to effectively detect and delineate isolates of *C. striatum* and *C. vulpis* from gastropod samples. A critical assessment of different slug tissue and nematode extraction methods, and epidemiological factors, is necessary for the improvement and development of the method described here. This method could provide significant support for monitoring and conservation efforts in hedgehogs, and could pave the way for similar methods to be employed for monitoring of other terrestrial species whose conservation is of concern.

Supplementary Materials: The following are available online at http://www.mdpi.com/2076-2615/10/8/1420/s1, Figure S1: Sequence alignment of ITS-2 sequences showing positions of discriminatory primers resulting in specific PCR products differing in length by 50 bp. Figure S2: PCR of extracted nematode DNA with CS/CV multiplexed primer set illustrating specific amplification.

Author Contributions: Conceptualization, S.A.; Formal analysis, C.G., M.W. and E.R.M.; Funding acquisition, S.A., D.F. and E.R.M.; Investigation, S.A., C.G. and B.R.; Methodology, S.A., C.G., R.B.G., A.J. and E.R.M.; Resources, C.G., D.F., R.B.G., A.J., S.M. and E.R.M.; Supervision, C.G. and E.R.M.; Writing—original draft, S.A., C.G., M.W. and E.R.M.; Writing—review & editing, B.R., D.F., R.B.G., A.J., S.M., M.W. and E.R.M. All authors have read and agreed to the published version of the manuscript.

Funding: Joint funded by a grant from the People's Trust for Endangered Species, the British Hedgehog Preservation Society, the Gower Society and the Welsh Institute for Sustainable Environments.

Acknowledgments: We would like to thank the Gower Society, the People's Trust for Endangered Species, the British Hedgehog Preservation Society and the Welsh Institute for Sustainable Environments for financial support. We would also like to thank the Wildlife Trust of South and West Wales for their help and hospitality on Skomer Island.

Conflicts of Interest: The authors declare no conflict of interest.

Abbreviations

rDNA	ribosomal deoxyribonucleic acid
PCR	Polymerase chain reaction
dNTPs	deoxyribonucleotide triphosphate
mM	millimolar
bp	base pair
agg	aggregate
ha	hectare

References

1. Thompson, W.L. *Sampling Rare or Elusive Species: Concepts, Designs and Techniques for Estimating Population Parameters*; Island Press: Washington, DC, USA, 2004.
2. Kendall, K.C.; Metzgar, L.H.; Patterson, D.A.; Steele, B.M. Power of Sign Surveys To Monitor Population Trends. *Ecol. Appl.* **1992**, *2*, 422–430. [CrossRef]
3. Linnell, J.D.C.; Swenson, J.E.; Kvam, T.; Nikus, N.; Fagrapport, N.; Oppdragsmelding, N. *Methods for Monitoring European Large Carnivores—A Worldwide Review of Relevant Experience*; Institutt Norsk for Naturforskning: Trondheim, Norway, 1998; Volume 549, ISBN 8242609519.
4. Eggert, L.S.; Eggert, J.A.; Woodruff, D.S. Estimating population sizes for elusive animals: The forest elephants of Kakum National Park, Ghana. *Mol. Ecol.* **2003**, *12*, 1389–1402. [CrossRef] [PubMed]
5. Amori, G. *Erinaceus Europaeus The IUCN Red List of Threatened Species 2016: E.T29650A2791303*; IUCN: Gland, Switzerland, 2016. [CrossRef]
6. Holsbeek, L.; Rodts, J.; Muyldermans, S. Hedgehog and other animal traffic victims in Belgium: Results of a countrywide survey. *Lutra* **1999**, *42*, 111–119.
7. Huijser, M.P.; Bergers, P.J.M. The effect of roads and traffic on hedgehog (Erinaceus europaeus) populations. *Biol. Conserv.* **2000**, *95*, 111–116. [CrossRef]
8. Macdonald, D.; Burnham, D. *The State of Britains Mammals—People's Trust for Endangered Species*; The Wildlife Conservation Research Unit, Oxford: London, UK, 2011.

9. Roos, S.; Johnston, A.; Noble, D. *UK Hedgehog Datasets and Their Potential for Long-Term Monitoring*; British Trust for Ornithology: Setford, UK, 2012; pp. 1–63.

10. Krange, M. Change in the Occurrence of the West European Hedgehog (Erinaceus Europaeus) in Western Sweden during 1950–2010. Ph.D. Thesis, Karlstad University, Karlstad, Sweden, 2015.

11. Hof, A.R.; Bright, P.W. Quantifying the long-term decline of the West European hedgehog in England by subsampling citizen-science datasets. *Eur. J. Wildl. Res.* **2016**, *62*, 407–413. [CrossRef]

12. Wembridge, D. *The State of Britain's Hedgehogs 2011*; British Hedgehog Preservation Society & Peoples Trust for Endangered Species: Ludlow, UK, 2011.

13. Cavallini, P. Faeces count as an index of fox abundance. *Acta* **1994**, *39*, 417–424. [CrossRef]

14. Silveira, L.; Jácomo, A.T.A.; Diniz-Filho, J.A.F. Camera trap, line transect census and track surveys: A comparative evaluation. *Biol. Conserv.* **2003**, *114*, 351–355. [CrossRef]

15. McKelvey, K.S.; Kienast, J.V.; Aubry, K.B.; Koehler, G.M.; Maletzke, B.T.; Squires, J.R.; Linduist, E.L.; Loch, S.; Schwartz, M.K. DNA Analysis of Hair and Scat Collected Along Snow Tracks to Document the Presence of Canada Lynx. *Wildl. Soc. Bull.* **2006**, *34*, 451–455. [CrossRef]

16. Yarnell, R.W.; Pacheco, M.; Williams, B.; Neumann, J.L.; Rymer, D.J.; Baker, P.J. Using occupancy analysis to validate the use of footprint tunnels as a method for monitoring the hedgehog Erinaceus europaeus. *Mammal Rev.* **2014**, *44*, 234–238. [CrossRef]

17. McKelvey, K.S.; Aubry, K.B.; Schwartz, M.K. Using Anecdotal Occurrence Data for Rare or Elusive Species: The Illusion of Reality and a Call for Evidentiary Standards. *Bioscience* **2008**, *58*, 549. [CrossRef]

18. Dogiel, V.; Bychowsky, B. Parasites of fisheries of the Caspian Sea. *Tr. Kompleks. Izucheniyu Kaspiiskogo Morya* **1939**, *7*, 150.

19. Herrington, W.C.; Bearse, H.; Firth, F.E. Observations on the life history, occurence, and distribution of the redfish parasite Sphyrion lumpium. *U. S. Bur. Fish. Spec. Rep.* **1939**, *5*, 1–18.

20. MacKenzie, K. Parasites as indicators of host populations. *Int. J. Parasitol.* **1987**, *17*, 345–352. [CrossRef]

21. Williams, H.H.; MacKenzie, K.; McCarthy, A.M. Parasites as biological indicators of the population biology, migrations, diet, and phylogenetics of fish. *Rev. Fish Biol. Fish.* **1992**, *2*, 144–176. [CrossRef]

22. Byers, J.E.; Altman, I.; Grosse, A.M.; Huspeni, T.C.; Maerz, J.C. Using parasitic trematode larvae to quantify an elusive vertebrate host. *Conserv. Biol.* **2011**, *25*, 85–93. [CrossRef]

23. Skrjabin, K.I.; Petrow, A.M. A description of the genus Crenosoma Molin, 1861 (Metastrongylidae, Nematota). *Parasitology* **1928**, *20*, 329–335. [CrossRef]

24. Dougherty, E.C. A review of the genus Crenosoma Molin, 1861 (Nematoda: Trichostrongylidae); its history, taxonomy, adult morphology, and distribution. *J. Helminthol. Soc. Wash.* **1945**, *12*, 44–62.

25. Lammler, G.; Saupe, E. Infection experiments with the lungworm of the hedgehog, Crenosoma striatum. *Z. Parasitenkd.* **1968**, *31*, 87–100.

26. Barus, V.; Blazek, K. The life cycle and the pathogenicity of the nematode Crenosoma-striatum. *Folia Parasitol.* **1971**, *18*, 215–226.

27. Reeve, N. *Hedgehogs*; T. & A.D. Poyser: London, UK, 1994; ISBN1 085661081X. ISBN2 9780856610813.

28. Mirzaei, M. Infection with Crenosoma striatum lungworm in Long-eared Hedgehog (Hemiechinus auritus) in Kerman province southeast of Iran. *Turk. J. Parasitol.* **2015**, *38*, 255–257. [CrossRef]

29. Gaglio, G.; Allen, S.; Bowden, L.; Bryant, M.; Morgan, E. Parasites of European hedgehogs (Erinaceus europaeus) in Britain: Epidemiological study and coprological test evaluation. *Eur. J. Wildl. Res.* **2010**, *56*, 839–844. [CrossRef]

30. Grewal, P.S.; Grewal, S.K.; Tan, L.; Adams, B.J. Parasitism of molluscs by nematodes: Types of associations and evolutionary trends. *J. Nematol.* **2003**, *35*, 146–156. [PubMed]

31. Mason, C.F. Snail populations, beech litter production, and role of snails in litter decomposition. *Oecologia* **1970**, *5*, 213–239. [CrossRef] [PubMed]

32. South, A. *Terrestrial Slugs, Biology, Ecology and Control*; Chapman and Hall: London, UK, 1992; ISBN 0-412-36810-2.

33. Kennedy, C.R. *Ecology of the Acanthocephala*; Cambridge University Press: Cambridge, UK, 2006; ISBN 9780521850087.

34. Chilton, N.B.; Gasser, R.B.; Beveridge, I. Differences in a ribosomal DNA sequence of morphologically indistinguishable species within the Hypodontus macropi complex (Nematoda: Strongyloidea). *Int. J. Parasitol.* **1995**, *25*, 647–651. [CrossRef]

35. Félix, M.A.; Braendle, C.; Cutter, A.D. A streamlined system for species diagnosis in caenorhabditis (Nematoda: Rhabditidae) with name designations for 15 distinct biological species. *PLoS ONE* **2014**, *9*, e0118327. [CrossRef]

36. Gasser, R.B.; Monti, J.R. Identification of parasitic nematodes by PCR-SSCP of ITS-2 rDNA. *Mol. Cell. Probes* **1997**, *11*, 201–209. [CrossRef]

37. Sinclair, R.; Melville, L.; Sargison, F.; Kenyon, F.; Nussey, D.; Watt, K.; Sargison, N. Gastrointestinal nematode species diversity in Soay sheep kept in a natural environment without active parasite control. *Vet. Parasitol.* **2016**, *227*, 1–7. [CrossRef]

38. Wimmer, B.; Craig, B.H.; Pilkington, J.G.; Pemberton, J.M. Non-invasive assessment of parasitic nematode species diversity in wild Soay sheep using molecular markers. *Int. J. Parasitol.* **2004**, *34*, 625–631. [CrossRef]

39. Gasser, R.B. PCR-based technology in veterinary parasitology. *Vet. Parasitol.* **1999**, *84*, 229–258. [CrossRef]

40. Gasser, R.B. Molecular tools—Advances, opportunities and prospects. *Vet. Parasitol.* **2006**, *136*, 69–89. [CrossRef]

41. Gasser, R.B.; Chilton, N.B.; Hoste, H.; Beveridge, I. Rapid sequencing of rDNA from single worms and eggs of parasitic helminths. *Nucleic Acids Res.* **1993**, *21*, 2525–2526. [CrossRef] [PubMed]

42. Hall, T.A. BioEdit: A user-friendly biological sequence alignment editor and analysis program for Windows 95/98/NT. *Nucleic Acids Symp. Ser.* **1999**, *45*, 95–98.

43. Aziz, N.A.A.; Daly, E.; Allen, S.; Rowson, B.; Greig, C.; Forman, D.; Morgan, E.R. Distribution of Angiostrongylus vasorum and its gastropod intermediate hosts along the rural–urban gradient in two cities in the United Kingdom, using real time PCR. *Parasit. Vectors* **2016**, *9*, 56. [CrossRef] [PubMed]

44. Folmer, O.; Black, M.; Hoeh, W.; Lutz, R.; Vrijenhoek, R. DNA primers for amplification of mitochondrial cytochrome c oxidase subunit I from diverse metazoan invertebrates. *Mol. Mar. Biol. Biotechnol.* **1994**, *3*, 294–299. [PubMed]

45. Hilborn, R.; Mangel, M. The Ecological Detective: Confronting models with data. In *The Ecological Detective*; Princeton University Press: Princeton, NJ, USA, 1997; p. 315. ISBN 978-0691034966.

46. Piaggio, A.J.; Engeman, R.M.; Hopken, M.W.; Humphrey, J.S.; Keacher, K.L.; Bruce, W.E.; Avery, M.L. Detecting an elusive invasive species: A diagnostic PCR to detect Burmese python in Florida waters and an assessment of persistence of environmental DNA. *Mol. Ecol. Resour.* **2014**, *14*, 374–380. [CrossRef]

47. Mills, L.S.; Citta, J.J.; Lair, K.P.; Schwartz, M.K.; Tallmon, D.A. Estimating Animal Abundance Using Noninvasive DNA Sampling: Promise and Pitfalls. *Ecol. Appl.* **2000**, *10*, 283–294. [CrossRef]

48. Jones, R.A.; Brophy, P.M.; Davis, C.N.; Davies, T.E.; Emberson, H.; Stevens, P.R.; Williams, H.W. Detection of Galba truncatula, Fasciola hepatica and Calicophoron daubneyi environmental DNA within water sources on pasture land, a future tool for fluke control? *Parasit. Vectors* **2018**, *11*, 1–9. [CrossRef]

49. Morgan, E.R.; Shaw, S.E.; Brennan, S.F.; De Waal, T.D.; Jones, B.R.; Mulcahy, G. Angiostrongylus vasorum: A real heartbreaker. *Trends Parasitol.* **2005**, *21*, 49–51. [CrossRef]

50. Barnes, H.F.; Weil, J.W. Slugs in Gardens: Their Numbers, Activities and Distribution. Part 1. *J. Anim. Ecol.* **1944**, *13*, 140–175. [CrossRef]

51. Poulin, R.; Keeney, D.B. Host specificity under molecular and experimental scrutiny. *Trends Parasitol.* **2007**. [CrossRef]

52. Desdevises, Y.; Morand, S.; Legendre, P. Evolution and determinants of host specificity in the genus Lamellodiscus (Monogenea). *Biol. J. Linn. Soc.* **2002**, *77*, 431–443. [CrossRef]

53. Cleaveland, S.; Laurenson, M.K.; Taylor, L.H. Diseases of humans and their domestic mammals: Pathogen characteristics, host range and the risk of emergence. *Philos. Trans. R. Soc. Lond.* **2001**, *356*, 991–999. [CrossRef] [PubMed]

54. Pedersen, A.B.; Altizer, S.; Poss, M.; Cunningham, A.A.; Nunn, C.L. Patterns of host specificity and transmission among parasites of wild primates. *Int. J. Parasitol.* **2005**, *35*, 647–657. [CrossRef] [PubMed]

55. Woolhouse, M.E.J.; Taylor, L.H.; Haydon, D.T. Population Biology of Multihost Pathogens. *Science* **2001**, *1828*, 1–5. [CrossRef] [PubMed]

56. Poulin, R.; Krasnov, B.R.; Mouillot, D. Host specificity in phylogenetic and geographic space. *Trends Parasitol.* **2011**, *27*, 355–361. [CrossRef]

57. Tkach, V.V.; Kuzmin, Y.; Pulis, E.E. A New Species of Rhabdias From Lungs of the Wood Frog, Rana Sylvatica, in North America: The Last Sibling of Rhabdias Ranae? *J. Parasitol.* **2006**, *92*, 631–636. [CrossRef]

58. Cable, J.; Barber, I.; Boag, B.; Ellison, A.R.; Morgan, E.R.; Murray, K.; Pascoe, E.L.; Sait, S.M.; Wilson, A.J.; Booth, M. Global change, parasite transmission and disease control: Lessons from ecology. *Philos. Trans. R. Soc. B Biol. Sci.* **2017**, *372*. [CrossRef]

59. Feng, K.; Li, W.; Guo, Z.; Duo, H.; Fu, Y.; Shen, X.; Tie, C.; Rijie, E.; Changqin, X.; Luo, Y.; et al. Development of LAMP assays for the molecular detection of taeniid infection in canine in Tibetan rural area. *J. Vet. Med. Sci.* **2017**, *79*, 1986–1993. [CrossRef]

60. Abbasi, I.; King, C.H.; Muchiri, E.M.; Hamburger, J. Detection of Schistosoma mansoni and Schistosoma haematobium DNA by loop-mediated isothermal amplification: Identification of infected snails from early prepatency. *Am. J. Trop. Med. Hyg.* **2010**, *83*, 427–432. [CrossRef]

61. Jones, H.I. Physalopterine nematodes in Australian reptiles: Interactions and patterns of infection. *Aust. J. Zool.* **2014**, *62*, 180–194. [CrossRef]

62. Zaffaroni, E.; Manfredi, M.T.; Citterio, C.; Sala, M.; Piccolo, G.; Lanfranchi, P. Host specificity of abomasal nematodes in free ranging alpine ruminants. *Vet. Parasitol.* **2000**, *90*, 221–230. [CrossRef]

63. Seville, R.S.; Addison, E.M. Nongastrointestinal helminths in marten (Martes americana) from Ontario, Canada. *J. Wildl. Dis.* **1995**, *31*, 529–533. [CrossRef] [PubMed]

64. Deak, G.; Mihalca, A.D.; Hirzmann, J.; Colella, V.; Alexandru, F.; Cavalera, M.A.; Gheorghe, F.; Bauer, C.; Monica, A.; Ali, A.; et al. Validity of genus Perostrongylus Schlegel, 1934 with new data on Perostrongylus falciformis (Schlegel, 1933) in European badgers, Meles meles (Linnaeus, 1758): Distribution, life-cycle and pathology. *Parasite Vector* **2018**, *56*, 1–16. [CrossRef] [PubMed]

65. Kuzmin, Y.; Tkach, V.V.; Snyder, S.D. Rhabdias ambystomae sp. n. (Nematoda: Rhabdiasidae) from the north American spotted salamander Ambystoma maculatum (Amphibia: Ambystomatidae). *Comp. Parasitol.* **2001**, *68*, 228–235.

66. Santoro, M.; Kinsella, J.M.; Galiero, G.; degli Uberti, B.; Aznar, F.J. Helminth Community Structure in Birds of Prey (Accipitriformes and Falconiformes) in Southern Italy. *J. Parasitol.* **2012**, *98*, 22–29. [CrossRef]

67. Bundy, D.A.P.; Harris, E.A. Helminth parasites of jamaican anoles (Reptilia: Iguanidae): A comparison of the helminth fauna of 6 anolis species. *J. Helminthol.* **1987**, *61*, 77–83. [CrossRef]

68. Gomes, T.F.F.; Melo, F.T.V.; Giese, E.G.; Furtado, A.P.; Gonçalves, E.C.; Santos, J.N. A new species of Mesocoelium (Digenea: Mesocoeliidae) found in Rhinella marina (Amphibia: Bufonidae) from Brazilian Amazonia. *Mem. Inst. Oswaldo Cruz* **2013**, *108*, 186–191. [CrossRef]

69. Stock, T.M.; Holmes, J.C. Host specificity and exchange of intestinal helminths among four species of grebes (Podicipedidae). *Can. J. Zool.* **1987**, *65*, 669–676. [CrossRef]

Article

3D Geometric Morphometrics Reveals Convergent Character Displacement in the Central European Contact Zone between Two Species of Hedgehogs (Genus *Erinaceus*)

Barbora Černá Bolfíková [1,*,†], Allowen Evin [2,†], Markéta Rozkošná Knitlová [3], Miroslava Loudová [3], Anna Sztencel-Jabłonka [4], Wiesław Bogdanowicz [4] and Pavel Hulva [3,5]

[1] Faculty of Tropical AgriSciences, Czech University of Life Sciences Prague, Kamýcká 129, 165 21 Prague, Czech Republic
[2] Institut des Sciences de l'Evolution—Montpellier (ISEM), Univ Montpellier, CNRS, EPHE, IRD, 2 place Eugène Bataillon, CC065, CEDEX 5, 34095 Montpellier, France; allowen.evin@umontpellier.fr
[3] Department of Zoology, Faculty of Science, Charles University in Prague, Viničná 7, 128 00 Prague, Czech Republic; knitlova@natur.cuni.cz (M.R.K.); miroslava.loudova@natur.cuni.cz (M.L.); hulva@natur.cuni.cz (P.H.)
[4] Museum and Institute of Zoology, Polish Academy of Sciences, Wilcza 64, 00-679 Warszawa, Poland; brysio@miiz.waw.pl (A.S.-J.); wieslawb@miiz.waw.pl (W.B.)
[5] Faculty of Science, University of Ostrava, Chittussiho 10, 710 00 Ostrava, Czech Republic
* Correspondence: bolfikova@ftz.czu.cz; Tel.: +420-22438-2497
† These authors contributed equally.

Received: 25 August 2020; Accepted: 30 September 2020; Published: 4 October 2020

Simple Summary: Hedgehogs, being insectivores with slow metabolisms, are quite sensitive to temperature and food availability. As a consequence, their ranges have oscillated in relation to past climate changes. Species that have evolved in different regions, but their ranges have shifted and overlapped subsequently, often represent intense competitors as a result of ecological similarities. The present study focuses on this phenomenon in the contact zone in central Europe and adjacent regions, using genetic determination of species and description of size and shape of skull, the morphological structure mirroring many selection pressures related to ecology. While animals living outside of the contact zone show marked differences between the two species, individuals within the contact zone are more alike with a smaller skull size and a convergent jawbone shape. Changes in skull size can be related to inter-species competition and also facilitated by selection pressure, mediated by overpopulated medium-sized predators such as foxes or badgers. Since the function of the lower jaw is mainly connected to feeding, we hypothesize that this pattern is due to the selection to size and shape related to competition for food resources. The present study helps to describe general patterns related to species formation, as well as species responses to anthropogenic environmental changes.

Abstract: Hedgehogs, as medium-sized plantigrade insectivores with low basal metabolic rates and related defensive anti-predator strategies, are quite sensitive to temperature and ecosystem productivity. Their ranges therefore changed dramatically due to Pleistocene climate oscillations, resulting in allopatric speciation and the subsequent formation of secondary contact zones. Such interactions between closely related species are known to generate strong evolutionary forces responsible for niche differentiation. In this connection, here, we detail the results of research on the phenotypic evolution in the two species of hedgehog present in central Europe, as based on genetics and geometric morphometrics in samples along a longitudinal transect that includes the contact zone between the species. While in allopatry, *Erinaceus europaeus* is found to have a larger skull than *E. roumanicus* and distinct cranial and mandibular shapes; the members of the two species in sympatry are smaller and more similar to each other, with a convergent shape of the mandible.

The relevant data fail to reveal any major role for either hybridisation or clinal variation. We, therefore, hypothesise that competitive pressure exerted on the studied species does not generate divergent selection sufficient for divergent character displacement to evolve, instead giving rise to convergent selection in the face of resource limitation in the direction of smaller skull size. Considering the multi-factorial constraints present in the relevant adaptive landscape, reduction in size could also be facilitated by predator pressure in ecosystems characterised by mesopredator release and other anthropogenic factors. As the function of the animals' lower jaw is mainly connected with feeding (in contrast to the cranium whose functions are obviously more complex), we interpret the similarity in shape as reflecting local adaptations to overlapping dietary resources in the two species and hence as convergent character displacement.

Keywords: convergent character displacement; *Erinaceus*; geometric morphometrics; species interactions

1. Introduction

Hedgehogs from the Western Palearctic play a key role as model organisms in the field of phylogeography and speciation studies (e.g., [1–3]). As medium-sized insectivores, they are sensitive to temperature and thus, to the productivity of ecosystems, to the extent that their ranges changed substantially during the Pleistocene climatic oscillations [2]. Indeed, isolation in southern refugia facilitated allopatric speciation scenarios, resulting in a recent pattern of east–west parapatry, with the northern white-breasted hedgehog (*Erinaceus roumanicus*) present in eastern Europe, and the European hedgehog (*E. europaeus*) occurring in western Europe. The two species form a secondary contact zone in central Europe (Italy, Austria, Czech Republic and Poland), with a relatively broad area of sympatry in the center of the zone (Czech Republic), possibly in relation to Neolithic deforestation [3]. This region therefore provides conditions suitable for studying species interactions in relation to genomic and ecological niche-differentiation in the context of anthropogenic environmental changes.

There has been little gene flow between *E. europaeus* and *E. roumanicus* [3–5]. Alongside genetic differentiation, there is a divergence of phenotypic traits in the closely related species which is an important source of information about adaptive processes, i.a. indicating nascent niche diversification [6]. In general, phenotypes respond to both abiotic and biotic factors. In sympatry, environmental factors are identical, but ecological, microallopatric or trophic differentiation may occur [7]. The integration of originating species into ecological networks also varies during the speciation process, in line with an increasing role for competition with the sister species that may facilitate the niche differentiation [8].

However, phenotypic variation in species with extensive ranges usually shows pronounced geographical variation, which complicates the comparison of allopatric populations. For example, the body size of *E. roumanicus* increases linearly from north to south in Europe and is thus shown to correlate positively with temperature, as well as negatively with precipitation in the summer [9]. It is hypothesised that size is in this case determined by seasonality of the resource availability [9].

Comparisons between the two species in allopatry thus tend to be lacking, given the requirement of sampling across the entire range, and/or controlled design of the study and homologous comparison, as regards environmental impact. Reeve [10] states that the species are of approximately the same size and weight, while comparisons between populations from Great Britain, Italy, Switzerland, Germany, Poland, Russia and Ukraine by Ruprecht [11] failed to find a key character allowing the species to be distinguished. The above author only sums up that "the western species is more differentiated in skull dimensions than the eastern species". Phenotypic differences between the white-breasted and European hedgehogs by reference to traditional morphometrics, or non-metric, discrete characters in the cranial phenotype prove relatively difficult. Hedgehogs are traditionally distinguished by cranial indices and the length of naso-maxillary sutures [12–14]; these characters in fact show marked intraspecific variability and interspecific overlap. Equally, sexual dimorphism related to size has never been found in either *E. roumanicus* or *E. europaeus* [15,16].

In sympatry, *E. europaeus* has a higher mean body mass, mean body length, mean hindfoot length and mean ear length than *E. roumanicus* in all adult age classes, as well as higher values for cranial indexes [15,17,18]. *Erinaceus roumanicus* has a longer tail in adult categories [17]. Body mass and neurocranial capacity in turn increase at a higher rate in *E. roumanicus* [17,19]. Deciduous dentition is replaced earlier in *E. roumanicus* than in *E. europaeus*, while also showing fewer deviations from the normal dental formula [20].

The objective of this research was to compare phenotypic differences in allopatric and sympatric populations of the above-mentioned species in terms of size and shape of their skulls, by reference to geometric morphometrics and genetic determination of species. The skull, as a complex morphological structure integrating traits associated with cognitive, sensory and food-processing functions, reflects the diverse selection pressures associated with the ecological niche [6,21]. In order to eliminate the effect of environmental adaptation (position in relation to the north–south and oceanic–continental gradients), we used longitudinal transect sampling spatially crossing the relatively bounded region in central Europe. We expect that the unified environmental context will allow us to describe in detail the interspecific differentiation, resulting from the speciation process. We hypothesise that sympatric populations will be affected by character displacement. We aim to discuss the observed patterns also in the perspective of environmental changes of the Anthropocene.

2. Materials and Methods

2.1. Specimens

Only adult specimens were used in our analyses, the age was estimated based on the date of death, dental abrasion [22], presence of milk teeth [20] and skull proportions [15,17]. A total of 69 skulls were examined (Table S1 in Supplementary Material), 29 of them represented *E. europaeus*, 25 *E. roumanicus* and 15 interspecific hybrids. The specimens originated from different localities (Figure 1) within the zone of sympatry (Czech Republic—14 *E. europaeus*, 13 *E. roumanicus*) and from allopatric localities in central Europe (Germany—15 *E. europaeus*, Slovakia—12 *E. roumanicus*). Samples from areas of sympatry were obtained from individuals that died in rescue centers. Skulls from areas of allopatry were borrowed from the Museum für Naturkunde, Berlin (Germany; MFN) and the Institute of Vertebrate Biology, Brno (Czech Republic; IVB). The skulls of interspecific hybrids deposited at MFN have also been examined. Seven such specimens originated from hybridisation experiments carried out in captivity by Herter [23], while eight hybrids were collected from the wild (w-hybrid) in Germany and the Czech Republic. Hybrid individuals show morphological traits of both species regarding classical morphometric indices as mentioned above [23].

2.2. Genetic Analyses

Considering the low frequency of interspecific hybridisation (one backcrossed individual out of 210 tested in Czechia and Slovakia) [3,4], species determination based on phenotypes has been confirmed by sequencing the mitochondrial control region in the samples originating from the zone of sympatry (N = 27). The tissue was collected in the course of skull preparation, and fixed in 96% ethanol, prior to storage at −20 °C. DNA was extracted using a DNA Blood and Tissue Kit (Qiagen, Prague, Czech Republic). Amplification of the mtDNA control region was achieved in line with the protocol used in Bolfíková and Hulva [3]. Sequences obtained were compared with those available in the GenBank® database.

Figure 1. Species ranges and localities of samples used in the present study. The species are marked in different colours: blue represents *E. europaeus*, red represents *E. roumanicus* and orange indicates wild hybrids of the two species. The contact zone position (based on available data) is highlighted in violet. The map was created using ArcGIS Online.

2.3. 3D Geometric Morphometrics Approach and Statistical Analyses

Skull sizes and shapes were assessed using three-dimensional landmark-based geometric morphometrics. A total of 47 three-dimensional coordinates were recorded corresponding to 13 landmarks on the mandible, and 34 landmarks on the cranium divided into 13 and 21 landmarks on the ventral and dorsal side, respectively (Figure 2, Table S2). Landmark coordinates were acquired using the Reflex microscope and Axel software (Reflex Measurement Ltd., Butleigh, Somerset BA6 8SP, UK) at the Museum and Institute of Zoology of the Polish Academy of Sciences with headquarters in Warsaw. All measurements were taken by the same person (MK).

The coordinates were superimposed using a generalised Procrustes analysis algorithm [24,25]. During this procedure, all specimens are translated, so the center of gravity of their landmark configuration coincides, normalised to the unit centroid size and rotated to minimise the squared summed distances between corresponding landmarks. Centroid size (CS) corresponds to the square root of the sum of squared distances of the landmarks from their centroid [26]. The coordinates after superimposition correspond to the shape data.

Differences in the logarithm of CS between groups were depicted using a boxplot. Because of the small number of specimens in some of the comparisons, significance of the between-group differences was tested using the non-parametric Wilcoxon rank tests for two groups, and the Kruskal–Wallis tests when more than two groups were compared. Shape differences were tested using one-way multivariate analysis of variance (MANOVA) and canonical variate analyses (CVA) in combination with leave-one-out cross validation percentages (CVP), following Evin et al. [27], taking into account the unbalanced sample size between groups. Mean CVP values are provided with a 90% confidence interval obtained with 100 resamples [27]. Shape changes along the CVA axes were visualised by calculating shape changes along the factorial axes using multivariate regression [28] using the 'Morpho v2.8' R package [29]. Because sample sizes were relatively small compared to the large number of variables, we applied dimensionality reduction of the data prior to MANOVAs and CVAs by substitution of the primary data by the first scores of the principal component analyses (PCA) maximizing the leave-one-out cross validation between groups [27,30].

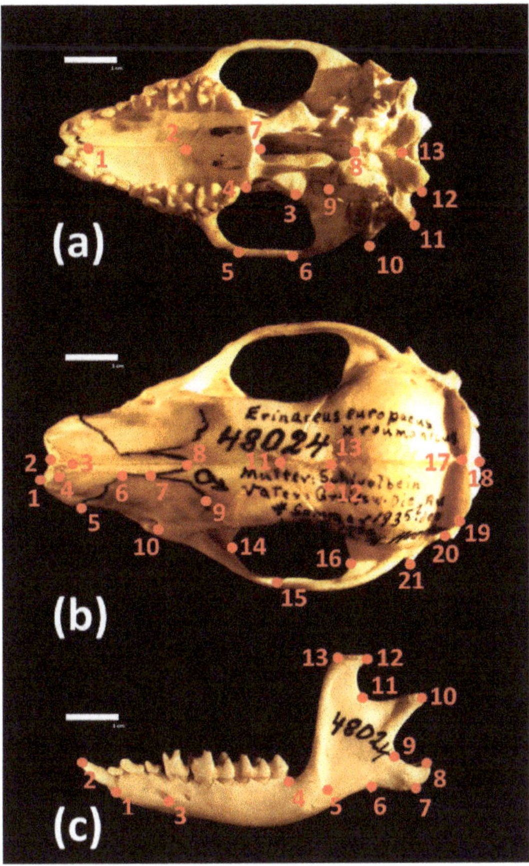

Figure 2. Location of the three-dimensional landmarks measured on (**a**) ventral and (**b**) dorsal sides of the cranium, and (**c**) the mandible.

To test for possible common differences in size and shape between allopatric and sympatric populations of the two species, we applied two-way ANOVA and MANOVA using size or shape as variables, species as the main classifier, and allopatric versus sympatric distribution as a sub-classifier factor. When necessary, *p*-values were adjusted for multiple comparisons after Benjamini and Hochberg [31]. Overall phenotypic similarities between groups were depicted using neighbour-joining networks, computed based on the Mahalanobis D^2 distances [32]. When allopatric and sympatric populations of the two species were compared, CVP were calculated for the main branches of the resulting neighbour-joining networks. The relationship between phenotypic data and geographic origin (latitude and longitude) was explored using the Mantel-test [33] for shape (Procrustes distances were used, and a randomized approach with 999 replicates), and regression [34] for centroid size for each structure and each species separately. To test whether the observed patterns of similarity can be explained by convergence [35], we adopted measures of the multidimensional convergence index (MCI) [36] calculated as the ratio of the Procrustes variance [37] within the putatively convergent lineages (i.e., sympatric populations) and within their sister lineages (i.e., allopatric populations). The obtained MCI values were compared to the distribution of 999 randomised MCI values (allopatric/sympatric attribution was randomised). Analyses were performed using the geomorph v3.2.0 [38] and ade4 v1.7.13 [39] for packages for R v3.6.3 [40].

3. Results

3.1. Differences between the Species

Morphologically related visual determinations of the species proved consistent with genetic assignment of animals within the sympatric population. On average, specimens of *E. europaeus* have a larger ventral side to the cranium and mandibles than those of *E. roumanicus* (Figure 3, Table 1), as well as a distinct cranial and mandibular shape (Table 1, Figure 4). When sympatric and allopatric populations are analysed jointly, 89.0% (CI: 86–92%), 94.8% (CI: 90–98%) and 84.0% (CI: 80–90%) of ventral, dorsal and mandibular shapes support the correct assignment of individuals to species (Table 1, Figure 4). If size alone is taken into account, the levels of correct cross-validation will drop respectively to 85.5% (CI: 84–88%), 78.2% (CI: 72–82%), and 72.7% (CI: 70–74%) (Table 1).

In terms of shape differences, *E. roumanicus* differs from *E. europaeus* in possessing a ventral size of the brain case that is proportionally more rounded (Figure S1a), as well as a dorsal side of the skull that is proportionally narrower (Figure S1b), and a mandible that is proportionally thinner, with the most anterior part shifted backward (Figure S1c).

3.2. Contrasted Species Differentiation in Allopatry and Sympatry

When allopatric and sympatric populations are analysed separately, as opposed to via pooled analysis, the results are seen to differ. In allopatry, the species exhibit significant differences in sizes and shapes of crania and mandibles (Table 1), to the extent that the minimal mean cross-validation between them is of 84.2% for the size of the dorsal side of the cranium (Table 1). *Erinaceus roumanicus* shows a smaller skull size than *E. europaeus* in all comparisons (Figure 3).

Table 1. Test of differences between species: overall (pooling sympatric and allopatric specimens) and separately sympatric or allopatric populations only. Differences in shape and size were tested using one-way multivariate analysis of variance (MANOVA) and Wilcoxon's tests, respectively, for ventral and dorsal sides of the cranium and the mandible. P: *p*-value, numDf and denDf: Numerator and denominator degrees of freedom, W: statistic of the Wilcoxon's tests. The numerator degree of freedom corresponds to the number of Principal Component (PC) scores included in the analyses. Cross-validation percentages (CVP) are provided as the mean and the 90% confidence interval of the distribution.

Trait		Shape			Size		
		F(numDf, denDf)	p	CVP	W	p	CVP
Mandible	Overall	$F(22,30) = 5.94$	5×10^{-6}	83.3% (80–88.1%)	16.648	4×10^{-5}	72.5% (70–76%)
	Sympatry	$F(15,10) = 2.16$	0.11	-	134	0.01	69.2% (69.2–69.2%)
	Allopatry	$F(13,13) = 11.4$	4×10^{-5}	89.8% (79–95.8%)	164	1×10^{-4}	85.7% (79–87.7%)
Ventral	Overall	$F(23,27) = 8.68$	2×10^{-7}	88.7% (84–92.1%)	25.12	5×10^{-7}	85.3% (84–88%)
	Sympatry	$F(19,5) = 4.47$	0.05	-	133	0.04	66.6% (61.5–69.2%)
	Allopatry	$F(10,15) = 7.32$	3×10^{-4}	92.3% (91.7–95.8%)	155	7×10^{-5}	92.2% (91.7–95.8%)
Dorsal	Overall	$F(29,24) = 12.41$	1×10^{-8}	95.0% (92–98%)	15.713	7×10^{-5}	78.2% (72–82%)
	Sympatry	$F(19,7) = 17$	4×10^{-4}	90.8% (84.6–100%)	145	7×10^{-5}	79.5% (79.2–83.3%)
	Allopatry	$F(15,11) = 6.6$	1×10^{-3}	84.25% (83.3–87.5%)	167	4×10^{-5}	84.7% (83.3–87.5%)

In relation to both size and shape, mean cross-validation percentages for the sympatric populations are always lower than those obtained in allopatry (Table 1). The sympatric populations of *E. europaeus* and *E. roumanicus* only differ significantly in dorsal skull shape, as well as the size of the mandibular and ventral skull view, with *E. roumanicus* characterised by smaller ventral size (Figure 3).

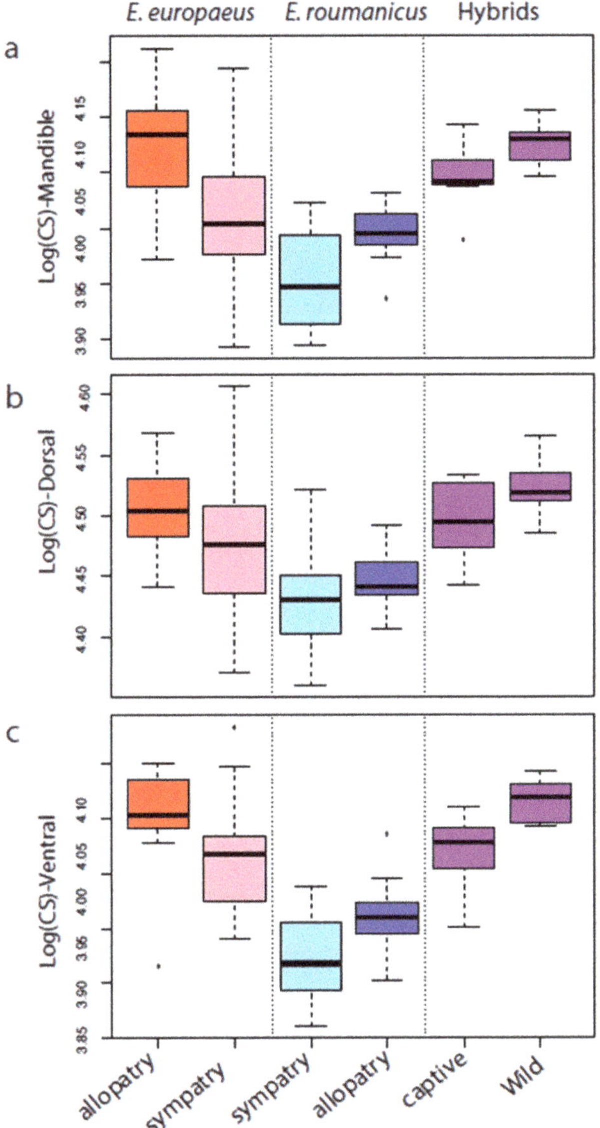

Figure 3. Cranial and mandibular size differences and variability between allopatric, sympatric and hybrid (captive and wild) populations of *E. europaeus* and *E. roumanicus*. Boxplots of the log centroid sizes (CS) of the mandible (**a**), dorsal (**b**) and ventral (**c**) sides of the cranium.

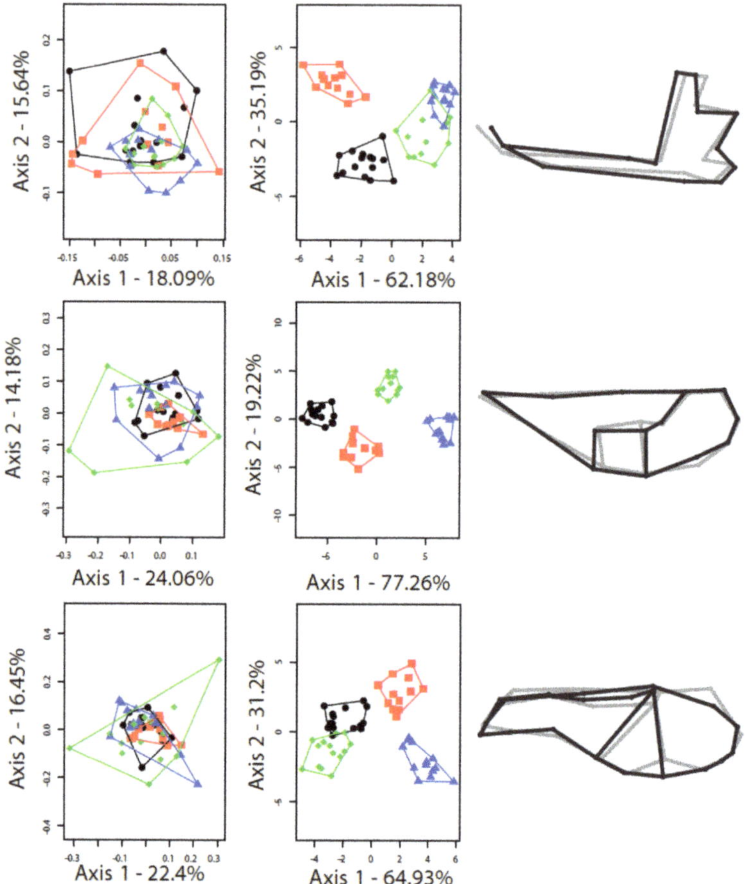

Figure 4. Principal component analysis (PCA; left section), discriminant analysis (DA; central section) and visualisation of the differences in shape between the two species (right section) in relation to mandible (top), ventral side (middle) and dorsal side (bottom). For PCA and DA, black stands for *E. europaeus* in allopatry, green for *E. europaeus* in sympatry, red for *E. roumanicus* in allopatry, blue for *E. roumanicus* in sympatry. For shape comparison, black stands for *E. roumanicus* and grey for *E. europaeus*; shape differences are amplified by a factor of 5.

3.3. Comparison of Allopatric and Sympatric Populations

Both species are of smaller size in sympatry (Figure 3), with a further effect being that the two are also more similar in size than they are in allopatry. CVA analyses thus reveal marked differences between allopatric and sympatric populations, as well as between the two species (Figure S1).

The overall phenotypic dissimilarity between the populations (Figure 5) shows a greater similarity between the sympatric populations of the two species than between the two (sympatric and allopatric) populations of the same species, when it comes to the ventral side of the cranium and the mandible (Figure 5a′,c′). In the case of these two structures, morphometric differentiation is more affected by allopatric/sympatric status than by taxonomy. Conversely, for the dorsal side of the cranium (Figure 5b′), the main differentiation is between the two species. The interaction term of the two-way analyses of variance reveals homogeneous patterns of differentiation between allopatric and sympatric populations of the two species (size and shape, all $p > 0.5$; Figure 5), with the one exception relating to mandible shape ($F_{14,49} = 3.27$, $p < 0.01$).

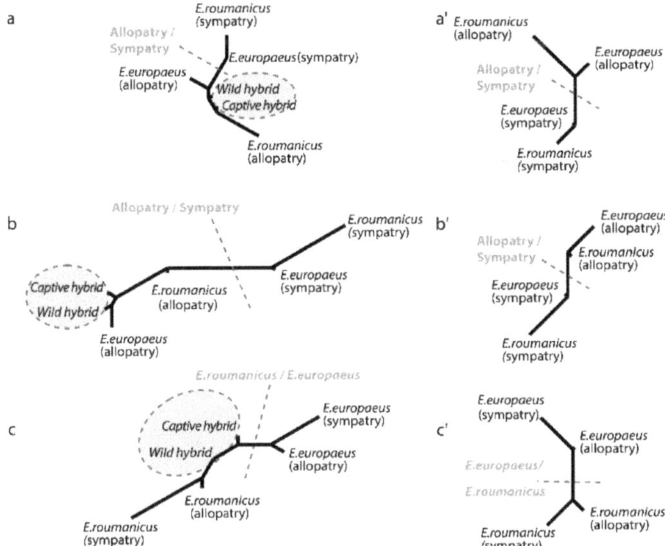

Figure 5. Overall shape differences among examined groups of *E. europaeus* and *E. roumanicus*: Right—between sympatric and allopatric populations of the two species; left—among all groups (allopatric, sympatric and hybrid populations). Neighbour-joining networks of the Mahalanobis distances for the mandible (**a**,**a′**), and the ventral (**b**,**b′**) and dorsal (**c**,**c′**) sides of the cranium.

MCI yielded values of 0.426 for the ventral and 0.441 for the dorsal side of the cranium, indicating the absence of convergence, while for the mandible, a convergence pattern with the MCI value of 1.541—well above one—and above 95% of the random MCI distribution (Figure 6), was detected.

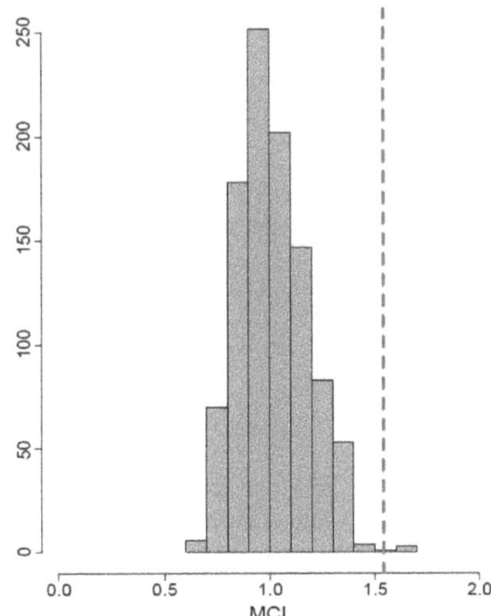

Figure 6. Multidimensional convergence index (MCI) of the sympatric and allopatric populations of both species obtained for the mandible (vertical dotted line) compared with a randomized distribution of MCI values.

3.4. Differences between the Two Hybrid Populations

Captive and wild hybrids do not differ in shape, in relation to any of the structures analysed (Table 2), though wild hybrids have a larger ventral side of the cranium than their captive counterparts (Figure 3). Neighbour-joining networks reveal closer proximity of the hybrids to the allopatric populations of the two species where data for the ventral side of the skull are concerned, as well as the mandible (Figure 5a,b). In relation to the dorsal side of the cranium, hybrids are seen to cluster between the two parent species (Figure 5c).

Table 2. Test of differences between captive and wild hybrids. Differences in shape and size were tested using MANOVA and Wilcoxon's tests, respectively, for the ventral and dorsal sides of the cranium and the mandible. P: *p*-value, numDf and denDf: Numerator and denominator degree of freedom. Cross-validation percentages (CVP) are provided as the mean and the 90% confidence interval (CI) of the distribution.

Trait	Mandible		Ventral		Dorsal	
	Tests	**CVP**	**Tests**	**CVP**	**Tests**	**CVP**
Shape	$F(2,12) = 0.6$, $p = 0.56$	-	$F(2, 12) = 2.8$, $p = 0.09$	-	$F(10,4) = 4.08$, $p = 0.09$	-
Size	$W = 46$, $p = 0.04$	73.9% (64.3–85.7%)	$W = 52$, $p = 0.004$	79.4% (71.4–92.9%)	$W = 40$, $p = 0.19$	-

3.5. Geographical Structure

Unlike in the case of the dorsal side of the skull—for which no geographical structure could be detected, data for the ventral side and the mandible are seen to vary geographically (Table 3), with differences between the two species noted. The population of *E. europaeus* shows a geographical structure, as regards data on mandibular shape, while specimens of *E. roumanicus* also vary geographically in this respect, as well as in the size and shape of the ventral cranial side. However, the sampling on this remains limited and this conclusion is awaiting better support.

Table 3. Geographic structure of the data. Results of Mantel tests for shape and regression for size for ventral and dorsal sides of the cranium and the mandible.

Trait	*E. europaeus*				*E. roumanicus*			
	obs	*p*-**Value**	**adj. R2**	*p*-**value**	**obs**	*p*-**Value**	**adj. R2**	*p*-**Value**
Mandible	0.283	0.036	0.167	0.039	0.221	0.005	0.274	0.011
Ventral	−0.09	0.775	0.04	0.232	−0.066	0.762	0.18	0.042
Dorsal	−0.070	0.626	−0.01	0.44	−0.014	0.53	−0.02	0.47

4. Discussion

4.1. Convergent Character Displacement in Size and Shape

Our first-ever reference to high-resolution three-dimensional geometric morphometrics in the case of hedgehogs helped reveal fine-scale differences between the two species in allopatry, while showing surprising similarity in circumstances of sympatry. Where matters of size are concerned, specimens of *E. europaeus* allopatric from *E. roumanicus* have larger crania and mandibles. In contrast, when present sympatrically, populations of the two species show a similar reduction in size, resulting in a much greater overlap than in allopatry, to the extent that cross-validation percentages are always lower in sympatric situations.

As regards shape, marked differences were detected for both species, between allopatric and sympatric populations. It emerged that most of the shape variation is clustered by allopatric/sympatric differences in ventral skull shape, as well as mandibles; while data for dorsal skull shape appear to be structured primarily by reference to inter-species differences. The overlap in the mandible variation, characterising sympatric populations in the discriminant analysis (Figure 4) confirmed by: their close proximity on the network (Figure 5); their lower percentage in cross-validation than their allopatric counterparts (Table 1); and the high MCI value, can be interpreted as a case of convergence in line with the pattern-based definition (Stayton 2015). However, interpretation of these results in accordance with the process-based definition of convergence (i.e., as bilateral convergent character-displacement) requires the examination of possible microevolutionary factors responsible for the presence of more-unified phenotypic traits under circumstances of sympatry.

4.2. Hybridisation and Introgression

Hybridisation might be the simplest explanation for the observed patterns, with hybrids in this case known to be large, and more similar in size to *E. europaeus* (perhaps as a reflection of the genome dominance of this species when it comes to the determination of size; or else in line with asymmetry in the ability to produce back-crosses). The experiments of Poduschka and Poduschka [41], combined with our data, suggest that asymmetrical introgression is to be anticipated (and therefore also asymmetrical changes in morphology).

In terms of shape, hybrids fall between the two parental allopatric morphologies when it comes to mandibles, and the ventral side of the skull. This is consistent with the renewal of the plesiomorphic state in hybrids. Interestingly, as captive and wild hybrids do not differ in skull shape (while the former have larger crania and mandibles than their wild counterparts), the suggestion is that populations held in isolation may develop different morphological traits [42,43].

Although past introgression events may have had a role to play in hedgehogs [4], the recent level of hybridisation is very low [3,5]. However, a genomic approach and ascertainment of the level of ancient introgression will be necessary if the potential role of hybridisation in phenotypic evolution is to be investigated fully.

4.3. Clinal Variation

A second potential explanation would involve general trends to the clinal variation characteristics of both species, possibly affected by the same selective environmental gradient that overrides the effect of competition [44]. However, given the limited width of the contact zone between the hedgehog species, it is difficult to imagine a longitudinal gradient taking on extreme values in central Europe. For example, Škoudlín [45] measured 23 metric characters and four proportional indices related to the skull, in order to compare specimens of *E. roumanicus* from the Czech Republic, Poland and Belarus. Those from Poland and Belarus were characterised by higher values for a majority of the characters studied, but the latitudinal pattern was rather in line with Bergmann's rule. A longitudinal cline has never been referred to in hedgehogs.

4.4. Ecological Species Interactions—Competitive and Predator Pressure

A third hypothesis would involve biotic ecological interactions, i.e., competitive pressure, as a possible causal factor accounting for patterns observed. The general expectation of divergent character displacement is based on a presumption of sufficiently narrow ecological valence, with resource competition generating selective pressure intense enough to promote niche diversification. It is well known that much of the diversity among terrestrial vertebrate skulls is associated with feeding [6,46,47] and that the adaptive evolution associated with it can be rapid [48–50]. Hedgehogs are medium-sized plantigrade mammals whose low basal metabolic rates and related defensive anti-predator strategies reflect energetic constraints associated with the unpredictable distribution of resources and predation [51]. Therefore, when two hedgehog species form a sympatric zone, potentially characterised by higher population density than in allopatry, the resulting guild may approach the carrying capacity of the environment, with competition for resources generating a convergent reduction in body size, rather than divergent character displacement. This explanation is also consistent with convergence in the mandible shape. As the function of the lower jaw connects mainly with feeding [47,52], in contrast to the cranium reflecting additional functions related to sensory organs and the brain [53,54], it is reasonable to ascribe respective patterns to evolutionary forces within the trophic niche. We hypothesize that decrease in size and convergence in shape is a result of local adaptations to overlapping dietary resources in the two species, and hence an example of convergent character displacement.

Background mechanisms underpinning the observed patterns could also act at the community level and entail predator pressure. The disruptive selection this causes could be seen as a proximate mechanism behind a bimodal distribution of body size in mammals, in line with Cope's rule [55]. Mammals filling "the mid-size gap" often possess defensive weaponry apomorphies, such as the spines in hedgehogs [56], thereby pointing to the major evolutionary consequences of predation among such medium-sized mammals. Recent ecological research on hedgehogs also reported a marked effect of intraguild predation [57]. The associated hypothesis accounting for our results would therefore involve higher predator pressure in the sympatric zone, as a reflection of the higher population density among hedgehogs in general, and the recent mesopredator release [58], i.e., population increases in e.g., red foxes and European badgers in central Europe. That may in turn reflect the population declines that characterised large carnivores for a long period in the past. All of this leaves evolution in the direction of smaller body size in prey species as more suitable for achieving crypsis. These ecological mechanisms could all be facilitated by such anthropogenic changes in the environment as declines in numbers of invertebrates due to the use of pesticides, fragmentation of habitats, or increased traffic.

As detailed knowledge about ecological niche of hedgehogs is crucial in understanding their population decline [59] and proposing rational conservation management, we propound further attention to described phenomena. Considering the substantial regional variation and steep east–west gradients in socioeconomic variables in Europe, resulting in geographic patterns in landscape fragmentation [60] and anthropogenic pressure on ecosystems and hedgehog populations, as well as differences in conservation management in rescue centers [61], large-scale studies using a standardized methodology are needed to compensate for possible observation bias and investigate more closely the complex phenotypic variation patterns and ecological niche characteristics. Taking into account environmental variables will also allow for the study of the presence of complex phenomena caused by selection pressures related to urbanization, the development of industrialised agriculture and other human-mediated factors such as anthropogenic dwarfing of species.

5. Conclusions

In the present study, seldom observed sympatric character displacement was ascertained in the contact zone of two sympatric hedgehog species in central Europe based on the analysis of skull size and shape. Considering almost complete formation of reproductive isolating barriers, we presume hybridisation and introgression are not the major processes behind the observed pattern. We hypothesise that unidirectional selection to size and shape related to competition in the trophic niche is responsible for that pattern, and that size changes of the skull could also be facilitated by a reduction in body size related to selection pressures, mediated by the Anthropocene mesopredator release.

Supplementary Materials: The following are available online at http://www.mdpi.com/2076-2615/10/10/1803/s1, Table S1: Samples used in the study, Table S2: Landmark coordinates, Figure S1: Cranial and mandibular shape differences.

Author Contributions: B.Č.B. and P.H. conceived and designed the study; B.Č.B. and M.R.K. collected and prepared the specimens for analyses; M.R.K. made morphometric measurements under the supervision of W.B. and A.S.-J. A.E. analysed the morphometric data; M.L. contributed to genetic characterization of the samples; B.Č.B., A.E., M.R.K., W.B. and P.H. drafted the manuscript and all authors contributed to the final version. All authors have read and agreed to the published version of the manuscript.

Funding: This study was funded by the Grant Agency of Charles University (GAUK 538218) and the Grant Agency of Czech University of Life Sciences Prague (IGA 20205007 and CIGA 20185006). MRK was supported by the Charles University Research Centre program No. 204069.

Acknowledgments: We would like to thank the Museum für Naturkunde, Berlin (Germany) and the Institute of Vertebrate Biology, Brno (Czech Republic) for a loan of the skulls. We also thank the rescue centers in the Czech Republic, namely Vlašim, Praha, Bruntál, Jaroměř and Litoměřice, for their contribution to the study. For valuable suggestions we thank to Dr. Vladimír Vohralík.

Conflicts of Interest: The authors declare no conflict of interest.

References

1. Hewitt, G.M. Post-Glacial Re-Colonization of European Biota. *Biol. J. Linn. Soc.* **1999**, *68*, 87–112. [CrossRef]
2. Seddon, J.M.; Santucci, F.; Reeve, N.J.; Hewitt, G.M. DNA Footprints of European Hedgehogs, *Erinaceus europaeus* and *E. concolor*: Pleistocene Refugia, Postglacial Expansion and Colonization Routes. *Mol. Ecol.* **2001**, *10*, 2187–2198. [CrossRef] [PubMed]
3. Bolfíková, B.; Hulva, P. Microevolution of Sympatry: Landscape Genetics of Hedgehogs *Erinaceus europaeus* and *E. roumanicus* in Central Europe. *Heredity* **2012**, *108*, 248–255. [CrossRef] [PubMed]
4. Černá Bolfíková, B.; Eliášová, K.; Loudová, M.; Kryštufek, B.; Lymberakis, P.; Sándor, A.D.; Hulva, P. Glacial Allopatry vs. Postglacial Parapatry and Peripatry: The Case of Hedgehogs. *PeerJ* **2017**, *5*, e3163:1–e3163:21. [CrossRef]
5. Curto, M.; Winter, S.; Seiter, A.; Schmid, L.; Scheicher, K.; Barthel, L.M.F.; Plass, J.; Meimberg, H. Application of a SSR-GBS Marker System on Investigation of European Hedgehog Species and Their Hybrid Zone Dynamics. *Ecol. Evol.* **2019**, *9*, 2814–2832. [CrossRef]
6. Evin, A.; Horáček, I.; Hulva, P. Phenotypic Diversification and Island Evolution of Pipistrelle Bats (*Pipistrellus pipistrellus* Group) in the Mediterranean Region Inferred from Geometric Morphometrics and Molecular Phylogenetics. *J. Biogeogr.* **2011**, *38*, 2091–2105. [CrossRef]
7. Coyne, J.; Orr, H. *Speciation*; Sinauer Associates: Sunderland, MA, USA, 2004.
8. Reifová, R.; Reif, J.; Antczak, M.; Nachman, M.W. Ecological Character Displacement in the Face of Gene Flow: Evidence from Two Species of Nightingales. *BMC Evol. Biol.* **2011**, *11*, 138:1–138:11. [CrossRef]
9. Kryštufek, B.; Tvrtković, N.; Paunović, M.; Özkan, B. Size Variation in the Northern White-Breasted Hedgehog *Erinaceus roumanicus*: Latitudinal Cline and the Island Rule. *Mammalia* **2009**, *73*, 299–306. [CrossRef]
10. Reeve, N. *Hedgehogs*; T & AD Poyser: London, UK, 1994.
11. Ruprecht, A. Correlation Structure of Skull Dimensions in European Hedgehogs. *Acta Theriol.* **1972**, *17*, 419–442. [CrossRef]
12. Holz, H.; Niethammer, J. *Erianceus concolor* Martin, 1838—Weissbrustigel, Ostigel. In *Handbuch der Säugetiere Europas 3/1*; Niethammer, J., Krapp, F., Eds.; Aula-Verlag: Wiesbaden, Germany, 1990; pp. 50–64.
13. Holz, H.; Niethammer, J. *Erinaceus europaeus* Linnaeus, 1758—Braunbrustigel, Westigel. In *Handbuch der Säugetiere Europas 3/1*; Niethammer, J., Krapp, F., Eds.; Aula-Verlag: Wiesbaden, Germany, 1990; pp. 26–49.
14. Kryštufek, B. Cranial Variability in the Eastern Hedgehog *Erinaceus concolor* (Mammalia: Insectivora). *J. Zool.* **2002**, *258*, 365–373. [CrossRef]
15. Hrabě, V. Variation in Cranial Measurement of *Erinaceus concolor roumanicus* (Insectivora, Mammalia). *Zool. List.* **1976**, *25*, 315–326.
16. Niethammer, J.; Krapp, F. *Handbuch Der Säugetiere Europas. Bd. 3/1, Insektenfresser-Insectivora, Herrentiere-Primates*; Niethammer, J., Krapp, F., Eds.; Aula-Verlag: Wiesbaden, Germany, 1990.
17. Hrabě, V. Variation in Somatic Characters of Two Species of *Erinaceus* (Insectivora, Mammalia) in Relation to Individual Age. *Zool. List.* **1975**, *24*, 335–351.
18. Wolff, P. Unterscheidungsmerkmale am Unterkiefer von *Erinaceus europaeus* L. und *Erinaceus concolor* Martin. *Ann. Naturhist. Mus. Wien* **1976**, *80*, 337–341. [CrossRef]
19. Kratochvíl, J. Die Hirnmasse Der Mitteleuropäischen Arten Der Gattung *Erinaceus* (Insectivora, Mamm.). *Folia Zool.* **1980**, *29*, 1–20.
20. Hrabě, V. Notes on the Dentition of Two *Erinaceus* Spp. from Czechoslovakia (Insectivora, Mammalia). *Folia Zool.* **1981**, *30*, 311–316.
21. Hanken, J.; Hall, B.K. *The Skull, Volume 3: Functional and Evolutionary Mechanisms*; University of Chicago Press: Chicago, IL, USA, 1993.
22. Škoudlín, J. Zur Altersbestimmung Bei *Erinaceus europaeus* Und *Erinaceus concolor* (Insectivora: Erinaceidae). *Věst. Československ. Spol. Zool.* **1976**, *40*, 300–306.

23. Herter, K. Studien Zur Verbreitung Der Europäischen Igel. *Arch. Naturgesch.* **1934**, *3*, 21–382.
24. Rohlf, F.J.; Slice, D. Extensions of the Procrustes Method for the Optimal Superimposition of Landmarks. *Syst. Zool.* **1990**, *39*, 40–59. [CrossRef]
25. Goodall, C.R. Procrustes Methods in the Statistical Analysis of Shape Revisited. In *Current Issues in Statistical Shape Analysis*; Mardia, K.V., Gill, C.A., Eds.; Leeds University Press: Leeds, UK, 1995; pp. 18–33.
26. Bookstein, F. *Morphometric Tools for Landmark Data: Geometry and Biology*; Cambridge University Press: New York, NY, USA, 1991. [CrossRef]
27. Evin, A.; Cucchi, T.; Cardini, A.; Strand Vidarsdottir, U.; Larson, G.; Dobney, K. The Long and Winding Road: Identifying Pig Domestication through Molar Size and Shape. *J. Archaeol. Sci.* **2013**, *40*, 735–743. [CrossRef]
28. Monteiro, L.R. Multivariate Regression Models and Geometric Morphometrics: The Search for Causal Factors in the Analysis of Shape. *Syst. Biol.* **1999**, *48*, 192–199. [CrossRef]
29. Schlager, S. Morpho and Rvcg—Shape Analysis in R. In *Statistical Shape and Deformation Analysis: Methods, Implementation and Applications*; Zheng, G., Li, S., Szekely, G., Eds.; Academic Press: San Diego, CA, USA, 2017; pp. 217–256.
30. Baylac, M.; Frieß, M. Fourier Descriptors, Procrustes Superimposition, and Data Dimensionality: An Example of Cranial Shape Analysis in Modern Human Populations. In *Modern Morphometrics in Physical Anthropology*; Slice, D.E., Ed.; Springer: Boston, MA, USA, 2005; pp. 145–162. [CrossRef]
31. Benjamini, Y.; Hochberg, Y. Controlling the False Discovery Rate: A Practical and Powerful Approach to Multiple Testing. *J. R. Stat. Soc. Ser. B* **1995**, *57*, 289–300. [CrossRef]
32. Mahalanobis, P. On the Generalized Distance in Statistics. *Proc. Natl. Acad. Sci. India* **1936**, *2*, 49–55.
33. Mantel, N. The Detection of Disease Clustering and a Generalized Regression Approach. *Cancer Res.* **1967**, *27*, 209–220. [PubMed]
34. Sokal, R.R.; Rohlf, F.J. *Biometry: The Principles and Practice of Statistics in Biological Research*, 3rd ed.; W.H. Freeman and Co.: New York, NY, USA, 1995.
35. Stayton, C.T. The Definition, Recognition, and Interpretation of Convergent Evolution, and Two New Measures for Quantifying and Assessing the Significance of Convergence. *Evolution* **2015**, *69*, 2140–2153. [CrossRef] [PubMed]
36. Stayton, C.T. Testing Hypotheses of Convergence with Multivariate Data: Morphological and Functional Convergence among Herbivorous Lizards. *Evolution* **2006**, *60*, 824–841. [CrossRef]
37. Zelditch, M.L.; Swiderski, D.L.; Sheets, H.D. *Geometric Morphometrics for Biologists: A Primer*, 2nd ed.; Elsevier, Academic Press: Amsterdam, The Netherlands, 2012.
38. Adams, D.; Collyer, M.; Kaliontzopoulou, A.; Sherratt, E. Geomorph: Software for Geometric Morphometric Analyses. R Package Version 3.1.0. 2019. Available online: https://github.com/geomorphR/geomorph (accessed on 20 December 2019).
39. Dray, S.; Dufour, A.-B. The Ade4 Package: Implementing the Duality Diagram for Ecologists. *J. Stat. Softw.* **2007**, *22*, 1–20. [CrossRef]
40. R Core Team. R: A Language and Environment for Statistical Computing. R Foundation for Statistical Computing. 2020. Available online: https://www.r-project.org (accessed on 29 February 2020).
41. Poduschka, W.; Poduschka, C. Kreuzungsversuche an Mitteleuropäischen Igeln. *Säugetierk. Mitt.* **1983**, *31*, 1–12.
42. O'Regan, H.J.; Kitchener, A.C. The Effects of Captivity on the Morphology of Captive, Domesticated and Feral Mammals. *Mamm. Rev.* **2005**, *35*, 215–230. [CrossRef]
43. Evin, A.; Dobney, K.; Schafberg, R.; Owen, J.; Strand Vidarsdottir, U.; Larson, G.; Cucchi, T. Phenotype and Animal Domestication: A Study of Dental Variation between Domestic, Wild, Captive, Hybrid and Insular *Sus scrofa*. *BMC Evol. Biol.* **2015**, *15*, 6. [CrossRef]
44. Meiri, S.; Simberloff, D.; Dayan, T. Community-Wide Character Displacement in the Presence of Clines: A Test of Holarctic Weasel Guilds. *J. Anim. Ecol.* **2011**, *80*, 824–834. [CrossRef]
45. Škoudlín, J. Craniometric Analysis of a Czechoslovak and a Polish Population of *Erinaceus concolor* (Mammalia: Erinaceidae). *Věst. Českosl. Spol. Zool.* **1982**, *46*, 304–316.
46. Smith, K.K. The Form of the Feeding Apparatus in Terrestrial Vertebrates: Studies of Adaptation and Constraint. In *The Skull*; Hanken, J., Hall, B.K., Eds.; University of Chicago Press: Chicago, IL, USA, 1993; pp. 150–196.

47. Zelditch, M.L.; Ye, J.; Mitchell, J.S.; Swiderski, D.L. Rare Ecomorphological Convergence on a Complex Adaptive Landscape: Body Size and Diet Mediate Evolution of Jaw Shape in Squirrels (Sciuridae). *Evolution* **2017**, *71*, 633–649. [CrossRef] [PubMed]
48. Hendry, A.P.; Kinnison, M.T. Perspective: The Pace of Modern Life: Measuring Rates of Contemporary Microevolution. *Evolution* **1999**, *53*, 1637–1653. [CrossRef] [PubMed]
49. Carroll, S.P.; Hendry, A.P.; Reznick, D.N.; Fox, C.W. Evolution on Ecological Time-Scales. *Funct. Ecol.* **2007**, *21*, 387–393. [CrossRef]
50. Herrel, A.; Huyghe, K.; Vanhooydonck, B.; Backeljau, T.; Breugelmans, K.; Grbac, I.; Van Damme, R.; Irschick, D.J. Rapid Large-Scale Evolutionary Divergence in Morphology and Performance Associated with Exploitation of a Different Dietary Resource. *Proc. Natl. Acad. Sci. USA* **2008**, *105*, 4792–4795. [CrossRef]
51. Lovegrove, B.G. The Evolution of Body Armor in Mammals: Plantigrade Constraints of Large Body Size. *Evolution* **2001**, *55*, 1464–1473. [CrossRef]
52. Morris, P.J.R.; Cobb, S.N.F.; Cox, P.G. Convergent Evolution in the Euarchontoglires. *Biol. Lett.* **2018**, *14*, 20180366:1–20180366:4. [CrossRef]
53. Cheverud, J.M. Phenotypic, Genetic, and Environmental Morphological Integration in the Cranium. *Evolution* **1982**, *36*, 499–516. [CrossRef]
54. Goswami, A. Cranial Modularity Shifts during Mammalian Evolution. *Am. Nat.* **2006**, *168*, 270–280. [CrossRef]
55. Alroy, J. Cope's Rule and the Dynamics of Body Mass Evolution in North American Fossil Mammals. *Science* **1998**, *280*, 731–734. [CrossRef]
56. Stankowich, T.; Campbell, L.A. Living in the Danger Zone: Exposure to Predators and the Evolution of Spines and Body Armor in Mammals. *Evolution* **2016**, *70*, 1501–1511. [CrossRef] [PubMed]
57. Williams, B.M.; Baker, P.J.; Thomas, E.; Wilson, G.; Judge, J.; Yarnell, R.W. Reduced Occupancy of Hedgehogs (*Erinaceus europaeus*) in Rural England and Wales: The Influence of Habitat and an Asymmetric Intra-Guild Predator. *Sci. Rep.* **2018**, *8*, 1–10. [CrossRef] [PubMed]
58. Ritchie, E.G.; Johnson, C.N. Predator interactions, mesopredator release and biodiversity conservation. *Ecol. Lett.* **2009**, *12*, 982–998. [CrossRef] [PubMed]
59. Hof, A.R.; Bright, P.W. Quantifying the Long-Term Decline of the West European Hedgehog in England by Subsampling Citizen-Science Datasets. *Eur. J. Wildl. Res.* **2016**, *62*, 407–413. [CrossRef]
60. Hulva, P.; Černá Bolfíková, B.; Woznicová, V.; Jindřichová, M.; Benešová, M.; Mysłajek, R.W.; Nowak, S.; Szewczyk, M.; Niedźwiecka, N.; Figura, M.; et al. Wolves at the Crossroad: Fission–Fusion Range Biogeography in the Western Carpathians and Central Europe. *Divers. Distrib.* **2018**, *24*, 179–192. [CrossRef]
61. Ploi, K.; Curto, M.; Bolfíková, B.Č.; Loudová, M.; Hulva, P.; Seiter, A.; Fuhrmann, M.; Winter, S.; Meimberg, H. Evaluating the Impact of Wildlife Shelter Management on the Genetic Diversity of *Erinaceus europaeus* and *E. roumanicus* in Their Contact Zone. *Animals* **2020**, *10*, 1452. [CrossRef]

 *animals*

Article

Evaluating the Impact of Wildlife Shelter Management on the Genetic Diversity of *Erinaceus europaeus* and *E. roumanicus* in Their Contact Zone

Kerstin Ploi [1], Manuel Curto [1,2], Barbora Černá Bolfíková [3], Miroslava Loudová [4], Pavel Hulva [4,5], Anna Seiter [1], Marilene Fuhrmann [1], Silvia Winter [6] and Harald Meimberg [1,*]

[1] Institute for Integrative Nature Conservation Research, University of Natural Resources and Life Sciences (BOKU), 1180 Vienna, Austria; kerstin.ploi@students.boku.ac.at (K.P.); manuel.curto@boku.ac.at (M.C.); anna.seiter@posteo.de (A.S.); marilene.fuhrmann@students.boku.ac.at (M.F.)
[2] MARE–Marine and Environmental Sciences Centre, Universidade de Lisboa, 1649-004 Lisboa, Portugal
[3] Faculty of Tropical AgriSciences, Czech University of Life Sciences Prague, Kamýcká 129, 165 21 Prague, Czech Republic; bolfikova@ftz.czu.cz
[4] Department of Zoology, Faculty of Science, Charles University in Prague, Viničná 7, 116 36 Prague, Czech Republic; miroslava.loudova@natur.cuni.cz (M.L.); pavel.hulva@natur.cuni.cz (P.H.)
[5] Faculty of Science, University of Ostrava, Chittussiho 10, 710 00 Ostrava, Czech Republic
[6] Institute of Plant Protection, University of Natural Resources and Life Sciences (BOKU), 1180 Vienna, Austria; silvia.winter@boku.ac.at
* Correspondence: meimberg@boku.ac.at

Received: 30 June 2020; Accepted: 12 August 2020; Published: 20 August 2020

Simple Summary: Hedgehogs are regularly brought to wildlife shelters. Depending on the area from where animals are accepted, translocation can occur between different regions or populations. In this study, the genetic diversity of wild hedgehog populations was compared with "shelter populations" within central Europe focusing on the western contact zone between both European hedgehog species. Some shelters were hosting both species at the same time, in one this could be shown genetically. Generally, no difference in genetic diversity between shelter individuals and wild populations was found. Two shelters from Innsbruck hosted individuals that probably belong to two subpopulations. This indicates that shelter management-related translocations could facilitate gene flow across a dispersal barrier.

Abstract: Hedgehogs are among the most abundant species to be found within wildlife shelters and after successful rehabilitation they are frequently translocated. The effects and potential impact of these translocations on gene flow within wild populations are largely unknown. In this study, different wild hedgehog populations were compared with artificially created "shelter populations", with regard to their genetic diversity, in order to establish basic data for future inferences on the genetic impact of hedgehog translocations. Observed populations are located within central Europe, including the species *Erinaceus europaeus* and *E. roumanicus*. Shelters were mainly hosting one species; in one case, both species were present syntopically. Apart from one exception, the results did not show a higher genetic diversity within shelter populations, indicating that individuals did not originate from a wider geographical area than individuals grouped into one of the wild populations. Two shelters from Innsbruck hosted individuals that belonged to two potential clusters, as indicated in a distance analysis. When such a structure stems from the effects of landscape elements like large rivers, the shelter management-related translocations might lead to homogenization across the dispersal barrier.

Keywords: hedgehog; animal shelter; translocation; genotyping by amplicon sequencing

---

Animals **2020**, *10*, 1452; doi:10.3390/ani10091452

1. Introduction

The two European hedgehog species *Erinaceus europaeus* (Linne, 1758) and *E. roumanicus* (Barrett-Hamilton, 1900) are assessed as 'least concern' by the IUCN (International Union for Conservation of Nature) Red List of Threatened Species [1,2]. Nevertheless, in several western European countries like England, the Netherlands, and Belgium, road-kill and citizen science data suggested a decline of *E. europaeus* [3–6]. Investigations on the hedgehog population structure have been performed across a wide range of European countries [7,8], including their central European zone of sympatry and potential hybridization zone [8–10]. Because hedgehogs are mainly found in close proximity to human settlements, rural as well as urban, occupying diverse man-made habitats [1,2,11], the population structure might be influenced by human infrastructure and anthropogenic barriers (e.g., roads and major transportation axes) fragmenting the landscape. Limited migration possibilities might consequently reduce and disturb gene flow [4,12–14]. This can manifest itself in an influence on the isolation-by-distance pattern on small geographical scales, like shown in the United Kingdom (UK) [7]. As small mammal species, hedgehogs are naturally subjected to dispersal impediments that can result in a genetic structure [15,16]. Therefore, the effects of anthropogenic as well as geographical barriers (e.g., rivers, mountain ranges) might be reinforced by the per se limited dispersal capacities of this species [7,9,14].

As likeable and well-known garden species, hedgehogs are often used as flagship or umbrella species in urban nature conservation (e.g., the garden projects "Garten Charata" in Switzerland, "Natur im Garten" in Austria, "Super-Igel-Garten, NAJU Rostock" in Germany). Due to this medial presence, and their popularity and presence in gardens, they are often the center of attention when it comes to human-mediated rescue actions. Hedgehogs are frequently translocated and among one of the most abundant species found in wildlife shelters in western Europe [17]. Wildlife shelters rehabilitating hedgehogs can also be found across Austria (at least one in every federal province) [18]. The primary aim of these shelters lies in the successful rehabilitation and subsequent release of their temporary patients back into the wild. This, however, may result in (unintentional) translocation; if the individual's origin is unknown, a release at the original site is not possible or if individuals cannot individually be identified after their stay in the shelter [17,18]. It is also known that such release locations are often chosen because of their suitable habitat, while the actual origin of hedgehog individuals is rarely recorded by the animal shelters (in Austria). Therefore, the origin may lie anywhere within a radius of 300 km around the shelter and might as well originate from a different federal province. Moreover, shelters may lack the ability of species identification (*E. europaeus* or *E. roumanicus*) [18]. Despite the proven importance of rehabilitation and translocation as conservation management tools [19,20], the lack of centralized monitoring, regulation, and licensing can lead to uncoordinated translocations, which may have consequences for the wild population at the release site [21].

E. europaeus and *E. roumanicus* are not under legal protection by the FFH (Flora-Fauna-Habitat) directive [22] and are not referred to within Austrian hunting regulations [23]. Nature conservation and species protection are regulated on the province level within Austria; therefore, nine different versions of legal regulation apply for the two hedgehog species within Austria. They either categorize hedgehogs as "not protected" or "protected", with the latter prohibiting the capture, killing, disturbance, ownership, and trade with the corresponding animal. Rehabilitation of wild hedgehog individuals is only allowed under certain circumstances, which include the necessity of rehabilitating injured or underweight individuals as well as the approval of exceptions for the protection of wildlife [18].

Potential outcomes of uncontrolled translocations may be outbreeding depression, accompanied by the loss of local adaptions, or a reduction in genetic variation and changes in the population genetic composition [19,20,24,25]. Even translocations that do not directly lead to gene flow within the affected population may result in genetic consequences through a reduction of the population size, triggered by increased competition or potential transmission of diseases [21].

Because human-mediated translocations can also lead to increased incidents of hybridization between species [19,24], a special focus needs to be put on the central European contact zone of *E. europaeus* and *E. roumanicus*, which reaches from Poland and the Czech Republic down through Austria to the border between Italy and Slovenia [9]. It is already known that (un)intentional translocations of various species led to an increase in hybridization and introgression [20]. The degree of hybridization between the two parapatric hedgehog species within their central European contact zone has been indicated to be low so far [8,9,26,27]. Recent investigations relying on a denser sampling of the zone of secondary contact (within Austria), as well as the usage of multiple microsatellite markers, however, indicate that the extent of potential hybridization might be higher than assumed and the actual zone of sympatry and species overlap might be broader [10]. Ongoing work is investigating this further; however, translocations might play a role as an influencing factor for distribution as well as hybridization.

E. europaeus* and *E. roumanicus* are known to have overcome the most recent period of repeated glaciations during the Pleistocene era in three different southern refugia. While *E. roumanicus* survived these glaciations in the Balkan peninsula, *E. europaeus* was restricted to the regions of the Iberian and Apennine peninsula [26,28–30]. Provoked range shifts, resulting from recurrent restrictions to glacial refugia and expansion during interglacial warmings [28,29,31,32], are reflected in today's hedgehog genetic structure, as has been shown by various studies [26,30,33,34]. They detected a major genomic division between *E. europaeus* and *E. roumanicus*, and a contact zone separating this divergence between eastern and western Europe [26,30,33,34]. As well as this, a divergence between *E. europaeus* individuals originating from either the Iberian or Apennine peninsula was shown [26,30,33,34]. For *E. roumanicus*, it is suggested that interglacial continental refugia might have led to further genetic differentiation among populations [8]. Nuclear as well as mitochondrial (mt)DNA markers have been used to reconstruct glacial refugia and post-glacial colonization routes of various species worldwide, with European hedgehogs among them [35]. However, it is known that different genetic markers show divergent resolution when observing genetical (sub)structuring and the phylogeographic origin of hedgehogs. While mtDNA markers are proven to be proficient for analyzing the divergence between *E. europaeus*, *E. roumanicus*, and *E. concolor*, as well as intraspecific genetic divergence within *E. europaeus*, nuclear markers were solely able to resolve the major interspecific splits among the three European *Erinaceus* species [26,34]. However, genetical substructuring of existing (sub)species and populations might be affected in regions where individuals of different phylogeographic and genetic origin are intermixed through human-mediated translocations of hedgehogs. In the current study, we aimed to use a set of 55 microsatellite markers, which given their high mutation rate and number should provide a higher discriminatory power than the above mentioned studies.

Only few studies have focused on the effect of animal translocations on gene flow as a whole [19]. We aimed to generate the basic data necessary to draw solid conclusions on whether the translocation of hedgehog individuals among a certain geographic range would result in the possible disturbance of local genetic structures. Therefore, we raise the question if hedgehog individuals congregated at wildlife shelters display a broader genetical structure and are more diverse than wild hedgehog populations. We answer this question by contrasting different wild hedgehog populations to artificially created "shelter populations", with regard to their genetic diversity, through Bayesian clustering and the application of principal coordinate analysis (PCoA). The results will be linked with data from questionnaires of Austrian wildlife/animal shelters and shall allow for inferences on the potential impact of hedgehog translocations within the central European contact zone of *E. europaeus* and *E. roumanicus*. Additionally, results will be discussed in a phylogeographic context, as the sampling area is assumed to host both refugial lineages of *E. europaeus*.

2. Material and Methods

2.1. Sample Collection and DNA Isolation

A total of 221 samples were used in the present study. Samples were assigned to *E. europaeus* (*n* = 144) and *E. roumanicus* (*n* = 77) by morphological characters. Some samples were already investigated in previous phylogeographical/population genetical assessments [10]. Samples were collected from shelter individuals (*n* = 130) and from wild hedgehogs (*n* = 91). Individuals from one shelter were treated as one population. Corresponding population assignment, as well as the geographical background can be found in Table 1. Figure S1 gives an overview on the sample locations.

Table 1. Sample information and populations. The table gives an overview on the used population name, number of individuals within each population, species assignment, and geographical origin. Abbreviations given: AUT = Austria, CZE = Czech Republic, GER = Germany, POL = Poland.

Population	# Individuals	Species	Geographical Origin
wild_LinzD_eu	7	*E. europaeus*	Linz/AUT (wild)
wild_Linz2_eu	2	*E. europaeus*	Linz/AUT (wild)
wild_Linz3_eu	6	*E. europaeus*	Linz/AUT (wild)
wild_Linz4_eu	6	*E. europaeus*	Linz/AUT (wild)
shelter_Bludenz_eu	31	*E. europaeus*	Bludenz/AUT (shelter)
wild_Chomutov_eu	8	*E. europaeus*	Chomutov/CZE (wild)
wild_Praha_eu	15	*E. europaeus*	Prague/CZE (wild)
wild_Hamburg_eu	7	*E. europaeus*	Hamburg/GER (wild)
wild_Berlin_eu	9	*E. europaeus*	Berlin/GER (wild)
shelter_Mossautal_eu	9	*E. europaeus*	Mossautal/GER (shelter)
shelter_Innsbruck1_eu	7	*E. europaeus*	Innsbruck/AUT (shelter)
shelter_Lea_eu	20	*E. europaeus*	Bavaria/GER (shelter)
shelter_Innsbruck2_eu	17	*E. europaeus*	Innsbruck/AUT (shelter)
wild_Linz_rou	6	*E. roumanicus*	Linz/AUT (wild)
wild_Markt_rou	3	*E. roumanicus*	Linz/AUT (wild
wild_Praha1_rou	10	*E. roumanicus*	Prague/CZE (wild)
wild_Praha2_rou	6	*E. roumanicus*	Prague/CZE (wild)
wild_Gdansk_rou	6	*E. roumanicus*	Gdansk/POL (wild)
shelter_Carinthia1_rou	21	*E. roumanicus*	Carinthia/AUT (shelter)
shelter_Carinthia2_rou	7	*E. roumanicus*	Carinthia/AUT (shelter)
shelter_Marilene_rou	18	*E. roumanicus*	Graz/AUT (shelter)

Individuals were either sampled through buccal swabs (shelter animals) or muscle tissue (from road fatalities or samples that were taken from museum specimens) [10]. All samples were stored in ethanol until further preparation and analysis. Detailed shelter identity and locations were not given.

DNA isolation followed the methods and procedures described in [10]. For DNA isolation of muscle tissue samples, a small piece of tissue was placed in 500 µL of lysis buffer (2% SDS, 2% PVP–40, 250 mM NaCl, 200 mM Tris-HCl, and 5 mM EDTA, pH8), to which 16.67 µL of proteinase K (10 mg/mL) were added, following a 2.5-h period of incubation at 56 °C. Then, 16.67 µL of RNase (10mg/mL) were subsequently added, followed by an incubation period of 20 min at 37 °C. After the second incubation, 125 µL of 3 M KOAc (pH 4.7) were added and samples were put on ice for 20 min. After a series of centrifugation steps (100 rcf for 1 min, 400 rcf for 1 min, 1700 rcf for 1 min, 7000 rcf for 1 min, and 20,000 rcf for 11 min), 400 μL of the supernatant were mixed with 15 μL of MagSi-DNA beads (size 300 nm, MagSi-DNA beads from MagnaMedics, Geleen, The Netherlands), as well as 600 μL of binding buffer (2 M GuHCl in 95% ethanol) and incubated at room temperature (5 min). To separate the resulting supernatant from the beads, the samples were placed on the magnetic separator SL-MagSep 96 (Steinbrenner, Germany) for 1 min. Two washing steps with 80% ethanol (600 μL each) followed. After discarding the supernatants, magnetic beads were air-dried at room temperature for 10 min. Two elutions were acquired, with 30 and 50 μL of preheated (65 °C) elution buffer (10 mM TrisHCl, pH 8), mixed with the beads, and incubated for 5 min at room temperature. DNA isolation of buccal swab samples followed the same procedure, with the exception that the product of lysis was filtered through NucleoSpin filter columns and centrifuged for 1 min at 562× *g*. Buccal swab samples were eluted in 20 and 25 μL of elution buffer [10].

2.2. Molecular Marker Enrichment and Amplification

A short sequence repeats genotyping by sequence approach (SSR-GBS), developed by [10], was used within the present study. Amplification of the microsatellite regions within the genomes of *E. europaeus* and *E. roumanicus* was conducted through multiplex PCR. In total, 55 different primers were available for this approach, 25 and 30 primer pairs for the species *E. europaeus* and *E. roumanicus*, respectively. A list of all primers that have been used for multiplex PCR can be found in the Supplementary Materials (Table S1). These primer sets were developed and improved by [10]. Amplification primers are built up by specific sequences that are elongated by Illumina P5/P7 sequences, which correspond to the Illumina adapter in the sequencing process. The forward primers were elongated with a part of the P5 motif (TCTTTCCCTACACGACGCTCTTCCGATCT) and the reverse primers with a part of the P7 motif (CTGGAGTTCAGACGTGTGCTCTTCCGATCT). The Multiplex PCR Kit from QIAGEN was used for the performance of multiplex PCR. Therefore, 1 μL of template was added to a prepared mix of 5 μL of QIAGEN Multiplex PCR Master Mix (Qiagen, CA, USA), 3.5 μL of H20, and 0.5 μL of a specific primer mix (1 μM). PCR was conducted using the following temperature profile: 95 °C for 15 min; 30 cycles of 95 °C for 30 s, 55 °C for 1 min, and 72 °C for 1 min; with a final extension at 72 °C for 10 min. For the evaluation of successful amplification and quality of the PCR products, agarose gel electrophoresis was conducted to visualize PCR results.

Purification of PCR products in preparation of index PCR was conducted using an inverse magnetic separator. A total 6-μL sample volume was mixed with 4.3 μL of magnetic beads and incubated at room temperature for 5 min. The DNA, bound to magnetic beads, was washed twice with 200 μL of 80% ethanol for 45 s. After discarding the ethanol, the magnetic particles were air-dried for 5 min and DNA was eluted with 17 μL of preheated (65 °C) elution buffer (10 mM TrisHCl, pH 8).

Index PCR was used to allow for the individual identification of the pooled samples with specific forward and revers primers through specific assignment of indices. On the one hand, the used primers help binding to the before amplified P5/P7 part of the primers used in multiplex PCR. On the other hand, they allow for binding to the flow cell in Illumina Sequencing and label the sample with a unique eight-bp index information, which helps to assign sequenced genotypes to single samples (P5: AATGATACGGCGACCACCGAGATCTACAC [Index] ACACTCTTTCCCTACACGACG; and P7: CAAGCAGAAGACGGCATACGAGAT [Index] GTGACTGGAGTTCAGACGTGT). For index PCR, 1 μL of clean PCR product was mixed with 5 μL of Multimix (QIAGEN), 2 μL of specific P5 forward primer, and 2 μL of specific P7 reverse primer. The following temperature profile was used for the performance of index PCR: 95 °C for 15 min; 10 cycles of 95 °C for 30 s, 58 °C for 1 min, and 72 °C for 1 min; with a final extension at 72 °C for 5 min [10].

2.3. Illumina Sequencing and Sequence Data Analysis

Following index PCR, samples were pooled and sent to the Biozentrum LMU in Munich, Germany, for sequencing in an Illumina MiSeq machine in both directions/through paired-end sequencing in a sequencing by synthesis approach.

The short sequence repeats genotyping by amplicon sequence (SSR-GBAS) method used in this paper follows the pipeline described in [10,36], available at [37]. Amplicon sequences are being used for the determination of genotypes. In this context, alleles are defined according to their length and the occurrence of single nucleotide polymorphisms (SNPs).

Sequences resulting from the Illumina run were supplied in two FASTQ files (read 1 and read 2), containing all sequences per index. The following processing and treatment of the samples was based on a combination of custom-made scripts, as well as third party programs [10]. This included the quality control of single bases, as well as each read, followed by the trimming of low-quality regions (Phred < 20, according to [10]). Sequences were aligned using the program PEAR [38], meaning that paired reads (read 1 and read 2) were merged. The necessary overlapping range was set to a minimum of 10 bp, with "a *p*-value below 0.01 for the highest observed expected alignment scores" [10,38]. Markers were explicitly designed to generate overlapping fragments of approximately 300 bp. Overlapping sequences below 250 bp and non-overlapping reads were not considered during further steps. Through a "demultiplexing" step, it was possible to identify primer sequences on each side of the merged reads and separate them per locus using the script primer_demultiplex.py [10,37], as merged reads were supposed to begin with the forward primer and end with the reverse primer sequence, based on library preparation. Finally, sequences were sorted by locus and sample, resulting in corresponding files, which were used for further genotyping analysis [10].

Allele definition was largely based on two major steps: (1) Sequence length, as well as (2) the possible occurrence of SNPs in each separate length class. Within each sample, each marker was examined for its most frequent sequence length class. This was done through a custom-made script (Rscript_Markerlength_develop_Color.R, [10,37]) and manually controlled based on length histograms. Loci were considered to be homozygous if they comprised a certain length with a frequency equal to or above 90% among all reads for the respective marker. If a locus showed two lengths with a frequency greater than 90% of all reads (and those frequencies differed by less than 20%), the genotype was considered heterozygous. As well as the calling for alleles based on length frequencies, the employed script (Rscript_Markerlength_develop_Color.R) checked for possible stutter within the selected alleles [10,37]. The remaining steps were preformed using the script Sequence_Allele_Call.py [37]. The various reads within the most frequent length class(es) of a (homozygous) locus were merged into one consensus sequence. Therefore, nucleotide positions were considered to be homozygous if they showed a frequency above 70% for a single position, and to be heterozygous if the frequency of a nucleotide within a single position was below 70%. Loci, within a specific sample, that had already been defined as homozygous based on their sequence length class could be considered heterozygous based on the two most frequent nucleotides for a position. Nucleotide positions were considered to be linked if more than one potential SNP (single nucleotid polymorphism) occurred in a sequence. For samples already defined as heterozygous based on length class, it was decided to choose the most frequent SNP combination. Based on the called alleles from all samples, a codominant matrix was set up, as input for subsequent population genetic analyses within different standard programs [10]. This matrix consists of two specific numbers, corresponding to unique sequences (i.e., specific alleles), for each investigated locus, of every sample.

2.4. Population Genetic Analysis

After the initial sequence analysis and exclusion of markers with too much missing data, 41 microsatellite markers were valid to be used (see Figure S1) for population genetic analysis of the 221 hedgehog samples.

Specific analyses were conducted in the following different approaches: (1) Analysis of all 221 hedgehog samples; (2) intraspecific analysis for the two respective species *E. europaeus* and *E. roumanicus*; as well as (3) a separate analysis of shelter and wild populations (on an inter- as well as intraspecific level).

The program STRUCTURE v.2.3.4 [39] was used for the detection of the underlying population structure within the investigated set of hedgehog individuals, based on Bayesian clustering (Markov Chain Monte Carlo, MCMC) of multilocus genotype data [39]. For all conducted approaches, the "Length of Burnin Period", as well as the "Number of MCMC Reps after Burnin" were selected as 10.000. An admixture model, with correlated allele frequencies was chosen for calculation. Calculations were conducted with 5 iterations.

To choose the most likely value for all observed Ks and thereby determine the most likely number of populations (genetic groups) to be found within the underlying set of samples, the "STRUCTURE HARVESTER" [40,41] was used. To summaries all iterations of each K into one single summary output and graphically represent the results calculated within STRUCTURE v.2.3.4, the web portal CLUMPAK (Cluster Markov Packager Across K) was used [42,43].

The Excel Add-In GenAlEx (Genetic Analysis in Excel) [44,45] was used for principal coordinate analysis (PCoA) using the covariance-standardized setting, and calculation of the standard population genetic parameters, like the observed and expected heterozygosity (He, Ho), fixation index (F), percentage of polymorphic loci, and number of alleles (Na).

3. Results

The data included in this study were obtained from different Illumina runs. Files corresponding to one sample containing all markers were used as input for quality control, merging of reads, and the allele call pipeline. After the exclusion of markers with more than 50% missing data, the matrix contained 41 primers for 221 samples. Primer characteristics were largely congruent with our earlier analysis [10] and are therefore not reported here. As expected, strong differentiation between *E. roumanicus* and *E. europaeus* was obvious, and intermediate specimens between the species recognized previously were not included in this study. No early generation hybrid was detected in this dataset and no individuals were positioned between the main clusters characterizing the two species (Figure 1). Within the single species, moderate genetic structure was indicated. In the analysis of all wild populations (K = 10), three clusters in *E. europaeus* (one for the populations in Linz, one for the Czech Republic (Prague and Chomutov), and one for the northern German populations of Berlin and Hamburg) are opposed to one cluster for all *E. roumanicus* populations (Figure 2). When looking only at clusters within one species, *E. roumanicus* was divided into three groups, one for Linz and Markt, the second for the populations in Prague, and the third for the population in Gdansk (Figure 2). When the populations from shelters are included, subdivision becomes clearer, as some of the shelters are from areas where no wild population was collected (Figure 2, Figure 3). For *E. europaeus*, despite the optimal K according to the Evanno method being 2, at K = 6, a second optimum is indicated and three clusters of the wild populations, as well as the shelter from Bludenz (Vorarlberg, western Austria), Innsbruck (Tyrol, western Austria), and Bavaria (shelter "Lea"; southern Germany) form their own clusters. The shelter from Mossautal was assigned to the same cluster as the northern German populations. In *E. roumanicus*, the analysis of all samples results in one additional cluster for the shelters in Carinthia and the shelter in Graz (shelter "Marilene"), which appears very admixed. In the shelter of Graz, we have some individuals that cluster together with *E. europaeus* (Figure 1), indicating that this population is not homogeneous.

Genetic differentiation between populations seems to be rather high with the marker system used. While the pairwise F_{st} between populations of the two species ranges from 0.3 and 0.4, intra-species values are between 0.02 and 0.21 in *E. europaeus* and between 0.13 and 0.3 in *E. roumanicus* (Figure 4). Low inter-species values (between 0.19 and 0.24) are attributed to the differences of admixed shelter populations where both species were included.

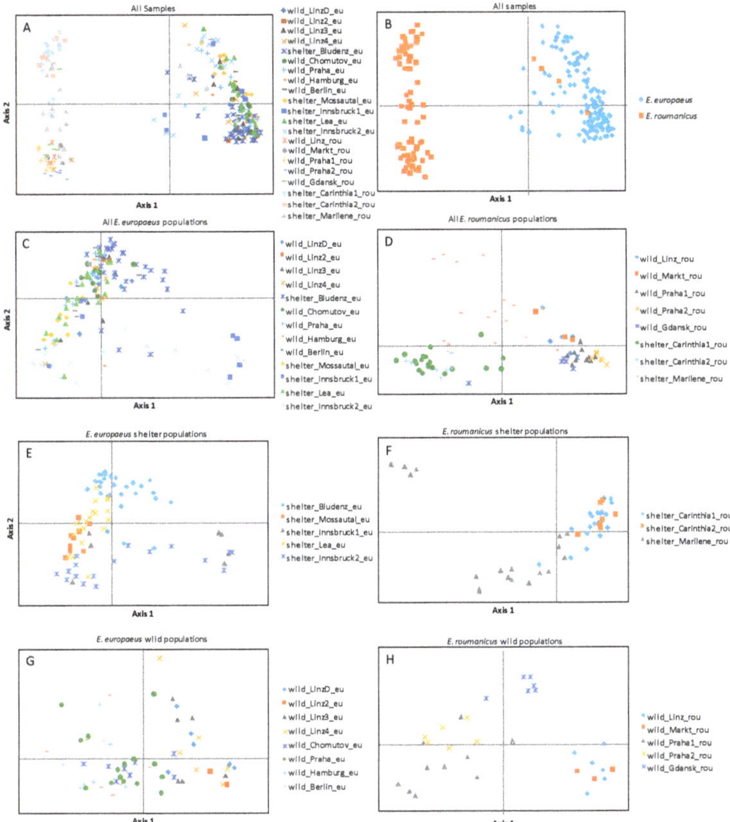

Figure 1. Principal coordinate analysis from genetic distances of the whole dataset ((**A**) for Populations, (**B**) for species), All populations divided per species (**C,D**), shelter populations (**E,F**), and wild populations (**G,H**). C through H are divided in *E. europaeus* (left) and *E. roumanicus* (right).

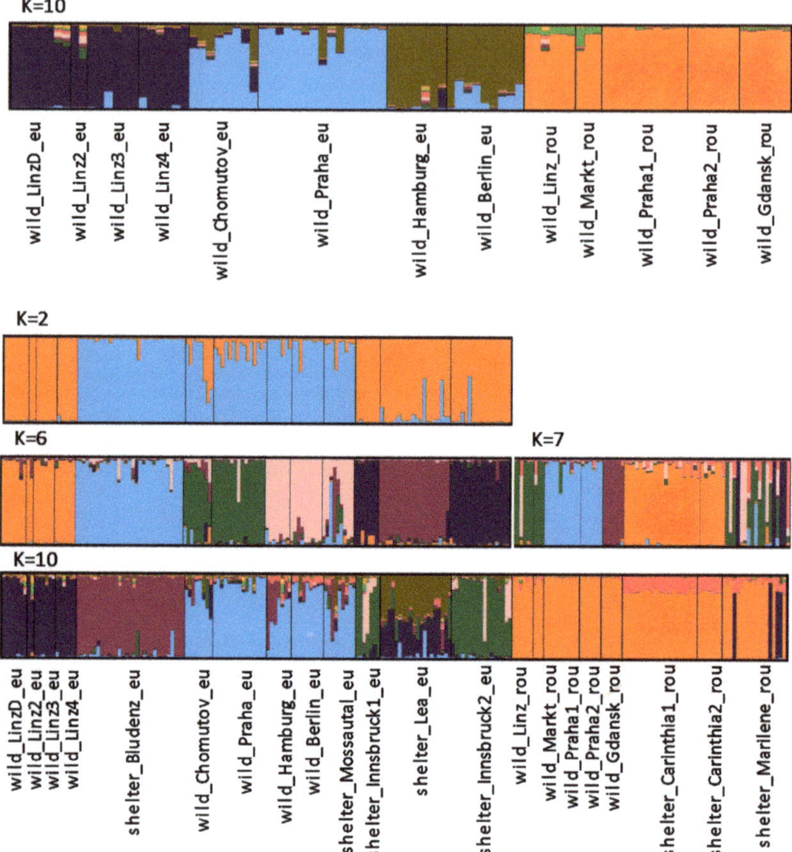

Figure 2. Structure analysis for the datasets of all wild (upper panel), all *E. europaeus* (K = 2, upper middle panel and K = 6 lower middle panel), all *E. roumanicus* (K = 7 lower middle panel, right), and all populations (K = 10 lower panel). The shown K corresponds to delta K_{max} (*E. europaeus* K = 2 and *E. roumanicus* K = 7) or suboptimal delta K (all populations K = 10; all wild populations K = 10, *E. europaeus* K = 6).

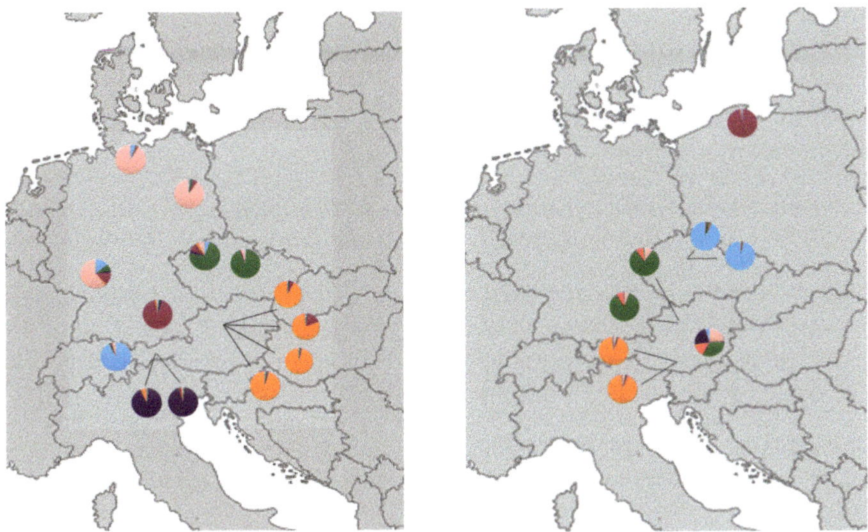

Figure 3. Spatial patterns of the genetic structure within *E. europaeus* (left panel) and *E. roumanicus* (right panel) shelter and wild populations. Raw data used and color-coding refer to structure analysis within the single species like shown in Figure 2 (K = 6 for *E. europaeus* and K = 7 for *E. roumanicus*). Basic map was taken from [46]. Localities are listed in Supplementary Materials Figure S1.

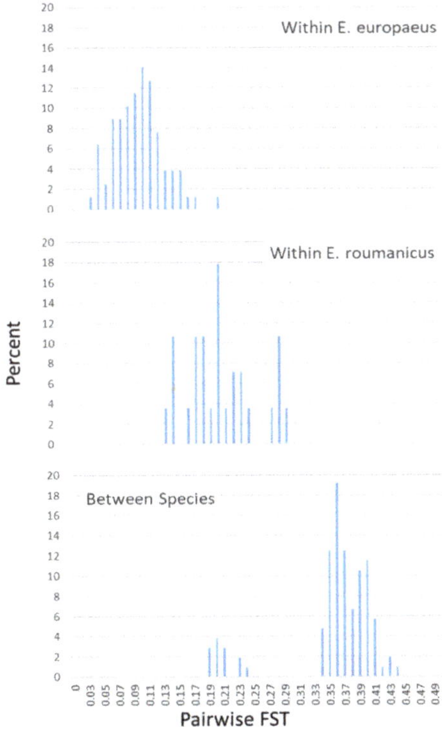

Figure 4. Frequency distribution of pairwise F_{st} between populations of *E. europaeus*, *E. roumanicus*, and between populations of both species.

4. Discussion

In concordance with previous studies on hedgehog genetic divergence and phylogeography [9,10,26,30], the results support a clear delimitation of the two species *E. europaeus* and *E. roumanicus*. The data allowed for phylogeographic inferences on factors shaping the genetic diversity of hedgehog populations.

Bayesian clustering revealed a most likely K = 2 for *E. europaeus* individuals within our dataset. This is corresponding to existing phylogeographic lineages of *E. europaeus* [26,30], based on analysis of mitochondrial markers. Hedgehogs survived the last glacial maximum in Mediterranean refugia and later colonized Europe northward in connection to spreading forests [28]. One of the lineages has Iberian origin [28,29] and recently is also occupying France, England, Switzerland, and southwest Germany [26] For the first time, we might have detected this lineage in Austria, which could be attributed to sampling gaps in previous studies. Due to its proximity to Switzerland and Germany, one could propose a possible existence of this lineage within Austria´s most western range. This is furthermore strengthened by the congruent assignment of the Bludenz and German (Berlin, Hamburg, Mossautal) populations within STRUCTURE analysis (K = 2), despite their relative geographic distance and the genetic distinctiveness of the Innsbruck population, despite its relative geographic proximity. Investigation on mitochondrial markers to verify the origin of this lineage is currently being undertaken.

The Apennine lineage of *E. europaeus* expanded out of the Italian Peninsula northwards through Austria, Switzerland, Germany, the Netherlands, Scandinavia, and Estonia [26]. In contrast to many other European species, hedgehogs have been shown to overcome the mountain barrier of the Alps and colonize Europe out from Italy [28,29]. The distinct pattern of the Bludenz population indicates the possibility that lineages of *E. europaeus* originating from the Iberian and Apennine Pleistocene refugia both occur in Austria. The separation of Linz and also Czech samples of K = 3 might be explained on their proximity to the zone of secondary contact and underlying genetic divergence or introgression [8]. Their genetic set-up might be established on possible introgression with *E. roumanicus* individuals, as they are situated in the potential hybridization zone. A possible influence of individuals from the secondary contact zone on the analysis of genetic divergence within the species has already been indicated by [8]. They also concluded that populations from the contact zone are different from the remaining study area, which they attributed to processes acting at parapatric range edges.

E. roumanicus showed a more pronounced genetic structure when looked at pairwise F_{st} than *E. europaeus*. It is suggested that only one lineage of *E. roumanicus* emerged from the southern refugium of the Balkan Peninsula [26,29,30]. Intraspecific structure analysis of *E. roumanicus* indicated a differentiation of all populations at K = 7 and populations of Linz, Prague, Gdansk, and Carinthia clearly differentiated from each other and stayed in a uniform pattern corresponding to their geographic origin. This pattern is in contradiction to previous shallow population structure found in this species [8,9]. It may be explained as a result of the hybrid zone dynamic with two populations (Linz and Praha) within the zone of sympatry [8]. Populations found within the cities of Linz (federal province of Upper Austria) and Prague are comprising both species. Individuals that were indicated as hybrids in the previous study with the same marker set [10] were not part of the populations included in this study. The division between the species is supported by a high F_{st}; however, also within one species populations were divided by F_{st}, indicating a moderate to high level of genetic structure. Moreover, their intraspecific F_{st} distribution is multimodal with some comparisons on the same range of the interspecific F_{st} values. Neutral markers exchanged between the two species could increase the signature of genetic differentiation between populations close and distant to the contact zone. For this study, only two areas with sympatric populations could be included so no specific analysis on introgression was performed. This might apply to *E. roumanicus* to a higher extent than to *E. europaeus* given their F_{st} distribution patterns. Ongoing work will investigate this further.

Landscapes across Europe appear to be highly fragmented due to anthropogenic infrastructure and various forms of human land use [4,12,13]. Hedgehog populations subjected to restricted gene flow due to other factors than geographic distance alone have been found in the UK [7] and in

Switzerland [14]. Especially within areas where the hedgehog´s distribution range is structured by major geographic barriers, like rivers or mountain ranges (as seen in the shelter populations of Innsbruck), these impediments might be circumvented through human-mediated infrastructure (such as bridges) or translocations. This artificial enhancement of species distribution may lead to a promotion of gene flow in specific areas. Whether or not this enhancement in gene flow has positive (enrichment of locally isolated populations in fragmented environment through an increase of genetic variation, [19]), negative (disturbance of locally adapted populations and outbreeding depression, [25]), or neutral effects on natural wild hedgehog populations remains the subject of future investigations.

The shelter populations observed within this study did not generally comprise a higher genetic diversity as could be expected because they comprise a mixture of individuals from a wide geographical area. While the shelter populations of Carinthia, Bludenz, and Bavaria appeared homogenous, the shelter population of Graz was highly diverse, with the appearance of both species. Personal information given in an interview of the shelter indicated that no differentiation of the two species is considered during sheltering. The two shelter populations from Innsbruck show a certain level of subdivision into two subpopulations in the PCoA. The exact origin of these samples is unknown; however, it is likely that these individuals could stem from both sides of the Inn, a river running through the federal province of Tyrol in the direction of south-west to north-east, which could constitute as a natural barrier [14].

The *E. europaeus* and *E. roumanicus* distribution within their central European contact zone is already known to be syntopic [9]. While increased landscape fragmentation might promote the limitation of gene flow [4,13,14] as well as a varying abundance and density of both hedgehog species in different regions [9], translocations have the possibility to interfere with these structures. Moreover, translocations might broaden the zone of overlap and might enhance the potential for hybridization. Using questionnaires, it was assessed that most shelters can differentiate between the species but regard species delimitation as irrelevant in practice [18]. The discovery of both species, *E. europaeus* and *E. roumanicus*, within a single shelter population (shelter Graz) shows shelters potential to interfere with the two species distribution range. In the thesis of [18], 8 out of 12 wildlife shelters (corresponding to 6 out of the 9 federal provinces) in Austria hosted both species. This is reinforced by shelter habits in animal translocations. Since wildlife shelters hold most of their hedgehogs in groups, the exact identification of single individuals is often neglected. The majority of wildlife shelters in Austria give rehabilitated hedgehogs back to the people that brought them (11 out of 13 shelters; [18]); however the individual that is handed over might be confused. A release at the place of origin is often not possible (due to a lack in individual identification; an unknown place of origin; or the place of origin being considered inappropriate; 8 out of 13 shelters in [18]), wherefore 6 out of 13 shelters turn to people voluntarily offering releases at places considered suitable [18]. Eventually, distances between the place of origin and release site are known to vary from 5–300 km in different shelters [18].

Studies mostly discuss the effect of translocation as the effect on the survival of individuals with lower fitness, which could interfere with natural selection [25]. Potential results might be relaxed selection pressure on survival during winter and a reduction of the adaptive response. In addition, more virulent forms of parasites and diseases could be kept in populations, allowing more damaging forms of diseases to develop [25]. This means that even releases that do not directly result in gene flow can have genetic consequences if they reduce the local population size by disease transmission or increased competition [21]. Various studies showed the capability of rehabilitated juvenile and adult hedgehogs to survive after their release [47,48]. Additionally, an improvement in the fitness of hedgehogs in Great Britain following temporary captivity was found [17]. The same study showed that rehabilitated and in turn released hedgehogs did not affect wild individuals through an increase in competition. This was attributed to suburban gardens serving as highly productive feeding sites [49].

Human-caused augmentation of hybridization pressure due to the translocation of individuals beyond the species range had been considered [19]. For the two parapatric hedgehog species *E. europaeus* and *E. roumanicus*, hybridization is recognized, but the degree seems to be very low in terms of early generation hybrids [9,36,50]. Long-term patterns of continuous introgression and its effect on genetic structure, like outlined above, could, however, very well be impactful. In the future, the data and marker system presented here, allowing for easy reproduction of genotyping data, will allow to monitor and investigate hybridization likelihood in detail.

Only few studies or data can be found about how translocating animals from wildlife shelters affects gene flow [19]. In our study, we found indications that genetic structure formed by landscape elements could be impacted over longer time scales when individuals from both populations are included in shelters. The example of the shelters from Innsbruck shows that this is possible. Even though, with a few exceptions, we did not find genetically strongly deviating individuals in the shelter populations, personal observation shows that hedgehogs can be transported via several hundred km for care and subsequent release. Even if this is unlikely to have a pronounced effect on the population structure [51], the consideration of the source for release is important, especially if the region is subdivided by migration barriers for the species.

5. Conclusions

Hosting hedgehog individuals in shelters can promote translocation of individuals because it can bring together animals that are isolated by dispersal barriers. In the zone of sympatry, the two different hedgehog species *E. europaeus* and *E. roumanicus* are sheltered together and are not always differentiated. Our data show that such cases occur, even if in most shelters a genetic effect is not seen. Shelter individuals should therefore be released under the consideration of natural barriers and treated as a species that can express small-scale genetic structure. Future research will investigate this further. Because the main consideration of shelters is animal welfare, also individuals from distant locations can be taken in, but for the individuals tested, this did not apply.

Supplementary Materials: The following are available online at http://www.mdpi.com/2076-2615/10/9/1452/s1. Figure S1: Map with locations of populations included in the study; Table S1: Primers used in the Multiplex PCR approach for sequence analysis of *E. europaeus* and *E. roumanicus*.

Author Contributions: Conceptualization, K.P., M.C. and H.M.; data curation, M.C., M.L., A.S., M.F. and S.W.; formal analysis, K.P., M.C., B.Č.B. and H.M.; investigation, K.P., M.C., B.Č.B., M.L., P.H., A.S., M.F. and H.M.; software, M.C. and S.W.; supervision, B.Č.B., P.H. and H.M.; writing—original draft, K.P. and H.M.; writing—review & editing, M.C., B.Č.B., M.L., P.H. and S.W. All authors have read and agreed to the published version of the manuscript.

Funding: Parts of the work was funded by Grant Agency of the Czech University of Life Sciences Prague projects IGA 20205007 and CIGA 20185006 to BCB and by the research stimulation programme of the Department of Integrative Biology and Biodiversity Research.

Acknowledgments: We thank the staff of the shelters for providing material and their help for taking the mouth swabs. We thank J. Plass from the Linzer Landesmuseum and Leon Barthel providing samples for analysis. Polish samples were provided by Pomorski Ośrodek Rehabilitacji Dzikich Zwierząt "Ostoja". Eva Dornstauder-Schrammel is acknowledged for technical assistance, David Lang for graphical and formatting changes and Lea Ficker for organizing the mouth swabs in Germany.

Conflicts of Interest: The authors declare no conflict of interest.

Ethical Statement: Material was obtained exclusively from traffic fatalities or as mouth swabs from individuals in shelters that are regularly handled by staff for examination. No animals were harmed and no wild animal was trapped for the investigation. Personal information of shelter owner and staff is not disclosed.

References

1. Amori, G. Erinaceus Europaeus. The IUCN Red List of Threatened Species 2016: E.T29650A2791303. In *IUCN Red List of Threatened Species*; IUCN: Gland, Switzerland, 2016. [CrossRef]

2. Amori, G.; Hutterer, R.; Kryštufek, B.; Yigit, N.; Mitsain, G.; Palomo, L.J. Erinaceus Roumanicus. The IUCN Red List of Threatened Species 2016: E.T136344A115206348. In *IUCN Red List of Threatened Species*; IUCN: Gland, Switzerland, 2016. [CrossRef]

3. Holsbeek, L.; Rodts, J.; Muyldermans, S. Hedgehog and other animal traffic victims in Belgium: Results of a countrywide survey. *Lutra* **1999**, *42*, 111–119.

4.   Huijser, M.P.; Bergers, P.J.M. The effect of roads and traffic on hedgehog (*Erinaceus europaeus*) populations. *Biol. Conserv.* **2000**, *95*, 111–116. [CrossRef]

5.   Hof, A.R.; Bright, P.W. Quantifying the long-term decline of the West European hedgehog in England by subsampling citizen-science datasets. *Eur. J. Wildl. Res.* **2016**, *62*, 407–413. [CrossRef]

6.   Hof, A.R. A Study of the Current Status of the Hedgehog (*Erinaceus europaeus*) and its Decline in Great Britain Since 1960. Ph.D. Thesis, Royal Holloway, University of London, Egham, UK, 2009.

7.   Becher, S.A.; Griffiths, R. Genetic differentiation among local populations of the European hedgehog (*Erinaceus europaeus*) in mosaic habitats. *Mol. Ecol.* **1998**, *7*, 1599–1604. [CrossRef] [PubMed]

8.   Černá Bolfíková, B.; Eliášová, K.; Loudová, M.; Kryštufek, B.; Lymberakis, P.; Sándor, A.D.; Hulva, P. Glacial allopatry vs. postglacial parapatry and peripatry: The case of hedgehogs. *PeerJ* **2017**, *5*, e3163. [CrossRef] [PubMed]

9.   Bolfíková, B.; Hulva, P. Microevolution of sympatry: Landscape genetics of hedgehogs *Erinaceus europaeus* and *E. roumanicus* in Central Europe. *Heredity* **2012**, *108*, 248–255. [CrossRef]

10.   Curto, M.; Winter, S.; Seiter, A.; Schmid, L.; Scheicher, K.; Barthel, L.M.F.; Plass, J.; Meimberg, H. Application of a SSR-GBS marker system on investigation of European Hedgehog species and their hybrid zone dynamics. *Ecol. Evol.* **2019**, *9*, 2814–2832. [CrossRef]

11.   Morris, P. *Hedgehog*; New Naturalist; Harper Collins: London, UK, 2018; Volume 137, ISBN 978-0-00-823573-4.

12.   Rondinini, C.; Doncaster, C.P. Roads as barriers to movement for hedgehogs. *Funct. Ecol.* **2002**, *16*, 504–509. [CrossRef]

13.   Orlowski, G.; Nowak, L. Road mortality of hedgehogs Erinaceus spp. in farmland in Lower Silesia (south-western Poland). *Pol. J. Ecol.* **2004**, *52*, 377–382.

14.   Braaker, S.; Kormann, U.; Bontadina, F.; Obrist, M.K. Prediction of genetic connectivity in urban ecosystems by combining detailed movement data, genetic data and multi-path modelling. *Landsc. Urban Plan.* **2017**, *160*, 107–114. [CrossRef]

15.   Avise, J.C. *Phylogeography: The History and Formation of Species*; Harvard University Press: Cambridge, MA, USA, 2000; ISBN 978-0-674-66638-2.

16.   Avise, J.C. Phylogeography: Retrospect and prospect. *J. Biogeogr.* **2009**, *36*, 3–15. [CrossRef]

17.   Molony, S.E.; Dowding, C.V.; Baker, P.J.; Cuthill, I.C.; Harris, S. The effect of translocation and temporary captivity on wildlife rehabilitation success: An experimental study using European hedgehogs (*Erinaceus europaeus*). *Biol. Conserv.* **2006**, *130*, 530–537. [CrossRef]

18.   Fuhrmann, M. Untersuchungen zur Wildtierrehabilitation und Wiederauswilderung von *Erinaceus europaeus* (Braunbrustigel) und Erinaceus roumanicus (Nördlicher Weißbrustigel) in Österreich. Master's Thesis, University of Natural Resources and Life Sciences Vienna, Vienna, Austria, 2018.

19.   Weeks, A.R.; Sgro, C.M.; Young, A.G.; Frankham, R.; Mitchell, N.J.; Miller, K.A.; Byrne, M.; Coates, D.J.; Eldridge, M.D.B.; Sunnucks, P.; et al. Assessing the benefits and risks of translocations in changing environments: A genetic perspective. *Evol. Appl.* **2011**, *4*, 709–725. [CrossRef] [PubMed]

20.   Allendorf, F.W.; Luikart, G.H.; Aitken, S.N. *Conservation and the Genetics of Populations*, 2nd ed.; Wiley–Blackwell: Hoboken, NJ, USA, 2013.

21.   Laikre, L.; Schwartz, M.K.; Waples, R.S.; Ryman, N. Compromising genetic diversity in the wild: Unmonitored large-scale release of plants and animals. *Trends Ecol. Evol.* **2010**, *25*, 520–529. [CrossRef] [PubMed]

22.   EUR-Lex: Access to European Union Law. Document L:1992:206:TOC. Available online: https://eur-lex.europa.eu/legal-content/DE/TXT/?uri=OJ:L:1992:206:TOC (accessed on 29 July 2020).

23.   Rechtsinformationssystem des Bundes RIS–Landesrecht. Available online: https://www.ris.bka.gv.at/Land/ (accessed on 31 July 2020).

24.   Edmands, S. Between a rock and a hard place: Evaluating the relative risks of inbreeding and outbreeding for conservation and management. *Mol. Ecol.* **2007**, *16*, 463–475. [CrossRef] [PubMed]

25.   Moore, M.; Early, G.; Touhey, K.; Barco, S.; Gulland, F.; Wells, R. Rehabilitation and Release of Marine Mammals in the United States: Risks and Benefits. *Mar. Mamm. Sci.* **2007**, *23*, 731–750. [CrossRef]

26. Seddon, J.M.; Santucci, F.; Reeve, N.J.; Hewitt, G.M. DNA footprints of European hedgehogs, *Erinaceus europaeus* and *E. concolor*: Pleistocene refugia, postglacial expansion and colonization routes. *Mol. Ecol.* **2001**, *10*, 2187–2198. [CrossRef]

27. Suchentrunk, F.; Haiden, A.; Hartl, G.B. On biochemical genetic variability and divergence of the two hedgehog species *Erinaceus europaeus* and *E. concolor* in central Europe. *Z. Säugetierkd.* **1998**, *63*, 257–265.

28. Hewitt, G.M. Post-glacial re-colonization of European biota. *Biol. J. Linn. Soc.* **1999**, *68*, 87–112. [CrossRef]

29. Hewitt, G. The genetic legacy of the Quaternary ice ages. *Nature* **2000**, *405*, 907–913. [CrossRef]

30. Santucci, F.; Emerson, B.C.; Hewitt, G.M. Mitochondrial DNA phylogeography of European hedgehogs. *Mol. Ecol.* **1998**, *7*, 1163–1172. [CrossRef] [PubMed]

31. Hewitt, G.M. Some genetic consequences of ice ages, and their role in divergence and speciation. *Biol. J. Linn. Soc.* **1996**, *58*, 247–276. [CrossRef]

32. Taberlet, P.; Fumagalli, L.; Wust-Saucy, A.-G.; Cosson, J.-F. Comparative phylogeography and postglacial colonization routes in Europe. *Mol. Ecol.* **1998**, *7*, 453–464. [CrossRef] [PubMed]

33. Filippucci, M.G.; Simson, S. Allozyme Variation and Divergence in Erinaceidae (Mammalia, Insectivora). *Isr. J. Zool.* **1996**, *42*, 335–345. [CrossRef]

34. Berggren, K.T.; Ellegren, H.; Hewitt, G.M.; Seddon, J.M. Understanding the phylogeographic patterns of European hedgehogs, *Erinaceus concolor* and *E. europaeus* using the MHC. *Heredity* **2005**, *95*, 84–90. [CrossRef] [PubMed]

35. Sommer, R.S. When east met west: The sub-fossil footprints of the west European hedgehog and the northern white-breasted hedgehog during the Late Quaternary in Europe. *J. Zool.* **2007**, *273*, 82–89. [CrossRef]

36. Tibihika, P.D.; Curto, M.; Dornstauder-Schrammel, E.; Winter, S.; Alemayehu, E.; Waidbacher, H.; Meimberg, H. Application of microsatellite genotyping by sequencing (SSR-GBS) to measure genetic diversity of the East African *Oreochromis niloticus*. *Conserv. Genet.* **2019**, *20*, 357–372. [CrossRef]

37. Mcurto/SSR-GBS-Pipeline. Available online: https://github.com/mcurto/SSR-GBS-pipeline (accessed on 26 July 2020).

38. Zhang, J.; Kobert, K.; Flouri, T.; Stamatakis, A. PEAR: A fast and accurate Illumina Paired-End reAd mergeR. *Bioinformatics* **2014**, *30*, 614–620. [CrossRef]

39. Pritchard, J.K.; Stephens, M.; Donnelly, P. Inference of Population Structure Using Multilocus Genotype Data. *Genetics* **2000**, *155*, 945–959.

40. Earl, D.A.; vonHoldt, B.M. Structure harvester: A website and program for visualizing structure output and implementing the Evanno method. *Conserv. Genet. Resour.* **2012**, *4*, 359–361. [CrossRef]

41. Structure Harvester. Available online: http://taylor0.biology.ucla.edu/structureHarvester/ (accessed on 26 July 2020).

42. Kopelman, N.M.; Mayzel, J.; Jakobsson, M.; Rosenberg, N.A.; Mayrose, I. Clumpak: A program for identifying clustering modes and packaging population structure inferences across K. *Mol. Ecol. Resour.* **2015**, *15*, 1179–1191. [CrossRef]

43. CLUMPAK. Cluster Markov Packager Across K. Available online: http://clumpak.tau.ac.il/ (accessed on 26 July 2020).

44. Peakall, R.; Smouse, P.E. GENALEX 6: Genetic analysis in Excel. Population genetic software for teaching and research. *Mol. Ecol. Notes* **2006**, *6*, 288–295. [CrossRef]

45. Peakall, R.; Smouse, P.E. GenAlEx 6.5: Genetic analysis in Excel. Population genetic software for teaching and research—An update. *Bioinformatics* **2012**, *28*, 2537–2539. [CrossRef]

46. Europakarte—Die Karte von Europa. Available online: https://www.europakarte.org/ (accessed on 30 July 2020).

47. Morris, P.A.; Meakin, K.; Sharafi, S. The Behaviour and Survival of Rehabilitated Hedgehogs (*Erinaceus europaeus*). *Anim. Welf.* **1993**, *2*, 53–66.

48. Morris, P.A.; Warwick, H. A Study of Rehabilitated Juvenile Hedgehogs After Release into the Wild. *Anim. Welf.* **1994**, *3*, 163–177.

49. Reeve, N.J. The Survival and Welfare of Hedgehogs (*Erinaceus europaeus*) After Release Back into the Wild. *Anim. Welf.* **1998**, *7*, 189–202.

50. Bogdanov, A.S.; Bannikova, A.A.; Pirusskii, Y.M.; Formozov, N.A. The first genetic evidence of hybridization between West European and Northern White-breasted hedgehogs (*Erinaceus europaeus* and *E. roumanicus*) in Moscow Region. *Biol. Bull. Russ. Acad. Sci.* **2009**, *36*, 647–651. [CrossRef]

51. Bolfíková, B.; Konečný, A.; Pfäffle, M.; Skuballa, J.; Hulva, P. Population biology of establishment in New Zealand hedgehogs inferred from genetic and historical data: Conflict or compromise? *Mol. Ecol.* **2013**, *22*, 3709–3720. [CrossRef]

Article

Unexpected Gene-Flow in Urban Environments: The Example of the European Hedgehog

Leon M. F. Barthel [1,2,*], Dana Wehner [3], Anke Schmidt [4], Anne Berger [1,2], Heribert Hofer [2,3,5] and Jörns Fickel [4,6,*]

[1] Department of Evolutionary Ecology, Leibniz Institute for Zoo and Wildlife Research (IZW), Alfred-Kowalke-Straße 17, 10315 Berlin, Germany; Berger@izw-berlin.de

[2] Berlin Brandenburg Institute of Advanced Biodiversity Research (BBIB), 14195 Berlin, Germany; hofer@izw-berlin.de

[3] Department of Biology, Chemistry, Pharmacy, Freie Universität Berlin, Takustrasse 3, 14195 Berlin, Germany; Wehner.Dana@gmail.com

[4] Department of Evolutionary Genetics, Leibniz Institute for Zoo and Wildlife Research (IZW), Alfred-Kowalke-Straße 17, 10315 Berlin, Germany; Aschmidt@IZW-Berlin.de

[5] Department of Ecological Dynamics, Leibniz Institute for Zoo and Wildlife Research, Alfred-Kowalke-Straße 17, 10315 Berlin, Germany

[6] Institute for Biochemistry & Biology, University of Potsdam, 14476 Potsdam-Golm, Germany

* Correspondence: Barthel.Leon@gmx.de (L.M.F.B.); Fickel@IZW-Berlin.de (J.F.)

Received: 19 November 2020; Accepted: 3 December 2020; Published: 7 December 2020

Simple Summary: An urban environment holds many barriers for mammals with limited mobility such as hedgehogs. These barriers appear often unsurmountable (e.g., rivers, highways, fences) and thus hinder contact between hedgehogs, leading to genetic isolation. In our study we tested whether these barriers affect the hedgehog population of urban Berlin, Germany. As Berlin has many of these barriers, we were expecting a strong genetic differentiation among hedgehog populations. However, when we looked at unrelated individuals, we did not see genetic differentiation among populations. The latter was only detected when we included related individuals too, a 'family clan' structure that is referred to as gamodemes. We conclude that the high percentage of greenery in Berlin provides sufficient habitat for hedgehogs to maintain connectivity across the city.

Abstract: We use the European hedgehog (*Erinaceus europaeus*), a mammal with limited mobility, as a model species to study whether the structural matrix of the urban environment has an influence on population genetic structure of such species in the city of Berlin (Germany). Using ten established microsatellite loci we genotyped 143 hedgehogs from numerous sites throughout Berlin. Inclusion of all individuals in the cluster analysis yielded three genetic clusters, likely reflecting spatial associations of kin (larger family groups, known as gamodemes). To examine the potential bias in the cluster analysis caused by closely related individuals, we determined all pairwise relationships and excluded close relatives before repeating the cluster analysis. For this data subset ($N = 65$) both clustering algorithms applied (Structure, Baps) indicated the presence of a single genetic cluster. These results suggest that the high proportion of green patches in the city of Berlin provides numerous steppingstone habitats potentially linking local subpopulations. Alternatively, translocation of individuals across the city by hedgehog rescue facilities may also explain the existence of only a single cluster. We therefore propose that information about management activities such as releases by animal rescue centres should include location data (as exactly as possible) regarding both the collection and the release site, which can then be used in population genetic studies.

Keywords: urban; hedgehog; genetic cluster; barrier

---

1. Introduction

Urbanisation involves some of the most rapid and intense human-induced transformation processes. Structures such as impervious surfaces, roads and buildings have fragmented the environment for many species. Semi-penetrable or impenetrable barriers now separate smaller patches of the former landscape, particularly in the urban conurbation. In order to gain access to adequate resources, animals living in such patches often have to cross these barriers to move between patches. Some wildlife species easily surmount barriers and cope with urban conditions (urban utilisers, urban dwellers; [1]) and the close proximity to people (e.g., wild boar *Sus scrofa*; [2]), whereas others cannot (e.g., great bustard *Otis tarda*; [3]). In addition, some species benefit from structures in urban spaces that mimic their original habitat (e.g., common swift *Apus apus*; [4]). Thus, for behaviourally flexible wildlife species urban habitats may provide a novel living environment with the opportunity to exploit novel resources [2,5,6].

Geographic separation of populations by barriers reduces gene-flow among them and thus increases genetic differentiation among populations. It also decreases the genetic variation within populations both by genetic drift and by reducing the availability of genetically different breeding partners, thereby increasing the risk of inbreeding and subsequent inbreeding depression as well as a higher incidence of infectious diseases and thus elevated mortality [7]. Thus, a consequence of habitat fragmentation may be local population extinction [8–11].

Urban landscapes also often contain large green patches such as parks, residential gardens, cemeteries, currently unused former industrial sites (brownfield sites) and other habitats that provide a relatively undisturbed living space for wildlife species. These patches may serve as stepping stone habitats, allowing gene flow between otherwise separated local populations [12]. Whether gene flow occurs depends on the mobility and dispersal capacity of each species in relation to the distances between suitable habitat patches and the distribution of the latter within the urban matrix. Thus, we expect species with high mobility and high dispersal capacity to be less affected by a strongly structured urban landscape (e.g., the red fox *Vulpes vulpes* [13]) than species with small home ranges and limited dispersal capacity. For the latter we therefore expect a fragmented urban landscape to promote genetic isolation of clusters of individuals, causing a highly structured meta-population [14–17].

The purpose of this study is to test these expectations by genotyping European hedgehogs (*Erinaceus europaeus*) across the highly structured and fragmented urban matrix of the city of Berlin, Germany. Although hedgehogs are widely distributed across Europe [18], we used this species as a model species because of its limited dispersal capacity and its relatively small home range [19,20]. The size of the latter may range from 0.8 ha (England; [21]), over 10 to 40 ha (England; [22]) up to 98 ha (Finland; [23]). Whereas female hedgehogs mostly stay within their habitat patches, male hedgehogs occasionally cover distances of up to 7 km per night [24]. Because the European hedgehog can use the urban matrix and cope with its structural characteristics [17,19], population densities in urban areas may actually be higher than in rural habitats [25,26]. However, despite their broad geographical distribution and their ability to utilise urban matrices, hedgehog populations have been declining in size and numbers across Europe [27–31]. Understanding the long-term consequences of progressive spatial fragmentation by urbanisation on hedgehog genetic population structure might become increasingly important for developing conservation strategies for this species [19,32,33].

2. Materials and Methods

2.1. Sample Collection and Sites

Over a period of five years (2013–2017), we collected mouth mucosal cells from free-ranging European hedgehogs (*N* = 250) using nylon swabs (FLOQSwabs, COPAN, Brescia, Italy) and Forensic cotton Swaps (Sarstedt, Nümbrecht, Germany) in the city of Berlin and its suburbs (~876 km², or 87,594 ha). Sampling was carried out between 10 p.m. and 4 a.m. during torchlight transect walks in different public parks, cemeteries and green areas in Berlin. Due to the very low sampling success in the

South-Western part of Berlin, we shifted the sampling effort during 2016 and 2017 to the North-eastern part of the city, mainly to two large parks: (a) the "Treptower Park", a 20 ha public park open to and accessible by the general public throughout day or night in south-central Berlin and (b) the ~160 ha large "Tierpark Berlin", Europe's largest landscape zoological garden and inaccessible to the general public after dusk. For each individual, we recorded the GPS coordinates of its sampling location. Additional samples (N = 56) were provided by animal rescue facilities and local veterinary surgeries in Berlin (Figure 1, Table S1). For these samples approximate locations were provided by staff members. Coordinates, which may have had an error margin of a few hundred metres, where recorded using online maps. We also asked staff working at these facilities whether they had implemented particular rules on how release sites were chosen after the rehabilitation of hedgehogs. All procedures in this study involving animals were performed in accordance with the ethical standards of the institution (IZW permit 2016-02-01) and German federal law (permission numbers Reg0115/15 and G0104/14).

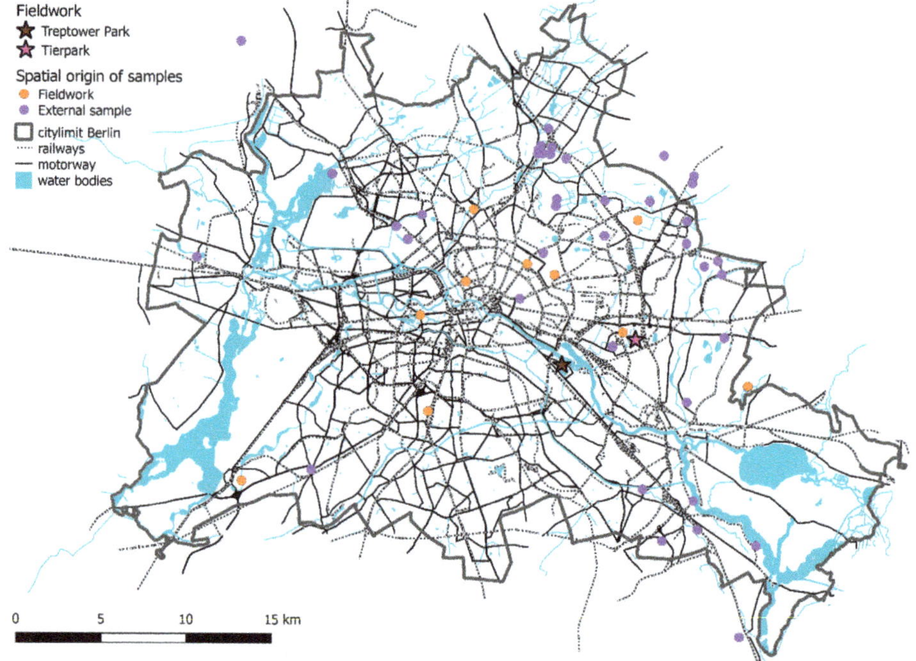

Figure 1. Map of Berlin and its surroundings and showing the locations from 139 out of 143 samples (four samples not shown because locations are outside of map). External samples: samples received from rescue centres. Their locations need to be viewed with caution. Samples from "Tierpark" and from, "Treptower Park" are lumped under pink and dark stars, respectively.

2.2. DNA Extraction and Analysis of Microsatellite Loci

DNA was extracted from all 306 samples using the DNeasy kit (Qiagen, Hilden, Germany) following the manufacturer's instructions, with a final DNA-elution in 80 μL distilled water (sterile). DNA concentrations were measured spectrophotometrically using a NanoDrop1000 (PeqLab GmbH, Erlangen, Germany). Individuals were genotyped at 10 microsatellite loci using a panel of nine loci from a previous landscape genetics study [34], with locus EEU1 added. The panel consisted of the following loci: EEU1, EEU2, EEU3, EEU4, EEU5 and EEU6 [35], EEU12H, EEU37H, EEU43H, and EEU54H [36]. One primer per pair was 5'- labelled with a fluorescent dye (6-FAM or HEX). To save time and costs, we prepared (after optimization) four primer master-mixes (Mix-A to Mix-D, 50 μL each). Mix-A contained the primers for loci EEU1, EEU2, and EEU54H (all 1 µM), Mix-B consisted of primers for loci EEU6 (1 µM) and EEU12H (2 µM), Mix-C of primers for loci EEU3 (1 µM) and EEU37H (2 µM), and Mix-D included the primers for loci EEU4 (4 µM) and EEU5 (2 µM). Primer pair EEU43H (3 µM) was run separately. The genotyping PCR mixture (10 µL) consisted of 5 µL 2′Type-it™ multiplex mix (Qiagen, Hilden, Germany), 1 µL primer mix, 3 µL H_2O and 1 µL DNA (50–120 ng). Cycling conditions were equal for all four master-mixes and locus EEU43H and were performed as touchdown-PCR: 95 °C 5 min, 4′ {94 °C 30 s, 63 °C down to 57 °C in 2 °C increments of 90 s each, 72 °C 30 s}, 31′ {94 °C 30 s, 55 °C 90 s, 72 °C 30 s}, 60 °C 30 min final elongation. Amplification products were analysed by capillary electrophoresis on an A3130*xl* automated sequencer (Thermo Fisher, Waltham, MA, USA) using POP7 and sized by comparison to a Genescan™ 500 ROX™ Size Standard (ABI) using the software Genemapper v.3.7 following the manufacturer's instructions. To avoid misleading results by allelic dropouts and false alleles, we applied a maximum likelihood approach [37] and genotyped each sample twice (in duplicates). The following quality filters were applied: (i) we did not allow for any allele mismatch between duplicates. If there was a mismatch, the sample was removed and genotyped again in duplicates from freshly extracted DNA. Genotypes were only scored if no mismatch was detected; otherwise, the sample was excluded from further analysis; (ii) we also excluded all individuals for which more than one locus had missing data.

2.3. Data Analysis

We calculated observed (H_O) and expected heterozygosities (H_E), number of alleles (N_A), as well as potential deviations from Hardy-Weinberg equilibrium (HWE) using the program Cervus v.3.0.7 [38–40]). We also used Cervus to search for matching genotypes across all samples. Tests for the presence of genotypic disequilibria among loci were performed using the software package Arlequin v.3.5.2.2 [41,42]. The significance level α was Bonferroni-corrected and set at 0.001 (0.05: 45 pairwise comparisons). Potential presence of null alleles was assessed using MicroChecker v.2.2.3 [43].

Although hedgehogs are solitary animals, their limited dispersal capacity (compared with larger mammals) may cause a population genetic structure by which closely related individuals may be living in closer proximity to each other than to more unrelated individuals. Because some clustering algorithms are affected by such associations of kin [44], we determined pairwise relatedness (r, [45] among all samples using the software package Coancestry v.1.0.1.9 [46]. Pairs with $r > 0.5$ were marked and subsequent cluster analysis (see below) was performed with and without these pairs (Figure S2).

The possible presence of genotypic clusters was evaluated both for the subset of only unrelated individuals and for the whole data set. For this purpose, we used two software packages with a Bayesian clustering approach: Structure v.2.3.4 [47–49] and Baps v.6.0 [50,51]. As priors for Structure, we applied the admixture model in conjunction with the correlated allele frequency model, because it is better suited to detect a subtle population structure, although this makes it more likely to overestimate the number of clusters K [48]. The model was applied to K-values ranging from $K = 1$ to 8. The required allele frequency distribution parameter λ was estimated for each run. To determine both appropriate burn-in and Markov chain lengths for parameter estimates of allele frequencies and membership coefficients per genotype in each genotypic cluster (Q), we set $K = 1$ and watched for the likelihoods to converge under various burn-in and run lengths. The final burn-in length was set at 20,000 iterations and Markov chains were run with a length of 200,000 iterations each. Each K was assessed independently 10 times to verify the consistency of estimates across runs. The most likely K was determined using both the log likelihood values (as ΔK cannot be applied if $K = 1$) and by following the ΔK method [52] using Structure Harvester [53]. For Baps the K prior ranged from 2 to 8 (as Baps cannot detect $K = 1$), whereby each K was also independently assessed 10 times. In addition, we used a location prior by providing the GPS coordinates of each sample's origin. We applied the algorithm using both the 'admixture' and the 'no admixture' prior.

2.4. Assignment

The threshold for the Q-value above which an individual will be assigned to a cluster is of importance. If the threshold is set too high it may underestimate a structure that in reality exists, whereas a threshold which is set too low will overemphasise a structure that in reality is not as pronounced as assumed. Here we chose a relatively conservative value of $Q \geq 0.85$ as the threshold for the assignment of individuals [54,55], thus allowing for some gene flow to have occurred among the inferred ancestral populations (at least three generations ago). Genetic distances between clusters of assigned individuals as well as the number of migrants (N_m) among clusters were estimated using ARLEQUIN. Input files for the different programmes were generated using the software CREATE [56]. Assignment results were averaged over ten runs using R [57] as means ± standard deviations (SD) unless stated otherwise.

3. Results

3.1. Genotyping

3.1.1. All Individuals

From the original dataset of 306 hedgehogs, data from 156 individuals had to be excluded. Out of those 156, 154 individuals were excluded because genotyping of their samples failed at more than one locus and two individuals were removed because the alleles of their duplicate sample genotypes did not match at all loci. In total, 150 individuals (49%) were successfully genotyped at all ten loci. Out of these genotypic profiles, one profile occurred three times and five others twice, leaving 143 unique genotypes (Table S2). These 143 genotypes, however, were not evenly distributed across the city of Berlin. The reasons were the low sampling success in the southern parts of the city, the shifted focus of sampling efforts in 2016 and 2017, and the fact that many samples from the southern part of Berlin did not pass the quality control filters.

The number of alleles per locus (N_A) ranged from four (locus EEU12H) to 16 (EEU37H), with a mean of 10.9 ± 4.1 (Table 1). H_O across all 143 unique genotypes ranged from 0.350 at locus EEU6 to 0.754 at locus EEU3, with a mean of H_O = 0.621 ± 0.133 (Table 1). Across all loci and individuals, one locus (EEH37H) deviated significantly from HWE (Table 1). Although several loci indicated the potential presence of null alleles, the probability was generally low. Pairwise relatedness analysis revealed numerous pairs of individuals with a high relatedness index ($r \geq 0.5$). Removal of these related individuals reduced the data set to 65 unrelated hedgehogs (Table S3). Presence of linkage disequilibria (LD) among the ten loci was tested both for the unrelated individuals ($N = 65$) and for 143 hedgehogs (all individuals). Among the unrelated individuals one out of 45 pairwise comparisons among the ten loci and among all hedgehogs 12 pairwise comparisons showed LD, although all loci had previously been declared to be independently inherited [36,58]; the former also included the loci from [34,58]. In our study, hedgehogs were sampled over a very large area (Figure 1), likely violating the assumption of an unstructured population, as expected for a small mammal in a highly fragmented landscape. The deviation from HWE at locus EEU37H and the linkage disequilibria may thus have been due to the Wahlund effect [59]. We therefore searched for an underlying population structure, first among the unrelated individuals and then among all individuals.

Table 1. Indices of ten microsatellite loci across the 143 unique genotypes (upper part) and averaged for the clusters (lower part).

Locus	N_typed	N_A	Allele Size Range (bp)	H_O	H_E	HWE	f_Null
EEU1	143	8	129–143	0.671	0.773	+	0.062
EEU2	141	13	257–281	0.752	0.863	+	0.064
EEU3	142	15	131–181	0.754	0.868	+	0.064

Table 1. *Cont.*

Locus	N_{typed}	N_A	Allele Size Range (bp)	H_O	H_E	HWE	f_{Null}
EEU4	143	14	144–170	0.699	0.785	+	<u>0.052</u>
EEU5	143	13	107–139	0.678	0.711	+	<u>0.011</u>
EEU6	143	6	145–159	0.350	0.331	+	−0.049
EEU12H	143	4	91–97	0.497	0.615	+	<u>0.098</u>
EEU37H	142	16	236–280	0.676	0.839	-	<u>0.095</u>
EEU43H	143	12	146–172	0.657	0.730	+	<u>0.047</u>
EEU54H	142	8	276–296	0.479	0.551	+	<u>0.067</u>
Mean	142.5	10.9		0.621	0.707		
SD	0.71	4.09		0.133	0.167		
Cluster assignment ($Q > 0.85$)		mean N_A		H_O	H_E	HWE	
1	29	8.6		0.623	0.685	+	
2 ("Tierpark")	14	3.3		0.557	0.524	+	
3 ("Trepower Park")	31	6.0		0.578	0.651	+	
single cluster ($N = 65$)	65	10.6		0.672	0.731	+	
no cluster, all ($N = 143$)	143	10.9		0.621	0.706	- *	

N_{typed}: number of individuals successfully genotyped at that locus, N_A: number of alleles per locus, bp.: base pairs, H_O: observed heterozygosity, H_E: heterozygosity expected under HWE, HWE: Hardy-Weinberg equilibrium. (+): locus was at HWE, (-): locus deviated from HWE, f_{Null}: probability for the presence of null-alleles (underlined values indicate an increased probability for the presence of null-alleles), SD: standard deviation. *: five out of 10 loci deviated from HWE.

3.1.2. Unrelated and Related Individuals

For the unrelated individuals data set, all individuals were assigned to a single cluster (mean LnP(K) = −2268.81 ± 0.481; results from STRUCTURE). For the entire data set, both clustering algorithms (STRUCTURE, BAPS) indicated the presence of three to four genotypic clusters (Table 2). The ΔK estimate (STRUCTURE HARVESTER) favoured three clusters over four (ΔK for $K = 3$ was 52.25; ΔK for $K = 4$ was 51.4), whereas BAPS favoured the presence of four clusters, with the fourth cluster being represented by two individuals (sampled at the same location). The likelihood for the number of genotypic clusters (K) to reflect the true number of ancestral populations had the following values (derived from BAPS): for $K = 3$: 0.00136, $K = 4$: 0.98883, and for $K = 5$: 0.0098).

Table 2. Animal ID, sampling location, Q-values (STRUCTURE) and cluster assignment for 143 hedgehogs of Berlin.

Internal ID	Locality	STRUCTURE Q-Value for Cluster 1	STRUCTURE Q-Value for Cluster 2	STRUCTURE Q-Value for Cluster 3	STRUCTURE Cluster Assignment ($Q \geq 0.85$)	BAPS Cluster Assignment *
147	Tiergarten, Berlin	0.87559	0.10548	0.01895	1	4
167	Tiergarten, Berlin	0.88068	0.10901	0.01031	1	4
176	Eisenhuettenstadt	0.85406	0.11966	0.02629	1	1
161	Hans-Baluschek-Park,	0.86127	0.11218	0.02655	1	1
311	Tierpark, Berlin	0.87213	0.10962	0.01824	1	1
334	12623 Berlin	0.86935	0.11628	0.01438	1	1
335	12623 Berlin	0.85235	0.11496	0.03271	1	1
220	Friedenstr., Berlin Near Volkspark Friedrichshain	0.87103	0.11977	0.00921	1	1

Table 2. *Cont.*

Internal ID	Locality	STRUCTURE Q-Value for Cluster 1	STRUCTURE Q-Value for Cluster 2	STRUCTURE Q-Value for Cluster 3	STRUCTURE Cluster Assignment ($Q \geq 0.85$)	BAPS Cluster Assignment *
175	Volkspark Prenzlauerberg, Berlin	0.86135	0.105	0.03368	1	1
199	Hellersdorf, Berlin	0.86134	0.12691	0.01175	1	1
159	Prenzlauerberg	0.85732	0.10238	0.04031	1	1
338	Park am Weidengrund	0.87595	0.11224	0.01182	1	1
117	Near graveyard	0.88408	0.10591	0.01	1	1
326	Zum Erlenbruch, 15344 Strausberg	0.86423	0.11255	0.02321	1	1
156	Buergerpark Pankow-Berlin	0.86845	0.11301	0.01854	1	1
341	Kleingartenanlage 750 Jahre Berlin, 13057 Berlin	0.85329	0.10368	0.04302	1	1
328	Warnemuender Str. 18, 13059 Berlin	0.86432	0.11747	0.01821	1	1
329	Warnemuender Str. 18, 13059 Berlin	0.86724	0.1151	0.01767	1	1
330	Warnemuender Str. 18, 13059 Berlin	0.87485	0.1077	0.01746	1	1
337	KGA Maerchenland, 13089 Berlin	0.87728	0.10819	0.01456	1	1
231	Friedenstr. 8, 16356 Ahrensfelde	0.87273	0.10608	0.02116	1	1
257	Dietrichstr. 5, 16356 Ahrensfelde	0.87167	0.10645	0.0219	1	1
193	Jungbornstr., 13129 Berlin	0.85709	0.12218	0.02072	1	1
189	Strasse 7, 13129 Berlin	0.85836	0.11349	0.02814	1	1
194	Schwarzwaldstr./Ilsenstr., 13129 Berlin	0.85756	0.11789	0.02457	1	1
185	Gutenfelsstr. 14, 13129 Berlin	0.8736	0.11369	0.01271	1	1
187	Gutenfelsstr. 14, 13129 Berlin	0.8837	0.1066	0.0097	1	1
113	Choise-le-Roi-Str. 3, Berlin	0.88496	0.10623	0.00881	1	1
118	Vielitzsee Ortsteil Strubensee, 16835 green area	0.85436	0.13413	0.01153	1	1
179	Eisenhuettenstadt	0.84074	0.10226	0.05701	admixed	1
243	Zeuthen	0.6249	0.20766	0.16744	admixed	1
129	Rohrwallallee 10. 12527 Berlin	0.83142	0.10412	0.06446	admixed	1
120	Altglienike Feldweg	0.84766	0.10475	0.0476	admixed	1
137	Kablower Weg 89, 12526 Berlin	0.78369	0.0984	0.1179	admixed	1
138	Kablower Weg 89, 12526 Berlin	0.8267	0.1031	0.0702	admixed	1
125	Riesserseestr. 10. 12527 Berlin	0.84665	0.11413	0.03923	admixed	1
135	Korkedamm 73, 12524 Berlin	0.69588	0.08972	0.21436	admixed	1
127	Rehwiese, Gerkrathstraße 2 Park	0.72131	0.14963	0.12906	admixed	1
182	Zehlendorf, Berlin	0.40151	0.09898	0.49948	admixed	1
158	Hans-Baluschek-Park, 10829 Berlin	0.56664	0.07147	0.36189	admixed	1

Table 2. *Cont.*

Internal ID	Locality	STRUCTURE Q-Value for Cluster 1	STRUCTURE Q-Value for Cluster 2	STRUCTURE Q-Value for Cluster 3	STRUCTURE Cluster Assignment ($Q \geq 0.85$)	BAPS Cluster Assignment *
169	Hans-Baluschek-Park, 10829 Berlin	0.75432	0.10163	0.14405	admixed	1
235	Glasberger Str. 43, 12555 Berlin	0.77728	0.10906	0.11368	admixed	1
110	Trainierbahn Hoppegarten	0.55419	0.42731	0.01849	admixed	1
A35_088	Treptower Park	0.66656	0.14738	0.18608	admixed	1
A4_317	Treptower Park	0.81876	0.16614	0.01512	admixed	1
A61_108	Treptower Park	0.67049	0.09899	0.23051	admixed	1
A68_108	Treptower Park	0.71856	0.17835	0.10308	admixed	1
126	Moldaustr. 30. 10319 Berlin near Tierpark	0.68485	0.2057	0.10944	admixed	1
128	Moldaustr. 24, 10319 Berlin near Tierpark	0.13048	0.18259	0.68692	admixed	1
136	Moldaustr. 24, 10319 Berlin near Tierpark	0.2581	0.25296	0.48897	admixed	1
174	Tierpark, Berlin	0.48697	0.40986	0.10316	admixed	1
308	Tierpark, Berlin	0.8438	0.13492	0.02127	admixed	1
310	Tierpark, Berlin	0.45013	0.09578	0.45409	admixed	1
314	Tierpark, Berlin	0.50102	0.45028	0.0487	admixed	1
317	Tierpark, Berlin	0.66602	0.15881	0.17519	admixed	1
320	Tierpark, Berlin	0.22576	0.45684	0.3174	admixed	1
333	12623 Berlin	0.76214	0.22126	0.01663	admixed	1
143	Tiergarten, Berlin	0.82986	0.15939	0.01076	admixed	1
152	Tiergarten, Berlin	0.79801	0.14577	0.05625	admixed	1
166	Tiergarten, Berlin	0.78003	0.16683	0.05316	admixed	1
309	Nordbahnhof park	0.72988	0.1093	0.16079	admixed	1
134	Volkspark Prenzlauerberg, Berlin	0.84225	0.1053	0.05244	admixed	1
142	Volkspark Prenzlauerberg, Berlin	0.62693	0.08785	0.28521	admixed	1
153	Volkspark Prenzlauerberg, Berlin	0.77372	0.0954	0.13087	admixed	1
168	Volkspark Prenzlauerberg, Berlin	0.75805	0.0945	0.14747	admixed	1
170	Volkspark Prenzlauerberg, Berlin	0.35675	0.0564	0.58687	admixed	1
172	Volkspark Prenzlauerberg, Berlin	0.83185	0.10633	0.06181	admixed	1
324	Eisenacher Str.,12629 Berlin near park Kastanienallee	0.71554	0.09224	0.19223	admixed	1
300	122/126, 12627 Berlin near Teupitzer Park	0.83854	0.12196	0.03952	admixed	1
261	Wolfshorststr. 25, 13591 Berlin	0.20168	0.04025	0.75806	admixed	1
340	Mahlerstraße, 13088 Berlin	0.75765	0.09987	0.14248	admixed	1
241	Glambecker Ring 4, 12679 Berlin	0.66759	0.16157	0.17085	admixed	1
114	Togostr. 45, 13351 Berlin near Volkspark Rehberge	0.61322	0.10559	0.28119	admixed	1
248	13053 Berlin	0.58362	0.0891	0.32725	admixed	1
119	Ghanastr. 27, 13351 Berlin near Volkspark Rehberge	0.7999	0.09627	0.10382	admixed	1
139	Falkenberger Krugwiesen, 13057 Berlin	0.78143	0.11325	0.10533	admixed	1

Table 2. *Cont.*

Internal ID	Locality	STRUCTURE Q-Value for Cluster 1	STRUCTURE Q-Value for Cluster 2	STRUCTURE Q-Value for Cluster 3	STRUCTURE Cluster Assignment ($Q \geq 0.85$)	BAPS Cluster Assignment *
140	Falkenberger Krugwiesen, 13057 Berlin	0.73941	0.18507	0.0755	admixed	1
150	Buergerpark Pankow-Berlin	0.66583	0.13902	0.19514	admixed	1
188	Schwarzelfenweg 19, 13088 Berlin	0.59525	0.23319	0.17157	admixed	1
116	Alt-Tegel 47c, 13507 Berlin	0.82319	0.14861	0.0282	admixed	1
191	Strasse 26 Nr. 30. 13129 Berlin near green area	0.65149	0.31845	0.03009	admixed	1
196	Schwarzwaldstr., 13129 Berlin	0.67859	0.10107	0.22037	admixed	1
186	Gutenfelsstr. 14, 13129 Berlin	0.81717	0.1626	0.02022	admixed	1
198	Gutenfelsstr. 14, 13129 Berlin	0.76845	0.1193	0.11227	admixed	1
200	Gutenfelsstr. 14, 13129 Berlin	0.8493	0.10802	0.04267	admixed	1
192	Urbacher Str., 13129 Berlin	0.54983	0.25383	0.19634	admixed	1
184	Freischuetzstr., 13129 Berlin	0.81933	0.14711	0.03358	admixed	1
203	Freischuetzstr., 13129 Berlin	0.8327	0.13544	0.03187	admixed	1
197	Krontalerstr., 13125 Berlin	0.84704	0.10899	0.04399	admixed	1
A3_317	Treptower Park	0.77842	0.11654	0.10505	admixed	admixed
A34_078	Treptower Park	0.67359	0.09012	0.23627	admixed	admixed
144	Tierpark, Berlin	0.10456	0.88882	0.00662	2	2
146	Tierpark, Berlin	0.10652	0.8851	0.00836	2	2
154	Tierpark, Berlin	0.10959	0.88179	0.00862	2	2
165	Tierpark, Berlin	0.10523	0.88487	0.00991	2	2
305	Tierpark, Berlin	0.10535	0.88925	0.00539	2	2
306	Tierpark, Berlin	0.10434	0.89067	0.005	2	2
307	Tierpark, Berlin	0.12173	0.85912	0.01916	2	2
312	Tierpark, Berlin	0.10548	0.88768	0.00684	2	2
313	Tierpark, Berlin	0.10518	0.88331	0.0115	2	2
315	Tierpark, Berlin	0.11634	0.86767	0.01601	2	2
318	Tierpark, Berlin	0.11617	0.87305	0.01078	2	2
319	Tierpark, Berlin	0.11028	0.87549	0.01424	2	2
342	Tierpark, Berlin	0.11259	0.87716	0.01028	2	2
344	IZW Garten, Berlin (bordering with Tierpark)	0.10948	0.87852	0.01199	2	2
141	Tierpark, Berlin	0.14033	0.79122	0.06844	admixed	2
149	Tierpark, Berlin	0.10468	0.79688	0.09844	admixed	2
321	Tierpark, Berlin	0.15174	0.82023	0.02802	admixed	2
322	Tierpark, Berlin	0.1618	0.8123	0.02591	admixed	2
157	Treptower Park	0.02233	0.01133	0.96636	3	3
345	Treptower Park	0.01293	0.01151	0.97556	3	3
346	Treptower Park	0.02053	0.00877	0.97068	3	3
348	Treptower Park	0.07228	0.0132	0.9145	3	3
349	Treptower Park	0.0197	0.00997	0.97034	3	3
350	Treptower Park	0.01276	0.01003	0.97721	3	3
A1_317	Treptower Park	0.01622	0.00786	0.97591	3	3
A10_028	Treptower Park	0.02804	0.02522	0.94674	3	3
A11_028	Treptower Park	0.06713	0.02515	0.90771	3	3
A12_028	Treptower Park	0.03142	0.0546	0.914	3	3
A13_028	Treptower Park	0.04447	0.01617	0.93935	3	3

Table 2. *Cont.*

Internal ID	Locality	STRUCTURE Q-Value for Cluster 1	STRUCTURE Q-Value for Cluster 2	STRUCTURE Q-Value for Cluster 3	STRUCTURE Cluster Assignment ($Q \geq 0.85$)	BAPS Cluster Assignment *
A14_028	Treptower Park	0.01659	0.00572	0.97765	3	3
A15_028	Treptower Park	0.01453	0.01432	0.97112	3	3
A16_028	Treptower Park	0.04624	0.01015	0.94359	3	3
A2_317	Treptower Park	0.03028	0.01088	0.95885	3	3
A20_038	Treptower Park	0.01454	0.01811	0.96735	3	3
A21_038	Treptower Park	0.05401	0.01349	0.9325	3	3
A22_038	Treptower Park	0.04167	0.01371	0.94463	3	3
A25_078	Treptower Park	0.01262	0.02414	0.96325	3	3
A27_078	Treptower Park	0.02487	0.03366	0.94146	3	3
A28_078	Treptower Park	0.01631	0.01948	0.96424	3	3
A30_078	Treptower Park	0.02939	0.01268	0.95794	3	3
A31_078	Treptower Park	0.01652	0.04733	0.93614	3	3
A32_078	Treptower Park	0.01376	0.01006	0.97617	3	3
A37_088	Treptower Park	0.01249	0.01269	0.97483	3	3
A43_088	Treptower Park	0.02567	0.02771	0.94662	3	3
A47_098	Treptower Park	0.01487	0.01031	0.97481	3	3
A5_317	Treptower Park	0.02926	0.0273	0.94345	3	3
A59_108	Treptower Park	0.0119	0.00944	0.97865	3	3
A9_028	Treptower Park	0.00902	0.00582	0.98515	3	3
252	Friedenstr., 16356 Ahrensfelde	0.04766	0.01531	0.93704	3	3
A56_098	Treptower Park	0.12642	0.03009	0.84349	admixed	3
A62_108	Treptower Park	0.13874	0.02767	0.83359	admixed	3
343	Tierpark, Berlin	0.16607	0.05859	0.77535	admixed	admixed

Using a value of $Q \geq 0.85$ (Structure), 74 out of 143 genotypes (51.7%) were assigned to either one of three genotypic clusters: cluster 1 with 29 genotypes, cluster 2 with 14 genotypes (all individuals but one were from "Tierpark"), and cluster 3 with 31 genotypes (all but one from "Treptower Park"). The 69 remaining genotypes were admixed, with admixture occurring across all clusters (Table 2, Figure S1). Each cluster was at HWE. Observed (H_O) and expected heterozygosities (H_E) were $H_O = 0.623$ and $H_E = 0.685$ for cluster 1 ($N = 29$), for cluster 2 ($N = 14$) they were $H_O = 0.557$ and $H_E = 0.524$, and for cluster 3 ($N = 31$) they were $H_O = 0.578$ and $H_E = 0.651$. Pairwise genetic distances (F_{ST}) among all clusters were significant ($p < 0.05$) with $F_{ST} = 0.169$ between clusters 1 and 2, $F_{ST} = 0.11$ between clusters 1 and 3, and $F_{ST} = 0.192$ between clusters 2 and 3.

Using the assignment threshold of $Q \geq 0.65$ [60], the number of hedgehogs per cluster would increase by 45 for cluster 1 ($N_{(Q=0.65)} = 74$), by four for cluster 2 ($N_{(Q=0.65)} = 18$), and by three for cluster 3 ($N_{(Q=0.65)} = 34$). Increasing the number of individuals per cluster in such a way would at the same time reduce the number of admixed individuals considerably ($N_{(Q=0.65)} = 17$).

3.1.3. Migrants

The number of migrants (N_m) per generation also differed among the three clusters. They were $N_m = 1.22$ between clusters 1 (wide-spread) and 2 ("Tierpark"), $N_m = 2.02$ between clusters 1 and 3 ("Treptower Park") and $N_m = 1.05$ between clusters 2 and 3. Applying the Baps clustering algorithm led to results very similar to the ones obtained from the Structure analysis, except for the introduction of a fourth cluster (2 individuals only) and an increase in the number of hedgehogs assigned to any cluster (Table 2). This increase in the number of individuals assigned to a cluster was particularly pronounced in cluster 1, into which Structure had only assigned 29 hedgehogs, whereas the Baps algorithm assigned three times as many individuals to that cluster ($N = 87$). Following the Baps assignment, hedgehogs from cluster 1 were also present in the "Tierpark" and the "Treptower Park".

3.2. Release of Hedgehogs after Rehabilitation

Although rescue facilities had no particular rules regarding the selection of release sites for rehabilitated hedgehogs, general policy was to release hedgehogs into favourable habitats, independent of their point of geographic origin. This policy led to the release of hedgehogs at distances far from the facilities and far from their previous pick-up points, in some cases at distances of >100 km.

4. Discussion

Considering only unrelated individuals, hedgehogs were assigned to a single cluster, whose members were spread across the city (Figure 2). Such a lack of genetic population structure was surprising in light of the presence of many potential barriers, and it contrasts with results from a genetic study on 149 urban hedgehogs in the city of Zurich (Switzerland), where a strong differentiation had been observed across an area of ~10,000 ha [60]. There, despite an eight times smaller spatial scale than in Berlin, three genotypic clusters had been inferred. The Zurich hedgehog clusters were well delineated by a major inner-city transportation axis as an anthropogenic barrier and two rivers as natural barriers [60]. The authors concluded that urban green areas were the most suitable habitat type to facilitate gene flow, whereas all other land cover types were more likely to impede gene flow [60].

The Zurich study differed from ours in several aspects: Their threshold for assigning individuals to a genetic cluster was considerably lower ($Q \geq 0.65$ instead of $Q \geq 0.85$), and they did not consider the potential effect of association of kin on genetic population structure. In our study, unrelated individuals did not demonstrate any obvious population genetic structure, although the city of Berlin is much larger than Zurich and even more divided by several highways and large rivers or canals.

The inclusion of all individuals in our study indicated the presence of at least three genotypic clusters ($Q \geq 0.85$), two of which were spatially well delineated (STRUCTURE: clusters 2 and 3, the "Tierpark" and the "Treptower Park"). If we used the threshold of $Q \geq 0.65$ as in the Zurich study [60], the relatively low assignment threshold and the inclusion of all 143 individuals (without removal of closely related individuals) would have led us to the conclusion of a strong population genetic structure consisting of three well-delineated clusters. Using a higher assignment threshold and removal of related individuals will reduce the number of individuals that can be assigned to a genotypic cluster and thus has a strong influence on the number of individuals per cluster and on the number of potential clusters to be detected.

As the genetic structure in Berlin hedgehogs only appeared if related individuals were included in the cluster analysis, we suggest the differentiation detected here to be a reflection of an underlying kinship network of gamodemes rather than to be a reflection of allele frequencies of three ancestral populations. The emergence of such gamodemes may be facilitated by the fact that hedgehogs are promiscuous as well as philopatric and that they have hetero-paternal superfecundation [61]. We do not know whether hedgehogs differentiate between kin and non-kin during mating season [61], but a lack of such differentiation may also contribute to the emergence and maintenance of "gamodemes". Such a gamodeme structure would also explain the local concentration of 'cluster 2 individuals' in the "Tierpark" and of "cluster 3 individuals" in the "Treptower Park". Although the "Tierpark" is a large park-like area (~160 ha) that was preserved after World War II and established as a zoological garden in 1954, it is almost fully fenced and surrounded by big streets both in the north and the west and by railway tracks in the east and the south. Thus, gene flow between hedgehogs from the "Tierpark" and the surrounding areas is clearly restricted, explaining the confinement from hedgehogs of cluster 2 to the "Tierpark". This is also shown by the significant pairwise F_{ST} values, which were the highest between clusters 2 and 3 ($F_{ST} = 0.192$) and clusters 2 and 1 ($F_{ST} = 0.169$). Interestingly, the hedgehogs inhabiting the "Treptower Park" (cluster 3) are strongly differentiated from the ones living in the "Tierpark" ($F_{ST} = 0.192$, lowest migration rate with $N_m = 1.05$), but not as strongly from the wide-spread cluster 1 ($F_{ST} = 0.11$, highest migration rate with $N_m = 2.02$). The main difference to the location "Tierpark" is that the "Treptower Park" is not fenced in and always accessible. However, it is bordered on one side by the river Spree (Berlin's main river) and on its three other sides by heavy-traffic roads.

An additional heavy-traffic road crosses the park longitudinally. Yet these barriers appear still to be more penetrable for hedgehogs than those in the "Tierpark". The reason why these "park gamodemes" have become so large may be the low landscape resistance within the parks, whereas at the borders of the parks landscape resistance increases drastically.

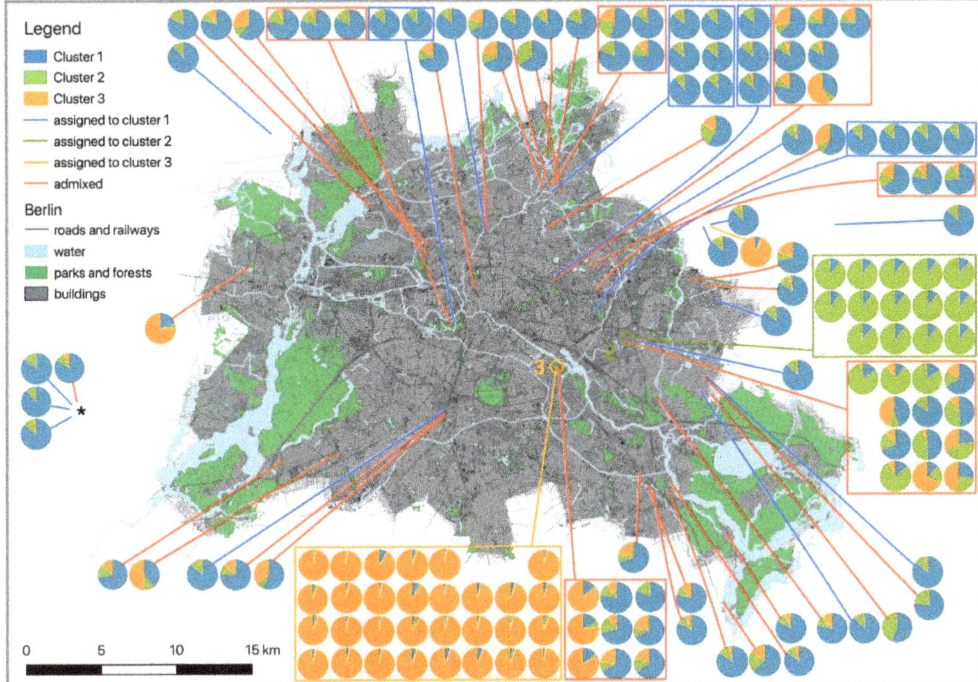

Figure 2. Map of Berlin and its surroundings indicating the locations of all samples displayed as pie charts showing the proportional membership of each individual in either one of three clusters (determined by STRUCTURE, including related animals). Different colours indicate membership to different clusters. Q-values were taken from Table 2. Individuals that belong to a cluster ($Q \geq 0.85$) are grouped within squares framed in the colour of their cluster. Admixed individuals are framed in red squares. Solid lines connect groups and individuals with their sampling location. For individuals assigned to either one of the three clusters, colours of lines correspond to cluster membership (cluster 1: blue, cluster 2: bright green, cluster 3: bright orange). Red lines connect admixed individuals to their sampling site. Pie charts of the four samples that were collected outside of the displayed city area are given on the left side, indicated by *. Locations "Tierpark" and "Treptower Park" are indicated by coloured circles and cluster number (2, green: Tierpark; 3, orange: Treptower Park).

Expected heterozygosity (H_E) for individuals of the most wide-spread cluster (cluster 1: $H_E = 0.685$; $N = 29$) was even slightly higher than the value of $H_E = 0.68$ measured in a country-wide study in the Czech Republic (average sampling site distances >450 km; [34]), indicating that in Berlin the urban environment does not lead to a reduction of genetic variability in hedgehogs. Because individuals from clusters 2 and 3 of our study were confined to single parks, either to the "Tierpark" or to the "Treptower Park", we expected to detect low observed (H_O) and expected heterozygosities there. Even though the values were indeed lower than in the widespread cluster 1, they were only lower by a small margin (cluster 2: $H_O = 0.557/H_E = 0.524$; cluster 3: $H_O = 0.578/H_E = 0.651$) and even slightly higher than those measured in free ranging hedgehog populations from New Zealand affected by a founder effect ($H_O = 0.42–0.52/H_E = 0.51–0.57$; [62]). These values were also in the range of values for the three urban clusters detected in Zurich ($H_O/H_E = 0.6/0.605$; $0.523/0.568$; $0.631/0.627$), although pairwise population

F_{ST}-values among the Zurich clusters ranged only between 0.059 and 0.082 [60], indicating a higher proportion of shared alleles among the Zurich hedgehog clusters than among the gamodeme clusters of Berlin. The values measured in the urban hedgehogs of Berlin were also much higher than the ones found in a recent study on rural hedgehogs in Denmark [61]. Here the population ($N = 14$) of Bornholm Island had values of only $H_O = 0.124$ and of $H_E = 126$, while the highest values were recorded in a subpopulation of Jutland (south of the Limfjord, $N = 71$) with $H_O = 0.293$ and of $H_E = 0.318$.

In contrast to many other mammalian species, hedgehogs lack a clear dispersal phase [63,64]. They rarely cover distances larger than 4 km [63] and are restricted in their movements by roads and other barrier-like structural elements [65,66]. Because the city of Berlin is approximately eight times larger than Zurich, the emergence of a genetic population structure due to restricted gene flow as seen in Zurich (but see our comments above on Q) appears to be inevitable. Yet hedgehogs in Berlin did not differentiate into a clear population genetic structure (if related animals were excluded), although the city of Berlin is much larger than Zurich. We thus expected dispersal over these large within-city distances to be even less likely (than in Zurich) and therefore a genetic population structure to be even more pronounced and clearly delineated by space. This was, however, not the case. Our results and observations would be compatible with the idea that all Berlin hedgehogs derived from a single ancestral population.

Because our results provide only a temporal snapshot, we do not know whether the spatial discrimination of clusters 2 ("Tierpark") and 3 ("Treptower Park") is the beginning of a process leading either to population differentiation or to complete admixture (as we found numerous admixed individuals), or whether it may represent a stable genotypic equilibrium.

Although we currently do not have a detailed knowledge about the ancestry of hedgehogs in Berlin, it is well known that hedgehogs have lived in Berlin for centuries and have experienced Berlin's increasing urbanisation throughout this period [67]. This raises the question as to what could be the reasons for the lack of a clear, spatially derived population genetic structure in a species that is considered to be substantially constrained by physical urban structures such as waterways, motorways, railways, and built-up areas [20,63,65], structures that characterise Berlin.

We argue that the main reason for our finding is the large proportion of green areas in Berlin. The city of Berlin is covered by 15,752 ha of forests (18%) and 10,885 ha of public green sites (12.4%) such as cemeteries, parks and gardens [2,68]. These areas provide a connective web of suitable habitats within the urban matrix, improving the opportunities for hedgehogs to maintain some amount of gene flow across the city. In addition, or alternatively, other factors may increase admixture. Given home ranges of 10–40 ha [22], the distances that needed to be covered to establish gene flow between "gamodeme" clusters are quite large for a short-legged ground-dwelling species, but numerous small and larger green areas can be stepping stones to link distant parts of the city. We also suggest that admixture had been enhanced by animals released by hedgehog rescue facilities [69]. These events are not fully quantified at present, but our interviews with personnel from rescue facilities confirmed that they are a regular occurrence, some estimates suggesting several hundred per year. Such rescue related translocations have also been observed in other studies [60,69,70].

5. Conclusions

We originally hypothesised urban hedgehogs, a species with relatively low mobility and low dispersal capacity, to be highly influenced by fragmented urban landscapes leading to genetic isolation of populations and thus a highly structured meta-population. Yet the hedgehog population in the city of Berlin is not genetically structured, if only unrelated individuals are being taken into account. A genetic structure becomes only visible if related individuals are also included in the analysis. Gene flow between these gamodeme-clusters is probably realised through natural means across the numerous green patches of Berlin's urban matrix and complemented by anthropogenic translocations. To maintain the currently existing genetic diversity in Berlin's hedgehog population, we suggest its repeated monitoring by census measures and population wide genetic analysis to determine whether current clusters (gamodemes) are at risk of becoming isolated.

Supplementary Materials: The following tables and figures are available online at http://www.mdpi.com/2076-2615/10/12/2315/s1, Figure S1: Map of all sampling locations (*N* = 143), displaying the distribution of individual genotype *Q*-values (Structure analysis results), in-between values are interpolated. Left: data for cluster 1, center: data for cluster 2, right: data for cluster 3, Figure S2: Pairwise relatedness after Queller and Goodman (QGEst) of sampled genotypes (Ind1 and Ind2) before (left) and after (right) removing related genotypes *r* > 0.5 (darker red), Table S1: Origin of samples (*N* = 143), Table S2: Unique genotypes (*N* = 143) of hedgehogs in Berlin across ten microsatellite loci, Table S3: Unique genotypes (*N* = 65) of unrelated (*r* < 0.5) hedgehogs in Berlin across ten microsatellite loci.

Author Contributions: Conceptualization, A.B., H.H. and J.F.; methodology, J.F.; software, J.F. and L.M.F.B.; validation, D.W. and A.S.; formal analysis, D.W., A.S.; investigation, D.W., A.S. and L.M.F.B.; data curation, D.W. and L.M.F.B.; writing—original draft preparation, L.M.F.B. and J.F.; writing—review and editing, J.F., A.B. and H.H.; visualization, L.M.F.B., J.F. and A.B.; supervision, A.B., and J.F.; project administration, A.B.; funding acquisition, A.B. and H.H. All authors have read and agreed to the published version of the manuscript.

Funding: This work as well as the publication costs was partly funded by the German Federal Ministry of Education and Research (BMBF) within the Collaborative Project "Bridging in Biodiversity Science-BIBS" (grant number 01LC1501).

Acknowledgments: We thank D. Müller and W. Rast for their assistance during field work and for organizing the database, P. Gras for his help with the GIS software and S. Kimmig, I. Möllenkotte and C. Scholz and many others for suggestions to improve former versions of the manuscript. We particularly acknowledge all rescue facilities, animal shelters and veterinary surgeries for their support by providing samples and sharing information on the origin of these samples.

Conflicts of Interest: The authors declare no conflict of interest.

References

1. Fischer, J.D.; Schneider, S.C.; Ahlers, A.A.; Miller, J.R. Categorizing wildlife responses to urbanization and conservation implications of terminology. *Conserv. Biol.* **2015**, *29*, 1246–1248. [CrossRef] [PubMed]

2. Stillfried, M.; Fickel, J.; Börner, K.; Wittstatt, U.; Heddergott, M.; Ortmann, S.; Kramer-Schadt, S.; Frantz, A.C. Do cities represent sources, sinks or isolated islands for urban wild boar population structure? *J. Appl. Ecol.* **2017**, *54*, 272–281. [CrossRef]

3. Pitra, C.; Lieckfeldt, D.; Frahnert, S.; Fickel, J. Phylogenetic relationships and ancestral areas of the bustards (Gruiformes: Otididae), inferred from mitochondrial DNA and nuclear intron sequences. *Mol. Phylogenetics Evol.* **2002**, *23*, 63–74. [CrossRef] [PubMed]

4. Schaub, T.; Meffert, P.J.; Kerth, G. Nest-boxes for Common Swifts Apus apus as compensatory measures in the context of building renovation: Efficacy and predictors of occupancy. *Bird Conserv. Int.* **2015**, *26*, 164–176. [CrossRef]

5. Johnson, M.T.J.; Munshi-South, J. Evolution of life in urban environments. *Science* **2017**, *358*, eaam8327. [CrossRef]

6. Maclagan, S.J.; Coates, T.; Ritchie, E.G. Don't judge habitat on its novelty: Assessing the value of novel habitats for an endangered mammal in a peri-urban landscape. *Biol. Conserv.* **2018**, *223*, 11–18. [CrossRef]

7. Belasen, A.H.; Bletz, M.C.; da Silva Leite, D.; Toledo, L.F.; James, T.Y. Long-Term Habitat Fragmentation Is Associated With Reduced MHC IIB Diversity and Increased Infections in Amphibian Hosts. *Front. Ecol. Evol.* **2019**, *6*, 236. [CrossRef]

8. Wilcox, B.A.; Murphy, D.D. Conservation Strategy: The Effects of Fragmentation on Extinction. *Am. Nat.* **1985**, *125*, 879–887. [CrossRef]

9. Munshi-South, J.; Nagy, C. Urban park characteristics, genetic variation, and historical demography of white-footed mouse (*Peromyscus leucopus*) populations in New York City. *PeerJ* **2014**, *2*, e310. [CrossRef]

10. Lourenço, A.; Álvarez, D.; Wang, I.J.; Velo-Antón, G. Trapped within the city: Integrating demography, time since isolation and population-specific traits to assess the genetic effects of urbanization. *Mol. Ecol.* **2017**, *26*, 1498–1514. [CrossRef]

11. Martins, R.F.; Fickel, J.; Le, M.; van Nguyen, T.; Nguyen, H.M.; Timmins, R.; Gan, H.M.; Rovie-Ryan, J.J.; Lenz, D.; Förster, D.W.; et al. Phylogeography of red muntjacs reveals three distinct mitochondrial lineages. *BMC Evol. Biol.* **2017**, *17*, 34. [CrossRef] [PubMed]

12. Kimura, M.; Weiss, G.H. The stepping stone model of population structure and the decrease of genetic correlation with distance. *Genetics* **1964**, *49*, 561–576. [PubMed]

13. Opdam, P. Metapopulation theory and habitat fragmentation: A review of holarctic breeding bird studies. *Landsc. Ecol.* **1991**, *5*, 93–106. [CrossRef]

14. Kimmig, S.E.; Beninde, J.; Brandt, M.; Schleimer, A.; Kramer-Schadt, S.; Hofer, H.; Börner, K.; Schulze, C.; Wittstatt, U.; Heddergott, M.; et al. Beyond the landscape: Resistance modelling infers physical and behavioural gene flow barriers to a mobile carnivore across a metropolitan area. *Mol. Ecol.* **2020**, *29*, 466–484. [CrossRef] [PubMed]

15. Andrén, H. Effects of Habitat Fragmentation on Birds and Mammals in Landscapes with Different Proportions of Suitable Habitat: A Review. *Oikos* **1994**, *71*, 355. [CrossRef]

16. Taylor, A.C.; Walker, F.M.; Goldingay, R.L.; Ball, T.; van der Ree, R. Degree of Landscape Fragmentation Influences Genetic Isolation among Populations of a Gliding Mammal. *PLoS ONE* **2011**, *6*, e26651. [CrossRef]

17. Braaker, S.; Moretti, M.; Boesch, R.; Ghazoul, J.; Obrist, M.K.; Bontadina, F. Assessing habitat connectivity for ground-dwelling animals in an urban environment. *Ecol. Appl.* **2014**, *24*, 1583–1595. [CrossRef]

18. Amori, G. Erinaceus europaeus. *IUCN Red List Threat. Species* **2016**, e.T29650A2791303. [CrossRef]

19. Doncaster, C.P.; Dickman, C.R. The ecology of small mammals in urban habitats. I. Population in a patchy environment. *J. Anim. Ecol.* **1987**, *56*, 629–640.

20. Baker, P.J.; Harris, S. Urban mammals: What does the future hold? An analysis of the factors affecting patterns of use of residential gardens in Great Britain. *Mamm. Rev.* **2007**, *37*, 297–315. [CrossRef]

21. Dowding, C.V.; Harris, S.; Poulton, S.; Baker, P.J. Nocturnal ranging behaviour of urban hedgehogs, *Erinaceus europaeus*, in relation to risk and reward. *Anim. Behav.* **2010**, *80*, 13–21. [CrossRef]

22. Morris, P.A. A study of home range and movements in the hedgehog (*Erinaceus europaeus*). *J. Zool.* **1988**, *214*, 433–449. [CrossRef]

23. Rautio, A.; Valtonen, A.; Kunnasranta, M. The effects of sex and season on home range in European hedgehogs at the northern edge of the species range. *Ann. Zool. Fenn.* **2013**, *50*, 107–123. [CrossRef]

24. Zingg, R. Aktivität sowie Habitat- und Raumnutzung von Igeln (*Erinaceus europaeus*) in einem ländlichen Siedlungsgebiet. Ph.D. Thesis, Universität Zürich, Zurich, Swiss, 1994. (In German).

25. Hubert, P.; Julliard, R.; Biagianti, S.; Poulle, M.-L.L.M. Ecological factors driving the higher hedgehog (*Erinaceus europaeus*) density in an urban area compared to the adjacent rural area. *Landsc. Urban Plan.* **2011**, *103*, 34–43. [CrossRef]

26. Hof, A.R.; Snellenberg, J.; Bright, P.W. Food or fear? Predation risk mediates edge refuging in an insectivorous mammal. *Anim. Behav.* **2012**, *83*, 1099–1106. [CrossRef]

27. Huijser, M.P.; Bergers, P.J.M. The effect of roads and traffic on hedgehog (*Erinaceus europaeus*) populations. *Biol. Conserv.* **2000**, *95*, 111–116. [CrossRef]

28. Krange, M. *Change in the occurrence of the West European Hedgehog (Erinaceus europaeus) in Western Sweden during 1950–2010*; Karlstad University: Karlstad, Sweden, 2015.

29. Hof, A.R.; Bright, P.W. Quantifying the long-term decline of the West European hedgehog in England by subsampling citizen-science datasets. *Eur. J. Wildl. Res.* **2016**, *62*, 407–413. [CrossRef]

30. Taucher, A.L.; Gloor, S.; Dietrich, A.; Geiger, M.; Hegglin, D.; Bontadina, F. Decline in distribution and abundance: Urban hedgehogs under pressure. *Animals* **2020**, *10*, 1606. [CrossRef]

31. Müller, F. Langzeit-Monitoring der Straßenverkehrsopfer beim Igel (*Erinaceus europaeus* L.) zur Indikation von Populationsdichteveränderungen entlang zweier Teststrecken im Landkreis Fulda. *Beiträge zur Naturkunde Osthessen* **2018**, *54*, 21–26. (In German)

32. Hof, A.R.A.; Bright, P.P.W. The value of green-spaces in built-up areas for western hedgehogs. *Lutra* **2009**, *52*, 69–82.

33. Beninde, J.; Veith, M.; Hochkirch, A. Biodiversity in cities needs space: A meta-analysis of factors determining intra-urban biodiversity variation. *Ecol. Lett.* **2015**, *18*, 581–592. [CrossRef]

34. Bolfíková, B.Č.; Hulva, P. Microevolution of sympatry: Landscape genetics of hedgehogs *Erinaceus europaeus* and *E. roumanicus* in Central Europe. *Heredity (Edinb.)* **2012**, *108*, 248–255. [CrossRef] [PubMed]

35. Becher, S.A.; Griffiths, R. Isolation and characterization of six polymorphic microsatellite loci in the European hedgehog *Erinaceus europaeus*. *Mol. Ecol.* **1997**, *6*, 89–90. [CrossRef] [PubMed]

36. Henderson, M.; Becher, S.A.; Doncaster, C.P.; Maclean, N. Five new polymorphic microsatellite loci in the European hedgehog *Erinaceus europaeus*. *Mol. Ecol.* **2000**, *9*, 1949–1951. [CrossRef]

37. Miller, C.R.; Joyce, P.; Waits, L.P. Assessing allelic dropout and genotype reliability using maximum likelihood. *Genetics* **2002**, *160*, 357–366.

38. Guo, S.W.; Thompson, E.A. Performing the Exact Test of Hardy-Weinberg Proportion for Multiple Alleles. *Biometrics* **1992**, *48*, 361. [CrossRef] [PubMed]

39. Marshall, T.C.; Slate, J.; Kruuk, L.E.B.; Pemberton, J.M. Statistical confidence for likelihood-based paternity inference in natural populations. *Mol. Ecol.* **1998**, *7*, 639–655. [CrossRef] [PubMed]

40. Kalinowski, S.T.; Taper, M.L.; Marshall, T.C. Revising how the computer program CERVUS accommodates genotyping error increases success in paternity assignment. *Mol. Ecol.* **2007**, *16*, 1099–1106. [CrossRef]

41. Excoffier, L.; Laval, G.; Schneider, S. Arlequin (version 3.0): An integrated software package for population genetics data analysis. *Evol. Bioinform. Online* **2005**, *1*, 47–50. [CrossRef]

42. Excoffier, L.; Lischer, H.E.L. Arlequin suite ver 3.5: A new series of programs to perform population genetics analyses under Linux and Windows. *Mol. Ecol. Resour.* **2010**, *10*, 564–567. [CrossRef]

43. Van Oosterhout, C.; Hutchinson, W.F.; Wills, D.P.M.; Shipley, P. Micro-Checker: Software for identifying and correcting genotyping errors in microsatellite data. *Mol. Ecol. Notes* **2004**, *4*, 535–538. [CrossRef]

44. Rodríguez-Ramilo, S.T.; Wang, J. The effect of close relatives on unsupervised Bayesian clustering algorithms in population genetic structure analysis. *Mol. Ecol. Resour.* **2012**, *12*, 873–884. [CrossRef] [PubMed]

45. Queller, D.C.; Goodnight, K.F. Estimating Relatedness Using Genetic Markers. *Evolution* **1989**, *43*, 258–275. [CrossRef]

46. Wang, J. Coancestry: A program for simulating, estimating and analysing relatedness and inbreeding coefficients. *Mol. Ecol. Resour.* **2011**, *11*, 141–145. [CrossRef] [PubMed]

47. Pritchard, J.K.; Stephens, M.; Donnelly, P. Inference of population structure using multilocus genotype data. *Genetics* **2000**, *155*, 945–959. [PubMed]

48. Falush, D.; Stephens, M.; Pritchard, J.K. Inference of population structure using multilocus genotype data: Linked loci and correlated allele frequencies. *Genetics* **2003**, *164*, 1567–1587. [PubMed]

49. Hubisz, M.J.; Falush, D.; Stephens, M.; Pritchard, J.K. Inferring weak population structure with the assistance of sample group information. *Mol. Ecol. Resour.* **2009**, *9*, 1322–1332. [CrossRef]

50. Corander, J.; Sirén, J.; Arjas, E. Bayesian Spatial Modelling of Genetic Population Structure. *Comput. Stat.* **2008**, *23*, 111–129. [CrossRef]

51. Cheng, L.; Connor, T.R.; Sirén, J.; Aanensen, D.M.; Corander, J. Hierarchical and spatially explicit clustering of DNA sequences with Baps software. *Mol. Biol. Evol.* **2013**, *30*, 1224–1228. [CrossRef]

52. Evanno, G.; Regnaut, S.; Goudet, J. Detecting the number of clusters of individuals using the software STRUCTURE: A simulation study. *Mol. Ecol.* **2005**, *14*, 2611–2620. [CrossRef]

53. Earl, D.A.; von Holdt, B.M. STRUCTURE HARVESTER: A website and program for visualizing STRUCTURE output and implementing the Evanno method. *Conserv. Genet. Resour.* **2012**, *4*, 359–361. [CrossRef]

54. Moßbrucker, A.M.; Apriyana, I.; Fickel, J.; Imron, M.A.; Pudyatmoki, S.; Sumardi; Suryadi, H. Non-invasive genotyping of Sumatran elephants: Implications for conservation. *Trop. Conserv. Sci.* **2015**, *8*, 745–759.

55. Schmidt, K.; Kowalczyk, R.; Ozolins, J.; Männil, P.; Fickel, J. Genetic structure of the Eurasian lynx population in north-eastern Poland and the Baltic states. *Cons. Genet.* **2009**, *10*, 497–501. [CrossRef]

56. Coombs, J.A.; Letcher, B.H.; Nislow, K.H. Create: A software to create input files from diploid genotypic data for 52 genetic software programs. *Mol. Ecol. Resour.* **2008**, *8*, 578–580. [CrossRef] [PubMed]

57. R Core Team. *R: A Language and Environment for Statistical Computing*; R Foundation for Statistical Computing: Vienna, Austria, 2018. Available online: https://www.R-project.org/ (accessed on 1 April 2017).

58. Becher, S.A.; Griffiths, R. Genetic differentiation among local populations of the European hedgehog (*Erinaceus europaeus*) in mosaic habitat. *Mol. Ecol.* **1998**, *7*, 1599–1604. [CrossRef] [PubMed]

59. Wahlund, S. Zusammensetzung von Populationen und Korrelationserscheinungen vom Standpunkt der Vererbungslehre aus betrachtet. *Hereditas* **1928**, *11*, 65–106. (In German) [CrossRef]

60. Braaker, S.; Kormann, U.; Bontadina, F.; Obrist, M.K.K. Prediction of genetic connectivity in urban ecosystems by combining detailed movement data, genetic data and multi-path modelling. *Landsc. Urban Plan* **2017**, *160*, 107–114. [CrossRef]

61. Rasmussen, S.L.; Nielsen, J.L.; Jones, O.R.; Berg, T.B.; Pertoldi, G. Genetic structure of the European hedgehog (*Erinaceus europaeus*) in Denmark. *PLoS ONE* **2020**, *15*, e0227205. [CrossRef]

62. Bolfíková, B.; Konečný, A.; Pfäffle, M.; Skuballa, J.; Hulva, P. Population biology of establishment in New Zealand hedgehogs inferred from genetic and historical data: Conflict or compromise? *Mol. Ecol.* **2013**, *22*, 3709–3720. [CrossRef]

63. Morris, P.A.; Reeve, N.J. Hedgehog *Erinaceus europaeus*. In *Mammals of the British Isles: Handbook*, 4th ed.; Harris, S., Yalden, D.W., Eds.; The Mammal Society: Southampton, UK, 2008; pp. 241–248.

64. Rasmussen, S.L.; Berg, T.B.; Dabelsteen, T.; Jones, O.R. The ecology of suburban juvenile European hedgehogs (*Erinaceus erinaceus*) in Denmark. *Ecol. Evol.* **2019**, *9*, 13174–13187. [CrossRef]

65. Doncaster, C.P.; Rondinini, C.; Johnson, P.C.D. Field test for environmental correlates of dispersal in hedgehogs *Erinaceus europaeus*. *J. Anim. Ecol.* **2001**, *70*, 33–46.

66. Rondinini, C.; Doncaster, C.P. Roads as barriers to movement for hedgehogs. *Funct. Ecol.* **2002**, *16*, 504–509. [CrossRef]

67. Herter, K. Gefangenschaftsbeobachtungen an europäischen Igeln II. *Mamm. Biol. (formerly Zeitschrift für Säugetierkunde)* **1933**, *8*, 195–218.

68. SenUVK: Senatsverwaltung für Umwelt, Verkehr und Klimaschutz Berlin, Referat Freiraumplanung und Stadtgrün, Grünflacheninformationssystem GRIS. 2018. Available online: https://www.berlin.de/senuvk/umwelt/stadtgruen/gruenanlagen/de/daten_fakten/downloads/ausw_5.pdf (accessed on 31 January 2019).

69. Ploi, K.; Curto, M.; Bolfíková, B.Č.; Loudová, M.; Hulva, P.; Seiter, A.; Fuhrmann, M.; Winter, S.; Meimberg, H. Evaluating the impact of wildlife shelter management on the genetic diversity of *Erinaceus europaeus* and *E.roumanicus* in their contact zone. *Animals* **2020**, *10*, 1452. [CrossRef] [PubMed]

70. Molony, S.E.; Dowding, C.V.; Baker, P.J.; Cuthill, I.C.; Harris, S. The effect of translocation and temporary captivity on wildlife rehabilitation success: An experimental study using European hedgehogs (*Erinaceus europaeus*). *Biol. Conserv.* **2006**, *130*, 530–537. [CrossRef]

Publisher's Note: MDPI stays neutral with regard to jurisdictional claims in published maps and institutional affiliations.

Article

An Estimate of the Scale and Composition of the Hedgehog (*Erinaceus europeaus*) Rehabilitation Community in Britain and the Channel Islands

Lucy E. Bearman-Brown [1,*] and Philip J. Baker [2]

[1] Department of Animal & Agriculture, Hartpury University, Gloucestershire GL19 3BE, UK
[2] School of Biological Sciences, University of Reading, Reading RG6 6AH, UK
* Correspondence: lucy.bearman-brown@hartpury.ac.uk; Tel.: +44-(0)1452702465

Simple Summary: Large numbers of animals enter wildlife hospitals/centres each year around the globe, but it is unclear whether the efforts of wildlife rehabilitators have significant impacts on the conservation of the species involved. In this study, we used a questionnaire survey to estimate the number and characteristics of practitioners helping to rehabilitate injured and orphaned hedgehogs (*Erinaceus europaeus*) in Britain, and the number of hedgehogs admitted in one benchmark year (2016). Overall, 304 rehabilitators were identified: 148 supplied data on their structure, and 174 outlined the number of hedgehogs admitted in 2016. Most hospitals (62.6%) were small (admitting <50 hedgehogs each year), but most hedgehogs (82.8%) were admitted to large hospitals (>250 hedgehogs each year). We estimated that this rehabilitation community collectively admitted >40,000 hedgehogs in the benchmark year, of which approximately half could have survived to be released. Assuming that most hedgehogs originated from urban areas, we estimate that >3% of the post-breeding population of hedgehogs entered wildlife hospitals in 2016. In contrast, the urban hedgehog population in Britain is estimated to have declined by approximately 2% per year during 2003–2017. These figures suggest, therefore, that wildlife rehabilitation has potentially been an important factor in the dynamics of hedgehog populations in this country during this period.

Abstract: The conservation benefits of wildlife rehabilitation are equivocal, but could be substantial for formerly common species that are declining rapidly but are still commonly admitted to wildlife centres. We used a questionnaire survey to estimate the number of practitioners rehabilitating West European hedgehogs (*Erinaceus europaeus*) in Britain and the numbers entering hospitals/centres in one benchmark year (2016); practitioners were identified using an internet search and snowball sampling. Overall, 304 rehabilitators were identified: 148 supplied data on their structure, and 174 outlined the number of hedgehogs admitted in 2016. The former comprised 62.6% small ($\leq$50 hedgehogs admitted year^{-1}), 16.7% medium-sized (51–250 yr^{-1}), and 20.7% large (>250 yr^{-1}) hospitals; however, these accounted for 4.8%, 12.4%, and 82.8% of hedgehog admissions, respectively. Small hospitals were less likely to be registered as a charity, have paid staff, have a social media account, to record admissions electronically, or to conduct post-release monitoring. However, they were more likely to operate from their home address and to have been established for $\leq$5 years. Extrapolations indicate that this rehabilitation community admitted >40,000 hedgehogs in 2016, of which approximately 50% could have been released. These figures suggest that wildlife rehabilitation has potentially been an important factor in the dynamics of hedgehog populations in Britain in the last two decades.

Keywords: animal welfare; conservation; *Erinaceus europaeus*; European hedgehog; wildlife hospital; wildlife rehabilitation

Citation: Bearman-Brown, L.E.; Baker, P.J. An Estimate of the Scale and Composition of the Hedgehog (*Erinaceus europeaus*) Rehabilitation Community in Britain and the Channel Islands. *Animals* **2022**, *12*, 3139. https://doi.org/10.3390/ani12223139

Academic Editors: Anne Berger and Sophie Lund Rasmussen

Received: 29 July 2022
Accepted: 4 November 2022
Published: 14 November 2022

Publisher's Note: MDPI stays neutral with regard to jurisdictional claims in published maps and institutional affiliations.

1. Introduction

The International Wildlife Rehabilitation Council defines wildlife rehabilitation (WR) as 'the treatment and temporary care of injured, diseased, and displaced indigenous animals, and the subsequent release of healthy animals to appropriate habitats in the wild' [1]. Although specific data are lacking, it is reasonable to assume that the practice of rehabilitating wild animals has increased at an international level as a result of the increasing negative impact of humans on natural ecosystems [2–9]. Although widely perceived as helping wildlife, the role of WR as a conservation tool is contested [10]. For example, it can be argued that most animals that enter wildlife hospitals are from common and widespread species [11,12], and that the money spent on their care cannot then be spent on wider conservation actions such as habitat preservation [13]. Similarly, unless released individuals have survival and/or reproduction rates comparable to those which have not required treatment, the cost-effectiveness of WR could be questioned [1,14–16]. Consequently, WR has more often been portrayed as an animal welfare issue or for the "benefit of the individual" [14,17–19].

However, WR can aid conservation in a more general context, such as aiding education [12,14,16,19,20], disease surveillance [21–24], monitoring of environmental pollutants [25], and the development of capture, treatment, and release protocols that can subsequently be applied to species of conservation concern [26,27]. In addition, it can help to reduce the welfare and population-level impacts of catastrophic events such as oil spills and wildfires, where large numbers of individuals may be affected in a very short space of time [28–31], by, e.g., reducing mortality rates and by euthanising individuals considered likely to die of their injuries. Furthermore, although high conservation status is often perceived in the context of rarity, it can also result from rapid declines of species that were formerly abundant and widespread [32,33]; in this context, wildlife rehabilitators may receive large numbers of individuals whose care could potentially contribute to the conservation of that species. The magnitude of this benefit is, therefore, partially dependent on the number of animals rehabilitated and released in relation to population size [14,34].

Quantifying the numbers of different species which are successfully rehabilitated and released can, however, be challenging because of the way that WR is often practised. In some countries, the wildlife rehabilitation community may encompass single individuals, charitable and non-charitable NGOs, and/or government agencies, some of which may focus on wildlife generally whilst others focus on just one or a few species [11,14,20]. Consequently, some rehabilitators may only treat a handful of animals each year, whereas larger organisations may treat thousands. As such, there may be substantive differences between organisations with respect to size, facilities, operating protocols, and experience which could affect patterns of care and, ultimately, release rates. In addition, not all countries require that wildlife rehabilitators are licenced or registered [20,35], such that even identifying the number of rehabilitators operating at any given time is problematic. This issue is currently of interest in Britain in the context of the rehabilitation of West European hedgehogs (*Erinaceus europeaus*; hereafter 'hedgehog'), a species of increasing conservation concern.

The hedgehog is a small (<1.5 kg), insectivorous mammal found throughout western Europe [36]. In Britain, it is found in a wide range of human-dominated landscapes including arable and pastoral farmland as well as urban areas [37–42]. Evidence from a range of different monitoring programs suggests that populations may have declined by up to 40% in some habitats in the last few decades [40,42–44], with declines to varying degrees also present throughout Europe [39,45–47]. Factors likely to be associated with this decline include: habitat loss, fragmentation, and degradation; the application of chemical biocides; an increase in the size of road networks and associated traffic volume; the increased abundance of an intra-guild predator, the Eurasian badger (*Meles meles*); and climate change [26,46,48–55]. Most recently, the British hedgehog population was estimated to number approximately 0.88 million individuals [44,56], down from 1.56m in the mid 1990s [57]; although the veracity of both estimates is equivocal, these, in combination with the trends outlined above [40,42–44], triggered the species' status to be upgraded to Vulnerable in these countries [56]. Despite this substantial decline, hedgehogs are frequently the most common mammal species admitted to wildlife hospitals in Britain [10,11], and are also commonly taken to veterinary surgeons by members of the public for treatment [58].

Hedgehog Rehabilitation in Britain

Whilst the specific details of wildlife legislation in Britain are complex, in general terms, members of the public are allowed to take any injured (or orphaned) wild animal into captivity for the purposes of treatment or care prior to its subsequent release. Whilst in captivity, the animal must receive appropriate husbandry and be taken to a veterinary surgeon for examination if necessary [35,59]; as part of this care, hedgehogs may be humanely killed (euthanised) in order to prevent suffering and/or where they are unsuitable for release. At the point the animal is deemed fit enough to survive in the wild, it should be released [1]. For those hedgehogs that cannot be released because they are unlikely to survive, many rehabilitators recommend euthanasia. Conversely, others consider retention in captivity an acceptable option, although some oppose this on welfare, and perhaps legal, grounds [1,4,59,60].

The wildlife rehabilitation community in Britain is large, diverse, and, in some respects, disjointed [10]. At one end of the spectrum is the Royal Society for the Protection of Animals (RSPCA), the largest animal welfare organisation in England and Wales, and the Scottish Society for the Protection of Animals (SSPCA), which operates in Scotland. Both organisations investigate and enforce cases associated with animal welfare and animal cruelty, including wild animals, but also rehabilitate injured wild animals; the RSPCA has four wildlife centres based in England, and the SSPCA has one based in Clackmannanshire. Similarly, Tiggywinkles Wildlife Hospital in Oxfordshire, England is considered the largest purpose-built wildlife hospital in Europe. Individually, these three organisations may each admit >1000 hedgehogs each year.

However, there are a substantial number of smaller organisations and individuals who also rehabilitate hedgehogs in Britain. For example, the British Hedgehog Preservation Society (BHPS) (http://www.britishhedgehogs.org.uk, accessed on 10 November 2022), a charitable organisation that has a specific focus on hedgehog conservation and rehabilitation, maintains a service whereby members of the public can call them to ask for the contact details of their nearest hedgehog carer/rehabilitator. Under UK data protection rules, this information cannot be disseminated to researchers, but it is estimated that this list may contain up to 600 different individuals and organisations (Fay Vass, President of the BHPS, pers. comm.; [36]). These are often individuals working from their own private residence, with hedgehogs housed within their home itself or in a shed or purpose-built unit in their garden. Given the widespread interest in the plight of hedgehogs in Britain, and the availability of training courses associated with treating and rehabilitating hedgehogs (e.g., http://www.valewildlife.org.uk/courses/, accessed on 10 November 2022), the numbers of these individuals/smaller organisations is likely to have grown in recent years.

Despite earlier recommendations about the scientific merits of collecting and collating data about animals entering wildlife hospitals in Britain [4,61], relatively few data exist on numbers admitted annually. For example, [11] reported an estimated 30,000–40,000 casualties per annum, the most common species being red fox (*Vulpes vulpes*), Eurasian badger (*Meles meles*), hedgehog and blackbird (*Turdus merula*). More recently, [4] reported 71,000 animals were admitted to the RSPCA's four wildlife centres and 23 other wildlife hospitals in 2011. Given this paucity of information, and the potentially increasing importance of rehabilitation as hedgehog numbers continue to decline, we used a structured internet search and questionnaire survey to estimate (1) the number of individuals and organisations rehabilitating hedgehogs in Britain, and (2) the number of hedgehogs admitted to wildlife hospitals in a single year (2016). The questionnaire also (3) requested information relating to the structure and practices of these individuals and organisations to identify how the British hedgehog rehabilitation community is structured at the current time. In addition, we (4) highlight the challenges associated with deriving estimates of the numb of rehabilitators practising and the numbers of hedgehogs admitted in the context of simil future studies.

2. Materials and Methods

For brevity, we use the term 'rehabilitator' to refer to any individual or organisatic that treats and releases hedgehogs; in Britain, this is also synonymous with the ter "carer." The terms 'hospital' or 'centre' refer to any building or structure from which rehabilitator operates (including private households, buildings on private premises such a garden shed and/or large purpose-build facilities). The terms 'hospital' or 'centre' do n necessarily reflect the size of an establishment nor the range of medical facilities present

A systematic internet search was undertaken from September 2016–January 2017 usir the online search engines Google and Bing, websites (e.g., www.britishhedgehogs.org.u accessed on 28 January 2017), online databases (e.g., www.helpwildlife.co.uk/director accessed on 28 January 2017), and the social media platforms Facebook and Twitter to crea a database of hedgehog rehabilitators. Search terms included: "wildlife hospital," "wildli rescue," "hedgehog hospital," "hogspital" and "hedgehog rescue." Where available, th name and location of each rehabilitator was provisionally recorded if they had an onlir presence in any form (e.g., social media profile, if they had been mentioned in local national media, if they had a fundraising campaign). However, because online informatic may be erroneous (e.g., they had ceased practising), rehabilitators were only classified active if their online information indicated that they were still operating in 2016; where th information was not immediately evident, the rehabilitator was contacted directly by ema or via social media to confirm.

Following this online search, snowball sampling (chain-referral sampling) was use to help identify additional rehabilitators. This was done by searching the social med associates of each provisionally identified rehabilitator, and by also asking them to fc ward/advertise our survey questionnaire to their contacts. This approach helped to identi rehabilitators that had no obvious online social media presence and individuals who ha only recently started practising.

2.1. Questionnaire Survey

A self-administered questionnaire was distributed via SmartSurvey® from Janua 2017–December 2017. The questionnaire was publicised through social media using w pages associated with wildlife and hedgehog rehabilitation, and through contacting th rehabilitators identified above directly. The questionnaire was further publicised v newsletters published by the BHPS, British Wildlife Rehabilitation Council, and the People Trust for Endangered Species. The BHPS promoted the questionnaire to all ~600 carers their database. All centres were contacted at least once via email or a social media messag depending on their preferred mode of communication advertised on their website or soci media. Information requested included: their name; whether they were a registered chari or not; how many paid personnel they employed; how many unpaid staff (volunteer worked at their hospital; whether they had a full-time veterinary surgeon on staff or worke in conjunction with an external veterinary practice; the year they had started rehabilitatir hedgehogs; whether their hospital was run from their personal residence or from a purpos built rehabilitation centre; whether they had a personal and/or business social med account for advertising their hospital to the general public; and whether they used pap records or a computer to record information about the animals they have cared for.

Respondents were then asked to indicate how many hedgehogs they had receive each year for the 5-year period from 2012–2016, inclusive; as 2016 was the most rece year for which respondents would have had complete information, this was taken as th benchmark year for estimating the number of animals admitted. Finally, respondents we asked to indicate whether they undertook post-release monitoring of any sort and, if s what methods they used (radio-tags, GPS-tags, injected microchips (PIT tags), ear tag marking spines with numbered tags (e.g., [62] or nail varnish). Such monitoring is typically passive, as it is focused on identifying animals that are readmitted to hospitals after their release, individuals that have been killed on roads near to where they were released, and/or the use of motion-activated video cameras at release sites.

2.2. Estimating the Number of Hedgehog Admissions

In order to estimate the total number of hedgehogs admitted in 2016 by all active hospitals identified, we first categorised those hospitals for which we had data on the number of admissions into three size classes: small, medium, and large. These divisions were estimated retrospectively based upon the frequency distribution of the numbers of hedgehogs admitted (see Results). These were assigned both to reflect the pattern of admissions but also to ensure that a sufficient number of hospitals (both from the original searches and from snowball sampling) was in each division to enable statistical analysis. Differences in the relative numbers of hospitals in each size class identified in the online searches versus the snowball sampling were compared using a chi-squared test. Differences in the median number of hedgehogs admitted in 2016 within each size class in the online searches versus snowball sampling were compared using a series of Mann–Whitney tests.

Two models were used to estimate the numbers of hedgehogs admitted by those hospitals for which there were no data available. In Model 1, the data from both the online searches and snowball sampling were merged and treated as a homogenous sample. These combined data were used to estimate the proportion of small (PS), medium (PM), and large (PL) hospitals in the overall community, as well as the median number of hedgehogs admitted by each hospital in each size class (NS, NM, and NL, respectively). These values were then used to estimate the number of hedgehogs (TX) admitted to those hospitals with missing data (H): TX = H × PM × NS + H × PM × NM + H × PL × NL. These were then added to the number known to have been admitted to those hospitals for which data were available (TY) in order to estimate the total number of hedgehogs admitted in 2016 (T): T = TX + TY.

Model 2 followed a similar approach, except that the data from the online searches and snowball sampling were treated separately, as there was evidence that the composition of each sample varied with respect to the proportion of hospitals in each size class and the median number of hedgehogs admitted within each size class (see Results).

2.3. Structure of the Rehabilitation Community

Differences in the characteristics of hospital size classes were quantified using data from those rehabilitators where we had both an estimate of their size and who had completed the questionnaire survey; rehabilitators who had completed the questionnaire but who had not indicated the numbers of hedgehogs admitted in 2016 were excluded. Similarly, rehabilitators who had failed to answer a specific question were excluded from the analysis relating to that variable.

A series of chi-squared tests were used to compare differences between the three hospital classes with respect to: (i) whether they were a registered charity or not; (ii) the type of veterinary care they had (five categories: none, work with an external veterinary practice, work with external wildlife hospital, they themselves are a veterinary nurse or veterinary surgeon, onsite veterinary surgeon); (iii) how long they had been established (data merged into two categories: ≤5 years and >5 years); (iv) where hedgehogs were housed during rehabilitation (three categories: building in private grounds, e.g., garden shed; in their private residence; a purpose-built facility); (v) the type of social media account(s) that they had (three categories: none; only one or more personal social media accounts; one or more business accounts, with or without personal accounts as well); (vi) how they kept records of the hedgehogs admitted (two categories: fully or partly on paper; fully or partly on computer. NB: the option "partly paper" indicates that the majority of records were recorded on paper with a minority on computer, whereas "partly computer" indicates the reverse); and (vii) whether they did or did not undertake any form of post-release monitoring.

In addition, we quantified the number of paid and unpaid (volunteers) staff working in each hospital size class. The latter was divided into three categories (1 volunteer, 2 volunteers, ≥3 volunteers) and analysed using a chi-squared test; where hospitals were run by just one unpaid person, this would be the person in charge of that hospital who was running it on a voluntary basis. It was not possible to statistically compare the number of paid staff, as very few hospitals employed any paid personnel. Therefore, we have simply summarised these data by indicating: the percentage of hospitals within each size class that employed one or more paid staff; and the mean number of paid staff in those hospitals where they were present.

Temporal trends in the annual number of admissions were investigated in order to identify whether 2016 was potentially an atypical year using two approaches. First, data from the current questionnaire were used to compare the median number of hedgehogs admitted in each hospital size class for the five-year period from 2012–2016, inclusive. Median (±IQR) numbers of admissions are presented for small, medium, and large hospitals separately, utilising all data available. However, not all hospitals supplied data for all five years; therefore, separate Friedman tests were used to identify between-year differences using those subsets of hospitals within each size class that did supply data for all five years. Second, data on the number of juveniles, adults, and hedgehogs of unknown age admitted to the four RSPCA wildlife centres in England (East Winch, Norfolk; Mallydams Wood, East Sussex; Stapeley Grange, Cheshire; West Hatch, Somerset) were collated for the period from January 2006 to December 2018, inclusive.

All statistical analyses were conducted using MINITAB version 19.1.1 and SPSS version 25. Nonparametric tests were used throughout, as the data were not normally distributed. Data are therefore presented as medians ± inter-quartile ranges unless otherwise specified.

3. Results

Overall, 239 rehabilitators were provisionally identified through online searches; 179 were considered active in 2016, 47 were considered not active, and 13 were of unknown status. Of the 179 that were active, information on the number of hedgehogs admitted was available from 59 (33.0%). A further 125 rehabilitators were identified by snowball sampling, all of which were considered active; 115 (92.0%) provided information on admissions. Data on the number of admissions in the benchmark year were available for 174 of 304 hospitals (57.2%: $n = 59 + 115$), but with data missing from a further 130 known establishments ($n = 120 + 10$).

Based on the pattern of admissions in 2016 (Figure 1), small, medium, and large hospitals were defined as those which admitted ≤50 ($n = 109$: 62.6%), 51–250 ($n = 29$: 16.7%), and >250 ($n = 36$: 20.7%) hedgehogs, respectively. Significantly more small hospitals and fewer large hospitals were detected by snowball sampling compared to the original online search (Chi-squared test: $X^2_2 = 67.18$, $p < 0.001$; Figure 1).

3.1. Estimating Total Hedgehog Admissions in Benchmark Year

Overall, the 174 rehabilitators for which data were available admitted 25,540 hedgehogs in 2016, with large hospitals collectively dealing with substantially larger numbers ($n = 21,145$; 82.8%), than medium-sized ($n = 3169$; 12.4%) or small ($n = 1226$; 4.8%) organisations. Significantly fewer hedgehogs were admitted to small hospitals depending on whether they had been identified by snowball sampling versus those that had been identified in the original online search (Mann–Whitney test: $W = 1124.00$, $p < 0.001$; Table 1); no significant differences were evident for medium-sized ($W = 256.50$, $p = 0.965$) or large ($W = 578.50$, $P = 0.097$) hospitals. Extrapolating from these data, the number of hedgehogs admitted by all 304 active hospitals in the combined sample was estimated to range from 40,991 (Model 1) to 59,308 (Model 2) individuals (Table 2).

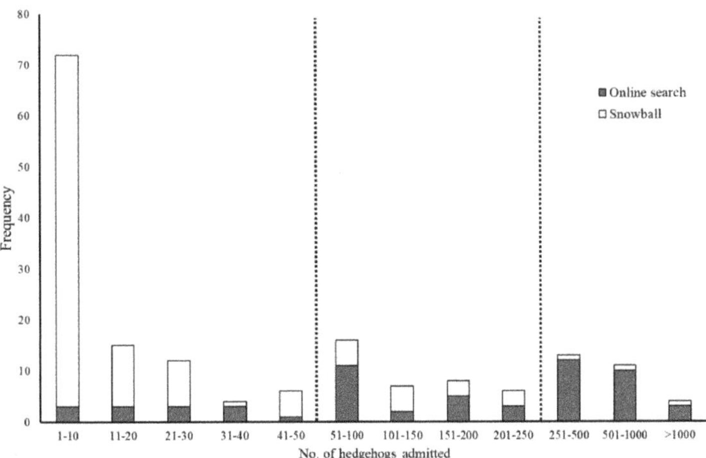

Figure 1. Frequency distribution of small, medium-sized, and large wildlife hospitals/rehabilitation organisations identified from original online searches (n = 59) versus snowball sampling (n = 115). Vertical lines indicate size categories defined retrospectively on the basis of the number of hedgehogs admitted in 2016: small $\leq$ 50 admissions; medium-sized = 51–250 admissions; and large $\geq$250 admissions.

Table 1. Summary of the median ($\pm$IQR) number of hedgehogs admitted to small ($\leq$50 admissions per annum), medium-sized (51–250 admissions), and large (>250 admissions) hospitals in 2016 based on whether they had been identified in the original online search (n = 59) versus those identified by snowball sampling (n = 115). Figures are also presented for all hospitals combined.

	Small	**Medium**	**Large**
Online	25.0 (10.0–32.0) (n = 13)	97.0 (67.5–156.5) (n = 17)	500.0 (346.0–701.0) (n = 29)
Snowball	6.0 (2.0–12.8) (n = 96)	121.0 (59.0–143.0) (n = 12)	235.0 (201.0–582.0) (n = 7)
Combined	6.0 (2.5–16.0) (n = 109)	97.0 (63.5–145.0) (n = 29)	478.0 (261.0–645.0) (n = 36)

3.2. Structure of the Hedgehog Rehabilitation Community

Overall, 148 of the 304 active rehabilitators contacted (48.7%) supplied information about both their structure and the number of hedgehogs they admitted in 2016. However, not all respondents answered all questions about their composition, so sample sizes varied between analyses.

Hospitals varied significantly with respect to their charitable status, the number of unpaid staff working at the hospital, the length of time they had been established, where hedgehogs were housed during the rehabilitation process, their social media presence, patterns of record-keeping but not patterns of veterinary care, and whether they conducted post-release monitoring (Table 3). In general terms, small and medium-sized hospitals were less likely to be registered as charities, more likely to have been established within the five years prior to 2016 (Figure 2), and more likely to operate out of the rehabilitator's private residence (Table 3). Furthermore, smaller hospitals were most commonly staffed by just one unpaid person (Figure 3), less likely to have a business social media presence, more likely to rely on paper records, and less likely to carry out post-release monitoring (Table 3). Post-release monitoring by all hospitals was predominantly via the use of spinal tags or nail varnish (n = 65 of 70 hospitals that conducted post-release monitoring). All three categories of hospital relied extensively on support from an external veterinary practice. Paid staff were present in <5% of small and medium-sized hospitals, but >40% of large hospitals (Table 3).

Table 2. Estimated number of hedgehogs admitted to all wildlife hospitals active in 2016 (n = 304); extrapolations are based upon data supplied by 59 of 159 hospitals identified from an online internet search and 115 of 125 hospitals identified from subsequent snowball searching. Hospital size was categorised on the basis of the number of hedgehogs admitted in 2016: small $\leq$50 admissions; medium = 51–250 admissions; and large $\geq$250 admissions. Model 1 assumed that there were no significant differences in (i) the relative distribution of hospitals of different sizes nor (ii) the median number of hedgehogs admitted annually between hospitals identified in the original online search versus snowball sampling; estimates were, therefore, based on the combined data. Model 2 assumed that there were significant differences in both of these components between hospitals identified in the original online search versus snowball sampling; estimates were, therefore, based on each subset of data separately.

Model 1: Online Internet Search and Snowball Sampling Data Combined			
	Small	**Medium**	**Large**
Total number of hospitals identified (N_O = hospitals identified by online search; N_S = hospitals identified by snowball search)		304 (N_0 = 179, N_S = 125)	
Total number of hospitals where number of hedgehogs admitted was known		174 (N_0 = 59, N_S = 115)	
Number (%) of hospitals in each size category where the number of hedgehogs admitted was known	109 (62.6%)	29 (16.7%)	36 (20.7%)
Absolute number of hedgehogs known to have been admitted in 2016 within each hospital size category	1226	3169	21,145
Number of hospitals where number of hedgehogs admitted in 2016 was not known		130 (N_0 = 120, NS = 10)	
Estimated number of hospitals in each size division where data on the number of hedgehogs admitted was unknown	81.4 (130 $\times$ 0.626)	21.7 (130 $\times$ 0.167)	26.9 (130 $\times$ 0.207)
Median number of hedgehogs admitted per hospital within each hospital size category (see Table 1)	6	97	478
Estimated number of additional hedgehogs admitted in 2016 within each hospital size category	488 (81.4 $\times$ 6)	2105 (21.7 $\times$ 97)	12,858 (26.9 $\times$ 478)
Estimated total number of hedgehogs admitted in 2016 within each hospital size category	1714	5274	34,003
Estimated total number of hedgehogs admitted in 2016		40,991	

Table 2. *Cont.*

Model 2: Online Internet Search and Snowball Sampling Data Treated Independently			
	Small	**Medium**	**Large**
(a) Hospitals identified through online search			
Total number of hospitals identified		179	
Total number of hospitals where number of hedgehogs admitted was known		59	
Number (%) of hospitals where number of hedgehogs admitted was known	13 (22.0%)	17 (28.8%)	29 (49.2%)
Known number of hedgehogs admitted within each hospital size category	299	1833	18,070
Median number of hedgehogs admitted per hospital within each hospital size category (see Table 1)	25	97	500
Number of hospitals where number of hedgehogs admitted was not known		120	
Estimated number of hospitals in each size division where data on the number of hedgehogs admitted was unknown	26.4 (120×0.220)	34.6 (120×0.288)	59.0 (120×0.492)
Estimated number of additional hedgehogs admitted in 2016 within each hospital size category	660 (26.4×25)	3356 (34.6×97)	29,500 (59.0×500)
Estimated total number of hedgehogs admitted in 2016 within each hospital size category	959	5189	47,570
Estimated total number of hedgehogs admitted in 2016		53,718	
(b) Hospitals identified through snowball sampling			
Total number of hospitals identified		125	
Total number of hospitals where number of hedgehogs admitted was known		115	
Number (%) of hospitals where number of hedgehogs admitted was known	96 (83.5%)	12 (10.4%)	7 (6.1%)
Known number of hedgehogs admitted within each hospital size category	927	1336	3075
Median number of hedgehogs admitted per hospital within each hospital size category (see Table 1)	6	121	135
Number of hospitals where number of hedgehogs admitted was not known		10	
Estimated number of hospitals in each size division where data on the number of hedgehogs admitted was unknown	8.4 (10×0.835)	1.0 (10×0.104)	0.6 (10×0.061)
Estimated number of additional hedgehogs admitted in 2016 within each hospital size category	50 (8.4×6)	121 (1.0×121)	81 (0.6×135)
Estimated total number of hedgehogs admitted in 2016 within each hospital size category	977	1457	3156
Estimated total number of hedgehogs admitted in 2016		5590	
(c) All hospitals combined			
Estimated total number of hedgehogs admitted in 2016 within each hospital size category	1936	6646	50,726
Estimated total number of hedgehogs admitted in 2016		59,308	

Table 3. Summary of the characteristics of small (≤50 hedgehogs admitted in 2016), medium-sized (51–250 hedgehogs admitted), and large (>250 hedgehogs admitted) wildlife hospitals (n = 148). Sample sizes vary between individual analyses if respondents had not answered that question.

Characteristics		Small (n = 108)	Medium (n = 22)	Large (n = 18)	Chi-Squared Results
Registered charity (n = 148)	No	96.3%	86.4%	27.8%	$X^2_2 = 61.98$, $p < 0.001$
	Yes	3.7%	13.6%	72.2%	
No. of paid staff (n = 146)	% of hospitals with paid staff	4.7%	0.0%	44.4%	-
	Mean no. of paid staff (range)	1.2 (1–2) [1]	-	6.4 (1–30)	
No. of unpaid staff (n = 148)	1 volunteer	83.3%	45.5%	22.2%	$X^2_4 = 74.70$, $p < 0.001$
	2 volunteers	14.8%	18.2%	-	
	3 or more volunteers	1.9%	36.4%	77.8%	
Veterinary care (n = 148)	None	2.8%	-	-	$X^2_8 = 13.03$, $p = 0.111$
	Work with external vet practice	81.5%	100.0%	88.9%	
	Work with external rescue/hospital	12.0%	-	-	
	I am a veterinary nurse/vet	1.9%	-	-	
	Have an onsite vet	1.9%	-	11.1%	
Length of time established (n = 137)	≤5 years	78.6%	68.2%	11.8%	$X^2_2 = 30.03$, $p < 0.001$
	>5 years	21.4%	31.8%	88.2%	
Housing (n = 148)	Building in private grounds	3.7%	18.2%	11.1%	$X^2_4 = 41.53$, $p < 0.001$
	Personal residence	94.4%	77.3%	50.0%	
	Purpose-built facility	1.9%	4.5%	38.9%	
Social Media (n = 148)	No social media account(s)	56.5%	4.5%	11.1%	$X^2_4 = 52.46$, $p < 0.001$
	Only personal account(s)	25.9%	18.2%	5.6%	
	Business and/or personal account(s)	17.6%	77.3%	83.3%	
Record-keeping (n = 132)	Paper (partly or fully)	78.3%	45.5%	50.0%	$X^2_2 = 12.42$, $p = 0.002$
	Computer (partly or fully)	21.7%	55.5%	50.0%	
Post-release monitoring (n = 145)	No	58.9%	38.1%	23.5%	$X^2_2 = 9.17$, $p = 0.010$
	Yes	41.1%	61.9%	76.5%	

[1] Two small hospitals based at higher education establishments were excluded from these figures as they listed the number of paid staff as "lots" and "4000" which presumably refers to the students at these establishment.

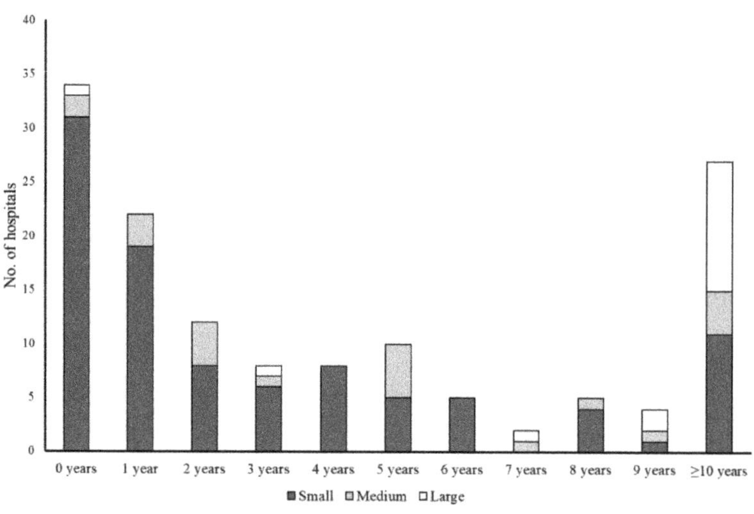

Figure 2. Number of years that small (≤50 admissions; n = 98), medium-sized (51–250 admissions; n = 22), and large (>250 admissions; n = 17) hospitals had been established in 2016.

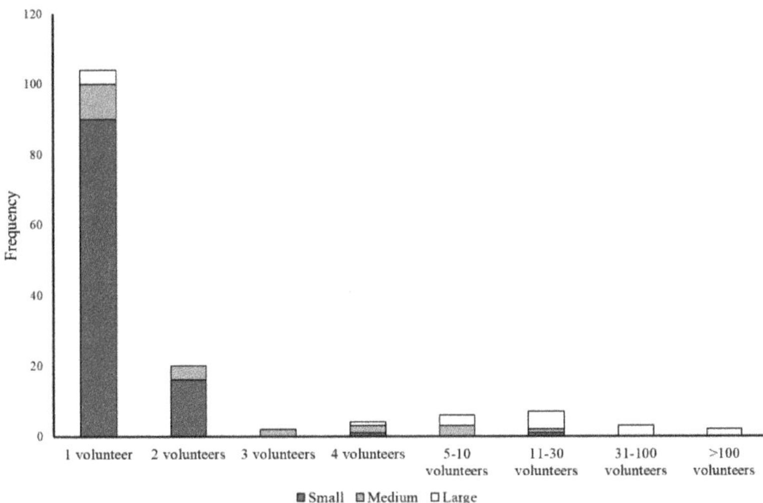

Figure 3. Frequency plot of the number of unpaid staff (volunteers) working at small ($\leq$50 admissions; n = 108), medium-sized (51–250 admissions; n = 22), and large (>250 admissions; n = 17) hospitals in 2016.

The median number of hedgehogs submitted annually throughout the period from 2012–2016 appeared to increase for small (Figure 4a) and medium-sized (Figure 4b) hospitals, and to a lesser degree for large hospitals (Figure 4c). Considering only those hospitals where there were 5 years' worth of data (n = 28), the median number of hedgehogs submitted in 2016 was significantly higher than in both 2012 and 2013 (Friedman test: H = 22.94, DF = 4, $p < 0.001$; Figure 4d). There was a significant positive correlation between the number of juveniles and adult hedgehogs admitted to the four RSPCA wildlife centres annually (Pearson correlation coefficient: r = 0.753, $p < 0.001$), with 2016 being associated with a particularly large number of admissions (Figure 5).

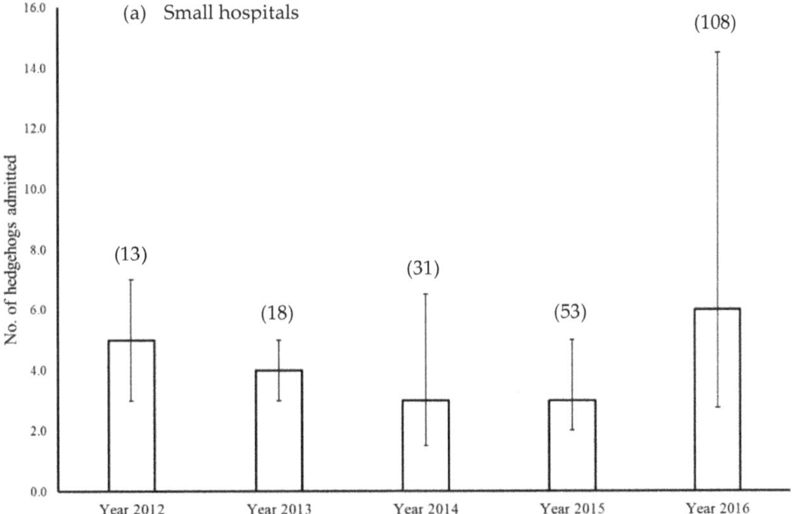

Figure 4. *Cont.*

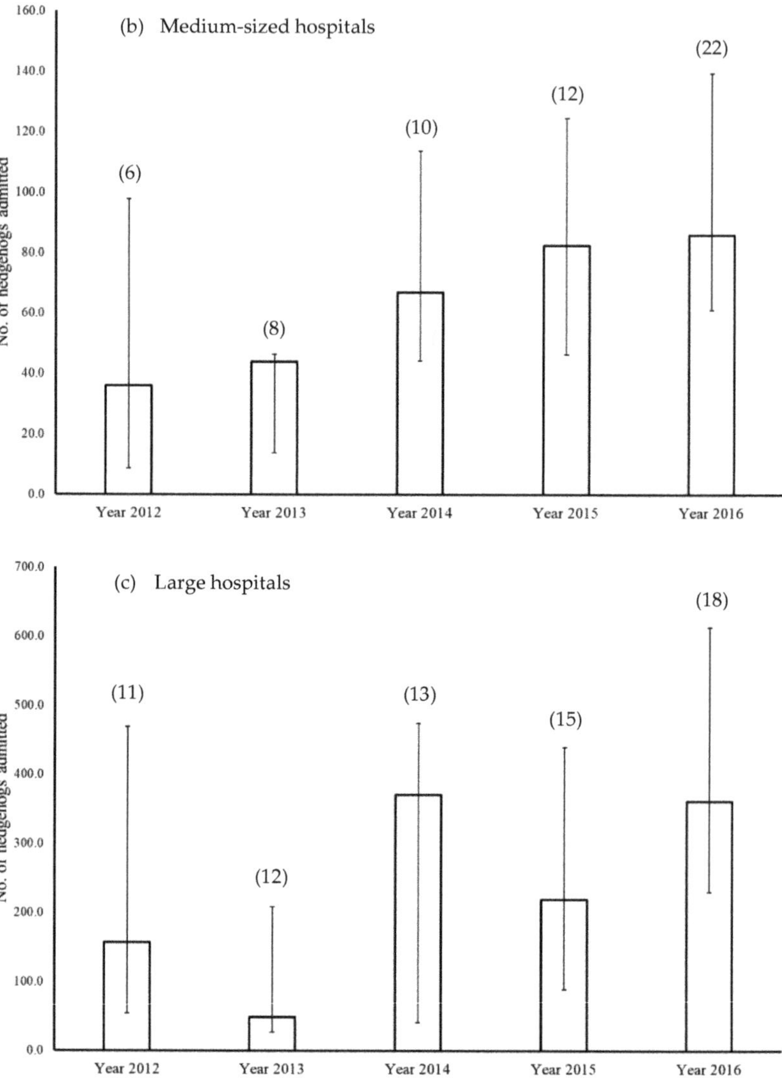

Figure 4. *Cont.*

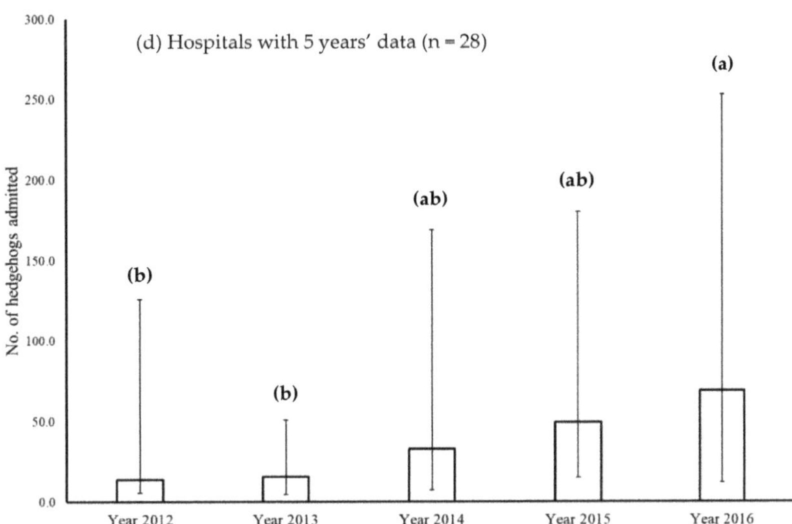

Figure 4. Median (±IQR) number of hedgehogs admitted annually to (**a**) small (≤50 admissions), (**b**) medium (51–250 admissions), and (**c**) large (>250 admissions) hospitals each year in the five-year period from 2012–2016, inclusive; figures above columns indicate sample sizes. (**d**) Number of hedgehogs admitted annually for those hospitals (n = 28) that provided data for all five years; letters above columns indicate post hoc groups from a Friedman test.

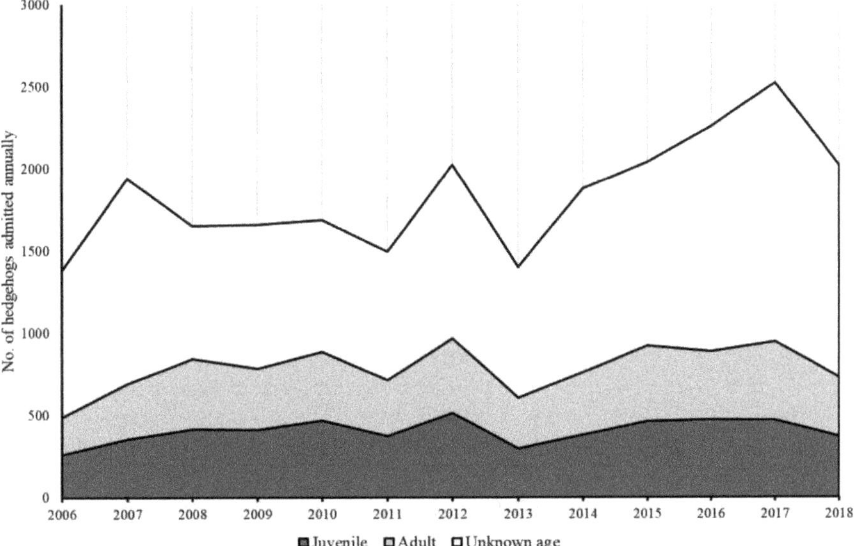

Figure 5. Number of juveniles, adults, and hedgehogs of unknown age admitted annually across the four wildlife hospitals in England run by the Royal Society for the Prevention of Cruelty to Animals during the period from 2006–2018, inclusive.

4. Discussion

This study is, to the best of our knowledge, the first to attempt to estimate the number of practitioners involved in the rehabilitation of hedgehogs in Britain and the number of hedgehogs admitted into their care. At one level, searching for rehabilitators via the internet should be straightforward: as members of the public need to be able to locate and contact individuals or organisations who take in and care for injured or orphaned hedgehogs, it would be expected that practitioners would maintain an active social media presence advertising their services. However, this did not seem to be the case. Overall, we identified 304 active rehabilitators, but only 58.9% were identified in the original online searches; the remainder were only identified by snowball sampling (i.e., relying on provisionally identified practitioners to further advertise our request for information to their personal contacts). This potentially indicates that a large proportion of hedgehog rehabilitators in Britain rely on indirect contact networks (e.g., referrals from other rehabilitators) or "word of mouth" in order to be found by members of the public. This increases the possibility that they will not be identified in studies like this one, such that the numbers presented below should be considered minimum estimates. In addition, this also means that some smaller hospitals may be difficult for members of the public to contact in the event that they discover an injured or orphaned hedgehog.

In terms of the number of hospitals, the hedgehog rehabilitation community in Britain is dominated by small hospitals (62.6%), with many fewer medium-sized (16.7%) and large (20.7%) establishments. This does, in part, reflect the approach we used to group hospitals into different size classes, but it is clear that a very large number of rehabilitators deal with relatively small numbers of admissions annually (Figure 1). This pattern is further reflected by a wide range of associated characteristics. For example, small hospitals were less likely to be registered charities, but more likely to consist of just one unpaid member of staff, to operate out of their house or associated building on their property, to rely on paper records rather than a computer, and not to carry out post-release monitoring. Collectively, these characteristics are consistent with the image of a passionate hedgehog enthusiast operating from their home address in conjunction with volunteer helpers whilst working full- or part-time.

Furthermore, most recent growth in the size and structure of this rehabilitation community was associated with small hospitals: of the 137 hospitals whose establishment time was known, 94 (68.6%) were <5 years old (77 small, 15 medium-sized, and 2 large). Despite their prevalence within the community, however, small hospitals only accounted for 4.8% of 25,540 hedgehogs admitted in 2016 (n = 174 rehabilitators); in comparison, medium-sized and large hospitals accounted for 12.4% and 82.8% of admissions, respectively. As such, it could be argued that small hospitals make a relatively minor contribution to the rehabilitation of hedgehogs in Britain. Yet, there are possible advantages associated with having a high density of small hedgehog hospitals operating throughout the country, but also limitations.

First, one factor that may increase the likelihood that an animal survives the rehabilitation process is the speed with which it receives care after having been discovered by a member of the public (hereafter 'finder'). Having a hospital close to the animal's initial location is likely to reduce transport times, especially if the finder was prepared to take it to the hospital themselves. In these circumstances, affected individuals could potentially be removed 'out of harm's way' more rapidly (e.g., out of direct sunlight, away from potential predators/scavengers, reduced risk of flystrike) and could receive appropriate first aid within a couple of hours of having been discovered, although it must be acknowledged that they may have been injured or unwell for considerably longer. At one extreme, this 'time to care' may be reduced even further if the finder was the person who caused the injury in the first place (e.g., with a garden strimmer) or who, for example, disturbed a nest of hoglets while gardening. However, these early advantages are potentially dependent on finders seeking advice from rehabilitators about the appropriate course of action; if these are difficult to contact, as may be the case for those hospitals that do not have a social media presence, then the affected animal may simply be left in an adverse location for some further period of time. In addition, there is anecdotal evidence that members of the public can become frustrated when rehabilitators ask them to deliver the animal to their hospital for treatment, especially if the distance required is large. At one extreme, this could potentially mean that the animal is subsequently ignored and never receives any treatment, or that the finder posts a negative comment on social media which could tarnish the reputation of the rehabilitator/hospital.

It is important to note that the number of hedgehog rehabilitators identified in this study is substantially lower than the >1700 veterinary practices in Britain [58]. However, there is often a degree of confusion relating to the obligation veterinary surgeons are under when it comes to treating wild animals in this country. The Royal College of Veterinary Surgeons code of professional conduct states that "all veterinary surgeons in practice must take steps to provide 24-hour emergency first aid and pain relief to animals according to their skills and the specific situation" (www.rcvs.org.uk/setting-standards/advice-and-guidance/code-of-professional-conduct-for-veterinary-surgeons/supporting-guidance/24-hour-emergency-first-aid-and-pain-relief/, accessed on 10 November 2022). This treatment may be administered free of charge, but they may also be able to claim payment for this emergency initial treatment through a memorandum of understanding between the RSPCA and the British Veterinary Association. After this initial 24-hour period, some practitioners may then expect to be paid for further treatment, whereas others may be working in partnership with local hedgehog rehabilitators, such that they may offer further care at a reduced price or for free. This collaboration enables hedgehogs to be cared for outside of veterinary practices where they would be surrounded by domestic animals and humans [58]. However, this distinction between free treatment in the initial 24-h emergency period, but not necessarily afterwards, has led to a degree of confusion by members of the public, including conflicting information appearing on social media that vets are obliged to treat hedgehogs and other wild animals for free regardless of the timescale involved. Consequently, this has led to a degree of antipathy towards vets from some quarters. In addition, there is also a perception that vets, and some rehabilitation organisations as well, are "too quick to euthanise." As a result, some rehabilitators recommend that members of the public should take injured hedgehogs to them in the first instance rather than to a vet/other rehabilitation organisation, as the latter are considered to be likely to euthanise an individual even if it presents with survivable injuries. Paradoxically, this would increase the amount of time that an animal requiring veterinary attention would have to suffer before actually being examined by a vet.

The second potential advantage associated with decreased distance between finding location and hospitals is that rehabilitated individuals may be more likely to be returned to their original site. However, even this is not always possible, as householders may refuse to have the animal returned to their garden and/or the rehabilitator may deem a site to be unsafe (e.g., due to the presence of badgers or busy roads) such that an alternative location is required. At present, there are insufficient data available on the specific release practices of different hospitals, although it is clear that hedgehogs are commonly released at alternative locations for practical or welfare reasons. Whilst this may not affect post-release survival rates significantly (*sensu* [26]), it may influence patterns of gene flow [63,64] and possibly disease transmission within and between populations [65,66].

Despite these putative advantages, small hospitals are also likely to have their limitations. As outlined above, 56.5% of the small hospitals that replied to the questionnaire survey stated that they did not have any social media accounts, and only 17.6% had a business social media account. Furthermore, 125 of the 304 rehabilitators identified, including 96 of 109 small hospitals, were only identified by snowball sampling. These data imply that members of the public may, therefore, not be able to easily find their nearest rehabilitator if they do not have a website/social media presence. In addition, small hospitals were associated with a range of characteristics that could affect their ability to maximise release rates. For example, only 4.9% of small hospitals had one or two paid staff, only 1.9% had three or more volunteers, 94.4% were located in the rehabilitator's private residence, and 2.8% stated they had no veterinary care. Furthermore, practitioners within small hospitals also need to carry out a wide range of ancillary activities beyond looking after the animals in their care, such as maintaining an online presence, training volunteers, and raising funds. As such, small hospitals may experience significantly greater challenges with, e.g., housing

hedgehogs in hygienic conditions, minimizing stress (*sensu* [67]), managing disease sprea
and may also have limited capacity at certain times of the year (e.g., when the number
orphaned juveniles is high or during heatwaves), although they are likely to be in conta
with other nearby rehabilitators who may have space.

At present, however, all of these putative benefits and concerns are conjecture, ar
may be unrealised and/or unwarranted, principally because of a lack of data on hedgeh
rehabilitation in Britain (either collectively or from individual rehabilitators) to substantia
or refute these claims. This is, in part, due to the difficulties associated with identifyir
practitioners, but also obtaining access to data for analysis. The latter is further exacerbate
by the fact that most hospitals/centres do not keep fully computerised records, which mea
that collating information about numbers admitted, underlying reasons for admissio
cause-specific release rates, and time in care before death, euthanasia, or release is difficu
Major reasons for this relate to the constraints associated with recording data electronical
where money and manpower are limited, meaning that those data which are accessib
tend to be from the larger centres, but also because hospitals/centres in Britain are n
required to collate such data or make it available to researchers. One potential, but high
contentious, option would be to make wildlife rehabilitation a licensed activity, with tl
submission of data compulsory. However, such a scheme would require funding, ar
smaller organisations argue that this additional cost would mean that they would have
cease operating. As such, additional methods for collecting such data within the curre
structure of the rehabilitation community need to be identified.

The Number of Admissions Relative to Population Size

Extrapolating from the data derived from the questionnaire survey, we estimate th
a total of 40,000–59,000 hedgehogs were admitted to the 304 active wildlife rehabilitato
identified in 2016. However, the disparity between these estimates suggests that they a
sensitive to the modelling approaches used. For example, Model 2 was particularly affecte
by applying the proportion of large hospitals identified in the online searches (49.2%)
order to estimate the corresponding number in the sample of 120 hospitals for which
data were available; this implies that we had missed 59.0 large hospitals in these initi
searches. We consider this to be unrealistic; however, large rehabilitators typically ha
business or personal social media accounts (88.9%), meaning that they were relative
straightforward to identify. This is also reflected in the fact that only three large hospita
were identified by snowball sampling. Furthermore, the number of active rehabilitato
used in deriving these estimates was substantially lower than the ~600 carers purporte
to be held on the BHPS's directory, although this organisation does not have detaile
information about their status (i.e., whether they are currently practicing, and if so, ho
many hedgehogs they take in each year). However, the fact that we were not able
identify such a large number of rehabilitators in this study does potentially suggest th
these undetected individuals/organisations may not actually be active, but that, if th
are practicing, they are likely to be dealing with small numbers of hedgehogs. For the
reasons, we consider the lower estimate of 40,000 is more plausible, but accept that it m
be higher.

Regardless, this estimate is substantially greater than those from previous studi
For example, [4,11] reported estimates of 30,000–40,000 and 71,000 admissions per annu
across the full range of bird (>200 species [68]) and mammal (>40 species [56]) speci
in Britain, respectively. It is also substantially larger than the number of West Europea
hedgehogs admitted to wildlife centres in other parts of its range (n = 490 in the 5-ye
period from 2009–2013 in three rescue centres in eastern Spain [69]; n = 16,967 in the 10-ye
period from 2010–2019 in 34 rescue centres in the Czech Republic [70]; n = 740 in tl
17-year period from 2002–2019 to the two main wildlife hospitals in northern Portugal [71
Furthermore, our estimate does not explicitly include the 31,000 hedgehogs admitted
veterinary surgeons each year [58], although there is likely some degree of overlap, sin
individual hedgehogs will be taken by rehabilitators to vets for treatment and/or vets w

pass on hedgehogs to rehabilitators for long-term care and for release. It does not, however, include those hedgehogs taken directly to vets which are then euthanised or die naturally in the practice.

The most recent estimate of the pre-breeding hedgehog population in Britain is 879,000 [44,56]. If we assume, based on the data from the RSPCA, which are the only data available at the current time, that approximately 50% of hedgehogs admitted to hospitals survive to be released [4], our results suggest that rehabilitators may collectively be saving 20,000 hedgehogs that would otherwise perish, a number equivalent to 2.3% of the national pre-breeding population. However, the causes for admission given by finders (Bearman-Brown and Baker, unpublished data) do suggest that most hedgehog casualties probably originate from urban areas, and that a large proportion of affected animals are juveniles [72]). If we assume that the pre-breeding population of hedgehogs associated with gardens and other urban green spaces is 200,000 [44], the adult sex ratio is 1:1, each adult female produces one litter a year, and mean litter size is 4.5 [36], this figure would equate to 10.0% of the pre-breeding or 3.1% of the post-breeding urban population. Given that presence–absence data from different habitats within urban areas suggest that this population has declined by approximately 2.0% per annum between 2003–2017 [73,74], it can be argued that hedgehog rehabilitation could have been an important component affecting the rate at which the urban population has been declining in the last two decades, rather than being merely a service related to animal welfare, as some authors have suggested previously [14,17–19]. It is important to acknowledge, however, that this estimate has been based on assumptions that require further examination once more detailed data are available from a much broader suite of hospitals/centres.

Furthermore, those data from the RSPCA and for those hospitals where we had a continuous set of data across the five-year survey period suggest that 2016 was a rather atypical year; in the former, the number admitted in 2016 was 22% higher than the average across the 13-year time period, whereas in the latter, the number admitted was significantly higher than in 2012 and 2013. At one level, the number of admissions would be expected to vary inter-annually, but, for a population that is assumed to be declining [44], we might expect to observe a general downward trajectory (assuming that the capacity of these hospitals has been approximately constant over time). In contrast, the pattern observed has been one for a general increase in admissions (Figures 4d and 5), with periodic peaks (Figure 5). The underlying reasons for these inter-annual fluctuations are unclear, but could be related to broad-scale changes associated with over-winter hibernation mortality, reproductive output, invertebrate prey availability, public awareness, risks related to anthropogenic activities, and/or inclement weather. As such, future research needs to focus on identifying how such ecological and human-mediated factors affect population dynamics and patterns of admission into wildlife hospitals.

5. Conclusions

This study was the first to attempt to directly estimate the number and characteristics of practitioners rehabilitating hedgehogs in Britain. A minimum of 304 rehabilitators were identified: most (n = 109) hospitals admitted less ≤50 hedgehogs annually, whereas most hedgehogs (82.8%) were admitted to the smaller number of large establishments (n = 36). However, the growth of the hedgehog rehabilitation community was mostly associated with the creation of these smaller hospitals. Overall, the collective number of hedgehogs admitted to (40,000–59,000) and potentially released from (20,000–29,500) these wildlife hospitals was large relative to the size of the pre-breeding population and the estimated annual rate of decline, especially if we assume that most individuals originated from urban areas. This implies that wildlife rehabilitation has potentially been an important factor in the dynamics of hedgehog populations in the last two decades. However, further information is still required regarding many facets of the rehabilitation process, as well as establishing how extrinsic ecological and anthropogenic factors affect the numbers of hedgehogs admitted to wildlife hospitals. Such future work would benefit particularly from facilitating means b which rehabilitators could store standardised information electronically.

Author Contributions: Conceptualization, L.E.B.-B. and P.J.B.; data curation, L.E.B.-B.; formal analy sis, L.E.B.-B. and P.J.B.; investigation, L.E.B.-B. and P.J.B.; methodology, L.E.B.-B. and P.J.B.; projec administration, L.E.B.-B.; supervision, P.J.B.; visualization, P.J.B.; writing—original draft, L.E.B.-B and P.J.B.; writing—review and editing, L.E.B.-B. and P.J.B. All authors have read and agreed to th published version of the manuscript.

Funding: This research received no external funding.

Institutional Review Board Statement: Ethical review and approval were waived for this study du to the use of information in the public domain and archive data relating to wildlife.

Informed Consent Statement: Informed consent was obtained from all subjects involved in the stud Participants were informed of time to completion, right to withdraw, and confidentiality and securit of data. As this study utilised retrospective/archive data from wildlife rehabilitation centres, ethica approval was not necessary. The data retrieved did not include any personal, protected, or sensitiv data; no identifying data regarding individual people were collected, and as animals involved wer wild, they are not owned by individuals.

Data Availability Statement: The data presented in this study are available on request from th corresponding author.

Acknowledgments: The authors would like to thank the participants who responded to the questior naire, who supported in promoting the questionnaire, and who provided feedback in the developmen of the questionnaire. The authors would like to specifically thank the RSPCA for provision of th comprehensive data set and to the BHPS, who advised on the development and promotion of th questionnaire. We would also like to thank two anonymous reviewers for their helpful comments.

Conflicts of Interest: The authors declare no conflict of interest.

References

1. Miller, E.A. (Ed.) *Minimum Standards for Wildlife Rehabilitation*, 4th ed.; National Wildlife Rehabilitators Association: St. Clou MN, USA, 2012.
2. Morner, T. Health Monitoring and Conservation of Wildlife in Sweden and Northern Europe. *Ann. N. Y. Acad. Sci.* **200** *969*, 34–38. [CrossRef] [PubMed]
3. Molina-López, R.A.; Casal, J.; Darwich, L. Causes of Morbidity in Wild Raptor Populations Admitted at a Wildlife Rehabilitatio Centre in Spain from 1995–2007: A Long Term Retrospective Study. *PLoS ONE* **2011**, *6*, e24603. [CrossRef] [PubMed]
4. Grogan, A.; Kelly, A. A Review of RSPCA Research into Wildlife Rehabilitation. *Vet. Rec.* **2013**, *172*, 211–215. [CrossRef] [PubMed]
5. Schenk, A.N.; Souza, M.J. Major Anthropogenic Causes for and Outcomes of Wild Animal Presentation to a Wildlife Clinic i East Tennessee, USA, 2000–2011. *PLoS ONE* **2014**, *9*, e93517. [CrossRef] [PubMed]
6. Mcruer, D.L.; Gray, L.C.; Horne, L.A.; Clark, E.E. Free-Roaming Cat Interactions with Wildlife Admitted to a Wildlife Hospital. *Wildl. Manag.* **2017**, *81*, 163–173. [CrossRef]
7. Montesdeoca, N.; Calabuig, P.; Corbera, J.A.; Oros, J. A Long-Term Retrospective Study on Rehabilitation of Seabirds in Grar Canaria Island, Spain (2003–2013). *PLoS ONE* **2017**, *12*, e0177366. [CrossRef]
8. Tejera, G.; Rodríguez, B.; Armas, C.; Rodríguez, A. Wildlife-Vehicle Collisions in Lanzarote Biosphere Reserve, Canary Island *PLoS ONE* **2018**, *13*, e0192731. [CrossRef]
9. Taylor-Brown, A.; Booth, R.; Gillett, A.; Mealy, E.; Ogbourne, S.M.; Polkinghorne, A.; Conroy, G.C. The Impact of Humar Activities on Australian Wildlife. *PLoS ONE* **2019**, *14*, e0206958. [CrossRef]
10. Kirkwood, J.K. Introduction: Wildlife Casualties and the Veterinary Surgeon. In *BSAVA Manual of Wildlife Casualties*; BSAVA Gloucester, UK, 2003; pp. 1–4. ISBN 0905214633.
11. Molony, S.E.; Baker, P.J.; Garland, L.; Cuthill, I.C.; Harris, S. Factors That Can Be Used to Predict Release Rates for Wildlif Casualties. *Anim. Welf.* **2007**, *16*, 361–367.
12. Wimberger, K.; Downs, C.T. Annual Intake Trends of a Large Urban Animal Rehabilitation Centre in South Africa: A Case Stud *Anim. Welf.* **2010**, 501–513.
13. Kirkwood, J.K. Wild Animal Welfare. In *Animal Welfare and the Environment*; Ryder, R.D., Ed.; Gerald Duckworth, Co., Ltd London, UK, 1992; pp. 139–154.
14. Guy, A.J.; Curnoe, D.; Banks, P.B.; Guy, A.J.; Curnoe, Á.D.; Curnoe, D.; Banks, P.B. A Survey of Current Mammal Rehabilitatio and Release Practices. *Biodivers. Conserv.* **2013**, *22*, 825–837. [CrossRef]

Guy, A.J.; Curnoe, D.; Banks, P.B. Welfare Based Primate Rehabilitation as a Potential Conservation Strategy: Does It Measure Up? *Primates* **2014**, *55*, 139–147. [CrossRef] [PubMed]

Pyke, G.H.; Szabo, J.K. Conservation and the Four Rs, Which Are Rescue, Rehabilitation, Release, and Research. *Conserv. Biol.* **2017**, *32*, 50–59. [CrossRef] [PubMed]

Kirkwood, J.K.; Sainsbury, A.W. Ethics of Interventions for the Welfare of Free-Living Wild Animals. *Anim. Welf.* **1996**, *5*, 235–243.

Kirkwood, J.K.; Best, R. Treatment and Rehabilitation of Wildlife Casualties: Legal and Ethical Aspects. *Practice* **1998**, *20*, 214–216. [CrossRef]

Dubois, S. A Survey of Wildlife Rehabilitation Goals, Impediments, Issues, and Success in British Columbia, Canada. Master's Thesis, The University of Victoria, Victoria, BC, Canada, 2003.

Wimberger, K.; Downs, C.T.; Boyes, R.S. A Survey of Wildlife Rehabilitation in South Africa: Is There a Need for Improved Management? *Anim. Welf.* **2010**, *19*, 481–499.

Trocini, S.; Pacioni, C.; Warren, K.; Butcher, J.; Robertson, I. Wildlife Disease Passive Surveillance: The Potential Role of Wildlife Rehabilitation Centres. In Proceedings of the Australian Wildlife Rehabilitation Conference, Canbarra, Australia, 22–24 July 2008; pp. 1–5.

Randall, N.J.; Blitvich, B.J.; Blanchong, J.A. Efficacy of Wildlife Rehabilitation Centers in Surveillance and Monitoring of Pathogen Activity: A Case Study with West Nile Virus. *J. Wildl. Dis.* **2012**, *48*, 646–653. [CrossRef]

Camacho, M.C.; Hernández, J.M.; Lima-Barbero, J.F.; Höfle, U. Use of Wildlife Rehabilitation Centres in Pathogen Surveillance: A Case Study in White Storks (Ciconia Ciconia). *Prev. Vet. Med.* **2016**, *130*, 106–111. [CrossRef]

Yabsley, M.J. The Role of Wildlife Rehabilitation in Wildlife Disease Research and Surveillance. In *Medical Management of Wildlife Species: A Guide for Practitioners*; Hernandez, S.M., Barron, H.W., Miller, E.A., Aguilar, R.F., Yabsley, M.J., Eds.; Wiley Blackwell: London, UK, 2020; pp. 159–165.

Jaspers, V.L.B.; Voorspoels, S.; Covaci, A.; Eens, M. Can Predatory Bird Feathers Be Used as a Non-Destructive Biomonitoring Tool of Organic Pollutants? *Biol. Lett.* **2006**, *2*, 283–285. [CrossRef]

Molony, S.E.; Dowding, C.V.; Baker, P.J.; Cuthill, I.C.; Harris, S. The Effect of Translocation and Temporary Captivity on Wildlife Rehabilitation Success: An Experimental Study Using European Hedgehogs (*Erinaceus europaeus*). *Biol. Conserv.* **2006**, *130*, 530–537. [CrossRef]

Soorae, P.S. *Global Re-Introduction Perspectives: 2013; Further Case Studies from around the Globe*; Soorae, P.S., Ed.; IUCN/SSC Reintroduction Specialist Group: Gland, Switzerland, 2013; ISBN 9782831716336.

Goldsworthy, S.D.; Giese, M.; Gales, R.P.; Brothers, N.; Hamill, J. Effects of the Iron Baron Oil Spill on Little Penguins (Eudyptula Minor). II. Post-Release Survival of Rehabilitated Oiled Birds. *Wildl. Res.* **2000**, *27*, 573–582. [CrossRef]

Newman, S.H.; Ziccardi, M.H.; Berkner, A.B.; Holcomb, J.; Clumpner, C.; Mazet, J.A.K. A Historical Account of Oiled Wildlife Care in California. *Mar. Ornithol.* **2003**, *31*, 59–64.

Lunney, D.; Gresser, S.M.; Mahon, P.S.; Matthews, A. Post-Fire Survival and Reproduction of Rehabilitated and Unburnt Koalas. *Biol. Conserv.* **2004**, *120*, 567–575. [CrossRef]

Griffith, J.E.; Dhand, N.K.; Krockenberger, M.B.; Higgins, D.P. A Retrospective Study of Admission Trends of Koalas to a Rehabilitation Facility over 30 Years. *J. Wildl. Dis.* **2013**, *49*, 18–28. [CrossRef] [PubMed]

Rodríguez, A.; Rodríguez, B.; Lucas, M.P. Trends in Numbers of Petrels Attracted to Artificial Lights Suggest Population Declines in Tenerife, Canary Islands. *Ibis* **2012**, *154*, 167–172. [CrossRef]

Monadjem, A.; Botha, A.; Murn, C. Survival of the African White-Backed Vulture Gyps Africanus in North-Eastern South Africa. *Afr. J. Ecol.* **2013**, *51*, 87–93. [CrossRef]

Pyke, G.H.; Szabo, J.K. What Can We Learn from Untapped Wildlife Rescue Databases? The Masked Lapwing as a Case Study. *Pacific Conserv. Biol.* **2018**, *24*, 148–156. [CrossRef]

Mullineaux, E. Veterinary Treatment and Rehabilitation of Indigenous Wildlife. *J. Small Anim. Pract.* **2014**, *55*, 293–300. [CrossRef]

Morris, P.A. *Hedgehog*; HarperCollins: London, UK, 2018.

Hof, A.R.; Bright, P.W. The Value of Green-Spaces in Built-up Areas for—Western Hedgehogs. *Lutra* **2009**, *52*, 69–82.

Hof, A.R.; Bright, P.W. Factors Affecting Hedgehog Presence on Farmland as Assessed by a Questionnaire Survey. *Acta Theriol.* **2012**, *57*, 79–88. [CrossRef]

Van de Poel, J.L.; Dekker, J.; Langevelde, F.V. Dutch Hedgehogs *Erinaceus europaeus* Are Nowadays Mainly Found in Urban Areas, Possibly Due to the Negative Effects of Badgers Meles Meles. *Wildl. Biol.* **2015**, *21*, 51–55. [CrossRef]

Pettett, C.E.; Moorhouse, T.P.; Johnson, P.J.; Macdonald, D.W. Factors Affecting Hedgehog (*Erinaceus europaeus*) Attraction to Rural Villages in Arable Landscapes. *Eur. J. Wildl. Res.* **2017**, *63*, 54. [CrossRef]

Pettett, C.E.; Johnson, P.J.; Moorhouse, T.P.; Macdonald, D.W. National Predictors of Hedgehog *Erinaceus europaeus* Distribution and Decline in Britain. *Mamm. Rev.* **2018**, *48*, 1–6. [CrossRef]

Williams, B.M.; Baker, P.J.; Thomas, E.; Wilson, G.J.; Judge, J.; Yarnell, R.W. Reduced Occupancy of Hedgehogs (*Erinaceus europaeus*) in Rural England and Wales: The Influence of Habitat and an Asymmetric Intra-Guild Predator. *Sci. Rep.* **2018**, *8*, 12156. [CrossRef]

Hof, A.R.; Bright, P.W. Quantifying the Long-Term Decline of the West European Hedgehog in England by Subsampling Citizen-Science Datasets. *Eur. J. Wildl. Res.* **2016**, *62*, 407–413. [CrossRef]

44. Wembridge, D.; Johnson, G.; Al-Fulaij, N.; Langton, S. *The State of Britain's Hedgehogs 2022*; People's Trust for Endangered Species: London, UK, 2022.
45. Holsbeek, L.; Rodts, J.; Muyldermans, S. Hedgehog and Other Animal Traffic Victims in Belgium: Results of a Countrywide Survey. *Lutra* **1999**, *42*, 111–119.
46. Huijser, M.P.; Bergers, P.J.M. The Effect of Roads and Traffic on Hedgehog (*Erinaceus europaeus*) Populations. *Biol. Conserv.* **2000**, *95*, 6–9. [CrossRef]
47. Müller, F. Langzeit-monitoring Der Strassenverkehrsopfer Beim Igel (*Erinaceus europaeus* L.) Zur Indikation von Populationsdichte Veränderungen Entlang Zweier Teststrecken Im Landkreis Fulda. *Beiträge Naturkd. Osthessen* **2018**, *54*, 21–26.
48. Doncaster, C.P. Factors Regulating Local Variations in Abundance: Field Tests on Hedgehogs. *Erinaceus europaeus. Oikos* **1994**, *69*, 182–192. [CrossRef]
49. Whalen, J.K.; Parmelee, R.W.; Edwards, C.A. Population Dynamics of Earthworm Communities in Corn Agroecosystems Receiving Organic or Inorganic Fertilizer Amendments. *Biol. Fertil. Soils* **1998**, *27*, 400–407. [CrossRef]
50. Rondinini, C.; Doncaster, C.P. Roads as Barriers to Movement for Hedgehogs. *Funct. Ecol.* **2002**, *16*, 504–509. [CrossRef]
51. Young, R.P.; Davison, J.; Trewby, I.D.; Wilson, G.J.; Delahay, R.J.; Doncaster, C.P. Abundance of Hedgehogs (*Erinaceus europaeus*) in Relation to the Density and Distribution of Badgers (Meles Meles). *J. Zool.* **2006**, *269*, 349–356. [CrossRef]
52. Dowding, C.V.; Harris, S.; Poulton, S.M.C.; Baker, P.J. Nocturnal Ranging Behaviour of Urban Hedgehogs, *Erinaceus europaeus*, in Relation to Risk and Reward. *Anim. Behav.* **2010**, *80*, 13–21. [CrossRef]
53. Geiger, F.; Bengtsson, J.; Berendse, F.; Weisser, W.W.; Emmerson, M.; Morales, M.B.; Ceryngier, P.; Liira, J.; Tscharntke, T.; Winqvist, C.; et al. Persistent Negative Effects of Pesticides on Biodiversity and Biological Control Potential on European Farmland. *Basic Appl. Ecol.* **2010**, *11*, 97–105. [CrossRef]
54. Moorhouse, T.P.; Palmer, S.C.F.; Travis, J.M.J.; Macdonald, D.W. Hugging the Hedges: Might Agri-Environment Manipulations Affect Landscape Permeability for Hedgehogs? *Biol. Conserv.* **2014**, *176*, 109–116. [CrossRef]
55. Trewby, I.D.; Young, R.P.; McDonald, R.A.; Wilson, G.J.; Davison, J.; Walker, N.; Robertson, P.A.; Doncaster, C.P.; Delahay, R. Impacts of Removing Badgers on Localised Counts of Hedgehogs. *PLoS ONE* **2014**, *9*, 2–5. [CrossRef]
56. Mathews, F.; Kubasiewicz, L.M.; Gurnell, J.; Harrower, C.A.; McDonald, R.A.; Shore, R.F. *A Review of the Population and Conservation Status of British Mammals. A Report by the Mammal Society under Contract to Natural England, Natural Resources Wales and Scottish Natural Heritage*; Natural England: Peterborough, UK, 2018; ISBN 9781783544943.
57. Harris, S.; Morris, P.A.; Wray, S.; Yalden, D.W. *A Review of British Mammals: Population Estimates and Conservation Status of British Mammals Other Than Cetaceans*; JNCC: Peterborough, UK, 1995; ISBN 1-873701-68-3.
58. Barnes, E.; Farnworth, M.J. Perceptions of Responsibility and Capability for Treating Wildlife Casualties in UK Veterinary Practices. *Vet. Rec.* **2016**, *108*, 197. [CrossRef]
59. Jones, S.A.; Chapman, S. The Ethics and Welfare Implications of Keeping Western European Hedgehogs (*Erinaceus europaeus*) in Captivity. *J. Appl. Anim. Welf. Sci.* **2019**, *23*, 467–483. [CrossRef]
60. Bullen, K. *Hedgehog Rehabilitation*; British Hedgehog Preservation Society: Ludlow, UK, 2010.
61. Grogan, A. The Importance of Research in Wildlife Rehabilitation. Available online: http://bwrc.org.uk/conference-proceedings/4549145806 (accessed on 25 July 2021).
62. Reeve, N.J.; Bowen, C.; Gurnell, J. An Improved Identification Marking Method for Hedgehogs. *Mammal Commun.* **2019**, *5*, 1–5.
63. Barthel, L.M.F.; Wehner, D.; Schmidt, A.; Berger, A.; Hofer, H.; Fickel, J. Unexpected Gene-Flow in Urban Environments: The Example of the European Hedgehog. *Animals* **2020**, *10*, 2315. [CrossRef]
64. Rasmussen, S.L.; Nielsen, J.L.; Jones, O.R.; Berg, T.B.; Pertoldi, C. Genetic Structure of the European Hedgehog (*Erinaceus europaeus*) in Denmark. *PLoS ONE* **2020**, *15*, e0227205. [CrossRef] [PubMed]
65. Rasmussen, S.L.; Larsen, J.; van Wijk, R.E.; Jones, O.R.; Berg, T.B.; Angen, Ø.; Larse, A.R. European hedgehogs (*Erinaceus europaeus*) as a natural reservoir of methicillin-resistant Staphylococcus aureus carrying mecC in Denmark. *PLoS ONE* **2019**, *14*, e0222031. [CrossRef] [PubMed]
66. Rasmussen, S.L.; Hallig, J.; van Wijk, R.E.; Huus Petersen, H. An investigation of endoparasites and the determinants of parasite infection in European hedgehogs (*Erinaceus europaeus*) from Denmark. *Int. J. Parasitol. Parasites Wildl.* **2021**, *16*, 217–227. [CrossRef] [PubMed]
67. Rasmussen, S.L.; Kalliokoski, O.; Dabelsteen, T.; Abelson, K. An exploratory investigation of glucocorticoids, personality and survival rates in wild and rehabilitated hedgehogs (*Erinaceus europaeus*) in Denmark. *BMC Ecol. Evol.* **2021**, *21*, 96. [CrossRef] [PubMed]
68. Harris, S.J.; Massimino, D.; Balmer, D.E.; Eaton, M.A.; Noble, D.G.; Pearce-Higgins, J.W.; Woodcock, P.; Gillings, S. *The Breeding Bird Survey 2019*; British Trust for Ornithology: Thetford, UK, 2020; ISBN 9781912642151.
69. Martínez, J.C.; Rosique, A.I.; Royo, M.S. Causes of Admission and Final Dispositions of Hedgehogs Admitted to Three Wildlife Rehabilitation Centers in Eastern Spain. *Hystrix* **2014**, *25*, 107–110. [CrossRef]
70. Lukešová, G.; Voslarova, E.; Vecerek, V.; Vucinic, M. Trends in Intake and Outcomes for European Hedgehog (*Erinaceus europaeus*) in the Czech Rescue Centers. *PLoS ONE* **2021**, *16*, e0248422. [CrossRef] [PubMed]
71. Garcês, A.; Soeiro, V.; Lóio, S.; Sargo, R.; Sousa, L.; Silva, F.; Pires, I. Outcomes, Mortality Causes, and Pathological Findings in European Hedgehogs (*Erinaceus europeus*, Linnaeus 1758): A Seventeen Year Retrospective Analysis in the North of Portugal. *Animals* **2020**, *10*, 1305. [CrossRef]

2. Burroughes, N.D.; Dowler, J.; Burroughes, G. Admission and Survival Trends in Hedgehogs Admitted to RSPCA Wildlife Rehabilitation Centres. *Proc. Zool. Soc.* **2021**, *74*, 198–204. [CrossRef]
3. Wembridge, D.; Langton, S. Living with Mammals: An Urban Study. *Br. Wildl.* **2016**, *27*, 188–195.
4. Wilson, E.; Wembridge, D. *State of Britain's Hedgehogs, 2018*; People's Trust for Endangered Species: London, UK, 2018.

Article

Differences in Mortality of Pre-Weaned and Post-Weaned Juvenile European Hedgehogs (*Erinaceus europaeus*) at Wildlife Rehabilitation Centres in the Czech Republic

Gabriela Kadlecova [1,*], Sophie Lund Rasmussen [2,3], Eva Voslarova [1] and Vladimir Vecerek [1]

[1] Department of Animal Protection and Welfare and Veterinary Public Health, Faculty of Veterinary Hygiene and Ecology, University of Veterinary Sciences Brno, 612 42 Brno, Czech Republic
[2] Wildlife Conservation Research Unit, The Recanati-Kaplan Centre, Department of Biology, University of Oxford, Tubney House, Abingdon Road, Tubney, Abingdon OX13 5QL, UK
[3] Department of Chemistry and Bioscience, Aalborg University, Fredrik Bajers Vej 7H, DK-9220 Aalborg, Denmark
* Correspondence: kadlecovag@vfu.cz

Simple Summary: Wildlife rehabilitation centres contribute to the conservation of wildlife by caring for sick, injured and orphaned animals that would not survive in the wild without human help and releasing healthy animals back into the wild in appropriate habitats. A total of 4388 European hedgehogs, identified as pre-weaned from normally timed litters (NL PRE), post-weaned from normally timed litters (NL POST), pre-weaned from late litters (LL PRE) and post-weaned from late litters (LL POST) were admitted to 27 wildlife rehabilitation centres in the Czech Republic in the period from 2011 to 2020. Where the outcome of rehabilitation care was known, young admitted before natural weaning were associated with a high mortality rate, especially in those from late litters. Among the four groups, the juveniles of the NL POST category experienced the lowest mortality (14%) with the highest release rate (86%). In contrast, LL PRE experienced the highest mortality (46%) with the lowest release rate (54%).

Citation: Kadlecova, G.; Rasmussen, S.L.; Voslarova, E.; Vecerek, V. Differences in Mortality of Pre-Weaned and Post-Weaned Juvenile European Hedgehogs (*Erinaceus europaeus*) at Wildlife Rehabilitation Centres in the Czech Republic. *Animals* **2023**, *13*, 337. https://doi.org/10.3390/ani13030337

Academic Editor: Emiliano Mori

Received: 21 December 2022
Revised: 12 January 2023
Accepted: 13 January 2023
Published: 17 January 2023

Abstract: Previous research from several European countries has indicated that the European hedgehog (*Erinaceus europaeus*) is in decline. Wildlife rehabilitation centres contribute toward the protection of debilitated hedgehogs, including the young. Based on data from 27 wildlife rehabilitation centres, the mortality rate and the release rate of juvenile hedgehogs were evaluated depending on whether they were from normally timed litters (admitted from April to September) or from late litters (admitted from October to March). A total of 4388 juvenile European hedgehogs were admitted to wildlife rehabilitation centres in the Czech Republic from 2011 to 2020. The number of post-weaned young from late litters admitted (28%) did not differ from the number of pre-weaned young from late litters (29%). Where the outcome was known, young from late litters had the highest mortality rate (46%) in the year of admission. The release rate was the highest in post-weaned young from normally timed litters (86%). Further research should focus on the definition of optimal care and treatment of the underlying causes for admission of juvenile hedgehogs. The reproductive strategy (the timing of litters) of European hedgehogs under the climatic conditions of the Czech Republic affects the chance of survival of young at wildlife rehabilitation centres and likely also in the wild.

Keywords: hoglet; weaning; rescue centre; release

1. Introduction

The western European hedgehog (*Erinaceus europaeus*) is an insectivore that inhabits a large part of Europe [1,2]. It is a species that displays true hibernation. It is, therefore, essential for hedgehogs to reach an optimal body condition to survive hibernation [3–5]. The length of hibernation and the timing of its ending may differ depending on how long the winter period lasts in a given area in which hedgehogs are found. Hibernation is known to be shorter in warmer areas [6], while hibernation is longer in the north due to the longer-lasting winter and low temperatures, therefore shortening the period of time during which hedgehogs are active [3,5]. Hibernation may also be affected by local fluctuations in the weather in the given period, as has been described in Denmark, for example, where the hibernation of hedgehogs was delayed by unusually warm weather in a particular year [7].

The timing of reproduction in European hedgehogs is also associated with the timing and length of hibernation. Hedgehogs begin to mate soon after awakening from hibernation if temperature conditions are optimal, leaving plenty of resources for the hedgehogs [8], and the body conditions of the females are sufficient. Female European hedgehogs appear to be seasonally polyoestrous [5], with a succession of oestrus cycles during spring and summer. Gestation in female European hedgehogs lasts approximately 34 days, after which usually 4–5 young are born [3], with previously documented examples of litter sizes of up to 11 individuals [9]. The young are dependent on their mother's care for a period of 5–6 weeks and are subsequently weaned at a weight of 250 g [10]. This is followed by the challenging time when the newly independent juveniles must manage to forage, with a diet comprised primarily of invertebrates, and obtain adequate day and night nests. The timing of birth influences the mortality rate in hedgehog young [11], which amounts to as much as 69% of offspring in the wild [3,12–14]. In the Czech Republic, hibernation ends in April and the first litters of European hedgehogs are usually born between May and August [15].

Usually, female hedgehogs have a single litter a year in areas with a colder climate, potentially enabling them to have a longer hibernation period [16]. When autumn has a high abundance of resources available, second litters may be produced in a given year [3–5]. Females that lose the young from the first litter may also have a second litter. They will go into oestrus again and mate, and wean their young later [10]. Offspring from second or later litters may be at a disadvantage over earlier-born young as they have less time to grow and obtain the necessary body condition before hibernation. However, Bunnell [10] studied juvenile hedgehogs taken into care—81 individuals from early litters and 38 from late litters—and demonstrated that young born in the late summer compensated for this delay by gaining weight faster than young from normally timed litters during their time in captivity with ad libitum food available [10]. It is, however, true of all young that they may find themselves in a situation in which they will not be able to survive without human help. This particularly applies to young that lose their mother before weaning. Such a scenario is sadly common as hedgehogs often live in the vicinity of towns and villages [17] and are exposed to many risks associated with anthropogenic activities, such as collisions with vehicles on the roads [18], being attacked by pets or dying as the result of injuries caused by garden tools and machinery, or from poisoning [3,4,19–23]. Wildlife rehabilitation centres can play an irreplaceable role in caring for individuals and therefore supporting the protection of this species [21,24]. Orphaned juvenile hedgehogs found by humans are often taken to wildlife rehabilitation centres where they are hand-raised with the aim of being subsequently released back into the wild. Previous research comparing the post-release survival of hand-raised and wild, juvenile hedgehogs has shown that hand-raised juveniles appear to have equal prospects as wild, suggesting that hand-raising of orphaned juvenile hedgehogs is an important contribution to the conservation of this species [13].

The aim of this study was to categorise the groups of juvenile western European hedgehogs (*Erinaceus europaeus*) admitted to wildlife rehabilitation centres in the Czech Republic before weaning and after weaning in the years 2011 to 2020, and to determine whether the timing of their birth (normally timed or late litters) influenced their survival chances in care. In addition, interpreting the results in a wider context to discuss how to optimise the conservation effort based on the hand-raising of orphaned, sick or injured juvenile hedgehogs.

2. Materials and Methods

Data on juvenile western European hedgehogs (*Erinaceus europaeus*) admitted wildlife rehabilitation centres in the Czech Republic were obtained from the Minist of the Environment, which is responsible for the work of wildlife rehabilitation centres ar keeps records on the animals for which these facilities care. As these rescue centres a part of the National Network of Rescue Centres, which cooperates with the Ministry of tł Environment, the recording system is unified. The data include information on the numb of admitted animals, the dates and reasons for admission, and voluntary information on tł weight or age of the animal and the place of finding if it is known. These records containe data on the numbers of juvenile European hedgehogs admitted from 2011 to 2020, tł reasons for their admission to wildlife rehabilitation centres, and the dates on which thes juveniles were admitted to and the outcome, e.g., whether they were released back into tł wild or died (naturally or euthanised). A total of 27 wildlife rehabilitation centres in tł Czech Republic provided data for the study. Identification of admitted species, includir distinguishing between the two native hedgehog species of the Czech Republic (*Erinace* *europaeus* and *E. roumanicus*), and recording characteristics of each individual, were tł responsibility of the rehabilitation centres' certified experts.

2.1. Categorising Data

We included data on reason for admission into care and weight at admission f juvenile European hedgehogs for the purpose of this study. The juvenile hedgehog were divided by weight at admission into the categories pre-weaned (1–250 g) and po weaned (251–400 g) young. Furthermore, the juvenile hedgehogs were divided by date admission into the categories "young from normally timed litters" (young admitted fro April to September) and "young from late litters" (young admitted from October to Marcł This division was based on calculation of the gestation period (5 weeks) and the age weaning (5–6 weeks) according to Bexton [22]: young from normally timed litters can from hedgehogs that mated during March and April at the earliest, while young from la litters came from litters for which hedgehogs had mated during August and September the earliest, i.e., in the late summer. A weight of 250 g is stated by Bunnell [10] as the weig of hedgehogs at weaning. The individuals were divided into four groups for the purpos of comparison: pre-weaned from normally timed litters (NL PRE), post-weaned fro normally timed litters (NL POST), pre-weaned from late litters (LL PRE) and post-weane from late litters (LL POST). In some cases, information on the outcome of the care (die naturally/euthanised or released back into the wild) was lacking, which resulted in tł exclusion of these records from the data analyses on survival and duration of time in ca For the young with known outcome, the proportion of young released after care ar young that died or were euthanised during care was calculated to establish the releas rate and mortality rate: the proportion of juveniles released and juveniles that died were euthanised in care compared to the total number of hedgehogs admitted in tł given group (NL PRE, NL POST, LL PRE, LL POST). Out of the 4388 juvenile hedgeho registered in the records, only 3441 individuals had information on the outcome of tł care (died naturally/euthanised or released back into the wild). Therefore, the remainir 947 individuals were excluded from the data analyses on survival and duration of tin in care.

The duration of care at the wildlife rehabilitation centres was analysed for individua that were subsequently released back into the wild, in the cases where date of admissic and date of release were stated in the records. The duration of care was calculated as tł number of days between these two dates.

2.2. Data Analysis

Statistical analysis was performed in the statistical program UNISTAT 6.5 for Exc (Unistat Ltd., London, UK). The data according to the normality test have a non-norm distribution, thus non-parametric tests were used. Spearman's coefficient was used to asse the trend in the number of young admitted in a given period, and a Fisher's exact test was used for evaluation of the difference in the numbers of young admitted in individual groups (NL PRE, NL POST, LL PRE and LL POST) and for comparison of the numbers of young released and young that died or were euthanised. Kruskal–Wallis ANOVA was used for the assessment of the duration of admission to the wildlife rehabilitation centre. In the tests used, the value of $p < 0.05$ was determined to be statistically significant.

3. Results

A total of 4388 juvenile European hedgehogs were admitted to 27 wildlife rehabilitation centres in the Czech Republic in the period from 2011 to 2020, with an increasing trend in the number of juveniles admitted per year during the entire period (rSp = 0.9879, $p< 0.001$). A significantly lower number ($p < 0.001$) of NL POST (post-weaned individuals from normally timed litters) young were admitted in comparison with all the other groups (Figure 1).

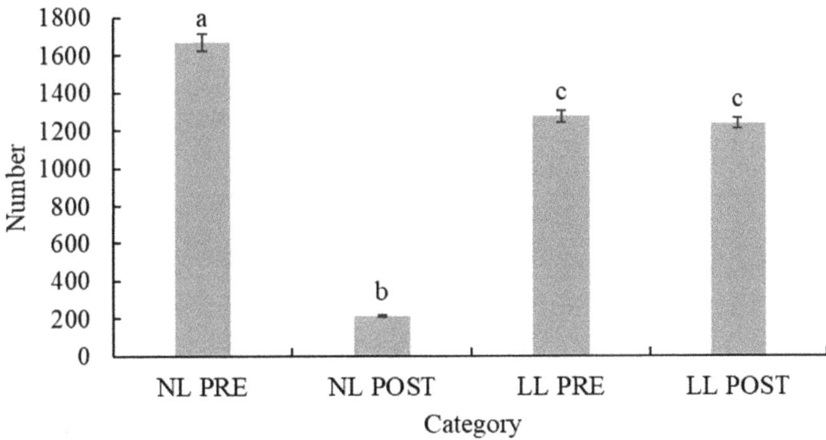

Figure 1. The number of juvenile European hedgehogs admitted to wildlife rehabilitation centres in the Czech Republic in the years 2011 to 2020, total 4388, divided into four categories by weight and timing of litter. Columns or bars with different letters (i.e., a versus b, a versus c, b versus c) indicate that they are statistically different from each other, while those with same letters (c versus c) indicate that they are not statistically different from each other. NL PRE = pre-weaned juvenile hedgehogs from normally timed litters; NL POST = post-weaned juvenile hedgehogs from normally timed litters; LL PRE = pre-weaned juvenile hedgehogs from late litters; LL POST = post-weaned juvenile hedgehogs from late litters.

The lowest mortality rate was found in NL POST juveniles (14% of the number of animals admitted in this category), for which the highest release rate was also found (86%) (Table 1.). In contrast, the highest mortality rate was seen in the category LL PRE (46%), which therefore also had the lowest release rate (54%). There was no significant difference ($p > 0.05$) in mortality rate and release rate between NL PRE and LL POST.

Table 1. Outcome of juvenile European hedgehogs admitted to wildlife rehabilitation centres from 2011 to 2020 divided into four categories by weight and timing of litter (data are from the 3441 individuals with known outcome).

Litter Timing	Category	Admitted (Number)	Mortality Rate		Release	
			Number	%	Number	%
Normally timed litter	pre-weaned	1518	453 [a]	30	1065 [x]	70
	post-weaned	210	30 [b]	14	180 [y]	86

Table 1. *Cont.*

Litter Timing	Category	Admitted (Number)	Mortality Rate		Release	
			Number	%	Number	%
Late litter	pre-weaned	983	455 [c]	46	528 [z]	54
	post-weaned	730	251 [a]	34	479 [x]	66

[a–c] different letters in a column indicate a statistically significant difference ($p < 0.01$), with fields marked with being statically significantly different from fields categorised as b and c, whereas fields indicated with the same letter represent data which are not statistically significantly different from each other. [x–z] different letters in column indicate a statistically significant difference ($p < 0.01$).

The length of stay in NL PRE (median 41 days) and LL PRE (median 41 days) young was significantly longer ($p < 0.001$) than in NL POST and LL POST young (median 21 and 24, respectively) (Figure 2).

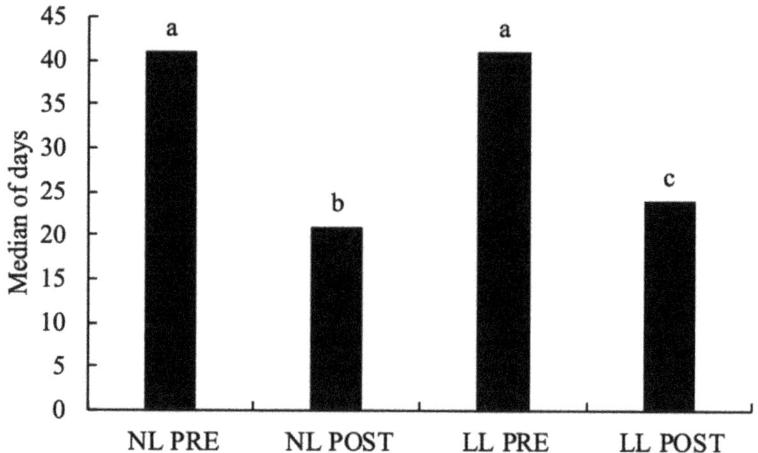

Figure 2. Period of time spent by released juvenile European hedgehogs at wildlife rehabilitation centres in the Czech Republic from 2011 to 2020 divided into four categories by weight and timing of litter (data are from the 3441 individuals with known outcome). Note: columns or bars with different letters (i.e., a versus b, a versus c, b versus c) indicate that they are statistically different from each other, while those with same letters (c versus c) indicate that they are not statistically different from each other. NL PRE = pre-weaned juvenile hedgehogs from normally timed litters (n = 806); NL POST = post-weaned juvenile hedgehogs from normally timed litters (n = 151); LL PRE = pre-weaned juvenile hedgehogs from late litters (n = 525); LL POST = post-weaned juvenile hedgehogs from late litters (n = 443).

4. Discussion

The number of European hedgehogs admitted to wildlife rehabilitation centres has been on the increase in recent years, with their admission to wildlife rehabilitation centres in the UK doubling in the period 2005–2017 [19]. The offspring of wildlife make up a large proportion of all sick, injured and orphaned individuals that are admitted to these facilities [24,25]. Young accounted for a proportion of almost 20% of all hedgehogs admitted to three wildlife rehabilitation centres in Spain in the years 2009–2013 [20] and almost 60% in 34 wildlife rehabilitation centres in the Czech Republic from 2010 to 2019 [21].

4.1. Distinguishing the Species in Care

The Czech Republic is inhabited by two species of hedgehogs, the European hedgehog (*Erinaceus europaeus*) and the northern white-breasted hedgehog (*E. roumanicus*), which can be challenging to distinguish from each other, as they share a range of similar features.

Our study focuses exclusively on *Erinaceus europaeus*. Our data derives from the records of 27 wildlife rehabilitation centres, where the carers have identified the individuals as European hedgehogs. Even though this categorisation cannot be validated, we acknowledge that all wildlife rescue centre staff in the Czech Republic must have a certificate from the Ministry of the Environment, which also includes an exam on identifying species living in the Czech Republic. In the case of *E. europaeus*, this includes distinguishing them from *E. roumanicus* based on morphology.

4.2. Categorising Individuals into Dependent and Independent Juveniles

Due to the lack of background knowledge of the individuals, our categorisation of juveniles into dependent and independent individuals was based on weight at admission, with pre-weaned weighing $\leq$250 g and post-weaned >250 g. There is a general consensus in the literature on the subject that juvenile hedgehogs tend to weigh around 250 g when they reach independence. However, there may have been cases where sickness or injury have caused independent individuals to weigh less than 250 g, which would ultimately have led them to become recorded as dependent juveniles. This potential bias should be considered when interpreting the results.

4.3. Independent Juveniles from Normally Timed Litters Have Better Prospects

Individuals belonging to the NL POST young category were admitted to wildlife rehabilitation centres least often and had a significantly lower mortality rate in care compared to the other categories. These individuals are adolescent hedgehogs weighing 251–400 g. They are already independent, beginning to live the solitary way of life typical of this species, and learning to forage for themselves and build nests [22]. They must also prepare themselves for the coming winter when hibernation awaits them, though in view of the appropriate timing of their birth in the summer months they have enough natural food and time to attain the necessary weight and body condition, if in otherwise good health. Therefore, their prospects are generally better compared to independent juveniles. This is also indicated by the fact that this group spent significantly shorter time in care compared to pre-weaned individuals from normally timed and late litters. It should, however, be taken into consideration that the lower sample size of this particular category of individuals, compared to the rest, may have influenced the results.

4.4. The Potentially Negative Consequences of Care and Alternative Solutions

Our results show that the highest proportions of juvenile hedgehogs admitted into care are dependent individuals (pre-weaned; NL PRE (38%) and LL PRE (29%)). These young are still dependent on their mother's care. However, every small (<250 g) hedgehog should not necessarily be considered orphaned and in need of care. These results suggest that dependent juvenile hedgehogs are often admitted to wildlife rehabilitation centres regardless of the timing of the litter. The ability to identify juvenile hedgehogs truly in need of human care is important, as they may otherwise be taken to wildlife rehabilitation centres needlessly. It has been shown that the capture of wild animals may itself lead to their death due to causes like capture myopathy or infection transmitted from the other patients [13,26]. Furthermore, wild animals placed in captivity, e.g., at a wildlife rehabilitation centre, encounter a novel, confined and unpredictable environment, which often includes handling and close proximity to humans [26]. These conditions cause physiological stress responses in a range of species and these increased stress levels may have severe effects on their health [13,27,28]. Rasmussen et al. [13] demonstrated that hedgehogs in care experience higher levels of faecal corticosterone metabolites and saliva corticosterone compared to wild individuals, suggesting that hedgehogs in care do experience higher levels of stress than their wild counterparts. Therefore, if a juvenile hedgehog seems to be healthy and mature enough to provide for itself, despite a small body size, human help in the form of supplementary feeding and provision of good nest sites in situ, should be recommended as an alternative to captivity.

4.5. When to Admit Juveniles into Care and When to Release Them

Several signs can be recognised in pre-weaned young indicating that they could nee care. These include the finding of an isolated juvenile or a litter of juveniles with individua weighing less than 200 g, being active during the day, vocalising, or having light-coloure and soft spines [23]. Such young may have lost their mother due to e.g. car collisions durir the summer, when lactating females are more active and may cross roads more frequent due to the increased rate of foraging needed to cover their intensified use of resourc caused by lactation [19]. They may also be less viable young that have been abandone by their mother [29], in which case any attempt to save them may be more complicate and perhaps even counterproductive. Caring for pre-weaned young is demanding ar is furthermore an economic burden for the wildlife rehabilitation centre. One princip problem lies in assuring an adequate milk diet, as is the case for the young of many oth wild animal species [30]. Hedgehog milk is extremely rich in fat and proteins and has a lo lactose content [31], for which reason it may be difficult to find a suitable milk substitu Gimmel et al. [32] have drawn attention to the shortcomings of commercially manufacture mixes for European hedgehogs, though their study did not consider milk substitutes. the milk of other mammals is to be used, Robinson and Routh [33] recommend using go milk, initially with a syringe, until the juveniles are mature enough to feed individual from a dish. Presently, the general practice at hedgehog wildlife rehabilitation centres to use commercially available puppy milk replacer combined with a careful monitorir of weight gain. Hand-raising of juvenile hedgehogs should only be carried out by traine hedgehog rehabilitators as it is a specialist's task.

The demands of hand-rearing pre-weaned hedgehog young are evidently also asso ated with the fact that young admitted to wildlife rehabilitation centres before weanir spent a significantly longer ($p < 0.001$) period of time at the centre than young admitted aft weaning. Pre-weaned young from both normally timed and late litters spent a median 41 days at wildlife rehabilitation centres before being released back into the wild. Hedgehc young in the wild are weaned at the age of 5–6 weeks [23]. Young are released at a simil age or an age several weeks older than the age of natural weaning depending on their a at the time of their admission and their weight before release.

Unfortunately, not all juvenile hedgehogs admitted to the wildlife rehabilitation centre could be released back into the wild. Some died while in care and some had to be euthanise due to severe injuries or poor prognosis. Age upon admission and the timing of the litt also influenced the outcome: Although more than a 70% of young from normally time litters admitted were released back into the wild in the same year (70% of pre-weaned ar 86% of post-weaned individuals), the numbers of pre-weaned animals released from la litters were significantly ($p < 0.001$) lower (53% individuals); the results may also be affecte by fewer number of admitted post-weaned hedgehogs form normally timed litters. Th smaller number of individuals from late litters released may have been influenced by th fact that the date of their release from the wildlife rehabilitation centre would have fallen the winter, i.e., during the hibernation period, leaving the registration as "unknown" in th record. Despite research demonstrating that hedgehogs can successfully be released durir winter when the conditions are suitable [34], and that hedgehogs in care experience high levels of stress which may decrease their welfare and chances of survival [13], opinior still differ between carers as to whether hibernating animals should be kept at wildli rehabilitation centres over the winter or released in the late autumn or during winter. Th question remains as to what impact stress has on animals during their time spent at wildlife rehabilitation centre, as it may have a negative effect on the health and norm behaviour of animals [35], and whether all animals at wildlife rehabilitation centres g into hibernation. External conditions, and temperature in particular, are important the commencement of hibernation [8]. Hibernation is often not possible, or disturbed, excessively small juvenile hedgehogs admitted before or during the winter when the require intensive care that includes frequent weighing and observation of the anima Nevertheless, research indicates that weighing of hedgehogs during hibernation and the possible wakening does not affect their chance of surviving hibernation as long as it is not done too frequently [36], particularly in cases in which the energy losses during such disturbance are compensated for by the provision of food that is available in sufficient quantities at wildlife rehabilitation centres. South et al. [36] also found that the duration of hibernation meant greater weight losses in smaller hedgehogs which lost weight more quickly during hibernation at the wildlife rehabilitation centre. The lower release rate in the group of LL POST individuals could also be caused by the condition of the individuals taken into care, as it is expected that individuals from later litters are generally less robust than individuals from normally timed litters due to the lack of food resources later in the season. The high release rate in post-weaned young from normally timed litters may also testify to the fact that rescue centres in the Czech Republic are in practice more likely to release young in summer and less likely to release them in winter for fear that they will not survive hibernation, even though previous studies have indicated that this is should not be a cause for concern [4,17,34,37].

4.6. Optimising the Care

In general, pre-weaned young from normally timed litters and post-weaned from late litters displayed similar mortality rate at wildlife rehabilitation centres in the Czech Republic; however, the highest mortality rate was found in pre-weaned hedgehogs from late litters. Mortality is generally high in animals at wildlife rehabilitation centres [24]. Furthermore, time spent at wildlife rehabilitation centres is a stress factor for hedgehogs increasing the level of glucocorticoids in the faeces and saliva in comparison with free-living hedgehogs [13] and the length of time they spend at such centres should be kept to the absolute minimum.

The results indicate that the timing of hedgehog births may also affect the chances of survival for juveniles admitted to wildlife rehabilitation centres and suggest that they may have different care requirements. Therefore, in addition to general care, the aim of wildlife rehabilitation centres should be to develop procedures and methods to care for hedgehogs at different ages and consider whether the effort should be adapted according to the period during which they were born and admitted to wildlife rehabilitation centres. Another task should be to optimise releases into the wild to ensure the highest possible survival rate after release and not to unnecessarily prolong the stressful stay in wildlife rehabilitation centres for these animals.

4.7. The Influence of Members of the Public on Admissions of Juvenile Hedgehogs

The public plays a large role in the admission of animals to wildlife rehabilitation centres by reporting findings of animals requiring human care to the staff of these centres or by bringing sick, injured or orphaned animals to the centres. Fewer NL POST than NL PRE juvenile hedgehogs were admitted to rescue centres, which may be due to public education stating that only young weighing less than 250 g that are found alone may need help, as opposed to young weighing more than 250 g. The situation may be different for young found in autumn, which raises concerns in people as to whether the juvenile hedgehog will be able to put on the necessary weight needed for hibernation. The results indicated a significantly ($p < 0.001$) lower proportion of NL POST hedgehogs admitted than LL POST. We suggest that even though these animals may have the same weight, the juveniles from late litters are considered more at risk by the public because of the approaching winter. This is causing members of the public to regard even large, healthy and independent juvenile hedgehogs as threatened and requiring human care simply because they remain active in the late autumn. The admission of seemingly healthy individuals belonging to the LL POST category could also be explained by the inability of members of the public and even hedgehog carers to distinguish between independent and dependent juvenile hedgehogs. The law in the Czech Republic requires rehabilitation centres to provide education to the public and thus participate in the protection of species in the wild including examples of the work of these centres and information materials.

The issue of an adequate weight before hibernation is well known, and many wildlife rehabilitation centres, along with other organisations contributing to the protection of European hedgehogs and other wildlife species, issue manuals containing information about how to identify individuals that require human help. Hibernation is a demanding period during which the animal uses a large amount of energy [36]. According to Robinson and Routh [33], hedgehogs weighing less than 450 g in the autumn, found out of the nest during the day, or showing signs of weakness or health problems may require help to survive the winter. Morris [38] suggested a hibernation weight of >450 g as a threshold for winter survival if a 25% weight loss during hibernation occurs. According to Bearman Brown et al. [39] hibernation is not a critical period for hedgehogs that hibernate at a weight of at least 600 g. Wildlife rehabilitation centres can help hedgehogs attain the necessary weight before hibernation during their time spent in captivity [40]. However, if a hedgehog is otherwise healthy, supplementary feeding and provision of suitable nest sites in situ is a much more desirable solution, avoiding taking the individual into care. Awareness of the risk of death in immature young hedgehogs during the winter may be the cause of the larger proportion (28%) of LL POST young admitted to rescue centres in the Czech Republic. Finding a young hedgehog in the autumn more often leads to its admission at a wildlife rehabilitation centre compared to the finding of a hedgehog of the same weight in the early summer. Rasmussen et al. [4] demonstrated that healthy juvenile hedgehogs are perfectly able to reach an adequate body condition to survive hibernation even when born later in the season, when resources are available. The same study also suggests directing the focus on body condition before hibernation instead of weight, as a small hedgehog of 600 g would be in good condition, while a large hedgehog of 600 g could be in a poor condition indicating weakness of some sort, making it less likely to survive hibernation. Furthermore, previous studies exploring the body mass change in hedgehogs before and after hibernation have shown that the individuals that can afford to lose most body mass do in fact lose most body mass [3,7,37,41]. This is also related to the timing of hedgehog releases into the wild, with Yarnell et al. [34] reporting that the survival rates of rehabilitated hedgehogs released during milder periods of the winter are equal to those of wild individuals.

5. Conclusions

Wildlife rehabilitation centres contribute towards maintaining populations of wild species of animal by caring for sick, injured or orphaned wildlife. The reproduction of European hedgehogs in the Czech Republic is influenced by climatic conditions, and young born in normally timed and late litters may have different care demands and survival chances. Although juvenile hedgehogs from normally timed litters can be released successfully back into the wild in most cases, the mortality rate is higher in pre-weaned young from late litters at the wildlife rehabilitation centres. As female hedgehogs often give birth to two litters per year in the Czech Republic, and the population of European hedgehogs is in decline, further efforts must be made to improve the care for juvenile hedgehogs, particularly before weaning, to increase the survival, and therefore also the release rate, of these animals.

Author Contributions: Conceptualization, G.K. and E.V.; Methodology, G.K.; Software, G.K.; Validation, S.L.R., V.V. and E.V.; Formal analysis, G.K.; Investigation, V.V.; Resources, G.K.; Data curation E.V.; Writing—original draft preparation, G.K.; Writing—review and editing, S.L.R.; Visualization G.K.; Supervision, V.V.; Project administration, E.V.; Funding acquisition, V.V. All authors have read and agreed to the published version of the manuscript.

Funding: This research was supported by ITA VFU Brno (Project No. FVHE/Vecerek/ITA2020).

Institutional Review Board Statement: Not applicable.

Informed Consent Statement: Not applicable.

Data Availability Statement: Data for the analysis was obtained from the Ministry of the Environment of the Czech Republic. The datasets generated and analysed during the current study are available from the corresponding author on reasonable request.

Acknowledgments: We would like to thank the Ministry of the Environment of the Czech Republic for providing data for this study.

Conflicts of Interest: The authors declare no conflict of interest. The funders had no role in the design of the study; in the collection, analyses, or interpretation of data; in the writing of the manuscript; or in the decision to publish the results.

References

IUCN Red List. 2019. Western European Hedgehog. Available online: https://www.iucnredlist.org/species/29650/2791303#text-fields (accessed on 3 May 2022).

Rautio, A.; Isomursu, M.; Valtonen, A.; Hirvelä-Koski, V.; Kunnasranta, M. Mortality, diseases and diet of European hedgehogs (*Erinaceus europaeus*) in an urban environment in Finland. *Mammal Res.* **2016**, *61*, 161–169. [CrossRef]

Morris, P. *Hedgehogs, Repr. with Rev.*; Whittet Books: London, UK, 2015.

Rasmussen, S.L.; Berg, T.B.; Dabelsteen, T.; Jones, O.R. The ecology of suburban juvenile European hedgehogs (*Erinaceus europaeus*) in Denmark. *Ecol. Evol.* **2019**, *9*, 13174–13187. [CrossRef]

Reeve, N.; Lindsay, R. *Hedgehogs, Poyser Natural History*, 1st ed.; Poyser: London, UK, 1994.

Haigh, A.; O'Riordan, R.M.; Butler, F. Nesting behaviour and seasonal body mass changes in a rural Irish population of the Western hedgehog (*Erinaceus europaeus*). *Acta Theriol.* **2012**, *57*, 321–331. [CrossRef]

Rasmussen, S.L. Personality, Survivability and Spatial Behavior of Juvenile Hedgehogs (*Erinaceus europaeus*): A Comparison between Rehabilitated and Wild Individuals. Master's Thesis, University of Copenhagen, Copenhagen, Denmark, 2013.

Fowler, P.A.; Racey, P.A. Daily and Seasonal Cycles of Body Temperature and Aspects of Heterothermy in the Hedgehog *Erinaceus europaeus*. *J. Comp. Physiol. B* **1990**, *160*, 299–307. [CrossRef]

Neumeier, M. *Wurfgrössen und Wurfzeiten der Igel in Deutschland, 3. Auflage, Neuausgabe*; Igel-Wissen Kompakt; Pro Igel: Lindau, Germany, 2016.

Bunnell, T. Growth rate in early and late litters of the European hedgehog (*Erinaceus europaeus*). *Lutra* **2009**, *52*, 15–22.

Morris, P. Pre-weaning mortality in hedgehog (*Erinacues europaeus*). *J. Zool.* **1977**, *182*, 162–164. [CrossRef]

Kristiansson, H. Population Variables and Causes of Mortality in a Hedgehog (*Erinaceus europaeus*) Population in Southern Sweden. *J. Zool.* **1990**, *220*, 391–404. [CrossRef]

Rasmussen, S.L.; Kalliokoski, O.; Dabelsteen, T.; Abelson, K. An Exploratory Investigation of Glucocorticoids, Personality and Survival Rates in Wild and Rehabilitated Hedgehogs (*Erinaceus europaeus*) in Denmark. *BMC Ecol. Evo.* **2021**, *21*, 96. [CrossRef]

Sæther, H.M. Overlevelse og Tidlig Spredning hos Juvenile Piggsvin (*Erinaceus europaeus*). Master's Thesis, Norges Teknisk-Naturvitenskapelige Universitet, Trondheim, Norway, 1997.

Anděra, M.; Gaisler, J. *SavciČeskéRepubliky: Popis, Rozšíření, Ekologie, Ochrana = Mammals of the Czech Republic: Description, Distribution, Ecology, and Protection*, 1st ed.; Academia: Praha, Czech Republic, 2012.

Jackson, D.B. The breeding biology of introduced hedgehogs (*Erinaceus europaeus*) on a Scottish Island: Lessons for population control and bird conservation. *J. Zool.* **2006**, *268*, 303–314. [CrossRef]

Pettett, C.E.; Moorhouse, T.P.; Johnson, P.J.; Macdonald, D.W. Factors Affecting Hedgehog (*Erinaceus europaeus*) Attraction to Rural Villages in Arable Landscapes. *Eur. J. Wildl. Res.* **2017**, *63*, 54. [CrossRef]

Wright, P.G.R.; Coomber, F.G.; Bellamy, C.C.; Perkins, S.E.; Mathews, F. Predicting Hedgehog Mortality Risks on British Roads Using Habitat Suitability Modelling. *PeerJ* **2020**, *7*, e8154. [CrossRef]

Burroughes, N.D.; Dowler, J.; Burroughes, G. Admission and survival trends in hedgehogs admitted to RSPCA Wildlife Wildlife rehabilitation centres. *Proc. Zool. Soc.* **2021**, *74*, 198–204. [CrossRef]

Crespo Martínez, J.; IzquierdoRosique, A.; SurrocaRoyo, M. Causes of admission and final dispositions of hedgehogs admitted to three wildlife rehabilitation centers in Eastern Spain. *Hystrix* **2014**, *25*, 107–110. [CrossRef]

Lukešová, G.; Voslarova, E.; Vecerek, V.; Vucinic, M. Trends in intake and outcomes for european hedgehog (*Erinaceus europaeus*) in the Czech rescue centers. *PLoS ONE* **2021**, *16*, e0248422. [CrossRef]

Mullineaux, E.; Keeble, E. *BSAVA Manual of Wildlife Casualties*, 2nd ed.; Bexton, S., Ed.; British Small Animal Veterinary Association: Quedgeley, UK, 2016.

Woods, M.; Mcdonald, R.A.; Harris, S. Predation of Wildlife by Domestic Cats Felis Catus in Great Britain: Predation of Wildlife by Domestic Cats. *Mamm. Rev.* **2003**, *33*, 174–188. [CrossRef]

Hanson, M.; Hollingshead, N.; Schuler, K.; Siemer, W.F.; Martin, P.; Bunting, E.M. Species, Causes, and Outcomes of Wildlife Rehabilitation in New York State. *PLoS ONE* **2021**, *16*, e0257675. [CrossRef]

Molina-López, R.A.; Mañosa, S.; Torres-Riera, A.; Pomarol, M.; Darwich, L. Morbidity, outcomes and cost-benefit analysis of wildlife rehabilitation in Catalonia (Spain). *PLoS ONE* **2017**, *12*, e0181331. [CrossRef]

26. Breed, D.; Meyer, L.C.R.; Steyl, J.C.A.; Goddard, A.; Burroughs, R.; Kohn, T.A. Conserving wildlife in a changing wor[l]. Understanding capture myopathy-a malignant outcome of stress during capture and translocation. *Conserv. Physiol.* **201**, 7, coz027. [CrossRef]

27. Stocker, L. *Practical Wildlife Care*, 2nd ed.; Wiley-Blackwell: Hoboken, NJ, USA, 2013.

28. Paci, G.; Bagliacca, M.; Lavazza, A. Stress Evaluation in Hares (*Lepus europaeus Pallas*) Captured for Traslocation. *Ital. J. Anim. S* **2006**, *5*, 175–181. [CrossRef]

29. Hrdy, S.B. Infanticide among animals: A review, classification, and examination of the implications for the reproductive strategi of females. *Ethol. Sociobiol.* **1979**, *1*, 13–40. [CrossRef]

30. Smith, A.T.; Johnston, C.H.; Alves, P.C.; Hackländer, K. *Lagomorphs: Pikas, Rabbits, and Hares of the World*; Smith, A.T., Ed.; Joh Hopkins University Press: Baltimore, MD, USA, 2018.

31. Landes, E.; Zentek, J.; Wole, P.; Kamphues, J. Investigation of Milk Composition in Hedgehogs. *J. Anim. Physiol. Anim. Nutr.* **199** 80, 179–184. [CrossRef]

32. Gimmel, A.; Eulenberger, U.; Liesegang, A. Feeding the European Hedgehog (*Erinaceus europaeus* L.)-Risks of Commercial Die for Wildlife. *J. Anim. Physiol. Anim. Nutr.* **2021**, *105*, 91–96. [CrossRef]

33. Robinson, I.; Routh, A. Veterinary care of the hedgehog. *Practice* **1999**, *21*, 128–137. [CrossRef]

34. Yarnell, R.W.; Surgey, J.; Grogan, A.; Thompson, R.; Davies, K.; Kimbrough, C.; Scott, D.M. Should Rehabilitated Hedgehogs Released in Winter? A Comparison of Survival, Nest Use and Weight Change in Wild and Rescued Animals. *Eur. J. Wildl. R* **2019**, *65*, 6. [CrossRef]

35. Morgan, K.N.; Tromborg, C.T. Sources of Stress in Captivity. *Appl. Anim. Behav. Sci.* **2007**, *102*, 262–302. [CrossRef]

36. South, K.E.; Haynes, K.; Jackson, A.C. Hibernation patterns of the european hedgehog, Erinaceus Europaeus, at a Cornish Resc Centre. *Animals* **2020**, *10*, 1418. [CrossRef]

37. Morris, P.A.; Warwick, H. A study of rehabilitated juvenile hedgehogs after release into the wild. *Anim. Welf.* **1994**, *3*, 163–1 [CrossRef]

38. Morris, P. An estimate of the minimum body weight necessary for hedgehogs (*Erinaceus europaeus*) to survive hibernation. *J. Zo* **1984**, *203*, 291–294.

39. Bearman-Brown, L.E.; Baker, P.J.; Scott, D.; Uzal, A.; Evans, L.; Yarnell, R.W. Over-winter survival and nest site selection of t West-European Hedgehog (*Erinaceus europaeus*) in arable dominated landscapes. *Animals* **2020**, *10*, 1449. [CrossRef]

40. Molony, S.E.; Dowding, C.V.; Baker, P.J.; Cuthill, I.C.; Harris, S. The effect of translocation and temporary captivity on wildli rehabilitation success: An experimental study using European hedgehogs (*Erinaceus europaeus*). *Biol. Conserv.* **2006**, *130*, 530–5; [CrossRef]

41. Jensen, A.B. Overwintering of European hedgehogs Erinaceus europaeus in a Danish rural area. *Acta Theriol.* **2004**, *49*, 145–1[[CrossRef]

 *animals*

rticle

Hematology, Biochemistry, and Protein Electrophoresis Reference Intervals of Western European Hedgehog (*Erinaceus europaeus*) from a Rehabilitation Center in Northern Portugal

Sofia Rosa [1], Ana C. Silvestre-Ferreira [2,3,4,5], Roberto Sargo [2,3,4,5], Filipe Silva [2,3,4,5] and Felisbina Luísa Queiroga [2,3,4,5,*]

[1] School of Life and Environmental Sciences (ECVA), University of Trás-os-Montes and Alto Douro (UTAD), Quinta dos Prados, 5001-801 Vila Real, Portugal
[2] Department of Veterinary Science, School of Agrarian and Veterinary Sciences (ECAV), University of Trás-os-Montes and Alto Douro (UTAD), Quinta dos Prados, 5001-801 Vila Real, Portugal
[3] Animal and Veterinary Research Centre (CECAV), University of Trás-os-Montes and Alto Douro (UTAD), Quinta dos Prados, 5001-801 Vila Real, Portugal
[4] Associated Laboratory for Animal and Veterinary Science (AL4AnimalS), University of Trás-os-Montes and Alto Douro (UTAD), Quinta dos Prados, 5001-801 Vila Real, Portugal
[5] Veterinary Teaching Hospital, University of Trás-os-Montes and Alto Douro (HVUTAD), Quinta dos Prados, 5001-801 Vila Real, Portugal
* Correspondence: fqueirog@utad.pt

Citation: Rosa, S.; Silvestre-Ferreira, A.C.; Sargo, R.; Silva, F.; Queiroga, F.L. Hematology, Biochemistry, and Protein Electrophoresis Reference Intervals of Western European Hedgehog (*Erinaceus europaeus*) from a Rehabilitation Center in Northern Portugal. *Animals* 2023, 13, 1009. https://doi.org/10.3390/ani13061009

Academic Editors: Anne Berger, Nigel Reeve and Sophie Lund Rasmussen

Received: 1 February 2023
Revised: 6 March 2023
Accepted: 7 March 2023
Published: 10 March 2023

Simple Summary: The Western European hedgehog (*Erinaceus europaeus*) is an insectivorous mammal with a wide geographic distribution. Owing mostly to climate changes and anthropogenic pressures, a considerable number of hedgehogs now live in urban areas close to humans, where they are exposed to contaminants and biological agents that may result in disease with the correspondent hematological and biochemical alterations. Hedgehogs can work as bioindicators to environmental pollution and host multiple zoonotic agents, making them relevant for One Health studies. Thus, it is essential to deepen the knowledge on this species and calculate reference intervals for the usual hematological and biochemical parameters. This would make it possible to recognize the "normal" and identify the "disease". In this study, some significant differences were evident, especially when comparing age groups (juveniles versus adults), showing the relevance of further investigations in this species.

Abstract: The Western European hedgehog (*Erinaceus europaeus*) can work as a bioindicator of environmental pollution and be a host for multiple zoonotic agents, making it relevant in terms of One Health studies. It is essential to deepen the knowledge on this species and calculate reference intervals (RIs) for the usual hematological and biochemical parameters. For this retrospective study (2017–2022), the archives of the Clinical Pathology Laboratory (LPC) of University of Trás-os-Montes and Alto Douro (UTAD) Veterinary Teaching Hospital were analyzed. Data of hematology, clinical biochemistry, and protein electrophoresis from 37 healthy hedgehogs of the Wild Animal Rehabilitation Center at UTAD, Northern Portugal, were included. It was possible to calculate RIs for almost all of the variables in the study, using Reference Value Advisor V2.1. Moreover, sex and age effects were investigated: alkaline phosphatase ($p = 0.012$, higher in males); total proteins ($p = 0.034$, higher in adults); mean cell volume ($p = 0.007$) and mean corpuscular hemoglobin ($p = 0.010$) (both higher in juveniles); and red blood cell distribution width ($p = 0.021$, higher in adults). Our study allowed for the first time to define RIs for a population of hedgehogs in Portugal, having a potentially relevant impact on species conservation and in the human–animal health interface.

Keywords: Western European hedgehog; *Erinaceus europaeus*; hematology; biochemistry; protein electrophoresis; reference intervals

1. Introduction

The Western European hedgehog, *Erinaceus europaeus*, belongs to the mammal order Eulipotyphla, family Erinaceidae, subfamily Erinaceinae, and genera *Erinaceus* [1]. There are two species of hedgehog in Europe, *Erinaceus europaeus* and *Erinaceus roumanicu*. *E. europaeus* may be found in western and central Europe, including Britain, the Mediterranean Islands, southern Scandinavia, and into Estonia and northern Russia [2,3]. It is also the most common one in Portugal [4,5].

The species *E. europaeus* is on the International Union for the Conservation of Nature (IUCN) and Red Book of Vertebrates of Portugal (LVVP) red list, as being "least concern" because it is common and abundant throughout its wide range [5,6]. However, in the last decades, there are registers of a decrease in the number of individuals of this species [7], which can be justified by several factors: the Western European hedgehog has some natural predators that pose a threat to the survival of the species (badgers (*Meles meles*) are the most significant ones) [8,9]; as the hedgehog occupies agricultural areas, it is frequently exposed to poisoning by pesticides and rodenticides [10]; the transformation and fragmentation of its habitat, as well as climate changes, affect its survival [11]; they are one of the vertebrates that frequently suffer mortality owing to road traffic [7]; and finally, they harbor a wide variety of different parasites and pathogens [8,12,13].

Owing to different pressure factors, hedgehogs are moving close to humans and urban centers, adapting to new habitats, food resources (as pet food), and refugia (as public and private gardens) [11,14–16], where they are exposed to contaminants and biological agents, which may result in disease with the correspondent hematological and biochemical alterations [17]. Furthermore, hedgehogs' ecological and feeding habits, as well as their high population densities and repeated contacts with wild and domestic animals and humans, make this species a possible sentinel for a One Health approach, mainly owing to its possible involvement in the ecology of potentially emerging pathogens [18,19], such as endoparasites (*Crenosoma striatum*, *Capillaria aerophila* (syn. *Eucoleus aerophilus*), *Capillaria* spp., coccidia, *Cryptosporidium* spp., *Brachylaemus* spp., and *Capillaria hepatica*) [12]. Published systematic reviews show that *E. europaeus* may harbor zoonotic pathogens and that the species can play an important role in the epidemiology of various zoonotic infections. The prevalence of zoonotic agents in hedgehogs, from both urban and rural habitats, is of major concern because there is a high probability of contact with humans and companion animals. Recently, several studies have shown that *E. europaeus* can harbor methicillin-resistant *Staphylococcus aureus* (MRSA) and that this resistance can predate the human discovery of antibiotics [20,21]. For this reason, studying local *E. europaeu* populations is highly relevant [22,23].

Reference intervals (RIs) are ranges calculated from a group or population of healthy individuals of a given species and are most widely used as a medical decision-making tool, serving as the basis of laboratory testing, differentiating whether or not a patient is healthy [24]. In order to determine RIs, the American Society of Veterinary Clinical Pathology (ASVCP) established guidelines with specific veterinary recommendations where it is necessary to define the population of interest as well as criteria to confirm the health status of selected individuals [25]. The studies on hematological and biochemical profiles for the species *E. europaeus* are scarce [17,26,27]. The most recent study, performed in a rehabilitation center in Italy (2014), evaluated hematological and biochemical profiles in the species, creating their own reference intervals [17].

In Portugal, as far as we know, there are no records on the subject. Therefore, it was our purpose to carry out a study on the hematological and biochemical (routine and electrophoretic) profiles of *E. europaeus* in a population of healthy individuals from a wildlife rehabilitation center in Northern Portugal, in order to create reference intervals and thus contributing to species conservation and a better understanding of the impact in the human–hedgehogs health interface.

2. Materials and Methods

This retrospective study was carried out at the Clinical Pathology Laboratory of the Veterinary Teaching Hospital of the University of Trás-os-Montes and Alto Douro (LPC-HVUTAD), located in the city of Vila Real, Portugal. A total of 94 registers of *E. europaeus* were identified from the LPC archive, corresponding to the routine post-quarantine evaluation of admitted individuals. The clinical files were reviewed to access the health status of animals on the date of blood collection. Animals that had normal physical examinations, no alterations on the diagnostic exams (whole body X-rays, blood collection, and coprological examination), and no signs of illness until the date of release were considered healthy. By reviewing clinical files, it was possible to identify 37 cases of healthy animals, thus 57 cases were discarded owing to several causes of disease ($n = 10$) or unavailable information ($n = 47$) (Figure 1). All of the hedgehogs included in this study originated from the region of northern Portugal, including the Douro River basin, and were received at the Wild Animal Rehabilitation Center of the Veterinary Teaching Hospital of UTAD (CRAS-HVUTAD) between 1 March 2017 and 31 August 2022. Age was evaluated at admission by a veterinarian, based on morphometrics, after Haigh et al. (2014) [28]. Only independent juveniles and adult animals were admitted to this study, being assigned on the day of blood collection to one of the two categories [28]. Sex was determined visually by the observation of the external genital organs. The entire process of animal capture and sample collection was carried out by veterinarians and qualified auxiliary personnel. Only animals considered healthy were enrolled. For this study, no ethical approvals were required, as all blood samples were routinely collected for official diagnostic and monitoring purposes and subsequently made available to this retrospective study.

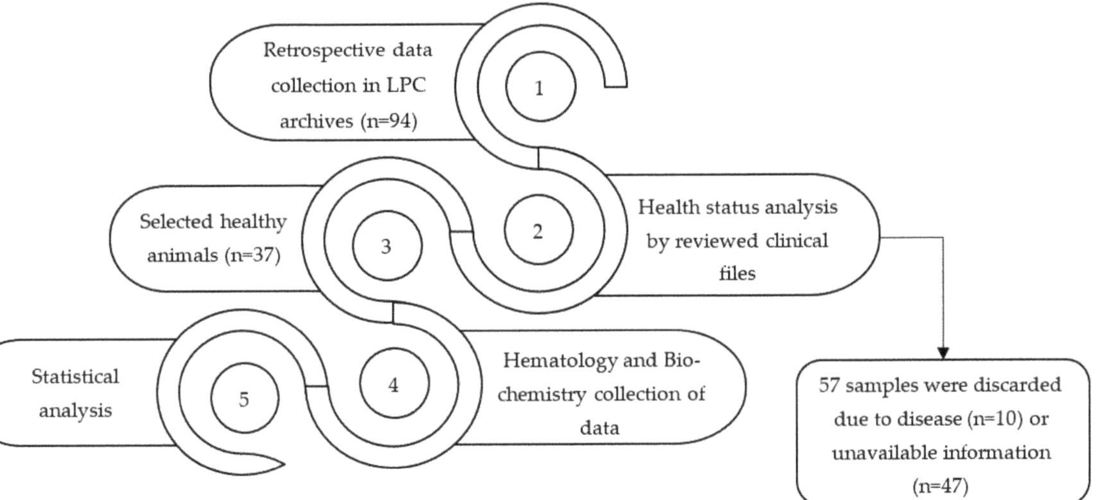

Figure 1. Flowchart of the retrospective analysis of this study.

2.1. Sample Collection

Animals were anesthetized by mask administration of 5% isoflurane in an oxygen flow of 2 L/min. After the loss of reflexes, the isoflurane concentration was diminished to 2% and maintained during the physical exam and sample collection. Blood samples were obtained through a cranial vein cava punction with 25 G needles and 1 mL syringes, always with the animal under anesthesia. Blood was transferred into 0.5 mL lithium heparin tubes (FL Medical). A complete record was kept for each animal, including sex, age group, and date of capture. After collection, all samples were transported to the LPC-HVUTAD for analysis.

2.2. Sample Treatment and Processing

For hematology, blood samples were analyzed in the ProCyte Dx (IDEXX) hematolog
analyzer and the following parameters were determined: red blood cells (M/μL), hemat
ocrit (%), hemoglobin (g/dL), mean cell volume (fL), mean corpuscular hemoglobin (pg
mean corpuscular hemoglobin concentration (g/dL), red blood cell distribution width (%
reticulocyte percent, reticulocyte count (K/μL), white blood cell (K/μL), neutrophil percer
lymphocyte percent, monocyte percent, eosinophil percent, basophil percent, neutroph
count (K/μL), lymphocyte count (K/μL), monocyte count (K/μL), eosinophil count (K/μl
basophil count (K/μL), platelet count (K/μL), mean platelet volume (fL), and plateletcr
(%). There is no predefined menu for hedgehogs in this analyzer; in addition, the optic
used was "others", with no associated reference ranges for any parameter.

Plasma samples were obtained through routine centrifugation at the SELECTA Centromi
BLT centrifuge, for 5 min at 2618× *g*. Then, it was transferred into 2 mL aliquots, proper.
identified, and kept at −20 °C until analysis. According to laboratory standards, strong
hemolyzed or lipemic samples were rejected. For the biochemical analysis, the DiaSy
Respons920 Clinical Chemistry Analyzer was used and the following parameters wei
determined: glucose (mg/dL), total proteins (g/dL), albumin (g/dL), alanine aminotrar
ferase (U/L), alkaline phosphatase (U/L), creatinine (mg/dL), urea (mg/dL), phosphort
(mg/dL), total calcium (mg/dL), cholesterol (mg/dL), triglycerides (mg/dL), gamma-G
(U/L), globulins (g/dL), aspartate aminotransferase (U/L), total bilirubin (mg/dL), sodiu.
(mmol/L), potassium (mmol/L), and chloride (mmol/L).

The remaining plasma was then used for protein electrophoresis on the Elephor&
automatic electrophoresis analyzer, following the manufacturer's instructions.

2.3. Data Collection Methodology, Statistical Analysis, and Reference Intervals

Data were registered in an Excel sheet and RIs were established according to th
ASVCP guidelines using the Reference Value Advisor V2.1 program [16].

To investigate the influence of sex and age in the analyzed parameters, a statistic.
analysis was performed using the SPSS program, version 27. The Shapiro–Wilk test wa
performed for testing samples' normality and then samples were subjected to analysis
variance (ANOVA) to investigate the influence of sex and age. A value of $p < 0.05$ wa
considered significant.

3. Results

3.1. Descriptive Analysis of the Total Number of Animals

Regarding sex, information was only available for 35 of the 37 animals: 13 (35.14%
males and 22 (59.46%) females; information was missing in 2 cases (5.41%).

For age, information was available in 34 cases of the 37 cases: 20 (54.05%) juvenil
and 14 (37.84%) adults; it was not possible to obtain the respective information for th
remaining 3 (8.11%) samples.

3.2. Reference Intervals

For hematology, Table 1 describes the RIs of the evaluated parameters in the tot.
sample (*n* = 37).

Table 1. Reference intervals of hematological parameters in 37 healthy animals of the speci
Erinaceus europaeus.

Parameters	N	Mean ± SD	Median	Min–Max	RI	LRL 90% CI	URL 90% CI
RBC (M/μL)	37	7.10 ± 1.50	7.20	4.07–11.01	4.1–10.2	3.5–4.8	9.5–10.9
HCT (%)	37	30.20 ± 5.30	29.40	17.30–41.40	19.4–41.1	16.7–21.9	38.6–43.6
HGB (g/dL)	37	10.20 ± 1.80	10.00	5.70–13.90	6.6–13.8	5.7–7.5	13.0–14.6

Table 1. *Cont.*

Parameters	N	Mean ± SD	Median	Min–Max	RI	LRL 90% CI	URL 90% CI
MCV (fL)	37	42.90 ± 5.40	40.90	35.50–57.50	34.3–55.8	33.7–35.2	50.5–59.7
MCH (pg)	37	14.40 ± 1.70	14.00	11.70–19.20	12.0–19.1	11.8–12.3	17.4–20.9
MCHC (g/dL)	37	33.60 ± 1.30	33.60	31.40–36.70	30.9–36.4	30.2–31.6	35.8–37.0
RDW (%)	37	29.00 ± 3.20	28.90	24.20–37.80	22.5–35.6	21.0–24.1	34.0–37.1
%RETIC	37	3.60 ± 3.40	2.40	0.20–14.00	*	*	*
RETIC (K/μL)	37	222.70 ± 168.30	183.10	10.60–645.30	11.5–705.7	2.6–29.0	549.2–892.2
WBC (K/μL)	37	9.10 ± 3.20	8.60	2.22–15.31	2.5–15.7	1.0–4.0	14.2–17.3
%NEU	37	52.70 ± 14.10	54.50	5.00–74.20	23.8–81.6	16.1–30.6	74.4–88.6
%LYM	37	37.20 ± 11.00	36.20	19.30–60.10	14.6–59.9	9.6–20.0	54.5–65.1
%MONO	37	7.90 ± 6.20	6.70	2.80–42.30	3.5–24.4	3.1–4.2	14.3–44.9
%EOS	37	1.80 ± 1.80	1.00	0.00–6.50	2.4–26.9	2.0–3.0	14.7–75.2
%BASO	37	0.30 ± 0.40	0.20	0.00–1.00	*	*	*
NEU (K/μL)	37	4.90 ± 2.30	4.50	0.11–9.89	0.8–10	0.2–1.5	8.2–11.5
LYM (K/μL)	37	3.30 ± 1.40	3.30	1.15–6.28	0.4–6.3	0.0–0.9	5.5–6.8
MONO (K/μL)	37	0.60 ± 0.30	0.60	0.22–1.40	0.2–1.4	0.2–0.3	1.1–1.6
EOS (K/μL)	37	0.20 ± 0.20	0.10	0.00–0.48	*	*	*
BASO (K/μL)	37	0.00 ± 0.00	0.00	0.00–0.11	*	*	*
PLT (K/μL)	37	268.30 ± 129.90	237.00	6.00–620.00	31.3–567.1	3.2–71.5	472.2–660.3
MPV (fL)	33	15.60 ± 1.10	15.50	12.40–17.80	13.2–17.9	12.7–13.9	17.4–18.4
PCT (%)	33	0.40 ± 0.20	0.40	0.05–0.91	0.1–0.8	0.0–0.2	0.7–0.9

SD, standard deviation; Min, minimum; Max, maximum; LRL, lower reference limit; URL, upper reference limit; CI, confidence interval. * Non-computable. RBC, red blood cells; HCT, hematocrit; HGB, hemoglobin; MCV, mean cell volume; MCH, mean corpuscular hemoglobin; MCHC, mean corpuscular hemoglobin concentration; RDW, red blood cell distribution width; %RETIC, reticulocyte percent; RETIC, reticulocyte count; WBC, white blood cell; %NEU, neutrophil percent; %LYM, lymphocyte percent; %MONO, monocyte percent; %EOS, eosinophil percent; %BASO, basophil percent; NEU, neutrophil count; LYM, lymphocyte count; MONO, monocyte count; EOS, eosinophil count; BASO, basophil count; PLT, platelet count; MPV, mean platelet volume; PCT, plateletcrit.

For routine plasma biochemistry and protein electrophoresis, the results are presented in Table 2 ($n = 37$).

Table 2. Reference intervals of biochemical parameters in 37 healthy animals of the species *Erinaceus europaeus*.

Parameters	N	Mean ± SD	Median	Min–Max	RI	LRL 90% CI	URL 90% CI
Glucose (mg/dL)	21	108.50 ± 24.10	109.00	44.20–141.91	57.0–160.0	43.0–72.2	144.0–175.3
TP (g/dL)	34	5.80 ± 1.20	5.90	3.55–8.59	3.4–8.3	2.8–3.9	7.6–8.9
Albumin (g/dL)	33	3.30 ± 0.60	3.20	2.29–4.46	2.1–4.5	1.8–2.4	4.2–4.8
ALT (U/L)	33	129.20 ± 61.20	125.00	36.40–298.00	*	*	*
ALP (U/L)	34	74.80 ± 48.00	59.50	18.00–203.10	19.2–217.8	17.8–22.7	166.3–281.4
Creatinine (mg/dL)	33	0.40 ± 0.20	0.30	0.10–0.77	0.1–0.9	0.1–0.1	0.7–1.0
Urea (mg/dL)	23	76.50 ± 42.70	68.00	16.20–215.40	*	*	*
Phosphorus (mg/dL)	33	7.70 ± 2.40	7.40	4.29–15.01	2.7–12.7	1.6–3.9	11.4–14.0
T. Calcium (mg/dL)	33	9.40 ± 1.60	9.90	2.55–11.35	5.6–11.5	1.4–7.4	11.0–11.7

Table 2. *Cont.*

Parameters	N	Mean ± SD	Median	Min–Max	RI	LRL 90% CI	URL 90% CI
Cholesterol (mg/dL)	15	134.90 ± 50.00	132.00	24.00–224.00	*	*	*
TG (mg/dL)	14	55.70 ± 14.70	56.00	18.00–79.00	22.9–88.5	10.6–35.7	75.1–100.9
Gamma-GT (U/L)	18	19.80 ± 17.90	8.70	2.60–56.00	1.8–151.8	*	*
Globulins (g/dL)	23	2.30 ± 1.00	2.10	0.00–3.97	0.4–4.7	0.1–0.9	3.6–5.9
AST (U/L)	9	19.60 ± 5.90	19.00	13.90–30.70	10.8–38.7	9.6–13.4	27.0–53.0
T. Bilirubin (mg/dL)	10	0.00 ± 0.00	0.00	0.01–0.03	*	*	*
Sodium (mmol/L)	20	142.90 ± 5.50	142.70	127.00–152.00	129.8–153.1	124.1–135.1	150.3–155.6
Potassium (mmol/L)	20	3.90 ± 0.70	3.80	2.15–5.44	2.5–5.4	2.0–2.9	4.9–5.9
Chloride (mmol/L)	20	110.20 ± 4.60	109.40	104.00–123.40	100.3–120.1	97.2–103.6	117.0–123.2
Protein electrophoresis in 12 cases							
Albumin (g/dL)	12	2.60 ± 0.40	2.50	1.91–3.14	1.7–3.5	1.4–2.1	3.1–3.8
α1-Globulin (g/dL)	12	0.60 ± 0.10	0.60	0.42–0.94	0.3–0.9	0.2–0.4	0.8–1.1
α2-Globulin (g/dL)	12	0.60 ± 0.10	0.60	0.32–0.76	0.2–0.9	0.1–0.4	0.8–1.0
β-Globulin (g/dL)	12	1.20 ± 0.50	1.10	0.70–2.37	0.6–3.4	0.5–0.8	1.8–8.9
γ-Globulin (g/dL)	12	0.20 ± 0.10	0.10	0.03–0.49	0.0–0.1	0.0–0.0	0.4–2.5

SD, standard deviation; Min, minimum; Max, maximum; LRL, lower reference limit; URL, upper reference limit; CI, confidence interval. * Non-computable. TP, total proteins; ALT, alanine aminotransferase; ALP, alkaline phosphatase; T. Calcium, total calcium; TG, triglycerides; AST, aspartate aminotransferase; T. Bilirubin, total bilirubin.

In Figure 2, an example of the electrophoretic pattern of a healthy *E. europaeus* individual is presented. Medium values for each category are expressed in Table 2.

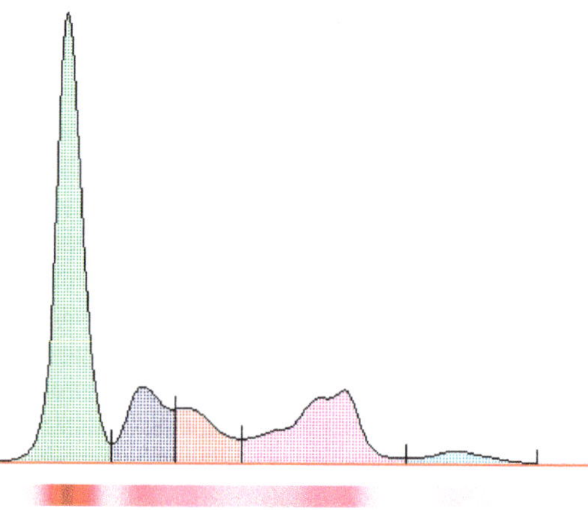

Figure 2. Electrophoretic pattern of a healthy individual of the species *Erinaceus europaeus*, respectively. Green—albumin; for globulins: dark blue—α1; red—α2; pink—β; light blue—γ.

3.3. Influence of Sex and Age

Based on the Shapiro–Wilk test, it was determined that the parameters followed a normal distribution ($p > 0.05$).

Considering the sex of the animals, only ALP revealed a statistically significant difference between groups, being higher in males. No other parameters revealed a statistically significant association with sex.

Considering the animals' age, regarding hematology, statistically significant differences were observed for MCV, MCH (both higher in juveniles), and RDW (higher in adults). Regarding clinical biochemistry, a statistically significant difference was observed for T.

that was higher in adults. No other parameters revealed a statistically significant association with the age.

The results described above are expressed in Table 3.

Table 3. Influence of sex and age on hematological and biochemical parameters in the species *Erinaceus europaeus*.

Category	Parameters	Mean ± SD	Mean ± SD	*p*
Hematology				
Age		Juveniles	Adults	
	MCV	44.95 ± 5.79	39.65 ± 3.88	0.007
	MCH	15.02 ± 1.72	13.49 ± 1.30	0.010
	RDW	27.99 ± 2.79	30.63 ± 3.41	0.021
Clinical Biochemistry				
Sex		Females	Males	
	ALP	58.50 ± 36.86	101.79 ± 55.84	0.012
Age		Juveniles	Adults	
	TP	5.47 ± 1.11	6.51 ± 1.34	0.034

4. Discussion

The Western European hedgehog (*Erinaceus europaeus*) is an insectivorous mammal widely distributed across Europe. A high number of hedgehogs live around urban areas, close to humans, where they are exposed to contaminants and biological agents that may affect their health status [17]. Moreover, hedgehogs can work as bioindicators and host multiple zoonotic agents, making them relevant in terms of a One Health approach [18,19]. The determination of RIs for usual hematological and biochemical parameters in wild healthy animals is the first step to identify potentially ill individuals. This is an important approach because those parameters, when altered, may be related to the presence of toxic or zoonotic agents in the environment [29,30]. That is the case of heavy metals such as zinc (Zn) and cadmium (Cd) that affect the liver and kidneys and can be detected by elevated levels of the biochemical parameters ALT and creatinine, respectively [31]. Identifying alterations in hedgehogs will allow the identification of other animal populations at risk of suffering the same type of exposure. In this way, the importance of the present study becomes clear in the context of One Health [24].

Existing works on RIs for hematology in this species are scarce. To the best of our knowledge, our study reports RIs for a wide range of parameters in comparison with previous investigations [17,27] and, to the best authors' knowledge, is the first to calculate RIs for RDW, MPV, and PCT in *E. europaeus*.

In our study, considering age, statistically significant differences were found in MCV, MCH, and RDW. In MCV and MCH, the mean value was higher in juveniles than in adults and, in RDW, it was higher in adults compared with juveniles. In the study by Lewis et al. (2002) [27], a statistically significant difference was found only for HCT. The fact that the present study was carried out 2 decades after that of Lewis et al. [27], and used different instruments, may help to justify the differences found. However, the authors cannot rule out the possibility that environmental changes, exposure to biological or toxic agents, as well as access to new food sources due to repeated contact with humans may also explain the differences found [18,19]. Another justification for the differences found has to do with the genetic variability that occurs in animals of the same species in different geographic locations, as recently confirmed in a study carried out in different regions of the Iberian Peninsula [32].

The study by Rossi et al. (2014) [17] only included juvenile animals, so it was not possible to make this type of comparison.

Considering sex, in our work, there were no statistically significant differences f[ ]
the analyzed parameters. The same observation was described by Lewis et al. (2002) [2[ ]
However, Rossi et al. (2014) [17] obtained a difference for MCV, but it was considered n[ ]
biologically relevant, and justified as a possible bias, a consequence of sex overlapping [ ]
their sample.

Studies describing RIs for biochemical parameters for *E. europaeus* are also scarce [17,2[ ]
as previous mentioned for hematology studies. The study by Rossi et al. (2014) [1[ ]
describes RI values for some biochemical parameters in this species. However, Rossi an[ ]
collaborators analyzed fewer parameters when compared with our study. Additional[ ]
Rossi and collaborators did not investigate the age effect (because only juveniles we[ ]
included in their study), nor the sex effect, which is comprehensible given the fact tha[ ]
using only juveniles, the animals did not reach sexual maturity. The study by Larsen an[ ]
Tönder (1967) describes values of electrophoresis, but without RI determination.

To the best of our knowledge, our study is the first to describe RI values for AST, tot[ ]
bilirubin, sodium, potassium, and chloride for the species *E. europaeus*. It is important [ ]
note that this is the first time, in the literature, that information about the electrolyte pan[ ]
is made available for this species.

Throughout our work, several important biochemical parameters were studied, pr[ ]
viding valid information at the level of different organs and systems such as the liver an[ ]
kidney, among others. Additionally, an investigation into the age and sex effect on the an[ ]
lyzed biochemical parameters was also performed. Considering age, statistically significa[ ]
differences were found in TP, where the mean value was higher in adults compared wi[ ]
juveniles. TP refers to all proteins in plasma that are made up of albumin and globulir[ ]
Higher TP values in adults may be related to a greater number of globulins owing to a mo[ ]
developed immune system in adults compared with juveniles [33]. Although, in our stud[ ]
we did not find a significant difference for globulin values between adults and juvenile[ ]
these values were in fact higher in adults, supporting this hypothesis.

Considering the sex effect, statistically significant differences were found in AL[ ]
where the mean value was higher in males than in females. Typically, a higher ALP valu[ ]
is related to the presence of liver, bone, and other diseases [34]. However, there are r[ ]
investigations in this species or others that indicate an increase or decrease in this paramet[ ]
associated with sex. Therefore, we can conclude that the fact that this value is higher i[ ]
males in our study is a coincidence related to the studied population and could be due [ ]
diet or some distinct environmental conditions [35].

Concerning protein electrophoresis, in a study in Bergen, Norway, by Larsen an[ ]
Tönder (1967) [26], a paper protein electrophoresis analysis was conducted, in a populatic[ ]
of 13 animals of the species *E. europaeus*. Contrary to our study, serum was used instea[ ]
of plasma. Because of the differences in sample type and methodology, it was impossib[ ]
for us to compare results. There are no more protein electrophoresis studies conducted i[ ]
this species.

Although all animals included in this study were wildlife casualties, the samples use[ ]
were collected only after a quarantine period, which would vary between 15 days an[ ]
3 months, depending on the cause of admittance. This time lapse between admittance an[ ]
blood collection would give animals time to recover from illness, but also subject the[ ]
to stress derived from captivity. Rasmussen et al. (2021) [36] showed that rehabilitate[ ]
hedgehogs had higher levels of endogenous corticosterone when compared with wil[ ]
caught individuals. Although endogenous steroids can impact physical responses and lea[ ]
to unbalances, the act of blood collection is considered a stressor and will contribute [ ]
this endogenous response [37]. Changes in the hematocrit and leucocyte count have bee[ ]
identified in a large set of species. This stress leukogram, however, varies from a transito[ ]
leukopenia, like in the rabbit, leukocytosis, as in camelids, or it can have no noticeab[ ]
changes, as in rats [38]. It is also expectable to find increased blood levels of glucos[ ]
and ALP with stress [39]. Isoflurane anesthesia can have an impact on measured bloo[ ]

parameters as it is known to increase serum glucose, AST, urea, nitrogen, and creatinine levels in rabbits [40] and to decrease erythrocyte parameters in ferrets [38].

Several of the results presented in both the Western European hedgehog and other species disagree with those observed in our work, supporting the existing differences between species and justifying the relevance of performing further studies. In addition, differences between methods and laboratories can also influence the results [41]. Moreover, the effects of stress in captivity and during handling on the hedgehogs investigated should be taken into consideration when interpreting the results, as it could influence the results and, thereby, the RIs determined [36].

It should be noted that the effects of sex and age on hematological and biochemical parameters are not necessarily uniform across geographic locations. So, for further studies, the inclusion of several different locations in an assessment or a meta-analysis should be noted. Therefore, it is really important that investigations are carried out in distinct countries/regions for a clear understanding of normal values, and thus cooperate for the improvement of the species' conservative status.

5. Conclusions

The Western European hedgehog is a mammal with an extensive geographic distribution that, because of climatic changes and other aspects, began to live in urban areas near humans, where it is exposed to several anthropogenic pressures, suffering a progressive reduction in population. Hematological and biochemical RIs are scarce and there is a need to continue performing such types of investigations.

In this study, it was possible to establish reference intervals for a wide range of hematological and biochemical parameters covering the majority of the values used in the clinical practice. Moreover, we found out that the RIs obtained in this study do not always fit those previously published in the literature, for populations with different geographic locations, concluding the necessity to create appropriated RIs for the target population.

Author Contributions: Conceptualization, A.C.S.-F. and F.L.Q.; formal analysis, F.L.Q.; investigation, S.R.; validation, A.C.S.-F. and R.S.; resources, F.S. and R.S.; writing—original draft preparation, S.R.; writing—review and editing, S.R., A.C.S.-F. and F.L.Q.; supervision, F.L.Q. All authors have read and agreed to the published version of the manuscript.

Funding: This work was financed by National Funds (FCT/MCTES, Fundação para a Ciência e a Tecnologia and Ministério da Ciência, Tecnologia e Ensino Superior) under the project UIDB/CVT/00772/2020. The authors also want to thank the support received by project UIDB/00211/2020 and UIDB/04033/2020, from FCT/MCTES.

Institutional Review Board Statement: Ethical review and approval were waived for this study owing to its retrospective nature.

Informed Consent Statement: Not applicable.

Data Availability Statement: The data that support the findings of this retrospective study are available upon reasonable request to the authors.

Conflicts of Interest: The authors declare no conflict of interest. The funders had no role in the design of the study; in the collection, analyses, or interpretation of data; in the writing of the manuscript; or in the decision to publish the results.

References

Morris, P. *Hedgehog*; Collins: London, UK, 2018; pp. 1–404.

Smith, A.J. Husbandry and Nutrition of Hedgehogs. *Veter Clin. N. Am. Exot. Anim. Pract.* **1999**, *2*, 127–141. [CrossRef] [PubMed]

Paupério, J.; Vale-Gonçalves, H.M.; Cabral, J.A.; Mira, A.; Bencatel, J. Insetívoros. In *Atlas de Mamíferos de Portugal*; Bencatel, J., Álvares, F., Moura, A.E., Barbosa, A.M., Eds.; Universidade de Évora: Évora, Portugal, 2017; pp. 42–43.

Seddon, J.M.; Santucci, F.; Reeve, N.J.; Hewitt, G.M. DNA footprints of *European hedgehogs, Erinaceus europaeus* and *E. concolor*: Pleistocene refugia, postglacial expansion and colonization routes. *Mol. Ecol.* **2001**, *10*, 2187–2198. [CrossRef] [PubMed]

Cabral, M. *Livro Vermelho dos Vertebrados de Portugal*; Instituto de Conservação da Natureza: Lisboa, Portugal, 2005.

Temple, H.J.; Cuttelod, A. *The Status and Distribution of Mediterranean Mammals*; IUCN: Gland, Switzerland; Cambridge, UK, 2009.

7. Hernández, M.C. Erizo común—*Erinaceus europaeus* Linnaeus, 1758. In *Enciclopedia Virtual de los Vertebrados Españoles*; López, I Martin, J., Barja, I., Eds.; Museo Nacional de Ciencias Naturales: Madrid, Spain, 2020.

8. Pfäffle, M. *Influence of Parasites on Fitness Parameters of the European Hedgehog (Erinaceus europaeus)*; Karlsruhe Institute c Technology: Karlsruhe, Germany, 2015.

9. Young, R.P.; Davison, J.; Trewby, I.D.; Wilson, G.J.; Delahay, R.J.; Doncaster, C.P. Abundance of hedgehogs (*Erinaceus europaeus*) i relation to the density and distribution of badgers (*Meles meles*). *J. Zool.* **2006**, *269*, 349–356. [CrossRef]

10. Rasmussen, S.L.; Berg, T.B.; Martens, H.J.; Jones, O.R. Anyone Can Get Old—All You Have to Do Is Live Long Enougl Understanding Mortality and Life Expectancy in European Hedgehogs (*Erinaceus europaeus*). *Animals* **2023**, *13*, 626. [CrossRef]

11. Rasmussen, S.L.; Berg, T.B.; Dabelsteen, T.; Jones, O.R. The ecology of suburban juvenile European hedgehogs (*Erinaceus europaeu* in Denmark. *Ecol. Evol.* **2019**, *9*, 13174–13187. [CrossRef]

12. Rasmussen, S.L.; Hallig, J.; van Wijk, R.E.; Petersen, H.H. An investigation of endoparasites and the determinants of parasit infection in European hedgehogs (*Erinaceus europaeus*) from Denmark. *Int. J. Parasitol. Parasites Wildl.* **2021**, *16*, 217–227. [CrossRef]

13. Mariacher, A.; Santini, A.; Del Lesto, I.; Tonon, S.; Cardini, E.; Barone, A.; Eleni, C.; Fichi, G.; Perrucci, S. Endoparasite Infection of the European Hedgehog (*Erinaceus europaeus*) in Central Italy. *Animals* **2021**, *11*, 3171. [CrossRef]

14. Hubert, P.; Julliard, R.; Biagianti, S.; Poulle, M.-L. Ecological factors driving the higher hedgehog (*Erinaceus europeaus*) density i an urban area compared to the adjacent rural area. *Landsc. Urban Plan.* **2011**, *103*, 34–43. [CrossRef]

15. Poel, J.L.; Dekker, J.; Langevelde, F. Dutch hedgehogs *Erinaceus europaeus* are nowadays mainly found in urban areas, possibl due to the negative Effects of badgers *Meles meles*. *Wildl. Biol.* **2015**, *21*, 51–55. [CrossRef]

16. Gazzard, A.; Yarnell, R.W.; Baker, P.J. Fine-scale habitat selection of a small mammalian urban adapter: The West Europea hedgehog (*Erinaceus europaeus*). *Mamm. Biol.* **2022**, *102*, 387–403. [CrossRef]

17. Rossi, G.; Mangiagalli, G.; Paracchini, G.; Paltrinieri, S. Hematologic and biochemical variables of hedgehogs (*Erinaceus europaeu* after overwintering in rehabilitation centers. *Veter Clin. Pathol.* **2014**, *43*, 6–14. [CrossRef]

18. Delogu, M.; Cotti, C.; Lelli, D.; Sozzi, E.; Trogu, T.; Lavazza, A.; Garuti, G.; Castrucci, M.R.; Vaccari, G.; De Marco, M.A.; et a Eco-Virological Preliminary Study of Potentially Emerging Pathogens in Hedgehogs (*Erinaceus europaeus*) Recovered at a Wildlif Treatment and Rehabilitation Center in Northern Italy. *Animals* **2020**, *10*, 407. [CrossRef] [PubMed]

19. Baptista, C.J.; Seixas, F.; Gonzalo-Orden, J.; Oliveira, P. Can the European Hedgehog (*Erinaceus europaeus*) Be a Sentinel for On Health Concerns? *Biologics* **2021**, *1*, 61–69. [CrossRef]

20. Rasmussen, S.L.; Larsen, J.; Van Wijk, R.E.; Jones, O.R.; Berg, T.B.; Angen, O.; Larsen, A.R. European hedgehog (*Erinaceus europaeus*) as a natural reservoir of methicillin-resistant Staphylococcus aureus carrying mecC in Denmark *PLoS ONE* **2019**, *14*, e0222031. [CrossRef]

21. Larsen, J.; Raisen, C.L.; Ba, X.; Sadgrove, N.J.; Padilla-González, G.F.; Simmonds, M.S.J.; Loncaric, I.; Kerschner, H.; Apfalter, P Hartl, R.; et al. Emergence of methicillin resistance predates the clinical use of antibiotics. *Nature* **2022**, *602*, 135–141. [CrossRef]

22. Ruszkowski, J.; Hetman, M.; Turlewicz-Podbielska, H.; Pomorska-Mól, M. Hedgehogs as a Potential Source of Zoonoti Pathogens—A Review and an Update of Knowledge. *Animals* **2021**, *11*, 1754. [CrossRef]

23. Baptista, C.J.; Oliveira, P.A.; Gonzalo-Orden, J.M.; Seixas, F. Do Urban Hedgehogs (*Erinaceus europaeus*) Represent a Relevar Source of Zoonotic Diseases? *Pathogens* **2023**, *12*, 268. [CrossRef] [PubMed]

24. Horn, P.S.; Pesce, A.J. Reference intervals: An update. *Clin. Chim. Acta* **2003**, *334*, 5–23. [CrossRef]

25. Friedrichs, K.R.; Harr, K.E.; Freeman, K.P.; Szladovits, B.; Walton, R.M.; Barnhart, K.F.; Blanco-Chavez, J. ASVCP reference interva guidelines: Determination of de novo reference intervals in veterinary species and other related topics. *Veter Clin. Pathol.* **2012**, *4* 441–453. [CrossRef]

26. Larsen, B.; Tonder, O. Serum Proteins in the Hedgehog. *Acta Physiol. Scand.* **1967**, *69*, 262–269. [CrossRef] [PubMed]

27. Lewis, J.C.M.; Norcott, M.R.; Frost, L.M.; Cusdin, P. Normal haematological values of European hedgehogs (*Erinaceus europaeu* from an English rehabilitation centre. *Veter Rec.* **2002**, *151*, 567–569. [CrossRef]

28. Haigh, A.; Kelly, M.; Butler, F.; O'Riordan, R.M. Non-invasive methods of separating hedgehog (*Erinaceus europaeus*) age classe and an investigation into the age structure of road kill. *Acta Thériol.* **2014**, *59*, 165–171. [CrossRef]

29. Dowding, C.V.; Shore, R.F.; Worgan, A.; Baker, P.J.; Harris, S. Accumulation of anticoagulant rodenticides in a non-targe insectivore, the European hedgehog (*Erinaceus europaeus*). *Environ. Pollut.* **2010**, *158*, 161–166. [CrossRef] [PubMed]

30. Vermeulen, F.; Covaci, A.; D'Havé, H.; Brink, N.W.V.D.; Blust, R.; De Coen, W.; Bervoets, L. Accumulation of background levels o persistent organochlorine and organobromine pollutants through the soil–earthworm–hedgehog food chain. *Environ. Int.* **201** *36*, 721–727. [CrossRef]

31. Al-Rikabi, Z.G.K.; Al-Saffar, M.A.; Abbas, A.H. The Accumulative Effect of Heavy Metals on Liver and Kidney Functions. *Me Leg. Update* **2021**, *21*, 1114–1119. [CrossRef]

32. Araguas, R.-M.; Vidal, O.; García, S.; Sanz, N. Genetic diversity and population structure of the Western European hedgeho *Erinaceus europaeus*: Conservation status of populations in the Iberian Peninsula. *Mamm. Biol.* **2022**, *102*, 375–386. [CrossRef]

33. Garner, B.C. Globulins. In *Clinical Veterinary Advisor*; Wiedmeyer, C., Ed.; Elsevier Inc.: Amsterdam, The Netherlands, 201 pp. 934–935. [CrossRef]

34. Sharma, U.; Pal, D.; Prasad, R. Alkaline phosphatase: An overview. *Indian J. Clin. Biochem.* **2014**, *29*, 269–278. [CrossRef]

35. Ness, R.D. Clinical Pathology and Sample Collection of Exotic Small Mammals. *Veter Clin. N. Am. Exot. Anim. Pract.* **1999**, *2* 591–620. [CrossRef]

Rasmussen, S.L.; Kalliokoski, O.; Dabelsteen, T.; Abelson, K. An exploratory investigation of glucocorticoids, personality and survival rates in wild and rehabilitated hedgehogs (*Erinaceus europaeus*) in Denmark. *BMC Ecol. Evol.* **2021**, *21*, 96. [CrossRef]

Romero, L.M.; Reed, J.M. Collecting baseline corticosterone samples in the field: Is under 3 min good enough? *Comp. Biochem. Physiol. Part A Mol. Integr. Physiol.* **2005**, *140*, 73–79. [CrossRef]

Campbell, T.W. Mammalian Hematology: Laboratory Animals and Miscellaneous Species. Section III: Hematology of Common Nondomestic Mammals, Birds, Reptiles, Fish, and Amphibians. In *Veterinary Hematology and Clinical Chemistry*, 3rd ed.; Thrall, M.A., Weiser, G., Allison, R.W., Campbell, T.W., Eds.; John Wiley & Sons: Hoboken, NJ, USA, 2022.

Allison, R.W. Laboratory Evaluation of the Liver. Section V: Clinical Chemistry of Common Nondomestic Mammals, Birds, Reptiles, Fish, and Amphibians. In *Veterinary Hematology and Clinical Chemistry*, 3rd ed.; Thrall, M.A., Weiser, G., Allison, R.W., Campbell, T.W., Eds.; John Wiley & Sons: Hoboken, NJ, USA, 2022.

Siegel, A.; Walton, R.M. Hematology and Biochemistry of small mammals. In *Ferrets, Rabbits, and Rodents*; Elsevier: Amsterdam, The Netherlands, 2020; pp. 569–582. [CrossRef]

Ferreira, A.F.; Queiroga, F.L.; Mota, R.A.; Rêgo, E.W.; Mota, S.M.; Teixeira, M.G.; Colaço, A. Hematological profile of captive bearded capuchin monkeys (*Sapajus libidinosus*) from Northeastern Brazil. *Cienc. Rural* **2018**, *48*, 1–7. [CrossRef]

Article

Outcomes, Mortality Causes, and Pathological Findings in European Hedgehogs (*Erinaceus europaeus*, Linnaeus 1758): A Seventeen Year Retrospective Analysis in the North of Portugal

Andreia Garcês [1,*], Vanessa Soeiro [2], Sara Lóio [2], Roberto Sargo [3], Luís Sousa [3], Filipe Silva [3,4] and Isabel Pires [4]

[1] Inno–Serviços Especializados em Veterinária, R. Cândido de Sousa 15, 4710-300 Braga, Portugal
[2] Wildlife Rehabilitation Centre of Parque Biológico de Gaia, Rua da Cunha, 4430-812 Avintes, Portugal; vanessasoeiro@cm-gaia.pt (V.S.); saraloio@cm-gaia.pt (S.L.)
[3] Wildlife Rehabilitation Centre of University of Trás-os-Montes and Alto Douro (CRAS-UTAD), University of Trás-os-Montes and Alto Douro, Quinta de Prados, 5000-801 Vila Real, Portugal; rsargo@utad.pt (R.S.); luissousa@utad.pt (L.S.); fsilva@utad.pt (F.S.)
[4] CECAV, University of Trás-os-Montes and Alto Douro, Quinta de Prados, 5000-801 Vila Real, Portugal; ipires@utad.pt
*   Correspondence: andreiamvg@gmail.com; Tel.: +351-918510495

Received: 30 June 2020; Accepted: 28 July 2020; Published: 30 July 2020

Simple Summary: The Western European hedgehog *Erinaceus europaeus* (Linnaeus, 1758) is one of the most common mammals in urban areas. We collected data over 17 years (2002–2019) regarding outcomes and causes of mortality on this species from two of the main wildlife rehabilitation centers in the north of Portugal. A total of 740 animals were admitted; the majority were juveniles, with the highest admission rate occurring during summer (36.8%). The main cause of admission was debilitation (30.7%). Of the total number of individuals admitted to these centers, 66.6% were released successfully. The main cause of death was trauma of unknown origin (32.7%).

Abstract: This study aimed to analyze the admission causes, outcomes, primary causes of death, and main lesions observed in the post mortem examinations of Western European hedgehogs, *Erinaceus europaeus* (Linnaeus, 1758), in the north of Portugal. The data were obtained by consulting the records from the two main wildlife rehabilitation centers located in the north of Portugal (Wildlife Rehabilitation Centre of Parque Biologico de Gaia and the Wildlife Rehabilitation Centre of the University of Trás-os-Montes and Alto Douro). Over 17 years (2002–2019) a total of 740 animals were admitted. Most of the animals were juveniles, with the highest number of admissions occurring during summer (36.8%) and spring (33.2%). The main reasons for admission were debilitation (30.7%) and random finds (28.4%). Of the total number of individuals admitted to these centers, 66.6% were successfully released back into the wild. The most relevant causes of death were trauma of unknown origin (32.7%), nontrauma causes of unknown origin (26.6%), and nutritional disorders (20.2%). The main lesions observed were related to trauma, including skeletal and skin lesions (fractures, hemorrhages, wounds) and organ damage, particularly to the lungs and liver. The hedgehog is a highly resilient and adaptable animal. The urban environment has many benefits for hedgehogs, yet the presence of humans can be harmful. In the future, the public needs to become even more involved in the activities of the wildlife centres, which will make a positive difference for these populations.

Keywords: *Erinaceus europaeus*; hedgehog; Portugal; mortality; trauma; pathology

---

Animals **2020**, *10*, 1305; doi:10.3390/ani10081305

1. Introduction

The Western European hedgehog, *Erinaceus europaeus* (Linnaeus, 1758), is a generalist nocturnal mammal, widely distributed across the European continent [1–3]. It hibernates from November to March, with some periods of awakening to forage or move in its nest [4]. Its diet consists mainly of macroinvertebrates, although, due to their great trophic adaptation potential, they can be generalist feeders [1,5]. This species has been classified as Least Concern (LC) in Portugal, according to the International Union for Conservation of Nature's (IUCN) Red List of Threatened Animals [6]. Recent monitoring data from the Netherlands, Sweden, United Kingdom (UK), Belgium, and Germany show that its population has suffered a decline in recent decades [7,8]. Agricultural intensification, habitat fragmentation, road traffic accidents, molluscicide and rodenticide poisoning, and predation (e.g., by foxes, badgers, dogs) have been suggested amongst the major causes of this decline [5,7–9].

E. europaeus is one of the species that seem to prefer urban areas as their current habitat. These animals are commonly found on green spaces in constructed areas such as gardens and parks [4,7]. Some studies in the UK have suggested that the hedgehog decline is more severe in rural than in urban areas. Urban areas have become a suitable habitat for hedgehogs due to the higher food densities associated with human occupation, the existence of additional nest sites, and a decreased risk of predation by their natural predators [4,6,7,10].

Hedgehogs are one of the most common mammal species admitted to wildlife rehabilitation centers, sanctuaries, or veterinary hospitals. The main reasons for admission described in the literature include skin, respiratory and gastrointestinal diseases, malnutrition, hypothermia, and traumatic injuries [4,5]. Due to their preference to inhabit urban areas, hedgehogs are subjected to a higher risk of human-related traumatic injuries, which can include drowning, injuries inflicted by domestic pets, poisoning, entrapment, and road injuries [4,5,7].

Even though data on causes of mortality and morbidity of hedgehogs have been published previously, long-term studies about admissions of these animals to rehabilitation centers are quite scarce, particularly in southern Europe.

The main purpose of this study was to collect data from hedgehog admittance records from the two major wildlife rehabilitation centers located in the north of Portugal, describing admission causes, outcomes, primary causes of death, and main lesions observed in the post mortem exams.

2. Materials and Methods

This study was conducted at the Wildlife Rehabilitation Centre of Parque Biológico de Gaia (WRC-PBG) (41°05′48.50″ N–8°33′21.34″ W), which is located on the northern Portuguese coastline, and at the Wildlife Rehabilitation Centre of University of Trás-os-Montes and Alto Douro (CRAS-UTAD) (41°17′18.13″ N–7°44′21.94″ W) in inland northern Portuguese.

For this study, we examined all hedgehog (*E. europaeus*) admittance records from these WRCs between 2002 and 2019. The variables analyzed were the arrival date, age (categorized as young or adult, based on their external characteristics), sex, the primary reason for admission, outcome, and post mortem examinations of the dead animals.

For each animal, the reason for admission was determined as one of the following: random find (healthy animals that were found in gardens, roads, or inside buildings), orphaned, debilitated, held in captivity (animals that were kept in captivity by people as pets for a period of time), dead, or injured, according to previous studies [11–13].

The outcome of each animal was categorized as follows: euthanized (EU), died during the recovery process (DE), and released into the wild (RE). The cause of death was determined based on clinical signs and post mortem examination, including a histopathological exam. The cause of death was categorized as infectious and parasitic disease, nutritional disorder, poisoning, nontraumatic death of unknown origin, collision with vehicles, predation (both from natural predators and dog/cat attacks), neoplasia, trapping, or trauma of unknown origin.

The post mortem examinations were performed in line with the established technique and safety procedures for this species [14]. The macroscopic exams were performed by I.P., A.G., V.S., and S.L., and the histopathological exams by I.P. All the macroscopic findings were recorded in a written protocol and using photographs. For the histopathological examinations, representative samples were collected from each animal. Samples were fixed in buffered 10% formalin solution, embedded in paraffin wax, sectioned at 3 μm, and stained with HE (hematoxylin-eosin).

All data collected were organized in Excel sheets, and the descriptive statistics, normality test, and inferential analyses were performed using SPSS Advanced Models TM 21.0 (SPSS Inc. 233 South Wacker Drive, 11th Floor Chicago, IL 60606-6412, USA). To study the differences between the observed and expected frequencies of categories of a field, one-sample nonparametric tests were used (binomial test or chi-square test, depending on the number of categories of the categorical field).

3. Results

3.1. Descriptive Data

A total of 740 animals were admitted to the WRC-PBG (n = 636) and CRAS-UTAD (n = 104), between 2002 and 2019. Figure 1 shows a map with the distribution of the admitted animals by the different district of origin in northern Portugal, based on the location from which they were collected. The majority came from Porto (n = 565) and Vila Real (n = 95).

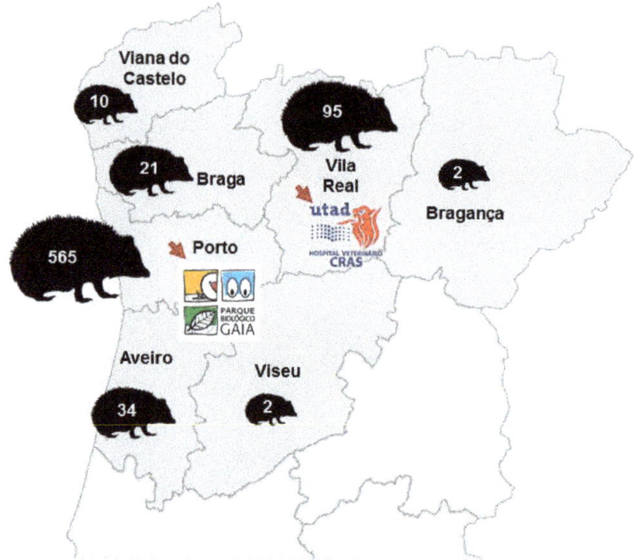

Figure 1. The number of animals distributed across different districts of origin in the north of Portugal, based on the locations from which they were collected. The locations of the Wildlife Rehabilitation Centre of Parque Biológico de Gaia (WRC-PBG) (41°05′48.50″ N–8°33′21.34″ W) and the Wildlife Rehabilitation Centre of University of Trás-os-Montes and Alto Douro (CRAS-UTAD) (41°17′18.13″ N–7°44′21.94″ W) are marked.

In only 90 out of the 740 animals was it possible to determine their sex, with 52% (n = 47) being females and 48% (n = 43) males. The differences observed between animal sexes were not significant ($p = 0.706$).

Regarding the age, 155 animals were juveniles, 49 were adults, and in the remaining 536 age was not identified. The difference between adults and juveniles was statistically significant ($p < 0.0001$).

The season with the highest hedgehog admission rate was the summer (36.8%) followed by spring (33.2%), autumn (22.3%), and winter (7.6%). Statistically significant differences were observed between seasons ($p < 0.001$).

The animals admitted to both WRCs and later released back to the wild represented 66.6% (n = 492) of the cases, and 33.3% (n = 248) died. Figure 2 represents the total number of animals admitted from 2002 to 2019, along with the number of dead and released animals per year.

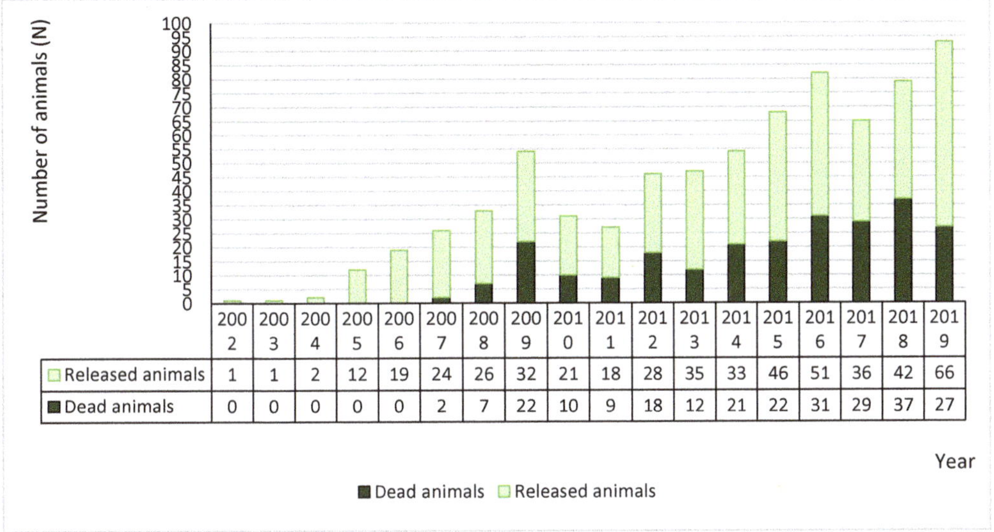

Figure 2. The number of animals admitted from 2002 to 2019, with the number of dead and released animals per year.

3.2. Reasons for Admission to the WRCs

The main reasons for admission were as follows: 30.7% (227) debilitated, 28.4% (n = 211) random finds, 21.4% (n = 158) injured, 17.3% (128) orphaned, and 2.2% (n = 16) held in captivity. There was a significant difference between the reasons for admission ($p < 0.01$). (Figure 3, Table S1) represents the distribution, in percentages, of the reasons for admission over the different years.

Table 1 displays the frequency and percentage of the reasons for admission by season, sex, age, and outcome.

As shown in Table 1, the seasons with the highest number of admissions were summer and spring for the categories random find, orphan, and debilitated ($p < 0.0001$). The injured and held in captivity categories had slightly higher numbers of admissions during autumn.

Regarding age and season, there were significant differences. During winter, most of the animals admitted were adults, while in the other seasons juveniles were the most common ($p < 0.001$). Although more females were admitted in spring and summer, and more males in winter and autumn, there were no significant differences ($p = 0.55$).

Except for injured animals, the predominant outcome across all categories was the release back to the wild.

There were statistically significant associations between cause of admission and the seasons ($p < 0.001$); in the winter and spring, most animals were admitted due to a debilitating physical condition, in the autumn due to injuries, and in the summer due to random finding.

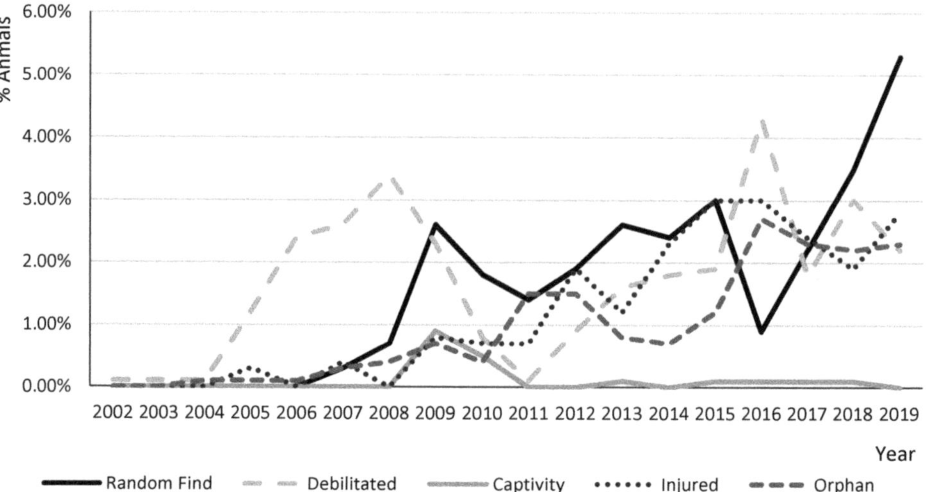

Figure 3. Distribution of the reasons for admission of *Erinaceus europaeus* to both wildlife rehabilitation centers from 2002 to 2019.

Table 1. Frequency and percentage of the reasons for admission by season, sex, age, and outcome.

	Random Find	**Debilitated**	**Captivity**	**Injured**	**Orphaned**
Season					
Winter	12 (21.4)	33 (58.8)	2 (3.6)	6 (10.7)	3 (5.4)
Autumn	41 (24.8)	44 (26.7)	6 (3.6)	52 (31.5)	22 (13.3)
Spring	68 (27.6)	75 (30.5)	5 (2.0)	50 (20.30)	48 (19.50)
Summer	90 (33.0)	75 (27.6)	3 (1.1)	50 (18.4)	55 (20.2)
Sex					
Female	20 (42.6)	7 (14.9)	1 (2.1)	11 (23.4)	8 (17.0)
Male	21 (48.8)	4 (9.30)	1 (2.3)	8 (16.8)	9 (20.9)
Age					
Adult	31 (63.3)	8 (16.3)	2 (4.1)	8(16.3)	0
Juvenile	22 (14.2)	3 (1.9)	0	8(5.2)	122(78.7)
Outcome					
Death	30 (14.3)	76 (33.0)	2 (12.5)	105 (65.8)	35 (27.4)
Released	180 (85.7)	152 (67.0)	14 (87.5)	54 (34.2)	92 (71.9)

3.3. Causes of Death

Of the 248 animals that died, 83.5% (n = 207) died during treatment and 16.5% (n = 41) were euthanized. Of those animals 7.7% (n = 19) were females, 9.2% (n = 23) were males, and in the remaining 83.1% (n = 206) sex was not identified. Pertaining to the age, 8.1% (n = 20) were adults, 21.7% (n = 54) were juveniles, and in the remaining 70.2% (n = 174) the age was not determined.

The highest percentage of death occurred during summer (n = 92, 37.1%), followed by spring (n = 74, 29.8%), autumn (n = 64, 25.8%), and winter (n = 18, 7.3%) (Table 2).

In decreasing order of frequency, the causes of death were as follows: 32.7% (n = 81) trauma of unknown origin, 26.6% (n = 66) nontraumatic death of unknown origin, 20.2% (n = 50) nutritional disorder, 8.1% (n = 20) infectious or parasitic disease, 4.8% (n = 12) predation, 4.8% (n = 12) collision with vehicles, 1.6% (n = 4) poisoning, 0.8% (n = 2) trapping and 0.4% (n = 1) neoplasia. (Figure 4, Table S2) represents the distribution of the different causes of death from 2002 to 2019. Table 3 presents the frequency and percentage of the causes of death by the type of death and cause of admission.

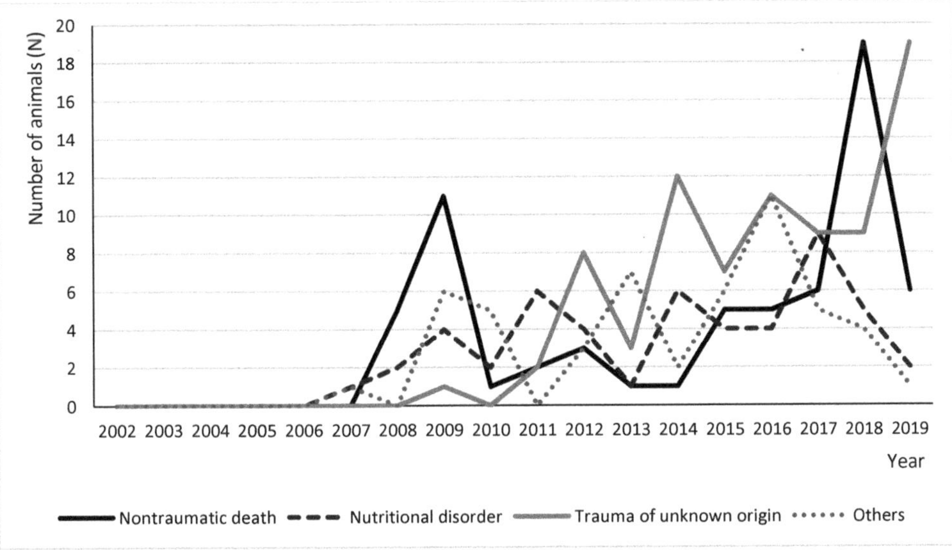

Figure 4. Distribution of the main causes of death of *Erinaceus europaeus* in both wildlife rehabilitation centers from 2002 to 2019. The category "others" includes predation, neoplasia, trapping, collision with vehicles, infectious and parasitic diseases, and poisoning.

3.4. Post Mortem Findings

Post mortem exams were performed on the 248 animals. Post mortem injuries were observed in different organs and systems. This section describes the main macroscopic lesions observed, complemented with some microscopic findings.

Upon external examination, 170 animals revealed signs of dehydration and emaciation. In all animals, the presence of ectoparasites such as fleas, mites (the most common being *Caparinia tripilis*), and ticks (mainly *Ixodes hexagonus*) was registered to various degrees. Three animals presented a high degree of parasitism, which led to emaciation, dehydration, and anemia and death. One animal presented a fungal infection due to dermatophytes (Figure 5).

Table 2. Frequency and percentage of the causes of death by season, sex, and age.

	Nontraumatic Death					Trauma			
	Nutritional Disorders	Poisoning	Infectious and Parasitic Diseases	Neoplasia	Nontraumatic Death of Unknown Origin	Predation	Trapping	Collision with Vehicles	Trauma of Unknown Origin
Season									
Winter	0	0	7 (38.8)	1 (5.5)	6 (33.3)	1 (5.5)	0	3 (16.6)	0
Autumn	6 (9.3)	1 (1.6)	4 (6.2)	0	14 (21.8)	5 (7.9)	1 (1.6)	2 (3.1)	31 (48.4)
Spring	20 (27.0)	1 (1.4)	3 (4.1)	0	21 (28.4)	4 (5.4)	1 (1.4)	4 (5.4)	20 (27.0)
Summer	24 (26.1)	2 (2.2)	6 (6.5)	0	25 (27.2)	2 (2.2)	0	3 (3.2)	30 (32.6)
Sex									
Female	6 (31.6)	2 (10.5)	4 (21.1)	1 (5.2)	1 (5.2)	2 (10.5)	0	2 (10.5)	1 (5.2)
Male	6 (26.1)	1 (4.3)	6 (26.1)	0	3 (13.0)	1 (4.3)	1 (4.3)	2 (8.7)	3 (13.0)
Age									
Adult	3 (15.0)	2 (10.0)	8 (40.0)	1 (5.0)	1 (5.0)	1 (5.0)	0	3 (15.0)	1 (5.0)
Juvenile	44 (81.4)	1 (1.8)	4 (7.4)	0	1 (1.8)	2 (3.7)	0	1 (1.8)	1 (1.8)

Table 3. Frequency and percentage of the causes of death by the type of death and reason for admission (euthanized—EU, died during the recovery process—DE).

	Nontraumatic Death					Trauma			
	Nutritional Disorders	Poisoning	Infectious and Parasitic Diseases	Neoplasia	Nontraumatic Death of Unknown Origin	Predation	Trapping	Collision with Vehicles	Trauma of Unknown Origin
Type of death									
DE	45 (21.7)	3 (1.4)	17 (8.2)	1 (0.5)	62 (30.0)	6 (2.8)	2 (1.0)	10 (4.8)	61 (29.5)
EU	5 (12.2)	1 (2.4)	3 (7.3)	0	4 (9.7)	6 (14.6)	0	2 (4.9)	20 (48.8)
Reason for admission									
Random find	10 (33.3)	1 (3.3)	5 (16.6)	1 (3.3)	8 (26.6)	0	0	4 (13.3)	1 (3.3)
Debilitated	4 (5.3)	2 (2.6)	12 (16.0)	0	55 (73.3)	0	0	1 (1.3)	1 (1.3)
Captivity	0	0	0	0	2 (100)	0	0	0	0
Injured	1 (0.9)	1 (0.9)	2 (1.9)	0	1 (0.9)	12 (11.5)	2 (1.9)	7 (6.7)	79 (75.2)
Orphan	35 (97.2)	0	1 (2.7)	0	0	0	0	0	0

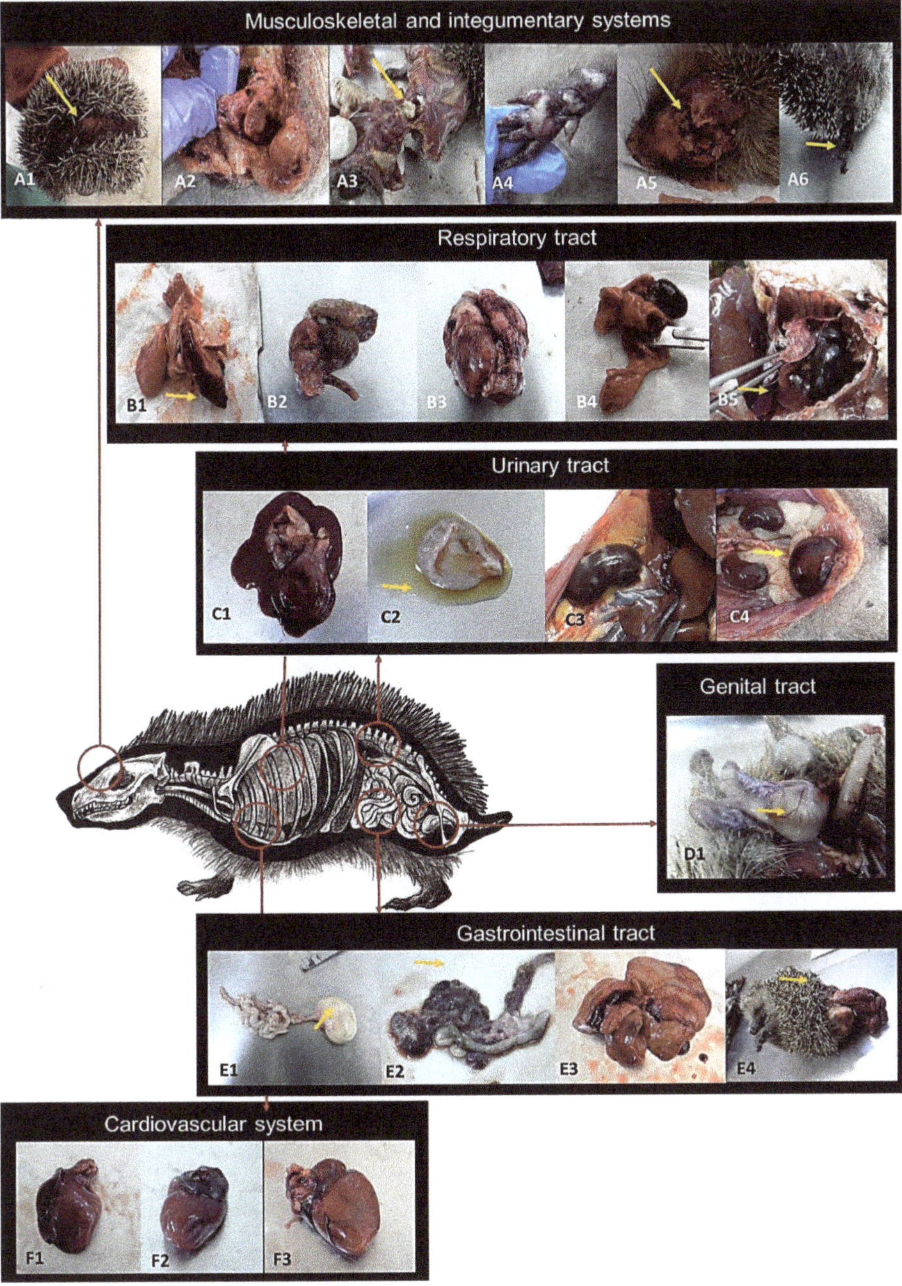

Figure 5. Schematic representation of lesion observed during the post mortem exams of *Erinaceus europaeus*. **A1**—skin wound; **A2**—pelvis fracture; **A3**—cervical abscess; **A4**—pododermatitis; **A5**—skull fracture; **A6**—posterior limb amputation; **B1**—lung hemorrhage; **B2**—fibrinous bronchopneumonia; **B3**—emphysema; **B4**—parasitic pneumonia; **B5**—hemothorax; **C1**—hemorrhagic cystitis; **C2**—altered urine pigmentation; **C3, C4**—kidney congestion; **D1**—pyometra; **E1**—gastric dilatation; **E2**—enteritis; **E3**—hepatic fracture and lipidosis; **E4**—evisceration; **F1**—myocardial discoloration; **F2, F3**—myocardial congestion.

One hundred and seven animals had skeletal and skin lesions (Figure 5A1–A6). They presented mild to severe traumatic injuries: skin lacerations (50), evisceration (2) (Figure 5E4), spine fractures (6) tibial fracture (4), fracture of the femur (2), sacrum fracture (6), limb amputation (1) (Figure 5A6), pododermatitis (2), and fracture of the skull (3) (Figure 5A5). Twelve animals presented polytrauma associated with a vehicle collision. Twelve animals had a cerebral concussion lesion. At least five animals presented eye evisceration prolapse or enucleation and ear hemorrhage. Among the muscular lesions, we found pallid muscles, hemorrhagic muscles, and subcutaneous hematomas associated with the fractured bones. In one animal, an abscess was observed in a cervical vertebra (Figure 5A3), causing an abnormal angulation of the neck (45°) in the caudal direction and causing severe respiratory distress.

At the level of internal organs, several animals presented congestion (115) and had free sanguineous liquid in the thoracic (47 hemothorax) (Figure 5B5) and abdominal (80 hemoperitoneum) cavities associated with traumatic lesions. One animal presented ascites and several ruptures of the diaphragm (7). The majority of these animals had been run over by vehicles.

Concerning the respiratory system, the lesion most commonly observed was congestion (50), followed by hemorrhage (44), edema (10), hydrothorax (1), and emphysema (2). Additionally, histological examination revealed bronchopneumonia (4), fibrinous bronchopneumonia (2), emphysematous areas of unknown etiology (3), interstitial pneumonia (3), aspiration pneumonia (15), hydrothorax (1), parasitic pneumonia (25), bronchitis (3), and emphysema (2) (Figure 5B1–B4). One animal presented a mucoid exudate in the nose and eyes.

In the cardiovascular system, observations included alterations in the shape of the heart (1), myocardial discoloration (2), hemopericardium (6), and myocardial hemorrhage (9) (Figure 5F1–F3). Histological examination of areas of myocardial discoloration also revealed vacuolization in isolated myocytes (1).

In the gastrointestinal tract, observations included split teeth (13), gingivitis, small wounds in the oropharynx and esophagus, hemorrhage (40), gastric ulcers (5), parasites (40), tympany (7) (Figure 5E1), enteritis of unknown origin (12) (Figure 5E2), hemorrhagic enteritis (2), intestinal rupture (5), and rectal prolapse (1). In the liver, nonspecific congestion (47), hepatic fractures (55), pale-discoloration lipidosis (116), subcapsular hemorrhages (1), and focal necrosis (5) (Figure 5E3) were observed. Additionally, in the histological exams, eosinophilic hepatitis, lipidosis, focal necrosis and hepatocellular atrophy were revealed (Figure 5).

Splenic fractures (43), splenomegaly (5), and, upon histological examination, the presence of megakaryocytes (2) and granulomatous splenitis (1) were detected in the spleen.

Two animals presented an enlargement of the adrenal gland with cortical hyperplasia.

In the urinary system, the main organ affected was the kidney, with nonspecific congestion and hemorrhage or circulatory lesions associated with trauma (97) (Figure 5C2–C4). Hemorrhagic cystitis (1) (Figure 5C1) and bladder rupture (3) were also observed. The histological examination exposed one focal non-purulent nephritis (1).

In the genital tract, two females presented lesions, one with a closed pyometra (Figure 5D) and other with neoplasia in the mammary gland.

In some individuals (n = 17), it was not possible to investigate any lesions of the internal organs because the carcasses were too autolytic. The majority of the animals' histopathological examinations were impaired, mostly by severe autolysis, and some were not performed due to lack of financial resources.

4. Discussion

The present study was the first investigation on outcomes, reasons for admission, mortality and pathological findings on wild hedgehogs in Portugal. Studies concerning wild hedgehogs' mortality are very scarce and almost nonexistent. To our knowledge, within the Mediterranean region, there was a similar study performed by Martinez et al. (2014) [5] in Spain. In most cases, information regarding outcomes or mortality is dispersed in the grey bibliography such as annual reports of the wildlife rehabilitation centers [11,12] or related to a specific cause of death such as collisions with road vehicles [10,13,14].

Our data were drawn from the records of a large number of cases admitted to two wildlife rehabilitation centers, WRC-PBG and CRAS-UTAD, over 17 years (2002–2019). These institutions are the main centers that receive and treat wild terrestrial vertebrates on the north of the country. These centers directly depend on the national government branch that regulates the wildlife services, ICNF. They are equipped with adequate facilities to accommodate wildlife and a team of technicians and veterinarians that rehabilitate the animals (Portaria n. ° 1112/2009, 2009). The injured animals are received from different official organizations (Serviço de Proteção da Natureza e do ambiente (SEPNA), police, etc.) and directly from the general public. Both possess a large database that registers the maximum possible information about the admissions.

A total of 740 hedgehogs were admitted during this period, with most of the animals being collected in the district of Porto and Vila Real. This was expected since these are the WRCs' locations, and there is therefore a higher probability of their receiving injured and debilitated animals from the neighboring area; this was also noticed in similar studies concerning other species in this region [15,16].

There was no significant difference between the sexes of the animals admitted, in contrast to what was perceived in other studies [1–3]. Regarding age, most of the animals were juveniles. The main seasons of admission were summer (36.8%) and spring (33.2%). Similar data were observed in other studies, and are expected based on the animals' natural behavior [4,6,7]. In spring, they emerge from the hibernation period and are more dispersed on the search of food. During the months of February–March and again in August–September, the males become more active during the reproductive season (in Portugal they can reproduce two times a year) [4,5]. In the months July–August (summer), females are more actively feeding and enlarging their home ranges after weaning their young, and the first juveniles start to disperse at the same time [4,6,7].

The main reasons for admission to the WRCs were debilitation, random finds, and injury. Admissions due to random finds and injury have been increasing in recent years. A similar trend has been observed in Spain [3] and the United Kingdom [2]. Nonetheless, in Spain, the main reasons for admission were random finds (40.8%) and orphaned young (19.4%). This difference could be related to many factors, such as differences in the habitat, human population, urbanization density, and how aware the public is of how to respond to an accidental find of wildlife.

Regarding the outcomes of the animals admitted to the two WRCs, 66.6% were released back to the wild, following the results described by Martinez et al. (2014) [5], with 69% of the admitted hedgehogs returned to the wild. Indeed, most of the admissions corresponded to healthy animals found accidentally or animals kept for a short period in the WRC when they were slightly underweight, dehydrated, or the weather was too cold [1]. Hedgehogs are resilient animals, which are relatively easy to keep in captivity, and recover well [5]. Most studies indicate that these animals, when released back in the wild, adapt and survive well [1,3,5].

Of the 248 dead animals, the vast majority, 83.4%, died during treatment. Several studies have suggested that there is no significant sex- or age-related specific mortality [1], and the same was observed in this study. The greatest peaks of mortality were in summer and spring, for the same reasons as explained previously. The main three causes of death were trauma of unknown origin (32.7%), nontraumatic death of unknown origin (26.6%), and nutritional disorder (20.2%). Unfortunately, one of the most common limitations of retrospective studies is the incomplete data. A large percentage of animals in this study presented unidentified cause of death, sex, or age. This could be related to the lack of information taken on the admittance form, scarce human and financial resources to perform complete examinations, and the absence of a digital system in which to store the data in the early years. The causes of death observed were oriented towards those described in the bibliography, where the main causes of death were associated with trauma, malnutrition, and dehydration [5,17–19]. The latter was present almost exclusively in juveniles, and is related to inexperience and early loss of their parents for diverse reasons [5,20]. Mortality in different years and regions can be related to variations in the surrounding environment. Some studies have shown that rigorous winters lead to greater mortality of individuals; this phenomenon has been observed in countries located in northern Europe such as Sweden, Denmark, Finland, and the United Kingdom [7,8,17,18]. Similarly, in urban areas, after hibernation, animals can become more susceptible to accidental finds by humans and pets (traps, predation, collision, and others) when foraging. The availability of resources can be lower and more disputed in certain areas, such as those that are more industrialized [21–23]. As a result of the curious nature of hedgehogs, it seems that they are somewhat accident-prone and death through misadventure is not uncommon. It is common to observe hedgehogs drowning in garden ponds, entangled in netting, falling into holes (particularly cattle grids), or licking poisonous substances [20]. Since hedgehogs seem to enjoy living in our gardens [3,18], it is common to observe these accidents. In our study, it was possible to identify some examples. Furthermore, direct contact with humans leads to a higher level of death related to anthropogenic sources, as traps, roadkill, and poisons, as was observed.

The pathological findings confirmed the main causes of mortality of these animals to be associated with trauma. The main lesions observed were related to trauma, with skeletal and skin lesions (fractures, hemorrhages, open wounds) and organ damage, particularly of the lungs and the liver. In this study, there was probably a greater number of animals with parasitic and infectious disease, but, due to the lack of complementary exams, it was not possible to confirm all diagnoses. In one case it was possible to identify the first case of a pyometra described in this species in a wild individual [24].

From 2002 to 2019, it was possible to observe an increase in the number of hedgehog admissions to the WRCs. This phenomenon could be related to the increase in the general public interest in wildlife and greater knowledge of the daily work of the WRCs [25]. An important part of the WRCs' mission is to educate the general public (particular youngsters) regarding wild fauna and flora, being a powerful tool for environmental awareness [26,27]. In terms of hedgehogs, the WRCs are making an effort to generate greater involvement of members of the public in the rehabilitation process of these animals. Due to the animal's peculiar aspect, small size, and friendly behavior, it is often used in environmental campaigns as a symbol, allowing the public, mostly children, to get involved in its recovery. These activities have great conservation value by allowing the public to understand more about WRCs' daily work and projects, acquiring knowledge about natural history, ecology, and threats on these wild species, particularly those severely affected by anthropogenic factors [1,4,28]. They also help to provide adequate information to people living in peri-urban and residential areas, allowing them to identify situations when rescue is necessary and to deliver animals to WRCs whenever special care is needed [4,5,18].

5. Conclusions

In conclusion, of the 740 animals admitted to two wildlife rehabilitation centers in the north of Portugal, more than half recovered and were released back to the wild. The main causes of admission were related to random finds and debilitating physical conditions. In this region, the main causes of death of these animals were associated with trauma, and in most cases linked to anthropogenic sources. The hedgehog is a highly resilient and adaptable animal. This is the first time that such a long study related to outcomes and mortality has been performed for this species. The urban environment has benefits for hedgehogs, offering supplementary sources of food and shelter, but the presence of humans may harm them. It is important that in the future, the public can become even more involved in the activities of wildlife centers and similar environmental associations, allowing them to make a positive difference to hedgehog populations.

Supplementary Materials: The following are available online at http://www.mdpi.com/2076-2615/10/8/1305/s1, Table S1: Distribution of the reasons for admission of *Erinaceus europaeus* in both wildlife rehabilitation centers from 2002 to 2019, Table S2: Distribution of the causes of death of *Erinaceus europaeus* in both wildlife rehabilitation centers from 2002 to 2019.

Author Contributions: Conceptualization, A.G. and I.P.; methodology, A.G.; software, I.P., F.S.; validation, I.P., F.S., R.S., V.S., S.L., formal analysis, A.G.; investigation, A.G.; resources, F.S., V.S., S.L., R.S., L.S.; data curation, F.S., V.S., S.L., R.S., L.S.; writing—original draft preparation, A.G.; writing—review and editing, A.G.; visualization, I.P., F.S., R.S., V.S., S.L., supervision, I.P., F.S., V.S., S.L.; project administration, A.G. All authors have read and agreed to the published version of the manuscript.

Funding: This research received no external funding.

Acknowledgments: We are thankful by the support of Wildlife Rehabilitation Centre of Parque Biologico de Gaia and the Wildlife Rehabilitation Centre of the University of Trás-os-Montes and Alto Douro.

Conflicts of Interest: The authors declare no conflict of interest.

Correction Statement: This article has been republished with a minor change. The change does not affect the scientific content of the article and further details are available within the backmatter of the website version of this article.

References

1. Reeve, N. *Hedgehogs*; Poyser: London, UK, 2002.
2. Mullineaux, E.; Keeble, E. *BSAVA Manual of Wildlife Casualties*; British Small Animal Veterinary Association: Gloucester, UK, 2003.
3. Pettett, C.E.; Moorhouse, T.P.; Johnson, P.J.; Macdonald, D.W. Factors affecting hedgehog (*Erinaceus europaeus*) attraction to rural villages in arable landscapes. *Eur. J. Wildlife. Res.* **2017**, *63*. [CrossRef]
4. Mullineaux, E.; Keeble, E. *BSAVA Manual of Wildlife Casualties*, 2nd ed.; British Small Animal Veterinary Association: Gloucester, UK, 2016; pp. 117–137.
5. Martínez, J.; Rosique, A.I.; Royo, M. Causes of admission and final dispositions of hedgehogs admitted to three Wildlife Rehabilitation Centers in eastern Spain. *Hystrix* **2014**, *25*, 107–110.
6. Amori, G. *Erinaceus europaeus*. In he IUCN Red List of Threatened Species 2016: e.T29650A2791303. Available online: https://dx.doi.org/10.2305/IUCN.UK.2016-3.RLTS.T29650A2791303.en (accessed on 15 April 2020).
7. Rasmussen, S.L.; Berg, T.B.; Dabelsteen, T.; Jones, O.R. The ecology of suburban juvenile European hedgehogs (*Erinaceus europaeus*) in Denmark. *Ecol. Evol.* **2019**, *9*, 13174–13187. [CrossRef]
8. The State of Britain's Hedgehogs. 2018. Available online: https://www.britishhedgehogs.org.uk/pdf/sobh-2018.pdf (accessed on 20 April 2020).
9. Huijser, M.P.; Bergers, P.J.M. The effect of roads and traffic on hedgehog (*Erinaceus europaeus*) populations. *Biol Conserv.* **2000**, *95*, 111–116. [CrossRef]
10. Wembridge, D.E.; Newman, M.R.; Bright, P.W.; Morris, P.A. An estimate of the annual number of hedgehog (*Erinaceus europaeus*) road casualties in Great Britain. *Mammal Commun.* **2016**, *2*, 8–14.
11. GREFA (Grupo de Rehabilitación de la Fauna Autóctona). Anuario 2012 [Internet]. 2012. Available online: http://www.grefa.org/multimedia/descargas/category/4-anuarios# (accessed on 20 April 2020).
12. 25 anos a Recuperar Animais Selvagens-Relatório 1985/2010 [Internet]. 2010. Available online: www.parquebiologico.pt/userdata/Relatorio-VET2010final.pdf (accessed on 20 April 2020).
13. Mikov, A.M.; Georgiev, D.G. On the road mortality of the Northern White-Breasted Hedgehog (*Erinaceus roumanicus* Barrett-Hamilton, 1900) in Bulgaria. *Ecol. Balk.* **2018**, *10*, 19–23.
14. Haigh, A.; O'Riordan, R.M.; Butler, F. Hedgehog *Erinaceus europaeus* mortality on Irish roads. *Wildlife Biol.* **2014**, *20*, 155–160. [CrossRef]
15. Garcês, A.; Pires, I.; Pacheco, F.; Fernandes, L.S.; Soeiro, V.; Lóio, S. Natural and anthropogenic causes of mortality in wild birds in a wildlife rehabilitation centre in Northern Portugal: A ten-year study. *Bird Study* **2019**, *66*, 484–493. [CrossRef]
16. Garcês, A.; Pires, I.; Pacheco, F.; Sanches, L.; Soeiro, V.; Lóio, S. Preservation of wild bird species in northern Portugal-Effects of anthropogenic pressures in wild bird population (2008–2017). *Sci. Total Environ.* **2018**, *650*. [CrossRef]
17. Kristiansson, H. Population variables and causes of mortality in a hedgehog (*Erinaceous europaeus*) population in southern Sweden. *J. Zool.* **1990**, *220*, 391–404. [CrossRef]
18. Rautio, A.; Isomursu, M.; Valtonen, A.; Hirvelä-Koski, V.; Kunnasranta, M. Mortality, diseases and diet of European hedgehogs (*Erinaceus europaeus*) in an urban environment in Finland. *Mammal. Res.* **2016**, *61*, 161–169. [CrossRef]

19. Bunnell, T. The incidence of disease and injury in displaced wild hedgehogs (*Erinaceus europaeus*). *Lutra* **2001**, *44*, 3–14.

20. Online, W. European Hedgehog Mortality-Introduction Wildlife Online [Internet]. Available online: https://www.wildlifeonline.me.uk/animals/article/european-hedgehog-mortality-introduction (accessed on 2 June 2020).

21. Ditchkoff, S.S.; Saalfeld, S.T.; Gibson, C.J. Animal behavior in urban ecosystems: Modifications due to human-induced stress. *Urban Ecosyst.* **2006**, *9*, 5–12. [CrossRef]

22. Rhythms, B. Stress and the city: Urbanization and its effects on the stress physiology in european blackbirds. *Ecology* **2006**, *87*, 1945–1952.

23. Magle, S.B.; Hunt, V.M.; Vernon, M.; Crooks, K.R. Urban wildlife research: Past, present, and future. *Biol. Conserv.* **2012**, *155*, 23–32. [CrossRef]

24. Garcês, A.; Poeta, P.; Soeiro, V.; Lóio, S.; Cardoso-Gomes, A.; Torres, C. Pyometra Caused by Staphylococcus lentus in a Wild European Hedgehog (*Erinaceus europaeus*). *J. Wildl. Dis.* **2019**, *55*. [CrossRef] [PubMed]

25. Tribe, A.; Brown, P.R. The role of wildlife rescue groups in the care and rehabilitation of Australian fauna. *Hum. Dimens. Wildl.* **2000**, *5*, 69–85. [CrossRef]

26. Mullineaux, E. Veterinary treatment and rehabilitation of indigenous wildlife. *J. Anim. Pract.* **2014**, *55*, 293–300. [CrossRef] [PubMed]

27. Trocini, S.; Pacioni, C.; Warren, K.; Butcher, J.; Robertson, I. Wildlife disease passive surveillance: The potential role of wildlife rehabilitation centres. In Proceedings of the 6th National Wildlife Rehabilitation Conference, Canberra, Australia, 9–13 August 2008.

28. Baker, P.J.; Harris, S. Urban mammals: What does the future hold? An analysis of the factors affecting patterns of use of residential gardens in Great Britain. *Mamm. Rev.* **2007**, *37*, 297–315. [CrossRef]

Article

Hibernation Patterns of the European Hedgehog, *Erinaceus europaeus*, at a Cornish Rescue Centre

Kathryn E. South [1,2,*], **Kelly Haynes** [2] and **Angus C. Jackson** [2]

1   Prickles and Paws Hedgehog Rescue, Cubert, Newquay, Cornwall TR8 5HD, UK
2   Centre for Applied Zoology, Cornwall College Newquay, Newquay, Cornwall TR7 2LZ, UK;
    kelly.haynes@cornwall.ac.uk (K.H.); angus.jackson@cornwall.ac.uk (A.C.J.)
*   Correspondence: katy@pricklesandpaws.org

Received: 30 June 2020; Accepted: 12 August 2020; Published: 14 August 2020

Simple Summary: Populations of the European hedgehog, *Erinaceus europaeus*, are declining in the UK. This small mammal is frequently admitted to rescue centres in the UK to be treated for a variety of illnesses or injuries. With many spending the winter in captivity, clear guidelines about how to look after hedgehogs during their hibernation would be very useful. We studied 35 hedgehogs over two winters to learn about their sleeping behaviour and how they change weight. We measured the total length of hibernation and the periods during hibernation when hedgehogs are more active (called spontaneous arousals). There were three main results. (1) The longer the hibernation, the more weight was lost. (2) Previous studies show that arousal is energetically expensive. Despite this, weight-loss was more related to the amount of time spent sleeping than to the number of times the hedgehog woke up, perhaps because they could easily feed each time they woke up. (3) Larger hedgehogs lost proportionally less weight per day, perhaps because they woke up and fed more often than did smaller hedgehogs. Behaviour by hibernating hedgehogs in captivity differs from that in the wild. Patterns revealed in this study are used to make some recommendations for guidelines that can be adapted for individual hedgehogs according to their size and behaviour during hibernation.

Abstract: The European hedgehog, *Erinaceus europaeus*, is frequently admitted to rescue centres in the UK. With many overwintering in captivity, there is cause to investigate hibernation patterns in order to inform and improve husbandry and monitoring protocols. Thirty-five hedgehogs were studied over two winters. Weight change during hibernation for the first winter was used to test for effects of disturbance on different aspects of hibernation, including total duration, frequency and duration of spontaneous arousals. There was no significant difference between the two winters for any of the four aspects studied. Significant positive correlations demonstrated that weight-loss increased with the duration of the hibernation period and with percent of nights spent asleep, but not with the number of arousal events. Thus, weight-loss appears more strongly associated with the proportion of time spent asleep than with the number of arousal events. This was surprising given the assumed energetic expense of repeated arousal and was potentially due to availability of food during arousals. In contrast with previous studies, larger hedgehogs lost less weight per day than did smaller hedgehogs. They also woke up more often (i.e., had more opportunities to feed), which may explain the unexpected pattern of weight-loss. Hibernatory behaviour in captivity differs from that in the wild, likely because of non-natural conditions in hutches and the immediate availability of food. This study provides a basis for further research into the monitoring and husbandry of hedgehogs such that it can be adapted for each individual according to pre-hibernation weight and behaviour during hibernation.

Keywords: hedgehog; hibernation; spontaneous arousal; metabolism; wildlife rehabilitation; rehabilitation protocols; wildlife rescue

---

1. Introduction

The Western European hedgehog (*Erinaceus europaeus*) is the most common mammal species admitted to wildlife rescue centres across the UK [1]. Their rehabilitation and release have the potential to contribute to conservation of this declining species. Whilst there are challenges with assessment of population sizes [2,3], the most recent estimate is 522,000, a 66% reduction from the 1995 estimate [3]. Reasons for rescue and rehabilitation are numerous, including: injuries from pets or gardening activities, entrapment, car-strikes, poisoning from pesticides and parasitic burdening [4,5]. Many hedgehogs undergoing or following rehabilitation are overwintered or hibernate at rescue centres [4,5].

1.1. Hibernation

In the UK, hibernation (a period of greatly reduced activity, temperature, respiration and metabolism) [6] by wild hedgehogs, commonly occurs in outdoor nests (or hibernacula) between November and March. It is triggered primarily by consistently low temperatures (<8 °C) [7], although photoperiod, body-condition and food availability may also be involved [8]. Date and duration of hibernation periods can be influenced by climate, individual condition and sex [7–9].

For wild hedgehogs, mortality during hibernation can be a major component of overall mortality [7,9] with as few as 30% of young surviving their first winter. Other estimates of survival probabilities during hibernation are more favourable, ranging from a mean of 0.66 in southern Sweden [10] to 0.89 ($n = 18$) in Denmark [9] and even 1.0 in Denmark ($n = 6$) [11]. Survival of winter may not occur if an individual is in poor condition and/or has small body weight [5,9,12]. Estimates of percent loss of weight during hibernation vary even within a country. For instance, in Britain, mean (±s.e.) weight-loss has been estimated as 14.11 (±3.08)% [13] and as 0.2% of original body-weight (or 0.8–2.0 g day^{-1}) being lost for each day spent in hibernation [14]. In southern Ireland, loss of body-weight was 17 (±0.53)% over hibernations lasting 148.9 (±0.5) days [15]. In Denmark, Jensen [11] recorded weight-loss during hibernation of 22.1 (±10.1)% for juveniles (during a mean of 79 days hibernation, $n = 6$) and 30.2 (±10.1)% for adult females (during a mean of 198 days, $n = 3$). Furthermore, also in Denmark, Rasmusson [9] measured juvenile weight-loss as 16 (±2.9)%. Weight-loss will depend on pre-hibernation weight and lighter individuals typically lose smaller percentages of their pre-hibernation mass than do larger individuals [11,16]. Where durations of hibernation vary among individuals, the % change in weight will also vary and so to make fair comparisons among individuals or studies, it may be more informative to consider change in weight day^{-1}.

1.2. Spontaneous Arousals

All hibernating mammals exhibit brief awakenings called spontaneous or periodic arousals during hibernation, where normal temperatures are regained and activity-levels increase [17]. Such spontaneous arousal is energetically expensive, with up to 75% of total energy requirement during hibernation being associated with arousals [18]. Considering how energetically expensive arousal during hibernation appears to be, the exact function and effects remain poorly understood. Possible functions are speculated to involve recovery from physiological costs accrued during metabolic depression, which may include oxidative stress, reduced immunocompetence and neuronal damage [17]. In the wild, hedgehog activity during these periods of arousal varies considerably. Walhovd [19] observed multiple arousals during hibernation (indicated by spontaneous increases in nest temperature), but no departure from the nest. Other studies report that arousals involve changing nest, sometimes several times [8,9,11]. Yarnell et al. [13] observed hibernation of 57 British hedgehogs in the wild, recording between one and seven nest changes per hedgehog. The types of nest and nesting material used (e.g., compost heap, twigs, under a bush, under a building) also vary among individuals and within a hibernation [9,20,21]. Arousal duration in outdoor hibernacula can last between 34 and 44 h [19] and frequency of spontaneous arousal has been recorded at 2.9 events per month ($n = 29$ individuals) [22]. Other similar figures are reported, yet demonstrate a large range, with 3–15 days between arousals and a total arousal duration during hibernation of 12–18 days [19]. Similarly, Buckle [23] reported a range of between 5 and 15 days between arousals whilst Kristoffersson and Soivio [24] reported that the longest periods between arousals lasted from 10–13 days.

Not all hedgehogs lose weight during hibernation. Of 10 radio-tracked hedgehogs in Copenhagen, five increased in weight during hibernation [9], almost certainly because of opportunities to feed during periods of arousal. If arousals are required regularly (for physiological reasons) then the number of arousals should increase with the duration of the total hibernation period. Different frequencies and durations of arousal may influence weight-loss, because feeding is possible during arousal and/or because arousal is energetically expensive. Evidence is contradictory for whether the number of arousals is positively or negatively correlated with weight-loss, as summarised nicely by Haigh et al. [15].

1.3. Threshold Mass to Survive Hibernation

Given all this variation, there is much discussion about the minimal body-weight that would allow a hedgehog to survive hibernation. From post-hibernation weights of 105 hedgehogs, Morris [25] estimated each individual's pre-hibernation weight, by assuming a likely loss of 25% of their body-weight during hibernation. With a buffer of 50 g to ensure sufficient fat reserves to survive the unpredictable British winters, he suggested a minimal weight for otherwise healthy individuals would be 450 g. In rural Denmark, a greater body-weight of 513 g was needed to survive hibernation, likely due to the harsher winter conditions [11]. In Ireland, late juveniles could survive the winter with a pre-hibernation weight of 475 g or more [15]. The minimal body-weight needed to survive hibernation at present may differ from these estimates, due to regional variation in climate and the passage of time since publication with the effects of climatic change on winter temperatures [9,26].

The health of a hedgehog and its probability of surviving winter in the wild is almost certainly down to more than just its absolute weight. A large, skinny hedgehog may weigh the same as a small fat hedgehog, but the former is in poor condition and may be more prone to mortality over winter than the latter. For those concerned with over-winter rehabilitation of hedgehogs in captivity, knowledge of hedgehog condition and likely patterns of behaviour will influence decisions about treatment. For example, body-weight prior to hibernation and anticipated change in weight may determine whether individuals are released to hibernate in the wild, or whether they are to hibernate in captivity or if they should be kept warm, awake, feeding and undergoing any necessary veterinary treatment. At present, in the UK, a minimal body-weight of 550 [7] or 600 g [5,12] is used as the threshold below which individuals are not released from rescue centres before winter and instead allowed to gain weight and hibernate in the centre.

1.4. Hibernation in Captivity

Conditions for hibernation in captivity are, however, clearly different to those in the wild. For instance, captive hedgehogs cannot move far during arousal, cannot change nest, have only one type of nest material and food is readily available in the enclosures. Handling during hibernation for weighing or health checks may also artificially induce arousal [8,27]. Whether or how captive conditions will affect hibernatory behaviours when compared to those in the wild is not certain.

Given the large numbers of hedgehogs hibernating in rescue centres over winter, knowledge of how weight will be lost and variation in behaviours among individuals is required because they have implications for husbandry (specifically mid-hibernation health-checks, weighing and artificial arousal). Thus, relationships between pre-hibernation mass, spontaneous arousal frequency and mass-loss during hibernation at rescue centres need further exploration. This will help ensure that appropriate intervals for weighing (and associated disturbance) are used, minimising the risk of fatalities during hibernation and improving the success of rehabilitation and release.

We used data on weight and activity of healthy juvenile hedgehogs available from records of existing husbandry protocols kept by a specialist hedgehog rescue centre, to test the following sets of hypotheses.

Hibernation with or without weighing: To test whether disturbance by weighing in mid-hibernation affected overall hibernation, we predicted that there would be a difference in mean total hibernation periods (nights), mean number of arousal events, mean duration of arousal events (nights) and mean number of nights sleep between arousal events between the two winters.

Associations with total hibernation period: We predicted that there would be correlations between the total hibernation period and the number of arousal events; or with percentage of time spent in arousal. More specifically, we expected that the longer the hibernation, the more arousal events there would be (a positive correlation) and that there could be more or less time spent asleep (a two-tailed correlation).

Associations with percent weight-loss: We tested hypotheses that weight-loss would be associated with: total hibernation period, number of arousal events, and percentage of nights asleep. Our expectation was that weight-loss would increase with the length of hibernation (a positive correlation), but we had no particular prediction for how it would change with number of arousal events or the proportion of time spent asleep (two-tailed correlations)

Associations with pre-hibernation weight: Finally, we tested whether pre-hibernation weight would be associated with: daily % weight-loss, actual weight-loss, total hibernation period, number of arousal events, mean number of nights awake per arousal event, and mean number of nights asleep between arousal events. In each case, we were uncertain about how the variables might change with pre-hibernation weight (i.e., all correlations were 2-tailed).

2. Materials and Methods

The study was conducted on 35 healthy, juvenile hedgehogs in good condition (17 males, 18 females) undergoing rehabilitation at Prickles and Paws Hedgehog Rescue Centre, Cornwall, UK, over the 2015/2016 and 2016/2017 winters. The centre, which works closely with a local veterinary practice, is a registered charity which operates to the standards of, and admits hedgehogs from, the Royal Society for the Prevention of Cruelty to Animals (RSPCA). The centre does not hold a UK Home Office licence under the Animal (Scientific Procedures) Act 1986, and data were collected only through the normal, approved operating protocols of the centre.

Individuals used in the study were admitted to the centre for a range of reasons and, if necessary, appropriate medical treatments, as prescribed by the centre's veterinarian, were administered (Supplementary Material 1). Historically, many rescue centres have opted to keep rehabilitated hedgehogs over winter, rather than releasing them before spring, even if in full health [7], and this has been the approach adopted by Prickles and Paws. No individual was considered for use in this study and allowed to hibernate unless it fulfilled the conditions of the centre's Hibernation Protocol (Supplementary Material 2) including having all medications completed, being >600 g in weight and in good body condition. To account for mass relative to size, a simple body-condition index specifically for hedgehogs was developed [28]. Testing the application of this index, Rasmussen et al. [9] recently found it to be of uncertain reliability due to the need for precise measurements, difficult to achieve on hedgehogs. In the present study, body condition was determined by visual assessment of body-shape; near-spherical shape when curled-up indicates good body-condition, a triangular rear-end indicates underweight condition [28]. Hedgehogs in the study were housed individually in outdoor hutches (floor area 0.31–0.49 m^2), where illumination, food and nesting material (shredded newspaper), were standardised as much as possible. Paper, whilst different to natural nesting materials, is readily available, dust-free, widely used in British rescue centres and recommended by other authors [7].

Between November 2015 and March 2016, weight-change and hibernatory behaviour were monitored for 21 healthy individuals. Prior to hibernation, individuals were weighed at least weekly to the nearest g using digital scales. When individuals displayed signs of entering hibernation (reduced food-intake and dark green faeces) all handling was ceased. There are several clinical reasons for green faeces, but in otherwise healthy hedgehogs, they may be an indicator of incipient hibernation and, therefore, a fair cue for changes in husbandry. If they had not entered hibernation within the next three days, they would then be weighed again. Thus, the time between last recorded weight and the start of hibernation varied from 1 to 6 days. When in hibernation, hedgehogs were checked daily by viewing through a mesh door and spontaneous arousals were noted when there was evidence of recent disturbance in the hutch (e.g., spread of nesting material, presence of faeces, food disturbed), which indicated the individual had left the nest area. After 6 weeks in hibernation, individuals were weighed; if their mass had fallen below 400 g, then arousal was induced artificially. This mass was considered a minimal threshold to ensure survival of the remainder of hibernation if there were no further spontaneous arousals. Dried food for small carnivores (e.g., cat biscuits) was available throughout, but food consumption during arousal was not recorded.

To assess potential effects of mid-hibernation weighing, frequencies and durations of spontaneous arousals, but not mid- or end-hibernation weights, were recorded for 14 individuals during a second winter (December 2016 to March 2017). The sample-size for 2016/17 was smaller, due to fewer healthy individuals of adequate mass being at the centre.

Entry into hibernation was defined as the first night that the individual did not leave the nest area (as determined by no disturbance in the hutch). The date of final arousal from hibernation was the first of five consecutive nights of activity outside the nest area. Total hibernation period was the number of nights between first entry into hibernation and final arousal (i.e., including nights in spontaneous arousal). A spontaneous arousal event constituted a series of consecutive nights spent active during the total hibernation period. Periods of spontaneous arousal began when an individual left the nest area and ended when the individual returned to the nest and no further disturbance was observed. The percentage time in spontaneous arousal was the total number of nights in spontaneous arousal as a percentage of the total hibernation period. The duration of an interval between arousals was the number of nights between the end of one arousal and the start of the next. Behaviours tending to indicate spontaneous rather than final arousal (e.g., minimal food intake and activity) meant that sometimes, several days elapsed before final arousal was certain. Thus, post-hibernation weight was recorded as soon as final arousal was confirmed (typically within 3 days and always within 6 days).

Statistical analysis: Weight change was measured as percentage or actual weight-loss relative to pre-hibernation weight, with positive or negative values representing decreases or increases in weight, respectively. Spontaneous arousal events were counted, measured for duration (number of nights per event) or total duration (sum of all events per hedgehog converted to a percentage of the total hibernation period). For correlations, Shapiro-Wilk normality tests were used to assess frequency distributions of data. Where distributions of data were non-normal or where at least one variable was a percentage, Spearman's rank correlation coefficients were used to establish significance of associations. When comparing variables between years, Fisher's F-tests were used to test for heteroscedasticity. Where variances were homogeneous, Student's *t*-test was used. All analyses were done in R version 4.0.2 [29].

3. Results

3.1. Hibernation with or without Weighing

Mean total hibernation periods were of similar duration in 2015/16 and in 2016/17 (Figure 1, Table 1). Maximal total hibernation periods for the two winters were 111 nights (106:5 nights asleep:awake) and 70 nights (68:2), respectively. Minimal durations were 13 nights (9:4) and 5 nights (5:0). Spontaneous arousals were observed in all but one individual (which hibernated for only 5 nights). The most spontaneous arousal events for a single individual in 2015/16 was 15 and in 2016/17, three individuals displayed six arousals. The mean number of arousals did not differ between the two winters (Figure 1, Table 1). Some spontaneous arousals lasted multiple nights, the longest being 9 nights, but neither the mean number of nights per event nor the number of nights between events differed between the two winters (Figure 1, Table 1). Variances were homogeneous for each variable (Table 1). The lack of differences between years meant that data from 2015/16 and 2016/17 were pooled.

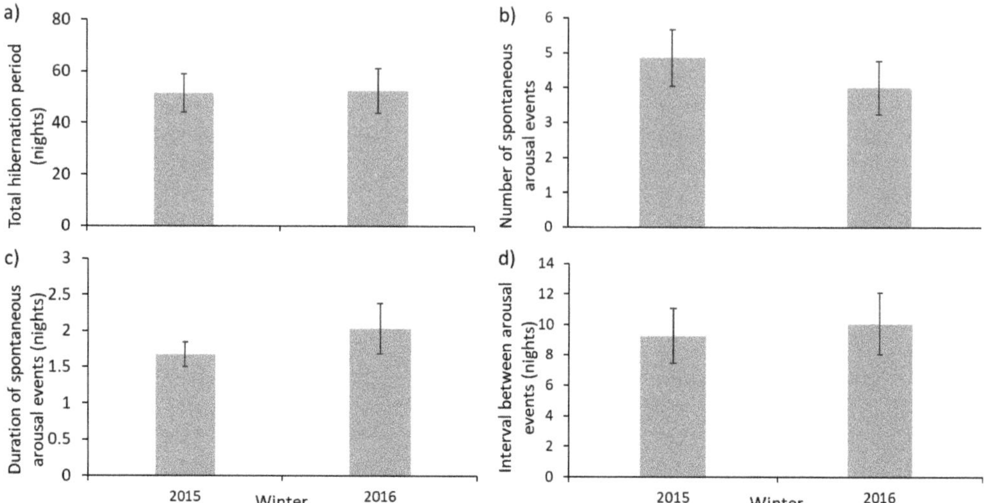

Figure 1. Mean (±s.e.) components of hibernation by hedgehogs in Cornwall between the winters of 2015–2016 (n = 21) and 2016–2017 (n = 13); (**a**) total hibernation period, (**b**) number of spontaneous arousal events, (**c**) duration of spontaneous arousal events, (**d**) interval between spontaneous arousal events i.e., number of consecutive days asleep.

Table 1. Fisher's F tests for heteroscedasticity and Student's *t*-tests to compare means for total hibernation period, number of arousal events, duration of arousal events (nights) and number of nights between arousal events between the winters of 2015/16 and 2016/17.

Variable	Fisher's *F* Test			Student's *t*-Test		
	F	d.f.	*p*	*t*	d.f.	*p*
Total hibernation period	1.20	20, 12	>0.70	−0.08	32	>0.90
No. of arousal events	1.80	20, 12	>0.20	0.72	32	>0.40
Duration of arousal events	0.39	20, 12	>0.05	−1.04	32	>0.30
Duration of interval between arousal events	1.31	20, 12	>0.60	−0.28	32	>0.70

3.2. Associations with Total Hibernation Period

Total hibernation period was positively correlated with the number of arousals (r_s = 0.40, n = 34, p < 0.01) and negatively correlated with the percentage of nights awake (r_s = −0.35, n = 34, p < 0.05, Figure 2). This meant that the longer the hibernation period, the greater the number of arousals and the smaller the proportion of time spent awake, which prompted a further analysis for which we had no a priori hypothesis. Because the number of arousals increased and the proportion of time awake decreased, the longer the total hibernation period, we predicted that arousals would on average be shorter, the longer the total hibernation period. There was a tendency towards this prediction, but the relationship was not significant (r_s = −0.22, n = 33, p > 0.1). None of the variables were normally distributed (Shapiro-Wilks, p < 0.05), so Spearman rank correlation coefficients were used.

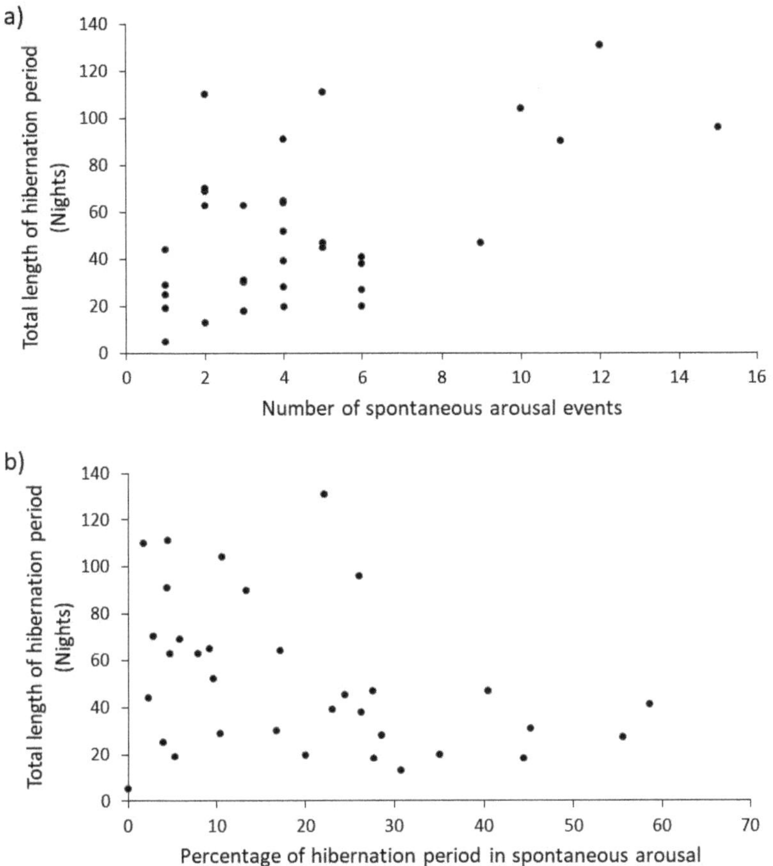

Figure 2. Correlations between total hibernation period and (**a**) number of spontaneous arousal events and (**b**) percentage of nights spontaneously aroused ($n = 34$).

3.3. Associations with Percent Weight-Loss

As predicted, the longer the total period of hibernation, the greater the percentage weight-loss ($r_s = 0.67$, $n = 21$, $p < 0.001$, Figure 3a). Weight-gain during hibernation, demonstrated by four individuals, shows that hedgehogs can eat during arousals. This pattern remained when hedgehogs that gained weight were removed from the analysis ($r_s = 0.69$, $n = 17$, $p < 0.001$), i.e., animals that gained weight by eating during arousals were not causing this pattern. Unexpectedly, percent weight-loss was not associated with the number of arousal events ($r_s = 0.12$, $n = 21$, $p > 0.60$, Figure 3b), but there was a significant positive correlation between percent weight-loss and percent of nights asleep ($r_s = 0.66$, $n = 21$, $p < 0.01$, Figure 3c). So, the proportion of time spent asleep in the hibernation period is more strongly associated with weight-loss during hibernation than is the number of arousal events. Variables were not normally distributed (Shapiro-Wilk, $p < 0.05$) or were percentages, so Spearman rank correlation coefficients were used in all cases.

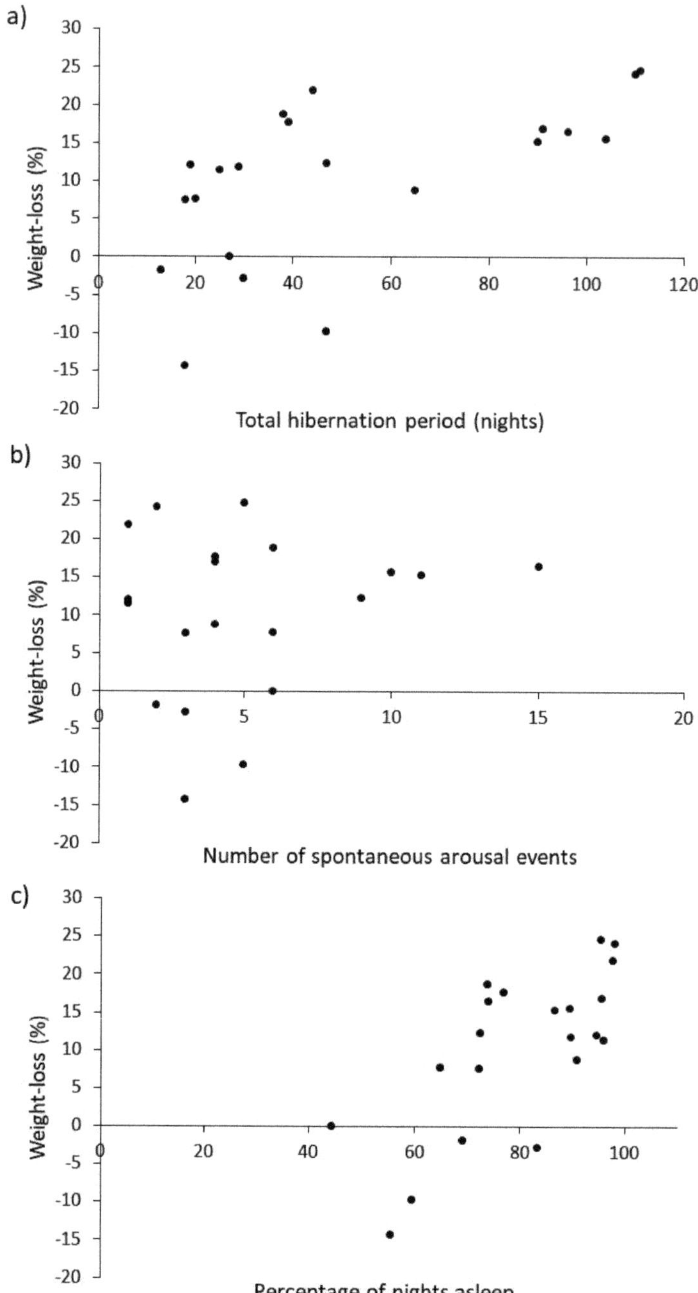

Figure 3. Correlations between percent weight-loss and (**a**) total hibernation period, (**b**) number of arousal events and (**c**) percentage of nights spent asleep (*n* = 34).

3.4. Associations with Pre-Hibernation Weight

There was a significant negative correlation between pre-hibernation weight of hedgehogs and the daily percentage weight-loss and with actual weight-loss; larger hedgehogs lost smaller proportions of their bodyweight each day than did smaller hedgehogs. This pattern was unaltered if hedgehogs that gained weight were excluded (Figure 4a, Table 2). There were no associations between pre-hibernation weight and total hibernation period, mean duration of arousal events or mean duration of intervals between arousal events (Table 2), but there was a significant positive correlation between pre-hibernation weight and the number of arousal events (Figure 4b, Table 2). So, the larger the hedgehog, the more often it wakes up during hibernation, but the less weight it loses. Variables were not normally distributed (Shapiro-Wilk, $p < 0.05$) or were percentages, so Spearman rank correlation coefficients were used in all cases.

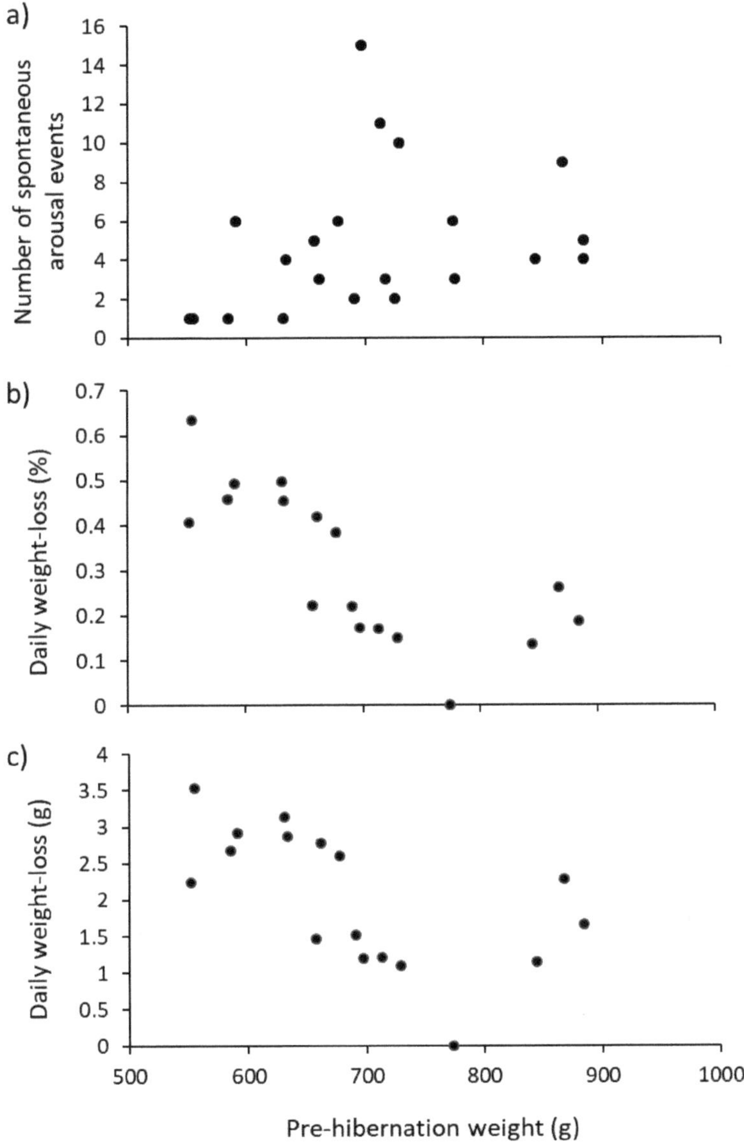

Figure 4. Correlations between pre-hibernation weight (g) of hedgehogs in the winter of 2015/16 ($n = 17$) and (**a**) number of arousal events, (**b**) percent daily weight-loss and (**c**) actual daily weight-loss (with (**b,c**) excluding individuals that gained weight).

Table 2. Correlations between pre-hibernation weights and daily weight-loss, total hibernation period, number and durations of arousal events and the duration of intervals between arousal events. Significant results are in italics.

Correlation of Pre-Hibernation Weight with	r_s	n	p
% daily weight-loss (incl hedgehogs that gained weight)	*−0.80*	*21*	*<0.001*
% daily weight-loss (excl hedgehogs that gained weight)	*−0.80*	*17*	*<0.001*
Actual daily weight-loss (excl hedgehogs that gained weight)	*−0.67*	*17*	*<0.01*
Total hibernation period	0.24	21	>0.2
No. of arousal events	*0.45*	*21*	*<0.05*
No. of nights per arousal event	0.23	21	>0.3
No. of nights between arousal events	−0.30	21	>0.1

4. Discussion

Patterns of weight-loss and activity by hedgehogs during hibernation described in this study can inform over-winter husbandry of hedgehogs at rescue centres, for which there are no set or widely-used guidelines or recommendations about frequency of weighing and handling, nor robust evidence for how to anticipate weight-loss. We recommend that monitoring (e.g., frequency of weighing) should be tailored to individuals, particularly those likely to lose weight faster based on their pre-hibernation weight and hibernatory behaviour.

4.1. Disturbance and Weighing

When comparing data from two winters, one with and one without weighing in mid-hibernation, there were no significant differences in the mean duration of total hibernation period, number of spontaneous arousal events, duration of these events or of the interval between these events (Table 1). This suggests that a single mid-hibernation weighing at the rescue centre did not influence important components of the hibernation. Of 21 hedgehogs weighed mid-hibernation in 2015–2016, only one woke soon afterwards and this may have coincided with a 'normal' spontaneous arousal. This suggests that concerns about effects of disturbance may be less serious than previously believed [8,27].

4.2. Spontaneous Arousals

The mean number of arousals per hedgehog fell within previously described ranges [19,24]. The mean and lower end of the range of number of nights between arousals was similar to those in Walhovd [19], but the upper end of the range (36 nights) was greater than elsewhere (15 nights [19]; 13 nights, [24]). The determinants of patterns of arousal are not yet known, but might include: differing methodologies used to observe these arousals (e.g., thermal measurements, visual observations, radiotracking); variation in environmental conditions; or variation in pre-hibernation mass or body-condition among studies. In the present non-invasive study, spontaneous arousal was only measured or recorded when an individual left their nest. This means that arousals that did not involve leaving the nest [8,19], could not be identified, and were not recorded. This could explain the apparently greater maximal number of nights between spontaneous arousals.

Although the mean duration of arousal events was typically <2 nights, some individuals woke for as long as nine consecutive nights, considerably longer than the 34–44 h reported by Walhovd [19] or the mean of 33.7 h ($n = 6$) observed by Kristoffersson and Soivio [24]. The latter did, however, observe one spontaneous arousal event lasting 121 h. Formal comparisons of published mean durations of spontaneous arousal with the present study cannot be made due to the different ways of measuring this variable. The long duration of some arousal events (up to 9 nights) could indicate that some individuals are not entering deep hibernation, perhaps due to conditions in the centre (food, nesting materials, disturbance) being different to those in the wild or to the relatively benign winter conditions in Cornwall.

4.3. Weight-Loss

Spontaneous arousal is energetically expensive [18,30] and so more frequent switching between metabolic states of hibernation and arousal should add to the total energy 'bill' for the hibernation. The expectation was that the number of spontaneous arousal events would increase linearly with the length of the total hibernation period (as demonstrated here, Figure 2a). If their purpose is to prevent oxidative stress, reduced immunocompetence and neuronal tissue damage [17], longer hibernation would increase these risks and so more arousals are 'required' to offset physiological costs. The percentage of time spent awake decreased (Figure 2b) as hibernations got longer. A logical inference from this is that mean duration of arousal event should also be shorter when hibernations are longer, but the physiological basis for this is not clear. Whilst our data tend towards this pattern, there was no significant negative correlation. Larger sample-sizes from data collected over more years would provide greater statistical power when testing this prediction.

As predicted, weight loss (a proxy for energetic cost) increased the longer the hibernation (Figure 3a). This strongly suggests that when making further comparisons between weight-loss and other variables (e.g., sex, body condition, starting weight, etc.) weight-loss should be presented as loss per day ($g \cdot d^{-1}$) rather than total weight-loss. In contrast with expectations, daily weight-loss was, however, associated more strongly with amount of sleep (Figure 3c) rather than the number of arousal events (Figure 3b). This may indicate that leaving and entering hibernation is not as energetically expensive as previously thought [18,30], but is perhaps more easily explained by the ready availability of food. In rescue centres, a small amount of food is provided throughout hibernation, so hedgehogs could feed each time they wake. The more often a hedgehog wakes and eats, the costs of arousal could be offset and less weight would be lost. Four individuals in this study even gained weight during their hibernation period (Figure 3). Such increases in weight may be explained by feeding during periods of arousal and/or taking on food in the time between final arousal and post-hibernation weighing. For reasons of certainty about final arousal, weighing post-hibernation can happen up to 6 days after waking, but is typically less than half this. Those that gained weight were also those that spent amongst the smallest proportion of their time asleep (Figure 3c), meaning that they had relatively more time available to feed during their hibernation. Although recording food consumption was not part of the original protocols, it merits further investigation as it may influence aspects of husbandry such as frequency of weighing required, and the amount of food provided during hibernation.

Body-weight is an important variable that helps determine types and amounts of medication prescribed by a veterinarian and administered by rescue centres to unwell hedgehogs. It also influences whether healthy hedgehogs should be allowed to enter hibernation and whether they should be released [5]. We expected there to be some association between pre-hibernation weight and five different characteristics of hibernation (Table 2). We present evidence that hedgehogs of different weight do hibernate differently. For instance, the number of arousal events increased with pre-hibernation weight (Figure 4a), but length of hibernation did not (Table 2). If larger hedgehogs need to wake more often during hibernation, then they may be developing relatively more physiological costs than smaller individuals and therefore need to off-set these costs with more regular arousals. This is in contrast with broad recognition that metabolic rates scale inversely with body-size [31]; however, it may be that hibernation creates different physiological costs than just calorific expenditure for arousal.

If weight is an underlying determinant of arousal, a regression (using a larger sample over multiple years) of number of arousals against pre-hibernation weight would allow prediction of how often hedgehogs will wake. Although not strictly statistically rigorous (due to values of the x-variable not being fixed), this could be used to anticipate the total (and different) amounts of food required for each animal (for potential consumption during arousal).

Daily weight-losses (mean ± s.e.; $1.18 \pm 0.48 \ g \cdot d^{-1}$) fell within Wroot's [14] values of 0.8–$2.0 \ g \cdot d^{-1}$, but our range (3.5 to $-6.2 \ g \cdot d^{-1}$) was much greater. Daily weight-loss in the present study (in terms of percentage and absolute amounts) were negatively related to pre-hibernation weight (Figure 4b,c). So, larger hedgehogs woke up more often (Figure 4a), but lost less weight than did smaller hedgehogs

(Figure 4b,c). This is the opposite to previous studies [9,11,32], which showed that individuals that had more mass to lose tended to lose more mass. This unexpected pattern is, again, perhaps best explained by the ready availability of food. Heavier individuals wake up more often and can, therefore, feed more often, meaning that they lose less weight, highlighting the need to monitor food consumption during captive hibernation.

4.4. Captive vs. Wild Hibernation

Conditions for hibernation in captivity are clearly rather different to those in the wild. Captive hedgehogs have no ability to move nests or select different nesting materials (as are observed in wild hedgehogs [9,15,21]. In the wild, animals have the opportunity to move around and forage during periods of arousal, but food may not be readily available because it is wintertime. Although captive hedgehogs can move around their hutch when awake, they cannot move the distances observed for wild hedgehogs. Decreases in food availability may well be one of the cues for hibernation [8], and moving nests may consume considerable energy. Thus, artificial conditions where food is plentiful and movement is restricted may cause hibernation by juveniles in captivity to be different from those in the wild. In particular, the duration and depth of hibernation may be less. Fruitful areas of research might include regional variation in duration and initiation of hibernation and how these may respond to climatic change. There is already evidence that the 'trigger' temperature for hibernation varies across Europe [8] and unseasonally mild conditions, potentially the consequence of a warming world [26], can delay the onset the hibernation [9,15]. For example, when the Autumn of 2014 was far milder than normal, juvenile hedgehogs in Denmark initiated hibernation a month later than shown in previous studies [9]. In the present study, durations of hibernation were similar between years (Figure 1a), but the maximal hibernation period was much shorter in the 2016/17 winter (70 nights) than in 2015/16 (111 nights). This may have been a consequence of the warmer mean minimal temperatures in February and March 2017 (6.1 °C and 8.0 °C) than in the previous year (5.1 °C and 5.1 °C; [33]). Unseasonably warm periods during hibernation may stimulate arousals and additional energy expenditure [34]. With increasingly mild winters [35], and unseasonably mild episodes in winter, the occurrence of multiple distinct periods of hibernation per winter, separated by abnormally long arousal events appear to be more evident. It would be useful to evaluate whether these behaviours occur more broadly. Long-term changes in the timing and duration of hibernation may require physiological and behavioural adaptation as well as phenological adjustments to altered availability of prey items. Such changes could have implications for rehabilitation methods and practice in order to maximise both welfare and its contribution to species conservation.

4.5. Recommendations for Overwinter Monitoring

The implications of these patterns observed here for over-winter husbandry of hedgehogs in rescue centres are several. We recommend closer monitoring of arousals and of food consumption during arousals.

Provision of food: Availability of food in captivity appears to reduce the amount of weight lost during hibernation; at least, those individuals that woke (and thus could feed) more often lost less weight per day than those that woke less frequently. If arousal is energetically expensive [18,30], but necessary for other physiological reasons [17], then these costs can be offset by eating. Food should certainly be made available for hedgehogs hibernating in captivity for consumption during periods of activity, to help offset weight-loss and improve likelihoods of surviving hibernation. The possibility that availability of food is encouraging shorter hibernations and more frequent, longer arousals requires greater investigation.

Monitoring weight and behaviour: The lack of evidence that mid-hibernation weighing causes disturbance could be interpreted as meaning that this level of monitoring should not be of major concern for the welfare of overwinter hedgehogs and supports the notion of weighing at intervals more frequent than 6 weeks. This could be particularly valuable for smaller hedgehogs (that lose weight faster than larger individuals) that may be approaching weights (e.g., ≤ 400 g) where the risk of not surviving hibernation is greater. This approach would benefit from further studies to test for any cumulative effects of multiple disturbances. Durations of hibernation vary amongst individuals, so weight-loss is better presented as $g \cdot d^{-1}$ or $\% \cdot d^{-1}$ rather than total loss.

Weights and duration of hibernation should not be the only variables to be monitored during over-winter care of hedgehogs at rescue centres. Frequency and duration of spontaneous arousals, indicated by disturbance outside the nest area, are easily and non-invasively observable, without disturbing the animal. Unfortunately, recording consumption of food was not part of the protocol that provided the data for the present study. Given the negative association between weight-loss and number of arousals, knowing how often and how much food is consumed would have been informative. Individuals waking regularly but not eating may well lose weight much faster than those that do eat and maybe even faster than those that wake infrequently. Such individuals may benefit from more frequent weighing to ensure they are artificially woken if they approach the minimal threshold for weight. Protocols for observations should be expanded to include the routine collection of data on frequency and duration of arousals and consumption of food, thereby allowing comparisons among locations, sex and prior conditions (weight, disease, medication, etc.). Further investigation is required to determine how these multiple variables influence hibernation.

5. Conclusions

Rescue centres can collect large amounts of data and have the potential to create evidence-based monitoring protocols intended to improve welfare of patients and success of rehabilitation. Key points that were demonstrated include: (i) mid-hibernation weighing did not seem to affect hibernation; (ii) weight-loss increased with duration of hibernation, but appeared more strongly associated with the proportion of time spent asleep than with the number of arousal events; (iii) in contrast with previous studies, larger hedgehogs lost less weight per day than did smaller hedgehogs; (iv) mean values for components of hibernation were similar to values recorded from individuals in the wild, but extremes for weight-loss (or gain) and duration of arousal events were much greater for animals being managed in a rescue centre. Much of (ii) and (iii) can be ascribed to plentiful availability of food. Differences in hibernatory behaviour between captive or wild hedgehogs may well be due to a range of non-natural conditions in hutches. There is increasing evidence that milder winters associated with climatic change are changing the ways in which hedgehogs hibernate. This has uncertain implications for the long-term future of declining populations. The patterns described here provide much-needed information for rescue centres caring for hedgehogs and highlight areas for further research. Hibernation protocols for rescue centres should be updated following the recommendations we make. We hope that this will allow more successful rehabilitation which will, in turn, help support populations of this charismatic, threatened species.

Supplementary Materials: The following are available online at http://www.mdpi.com/2076-2615/10/8/1418/s1, Table S1: Hedgehog husbandry records; Protocol S2: Hibernation protocol 2015–2016 winter.

Author Contributions: Conceptualization, K.E.S. and K.H.; methodology, K.E.S.; validation, K.E.S., K.H. and A.C.J.; formal analysis, A.C.J. and K.E.S.; investigation, K.E.S.; resources, K.E.S.; data curation, K.E.S.; writing—original draft preparation, K.E.S.; writing—review and editing, K.H. and A.C.J.; visualization, K.E.S. and A.C.J.; supervision, K.H. and A.C.J.; project administration, K.E.S. All authors have read and agreed to the published version of the manuscript.

Funding: This research received no external funding.

Acknowledgments: We thank all the staff, volunteers and trustees of Prickles and Paws Hedgehog Rescue for allowing us to carry out the study and supporting data collection. The authors are very grateful to Nigel Reeve for providing advice and direction to an early draft of this article and for the constructive comments of two anonymous reviewers.

Conflicts of Interest: The authors declare no conflict of interest.

References

1. Molony, S.E.; Baker, P.J.; Garland, L.; Cuthill, I.C.; Harris, S. Factors that can be used to predict release rates for wildlife casualties. *Anim. Welf.* **2007**, *16*, 361–367.
2. Croft, S.; Chauvenet, A.L.M.; Smith, G.C. A systematic approach to estimate the distribution and total abundance of British mammals. *PLoS ONE* **2017**, *12*, e0176339. [CrossRef] [PubMed]
3. Matthews, F.; Kubasiewicz, L.M.; Gurnell, J.; Harrower, C.A.; McDonald, R.A.; Shore, R.F. A Review of the Population and Conservation Status of British Mammals: Technical Summary. A Report by the Mammal Society under Contract to Natural England, Natural Resources Wales and Scottish Natural Heritage. Available online: https://www.mammal.org.uk/wp-content/uploads/2018/06/MAMMALS-Technical-Summary-FINALNE-Version-FM2.pdf (accessed on 24 August 2018).
4. Stocker, L. *Practical Wildlife Care*, 2nd ed.; Blackwell Publishing: Oxford, UK, 2005; pp. 200–215.
5. Bullen, K. *Hedgehog Rehabilitation*, 2nd ed.; British Hedgehog Preservation Society: Ludlow, UK, 2010; pp. 31–73.
6. Ruf, T.; Geiser, F. Daily torpor and hibernation in birds and mammals. *Biol. Rev.* **2014**, *90*, 891–926. [CrossRef] [PubMed]
7. Bexton, S. Hedgehogs. Chapter 12. In *BSAVA Manual of Wildlife Casualties*, 2nd ed.; Mullineaux, E., Keeble, E., Eds.; British Small Animal Veterinary Association: Gloucester, UK, 2016; pp. 117–136.
8. Reeve, N.J. *Hedgehogs*; T & A D Poyser Ltd.: London, UK, 1994; pp. 139–161.
9. Rasmussen, S.L.; Berg, T.B.; Dabelsteen, T.; Jones, O.R. The ecology of suburban juvenile European hedgehogs (*Erinaceus europaeus*) in Denmark. *Ecol. Evol.* **2019**, *9*, 13174–13187. [CrossRef]
10. Kristiansson, H. Population variables and causes of mortality in a hedgehog (*Erinaceus europaeus*) population in Southern Sweden. *J. Zool.* **1990**, *220*, 391–404. [CrossRef]
11. Jensen, A.B. Overwintering of European hedgehogs *Erinaceus europaeus* in a Danish rural area. *Acta Theriol. (Warsz)* **2004**, *49*, 145–155. [CrossRef]
12. Hibernation-Weight Hedgehog Hibernation Weight—A Collaborative View. Available online: http://www.britishhedgehogs.org.uk/pdf/Hibernation-Weight.pdf (accessed on 1 August 2020).
13. Yarnell, R.W.; Surgey, J.; Grogan, A.; Thompson, R.; Davies, K.; Kimbrough, C.; Scott, D.M. Should rehabilitated hedgehogs be released in winter? A comparison of survival, nest use and weight change in wild and rescued animals. *Eur. J. Wildl. Res.* **2019**, *65*, 6. [CrossRef]
14. Wroot, A.J. Feeding Ecology of the European Hedgehog *Erinaceus europaeus*, Royal Holloway, University of London, UK, 1984. Available online: http://ethos.bl.uk/OrderDetails.do?uin=uk.bl.ethos.363484 (accessed on 4 August 2017).
15. Haigh, A.; O'Riordan, R.M.; Butler, F. Nesting behaviour and seasonal body mass changes in a rural Irish population of the Western hedgehog (*Erinaceus europaeus*). *Acta Theriol. (Warsz)* **2012**, *57*, 321–331. [CrossRef]
16. Rasmussen, S.L. Personality, Survivability and Spatial Behaviour of Juvenile Hedgehogs (*Erinaceus europaeus*): A Comparison between Rehabilitated and Wild Individuals. Master's Thesis, University of Copenhagen, Copenhagen, Denmark, 2013.
17. Humphries, M.M.; Thomas, D.W.; Kramer, D.L. The role of energy availability in mammalian hibernation: A cost-benefit approach. *Physiol. Biochem. Zool.* **2003**, *76*, 165–179. [CrossRef]
18. Thomas, D.W.; Geiser, F. Periodic arousals in hibernating mammals: Is evaporative water loss involved? *Funct. Ecol.* **1997**, *11*, 585–591. [CrossRef]
19. Walhovd, H. Partial arousals from hibernation in hedgehogs in outdoor hibernacula. *Oecologia* **1979**, *40*, 141–153. [CrossRef] [PubMed]
20. Rautio, A.; Valtonen, A.; Auttila, M.; Kunnasranta, M. Nesting patterns of European hedgehogs (*Erinaceus europaeus*) under northern conditions. *Acta Theriol. (Warsz)* **2014**, *59*, 173–181. [CrossRef]
21. Huijser, M.P.; Bergers, P.J.M. The effect of roads and traffic on hedgehog (*Erinaceus europaeus*) populations. *Biol. Conserv.* **2000**, *95*, 111–116. [CrossRef]
22. Dmi'el, R.; Schwarz, M. Hibernation patterns and energy expenditure in hedgehogs from semi-arid and temperate habitats. *J. Comp. Physiol. B* **1984**, *155*, 117–123. [CrossRef]
23. Buckle, A.P. Some Aspects of the Relationship between Certain Epifaunistic Insects and Their Hosts. Available online: https://ethos.bl.uk/OrderDetails.do?uin=uk.bl.ethos.480373 (accessed on 9 July 2020).

24. Kristoffersson, R.; Soivio, A. Hibernation of the hedgehog (*Erinaceus europaeus*). The periodicity of hibernation of undisturbed animals during the winter in a constant ambient temperature. *Ann. Acad. Sci. Fenn. A IV Biol.* **1964**, *80*, 1–22.

25. Morris, P. An estimate of the minimum body-weight necessary for hedgehogs *Erinaceus europaeus* to survive hibernation. *J. Zool.* **1984**, *203*, 291–294.

26. IPCC Special Report: Global Warming of 1.5 °C. Chapter 1, p. 59. Available online: https://www.ipcc.ch/sr15/ (accessed on 23 March 2019).

27. Pajunen, I. Ambient-temperature dependence of the periodic respiratory pattern during long-term hibernation in the garden dormouse, *Eliomys quercinus* L. *Ann. Zool. Fennici.* **1984**, *21*, 143–148.

28. Bunnell, T. The assessment of British hedgehog (*Erinaceus europaeus*) casualties on arrival and determination of optimum release weights using a new index. *J. Wildl. Rehabil.* **2002**, *25*, 11–21.

29. R Core Team. *R: A Language and Environment for Statistical Computing*; R Foundation for Statistical Computing: Vienna, Austria, 2020. Available online: https://www.R-project.org/ (accessed on 14 August 2020).

30. Wang, L.C.H. Energetic and field aspects of mammalian torpor—The Richardson's ground squirrel. *J. Therm. Biol.* **1978**, *3*, 87. [CrossRef]

31. Glazier, D.S. Effects of metabolic level on the body size scaling of metabolic rate in birds and mammals. *Proc. R. Soc. Ser. B Biol. Sci.* **2008**, *275*, 1405–1410. [CrossRef]

32. Morris, P.A.; Warwick, H. A study of rehabilitated juvenile hedgehogs after release into the wild. *Anim. Welf.* **1994**, *3*, 163–177.

33. Newquay Weather Station. Newquay Weather Station Yearly Temperature Summaries. Available online: http://www.newquayweather.com/wxtempsummary.php (accessed on 6 June 2020).

34. Newman, C.; Macdonald, D.W. The Implications of Climate Change for Terrestrial UK Mammals; Biodiversity Climate Change Impacts. Report Card 2015; p. 20-1. Available online: https://nerc.ukri.org/research/partnerships/ride/lwec/report-cards/biodiversity/ (accessed on 20 July 2020).

35. Tandon, A.; Schultz, A. Met Office: The UK's Wet and Warm Winter of 2019–20. Available online: https://www.carbonbrief.org/met-office-the-uks-wet-and-warm-winter-of-2019-20 (accessed on 28 June 2020).

Review

A Review of the Occurrence of Metals and Xenobiotics in European Hedgehogs (*Erinaceus europaeus*)

Sophie Lund Rasmussen [1,2,3,*], Cino Pertoldi [2,4], Peter Roslev [2], Katrin Vorkamp [5] and Jeppe Lund Nielsen [2]

1. Wildlife Conservation Research Unit, The Recanati-Kaplan Centre, Department of Biology, University of Oxford, Tubney House, Tubney, Abingdon OX13 5QL, UK
2. Department of Chemistry and Bioscience, Aalborg University, 9220 Aalborg, Denmark; cp@bio.aau.dk (C.P.); pr@bio.aau.dk (P.R.); jln@bio.aau.dk (J.L.N.)
3. Linacre College, University of Oxford, St. Cross Road, Oxford OX1 3JA, UK
4. Aalborg Zoo, 9000 Aalborg, Denmark
5. Department of Environmental Science, Aarhus University, 4000 Roskilde, Denmark; kvo@envs.au.dk
* Correspondence: sophielundrasmussen@gmail.com or sophie.rasmussen@biology.ox.ac.uk

Simple Summary: The European hedgehog (*Erinaceus europaeus*) is a popular visitor in gardens and recreational areas all over Europe, but hedgehog populations are declining. Research exploring the causes of the decline, including exposure to potentially harmful pollutants and metals, may provide relevant information to improve conservation initiatives to protect this species in the wild. Hedgehogs are ground-dwelling mammals, feeding on a range of different food items such as insects, slugs, snails and earthworms but also eggs, live vertebrates, and carrion, and therefore come into close contact with pollutants present in their habitats and in their prey. This review investigated published research on the occurrence of metals and pollutants in hedgehogs and found that a vast range of different pesticides; rodenticides; persistent organic pollutants (POPs), including organochlorine compounds and brominated flame retardants (BFRs); as well as toxic heavy metals could be detected in samples from hedgehogs representing different European countries. Due to their ecology, combined with the opportunity to apply non-invasive sampling techniques through the collection of spines as sampling material, we suggest that the European hedgehog is a relevant bioindicator species for monitoring the exposure of omnivorous terrestrial wildlife to potential toxicants in urban and rural environments.

Citation: Rasmussen, S.L.; Pertoldi, C.; Roslev, P.; Vorkamp, K.; Nielsen, J.L. A Review of the Occurrence of Metals and Xenobiotics in European Hedgehogs (*Erinaceus europaeus*). *Animals* **2024**, *14*, 232. https://doi.org/10.3390/ani14020232

Academic Editor: D. N. Rao Veeramachaneni

Received: 23 November 2023
Revised: 23 December 2023
Accepted: 8 January 2024
Published: 11 January 2024

Abstract: Monitoring data from several European countries indicate that European hedgehog (*Erinaceus europaeus*) populations are declining, and research exploring the causes of the decline, including exposure to potentially harmful xenobiotics and metals, may inform conservation initiatives to protect this species in the wild. Hedgehogs are ground-dwelling mammals, feeding on a range of insects, slugs, snails, and earthworms, as well as eggs, live vertebrates, and carrion, including carcasses of apex predator species representing higher levels of the food chain. Consequently, hedgehogs come into close contact with contaminants present in their habitats and prey. This review investigated the studies available on the subject of the occurrence of metals and organic xenobiotics in hedgehogs. This study found that a vast range of different pesticides; persistent organic pollutants (POPs), including organochlorine compounds and brominated flame retardants (BFRs); as well as toxic heavy metals could be detected. Some compounds occurred in lethal concentrations, and some were associated with a potential adverse effect on hedgehog health and survival. Due to their ecology, combined with the opportunity to apply non-invasive sampling techniques using spines as sampling material, we suggest that the European hedgehog is a relevant bioindicator species for monitoring the exposure of terrestrial wildlife to potential toxicants in urban and rural environments.

Keywords: European hedgehogs; *Erinaceus europaeus*; xenobiotics; heavy metals; environmental pollution; toxicants; target screening; non-target screening; bioaccumulation; wildlife conservation

1. Introduction

According to the United Nations Environment Programme [1] and the European Environment Agency [2], an estimated 40,000–60,000 different industrial chemicals are globally used in commerce to produce a vast range of commodities and goods, including chemical-intensive products such as computers, mobile phones, furniture, and personal care products. The European Environment Agency also estimated for 2016 that 62% of the total volume of 345 million tonnes of chemicals consumed in the European Union were hazardous to human health [2]. Several programmes have been established to monitor the occurrence of hazardous chemicals in the environment, such as the European Union Water Framework Directive [3], the Arctic Monitoring and Assessment Programme [4], or the Partnership for the Assessment of Risks from Chemicals [5]. Where biota are included, measurements often focus on the aquatic environment—in particular, fish species—whereas relatively few monitoring initiatives exist for (terrestrial) wildlife [6].

Xenobiotics are chemical substances that do not occur naturally in the organism that is studied [7]. There are several origins of xenobiotics, including industrial, household, pharmaceutical, agricultural, and transportation sources [8]. They are used in a variety of products of modern-day society and include compounds such as pharmaceuticals, personal care products, food additives, pesticides and biocides, plastic additives, and detergents [9]. Some xenobiotic compounds may have problematic properties, including toxic effects on wildlife [8]. The xenobiotics represented in this review include organochlorine industrial chemicals (e.g., polychlorinated biphenyls), brominated flame retardants (BFRs), pesticides—i.e., insecticides (including phased-out persistent ones such as DDT), rodenticides, fungicides, herbicides, nematicides, and biocides.

Metals are defined as solid substances with high electrical and thermal conductivity. They occur naturally in the lithosphere, and their compositions and concentrations vary among different localities [10]. Heavy metals are metals with relatively high atomic weights and specific densities (e.g., $\geq 5 \text{ g/cm}^3$). In low concentrations, some metals play an essential role in maintaining various biochemical and physiological functions in living organisms (i.e., essential metals), but they become harmful when threshold concentrations are exceeded [10], similar to non-essential metals. This can lead to adverse effects on living organisms and the environment, specifically with exposure to lead, cadmium, mercury, and arsenic as the main threats [11]. Simultaneously, some metals are essential and therefore occur naturally in vertebrates such as hedgehogs [12]. These include sodium (Na), potassium (K), magnesium (Mg), calcium (Ca), iron (Fe), manganese (Mn), cobalt (Co), copper (Cu), zinc (Zn), and molybdenum (Mo), and it is currently also accepted that metal elements such as chromium (Cr) and nickel (Ni) should be included in that category, as vertebrates show certain deficiency symptoms when these metals are absent or in low concentrations [12]. Therefore, being mindful of this, distinguishing between naturally occurring low levels and elevated toxic levels remains important when interpreting results for chemically screening and detecting metals in hedgehogs.

The European hedgehog (*Erinaceus europaeus*), hereafter referred to as hedgehog, is widely distributed across Europe [13,14]. Nowadays, it primarily resides in habitats with human activity and occupation [15–19], including habitats with a potential exposure to xenobiotic chemicals, such as urban areas. As hedgehogs prey on a variety of insect species, earthworms, and slugs, occasionally supplementing their diet with carrion, eggs, and live vertebrate prey when available, they are potentially exposed to xenobiotic compounds from a variety of sources such as soil (topic absorption) and different types of prey species, including apex predators (carrion), by ingestion [20–24]. Hedgehogs have small home ranges and tend to stay in the same area throughout their lives [13–15]. It is therefore likely that xenobiotics in hedgehogs represent local pollution levels from the area from which the hedgehogs originated. Despite a mean suggested life expectancy of around two years (see Rasmussen et al. (2023) [25] Table 1 for an overview), hedgehogs have the potential to reach 16 years of age [25], which means they could experience long-term exposure to different xenobiotics and metals, potentially causing harmful effects in some individuals. Previous studies have found that insectivores have a greater risk of metal intoxication compared to other small mammal species like rodents [26,27]. Hedgehogs are likely exposed to metal during foraging, as they prey on a variety of insect species, earthworms, and slugs [22], all of which are known to accumulate high metal levels [28].

Hedgehogs are easy to catch and handle, and their spines and hair can be used for chemical analyses, which allows for non-destructive sampling methods [29,30]. Although these arguments make hedgehogs good candidates for monitoring programmes, few studies have focussed on their exposure to environmental xenobiotics and metals in urban and rural environments. Furthermore, substantial evidence, based on monitoring data from a range of European countries, indicates that hedgehog populations are declining [16,31–40]. The suspected causes for the decline include habitat loss; habitat fragmentation; inbreeding; intensified agricultural practices; road traffic accidents; a reduction in biodiversity and hence, food items; lack of suitable nest sites in residential gardens; accidents caused by garden tools; netting and other anthropogenic sources in residential gardens; badger predation; and infections with pathogens and endoparasites [15,17,32,41–53].

Research exploring the potential causes of the decline and conservation initiatives to protect this species in the wild should consider the role of potentially harmful chemicals. Consequently, we consider it relevant to review the existing data on the occurrence of xenobiotics and metals in hedgehogs. Thus, our objective was to provide an update on the occurrence of organic xenobiotics and metals in hedgehogs, which may inspire and inform future studies on exposure to xenobiotics and metals, including their effects in hedgehogs.

2. Materials and Methods

To produce this literature review, the Google Scholar and Web of Science (WOS) search tools were used with these keywords: Erinaceus europaeus OR hedgehog AND a combination of 19 different search words, each entered separately, e.g., *Erinaceus europaeus* OR hedgehog AND toxicology (toxicology OR ecotoxicology OR accumulation OR xenobiotic OR bioindicator OR chemicals OR target screening OR non-target screening OR metal OR pollutants OR rodenticides OR pesticides OR herbicides OR insecticides OR molluscicides OR acaricides OR brominated flame retardants OR persistent organic pollutants OR organochlorine compounds).

A total of 25 results were obtained, although some were unpublished conference abstracts or reports mentioning the poisoning of hedgehogs without presenting concentration levels or specific chemicals detected [29,30,54–76]. These studies used a variety of sample types from hedgehogs to detect chemical compounds such as rodenticides, persistent organochlorine compounds, brominated flame retardants, metalloids, and metals. Figure provides an overview of the different sample types and compounds studied.

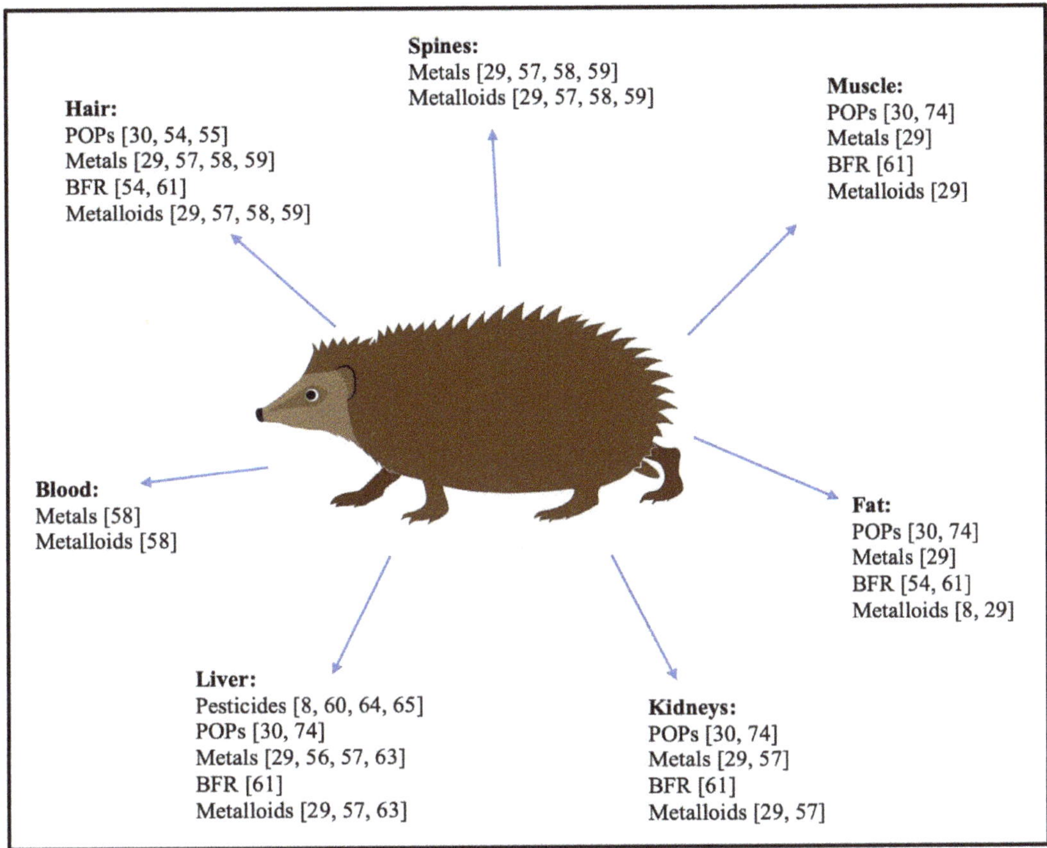

Figure 1. Overview of sample types used for research on xenobiotics in European hedgehogs. Abbreviations: brominated flame retardants (BFRs). Pesticides (rodenticides, insecticides, fungicides, herbicides, and nematicides) analysed: rodenticides: warfarin, coumatetralyl, difenacoum, bromadiolone, brodifacoum, flocoumafen. A total of 55 different insecticides, fungicides, herbicides, and nematicides [65]. Organochlorine compounds analysed: polychlorinated biphenyls (PCBs), dichlorodiphenyl-trichloroethanes (DDTs), hexachlorobenzene (HCB), octachlorostyrene (OCS), chlordane (CHL), hexachlorocyclohexanes (HCHs). BFRs analysed: polybrominated diphenyl ethers (PBDEs) and brominated biphenyl 153 (BB 153). Metalloids analysed: selenium and arsenic. Metals analysed: silver (Ag), aluminium (Al), cadmium (Cd), cobalt (Co), chromium (Cr), copper (Cu), iron (Fe), mercury (Hg), magnesium (Mg), manganese (Mn), molybdenum (Mo), nickel (Ni), lead (Pb), zinc (Zn).

3. Results

3.1. Insecticides, Fungicides, Herbicides, and Nematicides

In 2021, Schanzer et al. [65] published their research screening for 55 pesticides (insecticides, fungicides, herbicides, and nematicides) in livers from six hedgehogs, dying at a wildlife rehabilitation centre in Germany. In these liver samples, the fungicides fenpropimorph and tebuconazole, the insecticides dieldrin and permethrin, as well as the metabolites fipronil sulfone (originating from the insecticide fipronil) and p,p'-DDE (originating from the persistent organic pollutant insecticide p,p'-DDT) were detected [65]. A data summary is provided in Table 1. Adding to these investigations of pesticides, Luzardo et al. (2014) [75] detected six unspecified carbamate insecticides in liver samples from six hedgehogs collected from wildlife poisoning episodes in 2010–2012. Additionally, a study reported one incidence of poisoning with the herbicide paraquat in a hedgehog the UK and one case of poisoning of a hedgehog with the rodenticide chlorophacinone France [72]. Carbamate insecticides and organophosphate insecticides were detected in or hedgehog, and anticoagulant rodenticides were detected in two individuals in a study c wildlife dying from suspected poisoning in Italy [73]. Gemmeke (1995) [76] experimente with the dosage of metaldehyde in live hedgehogs to determine the risk of seconda poisoning with metaldehyde. The author served 200 slugs poisoned with metaldehyd to six adult hedgehogs. Of the six hedgehogs tested, four ate all, or close to all, of tl 200 slugs served, and two ate 0 and 12, respectively. None of the hedgehogs were reporte to show any adverse symptoms, behavioural differences, or signs of poisoning. Howeve Keymer et al. (1991) [77] diagnosed metaldehyde poisoning in three dead hedgehogs cc lected from the UK between 1976 and 1986, and detected concentrations of up to 80 mg/k of acetaldehyde (a by-product and metabolite of metaldehyde).

Table 1. Results from the screening for insecticides, fungicides, herbicides, and nematicides in liv samples from German hedgehogs by Schanzer et al. (2021) [65]. The sampling material was six live from German hedgehogs dying in care, and they were analysed by gas chromatography coupled tandem mass spectrometry (GC-MS/MS).

Compound	Pesticide Type	Levels Detected µg/g	Frequency of Positives %	Frequency of Positives N/n
p,p′-DDE	Metabolite of the insecticide p,p′-DDT	0.001–0.22	50	3/6
Fenpropimorph	Fungicide	0.0005–0.002	66.67	4/6
Fipronil sulfone	Insecticide	0–0.05	100	6/6
Tebuconazole	Fungicide	0–0.008	33.33	2/6
Dieldrin	Insecticide	0.003–0.0031	16.67	1/6
Permethrin	Insecticide	0.007–0.0073	16.67	1/6

3.2. Rodenticides

Rodenticides are widely used and are known to accumulate within food chain posing a threat to the survival of birds of prey and predatory mammals [78]. Given tha the hedgehogs' natural diet includes vertebrate cadavers [20–24], scavenging on poisone rats and mice is not an uncommon behaviour for hedgehogs, potentially causing seconda poisoning with rodenticides. Furthermore, hedgehogs may also scavenge on carcasse of predatory species preying on rodents or ingest rodenticide pellets directly, if these a accessible to the hedgehogs in, e.g., bait boxes with holes large enough to fit a hedgeho head or by spreading the pellets directly on the ground [79–81]. Rodenticides are als detected in non-target invertebrates such as beetles and slugs [82,83], which constitute considerable proportion of the natural diet of hedgehogs [22,23].

The occurrence of rodenticides in hedgehogs has been examined in a few studi described in this section. Dowding et al., 2010 [60] analysed 120 livers from hedgehog dying in care in the UK for first-generation anticoagulant rodenticides (FGAR) and secon generation anticoagulant rodenticides (SGAR). They detected rodenticides in a total 67% of the samples (Table 2), with a low detection frequency of flocoumafen (1/120) an the highest detection frequency of difenacoum (57/120). Detectable levels of rodenticid in liver samples from European hedgehogs ranged from 0.03 to 0.25 µg/g wet weigl (Table 2). Lopéz-Perea et al. (2015) [70] screened for six anticoagulant rodenticides i liver samples from 48 hedgehogs dying in care in Spain in 2011–2013. The results showe a detection frequency ranging from 0% (warfarin) to 50% (brodifacoum), with anticoa ulant rodenticides detected in 28 out of 48 individuals and a total mean concentratio of 0.122 µg/g anticoagulant rodenticides detected per individual [70]. A study of live from two hedgehogs from Spain, dying of suspected poisoning at a wildlife rehabilitatio centre in 2005–2010, screened for six different rodenticides [64]. The screening detecte bromadiolone (N = 2/2, mean 0.026 µg/g wet weight, range 0.013–0.049 µg/g wet weigh and brodifacoum (N = 1/2, mean 0.092 µg/g wet weight). A conference poster, presenting a study screening for anticoagulant rodenticides in six hedgehogs from Scotland collected in 2003–2013, showed a detection range of 33% and a median residue of 0.047 µg/g for unspecified anticoagulant rodenticides [69].

Based on the high levels of rodenticides detected in some of the hedgehogs, Dowding et al. (2010) [60] suggested that lethal poisoning by rodenticides was likely to occur in some hedgehogs. In the study by Sánchez-Barbudo et al. (2012) [64], it was described that, in the wildlife carcasses chosen for rodenticide screening, death was suspected to have occurred from poisoning by anticoagulant rodenticides due to discernible haemorrhages detected during necropsies, which presumably then also applied to the hedgehog carcasses included in the study, suggesting that the concentrations of rodenticides found in the two individuals were lethal. Rasmussen et al. (2019) [15] used radio tracking to monitor independent juvenile hedgehogs in the suburbs of Copenhagen (Denmark) and found one suspected case of lethal rodenticide poisoning, but the carcass was in an advanced state of decay when retrieved, preventing a chemical analysis of rodenticides. In New Zealand, where the European hedgehogs introduced are considered pests, publications have described how hedgehogs are efficiently controlled with rodenticides through the application of aerial baits and bait stations containing sodium fluoroacetate (1080) [66], which is not approved for use in European countries [84], and brodifacoum [67,68]. In one of these studies, 32 of the targeted hedgehogs were found dead and later confirmed as having been poisoned [68]. In a study conducted under laboratory conditions to investigate blood coagulation factors and the effect of warfarin, ten hedgehogs were injected with 0.4 mg/kg body weight warfarin on three consecutive days [71]. No individuals died, but the effect was a large decrease in coagulation factor activity after warfarin treatment [71]. However, even if the rodenticide doses analysed in the studies in Table 2 may not be lethal for the hedgehogs, repeated exposure, or a certain bioaccumulation, has the potential to cause different toxicological effects, which may compromise the fitness and survival of the hedgehogs. With detection frequencies of rodenticides reaching up to 99% in predatory species [81] and >90% in slugs [85], the secondary exposure to rodenticides in hedgehogs through the ingestion of poisoned prey could potentially lead to an even higher prevalence in hedgehogs than the 67% found by Dowding et al. (2010) [60].

Table 2. Results from the screening for rodenticides in liver samples from hedgehogs. Abbreviations used: first-generation anticoagulant rodenticides (FGAR), second-generation anticoagulant rodenticides (SGAR), United Kingdom (UK), Spain (S), not applicable (na). All the hedgehogs tested had died in care. All the analyses were based on liquid chromatography (LC), with either fluorescence or mass spectrometric detection.

Reference	Compound	Type	Sample Size	Sampling Year	Country	Median Levels Detected (µg/g Wet Weight)	Mean Levels Detected (µg/g Wet Weight)	Frequency %	Frequency N/n
Dowding et al., 2010 [60]	Warfarin	FGAR	120	2004–2006	UK	0.03		8	10/120
Dowding et al., 2010 [60]	Coumatetralyl	FGAR	120	2004–2006	UK	0.05		14	17/120
Dowding et al., 2010 [60]	Difenacoum	SGAR	120	2004–2006	UK	0.05		13	16/120
Dowding et al., 2010 [60]	Difenacoum	SGAR	120	2004–2006	UK	0.06		48	57/120
Dowding et al., 2010 [60]	Bromadiolone	SGAR	120	2004–2006	UK	0.25		11	13/120
Dowding et al., 2010 [60]	Bromadiolone	SGAR	120	2004–2006	UK	0.05		19	23/120
Dowding et al., 2010 [60]	Brodifacoum	SGAR	120	2004–2006	UK	0.06		5	6/120
Dowding et al., 2010 [60]	Brodifacoum	SGAR	120	2004–2006	UK	0.04		3	4/120
Dowding et al., 2010 [60]	Flocoumafen	SGAR	120	2004–2006	UK	na		1	1/120
Dowding et al., 2010 [60]	Flocoumafen	SGAR	120	2004–2006	UK	na		0	0/120
Dowding et al., 2010 [60]	Total SGAR	SGAR	120	2004–2006	UK	na		58	69/120
Dowding et al., 2010 [60]	Total SGAR	SGAR	120	2004–2006	UK	na		23	27/120
Dowding et al., 2010 [60]	Total FGAR and SGAR	FGAR + SGAR	120	2004–2006	UK	na		67	80/120
Sánchez-Barbudo et al., 2012 [64]	Bromadiolone	SGAR	2	2005–2010	S		0.026	100	2/2
Sánchez-Barbudo et al., 2012 [64]	Brodifacoum	SGAR	2	2005–2010	S		0.092	50	1/2
Lopéz-Perea et al., 2015 [70]	Warfarin	FGAR	48	2011–2013	S		ND	0	0/48
Lopéz-Perea et al., 2015 [70]	Coumatetralyl	FGAR	48	2011–2014	S		0.08	27	13/48
Lopéz-Perea et al., 2015 [70]	Difenacoum	SGAR	48	2011–2015	S		0.008	25	12/48
Lopéz-Perea et al., 2015 [70]	Bromadiolone	SGAR	48	2011–2016	S		0.0074	12.5	6/48
Lopéz-Perea et al., 2015 [70]	Brodifacoum	SGAR	48	2011–2017	S		0.043	50	24/48
Lopéz-Perea et al., 2015 [70]	Flocoumafen	SGAR	48	2011–2018	S		0.023	4	2/48

3.3. Organochlorine Compounds

Many of the organochlorine compounds that have been detected in hedgehogs are regulated by the UN Stockholm Convention on persistent organic pollutants (POPs), which covers chemicals that are persistent, bioaccumulative, toxic to humans and wildlife, and can be transported over long distances [86].

The compounds that have been analysed in hedgehogs include polychlorinated biphenyls (PCBs), dichloro-diphenyl-trichloroethane (DDT and its metabolites), hexachlorobenzene (HCB), octachlorostyrene (OCS), chlordane (CHL), and hexachlorocyclohexanes (HCHs). All of these compounds, except OCS, are considered POPs according to the UN Stockholm Convention.

They were originally synthesised for industrial purposes (e.g., PCBs,) or as agrochemicals (DDT, chlordane, lindane (an HCH isomer)), while octachlorostyrene mainly forms unintentionally [87]. A common characteristic of these organochlorine compounds is their hydrophobic and lipophilic nature, leading them to bind strongly to solids such as organic matter in soil and aquatic systems and accumulate in fatty tissues of the organisms exposed to the chemicals. Lipolysis triggered by exercise can release PCBs from adipose tissue into the bloodstream [88]. As a result, the concentration, distribution, and metabolism of PCBs in plasma can differ significantly among individuals due to variations in lifestyle and behaviour.

Polychlorinated biphenyls can biomagnify [89], which has caused concern about their impact on organisms, especially on those higher up in the food chain. Exposure to organochlorine compounds can result in severe health effects, including specific cancers [90,91], birth defects, compromised immune function [92], and reproductive system dysfunction [93–96]. Additionally, it may lead to increased susceptibility to diseases [97] and damage to both the central and peripheral nervous systems [86]. Due to potential synergistic effects, it is challenging to determine the ecotoxicological influence of POPs, as a range of different POPs typically co-occur and accumulate simultaneously in biota due to their omnipresence in the environment [89].

As hedgehogs are ground-dwelling mammals, feeding on earthworms [13], they frequently come into close contact with soil, where hydrophobic organochlorines tend to accumulate, and are therefore also potentially exposed to POPs from this particular source.

Hedgehog muscles, fat, hair, livers, and kidneys have been used to study the accumulation of these organochlorine compounds [30,54–56] (Table 3 and Figure 1).

With sample sizes ranging from 6 to 77 individuals and sampling taking place in Belgium, the Netherlands, and Italy in 1994–2008, the different studies found a high detection frequency of organochlorine compounds in hedgehog samples in general, reaching close to 100% or 100% in many cases (Table 3), with a few exceptions in the hair samples. The occurrence in hair samples appears to be lower compared to the other sample types such as the liver, kidneys, fat, and muscle, which may be explained by the fact that these compounds accumulate in fatty tissues. The levels varied between not detected up to 31,780 ng/g dry weight and 0.1–2.8 ng/mL wet weight in blood samples. Furthermore, but not included in Table 3, the organochlorine compound p,p′-DDE (a metabolite of p,p′-DDT) was detected in three out of six hedgehog livers from Germany at concentrations ranging from 1.03 to 22.23 ng/g [65].

D'Havé et al. (2006a) [30] targeted the PCB congeners 28, 31, 74, 95, 99, 101, 105, 110, 118, 128, 138, 149, 153, 156, 163, 170, 180, 183, 187, 194, and 199 in their analyses. They found that the majority of PCB congeners in all the tissues were PCBs 153, 138/163, and 180, with a joint mean concentration of 53–63% out of the total PCB concentration in each tissue analysed.

D'Havé et al. (2007) analysed the PCB congeners 28, 31, 74, 99, 101, 105, 110, 118, 128, 138, 149, 153, 156, 170, 180, 183, 187, 194, and 199 in the hair samples from hedgehogs tested in the study. They found that PCB 118, PCB 138, and PCB 153 were dominant in the hair samples.

Table 3. Results from studies investigating organochlorine compounds in European hedgehogs. Abbreviations used: Belgium (BE), the Netherlands (NL), Italy (I), not detected (ND), not applicable (na), polychlorinated biphenyls (PCBs), dichloro-diphenyl-trichloroethane and metabolites (DDTs), hexachlorobenzene (HCB), octachlorostyrene (OCS), chlordane (CHL), hexachlorocyclohexanes (HCHs). Unknown is written when the number of samples is unknown and when it is unknown whether only positive samples were represented in the dataset. All the samples were analysed with gas chromatograph (GC)–mass spectrometry (MS). * denotes a range of medians for samples representing seven study sites.

Reference	Compound	Sample Material	Sample Size	Sampling Year	Country	Levels Detected (ng/g Dry Weight)	Levels Detected (ng/mL Wet Weight)	Median Levels Detected (ng/g Dry Weight)	Hedgehogs Analysed	Frequency of Positives %	Frequency of Positives N/n
CHL											
D'Havé et al., 2006a [30]	CHL	Fat	6	2002–2003	BE + NL	1.9–14.1	na	4.7	Dead, roadkill, and dead in care	100	6/6
D'Havé et al., 2006a [30]	CHL	Hair	45	2002–2003	BE + NL	ND–1.2	na	ND	Dead, roadkill, and dead in care	33	15/45
D'Havé et al., 2006a [30]	CHL	Kidney	44	2002–2003	BE + NL	0.02–26.4	na	0.5	Dead, roadkill, and dead in care	100	44/44
D'Havé et al., 2006a [30]	CHL	Liver	43	2002–2003	BE + NL	0.2–75.7	na	5.1	Dead, roadkill, and dead in care	100	43/43
D'Havé et al., 2006a [30]	CHL	Muscle	44	2002–2003	BE + NL	0.01–20.7	na	0.5	Dead, roadkill, and dead in care	100	44/44
DDT											
Vermeulen et al., 2010 [54]	DDTs	Blood	13	2005–2007	BE	na	0.1–0.4	na	Wild, live hedgehogs	Unknown	Unknown
D'Havé et al., 2006a [30]	DDTs	Fat	6	2002–2003	BE + NL	5.51–194	na	18.1	Dead, roadkill, and dead in care	100	6/6
Alleva et al., 2006 [56]	DDTs	Fat	Unknown	1994–1995	I	0–27,680	na	1490	Dead, roadkill	na	na
Vermeulen et al., 2010 [54]	DDTs	Hair	18	2005–2006	BE	0.2–5.7	na	na	Wild, live hedgehogs	Unknown	Unknown
D'Havé et al., 2006a [30]	DDTs	Hair	45	2002–2003	BE + NL	ND–725	na	2.5	Dead, roadkill, and dead in care	91	41/45
D'Havé et al., 2007 [55]	DDTs	Hair	77	2002	BE	ND–84	na	ND–0.9	Wild, live hedgehogs	60	46/77
D'Havé et al., 2006a [30]	DDTs	Kidney	44	2002–2003	BE + NL	ND–1313	na	1.8	Dead, roadkill, and dead in care	98	43/44
D'Havé et al., 2006a [30]	DDTs	Liver	43	2002–2003	BE + NL	ND–750	na	1.4	Dead, roadkill, and dead in care	98	42/43
D'Havé et al., 2006a [30]	DDTs	Muscle	44	2002–2003	BE + NL	0.02–1444	na	2.3	Dead, roadkill, and dead in care	100	44/44

Table 3. *Cont.*

Reference	Compound	Sample Material	Sample Size	Sampling Year	Country	Levels Detected (ng/g Dry Weight)	Levels Detected (ng/mL Wet Weight)	Median Levels Detected (ng/g Dry Weight)	Hedgehogs Analysed	Frequency of Positives %	Frequency of Positives N/n
HCB											
D'Havé et al., 2006a [30]	HCB	Fat	6	2002–2003	BE + NL	1.6–67.2	na	4.42	Dead, roadkill, and dead in care	100	6/6
D'Havé et al., 2006a [30]	HCB	Hair	45	2002–2003	BE + NL	0.02–334.7	na	0.16	Dead, roadkill, and dead in care	78	35/45
D'Havé et al., 2007 [55]	HCB	Hair	77	2002	BE	ND–679	na	ND–40.7	Wild, live hedgehogs	55	42/77
D'Havé et al., 2006a [30]	HCB	Kidney	44	2002–2003	BE + NL	0.02–160.1	na	0.26	Dead, roadkill, and dead in care	100	44/44
D'Havé et al., 2006a [30]	HCB	Liver	43	2002–2003	BE + NL	0.02–247.6	na	0.28	Dead, roadkill, and dead in care	100	43/43
D'Havé et al., 2006a [30]	HCB	Muscle	44	2002–2003	BE + NL	0.04–135.3	na	0.31	Dead, roadkill, and dead in care	100	44/44
Chu et al., 2003 [74]	HCB	Fat	5	2001–2002	BE	1.61–82.54	na	20.08	Unknown	na	na
Chu et al., 2003 [74]	HCB	Liver	10	2001–2002	BE	0.11–4.49	na	1.27	Unknown	na	na
Chu et al., 2003 [74]	HCB	Muscle	11	2001–2002	BE	0.09–5.03	na	0.97	Unknown	na	na
Chu et al., 2003 [74]	HCB	Kidney	11	2001–2002	BE	0.09–4.65	na	0.96	Unknown	na	na
HCHs											
D'Havé et al., 2006a [30]	HCHs	Fat	6	2002–2003	BE + NL	1.1–2.4	na	1.4	Dead, roadkill, and dead in care	100	6/6
D'Havé et al., 2006a [30]	HCHs	Hair	45	2002–2003	BE + NL	ND–105.5	na	0.7	Dead, roadkill, and dead in care	93	42/45
D'Havé et al., 2007 [55]	HCHs	Hair	77	2002	BE	ND–134.8	na	ND–12.8	Wild, live hedgehogs	70	54/77
D'Havé et al., 2006a [30]	HCHs	Kidney	44	2002–2003	BE + NL	0.03–2.9	na	0.2	Dead, roadkill, and dead in care	100	44/44
D'Havé et al., 2006a [30]	HCHs	Liver	43	2002–2003	BE + NL	ND–8.8	na	0.1	Dead, roadkill, and dead in care	98	42/43
D'Havé et al., 2006a [30]	HCHs	Muscle	44	2002–2003	BE + NL	ND–11.5	na	0.2	Dead, roadkill, and dead in care	98	43/44

Table 3. *Cont.*

OCS

Reference	Compound	Sample Material	Sample Size	Sampling Year	Country	Levels Detected (ng/g Dry Weight)	Levels Detected (ng/mL Wet Weight)	Median Levels Detected (ng/g Dry Weight)	Hedgehogs Analysed	Frequency of Positives %	Frequency of Positives N/n
D'Havé et al., 2006a [30]	OCS	Fat	6	2002–2003	BE + NL	0.08–0.5	na	0.4	Dead, roadkill, and dead in care	100	6/6
D'Havé et al., 2006a [30]	OCS	Hair	45	2002–2003	BE + NL	ND–0.06	na	ND	Dead, roadkill, and dead in care	9	4/45
D'Havé et al., 2006a [30]	OCS	Kidney	44	2002–2003	BE + NL	0.01–1.1	na	0.1	Dead, roadkill, and dead in care	100	44/44
D'Havé et al., 2006a [30]	OCS	Liver	43	2002–2003	BE + NL	0.03–3.1	na	0.2	Dead, roadkill, and dead in care	100	43/43
D'Havé et al., 2006a [30]	OCS	Muscle	44	2002–2003	BE + NL	0.01–0.9	na	0.1	Dead, roadkill, and dead in care	100	44/44
Chu et al., 2003 [74]	OCS	Fat	5	2001–2002	BE	0.08–0.49	na	0.34	Unknown	na	na
Chu et al., 2003 [74]	OCS	Liver	10	2001–2002	BE	0.14–1.10	na	0.39	Unknown	na	na
Chu et al., 2003 [74]	OCS	Muscle	11	2001–2002	BE	0.01–0.29	na	0.08	Unknown	na	na
Chu et al., 2003 [74]	OCS	Kidney	11	2001–2002	BE	0.01–0.32	na	0.12	Unknown	na	na

PCB

Reference	Compound	Sample Material	Sample Size	Sampling Year	Country	Levels Detected (ng/g Dry Weight)	Levels Detected (ng/mL Wet Weight)	Median Levels Detected (ng/g Dry Weight)	Hedgehogs Analysed	Frequency of Positives %	Frequency of Positives N/n
Vermeulen et al., 2010 [54]	PCB	Blood	13	2005–2009	BE	na	0.2–2.8	na	Wild, live hedgehogs	Unknown	Unknown
D'Havé et al., 2006a [30]	PCB	Fat	6	2002–2003	BE + NL	89–739	na	273	Dead, roadkill, and dead in care	100	6/6
Alleva et al., 2006 [56]	PCB	Fat	Unknown	1994–1995	I	0–31,780	na	1800	Dead, roadkill	na	na
Vermeulen et al., 2010 [54]	PCB	Hair	18	2005–2008	BE	0.6–13.5	na	na	Wild, live hedgehogs	Unknown	Unknown
D'Havé et al., 2006a [30]	PCB	Hair	45	2002–2003	BE + NL	ND–789	na	10	Dead, roadkill, and dead in care	96	43/45
D'Havé et al., 2007 [55]	PCB	Hair	77	2002	BE	ND–65	na	1–5*	Wild, live hedgehogs	69	53/77
D'Havé et al., 2006a [30]	PCB	Kidney	44	2002–2003	BE + NL	3–5150	na	49	Dead, roadkill, and dead in care	100	44/44
D'Havé et al., 2006a [30]	PCB	Liver	43	2002–2003	BE + NL	2–5910	na	75	Dead, roadkill, and dead in care	100	45/45
D'Havé et al., 2006a [30]	PCB	Muscle	44	2002–2003	BE + NL	5–2940	na	50	Dead, roadkill, and dead in care	100	44/44

Vermeulen et al. (2010) [54] targeted the PCB congeners 99, 101, 118, 138, 153, 156, 170, 180, 183, and 187. Except for PCB 101, all the congeners were detected in the blood and hair samples used in the study, but PCB 118, PCB 138, PCB 153, and PCB 180 were dominant, which is in agreement with the studies by D'Havé et al. and generally presented a group of the most bioaccumulating PCB congeners. Alleva et al. (2006) [56] found that European hedgehogs had the highest levels of PCBs of the mammal species analysed in the study (mean ± SE: 6430 ± 4330 ng/g weight) and that this level was equivalent to those of insectivorous bird species, whereas the levels of fish- and small-mammal-eating bird species were considerably higher (see Table 1 in Alleva et al. (2006) [56] for an overview of the species included in the study). The authors suggested that the lower concentration of organochlorine compounds found in mammals compared to fish- and mammal-eating birds is due to the fact that mammals metabolise organochlorine compounds more readily than birds [98]. They also argued that the higher levels detected in the insectivorous species in general compared to, e.g., herbivorous species could be caused by the direct poisoning of their prey with organochlorine pollutants. Other factors to consider in the interpretation of POPs in wildlife is the sex differences found in species like the polar bear [99] and striped dolphin [100]. The POPs tend to accumulate in fat tissue, exhibiting a notable distinction between males and females in terms of a generally lower BMI index in males with more muscle mass compared to fat tissue [101]. In periods of stress, where an animal is starving, the fat is metabolised, and the accumulated POPs are then transferred from the fat to the blood stream [102,103]. However, females offload POPs via pregnancy and lactation, especially in species like the polar bear with high-nutrition milk (between 27.5 and 35.8% fat) [104,105]. However, the fat percentage in the mother's milk of European hedgehogs is only 10% [14], which may therefore not have a similarly strong effect on the levels of POPs detected in female versus male hedgehogs. In comparison, Chu et al. (2003) [74] found that OCS concentrations in the fat tissue of the hedgehogs (mean 0.34 ng/g ww, n = 5) were similar with levels in the liver (mean 0.39 ng/g wet weight, n = 10), whereas mean HCB levels in the fat tissue (mean 20.08 ng/g lipid weight, n = 5) were markedly higher than in the liver, kidney, and muscle tissue (means ranging from 0.09 to 5.03 ng/g wet weight, n = 10–11) analysed.

3.4. Brominated Flame Retardants (BFRs)

Several families of BFRs have been listed as POPs in the UN Stockholm Convention, including polybrominated diphenyl ethers (PBDEs) and hexabromobiphenyl (HBB) [86]. They were listed later than the most prominent organochlorine compounds, such as PCBs and DDT. HBB and two technical PBDE mixtures, Penta- and OctaBDE, were regulated in 2009, while the third technical PBDE product, DecaBDE, was added to the Stockholm Convention in 2017.

PBDEs and polybrominated biphenyls (PBBs) are two classes of BFRs, which have been used to improve the fire safety of synthetic polymers used in, e.g., electronic equipment, cars, building materials, and textiles [106,107]. The described toxic effects of these BRFs in vertebrates are developmental neurotoxicity, altered thyroid hormone homeostasis, liver conditions (hepatotoxicity), limb deformities in foetuses, and carcinogenic effects (tumours) [108,109]. Two studies so far have analysed the occurrence of PBDEs and the hexabrominated biphenyl BB 153 in hedgehogs [54,61] (Table 4).

Using fat, hair, kidney, liver, and muscle samples from individuals collected in Belgium and the Netherlands, with samples sizes ranging from 6 to 44, BB 153 and PBDEs were detected in all the samples, with median values of <0.10 ng/g wet weight for BB 153 and 1.2–9.5 ng/g wet weight for ∑PBDE.

Table 4. An overview of the results from studies on brominated flame retardants (BFRs) in European hedgehogs. Abbreviations used: polybrominated diphenyl ethers (PBDEs), brominated biphenyl 153 (BB 153), the Netherlands (NE), Belgium (BE), not applicable (na). All the samples were analysed with gas chromatography–mass spectrometry (GC-MS).

Reference	Compound	Sample Type	Sample Size	Sampling Year	Country	Levels Detected (ng/g Wet Weight)	Median Levels Detected (ng/g Wet Weight)	Hedgehogs Analysed
BB 153								
D'Havé et al., 2005b [61]	**BB 153**	Fat	6	2002–2003	BE + NL	<0.10–0.2	<0.10	Roadkill and dead in care
D'Havé et al., 2005b [61]	**BB 153**	Hair	32	2002–2003	BE + NL	<0.05–0.6	0.09	Roadkill and dead in care
D'Havé et al., 2005b [61]	**BB 153**	Kidney	44	2002–2003	BE + NL	<0.10–1.1	<0.10	Roadkill and dead in care
D'Havé et al., 2005b [61]	**BB 153**	Liver	43	2002–2003	BE + NL	<0.10–2.5	<0.10	Roadkill and dead in care
D'Havé et al., 2005b [61]	**BB 153**	Muscle	44	2002–2003	BE + NL	<0.10–1.1	<0.10	Roadkill and dead in care
PBDE								
D'Havé et al., 2005b [61]	**PBDE**	Fat	6	2002–2003	BE + NL	3.1–19.4	9.1	Roadkill and dead in care
Vermeulen et al., 2010 [54]	**PBDE**	Hair	18	2005–2010	BE	0.01–3.3	na	Wild, live hedgehogs
D'Havé et al., 2005b [61]	**PBDE**	Hair	32	2002–2003	BE + NL	0.8–11	1.5	Roadkill and dead in care
D'Havé et al., 2005b [61]	**PBDE**	Kidney	44	2002–2003	BE + NL	0.4–238.9	1.2	Roadkill and dead in care
D'Havé et al., 2005b [61]	**PBDE**	Liver	43	2002–2003	BE + NL	1–1177.5	9.5	Roadkill and dead in care
D'Havé et al., 2005b [61]	**PBDE**	Muscle	44	2002–2003	BE + NL	0.3–316.3	1.5	Roadkill and dead in care

D'Havé et al. (2005b) [61] reported a positive correlation between BFRs in hair and organs when considering the sum of PBDEs, concluding that hair can be used as a non-invasive alternative to organs for the monitoring of PBDE accumulation in hedgehogs. The chosen PBDE congeners 28, 47, 99, 100, 153, 154, and 183 were detected in all the sample types from the hedgehogs (hair, liver, kidney, muscle, and fat tissue) [61]. Except for the hair samples, the PBDE pattern in hedgehogs was dominated by the PBDE 47, followed by PBDEs 153 and 99. Compared with other species of wildlife, the most common PBDE congeners found in a selection of terrestrial herbivorous mammals (rabbits, moose, and reindeer) were BDEs 47, 99, and 100 [110], but, in predatory bird species, BDE 153 was the predominant congener [111,112]. These differences in the detection patterns of PDBEs between the terrestrial wildlife species may be explained by species-specific differences in PBDE metabolism and accumulation as well as food preferences, as hedgehogs and birds of prey are positioned at a higher trophic level than herbivorous species, causing diets and metabolisation to differ [112]. Furthermore, different studies were conducted in different decades, and the composition of PBDE mixtures may have changed during that period, exposing wildlife to different congeners.

3.5. Metals

Research has indicated that insectivores have a greater risk of metal intoxication compared to other small mammal species like rodents [26,27]. As hedgehogs prey on a variety of insect species, earthworms, and slugs [22], all of which are known to accumulate high metal levels [28], they are likely exposed to metals during foraging.

Some metals are essential and therefore occur naturally in vertebrates such as hedgehogs [12]. These include sodium (Na), potassium (K), magnesium (Mg), calcium (Ca), iron (Fe), manganese (Mn), cobalt (Co), copper (Cu), zinc (Zn), and molybdenum (Mo), and it is currently also accepted that metal elements such as chromium (Cr) and nickel (Ni) should be included in that category, as vertebrates show certain deficiency symptoms when these metals are absent or in low concentrations [12]. Therefore, being mindful of this, distinguishing between naturally occurring low levels and elevated toxic levels remains important when interpreting results from chemical screening and detection of metals in hedgehogs.

In our literature search, we found six studies that investigated the presence of metals in hedgehogs, using sample material ranging from hair; spines; and tissues such as kidney, liver, fat, and muscle to blood [29,56–59,63]. The metals tested were silver (Ag), aluminium (Al), cadmium (Cd), cobalt (Co), chromium (Cr), copper (Cu), iron (Fe), mercury (Hg), magnesium (Mg), manganese (Mn), molybdenum (Mo), nickel (Ni), lead (Pb), and zinc (Zn) (see Supplementary Materials for an overview).

The detection of metals was based on samples of blood, spines, livers, muscles, kidneys, hair, and fat, with sample sizes between 7 and 83. The levels detected (blood samples excluded) were Ag (ND–0.62 µg/g, mean ND–0.12 µg/g), Al (ND–230 µg/g, mean 8–76 µg/g), Cd (ND–337 µg/g, mean 0.04–45.17 µg/g), Co (ND–1366 µg/g, mean 0.01–0.99 µg/g), Cr (0–30.9 µg/g, mean 0–5.4 µg/g), Cu (0.2–200 µg/g, mean 1.8–64 µg/g), Fe (ND–2849.76 µg/g, mean 22.94–2339 µg/g), Hg (0.19 µg/g, mean 0.06 µg/g), Mg (46.33–1086.24 µg/g, mean 144.88–731.04 µg/g), Mn (ND–31.11 µg/g, mean 1.85–6.33 µg/g), Mo (ND–5.12 µg/g, mean ND–2.55 µg/g), Ni (ND–35 µg/g, mean 0.07–0.73 µg/g), Pb (ND–31.5 µg/g, mean 0.54–10.9 µg/g), and Zn (0.1–7.47 µg/g, mean 0.06–228.97µg/g).

Including information on the age of the individuals in their study, Rautio et al. (2010) [57] found significant increases in the levels of several metals (Cd, Mo, Cu, Fe, Mn) with increasing age, although this depended on the tissue types analysed, suggesting an age-related bioaccumulation of metals in hedgehogs. This was supported by Jota Baptista et al. (2023) [63], showing that concentrations of Cd and Co were significantly lower in juvenile compared to adult individuals. At present, it appears that no values describing physiologically normal concentrations of essential metals exist for European hedgehogs, which complicates an interpretation and discussion of the concentrations of essential metals in hedgehogs. However, Jota Baptista et al. (2023) [63] detected biliary hyperplasia in 16 of the 45 hedgehogs examined, concluded that concentrations of metals were higher in individuals with this condition, and suggested that heavy metals and metalloids may be the primary contributing factor causing biliary hyperplasia in hedgehogs.

3.6. Selenium and Arsenic Metalloids in Hedgehogs

Selenium is an essential trace element and has several important functions in the metabolism of animals, e.g., as an antioxidant constituting a component of glutathione peroxidase (GSHPx), assisting in intracellular defence mechanisms against oxidative damage [113]. However, selenium poisoning, or selenosis, has been described in production animals through conditions called "blind staggers" and "alkali disease" [114], causing impaired vision, ataxia, and deformities in nails, hooves, and hair [114–116]. However research also suggests that selenoproteins and other selenium metabolites are important in regulating immune function and reducing cancer risk [117]. Selenium deficiency is known to cause a range of health conditions in vertebrates [118,119], which is why selenium is used extensively in fertilizers, especially as an enrichment of livestock feed crops [120]. Natural sources of selenium include marls, gypsum, volcanic eruptions, sea spray, and the weathering of rocks and soils containing selenium. Anthropogenic sources, constituting the majority of the influx of selenium to the environment, include mining, agriculture, coal combustion, insecticide production, oil refining, photocells, and glass production [119]. Industrial and agricultural activities are the dominant anthropogenic sources of selenium pollution in, e.g., soil and drinking water [119].

Arsenic is a widespread element occurring worldwide [121], which originates from natural geogenic sources, as it is a major constituent of more than 245 minerals [122], and from anthropogenic sources. Anthropogenic activities contribute three times as much as natural sources to the accumulation of arsenic in the environment [122]. Out of these industrial effluents constitute the largest contribution. Most of the arsenic is used for the preservation of wood, but the manufacturing of paints, dyes, ceramics and glass, electronics, pigments, and antifouling agents also include arsenic. Agricultural inputs from chemicals such as insecticides, herbicides, desiccants, and fertilizers are a major source of arsenic in soils. Insecticidal products containing arsenic have previously been used extensively for the treatment of ectoparasites in livestock [123].

Arsenic appears in several chemical forms, all with different degrees of toxicity. In organic forms of arsenic (arsenite and arsenate) are more toxic, while methylated forms (methylarsonate (MMA) and dimethylarsinate (DMA)) are moderately toxic [121]. Other arsenic species, like trimethyl-arsine oxide (TMAO) and tetramethyl-arsonium (TETRA) are also considered moderately toxic. By contrast, the forms arsenobetaine (AsB), arsenocholine (AsC), and other arsenosugars (AsS) appear to have low or very low toxicity [124]. The toxicity caused by arsenic exposure is linked to an imbalance between pro-oxidant and antioxidant homeostasis, which results in oxidative stress [125]. The general mechanism behind the toxic effects of arsenic is the oxidative deterioration of polyunsaturated fatty acids, a process known as lipid peroxidation [126]. Research on the health aspects of arsenic exposure has revealed how chronic exposure may cause cancer in the skin, lungs, bladder and liver [127].

Given the potential toxicity of arsenic and selenium exposure, and the presence of these metalloids in the soil and water, it is relevant to explore the occurrence and bioaccumulation of these compounds in hedgehogs. Five different studies so far have addressed the prevalence of the specific metalloids selenium [57] and arsenic [29,57–59,63] in hedgehogs. Table 5 provides an overview of the findings.

Table 5. An overview of the prevalence and detection levels of the metalloids selenium and arsenic in European hedgehogs. Abbreviations used: Belgium (BE), the Netherlands (NL), Finland (FI), Portugal (PT), not detected (ND), not applicable (na), inductively coupled plasma mass spectrometry (ICP-MS), inductively coupled plasma optical emission spectrometry (ICP-OES). * denotes measures based on a range of means from seven study sites.

Reference	Compound	Method	Sample Material	Sample Size	Sampling Year	Country	Levels Detected (µg/g Dry Weight)	Levels Detected (µg/mL Wet Weight)	Mean Levels Detected (µg/g)	Median Levels Detected (µg/g or µg/mL)	Hedgehogs Analysed
Arsenic											
Vermeulen et al., 2009 [58]	Arsenic	ICP-MS	Blood	26	2005–2006	BE	na	0.1–155.5	na	0.2–17.8	Live, wild
D'Havé et al., 2006b [29]	Arsenic	ICP-MS	Fat	7	2002–2003	BE + NL	ND–0.13	na	0.08	na	Dead, roadkill + in care
Vermeulen et al., 2009 [58]	Arsenic	ICP-MS	Hair	26	2005–2006	BE	0.2–17.3	na	na	0.3–8.2	Live, wild
Rautio et al., 2010 [57]	Arsenic	ICP-OES	Hair	65	2004–2005	FI	ND–1.6	na	0.46	na	Dead, roadkill + starvation
D'Havé et al., 2006b [29]	Arsenic	ICP-MS	Hair	43	2002–2003	BE + NL	ND–2.33	na	0.69	na	Dead, roadkill + in care
D'Havé et al., 2005a [59]	Arsenic	ICP-OES	Hair	83	2002	BE	0.11–6.46 *	na	na	na	Live, wild
Rautio et al., 2010 [57]	Arsenic	ICP-OES	Kidney	64	2004–2005	FI	0.13–1.1	na	0.47	na	Dead, roadkill + starvation
D'Havé et al., 2006b [29]	Arsenic	ICP-MS	Kidney	44	2002–2003	BE + NL	ND–2.06	na	0.58	na	Dead, roadkill + in care
Rautio et al., 2010 [57]	Arsenic	ICP-OES	Liver	58	2004–2005	FI	0.26–1.06	na	0.45	na	Dead, roadkill + starvation
D'Havé et al., 2006b [29]	Arsenic	ICP-MS	Liver	43	2002–2003	BE + NL	ND–4.23	na	0.69	na	Dead, roadkill + in care
Jota Baptista et al., 2023 [63]	Arsenic	ICP-MS	Liver	41	2019–2021	PT	0–0.64	na	0.13	na	Dead, in care
D'Havé et al., 2006b [29]	Arsenic	ICP-MS	Muscle	44	2002–2003	BE + NL	ND–1.42	na	0.29	na	Dead, roadkill + in care
Vermeulen et al., 2009 [58]	Arsenic	ICP-MS	Spines	26	2005–2006	BE	0.2–23.6	na	na	0.4–6.3	Live, wild
Rautio et al., 2010 [57]	Arsenic	ICP-OES	Spines	63	2004–2005	FI	0.16–1.56	na	0.42	na	Dead, roadkill + starvation
D'Havé et al., 2006b [29]	Arsenic	ICP-MS	Spines	43	2002–2003	BE + NL	ND–5.66	na	1.24	na	Dead, roadkill + in care
D'Havé et al., 2005a [59]	Arsenic	ICP-OES	Spines	82	2002	BE	0.23–7.97 *	na	na	na	Live, wild
Selenium											
Rautio et al., 2010 [57]	Selenium	ICP-OES	Hair	65	2005–2006	FI	ND–1.02	na	0.18	na	Dead, roadkill + starvation
Rautio et al., 2010 [57]	Selenium	ICP-OES	Kidney	64	2004–2005	FI	2.05–11.36	na	4.63	na	Dead, roadkill + starvation
Rautio et al., 2010 [57]	Selenium	ICP-OES	Liver	58	2004–2005	FI	1.22–3.89	na	2.4	na	Dead, roadkill + starvation
Rautio et al., 2010 [57]	Selenium	ICP-OES	Spines	63	2004–2005	FI	0.05–1.65	na	0.69	na	Dead, roadkill + starvation

The sampling took place in Finland, Belgium, and the Netherlands in the yea
2002–2006 and in Portugal in 2019–2021, with sample sizes ranging from 7 to 83 hedgehog
representing the sample types of hair, kidney, liver, spine, muscle, blood, and fat. Fe
selenium, the detection levels ranged between not detected to 11.36 μg/g dry weight, wi
means ranging from 0.18 to 4.63 μg/g. Arsenic had detection levels from ND to 23.6 μg/
in the hair, kidneys, livers, spines, muscles, and fat, with means ranging from 0.08
1.24 μg/g. In the blood samples, levels of arsenic ranged from 0.1 to 155.5 μg/mL. Th
data presented in the studies unfortunately did not allow for a representation of detectic
frequencies for the two metalloids.

Rautio et al. (2010) [57] found that selenium concentrations increased significant
with increasing age in all the tissue types studied, suggesting a gradual accumulatic
of this compound in hedgehogs with age. Jota Baptista et al. (2023) [63] described ho
levels of arsenic were significantly lower in independent juveniles compared to adults ar
dependent juveniles.

4. Discussion

4.1. More Research on Exposure to Xenobiotics

Hedgehogs are increasingly inhabiting areas of human occupation [15–19], such i
residential gardens and urban parks, where they navigate through dense shrubberie
flower beds, vegetable gardens, and open green spaces. Facilitated by their short statui
rarely reaching > 15 cm in height, they may be exposed to many sources of herbicid
and insecticides, as they come into close contact with plants during foraging, in additic
to consuming prey items living and feeding on these plants, which are then targeted k
insecticides [20,22,23,128]. Dietary studies have also revealed remnants of plants and fru
in the stomachs of hedgehogs, although it is unknown whether they intentionally fed on th
plants or whether they were ingested during an attempt to catch prey items positioned c
the plant material [13,20]. The use of insecticides in residential gardens serving to elimina
species of, e.g., ants and aphids, is common. Hedgehogs may be exposed to insecticid
through foraging on poisoned prey items but also by moving through treated shrubs ar
areas (in the case of aerosol or liquid insecticidal products) or by ingesting the poisc
through oral intake, if the poison is placed in the open.

Hedgehog populations in rural areas appear to face the highest decline [37,39], i
cluding agricultural landscapes, with a possible pathway of exposure during foraging
herbicides and insecticides used in cultivated fields [129].

Hedgehogs admitted into care at a wildlife rehabilitation centre may furthermoi
become exposed to insecticides through flea treatments in cases where their ectoparasi
burdens have become extensive enough to cause a reduction in fitness, such as anaemi
requiring treatment. Evidence from wildlife rehabilitation centres has led to a gener;
consensus among hedgehog rehabilitators and veterinarians treating hedgehogs that pe
methrin is likely lethal to hedgehogs, as it is to cats [130], with its documented critical effe
concentration (PNEC) of 120 mg permethrin/kg food for small mammals [131] and a
oral lethal dose of 50 (LD_{50}) for rats of 480 mg permethrin/kg bodyweight [131], with a
estimated LD_{50} of 480 mg permethrin/kg bodyweight for mammals in general, in case c
primary poisoning [132]. But are hedgehogs otherwise exposed to insecticides intende
for the treatment of ectoparasites in pets? As these substances are excreted from dogs ar
cats through urine and faeces [133–135], there is a risk that hedgehogs may come into clo:
contact with the compounds, as they sometimes cover themselves in faeces from dogs ar
cats, likely as an attempt to disguise their own smell for predators. Furthermore, sharir
sources of fresh water with pets could also lead to an exposure of insecticides used fc
treatment against endoparasites, if a dog swims or rolls in a small puddle or a lake c
stream from which the hedgehog drinks, as a range of the products against ectoparasite
are nowadays "spot on" products, which are applied directly onto the skin and fur of th
pets [136]. Schanzer et al. (2021) [65] detected permethrin in one individual and a metab

lite of fipronil in all of the six hedgehogs analysed, which may indicate that exposure to insecticidal treatments for ectoparasites in cats and dogs is widespread in hedgehogs.

4.2. Food and Waterborne Contaminants

As hedgehogs are frequently offered supplementary feeding with commercial cat food in residential gardens [137], it would also be relevant to analyse cat food for different potentially toxic compounds, as recent research discovered that perfluoroalkylated substances (PFASs) were found, especially in organic chicken eggs, likely due to the addition of fish meal in commercial chicken feed [138,139]. Fish meal is also a common ingredient in commercial cat and dog food [140], which may cause PFASs to accumulate in hedgehogs feeding on these products, in addition to the exposure of PFAS from other sources, such as contaminated sites and wastewater [141]. Even though the toxic effects of PFASs are currently unknown for hedgehogs, previous research indicates negative health impacts of the bioaccumulation of PFASs in wildlife, e.g., a significant association between infectious diseases and elevated concentrations of perfluorooctanesulfonate (PFOS) and perfluorooctanoic acid (PFOA) in the livers of southern sea otters (*Enhydra lutris nereis*) [142].

Municipal wastewater is well known to contain a range of xenobiotics, including pesticides, PFAS, and flame retardants but also pharmaceutical products [143], excreted through urine and faeces from their human users. Sewage sludge is also a common fertiliser that contains a range of compounds such as heavy metals [144], pharmaceuticals [145], and pesticides [146]. The same applies to manure from livestock treated with different types of medical drugs. When sludge or manure is spread as fertilisers on the fields in which hedgehogs forage, a possible exposure pathway for hedgehogs is created. Furthermore, there may also be residues of pesticides, PFAS, pharmaceuticals, and other pollutants in the plain drinking water [147] provided for hedgehogs in people's gardens in, e.g., ponds and water bowls.

4.3. Non-Target Screening

Traditionally, analytical chemistry applies trace-level chemical analytical methods for a specific type of sample and group of substances. This form of targeted analysis is used for the identification and quantification of specific compounds, especially at low levels. However, target analyses only identify compounds that have been defined in advance, potentially overlooking other compounds with toxic potential. New analytical techniques, such as non-target screening based on high resolution mass spectrometry (HRMS), offer a possibility to scan for unknown compounds in a sample (e.g., Hollender et al. (2017) [148]). HRMS can be coupled with different types of chromatographic separation methods, i.e., liquid chromatography (LC) and gas chromatography (GC), for polar and non-polar compounds, respectively. Thus, a combination of both will be required to cover a broad spectrum of organic chemicals [149]. Following the recording of high-resolution mass spectra, bioinformatic analysis is applied to identify the compounds via comparisons with mass spectra libraries. Non-target screening would be a useful complementary approach for research on xenobiotics in hedgehogs, as it would allow for a more comprehensive screening of substances in hedgehog samples, providing insights into potentially overlooked compounds. If analytical standards are available for the identified compound, a target method for quantification can be developed as a second step. Due to the potentially high number of chemicals hedgehogs can be exposed to, we would like to advocate for the use of non-target screening in future studies on xenobiotics in hedgehogs.

4.4. The Health- and Age-Related Effects of the Occurrence of Contaminants in Hedgehogs

So far, research has primarily focused on quantifying the extent of contaminants in hedgehogs, detecting levels and frequencies of toxic, and potentially lethal, compounds but has, until recently (2023), not related these exposure levels to health effects in hedgehogs. However, Jota Baptista et al. (2023) [63] detected biliary hyperplasia in 16 of the 45 dead hedgehogs examined in their study and concluded that concentrations of metals were higher in individuals with biliary hyperplasia. We encourage future studies to investigate the potential toxicological effects of the widespread occurrence of rodenticides, organochlorine compounds, and BFRs in this declining species and expand this to currently understudied compounds, such as insecticides and PFAS. Since hedgehogs are exposed to multiple compounds at the same time, their combined effects will be relevant to address as well. Rautio et al. (2010) [57] found evidence of an age-related increase in concentrations of different metals (Cd, Se, Mo, Cu, Fe, and Mn) in hedgehogs, which is also a relevant subject in need of further study, including the health effects of bioaccumulation of multiple metals, especially given the fact that European hedgehogs have the potential to reach 16 years of age [25].

4.5. Selecting the Relevant Sample Types

The published studies investigated in this review used different approaches and sample types for studying the occurrence of xenobiotics and metals in hedgehogs, including spines, hair, muscles, fat, livers, kidneys, and blood (Figure 1). Depending on the compound in focus, its physical–chemical characteristics, and physiological processes, some sample types seemed more representative than others. As an example, Vermeulen et al. (2009) [58] found that the levels of As, Cd, and Pd were correlated in the hair, spines, and blood, but by contrast, this did not apply to Al, Cr, Cu, Fe, Mn, Ni, or Zn. D'Havé et al. (2006B) [29] discovered the highest concentrations of metals in internal tissues compared to the hair and spines, with Ag, Fe, Pb, and Zn concentrations being dominant in the livers, and Cd and Co measured in the highest levels in liver and kidney tissue. Furthermore, the authors concluded that external tissues, such as the hair and spines, may accumulate substantial concentrations of certain metals (Al, Cr, Cu, and Ni) and As. They recorded the highest concentrations for Al in spines, while As was predominant in the hair and spines. Rautio et al. (2010) [57] found that As, Cd, and Se concentrations were the highest in the kidneys, compared to Fe, Mg, Mn, Mo, Pb, and Zn, which were the highest in the livers, and Cu and Ni levels being the highest in the hair. In this study, there was a general tendency for the concentrations of the chemical compounds investigated to be lower in spine samples compared to samples from internal organs [57]. Lipophilic compounds typically accumulate in lipid-rich tissues, and the concentrations of liposoluble toxicants may vary due to morphological and behavioural differences between the sexes.

We encourage harmonised approaches for monitoring purposes, including an alignment of protocols regarding tissue types selected for analyses and sampling techniques, as well as quality control measures for the harmonisation of analytical methods. These combined efforts would improve the comparability of the results. However, while the standardisation of tissue types is important, there is also a need for analyses of different organs and tissues to improve the toxicokinetic understanding of xenobiotics and metals in hedgehogs.

4.6. Non-Destructive Measures and Hibernating Mammals as Bioindicators

Several of the research papers reviewed suggest that hedgehogs may serve as potential bioindicators for studies on the presence and accumulation of different environmental pollutants, as they share habitats with a wide range of vertebrates, and their spines appear to be a valuable and non-invasive sample type for the analysis of selected chemicals. However, for a correct interpretation of the detection of chemical compounds, a better understanding of the metabolism of contaminants in hedgehogs would be useful. It should also be considered that the direct causes of exposure to chemicals in humans and hedgehogs are not necessarily identical even though they share habitats, as humans generally do not tend to eat insects in Europe. Instead, humans may eat the same plants as the insects, which are then consumed by the hedgehogs. However, signals of potentially harmful compounds in hedgehog samples may indicate exposure sources in specific areas that would benefit from closer investigation to prevent or reduce exposure to other species.

Using spines from hedgehogs may serve as an important non-invasive alternative to traditional organ analyses of sacrificed animals [150]. The spines can be collected through a non-invasive method, as they do not contain any nerves [13], and can be sampled very rapidly with a minimum duration of handling, potentially only causing a low degree of acute stress to the hedgehog being sampled [151]. However, it should be considered that the concentrations and chemical compounds found in spines are not necessarily directly comparable to those found in organs [57,58]. Additionally, the use of dead hedgehogs for research collected by volunteers in the wild is also widely applied, and citizen science projects like The Danish Hedgehog Project have provided large numbers of samples from a wide range of habitat types for a variety of different research purposes [25,47,48,50,51]. The public adoration of hedgehogs makes large-scale citizen science projects possible, where the use of dead hedgehogs collected in the wild could also serve as a non-invasive sampling technique for future studies of xenobiotic exposure and ecotoxicology.

In contrast to actively wintering small mammals that are forced to increase their food intake during colder temperatures, potentially leading to a higher exposure of pollutants during the winter, hedgehogs hibernate for up to six months a year in most of their geographical distribution [13,14]. This may influence the accumulation of xenobiotics in their tissues. The potential lack of metabolisation of different chemical compounds during the state of torpor in hedgehogs could perhaps affect the levels detected in hedgehogs compared to non-hibernating species. Additionally, they are also likely to be affected by "delayed toxicity" through the metabolisation of adipose tissue with accumulated pollutants during hibernation.

Therefore, we advocate for research investigating these potential influences on the levels of xenobiotics and metals detected in hedgehogs compared to other small mammal species, enabling a more robust comparison between future studies with hedgehogs utilised as bioindicator species.

5. Conclusions

This review aimed to provide a comprehensive overview of the available studies screening for xenobiotics and metals in hedgehogs. Our findings revealed that a vast range of different pesticides, POPs, including organochlorine compounds and BFRs, metals, and metalloids, could be detected in samples from hedgehogs collected from different locations throughout Europe. In some cases, the compounds reached lethal concentrations, causing fatal poisoning in hedgehogs, and, in other cases, adverse health impacts, such as biliary hyperplasia, were described in the poisoned hedgehogs. Since some studies included animals that had died from poisoning, it is important to note that these might present a bias towards high concentrations, rather than representing general exposure levels. Moreover, given the lack of information on lethal doses for European hedgehogs, the interpretation of the concentrations of xenobiotics and metals present in the hedgehogs with regard to toxic effects is challenging and restricts us to drawing conclusions about the presence of these compounds in the hedgehogs.

Because we share habitats, toxicological screenings of hedgehogs could also indicate the potential exposure of xenobiotics to other terrestrial vertebrates. Hedgehogs are ground-dwelling mammals, feeding on a range of insects, slugs, snails, and earthworms and thereby come into close contact with contaminants present in the soil. They also feed on carrion, potentially accumulating compounds found in higher levels of the food chain from apex predator species. Combined with the opportunity to apply non-invasive sampling techniques through the collection of spines as sampling material, as well as the large potential for citizen science projects collecting dead hedgehogs in the wild, the European hedgehog should be regarded as a relevant bioindicator species. Furthermore, hedgehogs are declining in Europe, and insights gained through research on the role of xenobiotics and heavy metals in this decline will help inform future conservation actions directed at this species.

Due to this important potential, we advocate for more research into the exposure to and potential bioaccumulation of xenobiotics and metals in hedgehogs with a standardisation and harmonisation of sampling techniques, sample types, and methods of analysis in future studies, which would be imperative for facilitating robust comparisons. Additionally, incorporating non-target screening techniques will enable the detection of hitherto overlooked relevant and potentially toxic substances.

Supplementary Materials: The following supporting information can be downloaded at https://www.mdpi.com/article/10.3390/ani14020232/s1. An overview of the prevalence and detection levels of metals in European hedgehogs. Abbreviations used: Belgium (BE), the Netherlands (NL), Finland (FI), Portugal (PT), Italy (I), not detected (ND), not applicable (na), inductively coupled plasma mass spectrometry (ICP-MS), inductively coupled plasma optical emission spectrometer (ICP-OES), atomic absorption spectrometry (AAS). * denotes measures based on a range of means from seven study sites. An overview of the prevalence and detection levels of metals in European hedgehogs. Abbreviations used: Belgium (BE), the Netherlands (NL), Finland (FI), Portugal (PT), Italy (I), not detected (ND), not applicable (na), inductively coupled plasma mass spectrometry (ICP-MS), inductively coupled plasma optical emission spectrometer (ICP-OES), atomic absorption spectrometry (AAS). * denotes measures based on a range of means from seven study sites.

Author Contributions: Conceptualization, S.L.R., C.P., P.R., K.V. and J.L.N.; methodology, S.L.R., C.P., P.R., K.V. and J.L.N.; investigation, S.L.R.; resources, S.L.R.; data curation, S.L.R.; writing—original draft preparation, S.L.R.; writing—review and editing, S.L.R., C.P., P.R., K.V. and J.L.N.; visualization, S.L.R.; supervision, C.P., P.R., K.V. and J.L.N.; All authors have read and agreed to the published version of the manuscript.

Funding: This research was funded by the Carlsberg Foundation, grant number CF20_0443, and the Danish Environmental Protection Agency's Pesticide Research Programme J.nr. 2019-15128.

Institutional Review Board Statement: Not applicable.

Informed Consent Statement: Not applicable.

Data Availability Statement: All the data are available in the published studies used for this review.

Conflicts of Interest: The authors declare no conflicts of interest.

References

1. United Nations Environment Programme. *Global Chemicals Outlook Ii: Summary for Policymakers*; United Nations Environment Assembly of the United Nations Environment Programme, Ed.; United Nations: Nairobi, Kenya, 2019.
2. European Environment Agency. Consumption of Hazardous Chemicals. In *Environmental Indicator Reports, Environment and Health*; European Environment Agency: Luxembourg, 2018.
3. The European Commission. *Directive 2000/60/EC of the European Parliament and of the Council of 23 October 2000 Establishing a Framework for Community Action in the Field of Water Policy*; European Commission: Brussels, Belgium, 2014.
4. Reiersen, L.-O.; Vorkamp, K.; Kallenborn, R. The role of the Arctic Monitoring and Assessment Programme (AMAP) in reducing pollution of the Arctic and around the globe. *Environ. Sci. Ecotechnol.* **2024**, *17*, 100302. [CrossRef] [PubMed]
5. Marx-Stoelting, P.; Rivière, G.; Luijten, M.; Aiello-Holden, K.; Bandow, N.; Baken, K.; Cañas, A.; Castano, A.; Denys, S.; Fillol, C.; et al. A walk in the PARC: Developing and implementing 21st century chemical risk assessment in Europe. *Arch. Toxicol.* **2023**, *97*, 893–908. [CrossRef]
6. Badry, A.; Slobodnik, J.; Alygizakis, N.; Bunke, D.; Cincinelli, A.; Claßen, D.; Dekker, R.W.R.J.; Duke, G.; Dulio, V.; Göckener, B.; et al. Using environmental monitoring data from apex predators for chemicals management: Towards harmonised sampling and processing of archived wildlife samples to increase the regulatory uptake of monitoring data in chemicals management. *Environ. Sci. Eur.* **2022**, *34*, 81. [CrossRef]
7. Schlegel, H. Xenobiotics. In *General Microbiology*; Cambridge University Press: Cambridge, UK, 1986; p. 433.
8. Mishra, V.K.; Singh, G.; Shukla, R. Chapter 6—Impact of Xenobiotics Under a Changing Climate Scenario. In *Climate Change and Agricultural Ecosystems*; Choudhary, K.K., Kumar, A., Singh, A.K., Eds.; Woodhead Publishing: Sawston, UK, 2019; pp. 133–15.
9. Rieger, P.-G.; Meier, H.-M.; Gerle, M.; Vogt, U.; Groth, T.; Knackmuss, H.-J. Xenobiotics in the environment: Present and future strategies to obviate the problem of biological persistence. *J. Biotechnol.* **2002**, *94*, 101–123. [CrossRef] [PubMed]
10. Jaishankar, M.; Tseten, T.; Anbalagan, N.; Mathew, B.B.; Beeregowda, K.N. Toxicity, mechanism and health effects of some heavy metals. *Interdiscip. Toxicol.* **2014**, *7*, 60. [CrossRef] [PubMed]
11. Järup, L. Hazards of heavy metal contamination. *Br. Med. Bull.* **2003**, *68*, 167–182. [CrossRef]

2. Zoroddu, M.A.; Aaseth, J.; Crisponi, G.; Medici, S.; Peana, M.; Nurchi, V.M. The essential metals for humans: A brief overview. *J. Inorg. Biochem.* **2019**, *195*, 120–129. [CrossRef]

3. Morris, P. *Hedgehog, [New Naturalist Vol. 137]*; Collins: London, UK, 2018.

4. Reeve, N. *Hedgehogs*; Poyser: London, UK, 1994.

5. Rasmussen, S.L.; Berg, T.B.; Dabelsteen, T.; Jones, O.R. The ecology of suburban juvenile European hedgehogs (*Erinaceus europaeus*) in Denmark. *Ecol. Evol.* **2019**, *9*, 13174–13187. [CrossRef]

6. van de Poel, J.L.; Dekker, J.; van Langevelde, F. Dutch hedgehogs *Erinaceus europaeus* are nowadays mainly found in urban areas, possibly due to the negative effects of badgers Meles meles. *Wildl. Biol.* **2015**, *21*, 51–55. [CrossRef]

7. Hubert, P.; Julliard, R.; Biaganti, S.; Poulle, M.-L. Ecological factors driving the higher hedgehog (*Erinaceus europeaus*) density in an urban area compared to the adjacent rural area. *Landsc. Urban Plan.* **2011**, *103*, 34–43. [CrossRef]

8. Pettett, C.E.; Moorhouse, T.P.; Johnson, P.J.; Macdonald, D.W. Factors affecting hedgehog (*Erinaceus europaeus*) attraction to rural villages in arable landscapes. *Eur. J. Wildl. Res.* **2017**, *63*, 54. [CrossRef]

9. Gazzard, A.; Yarnell, R.W.; Baker, P.J. Fine-scale habitat selection of a small mammalian urban adapter: The West European hedgehog (*Erinaceus europaeus*). *Mamm. Biol.* **2022**, *102*, 387–403. [CrossRef]

10. Dickman, C.R. Age-related dietary change in the European hedgehog, *Erinaceus europaeus*. *J. Zool.* **1988**, *215*, 1–14. [CrossRef]

11. Rautio, A.; Isomursu, M.; Valtonen, A.; Hirvela-Koski, V.; Kunnasranta, M. Mortality, diseases and diet of European hedgehogs (*Erinaceus europaeus*) in an urban environment in Finland. *Mammal Res.* **2016**, *61*, 161–169. [CrossRef]

12. Wroot, A.J. Feeding Ecology of the European Hedgehog, Erinaceus Europeaus. Ph.D. Thesis, University of London, London, UK, 1984.

13. Yalden, D. The food of the hedgehog in England. *Acta Theriol.* **1976**, *21*, 401–424. [CrossRef]

14. Jones, C.; Moss, K.; Sanders, M. Diet of hedgehogs (*Erinaceus europaeus*) in the upper Waitaki Basin, New Zealand: Implications for conservation. *N. Z. J. Ecol.* **2005**, *29*, 29–35.

15. Rasmussen, S.L.; Berg, T.B.; Martens, H.J.; Jones, O.R. Anyone Can Get Old—All You Have to Do Is Live Long Enough: Understanding Mortality and Life Expectancy in European Hedgehogs (*Erinaceus europaeus*). *Animals* **2023**, *13*, 626. [CrossRef]

16. Nickelson, S.A.; West, S.D. Renal Cadmium Concentrations in Mice and Shrews Collected from Forest Lands Treated with Biosolids. *J. Environ. Qual.* **1996**, *25*, 86–91. [CrossRef]

17. Wijnhoven, S.; Leuven, R.S.E.W.; van der Velde, G.; Jungheim, G.; Koelemij, E.I.; de Vries, F.T.; Eijsackers, H.J.P.; Smits, A.J.M. Heavy-Metal Concentrations in Small Mammals from a Diffusely Polluted Floodplain: Importance of Species- and Location-Specific Characteristics. *Arch. Environ. Contam. Toxicol.* **2007**, *52*, 603–613. [CrossRef]

18. Šerić Jelaska, L.; Jurasović, J.; Brown, D.S.; Vaughan, I.P.; Symondson, W.O. Molecular field analysis of trophic relationships in soil-dwelling invertebrates to identify mercury, lead and cadmium transmission through forest ecosystems. *Mol. Ecol.* **2014**, *23*, 3755–3766. [CrossRef] [PubMed]

19. d'Havé, H.; Scheirs, J.; Mubiana, V.K.; Verhagen, R.; Blust, R.; De Coen, W. Non-destructive pollution exposure assessment in the European hedgehog (*Erinaceus europaeus*): II. Hair and spines as indicators of endogenous metal and As concentrations. *Environ. Pollut.* **2006**, *142*, 438–448. [CrossRef]

20. d'Havé, H.; Scheirs, J.; Covaci, A.; Schepens, P.; Verhagen, R.; De Coen, W. Nondestructive pollution exposure assessment in the European hedgehog (*Erinaceus europaeus*): III. Hair as an indicator of endogenous organochlorine compound concentrations. *Environ. Toxicol. Chem. Int. J.* **2006**, *25*, 158–167. [CrossRef]

21. Müller, F. Langzeit-Monitoring der Strassenverkehrsopfer beim Igel (*Erinaceus europaeus* L.) zur Indikation von Populations-dichteveränderungen entlang zweier Teststrecken im Landkreis Fulda. *Beiträge Zur Naturkunde Osthessen* **2018**, *54*, 21–26.

22. British Trust for Ornithology (BTO); Commissioned by British Hedgehog Preservation Society and People's Trust for Endangered Species. *The State of Britain's Hedgehogs 2011*; BTO: London, UK, 2011.

23. British Hedgehog Preservation Society; People's Trust for Endangered Species. *The State of Britain's Hedgehogs 2015*; British Hedgehog Preservation Society and People's Trust for Endangered Species: London, UK, 2015.

24. Wilson, E.; Wembridge, D. (Eds.) *The State of Britain's Hedgehogs 2018*; British Hedgehog Preservation Society and People's Trust for Endangered Species: London, UK, 2018.

25. Hof, A.R.; Bright, P.W. Quantifying the long-term decline of the West European hedgehog in England by subsampling citizen-science datasets. *Eur. J. Wildl. Res.* **2016**, *62*, 407–413. [CrossRef]

26. Krange, M. Change in the Occurrence of the West European Hedgehog (*Erinaceus europaeus*) in Western Sweden during 1950–2010. Master's Thesis, Karlstad University, Karlstad, Sweden, 2015.

27. Williams, B.M.; Baker, P.J.; Thomas, E.; Wilson, G.; Judge, J.; Yarnell, R.W. Reduced occupancy of hedgehogs (*Erinaceus europaeus*) in rural England and Wales: The influence of habitat and an asymmetric intra-guild predator. *Sci. Rep.* **2018**, *8*, 12156. [CrossRef]

28. Taucher, A.L.; Gloor, S.; Dietrich, A.; Geiger, M.; Hegglin, D.; Bontadina, F. Decline in Distribution and Abundance: Urban Hedgehogs under Pressure. *Animals* **2020**, *10*, 1606. [CrossRef] [PubMed]

29. Wembridge, D.; Johnson, G.; Al-Fulaij, N.; Langton, S. *The State of Britain's Hedgehogs 2022*; British Hedgehog Preservation Society and People's Trust for Endangered Species: London, UK, 2022.

30. Mathews, F.; Harrower, C. *IUCN—Compliant Red List for Britain's Terrestrial Mammals. Assessment by the Mammal Society under Contract to Natural England, Natural Resources Wales and Scottish Natural Heritage*; Natural England: Peterborough, UK, 2020.

41. Brakes, C.R.; Smith, R.H. Exposure of non-target small mammals to rodenticides: Short-term effects, recovery and implications for secondary poisoning. *J. Appl. Ecol.* **2005**, *42*, 118–128. [CrossRef]

42. Haigh, A.; O'Riordan, R.M.; Butler, F. Nesting behaviour and seasonal body mass changes in a rural Irish population of the Western hedgehog (*Erinaceus europaeus*). *Acta Theriol.* **2012**, *57*, 321–331. [CrossRef]

43. Hof, A.R.; Bright, P.W. The value of agri-environment schemes for macro-invertebrate feeders: Hedgehogs on arable farms in Britain. *Anim. Conserv.* **2010**, *13*, 467–473. [CrossRef]

44. Huijser, M.P.; Bergers, P.J.M. The effect of roads and traffic on hedgehog (*Erinaceus europaeus*) populations. *Biol. Conserv.* **2000**, *9*, 111–116. [CrossRef]

45. Young, R.P.; Davison, J.; Trewby, I.D.; Wilson, G.J.; Delahay, R.J.; Doncaster, C.P. Abundance of hedgehogs (*Erinaceus europaeus*) in relation to the density and distribution of badgers (*Meles meles*). *J. Zool.* **2006**, *269*, 349–356. [CrossRef]

46. Dowding, C.V.; Harris, S.; Poulton, S.; Baker, P.J. Nocturnal ranging behaviour of urban hedgehogs, *Erinaceus europaeus*, in relation to risk and reward. *Anim. Behav.* **2010**, *80*, 13–21. [CrossRef]

47. Rasmussen, S.L.; Nielsen, J.L.; Jones, O.R.; Berg, T.B.; Pertoldi, C. Genetic structure of the European hedgehog (*Erinaceus europaeus*) in Denmark. *PLoS ONE* **2020**, *15*, e0227205. [CrossRef]

48. Rasmussen, S.L.; Yashiro, E.; Sverrisdóttir, E.; Nielsen, K.L.; Lukassen, M.B.; Nielsen, J.L.; Asp, T.; Pertoldi, C. Applying the GBS Technique for the Genomic Characterization of a Danish Population of European Hedgehogs (*Erinaceus europaeus*). *Gene Biodivers. (GABJ)* **2019**, *3*, 78–86. [CrossRef]

49. Larsen, J.; Raisen, C.L.; Ba, X.; Sadgrove, N.J.; Padilla-González, G.F.; Simmonds, M.S.J.; Loncaric, I.; Kerschner, H.; Apfalter, P.; Hartl, R.; et al. Emergence of methicillin resistance predates the clinical use of antibiotics. *Nature* **2022**, *602*, 135–141. [CrossRef] [PubMed]

50. Rasmussen, S.L.; Larsen, J.; van Wijk, R.E.; Jones, O.R.; Berg, T.B.; Angen, O.; Larsen, A.R. European hedgehogs (*Erinaceus europaeus*) as a natural reservoir of methicillin-resistant *Staphylococcus aureus* carrying *mecC* in Denmark. *PLoS ONE* **2019**, *14*, e0222031. [CrossRef]

51. Rasmussen, S.L.; Hallig, J.; van Wijk, R.E.; Petersen, H.H. An investigation of endoparasites and the determinants of parasite infection in European hedgehogs (*Erinaceus europaeus*) from Denmark. *Int. J. Parasitol. Parasites Wildl.* **2021**, *16*, 217–227. [CrossRef]

52. Rasmussen, S.L.; Schrøder, A.E.; Mathiesen, R.; Nielsen, J.L.; Pertoldi, C.; Macdonald, D.W. Wildlife Conservation at a Garden Level: The Effect of Robotic Lawn Mowers on European Hedgehogs (*Erinaceus europaeus*). *Animals* **2021**, *11*, 1191. [CrossRef]

53. Moore, L.J.; Petrovan, S.O.; Baker, P.J.; Bates, A.J.; Hicks, H.L.; Perkins, S.E.; Yarnell, R.W. Impacts and potential mitigation of road mortality for hedgehogs in Europe. *Animals* **2020**, *10*, 1523. [CrossRef] [PubMed]

54. Vermeulen, F.; Covaci, A.; d'Havé, H.; Van den Brink, N.W.; Blust, R.; De Coen, W.; Bervoets, L. Accumulation of background levels of persistent organochlorine and organobromine pollutants through the soil–earthworm–hedgehog food chain. *Environ. Int.* **2010**, *36*, 721–727. [CrossRef]

55. d'Havé, H.; Scheirs, J.; Covaci, A.; van den Brink, N.W.; Verhagen, R.; De Coen, W. Non-destructive pollution exposure assessment in the European hedgehog (*Erinaceus europaeus*): IV. Hair versus soil analysis in exposure and risk assessment of organochlorine compounds. *Environ. Pollut.* **2007**, *145*, 861–868. [CrossRef] [PubMed]

56. Alleva, E.; Francia, N.; Pandolfi, M.; De Marinis, A.M.; Chiarotti, F.; Santucci, D. Organochlorine and Heavy-Metal Contaminants in Wild Mammals and Birds of Urbino-Pesaro Province, Italy: An Analytic Overview for Potential Bioindicators. *Arch. Environ. Contam. Toxicol.* **2006**, *51*, 123–134. [CrossRef]

57. Rautio, A.; Kunnasranta, M.; Valtonen, A.; Ikonen, M.; Hyvärinen, H.; Holopainen, I.J.; Kukkonen, J.V. Sex, age, and tissue specific accumulation of eight metals, arsenic, and selenium in the European hedgehog (*Erinaceus europaeus*). *Arch. Environ. Contam. Toxicol.* **2010**, *59*, 642–651. [CrossRef]

58. Vermeulen, F.; d'Havé, H.; Mubiana, V.K.; Van den Brink, N.W.; Blust, R.; Bervoets, L.; De Coen, W. Relevance of hair and spines of the European hedgehog (*Erinaceus europaeus*) as biomonitoring tissues for arsenic and metals in relation to blood. *Sci. Total Environ.* **2009**, *407*, 1775–1783. [CrossRef] [PubMed]

59. d'Havé, H.; Scheirs, J.; Mubiana, V.K.; Verhagen, R.; Blust, R.; De Coen, W. Nondestructive pollution exposure assessment in the European hedgehog (*Erinaceus europaeus*): I. Relationships between concentrations of metals and arsenic in hair, spines, and soil. *Environ. Toxicol. Chem. Int. J.* **2005**, *24*, 2356–2364. [CrossRef]

60. Dowding, C.V.; Shore, R.F.; Worgan, A.; Baker, P.J.; Harris, S. Accumulation of anticoagulant rodenticides in a non-target insectivore, the European hedgehog (*Erinaceus europaeus*). *Environ. Pollut.* **2010**, *158*, 161–166. [CrossRef] [PubMed]

61. d'Havé, H.; Covaci, A.; Scheirs, J.; Schepens, P.; Verhagen, R.; De Coen, W. Hair as an indicator of endogenous tissue levels of brominated flame retardants in mammals. *Environ. Sci. Technol.* **2005**, *39*, 6016–6020. [CrossRef]

62. D'Havé, H. De egel als bio-indicator. Monitoring van milieuvervuiling. *Zoogdier* **2009**, *20*, 12–15.

63. Jota Baptista, C.; Seixas, F.; Gonzalo-Orden, J.M.; Patinha, C.; Pato, P.; Ferreira da Silva, E.; Casero, M.; Brazio, E.; Brandão, R.; Costa, D.; et al. High Levels of Heavy Metal(loid)s Related to Biliary Hyperplasia in Hedgehogs (*Erinaceus europaeus*). *Animals* **2023**, *13*, 1359. [CrossRef]

64. Sánchez-Barbudo, I.S.; Camarero, P.R.; Mateo, R. Primary and secondary poisoning by anticoagulant rodenticides of non-target animals in Spain. *Sci. Total Environ.* **2012**, *420*, 280–288. [CrossRef]

Schanzer, S.; Kröner, E.; Wibbelt, G.; Koch, M.; Kiefer, A.; Bracher, F.; Müller, C. Miniaturized multiresidue method for the analysis of pesticides and persistent organic pollutants in non-target wildlife animal liver tissues using GC-MS/MS. *Chemosphere* **2021**, *279*, 130434. [CrossRef] [PubMed]

Dilks, P.; Sjoberg, T.; Murphy, E.C. Effectiveness of aerial 1080 for control of mammal pests in the Blue Mountains, New Zealand. *N. Z. J. Ecol.* **2020**, *44*, 1–7. [CrossRef]

Griffiths, R.; Buchanan, F.; Broome, K.; Neilsen, J.; Brown, D.; Weakley, M. Successful eradication of invasive vertebrates on Rangitoto and Motutapu Islands, New Zealand. *Biol. Invasions* **2015**, *17*, 1355–1369. [CrossRef]

Tearne, D.C. Eradication and Population Dynamics of the European Hedgehog (*Erinaceus europaeus*) on Rangitoto and Motutapu Islands. Ph.D. Thesis, University of Auckland, Auckland, New Zealand, 2010.

Hughes, J.; Sharp, E.; Taylor, M.; Melton, L. Monitoring Plant Protection Product and Rodenticide Use and Wildlife Exposure in Scotland. Available online: https://research-scotland.ac.uk/bitstream/handle/20.500.12594/24942/SETAC_2013_Monitoring_PPP_and_rodenticide_use_and_wildlife_exposure_in_Scotland.pdf?sequence=1 (accessed on 1 November 2023).

López-Perea, J.J.; Camarero, P.R.; Molina-López, R.A.; Parpal, L.; Obón, E.; Solá, J.; Mateo, R. Interspecific and geographical differences in anticoagulant rodenticide residues of predatory wildlife from the Mediterranean region of Spain. *Sci. Total Environ.* **2015**, *511*, 259–267. [CrossRef] [PubMed]

de Wit, C.A.; Persson, G.; Nilsson, I.M.; Johansson, B.W. Circannual changes in blood coagulation factors and the effect of warfarin on the hedgehog Erinaceus europaeus. *Comp. Biochem. Physiol. Part A Physiol.* **1985**, *80*, 43–47. [CrossRef] [PubMed]

de Snoo, G.R.; Scheidegger, N.M.; de Jong, F.M. Vertebrate wildlife incidents with pesticides: A European survey. *Pestic. Sci.* **1999**, *55*, 47–54. [CrossRef]

Bertero, A.; Chiari, M.; Vitale, N.; Zanoni, M.; Faggionato, E.; Biancardi, A.; Caloni, F. Types of pesticides involved in domestic and wild animal poisoning in Italy. *Sci. Total Environ.* **2020**, *707*, 136129. [CrossRef] [PubMed]

Chu, S.; Covaci, A.; Voorspoels, S.; Schepens, P. The distribution of octachlorostyrene (OCS) in environmental samples from Europe. *J. Environ. Monit.* **2003**, *5*, 619–625. [CrossRef]

Luzardo, O.P.; Ruiz-Suárez, N.; Valerón, P.F.; Camacho, M.; Zumbado, M.; Henríquez-Hernández, L.A.; Boada, L.D. Methodology for the Identification of 117 Pesticides Commonly Involved in the Poisoning of Wildlife Using GC–MS-MS and LC–MS-MS. *J. Anal. Toxicol.* **2014**, *38*, 155–163. [CrossRef]

Gemmeke, H. Untersuchungen uber die Gefahr der Sekundarvergiftung bei Iqeln (*Erinaceus europaeus* L.) durch metaldehyd-vergiftete Ackerschnecken. *Nachrichtenblatt Des Deutschen Pflanzenschutzdienstes* **1995**, *47*, 237–239.

Keymer, I.F.; Gibson, E.A.; Reynolds, D.J. Zoonoses and other findings in hedgehogs (*Erinaceus europaeus*): A survey of mortality and review of the literature. *Vet. Rec.* **1991**, *128*, 245–249. [CrossRef] [PubMed]

Van den Brink, N.W.; Elliott, J.E.; Shore, R.F.; Rattner, B.A. *Anticoagulant Rodenticides and Wildlife*; Springer: Berlin/Heidelberg, Germany, 2018; Volume 5.

Christensen, T.K.; Lassen, P.; Elmeros, M. High exposure rates of anticoagulant rodenticides in predatory bird species in intensively managed landscapes in Denmark. *Arch. Environ. Contam. Toxicol.* **2012**, *63*, 437–444. [CrossRef] [PubMed]

Topping, C.J.; Elmeros, M. Modeling exposure of mammalian predators to anticoagulant rodenticides. *Front. Environ. Sci.* **2016**, *4*, 80. [CrossRef]

Elmeros, M.; Topping, C.J.; Christensen, T.K.; Lassen, P.; Bossi, R. *Spredning af Antikoagulante Rodenticider med mus og Eksponeringsrisiko for Rovdyr*; Miljøstyrelsen, Ministry of Environment Denmark: Copenhagen, Denmark, 2015.

Spurr, E.; Drew, K. Invertebrates feeding on baits used for vertebrate pest control in New Zealand. *N. Z. J. Ecol.* **1999**, *23*, 167–173.

Elliott, J.E.; Hindmarch, S.; Albert, C.A.; Emery, J.; Mineau, P.; Maisonneuve, F. Exposure pathways of anticoagulant rodenticides to nontarget wildlife. *Environ. Monit. Assess.* **2014**, *186*, 895–906. [CrossRef] [PubMed]

University of Hertfordshire. Ppdb: Pesticide Properties Database. Available online: http://sitem.herts.ac.uk/aeru/ppdb/en/Reports/3160.htm (accessed on 21 December 2023).

Alomar, H.; Chabert, A.; Coeurdassier, M.; Vey, D.; Berny, P. Accumulation of anticoagulant rodenticides (chlorophacinone, bromadiolone and brodifacoum) in a non-target invertebrate, the slug, Deroceras reticulatum. *Sci. Total Environ.* **2018**, *610*, 576–582. [CrossRef] [PubMed]

United Nations Environment Programme. *The Stockholm Convention on Persistent Organic Pollutants*; United Nations: Stockholm, Sweden, 2001.

Vorkamp, K.; Rigét, F.F.; Bossi, R.; Sonne, C.; Dietz, R. Endosulfan, Short-Chain Chlorinated Paraffins (SCCPs) and Octachlorostyrene in Wildlife from Greenland: Levels, Trends and Methodological Challenges. *Arch. Environ. Contam. Toxicol.* **2017**, *73*, 542–551. [CrossRef]

Domazet, S.L.; Grøntved, A.; Jensen, T.K.; Wedderkopp, N.; Andersen, L.B. Higher circulating plasma polychlorinated biphenyls (PCBs) in fit and lean children: The European youth heart study. *Environ. Int.* **2020**, *136*, 105481. [CrossRef]

Jones, K.C.; De Voogt, P. Persistent organic pollutants (POPs): State of the science. *Environ. Pollut.* **1999**, *100*, 209–221. [CrossRef]

Qing Li, Q.; Loganath, A.; Seng Chong, Y.; Tan, J.; Philip Obbard, J. Persistent organic pollutants and adverse health effects in humans. *J. Toxicol. Environ. Health Part A* **2006**, *69*, 1987–2005. [CrossRef]

Greim, L.W.W.H. The toxicity of brominated and mixed-halogenated dibenzo-p-dioxins and dibenzofurans: An overview. *J. Toxicol. Environ. Health Part A* **1997**, *50*, 195–216. [CrossRef]

92. Scheringer, M. Persistence and spatial range as endpoints of an exposure-based assessment of organic chemicals. *Environ. S Technol.* **1996**, *30*, 1652–1659. [CrossRef]

93. Bergman, A.; Olsson, M. Pathology of Baltic grey seal and ringed seal females with special reference to adrenocortical hyperplas Is environmental pollution the cause of a widely distributed disease syndrome. *Finn. Game Res.* **1985**, *44*, 47–62.

94. Reijnders, P.J. Reproductive failure in common seals feeding on fish from polluted coastal waters. *Nature* **1986**, *324*, 456–4! [CrossRef]

95. Giesy, J.P.; Ludwig, J.P.; Tillitt, D.E. Deformities in birds of the Great Lakes region. *Environ. Sci. Technol.* **1994**, *28*, 128A–135 [CrossRef] [PubMed]

96. Ratcliffe, D. Decrease in eggshell weight in certain birds of prey. *Nature* **1967**, *215*, 208–210. [CrossRef] [PubMed]

97. de Swart, R.; Ross, P.; Vedder, L.; Timmerman, H.; Heisterkamp, S.; Van Loveren, H.; Vos, J.; Reijnders, P.J.; Osterhaus, Impairment of immune function in harbor seals (*Phoca vitulina*) feeding on fish from polluted waters. *Ambio* **1994**, *23*, 155–159

98. de Solla, S.R. Exposure, Bioaccumulation, Metabolism and Monitoring of Persistent Organic Pollutants in Terrestrial Wildlife. *Dioxin and Related Compounds: Special Volume in Honor of Otto Hutzinger*; Alaee, M., Ed.; Springer International Publishing: Cha Switzerland, 2016; pp. 203–252.

99. Daugaard-Petersen, T.; Langebæk, R.; Rigét, F.F.; Letcher, R.J.; Hyldstrup, L.; Jensen, J.-E.B.; Bechshoft, T.; Wiig, Ø.; Jenssen, B.N Pertoldi, C.; et al. Persistent organic pollutants, skull size and bone density of polar bears (*Ursus maritimus*) from East Greenlar 1892–2015 and Svalbard 1964–2004. *Environ. Res.* **2018**, *162*, 74–80. [CrossRef]

100. Marsili, L.; D'Agostino, A.; Bucalossi, D.; Malatesta, T.; Fossi, M.C. Theoretical models to evaluate hazard due to organochlorin compounds (OCs) in Mediterranean striped dolphin (*Stenella coeruleoalba*). *Chemosphere* **2004**, *56*, 791–801. [CrossRef]

101. Glucksmann, A. Sexual dimorphism in mammals. *Biol. Rev.* **1974**, *49*, 423–475. [CrossRef]

102. Henriksen, E.O.; Gabrielsen, G.W.; Skaare, J.U. Levels and congener pattern of polychlorinated biphenyls in kittiwakes (*Ris tridactyla*), in relation to mobilization of body-lipids associated with reproduction. *Environ. Pollut.* **1996**, *92*, 27–37. [CrossRe [PubMed]

103. Fenstad, A.A.; Jenssen, B.M.; Moe, B.; Hanssen, S.A.; Bingham, C.; Herzke, D.; Bustnes, J.O.; Krøkje, Å. DNA double-strand breal in relation to persistent organic pollutants in a fasting seabird. *Ecotoxicol. Environ. Saf.* **2014**, *106*, 68–75. [CrossRef] [PubMed]

104. Derocher, A.E.; Andriashek, D.; Arnould, J.P. Aspects of milk composition and lactation in polar bears. *Can. J. Zool.* **1993**, 7 561–567. [CrossRef]

105. Bernhoft, A.; Wiig, Ø.; Utne Skaare, J. Organochlorines in polar bears (*Ursus maritimus*) at Svalbard. *Environ. Pollut.* **1997**, 9 159–175. [CrossRef]

106. De Wit, C.A. An overview of brominated flame retardants in the environment. *Chemosphere* **2002**, *46*, 583–624. [CrossRe [PubMed]

107. de Wit, C.A.; Herzke, D.; Vorkamp, K. Brominated flame retardants in the Arctic environment—Trends and new candidates. *S Total Environ.* **2010**, *408*, 2885–2918. [CrossRef] [PubMed]

108. Darnerud, P.O. Toxic effects of brominated flame retardants in man and in wildlife. *Environ. Int.* **2003**, *29*, 841–853. [CrossRef]

109. Sun, Y.; Wang, Y.; Liang, B.; Chen, T.; Zheng, D.; Zhao, X.; Jing, L.; Zhou, X.; Sun, Z.; Shi, Z. Hepatotoxicity of decabromodiphen ethane (DBDPE) and decabromodiphenyl ether (BDE-209) in 28-day exposed Sprague-Dawley rats. *Sci. Total Environ.* **2020**, 7C 135783. [CrossRef]

110. Jansson, B.; Andersson, R.; Asplund, L.; Litzen, K.; Nylund, K.; Sellström, U.; Uvemo, U.B.; Wahlberg, C.; Wideqvist, U.; Ods T. Chlorinated and brominated persistent organic compounds in biological samples from the environment. *Environ. Toxicol. Che Int. J.* **1993**, *12*, 1163–1174. [CrossRef]

111. Lindberg, P.; Sellström, U.; Häggberg, L.; de Wit, C.A. Higher brominated diphenyl ethers and hexabromocyclododecane foun in eggs of peregrine falcons (*Falco peregrinus*) breeding in Sweden. *Environ. Sci. Technol.* **2004**, *38*, 93–96. [CrossRef]

112. Law, R.J.; Alaee, M.; Allchin, C.R.; Boon, J.P.; Lebeuf, M.; Lepom, P.; Stern, G.A. Levels and trends of polybrominate diphenylethers and other brominated flame retardants in wildlife. *Environ. Int.* **2003**, *29*, 757–770. [CrossRef] [PubMed]

113. Tinggi, U. Essentiality and toxicity of selenium and its status in Australia: A review. *Toxicol. Lett.* **2003**, *137*, 103–110. [CrossRe [PubMed]

114. Magg, D.D.; Glen, M.W. Toxicity of Selenium: Farm Animals. In *Selenium in Biomedicine*; Muth, O.H., Oldfield, J.E., Weswig, P.F Eds.; International Symposium on Selenium in Biomedicine; AVI Pub. Co.: Westport, CT, USA, 1967; pp. 127–140.

115. Hovda, L.R.; Benson, D.; Poppenga, R.H. (Eds.) Selenium. In *Blackwell's Five-Minute Veterinary Consult Clinical Companion: Equi Toxicology*; Wiley-Blackwell: Hoboken, NJ, USA, 2021; pp. 207–211.

116. Raisbeck, M.F. Selenosis in ruminants. *Vet. Clin. Food Anim. Pract.* **2020**, *36*, 775–789. [CrossRef] [PubMed]

117. Rayman, M.P. Selenium in cancer prevention: A review of the evidence and mechanism of action. *Proc. Nutr. Soc.* **2005**, 6 527–542. [CrossRef] [PubMed]

118. Reilly, C. *Selenium in Food and Health*; Blackie Academic and Professional, Chapman and Hall: London, UK, 1996.

119. He, Y.; Xiang, Y.; Zhou, Y.; Yang, Y.; Zhang, J.; Huang, H.; Shang, C.; Luo, L.; Gao, J.; Tang, L. Selenium contaminatio consequences and remediation techniques in water and soils: A review. *Environ. Res.* **2018**, *164*, 288–301. [CrossRef]

120. Gupta, U.C.; MacLeod, J. Effect of various sources of selenium fertilization on the selenium concentration of feed crops. *Can. Soil Sci.* **1994**, *74*, 285–290. [CrossRef]

21. Ventura-Lima, J.; Bogo, M.R.; Monserrat, J.M. Arsenic toxicity in mammals and aquatic animals: A comparative biochemical approach. *Ecotoxicol. Environ. Saf.* **2011**, *74*, 211–218. [CrossRef]
22. Akter, K.F.; Owens, G.; Davey, D.E.; Naidu, R. *Arsenic Speciation and Toxicity in Biological Systems*; Springer: Berlin/Heidelberg, Germany, 2005.
23. George, J.E. Present and future technologies for tick control. *Ann. N. Y. Acad. Sci.* **2000**, *916*, 583–588. [CrossRef]
24. Fattorini, D.; Notti, A.; Regoli, F. Characterization of arsenic content in marine organisms from temperate, tropical, and polar environments. *Chem. Ecol.* **2006**, *22*, 405–414. [CrossRef]
25. Samuel, S.; Kathirvel, R.; Jayavelu, T.; Chinnakkannu, P. Protein oxidative damage in arsenic induced rat brain: Influence of DL-α-lipoic acid. *Toxicol. Lett.* **2005**, *155*, 27–34. [CrossRef]
26. Flora, S.J.S.; Bhadauria, S.; Pant, S.C.; Dhaked, R.K. Arsenic induced blood and brain oxidative stress and its response to some thiol chelators in rats. *Life Sci.* **2005**, *77*, 2324–2337. [CrossRef] [PubMed]
27. Aposhian, H.V.; Zakharyan, R.A.; Avram, M.D.; Kopplin, M.J.; Wollenberg, M.L. Oxidation and detoxification of trivalent arsenic species. *Toxicol. Appl. Pharmacol.* **2003**, *193*, 1–8. [CrossRef] [PubMed]
28. Dimelow, E.J. Observations on the Feeding of the Hedgehog (*Erinaceus europaeus* L.). *Proc. Zool. Soc. Lond. (J. Zool.)* **1963**, *141*, 291–309. [CrossRef]
29. Tudi, M.; Daniel Ruan, H.; Wang, L.; Lyu, J.; Sadler, R.; Connell, D.; Chu, C.; Phung, D.T. Agriculture development, pesticide application and its impact on the environment. *Int. J. Environ. Res. Public Health* **2021**, *18*, 1112. [CrossRef] [PubMed]
30. Boland, L.A.; Angles, J.M. Feline permethrin toxicity: Retrospective study of 42 cases. *J. Feline Med. Surg.* **2010**, *12*, 61–71. [CrossRef] [PubMed]
31. European Chemicals Agency (ECHA). *Product Assessment Report of a Biocidal Product (Family) for National Authorisation Applications. Product Type: Pt 18. Pyriproxyfen and Permethrin. Case Number in R4bp: Bc-Gy023113-30*; European Chemicals Agency: Helsinki, Finland, 2020.
32. *Product Assessment Report of a Biocidal Product for National Authorisation Applications. Product Type 18: Permethrin. Bc-Yr023158-12*; European Chemicals Agency: Helsinki, Finland, 2019.
33. Lestremau, F.; Willemin, M.E.; Chatellier, C.; Desmots, S.; Brochot, C. Determination of cis-permethrin, trans-permethrin and associated metabolites in rat blood and organs by gas chromatography–ion trap mass spectrometry. *Anal. Bioanal. Chem.* **2014**, *406*, 3477–3487. [CrossRef]
34. Snodgrass, H.L., Jr.; Nelson, D.C. *Dermal Penetration and Distribution of 14c-Labeled Permethrin Isomers*; United States Army Environmental Hygiene Agency: Aberdeen, MD, USA, 1982.
35. Chrustek, A.; Hołyńska-Iwan, I.; Dziembowska, I.; Bogusiewicz, J.; Wróblewski, M.; Cwynar, A.; Olszewska-Słonina, D. Current Research on the Safety of Pyrethroids Used as Insecticides. *Medicina* **2018**, *54*, 61. [CrossRef]
36. Teerlink, J.; Hernandez, J.; Budd, R. Fipronil washoff to municipal wastewater from dogs treated with spot-on products. *Sci. Total Environ.* **2017**, *599–600*, 960–966. [CrossRef]
37. Gazzard, A.; Baker, P.J. Patterns of feeding by householders affect activity of hedgehogs (*Erinaceus europaeus*) during the hibernation period. *Animals* **2020**, *10*, 1344. [CrossRef]
38. DTU National Food Institute. *Indhold Af Pfas I Fiskemel Og Via Indhold I Økologisk Foder I Økologiske Æg*; DTU National Food Institute: Lyngby, Denmark, 2023.
39. Pfas Found in Organic Eggs in Denmark. Available online: https://www.food.dtu.dk/english/news/pfas-found-in-organic-eggs-in-denmark?id=789f9ba1-bdfc-4a7d-908b-fc6cccff4742 (accessed on 1 February 2023).
40. De Silva, S.S.; Turchini, G.M. Towards understanding the impacts of the pet food industry on world fish and seafood supplies. *J. Agric. Environ. Ethics* **2008**, *21*, 459–467. [CrossRef]
41. De Silva, A.O.; Armitage, J.M.; Bruton, T.A.; Dassuncao, C.; Heiger-Bernays, W.; Hu, X.C.; Kärrman, A.; Kelly, B.; Ng, C.; Robuck, A. PFAS exposure pathways for humans and wildlife: A synthesis of current knowledge and key gaps in understanding. *Environ. Toxicol. Chem.* **2021**, *40*, 631–657. [CrossRef] [PubMed]
42. Kannan, K.; Perrotta, E.; Thomas, N.J. Association between Perfluorinated Compounds and Pathological Conditions in Southern Sea Otters. *Environ. Sci. Technol.* **2006**, *40*, 4943–4948. [CrossRef]
43. Surra, E.; Correia, M.; Figueiredo, S.; Silva, J.G.; Vieira, J.; Jorge, S.; Pazos, M.; Sanromán, M.Á.; Lapa, N.; Delerue-Matos, C. Life Cycle and Economic Analyses of the Removal of Pesticides and Pharmaceuticals from Municipal Wastewater by Anodic Oxidation. *Sustainability* **2021**, *13*, 3669. [CrossRef]
44. Pathak, A.; Dastidar, M.G.; Sreekrishnan, T.R. Bioleaching of heavy metals from sewage sludge: A review. *J. Environ. Manag.* **2009**, *90*, 2343–2353. [CrossRef] [PubMed]
45. Verlicchi, P.; Zambello, E. Pharmaceuticals and personal care products in untreated and treated sewage sludge: Occurrence and environmental risk in the case of application on soil—A critical review. *Sci. Total Environ.* **2015**, *538*, 750–767. [CrossRef] [PubMed]
46. Tadeo, J.L.; Sánchez-Brunete, C.; Albero, B.; Garcí-Valcárcel, A.I. Determination of pesticide residues in sewage sludge: A review. *J. AOAC Int.* **2010**, *93*, 1692–1702. [CrossRef]
47. Chander, V.; Sharma, B.; Negi, V.; Aswal, R.S.; Singh, P.; Singh, R.; Dobhal, R. Pharmaceutical compounds in drinking water. *J. Xenobiotics* **2016**, *6*, 5774. [CrossRef]
48. Hollender, J.; Schymanski, E.L.; Singer, H.P.; Ferguson, P.L. Nontarget screening with high resolution mass spectrometry in the environment: Ready to go? *Environ. Sci. Technol.* **2017**, *51*, 11505–11512. [CrossRef]

149. Hollender, J.; Schymanski, E.L.; Ahrens, L.; Alygizakis, N.; Béen, F.; Bijlsma, L.; Brunner, A.M.; Celma, A.; Fildier, A.; Fu, Q.; et al. NORMAN guidance on suspect and non-target screening in environmental monitoring. *Environ. Sci. Eur.* **2023**, *35*, 75. [CrossRef]
150. Fossi, M.C. Nondestructive biomarkers in ecotoxicology. *Environ. Health Perspect.* **1994**, *102*, 49–54. [CrossRef]
151. Rasmussen, S.L.; Kalliokoski, O.; Dabelsteen, T.; Abelson, K. An exploratory investigation of glucocorticoids, personality and survival rates in wild and rehabilitated hedgehogs (*Erinaceus europaeus*) in Denmark. *BMC Ecol. Evol.* **2021**, *21*, 96. [CrossRef] [PubMed]

 **animals**

...ticle

Occurrence and Characteristics of Cut Injuries in Hedgehogs in Germany: A Collection of Individual Cases

Anne Berger

Leibniz Institute for Zoo and Wildlife Research, Alfred-Kowalke-Straße 17, 10315 Berlin, Germany; berger@izw-berlin.de

Simple Summary: The European hedgehog is a protected species, but its populations are declining across Europe. This decline has various causes, such as lack of food, habitat loss and fragmentation or life-threatening injuries caused by human activities. Hedgehog rescue centres increasingly report hedgehogs found with severe cuts, presumably caused by garden tools. Responsibility for hedgehog injuries caused by robotic lawnmowers and possible technical or political solutions to prevent such injuries are currently being discussed between wildlife conservationists, mower manufacturers and politicians. This discussion has so far lacked basic data on the extent of cutting injuries in hedgehogs. In this study, data on hedgehogs with cut injuries were collected throughout Germany in order to gain an impression of where, when and how frequently these injuries occur. The number of reporting hedgehog care centres and thus the number of hedgehogs reported per federal state varied highly. Out of the total of 370 injured hedgehogs reported, at least 60% were found over 12 h after the accident and at least 47% did not survive as a result of the injury. Overall, this study shows that cutting injuries caused by garden maintenance equipment pose an additional lethal danger to this declining, protected wildlife species.

Abstract: The number of European hedgehogs (*Erinaceus europaeus*) is in long-term decline across Europe. Recently, an additional threat to hedgehogs' lives has been cutting injuries caused by garden care equipment, but to date, there have been no reliable data on their spatial and temporal occurrence as well as characteristics such as mortality rate. Usually, found injured hedgehogs are admitted to care centres. In this study, data on hedgehogs with cutting injuries were collected from care centres throughout Germany. Over a period of 16 months, data on a total of 370 hedgehogs with cut injuries were reported by 71 care centres. At least 60% of these hedgehogs were found more than 12 h after the accident and at least 47% did not survive as a result of the injury. The comparatively high mortality rate coupled with a possible high number of unreported cases of hedgehogs with laceration injuries show that these accidents pose an additional, serious danger to hedgehogs, both impacting the welfare of individual animals and having a broader effect on the conservation potential of this species. Moreover, the data collected objectify the current discussion on the need for possible technical or political solutions to prevent such injuries.

Keywords: European hedgehog; *Erinaceus europaeus*; rehabilitation centre; cut injuries; mortality rate; wildlife conservation; human–wildlife conflict; animal suffer; animal welfare

Citation: Berger, A. Occurrence and Characteristics of Cut Injuries in Hedgehogs in Germany: A Collection of Individual Cases. *Animals* **2024**, *14*, 57. https://doi.org/10.3390/ani14010057

Academic Editor: Anna Nekaris

Received: 15 November 2023
Revised: 13 December 2023
Accepted: 19 December 2023
Published: 22 December 2023

1. Introduction

The West European hedgehog (*Erinaceus europaeus*, hereafter referred to as "hedgehog") is a solitary, nocturnal insectivore and is one of the most popular and well-known wild animals among the general population [1]. People's interest in these animals is probably due to the fact that they are often found in close proximity to humans. Human habitations, especially gardens, have structures that are particularly attractive to hedgehogs, such as a high proportion of green areas, bushes or additional sources of water and food [2]. Despite their high popularity, hedgehog populations have experienced a serious and continuous decline in recent decades [3–7], especially in rural areas [8,9]. Reasons for this decli[ne] include habitat loss and fragmentation [9–11], reduced food availability, partly due [to] the use of pesticides and insecticides [12,13], in some areas, also intra-guild predati[on] by badgers [8,14–16] and possible climate-related effects [17]. Moreover, there are ofte[n] various fatal accident hazards to which hedgehogs are exposed, such as road traffic [or] entanglement in garbage [18–20].

Hedgehogs are subject to protection in large parts of Europe; in Germany, they a[re] specially protected wild animals that cannot be hunted and may not be caught, injure[d] or killed [21]. However, the law does grant one exception to the ban on possession: [It] states that injured, helpless and/or sick animals may be taken into human care in ord[er] to nurse them back to health. As soon as the hedgehogs are able to maintain themselve[s] independently, they must be released into the wild immediately. Caring for sick, injure[d] or orphaned wild animals and their preparation for release back into the wild are usual[ly] carried out in wildlife rehabilitation centres. The rehabilitation of wild animals requir[e] large investments of time, personnel and money [22] and therefore it is often important [to] ask which animals can be taken in at all, that is, if they have a high chance of survivin[g] well during care and also after release [23]. Hedgehogs are among the European wildli[fe] species that are most frequently rehabilitated by humans [24], often not even in publ[ic] rehabilitation centres, but on a private, voluntary, non-profit basis.

Many of these hedgehog care centres report a significant increase in the incidence [of] cutting injuries caused by garden maintenance equipment (scythes, string trimmers, robot[ic] lawnmowers). As the global market for robotic lawnmowers is expanding at an annu[al] growth rate of more than 12% in the period 2019 to 2025 [25], and as the use of robot[ic] mowers is increasing significantly compared to other maintenance equipment, it could b[e] hypothesised that the increase in the number of cut injuries in hedgehogs is associate[d] with the use of these robotic mowers. An initial study has already shown that—contrary [to] the manufacturer's specifications—many models of robotic lawnmowers can cause serio[us] cutting injuries to hedgehogs [26]. A great attractiveness of robotic lawnmowers stems fro[m] the fact that, unlike other lawnmowers, these devices can be used legally for an unlimite[d] period of time (i.e., also at night and on public holidays and on Sundays) due to their lo[w] noise emissions. They can also work unattended (i.e., in the absence of humans). Thes[e] two characteristics in particular make it very likely that many collisions between robot[ic] lawnmowers and nocturnal hedgehogs occur and that these are often not even noticed b[y] humans, especially in cases where the hedgehog is only slightly injured and still able [to] run away from the scene after the collision.

In Germany, there have already been many petitions and political efforts by hedgeho[g] protection and care organisations calling for a general ban on the night-time use of mowin[g] robots [27–29]. So far, these have all failed due to a lack of public interest, a lack of dat[a] on the extent of cut injuries in hedgehogs or a lack of legislation to enforce these rules a[t] the regional level. The aim of this study is to collect and quantify concrete data on th[e] temporal and spatial distribution of these injuries, specific characteristics of the wound[s,] the probability of survival and the extent of the care required for hedgehogs found wit[h] cut injuries in Germany. These figures are intended to add an objective perspective [to] the emotionally charged debate on hedgehogs with cut wounds and are fundamental [to] the current discussion on the need for possible technical or political solutions to preve[nt] such injuries.

By statistically analysing the individual cases collected, the following question[s] were investigated:

(A) Are there days of the week when there are significantly more cases of cut injuri[es] in hedgehogs or when injured hedgehogs are found more frequently? On Sundays an[d] public holidays, and generally between 8 p.m. and 6 a.m. (at night, during the natur[al] activity period of hedgehogs), the use of garden maintenance devices is prohibited b[y] law for noise protection reasons and no maintenance work is carried out by public gree[n] maintenance authorities at weekends (Saturday and Sunday). An exception to this is robot[ic] lawnmowers, which are not subject to these time restrictions, as they are usually quieter than the noise limits to be observed [30] and are therefore also used on Sundays and at night and are preferable in private areas. An above-average incidence of hedgehogs injured by cutting on Sundays would suggest that they are predominantly injured by robotic lawnmowers, especially those for private use;

(B) How are the cases of hedgehogs found with cutting injuries distributed across different age and sex groups?

(C) Are there individual characteristics of the injuries that lead to an increased mortality risk?

2. Materials and Methods

2.1. Data Collection

All data on cases of hedgehogs with cut injuries in this study were collected via a Facebook group set up on 28 June 2022. Due to graphic nature of the photos of reported animals and to protect personal data, the Facebook group was not accessible to the general population but individually to people who had come into direct contact with cut hedgehogs and could provide information about them. Every report of a cut hedgehog contained the following *minimal information*:

- Date on which the animal was found;
- Place where the animal was found;
- One or more photos showing the injuries;
- Who treated or diagnosed the animal or who provided initial treatment and care for the injured hedgehog (this was asked to ensure that *additional* diagnostic *information* was provided by expert veterinarians or experienced hedgehog carers).

The following *additional information* was provided, where known:

- Sex of the animal;
- Estimated age of the animal;
- Fate of the animal (euthanasia/died while in care/recovered/released back into the wild).

All information from the reports that contained at least the *minimal information* was transferred to a table; thereby, information on the wound characteristics was taken from the photos or diagnostic descriptions sent in. The table contained the following columns with the following (categorical) content entries:

- Hedgehog identification number: xxx;
- Date (day of found): YYYYMMDD;
- Day of the week: Monday/Tuesday/Wednesday/Thursday/Friday/Saturday/Sunday;
- Location: exact address or at least the postcode;
- Sex: male/female/reproducing female (means pregnant, lactating)/unknown;
- Age: adult (from the survival of the first hibernation, note: all animals found from January to May were always considered adult here)/juvenile (heavier than 100 g and before the first hibernation)/nestling (less than 100 g)/unknown (age was not reported or cannot be estimated from the reported information);
- Fate: euthanasia/died (during treatment or rehabilitation)/survived or released into the wild/unknown (this includes cases where no information was provided, but also all cases that are still open with regard to survival, e.g., still undergoing treatment);
- Characteristics of the wound:
- Size of the cut surface larger than 2 × 2 cm: yes/no/unknown (all cases were listed as "unknown" for which the photos did not allow a clear size estimation, as reference size objects—such as the fingers or hands of the treating person—were missing);
- Presence of maggots: yes/no/unknown;
- Presence of necrosis: yes/no/unknown;
- Presence of abscesses: yes/no/unknown;
- Bone damage (e.g., fractures, splintering): yes/no/unknown;

- Loss of body parts: yes/no/unknown;
- Age of the wound: <12 h (the injured animal was noticed during the accident and taken to the vet or no maggots or necrosis were recognisable on the wound or mentioned in the reports)/>12 h/already healed (cut wound is already scarred and overgrown)/unknown (all cases were classified here in which an assessment could not be clearly made on the basis of the photos and reports). This wound age estimate could be made "remotely" from photos and reports based on the knowledge that fly maggots need 8–12 h to hatch from the fly eggs even under ideal conditions [31–33], thus wounds with maggots had to be at least 8–12 h old. According to textbooks on wound healing processes in wild animals [34,35] and to personal reports on the duration of successful wound healing in hedgehogs from care stations, the age of wounds that have already healed can be estimated at around 1 week up to several months.

Body parts affected:

- Head (in front of the imaginary line between the ears): yes/no/unknown;
- Neck and shoulders (starting behind the imaginary line between the ears): yes/no/unknown;
- Back (spines on the back up to the edge): yes/no/unknown;
- Extremities: yes/no/unknown;
- Flank/belly (from the edge of the spines towards the belly): yes/no/unknown.

2.2. Statistical Analysis

Analyses were performed using the basic package R Studio [36]. Pearson's chi-square tests were used to check whether (1) all animals were found and (b) all animals with a wound age <12 h were found at the expected frequency on each day of the week ($p = 1/7$).

For the following data analysis, the category "fate" was pooled into two definitions "died" (all animals that were euthanised or died during care) and "survived" (all animals that survived). All cases where the fate of the animals was not known (either not reported or the animals are still under treatment) were excluded from the following calculations.

For the following parameters (A) wound size, (B) maggots, (C) necrosis, (D) injured bones, (E) abscess, (F) severed body parts, (G) head, (H) neck, (I) back, (J) extremities, (K) flank/abdomen and (L) wound age, the number of "died" (g) versus "survived" (s) animals was counted and the mortality rate ($mr = g \times 100/(g + s)$) was calculated.

To investigate whether the age of the wound, the presence of certain wound characteristics (A–F) or the affected body part (G–K) influenced the fate of the animal "died"/"survived", the statistical significance for each of these frequency ratios was tested using chi-square tests.

3. Results

The Facebook group was set up on 28 June 2022. By 31 October 2023 (during 16 months), a total of 370 cases of hedgehogs with cut injuries were reported by 71 reporters (average: 5.2, median: 2.0). Figure 1 shows the locations where the hedgehogs were found on the map of Germany, separated by colour and by symbol according to the information on their fate (euthanised or died during care versus survived or still in care or fate is unknown). Figure 1 also shows the 17 hedgehog care centres that reported an above-average number (n > 5) of hedgehogs. The majority of the 71 reporters were small hedgehog care centres with only a few hedgehogs in their care; thus, more than half reported only one hedgehog (n = 27) or two hedgehogs (n = 14). Table 1 shows the number of hedgehogs found and the number of hedgehog care centres that reported more than five hedgehogs in each German federal state.

The earliest find data are from 2013; Figure 2 shows how many hedgehogs were found per year and Figure 3 shows the find data per month, also showing the proportion of animals found in 2023, in 2022 and in previous years.

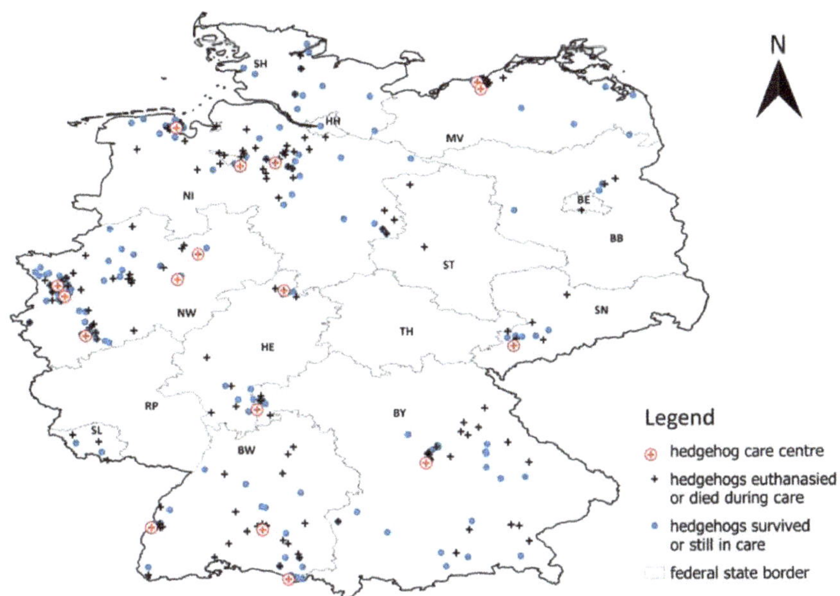

Figure 1. Map of the locations in Germany in which the 370 reported hedgehogs with cut injuries were found, divided according to their fate (black cross = did not survive, blue dot = did survive or fate is unknown), and of 17 hedgehog care centres that reported more than 5 of these hedgehogs. Uppercase letters give the German federal state abbreviations, see Table 1.

Table 1. Overview of the number of reported hedgehogs injured by cuts and the number of hedgehog care centres reporting more than 5 hedgehogs in several German federal states.

Federal State	Federal State Abbreviation	Number of Reported Hedgehogs	Number of Reporting Care Centres
Baden-Württemberg	BW	42	3
Bavaria	BY	54	2
Berlin	BE	1	0
Brandenburg	BB	8	0
Bremen	HB	5	1
Hamburg	HH	2	0
Hesse	HE	14	1
Mecklenburg-West Pomerania	MV	24	2
Lower Saxony	NI	80	2
Northrhine-Westphalia	NW	107	5
Rhineland Palatinate	RP	0	0
Saarland	SL	5	0
Saxony	SN	12	1
Saxony Anhalt	ST	2	0
Schleswig Holstein	SH	14	0
Thuringia	TH	0	0

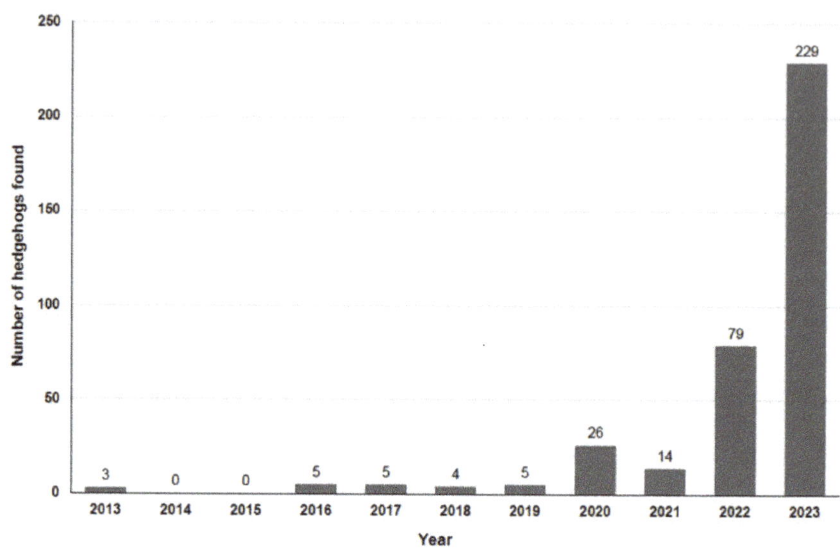

Figure 2. Number of 370 reported hedgehogs plotted according to the year in which they were foun

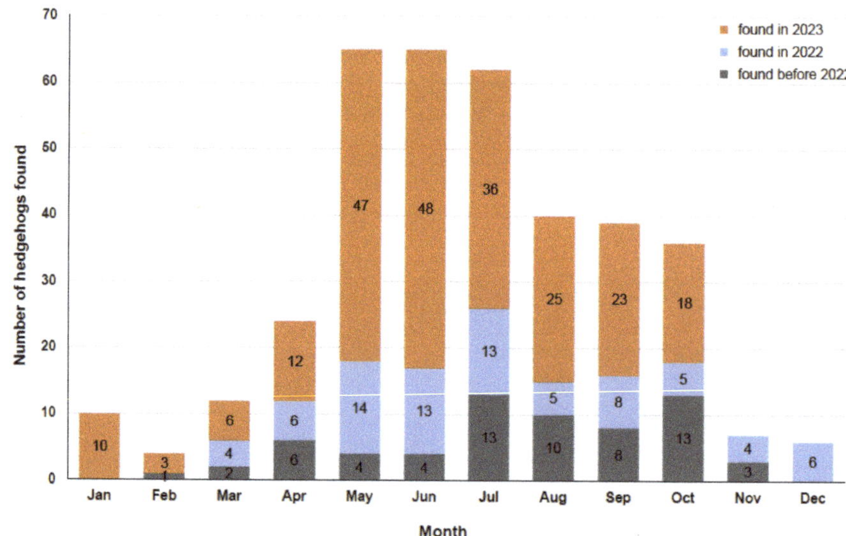

Figure 3. Number of 370 reported hedgehogs plotted according to the month in which they wei
found. The respective proportions of finds from 2023 (orange), 2022 (blue-grey) and the years befo
2022 (grey) are colour-coded.

Although the distribution of hedgehogs found by day of the week shows that fewe
animals were found on Fridays than on other days and that a particularly large number wei
found at weekends and on Tuesdays, these differences are not significant (chi-square te:
for given probabilities for hedgehogs found within 12 h of the accident: x-squared = 7.633
df = 6, *p*-value = 0.2662; for hedgehogs which were found later than 12 h after the accider
x-squared = 2.3457, df = 6, *p*-value = 0.8853) (Figure 4).

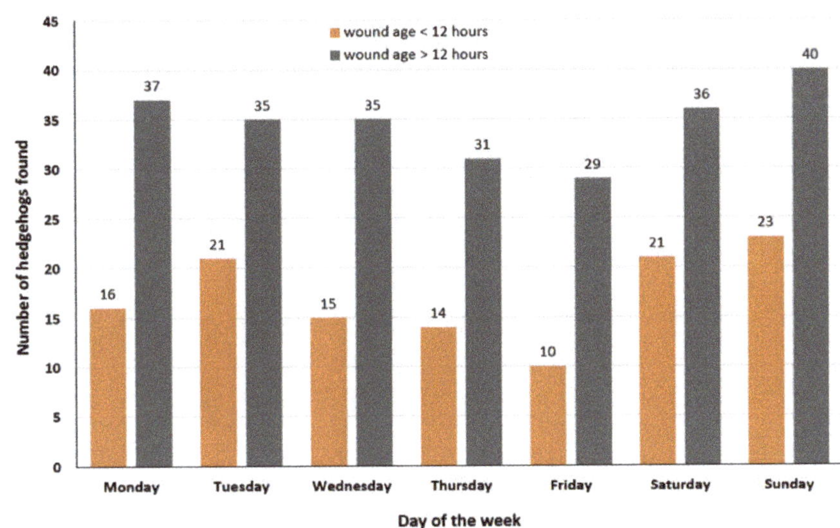

Figure 4. Number of 363 injured hedgehogs plotted according to the day of the week on which they were found: orange: hedgehogs found within 12 h of the accident (x-squared = 7.6333, df = 6, *p*-value = 0.2662), grey: hedgehogs whose wounds were older than 12 h on the date they were found (x-squared = 2.3457, df = 6, *p*-value = 0.8853). In 7 cases, the exact date and therefore the day of the week on which the injured hedgehog was found was not known.

Out of the 370 reported hedgehogs, 115 were euthanised due to the severity of the injury, 60 died during treatment or further care, 120 survived the treatment or were released back into the wild and there was no information on 75 or their care was still ongoing (Figure 5). A total of 32.7% (n = 121) of the hedgehogs were found within the first 12 h after the injury, 44.9 % (n = 166) of the hedgehogs had wounds older than 12 h and 14.6% (n = 54) had ones older than several weeks. A total of 33.2% (n = 123) of the hedgehogs were males, 28.4% (n = 105) were females (of which 3.5% (n = 13) were currently pregnant or lactating); the sex of 142 animals (38.4%) was unknown. Among the reported hedgehogs, 5 (1.4%) were dependent nestlings (<100 g), 35 (9.5%) were identified as juveniles and 191 (51.6%) were classified as adults (Figure 5).

Figure 6 shows the number of hedgehogs that "died" and "did not die" and the resulting mortality rate calculated for the respective wound characteristics or wounded body parts. If the mortality rate is >50%, this means that more than half of the animals exhibiting this wound characteristic or wounded body part died. The results of the chi-square test, which tested whether the certain wound characteristics or wounded body parts had an effect on the fate of the animal (died or survived), are shown as results with asterisks in Figure 6 and as specific *p*-values in Table 2.

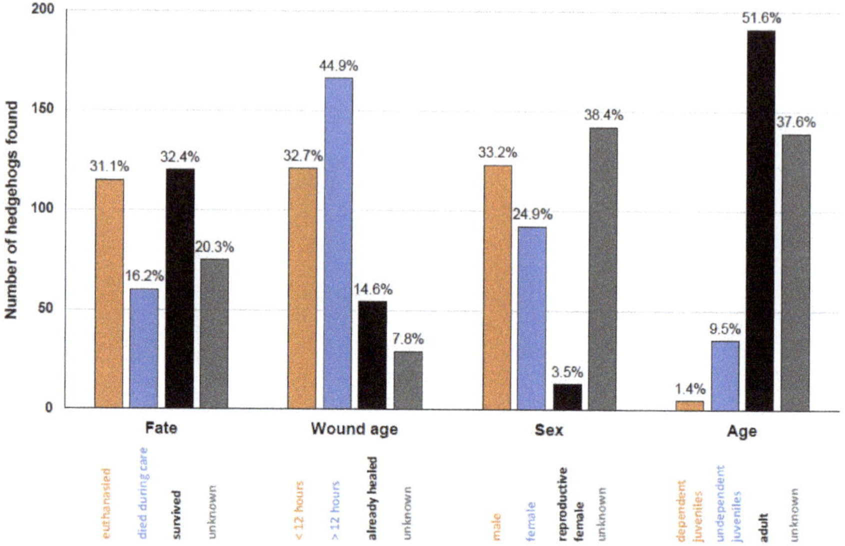

Figure 5. Number and percentages of the different fates (euthanised/died during care/survived or released/unknown), age of the wound when the hedgehog was found (<12 h/>12 h/already healed/unknown), sex (male, female, reproducing female/unknown) and age of the hedgehog (dependent nestling/independent juvenile/adult/unknown) of the 370 reported hedgehogs found with cut wounds.

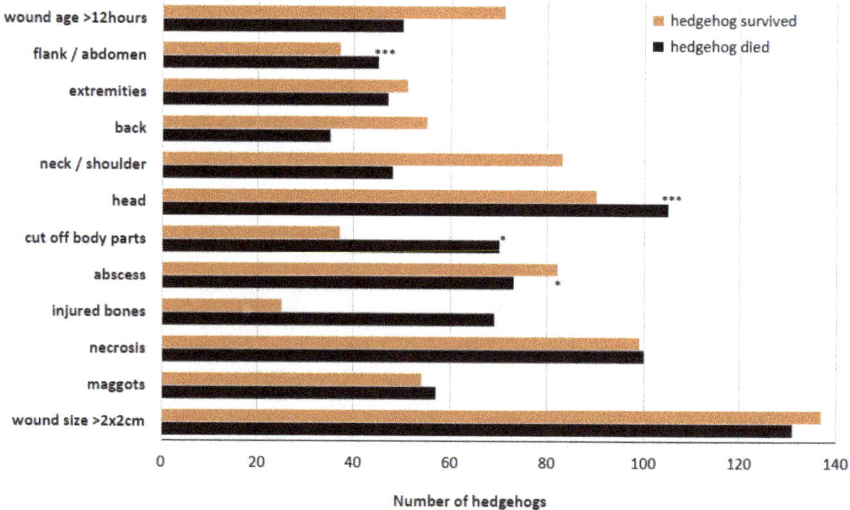

Figure 6. Number of hedgehogs with certain wound characteristics (wound is larger than 2 × 2 cm/presence of maggots/necrosis/bone injuries/body parts cut off/wound is older than 12 h) or whose wounds occurred on certain parts of the body (head/neck or shoulders/back/extremities/flank or abdomen) (multiple counts possible). The numbers are divided into hedgehogs that died by euthanasia or during treatment/care (black) and hedgehogs that survived. Cases where the fate of the animals was not known are not included. Significance results for the ratio of the parameters to the overall mortality of the 370 reported hedgehogs are shown symbolically (* *p*-value < 0.05, *** *p*-value < 0.001).

Table 2. Results of the chi-square test for the tested wound parameters with regard to their relationship to overall mortality (euthanasia or died during treatment/care) in 370 reported cut hedgehogs (* *p*-value < 0.05, *** *p*-value < 0.001).

	Mortality Rate [%]	X-Squared	*p*-Value
wound size > 2 × 2 cm	48.9	6.84	0.1447
maggots	51.4	11.73	0.06832
necrosis	50.3	1.67	0.796
injured bones	73.4	2.22	0.6945
abscess	47.1	11.67	0.0199 *
body parts cut off	65.4	11.09	0.02549 *
head	53.8	17.81	0.00013 ***
neck/shoulder	36.6	0.11	0.947
back	38.9	7.02	0.1348
extremities	48.0	1.29	0.5257
flank/abdomen	54.9	60.14	<0.0001 ***
wound age > 12 h	41.3	12.34	0.1947

4. Discussion

In this study, data on hedgehogs found with cut injuries throughout Germany were compiled for the first time and statistically analysed for their spatial and temporal occurrence and injury characteristics.

Although the spatial distribution of the hedgehogs found suggests that some parts of Germany (particularly North Rhine-Westphalia and Lower Saxony) have a particularly high incidence of cut injuries in hedgehogs (Figure 1), it should be considered that most of the larger hedgehog rescue centres that took part in the collection are located in these areas of high occurrence. When asking at other larger hedgehog centres in Germany that did not take part in the data collection of this study, we were told that they also have many hedgehogs with cuts in their care, but that they do not have the capacity to document or report this information as they are too busy caring for the hedgehogs, whose welfare comes before documentation. Moreover, most German hedgehog centres use an analogue protocol system, which means that queries about certain types of wounds require considerable effort for them. Apart from the fact that there is no complete and, above all, up-to-date list of all hedgehog care centres operating in Germany, many of the well-known and, above all, larger hedgehog centres did not participate in the collection of the data presented here, despite being asked. Therefore, the spatial distribution of the 370 hedgehogs found in this study only reflects a section of the overall German situation; areas without reports do not mean that there are no hedgehogs injured by cuts there.

Nevertheless, the map also shows that there is a very large number of hedgehogs with cut injuries, particularly around the larger hedgehog centres. Even if the area from which hedgehogs in need of help are brought to the (especially larger) hedgehog centres is very large [37], the willingness of people to take an injured hedgehog to a vet or care centre located far away is certainly limited at some point. Therefore, hedgehogs can of course only be helped where there are well-known and specialised hedgehog care centres or vets nearby. From the reports on the hedgehogs that could be released back into the wild after treatment (n = 120), it was clear that cut hedgehogs usually take many months to recover, and the treatment descriptions also suggest that several or complicated surgeries and expensive medication might be necessary. A study of 11,801 hedgehog patients from seven hedgehog centres in Germany showed a duration of stay of 0–359 days (mean 65.9 days, spread 77.2 days) [37]; hedgehogs with cut injuries therefore require above-average and likely more expensive treatment than other hedgehogs in need of care. Many centres also reported that the steadily increasing number and, above all, the severity of the injuries were pushing them to their spatial, financial, physical and mental capacity limits and that they would either have to close or would no longer be able to take in any more hedgehogs if this trend of rising number of hedgehogs with cut injuries continued.

The data in our study (Figures 2 and 3) show that cut injuries in hedgehogs a[re] becoming increasingly common. Though the collection of data only started on 28 June 202[2] and continued for 16 months, the 4 months (July–October) in which data were collecte[d] and reported for both 2022 and 2023 show that there was an increase in the number of case[s] from 2022 to 2023 (as in all other months too). The distribution of the findings by mon[th] (Figure 3) shows that in 8.6% of cases, cuts also occurred during the hedgehog hibernatio[n] period (January to March), a time when the animals generally do not leave their hibernatio[n] nest [17]. Studies on the time of admission of hedgehogs to rehabilitation centres show th[at] hedgehogs are brought in throughout the year with a bimodal distribution pattern, with [a] peak in summer with early litters (July–August) and one larger peak in autumn (August [to] November) with late litters and animals that are too weak to hibernate [24,38,39]. Most [of] the cut hedgehogs in this study were reported in May and June (Figure 3), a time when the[re] have always been very few new hedgehogs in need of care at hedgehog centres [38–41]. A[s] the majority of hedgehogs with cut injuries are also long-term patients, these high numbe[rs] in early summer (i.e., before the previous peak periods) mean that important inpatie[nt] capacities (e.g., hedgehog boxes) are already occupied resulting in no space for the man[y] animals that come in summer or autumn. The high number of hedgehogs injured by cu[ts] from May to July is due to both hedgehogs' way of life (hedgehogs are active and mal[es] travel longer distances and explore unknown terrain during the mating season [17]), b[ut] also to human activities in garden and green space maintenance. Even though this stud[y] shows a trend that a particularly high number of hedgehogs with fresh cuts were foun[d] on Tuesdays, Saturdays and Sundays and particularly few on Fridays (Figure 4), whic[h] gives an indication of the day of the week when the hedgehogs were injured by gard[en] maintenance works, these differences are not significant and may therefore have arisen b[y] chance. However, since robotic lawnmowers are the only gardening tools that can legal[ly] be used at any time, including Sundays and at night, the results also suggest that many [of] the injuries could have resulted from collisions with robotic lawnmowers; since the use [of] all other devices is not allowed on Sundays for noise protection reasons [30], there woul[d] also be fewer cuts on Sundays if these devices were mainly responsible for these cut injuri[es.] Hedgehogs with older cuts were found frequently (although not statistically significantl[y]) on Saturdays, Sundays and Mondays and rarely on Fridays, which gives an indication [of] when people are in green spaces/gardens and become aware of injured hedgehogs.

This study showed a mortality rate of at least 47.3% (in 20.3% of cases, the fate of th[e] animals is unknown) (Figure 5). Long-term studies in other European rehabilitation centr[es] report a mortality rate of approximately 1/3 of the total number of animals admitted [24,4[2] but it is apparent that the mortality rate is highly dependent on the reason for admissio[n,] the time of year (which in turn is related to the reason for admission) and the respectiv[e] care centre (some centres specialise in intensive care patients with an increased probabili[ty] of mortality) [24,37–39,42,43]. In addition, data from many rehabilitation centres show [a] decreasing mortality rate over the years, probably due to better treatment options and ca[re] practices [43]. In this respect, although the number of animals that died and, in particula[r,] were euthanised in our study is high, it is comparable with data from other studies.

In our study, 32.7% of the animals were found within 12 h of the cut injury (Figure 5[).] The majority of animals were not found until several days or even weeks after the accide[nt.] These animals were often no longer able to search for food or eat on their own or to lic[k] their wounds, so that their wounds became infected or flies laid their eggs there. Anima[ls] with such wounds have little chance of surviving in the wild. In 14.6% of the animals in ou[r] study, the wounds had nearly healed and some even appeared to be able to continue livin[g] with the injury, but here too there were individual cases with only a very low probabili[ty] of survival (they were severely dehydrated, emaciated or missing several limbs). In an[y] case, these injured animals suffer prolonged severe pain, suffering and harm which a[re] caused by human action and which according to European and German animal welfa[re] law can only be excused with a reasonable cause and must be avoided or limited as fa[r] as possible [44].

Out of the 370 animals reported, 33.2% were male and 28.4% female (Figure 5); in other studies, the sex ratio was approximately equal [37,38,42,45] or the injured males outnumbered the females due to their wider ranging behaviour and the resulting higher accident risk [43]. In this study, 3.5% of the reported hedgehogs were pregnant or lactating females at the time of the accident, thereby lowering the survival possibility their young, too. The vast majority of animals in this study were adults (Figure 5), but there were also dependent nestlings with lacerations which had sustained these injuries through the destruction of the litter nest and not through their own movement behaviour. The relatively high number of adult animals compared to other studies on hedgehog care centres [38,39,42,45] can be explained by the fact that subadult animals were also assessed as adults in this study, as a more precise age estimate is in general quite difficult and was not possible based on the reported information [46–48]. In this study, the proportion of animals for which the sex or age category was unknown is quite high, as this information was mostly not provided by the reporters as it was not demanded as necessary (minimal) information.

Even though the criteria for wound assessment in this study were chosen to be quite simple so that they could be made remotely based on photos and diagnostic reports, it must be mentioned that this methodology can be very error-prone since the diagnostic reports and photos are from many different veterinarians and hedgehog keepers, whose assessments and working methods can sometimes differ greatly from one another. A further limitation of the "remote" methodology when assessing wound characteristics or the sex or age of the animal is shown by the high proportion of "unknown" case assignments; in order to avoid incorrect assignments, the rating "unknown" was often given, which can falsify overall statistics or make them difficult to interpret.

This study showed that the animals that died (47.3%) were significantly more likely to have cuts to the head, flank and abdomen (Figure 6). There were also significant correlations between the occurrence of abscesses and removed body parts. The body parts that were cut off were noses, eyes, ears, snouts, toes, feet or whole legs. With these kinds of wounds, compared to other types of cut injuries (e.g., in the neck or very large wounds), there is virtually no chance of healing and survival. Other parameters, such as the age of the wound, had no influence on the probability of survival of the injured animal, which means an average probability of survival even if the injured animal was found days to weeks after the accident. However, due to their small body size, their hidden, nocturnal way of life and their danger-avoiding behaviour, it is rather coincidental that injured hedgehogs are found and the number of unknown cases of injured animals that are not found must be considered high. Hedgehogs try to behave as inconspicuously as possible when injured or try to find shelter in bushes in order to avoid attracting the attention of potential predators such as crows or foxes [49]. This behaviour also explains the high number of hedgehogs that were only found days or weeks after the injury. However, some individuals in this study, particularly those with extreme head wounds, were only found because they sought out human's vicinity on their own. But even hedgehogs that have died from cuts in the wild are not that easy to find: either they reached shelter before their death and are difficult to discover there, or other wild animals attacked and fed on them, and their carcasses disappeared relative quickly [50].

Various studies have examined the causes of death in hedgehogs in the wild and their possible impact on the population [20,51,52]; however, estimations of impact on the population always need solid figures about the population itself. Thanks to many years of citizen science monitoring, there is already a relatively solid database on the hedgehog population in Great Britain [7], which, for instance, made it possible to estimate the yearly number of hedgehogs which become life-threateningly entangled in garbage and the influence of this number on the development of the hedgehog population [20]. However, such figures on the number of hedgehogs in Germany are missing; thus, such estimates cannot be made with the data from this study. This will therefore be the subject of further investigations, because politicians and society first demand reliable information on the extent of hedgehogs with cut injuries and the influence of this phenomenon on their population development before they introduce restrictions (such as a ban on night-time us of robotic lawnmowers). Nevertheless, this study was able to prove that hedgehogs wit cuts caused by humans are not rare, isolated cases, but rather represent a problem that i relevant to animal welfare and requires technical (like devices that can be programmed fo a specific time of day and thus can only be used in daylight hours) or political solutions a soon as possible.

5. Conclusions

With the help of several hedgehog care centres, data on hedgehogs injured by cut across Germany were compiled. The analysis of these data showed that cut injuries increas from year to year, placing an enormous burden on many hedgehog care centres and usin up important resources, as these injuries often require above-average care and treatmen There is also a considerable animal welfare problem, as the majority of hedgehogs wit cut injuries are found days or weeks after the accident and therefore have to endur considerable suffering, pain and harm over a long period of time. Such animal suffering i prohibited by law, provided there are alternatives that do not cause animal suffering. At th very least, alternatives that do not cause that much animal suffering are certainly availabl through the technical or political implementation of a ban on night-time use of roboti mowers and these must be implemented immediately, which this study has attempted t contribute to.

Funding: This research received no external funding.

Institutional Review Board Statement: Not applicable.

Informed Consent Statement: Not applicable.

Data Availability Statement: All original data (photos and reports) can be found at https://www facebook.com/groups/igelmitschnittverletzungen/, (accessed on 10 December 2023). Due to graphi nature of the photos and sensitive data (such as the private addresses of gardens in which hedgehog with cuts were found), this page is only accessible by personal registration with the author.

Acknowledgments: I would like to take this opportunity to thank all the hedgehog carers who, i addition to their voluntary, self-sacrificing care for the injured hedgehogs, also found the time t report the data on their animals to me. This study would not exist without you! Analysing the photo of the cut hedgehogs and reading the treatment reports took a lot out of me because of the extent o the injuries, but that is nothing compared to the nerves of steel, patience and strength that you brin to the care of these animals. You rock!

Conflicts of Interest: The author declares no conflict of interest.

References

1. Sweet, F.S.T.; Noack, P.; Hauck, T.E.; Weisser, W.W. The Relationship between Knowing and Liking for 91 Urban Animal Specie among Students. *Animals* **2023**, *13*, 488. [CrossRef] [PubMed]
2. Hof, A.R.; Bright, P.W. The value of green-spaces in built-up areas for western hedgehogs. *Lutra* **2009**, *52*, 69–82.
3. Hof, A.R.; Bright, P.W. Quantifying the long-term decline of the West European hedgehog in England by subsampling citizer science datasets. *Eur. J. Wildl. Res.* **2016**, *62*, 407–413. [CrossRef]
4. Taucher, A.L.; Gloor, S.; Dietrich, A.; Geiger, M.; Hegglin, D.; Bontadina, F. Decline in Distribution and Abundance: Urba Hedgehogs under Pressure. *Animals* **2020**, *10*, 1606. [CrossRef] [PubMed]
5. Krange, M. Change in the Occurrence of the West European Hedgehog (*Erinaceus europaeus*) in Western Sweden during 1950–201(Master's Thesis, Karlstad University, Karlstad, Sweden, 2015.
6. Müller, F. Langzeit-Monitoring der Strassenverkehrsopfer beim Igel (*Erinaceus europaeus* L.) zur Indikation von Populations dichteveränderungen entlang zweier Teststrecken im Landkreis Fulda. *Beiträge Nat. Osthess.* **2018**, *54*, 21–26.
7. SoBH. The State of Britain's Hedgehogs 2022. British Hedgehog Preservation Society and People's Trust for Endangered Specie 2022. Available online: https://www.hedgehogstreet.org/wp-content/uploads/2022/02/SoBH-2022-Final.pdf (accessed on 1(December 2023).
8. van de Poel, J.L.; Dekker, J.; van Langevelde, F. Dutch hedgehogs Erinaceus europaeus are nowadays mainly found in urba areas, possibly due to the negative effects of badgers Meles meles. *Wildl. Biol.* **2015**, *21*, 51–55. [CrossRef]

Hubert, P.; Julliard, R.; Biagianti, S.; Poulle, M.L. Ecological factors driving the higher hedgehog (*Erinaceus europeaus*) density in an urban area compared to the adjacent rural area. *Landsc. Urban Plann.* **2011**, *103*, 34–43. [CrossRef]

Yarnell, R.W.; Pettett, C.E. Beneficial Land Management for Hedgehogs (*Erinaceus europaeus*) in the United Kingdom. *Animals* **2020**, *10*, 1566. [CrossRef]

Braaker, S.; Kormann, U.; Bontadina, F.; Obrist, M.K. Prediction of genetic connectivity in urban ecosystems by combining detailed movement data, genetic data and multi-path modelling. *Landsc. Urban Plann.* **2017**, *160*, 107–114. [CrossRef]

Dowding, C.V.; Shore, R.F.; Worgan, A.; Baker, P.J.; Harris, S. Accumulation of anticoagulant rodenticides in a non-target insectivore, the European hedgehog (*Erinaceus europaeus*). *Environ. Pollut.* **2010**, *158*, 161–166. [CrossRef]

Brakes, C.R.; Smith, R.H. Exposure of non-target small mammals to rodenticides: Short-term effects, recovery and implications for secondary poisoning. *J. Appl. Ecol.* **2005**, *42*, 118–128. [CrossRef]

Trewby, I.D.; Young, R.; McDonald, R.A.; Wilson, G.J.; Davison, J.; Walker, N.; Robertson, A.; Doncaster, C.P.; Delahay, R.J. Impacts of Removing Badgers on Localised Counts of Hedgehogs. *PLoS ONE* **2014**, *9*, e95477. [CrossRef] [PubMed]

Williams, B.M.; Baker, P.J.; Thomas, E.; Wilson, G.; Judge, J.; Yarnell, R.W. Reduced occupancy of hedgehogs (*Erinaceus europaeus*) in rural England and Wales: The influence of habitat and an asymmetric intra-guild predator. *Sci. Rep.* **2018**, *8*, 12156. [CrossRef] [PubMed]

Young, R.P.; Davison, J.; Trewby, I.D.; Wilson, G.J.; Delahay, R.J.; Doncaster, C.P. Abundance of hedgehogs (*Erinaceus europaeus*) in relation to the density and distribution of badgers (*Meles meles*). *J. Zool.* **2006**, *269*, 349–356. [CrossRef]

Morris, P. *Hedgehog*; Collins New Naturalist Library; Wiliam Collins: Glasgow, UK, 2018; pp. 1–404.

Huijser, M.P.; Bergers, P.J.M. The effect of roads and traffic on hedgehog (*Erinaceus europaeus*) populations. *Biol. Conserv.* **2000**, *95*, 6–9. [CrossRef]

Holsbeek, L.; Rodts, J.; Muyldermans, S. Hedgehog and other animal traffic victims in Belgium: Results of a countrywide survey. *Lutra* **1999**, *42*, 111–119.

Thrift, E.; Nouvellet, P.; Mathews, F. Plastic Entanglement Poses a Potential Hazard to European Hedgehogs *Erinaceus europaeus* in Great Britain. *Animals* **2023**, *13*, 2448. [CrossRef]

Bundesnaturschutzgesetz. Gesetz über Naturschutz und Landschaftspflege (Bundesnaturschutzgesetz-BNatSchG). Available online: https://www.gesetze-im-internet.de/bnatschg_2009/BJNR254210009.html (accessed on 1 October 2023).

Guy, A.J.; Curnoe, D.; Banks, P.B. A survey of current mammal rehabilitation and release practices. *Biodivers. Conserv.* **2013**, *22*, 825–837. [CrossRef]

Grogan, A.; Kelly, A. A review of RSPCA research into wildlife rehabilitation. *Vet. Rec.* **2013**, *172*, 211–214. [CrossRef]

Martínez, J.; Rosique, A.I.; Royo, M. Causes of admission and final dispositions of hedgehogs admitted to three Wildlife Rehabilitation Centers in eastern Spain. *Hystrix* **2014**, *25*, 107–110.

Research and Markets. *Robotic Lawn Mowers Market-Global Outlook and Forecast. 2020–2025*; Arizton Global: Chicago, IL, USA, 2020.

Rasmussen, S.L.; Schrøder, A.E.; Mathiesen, R.; Nielsen, J.L.; Pertoldi, C.; Macdonald, D.W. Wildlife Conservation at a Garden Level: The Effect of Robotic Lawn Mowers on European Hedgehogs (*Erinaceus europaeus*). *Animals* **2021**, *11*, 1191. [CrossRef]

Open Petition: Nachtfahrverbot für Mähroboter. Available online: https://www.openpetition.de/petition/online/nachtfahrverbot-fuer-maehroboter (accessed on 10 December 2023).

Open Petition: Nächtliche Einschränkung für Mähroboter. Available online: https://www.openpetition.de/petition/online/igelschutz-nachtverbot-fuer-maehroboter (accessed on 10 December 2023).

Kleine Anfrage Landtag Baden-Württemberg. Available online: https://www.landtag-bw.de/files/live/sites/LTBW/files/dokumente/WP17/Drucksachen/5000/17_5699_D.pdf (accessed on 10 December 2023).

Mähroboter und Ruhezeiten: Was Ist Erlaubt? Available online: https://www.maehroboter-guru.de/maehroboter-und-ruhezeiten-was-ist-erlaubt/ (accessed on 10 December 2023).

Scholl, P.J.; Catts, E.P.; Mullen, G.R. Myiasis (Muscoidea, Oestroidea). In *Medical and Veterinary Entomology*; Mullen, G., Durden, L., Eds.; Academic Press: Cambridge, MA, USA; Elsevier: Amsterdam, The Netherlands, 2009; pp. 309–338.

Bambaradeniya, Y.T.B.; Magni, P.A.; Dadour, I.R. Traumatic sheep myiasis: A review of the current understanding. *Vet. Parasitol.* **2023**, *314*, 109853. [CrossRef] [PubMed]

Francesconi, F.; Lupi, O. Myiasis. *Clin. Microbiol. Rev.* **2012**, *25*, 79–105. [CrossRef] [PubMed]

Lux, C.N. Wound healing in animals: A review of physiology and clinic evaluation. *Vet. Dermatol.* **2022**, *33*, 91-e27. [CrossRef] [PubMed]

Maxwell, E.A.; Bennett, R.A. Wound Management in Wildlife. In *Medical Management of Wildlife Species: A Guide for Practitioners*; Hernandez, S.M., Barron, H.W., Miller, E.A., Aguilar, R.F., Yabsley, M.J., Eds.; Wiley Online Library: Hoboken, NJ, USA, 2019; pp. 129–143.

R Studio. RStudio: Integrated Development Environment for R. RStudio: Boston. Available online: http://www.rstudio.com/ (accessed on 10 September 2023).

Kögel, B. Untersuchungen zu Igelpfleglingen Ausgewählter Deutscher Igelstationen und Erfolge der Therapie aus den Jahren 1984 bis 2006. Ph.D. Thesis, Tierärztliche Hochschule Hannover, Hannover, Germany, 4 May 2009.

Shelton, M. Can UK Wildlife Rehabilitation Centre Data Be Used to Help Monitor the Health of *Erinaceus europaeus* and Spot Trends Relating to Disease and Injury within the Species? Bachelor's Thesis, University Centre Sparsholt, Sparsholt, UK, 2017.

39. Burroughes, N.D.; Dowler, J.; Burroughes, G. Admission and Survival Trends in Hedgehogs Admitted to RSPCA Wildlife Rehabilitation Centres. *Proc. Zool. Soc.* **2021**, *74*, 198–204. [CrossRef]
40. Fuhrmann, M. Untersuchungen zur Wildtierrehabilitation und Wiederauswilderung von *Erinaceus europaeus* (Braunbrustigel) und *Erinaceus roumanicus* (Nördlicher Weißbrustigel) in Österreich. Master's Thesis, Universität für Bodenkultur, Wien, Austria, 20?
41. Lukešová, G.; Voslarova, E.; Vecerek, V.; Vucinic, M. Trends in intake and outcomes for European hedgehog (*Erinaceus europaeus*) in the Czech rescue centers. *PLoS ONE* **2021**, *16*, e0248422. [CrossRef] [PubMed]
42. Garcês, A.; Soeiro, V.; Lóio, S.; Sargo, R.; Sousa, L.; Silva, F.; Pires, I. Outcomes, Mortality Causes, and Pathological Findings in European Hedgehogs (*Erinaceus europeus*, Linnaeus 1758): A Seventeen Year Retrospective Analysis in the North of Portugal. *Animals* **2020**, *10*, 1305. [CrossRef] [PubMed]
43. Reeve, N.J.; Huijser, M.P. Mortality factors affecting wild hedgehogs: A study of records from wildlife rescue centres. *Lutra* **1999**, *42*, 7–24.
44. Tierschutzgesetz. Available online: https://dejure.org/gesetze/TierSchG/1.html (accessed on 2 October 2023).
45. Bunnell, T. The incidence of disease and injury in displaced wild hedgehogs (*Erinaceus europaeus*). *Lutra* **2001**, *44*, 3–14.
46. Morris, P.A. A method for determining absolute age in the hedgehog. *J. Zool.* **1970**, *161*, 277–281. [CrossRef]
47. Reeve, N.J.; Love, B.; Shore, R. X-ray measures of long bones as an aid to age-determination in hedgehogs (*Erinaceus europaeus*). Proceedings of the 1st European Hedgehog Research Group Meeting, Arendal, Norway, 1996.
48. Rasmussen, S.L.; Berg, T.B.; Martens, H.J.; Jones, O.R. Anyone Can Get Old-All You Have to Do Is Live Long Enough: Understanding Mortality and Life Expectancy in European Hedgehogs (*Erinaceus europaeus*). *Animals* **2023**, *13*, 62. [CrossRef] [PubMed]
49. Dowding, C.V.; Harris, S.; Poulton, S.; Baker, P.J. Nocturnal ranging behaviour of urban hedgehogs, *Erinaceus europaeus*, in relation to risk and reward. *Anim. Behav.* **2010**, *80*, 13–21. [CrossRef]
50. Santos, S.M.; Carvalho, F.; Mira, A. How Long Do the Dead Survive on the Road? Carcass Persistence Probability and Implications for Road-Kill Monitoring Surveys. *PLoS ONE* **2011**, *6*, e25383. [CrossRef] [PubMed]
51. Kristiansson, H. Population variables and causes of mortality in a hedgehog (*Erinaceus europaeus*) population in southern Sweden. *J. Zool.* **1990**, *220*, 391–404. [CrossRef]
52. Wembridge, D.E.; Newman, M.R.; Bright, P.W.; Morris, P.A. An estimate of the annual number of hedgehog (*Erinaceus europaeus*) road casualties in Great Britain. *Mammal Commun.* **2016**, *2*, 8–14.

Article

Wildlife Conservation at a Garden Level: The Effect of Robotic Lawn Mowers on European Hedgehogs (*Erinaceus europaeus*)

Sophie Lund Rasmussen [1,2,*], Ane Elise Schrøder [3,4], Ronja Mathiesen [5], Jeppe Lund Nielsen [2], Cino Pertoldi [2] and David W. Macdonald [1]

1   Wildlife Conservation Research Unit, The Recanati-Kaplan Centre, Department of Zoology, University of Oxford, Tubney House, Abingdon Road, Tubney, Abingdon OX13 5QL, UK; david.macdonald@zoo.ox.ac.uk
2   Department of Chemistry and Bioscience, Aalborg University, Fredrik Bajers Vej 7H, DK-9220 Aalborg, Denmark; jln@bio.aau.dk (J.L.N.); cp@bio.aau.dk (C.P.)
3   Natural History Museum of Denmark, University of Copenhagen, Universitetsparken 15, DK-2100 Copenhagen Ø, Denmark; aneelises@snm.ku.dk
4   Fossil and Moclay Museum, Museum Mors, Skarrehagevej 8, DK-7900 Nykøbing Mors, Denmark
5   Agilent Technologies Denmark ApS, Produktionsvej 42, DK-2600 Glostrup, Denmark; ronjamathiesen@gmail.com
*   Correspondence: sophie.rasmussen@zoo.ox.ac.uk or sophielundrasmussen@gmail.com; Tel.: +45-2211-7268 or +44-787-1837-510

Simple Summary: Injured European hedgehogs are frequently admitted to hedgehog rehabilitation centres with different types of cuts and injuries. Although not rigorously quantified, a growing concern is that an increasing number of cases may have been caused by robotic lawn mowers. Research indicates that European hedgehogs are in decline. It is therefore important to identify and investigate the factors responsible for this decline to improve the conservation initiatives directed at this species. Because hedgehogs are increasingly associated with human habitation, it seems likely that numerous individuals will encounter several robotic lawn mowers during their lifetimes. Consequently, this study aimed to describe and quantify the effects of robotic lawn mowers on hedgehogs, and we tested 18 robotic lawn mowers in collision with dead hedgehogs. Some models caused extensive damage to the dead hedgehogs, but there were noteworthy differences in the degree of harm inflicted, with some consistently causing no damage. None of the robotic lawn mowers tested was able to detect the presence of dead, dependent juvenile hedgehogs, and no models could detect the hedgehog cadavers without physical interaction. We therefore encourage future collaboration with the manufacturers of robotic lawn mowers to improve the safety for hedgehogs and other garden wildlife species.

Abstract: We tested the effects of 18 models of robotic lawn mowers in collision with dead European hedgehogs and quantified the results into six damage categories. All models were tested on four weight classes of hedgehogs, each placed in three different positions. None of the robotic lawn mowers tested was able to detect the presence of dependent juvenile hedgehogs (<200 g) and all models had to touch the hedgehogs to detect them. Some models caused extensive damage to the hedgehog cadavers, but there were noteworthy differences in the degree of harm inflicted, with some consistently causing no damage. Our results showed that the following technical features significantly increased the safety index of the robotic lawn mowers: pivoting blades, skid plates, and front wheel drive. Based on these findings, we encourage future collaboration with the manufacturers of robotic lawn mowers to improve the safety for hedgehogs and other garden wildlife species.

Keywords: animal behaviour; applied conservation biology; *Erinaceus europaeus*; human–wildlife conflicts; robotic lawn mowers; wildlife conservation

Citation: Rasmussen, S.L.; Schrøder, A.E.; Mathiesen, R.; Nielsen, J.L.; Pertoldi, C.; Macdonald, D.W. Wildlife Conservation at a Garden Level: The Effect of Robotic Lawn Mowers on European Hedgehogs (*Erinaceus europaeus*). *Animals* **2021**, *11*, 1191. https://doi.org/10.3390/ani11051191

Academic Editors: Krishna R. Balasubramaniam and Stefano S.K. Kaburu

Received: 5 March 2021
Accepted: 16 April 2021
Published: 21 April 2021

Publisher's Note: MDPI stays neutral with regard to jurisdictional claims in published maps and institutional affiliations.

1. Introduction

Research on both national and local scales has either documented, or expressed concern about the likelihood of, a decline in European hedgehog (*Erinaceus europaeus*) populations in several western European countries [1–10]. It is therefore a priority to identify and investigate the factors responsible for this decline to provide the evidence necessary to underpin remedial conservation interventions.

Injured hedgehogs are frequently admitted to hedgehog rehabilitation centres with different types of cuts and injuries. Some injuries are consistent with known risks to hedgehogs in the form of garden trimmers and dog bites [11–13]. However, although not rigorously quantified, a concern has arisen in several European countries that an increasing number of cases may have been caused by robotic lawn mowers. Although not previously investigated, these growing rumours have led to several articles in the media and on social media claiming that these mowers are lethal to hedgehogs. If the threat is real, then it would indeed be a cause for concern, as the global market for robotic lawn mowers is expanding dramatically and was expected to reach USD 1.3 billion in 2020, growing at an annual rate of more than 12 percent during the period 2019–2025 [14].

As research indicates that European hedgehogs are increasingly associated with human habitation [7,8,15–17] and are often seen foraging on grassy turf in the garden and green spaces of urban areas [18–22], it seems likely that numerous individuals will encounter several robotic lawn mowers during their lifetimes. To our knowledge, there has thus far been no systematic scientific research evaluating whether this risk of physical damage is mere hearsay or a real and present threat to be added to the already vulnerable species. Therefore, the aims of this study are to describe and quantify the physical effect of robotic lawn mowers on hedgehogs and provide information on potential technical features of the machines that could increase the safety index of the robotic lawn mowers. The main purpose of providing this information is to improve the conservation of European hedgehogs living in residential areas by reducing the plausible negative anthropogenic effects potentially caused by robotic lawn mowers.

2. Materials and Methods

In total, 18 designs of robotic lawn mowers were selected for the study. The selection was based on the advice of a product specialist in robotic lawn mowers and is considered to represent the spectrum of brands, models, and specifications of the products available on the European market (Table 1). The cutting height of the machines was adjusted to the highest setting to keep the grass at the test site intact to ensure equal test conditions for all trials.

Table 1. Overview of the models of robotic lawn mowers tested. In the column "Blades", Pivoting indicates "low energy pivoting blades" and Fixed indicates "heavy duty fixed blades". WMCC detection is short for "wheel motor current collision detection".

Test Number	Brand	Model	Blades	Collision Sensor	WMCC Detection	Wheels	Front/Rear Wheel Drive	Skid Plate	Headlights	Ultrasonic Sensors	Camera Vision
1	Husqvarna	Automower® 105	Pivoting	Yes		3	Front	Yes			
2	Husqvarna	Automower® 305	Pivoting		Yes	4	Rear	Yes			
3	Husqvarna	Automower® 315X	Pivoting	Yes		4	Rear	Yes	Yes		
4	Husqvarna	Automower® 450X	Pivoting	Yes		4	Rear	Yes	Yes	Yes	
5	Gardena	Sileno City	Pivoting		Yes	3	Front				
6	Gardena	Sileno Life	Pivoting		Yes	4	Front				
7	Worx	Landroid L (WR153E)	Pivoting		Yes	4	Rear				
8	Worx	Landroid M (WR143E)	Pivoting		Yes	4	Rear			Yes	
9	Kress	Mission KR111	Pivoting	Yes		4	Rear			Yes	
10	LandXcape	LX8212i	Pivoting		Yes	3	Rear			Yes	
11	Honda	Miimo HRM 40 Live	Pivoting		Yes	4	Rear				
12	Honda	Miimo HRM 3000	Pivoting	Yes		4	Rear				
13	Robomow	RS635 PRO	Fixed		Yes	3	Rear				Yes
14	AL-KO	Robolinho® 1150	Fixed		Yes	4	Rear				
15	Ambrogio Robot	4.0 Elite	Fixed	Yes		4	Rear				
16	Stiga	Autoclip 530 SG	Fixed		Yes	4	Rear				
17	Stihl	iMow® 422PC	Fixed	Yes		4	Rear				
18	DAYE	Grouw M900	Pivoting	Yes		4	Rear				

Of the 18 robotic lawn mowers tested, 5 had fixed blades (Figure 1A) and 13 had pivoting blades (Figure 1B).

Figure 1. Fixed or pivoting blades. (**A**) A robotic lawn mower with fixed blades. (**B**) A robotic lawn mower with pivoting blades. Pivoting blades will fold into a protective frame when hitting something harder than grass, as opposed to fixed blades, which are constantly exposed. Photographs by Petrus Ekbladh and Ronja Mathiesen.

The robotic lawn mower tests were performed on dead hedgehogs, henceforth referred to as "hedgehogs". These animals had died in, and were secured from, hedgehog rehabilitation centres in Denmark from June to August 2020. All hedgehogs chosen for this

study were intact with no visible injuries. The hedgehogs were stored in freezers at −20 °
and were thawed before the tests. The 70 selected hedgehog cadavers were divided in
four different weight classes to represent four stages of life (Table 2).

Each robotic lawn mower model was tested on four hedgehogs representing eac
of the four described weight classes. If an individual was injured by the mower durir
a test, it would be discarded to avoid confusion or interaction with previous injuries
subsequent tests (with one uncomplicated exception, where we reused a cadaver wit
superficial injuries in weight class 4, due to a shortage of individuals in this category).

Table 2. Weight classes. Graphical representation of the four weight classes of dead hedgehogs used in the study. The
pictures of live hedgehogs were provided to illustrate the sizes of individuals belonging to the four weight classes. No live
hedgehogs were tested in this study. The ruler on the pictures indicates length of the individuals in cm. Photographs by
Michela Dugar.

Weight Class	Weight (g)	No. of Individuals	Total No. of Individuals per Weight Class	Stages of Life	Representation
1	Up to 199	22	22	Dependent juveniles	Weight 46 g Length 7.5 cm
2	200	3	21	Independent juveniles	Weight 530 g Length 19.5 cm
	300	3			
	400	9			
	500	6			
3	600	8	20	Adults	Weight 860 g Length 23 cm
	700	8			
	800	4			
4	900	4	7	Large adults	Weight 1080 g Length 25 cm
	1000	2			
	1100	1			

Each individual was tested in three different positions (Figure 2), as an attempt
mimic the behaviour of a live individual:

1. Lying on the side with the back pointing towards the approaching robotic law
   mower, mimicking the curled up position a hedgehog often adopts as a defenc
   mechanism against approaching danger [19,23].
2. Lying on the side with the stomach pointing towards the approaching robotic law
   mower (somewhat unnatural, but extremely vulnerable position).

3. Standing upright on its feet with the head pointing towards the approaching robotic lawn mower (an expression of curiosity but not alarm).

To sum up, each robotic lawn mower was tested 12 times:

- Three times per individual (once in each of the three positions).
- One individual from each of the four weight classes.

The tests were filmed with a GoPro Hero 8 Black action camera placed on a tripod. If a hedgehog was injured during the tests, we recorded the injuries and documented them with the camera.

The tests of 17 out of 18 machines were carried out in a private garden in Hok, Sweden, with a flat and well-trimmed lawn, on 25 and 26 August 2020. The last machine (model: Grouw M900) was tested in a private garden with a flat and well-trimmed lawn in Aarhus, Denmark on 25 September 2020. All 216 tests were performed during daylight.

The setup for most of the tests was as follows (Figure 2): The hedgehog was placed on the lawn at a 3 m distance from the robotic lawn mower. The camera was placed next to the hedgehog on the left-hand side at a 1.5 m distance. The mower was then turned on and manually directed to move towards the hedgehog. The distance of 3 m was sufficient to ensure the machine was operating at maximum speed, and the blades were in action, before reaching the hedgehog. If the machine did not move in a straight line towards the hedgehog, it was then relocated back to the initial position and turned on again. This was done to standardise the tests and to ensure that the hedgehog was located to the centre of the front of each approaching machine.

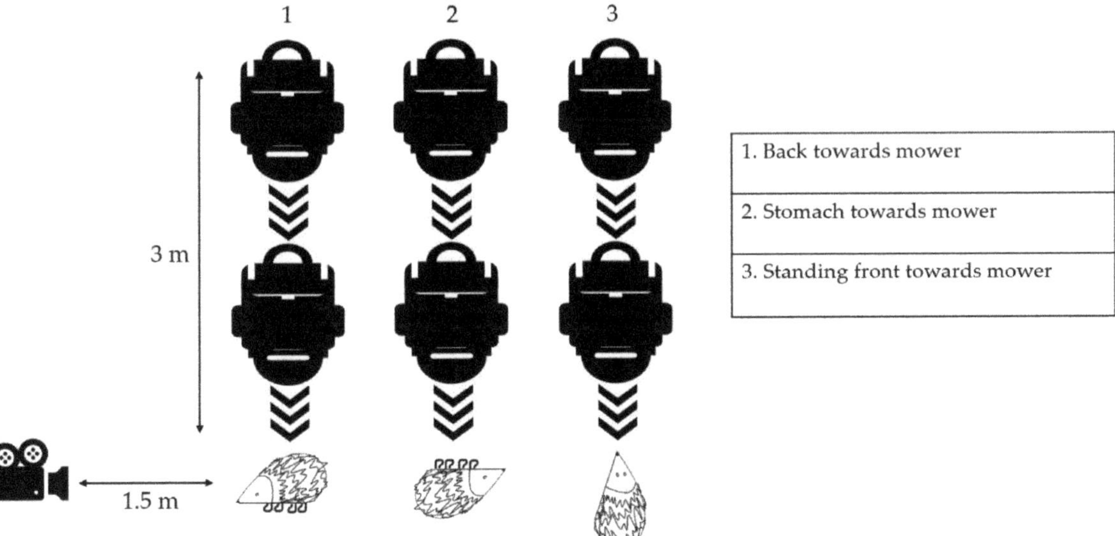

Figure 2. An overview of the test setup. Each robotic lawn mower was tested 12 times in total, 3 times per dead hedgehog representing 1 of 4 weight classes. A hedgehog from each of the four weight classes was placed in three different positions. The three positions were (**1**) Lying on the side with the back oriented towards the approaching robotic lawn mower; (**2**) Lying on the side with the stomach oriented towards the approaching robotic lawn mower; (**3**) Standing upright on its feet with the head oriented towards the approaching robotic lawn mower. The damage recorded from each test was categorised as 0-A according to the damage categories.

In the cases of two models (tests on Stiga Autoclip 530 SG (Stiga, Castelfranco Venetto, Iltaly) and Ambrogio Robot 4.0 Elite (Zucchetti Centro Sistemi Spa, Arezzo, Italy)) the machine was turned on at a greater distance than 3 m from the hedgehog cadaver, because these particular robotic lawn mowers took longer distances to gain momentum and for their blades to be functioning fully. Due to its specifications (causing erratic movements) the tests of the model Honda Miimo HRM 40 Live (Honda France Manufacturing, Ormes, France) were filmed with a mobile camera, with the hedgehog placed in front of the approaching machine once it had gained full speed.

2.1. Quantifying the Damage

We divided severity of damage caused by the robotic lawn mowers into six damage categories:

0. No physical contact between the machine and the hedgehog. The machine senses the hedgehog from a distance, changes direction, and drives on without touching the hedgehog. No damage is caused to the hedgehog cadaver.
1. The robotic lawn mower approaches the hedgehog and the front of the machine touches the hedgehog lightly (a "nudge") and thereby detects the corpse. Immediately the machine changes direction and drives on without touching the hedgehog further. No damage is caused to the hedgehog cadaver.
2. The robotic lawn mower approaches the hedgehog and the front of the machine touches the hedgehog (a "flip") to detect the hedgehog. The physical interaction causes the hedgehog to be moved into a different body position (flipped from lying on one side of the body to the other side of the body) or being lifted partly from the ground before settling in the same position again. Afterwards, the machine changes direction and drives on without touching the hedgehog further. The damage to the hedgehog is at most minimal and involves no contact with the blades (at worst this might cause a slight bruise).
3. The robotic lawn mower fails to detect the presence of the hedgehog and continues to drive across the hedgehog. The front panel of the machine is lifted as the machine drives over the cadaver, which causes the blades to stop running [24]. In some cases the machine withdraws and changes direction, so that only part of the dead hedgehog's body was situated underneath the machine. The blades of the robotic lawn mower may have come into contact with the dead hedgehog but have not punctured the skin. The damages observed ranged from undetectable to the cutting of a small number of spines, but might have involved minor bruising to a live hedgehog.
4. The robotic lawn mower fails to detect the presence of the hedgehog and continues to drive across it. The blades of the machine have come into contact with the dead hedgehog and have caused injuries to the cadaver. The severity of the injuries range from small puncture wounds on the skin (1 cm) to clipping of limbs or complete exposure of the entire abdominal region and decapitation.
A. The machine does not detect the juvenile hedgehog (<200 g, weight class 1) and continues to drive across it. As the body of the small hedgehog is situated below the blades of the robotic lawn mower, the juvenile hedgehog is left with no visible injuries. It is possible that in life this could have caused injury or bruising, perhaps by the wheels rather than the blades (and much would depend on the response of the juvenile hedgehog in life).

2.2. Data Analyses

The proportion (ratio) of "no damage" or "damage" during the tests, the safety index was calculated for the following features on the mowers: (1) blade type, (2) front or rear wheel drive, (3) wheel numbers, (4) skid plate, (5) ultrasonic sensor, (6) camera vison, (7) collision sensors, and (8) wheel motor current collision detection.

For the data analyses, the damage categories were divided into two definitions of "no damage" and "damage":

- "No damage":
    1. Pooled damage categories 0, 1, and 2.
    2. Pooled damage categories 0, 1, 2, and 3.

- "Damage":
    1. Pooled damage categories 3 and 4.
    2. Only damage category 4.

The ratios of "no damage"/"damage" were interpreted as an overall index of safety for the hedgehogs. The higher the number of "no damage" events compared to "damage", the higher the safety index, and hence the judgement that the robotic lawn mower was more "hedgehog friendly".

The statistical significance for each of the "no damage"/"damage" ratios was tested with 2×2 Chi square tests with Yate's correction ($Chi_{Yate's\ correction}$) to investigate if the presence or absence of a given technical feature on a robotic lawn mower significantly affected the ratios.

The 2×2 Chi square tests with Yate's correction were firstly conducted for each of the four weight classes and three positions of the hedgehogs, separately. However, due to the low statistical power caused by analysing weight and position separately, all weight classes and positions were combined. Subsequently, 2×2 Chi square tests with Yate's correction for all weight classes and positions combined were calculated for each of the two definitions of damage, testing the effect of each of the eight chosen technical features on the robotic lawn mowers on the safety index.

Lastly, we calculated the percentage distribution of damage to hedgehogs during the 12 tests on each mower based on the total number of cases where damage was recorded (either damage category 3 + 4 or damage category 4). Damage category A was omitted from the analyses, because including it resulted in different sample sizes of tests for different models of robotic lawn mowers. Therefore, the percentage distribution was chosen as a measure of safety.

3. Results

Regardless of brand, model, and specifications, none of the robotic lawn mowers detected the dependent juvenile hedgehogs (<200 g, weight class 1). Some machines did, however, move over the individuals resulting in no apparent damage, as the juveniles were sufficiently small, i.e., smaller than the minimum mowing height, thereby avoiding the running blades of the mowers (damage category A).

In all tests of weight category classes > 200 g (weight classes 2–4), the robotic lawn mowers had to physically interact with the hedgehog cadaver to detect it. None of the machines, not even models with camera vision and ultrasonic sensors, was able to detect the hedgehog in advance and change direction before touching the hedgehog. Therefore, we did not record any damage category 0. In many cases, the mowers would only touch the hedgehog (damage category 1 or 2), subsequently detect it, and change direction. However, some machines did not detect the hedgehogs and ran straight over them. In some cases the mandatory safety measures of the machine [24] caused the blades to stop rotating within seconds of contact, leaving the hedgehog undamaged or with slight cuts to the spines (damage category 3). In the event that the safety features of the machine failed to detect the hedgehog, the result was injury to the cadaver (damage category 4) ranging from lighter skin abrasions and puncture wounds, to the amputation of extremities like legs and penises, to complete disembowelment, and in one case a partial decapitation. The injuries appeared on all areas of the body in no particular pattern, as it depended on the position in which the hedgehog was caught under the robotic lawn mower, as well as the angles of the blades. Figure 3 provides an overview of the damage categories recorded for each of the 12 different tests performed on the 18 robotic lawn mowers.

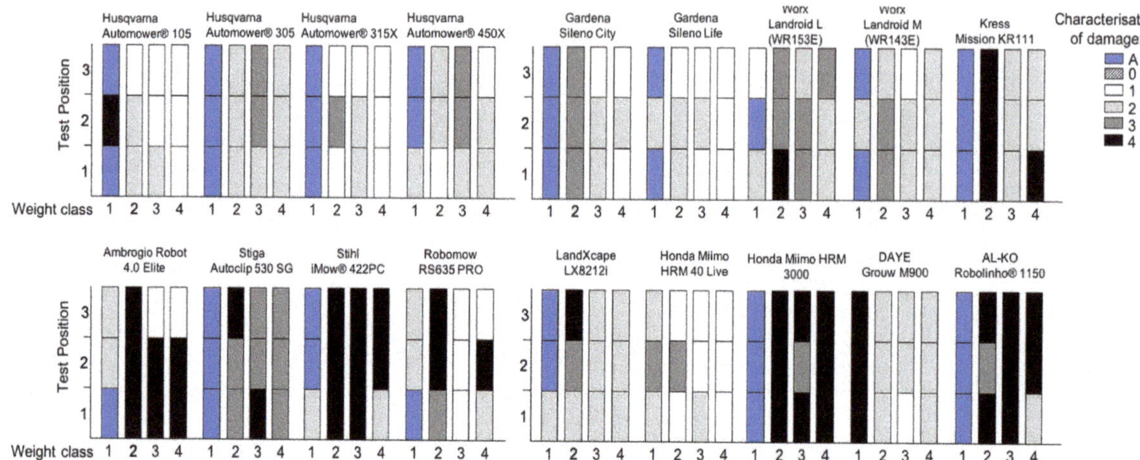

Figure 3. The test results for each of the 18 robotic lawn mowers tested. Every result for each of the four weight classes in each of the three positions have been described based on a categorisation of damage ranging from 0 to 4, with damage category 4 being the most severe. Damage category A represents the events where the machine does not detect the juvenile hedgehog (<200 g, weight class 1) and continues to drive across the juvenile hedgehog, but as the body of the small hedgehog is situated below the blades of the robotic lawn mower, the juvenile hedgehog is left with no visible injuries or bruises.

Comparing the effect of fixed and pivoting blades, the results showed that pivoting blades significantly reduced the number of damages during the tests, regardless of the definition of the category "damage" (either damage category 3 + 4 or damage category 4) ($Chi_{Yate's\ correction}$ = 28.95 and 26.62, p < 0.0001). The same applied to machines with front wheel drive compared to rear wheel drive ($Chi_{Yate's\ correction}$ (4) = 7.25, p = 0.007; $Chi_{Yate's\ correction}$ (3 + 4) = 8.99, p = 0.003) as well as the presence of skid plates on the machines ($Chi_{Yate's\ correction}$ = 11.39 and 10.99, p = 0.001). Robotic lawn mowers with three wheels instead of four had a significantly higher safety index, meaning that there were fewer cases of damage to the hedgehogs during the tests for the damage categorisation based on both damage category 3 and 4 ($Chi_{Yate's\ correction}$ = 4.37, p = 0.037), but not for the damage categorisation based only on damage category 4. Ultrasonic sensors also appeared to increase the safety index for the damage categorisation based on damage category 3 and 4 ($Chi_{Yate's\ correction}$ = 3.84, p = 0.05), but not for the damage categorisation based only on damage category 4. The presence of collision sensors, compared to wheel motor current collision detection, reduced the safety index for damage category 4 ($Chi_{Yate's\ correction}$ = 13.23, p = 0.0003). Table 3 provides a summary of the Chi square statistics.

Table 3. Results from the data analyses investigating if the presence or absence of a given technical feature on a robotic lawn mower significantly influenced the safety index. Damage category A was omitted from the analyses.

Features	Damage Categories Included	Type	No Damage	Damage	Safety Index (No Damage/Damage)	Safety Index	Chi Square with Yates Correction	*p*-Value
Fixed or pivoting blades	4	Fixed	23	27	23/27	0.85	28.95	<0.0001 ***
		Pivoting	112	18	112/18	6.22		
	3 + 4	Fixed	14	36	14/36	0.39	26.62	<0.0001 ***
		Pivoting	93	37	93/37	2.51		
Front or rear wheel drive	4	Front	28	1	28/1	28.00	7.25	0.007 **
		Rear	107	44	107/44	2.43		
	3 + 4	Front	25	4	25/4	6.25	8.99	0.003 **
		Rear	82	69	82/69	1.19		
3 or 4 wheels	4	3 wheels	35	5	35/5	7.00	3.47	0.062
		4 wheels	100	40	100/40	2.50		
	3 + 4	3 wheels	30	10	30/10	3.00	4.37	0.037 *
		4 wheels	77	63	72/63	1.22		
Skid plate	4	Yes	37	1	37/1	37.00	11.39	0.001 ***
		No	98	44	98/44	2.23		
	3 + 4	Yes	32	6	32/6	5.33	10.99	0.001 ***
		No	75	67	75/67	1.12		
Ultrasonic sensors	4	Yes	34	5	34/5	6.80	3.15	0.076
		No	101	40	101/40	2.53		
	3 + 4	Yes	29	10	29/10	2.90	3.84	0.050 *
		No	78	63	78/63	1.24		
Camera vision	4	Yes	8	3	8/3	2.67	0.03	0.857
		No	127	42	127/47	3.02		
	3 + 4	Yes	7	4	7/4	1.75	0.01	0.980
		No	100	69	100/69	1.45		
Collision sensors	4	Yes	49	31	49/31	1.58	13.23	0.0003 ***
		No	86	14	86/14	6.14		
	3 + 4	Yes	45	35	45/35	1.29	0.39	0.53
		No	62	38	62/38	1.63		
Wheel motor current collision detection	4	Yes	86	14	86/14	6.14	13.23	0.0003 ***
		No	49	31	49/31	1.58		
	3 + 4	Yes	62	38	62/38	1.63	0.39	0.53
		No	45	35	45/35	1.29		

* *p*-value ≤ 0.05, ** *p*-value ≤ 0.01, *** *p*-value ≤ 0.001.

The percentage distribution of damages to the hedgehogs during the tests of each robotic lawn mower (Table 4) provides an overview of the performance of each machine. The lower percentage of damages during the tests, the safer the mower is for hedgehogs, insofar as the injuries are a good approximation to what could be sustained on live hedgehogs. The percentage distribution of damages varied accordingly, as some models may have caused no or few category 4 damages but had a higher occurrence of damage category 3 during the tests.

Table 4. The percentage distribution of tests resulting in damage to the hedgehogs, defined either as damage category 4 or damage category 3 + 4. Damage category A was omitted from the analyses, leaving the total number of tests between 9 and 12 depending on the amount of damage category A results recorded per robotic lawn mower. The lower percentage of cases of damage during the tests, the safer the robotic lawn mower. The robotic lawn mower models have been listed in accordance with the percentage distribution of damage defined as damage category 3 + 4. Models showing the lowest damage percentage are listed first.

Robotic Lawn Mowers		Tests with Damage Category 4			Tests with Damage Category 3 + 4		
Brand	**Model**	**No Damage (0–3)**	**Damage (4)**	**Cases of Damage in Tests (%)**	**No Damage (0–2)**	**Damage (3–4)**	**Cases of Damage in Tests (%)**
Gardena	Sileno Life	10	0	**0**	10	0	**0**
Husqvarna	Automower® 105	9	1	**10**	9	1	**10**
Husqvarna	Automower® 315X	9	0	**0**	8	1	**11**
Honda	Miimo HRM 40 Live	12	0	**0**	10	2	**17**
Husqvarna	Automower® 450X	10	0	**0**	8	2	**20**
Worx	Landroid M (WR143E)	10	0	**0**	8	2	**20**
LandXcape	LX8212i	9	1	**10**	8	2	**20**
Husqvarna	Automower® 305	9	0	**0**	7	2	**22**
DAYE	Grouw M900	9	3	**25**	9	3	**25**
Gardena	Sileno City	9	0	**0**	6	3	**33**
Robomow	RS635 PRO	8	3	**27**	7	4	**36**
Kress	Mission KR111	5	4	**44**	5	4	**44**
Worx	Landroid L (WR153E)	10	1	**9**	5	6	**55**
Ambrogio Robot	4.0 Elite	4	7	**64**	4	7	**64**
Stihl	iMow® 422PC	2	8	**80**	2	8	**80**
AL-KO	Robolinho® 1150	2	7	**78**	1	8	**89**
Honda	Miimo HRM 3000	1	8	**89**	0	9	**100**
Stiga	Autoclip 530 SG	7	2	**22**	0	9	**100**

4. Discussion

As the results showed that none of the robotic lawn mowers tested was able to detect the hedgehogs without physical interaction and none detected the dependent juveniles, we cannot be confident that any of the robotic lawn mowers tested were entirely safe to hedgehogs. Preferably, the machines should not interact physically at all with the hedgehogs. However, the damages categorised as 1–2 do not appear to harm the hedgehogs and perhaps the hedgehogs may even learn to avoid robotic lawn mowers after such an encounter. Furthermore, there were obvious differences in the outcome on the hedgehogs depending on the machines tested, with some models consistently causing no damage on collision (Figure 3 and Table 4).

Some of the injuries recorded would have been immediately lethal, and all of the damages in category 4 would have had the potential to become lethal if left untreated. A small puncture wound, if untreated, might get infected and progress to balloon syndrome, a potentially lethal condition caused by subcutaneous emphysema, which makes the skin of the hedgehog blow up like a balloon [25], or a general systemic infection. As hedgehogs are considered quite elusive even when damaged and in pain, it must be assumed that a proportion of hedgehogs injured by robotic lawn mowers will not be found and helped in time and will likely die from their injuries in the wild.

In some cases the robotic lawn mowers failed to detect the hedgehog but met the safety regulations insofar as the blades stopped when the surface of the machine was lifted (activating a tilt-, lift- or obstruction sensor) leaving the skin of the hedgehog unbroken (damage category 3) [24]. However, there were situations where the mower continued to run over and hence injure the hedgehog cadaver. We reduced our recording of injuries to one category (damage category 4), as the outcome may be influenced by a range of different factors, such as the soil softness and type, height of the grass, position of the hedgehog as the robotic lawn mower runs over the individual, and how the collision with the hedgehog positions the individual underneath the blades. As these different factors may have influenced the results of the tests, causing uncertainty of the potential outcome in all scenarios where the robotic lawn mowers failed to detect the hedgehogs and continued to run over the individuals (damage categories 3 and 4), we decided to represent both types of damage categories in our analyses of the results (damage category 4 and damage category 3 + 4) as a precautionary measure.

During our experiments, there was a greater likelihood that robotic lawn mowers with fixed blades would fail to detect the dead hedgehogs, causing more extensive damage to them. These results may be explained by various factors. In contrast to fixed blades, which are constantly exposed, pivoting blades fold into a protective frame when they hit something harder than grass. Furthermore, robotic lawn mowers with fixed blades require more heavy-duty machine power to run the blades, and this greater power appeared to render the machines less controllable and less sensitive in their detection technology. The engineering of front- compared to rear-wheel drive, as well as the use of three compared to four wheels, influenced the safety index positively. This may also be because models with front-wheel drive and three wheels all had pivoting blades. The same explanation may apply to the significantly lower incidence of damage for tests on robotic lawn mowers with ultrasonic sensors, all of which were fitted with pivoting blades. Lastly, the presence of skid plates significantly reduced the number of tests causing damage to the hedgehogs. The skid plate is designed to protect the pivoting blades from hard objects and thereby also protects foreign objects, such as a hedgehog, from the blades. Only one of the models tested contained a combination of these beneficial features (except ultrasonic sensors). These should be the focus of future designs of robotic lawn mowers with hedgehog safety in mind.

We cannot rule out the possibility that the results were also influenced by the lift detection sensitivity of the robotic mowers. We could not test this, but presume that if lift detection sensitivity was sufficiently high, the machines would detect the hedgehogs and change direction or stop the blades rotating as soon as the surface of the machine was lifted, reducing the risk of injuries.

4.1. Using Dead Hedgehogs as Test Subjects

Working with dead hedgehogs as test subjects may not perfectly mimic the outcomes of natural collisions. Firstly, live hedgehogs might detect and evade the robotic lawn mower. Secondly, they might curl up, and their tightened muscles and raised spines could provide protection. We sought to mimic these behaviours in the positions we chose for the cadavers, but of course their muscle tone and reactions were different. Alternative insights would come from simulations using live hedgehogs with safely modified mowers.

4.2. Failed Detection of Dependent Juveniles and the Consequences

None of the tested robotic lawn mowers was able to detect the dependent juvenile hedgehogs (<200 g, weight class 1). In most cases, these small individuals passed beneath the rotating blades. We do not know how mother hedgehogs accompanied by their litters would react to an active robotic lawn mower, but reports from the public indicate that they generally tend to stay in the nests during ordinary human garden activity, although this would have to be investigated further in future work. An orphaned juvenile hedgehog is more likely to be exposed to running robotic lawn mowers. However, such an individual is already very vulnerable with a low chance of survival, regardless of the presence of a robotic lawn mower, unless found in good time and taken into care at a wildlife rehabilitation centre.

4.3. Results in Relation to Discussions with the Public

The public discourse has raised questions of whether hedgehogs can outrun robot lawn mowers and whether hedgehogs are able to detect them properly. As we use cadavers, we were not able to test this, but we do know that hedgehogs can run at up to 50 m per minute [26], whereas the maximum speed of Husvarna's robotic lawn mower ranges between 21 m per minute and 39 m per minute (pers. comm. Husqvarna). In terms of cues likely to alert the hedgehogs, these machines make characteristic sounds and smells detectable by human senses. We made no observations of the behavioural responses of live hedgehogs to the mowers, although this could be done at no risk to the hedgehogs using disarmed machines.

As hedgehogs are nocturnal, it has been widely recommended that any problem would be circumvented by running robotic mowers only by day. This might indeed largely obviate the problem, nonetheless being mindful that hedgehogs may be active during the daytime for several different reasons [19,23].

In the light of the results from the present study, we encourage manufacturers, distributors, and sellers of robotic lawn mowers to educate customers on the importance of refraining from using robotic lawn mowers at night time and to check the lawn for wildlife species that are potentially vulnerable to the machines, such as hedgehogs, leverets, fledglings, and amphibians, before mowing.

5. Conclusions

As hedgehogs are increasingly associated with human habitation, they are likely to encounter robotic lawn mowers, and our results show the encounters, depending on the model, could be injurious and even fatal. That said, while our study answers critical questions regarding the likely nature and extent of injuries, we cannot comment on the likelihood of these encounters or the hedgehogs' responses to them. However, a major step towards resolving the risk of robotic lawn mowers on hedgehog survival involves the design and purchase of hedgehog-friendly mowers, a topic of potentially fruitful collaboration between hedgehog conservationists, behavioural ecologists, and mower manufacturers.

Author Contributions: Conceptualization, S.L.R., A.E.S., R.M., J.L.N., C.P., and D.W.M.; Data curation, S.L.R., A.E.S., and R.M.; Formal analysis, S.L.R. and C.P.; Funding acquisition, S.L.R.; Investigation, S.L.R., A.E.S., and R.M.; Methodology, S.L.R., A.E.S., and R.M.; Project administration, S.L.R.; Resources, S.L.R.; Supervision, J.L.N., C.P., and D.W.M.; Visualization, S.L.R., A.E.S., and R.M.; Writing—original draft, S.L.R.; Writing—review and editing, S.L.R., A.E.S., R.M., J.L.N., C.P., and D.W.M. All authors have read and agreed to the published version of the manuscript.

Funding: This research was funded by the British Hedgehog Preservation Society. The APC was funded by the QATO Foundation and Dyrenes Beskyttelse (Animal Protection Denmark).

Institutional Review Board Statement: Ethical review and approval were waived for this study due to the fact that, regardless of the status of European hedgehogs as protected animals in the Danish legislation, the hedgehogs used in the study had already died in care at Danish hedgehog rehabilitation centres authorised and monitored by the Danish Nature Agency.

Data Availability Statement: Further data from the research is available from Zenodo (DOI: 10.5281/zenodo.4707658, https://doi.org/10.5281/zenodo.4707658).

Acknowledgments: The authors thank robotic lawn mower product specialist Petrus Ekbladh for advice and technical assistance. We thank the Danish hedgehog carers from Pindsvine Plejerne and Dyrenes Beskyttelse (Inge Tornkjær, Dorte Sigga Lund, Janne Montell, Tine Deth Christiansen, Leo and Maria Holm Kragh, and Dorthe Madsen) for collecting hedgehogs dying in care for the tests. We also thank Michela Dugar from Progetto Riccio Europeo—The Wild Hedgehog Project for providing pictures of hedgehogs for Table 2.

Conflicts of Interest: The authors declare no conflict of interest. The funders had no role in the design of the study; in the collection, analyses, or interpretation of data; in the writing of the manuscript; or in the decision to publish the results.

References

1. Hof, A.R.; Bright, P.W. Quantifying the long-term decline of the West European hedgehog in England by subsampling citizen-science datasets. *Eur. J. Wildl. Res.* **2016**, *62*, 407–413. [CrossRef]
2. Krange, M. Change in the Occurrence of the West European Hedgehog (*Erinaceus europaeus*) in Western Sweden during 1950–2010. Master's Thesis, Karlstad University, Karlstad, Sweden, 2015.
3. Müller, F. Langzeit-Monitoring der Strassenverkehrsopfer beim Igel (*Erinaceus europaeus L.*) zur Indikation von Populations-dichteveränderungen entlang zweier Teststrecken im Landkreis Fulda. *Beiträge zur Nat. Osthess.* **2018**, *54*, 21–26.
4. SoBH. *The State of Britain's Hedgehogs 2011.* British Trust for Ornithology (BTO) Commissioned by People's Trust for Endangered Species (PTES) and the British Hedgehog Preservation Society (BHPS). 2011. Available online: https://www.britishhedgehogs.org.uk/leaflets/sobh.pdf (accessed on 1 March 2021).
5. SoBH. *The State of Britain's Hedgehogs 2015.* British Hedgehog Preservation Society and People's Trust for Endangered Species. 2015. Available online: http://ptes.org/wp-content/uploads/2015/11/SoBH-2015.pdf (accessed on 1 March 2021).
6. SoBH. *The State of Britain's Hedgehogs 2018.* British Hedgehog Preservation Society and People's Trust for Endangered Species. 2018. Available online: https://www.hedgehogstreet.org/wp-content/uploads/2018/02/SoBH-2018_final.pdf (accessed on 1 March 2021).
7. van de Poel, J.L.; Dekker, J.; van Langevelde, F. Dutch hedgehogs *Erinaceus europaeus* are nowadays mainly found in urban areas, possibly due to the negative effects of badgers *Meles meles*. *Wildl. Biol.* **2015**, *21*, 51–55. [CrossRef]
8. Williams, B.M.; Baker, P.J.; Thomas, E.; Wilson, G.; Judge, J.; Yarnell, R.W. Reduced occupancy of hedgehogs (*Erinaceus europaeus*) in rural England and Wales: The influence of habitat and an asymmetric intra-guild predator. *Sci. Rep.* **2018**, *8*, 12156. [CrossRef] [PubMed]
9. Taucher, A.L.; Gloor, S.; Dietrich, A.; Geiger, M.; Hegglin, D.; Bontadina, F. Decline in Distribution and Abundance: Urban Hedgehogs under Pressure. *Animals* **2020**, *10*, 1606. [CrossRef] [PubMed]
10. Mathews, F.; Harrower, C. *IUCN–Compliant Red List for Britain's Terrestrial Mammals. Assessment by the Mammal Society under Contract to Natural England, Natural Resources Wales and Scottish Natural Heritage*; Natural England: Peterborough, UK, 2020.
11. Garcês, A.; Soeiro, V.; Lóio, S.; Sargo, R.; Sousa, L.; Silva, F.; Pires, I. Outcomes, Mortality Causes, and Pathological Findings in European Hedgehogs (*Erinaceus europeus*, Linnaeus 1758): A Seventeen Year Retrospective Analysis in the North of Portugal. *Animals* **2020**, *10*, 1305. [CrossRef] [PubMed]
12. Lukešová, G.; Voslarova, E.; Vecerek, V.; Vucinic, M. Trends in intake and outcomes for European hedgehog (*Erinaceus europaeus*) in the Czech rescue centers. *PLoS ONE* **2021**, *16*, e0248422. [CrossRef] [PubMed]
13. Martínez, J.C.; Rosique, A.I.; Royo, M.S. Causes of admission and final dispositions of hedgehogs admitted to three Wildlife Rehabilitation Centers in eastern Spain. *Hystrix Ital. J. Mammal.* **2014**, *25*, 107–110.
14. Research and Markets. *Robotic Lawn Mowers Market-Global Outlook and Forecast. 2020–2025*; Arizton Global: Chicago, IL, USA, 2020.
15. Pettett, C.E.; Moorhouse, T.P.; Johnson, P.J.; Macdonald, D.W. Factors affecting hedgehog (*Erinaceus europaeus*) attraction to rural villages in arable landscapes. *Eur. J. Wildl. Res.* **2017**, *63*, 12. [CrossRef]
16. Hubert, P.; Julliard, R.; Biagianti, S.; Poulle, M.L. Ecological factors driving the higher hedgehog (*Erinaceus europeaus*) density in an urban area compared to the adjacent rural area. *Landsc. Urban Plan.* **2011**, *103*, 34–43. [CrossRef]
17. Doncaster, C.P.; Rondinini, C.; Johnson, P.C.D. Field test for environmental correlates of dispersal in hedgehogs *Erinaceus europaeus*. *J. Anim. Ecol.* **2001**, *70*, 33–46. [CrossRef]
18. Rasmussen, S.L.; Berg, T.B.; Dabelsteen, T.; Jones, O.R. The ecology of suburban juvenile European hedgehogs (*Erinaceus europaeus*) in Denmark. *Ecol. Evol.* **2019**, *9*, 13174–13187. [CrossRef] [PubMed]
19. Morris, P. *Hedgehog*; William Collins: Glasgow, UK, 2018.
20. Berger, A.; Lozano, B.; Barthel, L.M.; Schubert, N. Moving in the Dark—Evidence for an Influence of Artificial Light at Night on the Movement Behaviour of European Hedgehogs (*Erinaceus europaeus*). *Animals* **2020**, *10*, 1306. [CrossRef] [PubMed]
21. Dowding, C.V.; Harris, S.; Poulton, S.; Baker, P.J. Nocturnal ranging behaviour of urban hedgehogs, *Erinaceus europaeus*, in relation to risk and reward. *Anim. Behav.* **2010**, *80*, 13–21. [CrossRef]
22. Gazzard, A.; Baker, P.J. Patterns of feeding by householders affect activity of hedgehogs (*Erinaceus europaeus*) during the hibernation period. *Animals* **2020**, *10*, 1344. [CrossRef] [PubMed]
23. Reeve, N. *Hedgehogs*; Poyser: London, UK, 1994.
24. CENELEC European Committee for Electrotechnical Standardization. *Safety of Household and Similar Appliances-Part 2-107: Particular Requirements for Robotic Battery Powered Electrical Lawnmowers*; EN 50636-2-107; European Committee for Electrotechnical Standardization, CENELEC Management Centre: Brussels, Belgium, 2015.
25. Bullen, K. *Hedgehog Rehabilitation*; British Hedgehog Preservation Society: Ludlow, UK, 2010.
26. Wroot, A.J. Feeding ecology of the European hedgehog *Erinaceus europaeus*. Ph.D. Thesis, Royal Holloway, University of London, Egham, UK, 1984.

Article

Testing the Impact of Robotic Lawn Mowers on European Hedgehogs (*Erinaceus europaeus*) and Designing a Safety Test

Sophie Lund Rasmussen [1,2,3,*], Bettina Thuland Schrøder [4], Anne Berger [5], Rahel Sollmann [6], David W. Macdonald [1], Cino Pertoldi [2,7] and Aage Kristian Olsen Alstrup [8,9]

1 Wildlife Conservation Research Unit, The Recanati-Kaplan Centre, Department of Biology, University of Oxford, Tubney House, Abingdon Road, Tubney, Abingdon OX13 5QL, UK; david.macdonald@biology.ox.ac.uk

2 Department of Chemistry and Bioscience, Aalborg University, Fredrik Bajers Vej 7H, 9220 Aalborg, Denmark; cp@bio.aau.dk

3 Linacre College, University of Oxford, St. Cross Road, Oxford OX1 3JA, UK

4 Behavioral Ecology Group, Section for Ecology and Evolution, Department of Biology, University of Copenhagen, Universitetsparken 15, 2100 Copenhagen Ø, Denmark; catsandconservation@gmail.com

5 Department of Evolutionary Ecology, Leibniz Institute for Zoo and Wildlife Research, Alfred-Kowalke-Straße 17, 10315 Berlin, Germany; berger@izw-berlin.de

6 Department of Ecological Dynamics, Leibniz Institute for Zoo and Wildlife Research, Alfred-Kowalke-Straße 17, 10315 Berlin, Germany; sollmann@izw-berlin.de

7 Aalborg Zoo, Mølleparkvej 63, 9000 Aalborg, Denmark

8 Department of Nuclear Medicine and PET, Aarhus University Hospital, Palle Juul-Jensens Boulevard 165, 8200 Aarhus, Denmark; aagealst@rm.dk

9 Department of Clinical Medicine, Aarhus University, Palle Juul-Jensens Boulevard 165, 8200 Aarhus, Denmark

* Correspondence: sophie.rasmussen@biology.ox.ac.uk

Simple Summary: The declining populations of European hedgehogs (*Erinaceus europaeus*) are increasingly inhabiting areas with human occupation. However, sharing habitats with humans comes at a cost: a residential garden holds many potential dangers for hedgehogs. Previous research has shown that certain models of robotic lawn mowers may harm hedgehogs. This study investigated the effects of 19 models of robotic lawn mowers on hedgehog cadavers. The insights gained from the current and previous research led to the design of a protocol for testing the safety of robotic lawn mowers on hedgehogs. The proposed standardised safety test will hopefully be implemented in the European Committee for Electrotechnical Standardization (CENELEC) protocol, potentially allowing for a labelling system indicating whether a robotic lawn mower is safe for hedgehogs, guiding the consumers to purchase hedgehog-friendly robotic lawn mowers in the future, thus reducing the negative impact some models of robotic lawn mowers may have on hedgehog conservation.

Citation: Rasmussen, S.L.; Schrøder, B.T.; Berger, A.; Sollmann, R.; Macdonald, D.W.; Pertoldi, C.; Alstrup, A.K.O. Testing the Impact of Robotic Lawn Mowers on European Hedgehogs (*Erinaceus europaeus*) and Designing a Safety Test. *Animals* **2024**, *14*, 122. https://doi.org/10.3390/ani14010122

Academic Editor: Brian L. Cypher

Received: 30 November 2023
Revised: 22 December 2023
Accepted: 27 December 2023
Published: 29 December 2023

Abstract: Previous research has established that some models of robotic lawn mowers are potentially harmful to hedgehogs. As the market for robotic lawn mowers is expanding rapidly and the populations of European hedgehogs (*Erinaceus europaeus*) are in decline, it is important to investigate this risk further to understand the potential threat which some robotic lawn mowers may pose to hedgehogs. We tested 19 models of robotic lawn mowers in collision with hedgehog cadavers to measure their effect on hedgehogs. Our results showed that some models of robotic lawn mowers may injure hedgehogs, whereas others are not harmful to them. Apart from one single incidence, all robotic lawn mowers had to physically touch the hedgehog carcasses to detect them. Large hedgehog cadavers were less likely to be "injured", with height being the most influential measure of size. The firmness of the tested hedgehog cadavers (frozen or thawed) did not influence the outcome of the collision tests. Neither the position of the hedgehog cadavers nor the selected technical features of the lawn mowers affected the probability of injury. Based on the results, we designed a standardised safety test to measure the effect of a specific model of robotic lawn mower on hedgehogs.

Keywords: European hedgehog; *Erinaceus europaeus*; robotic lawn mowers; wildlife conservation; safety tests; garden technology; anthropogenic disturbance; lawn care

1. Introduction

The European hedgehog (*Erinaceus europaeus*) is in documented decline in several western European countries [1–11]. Previous research has unravelled a variety of suspected causes for the decline, such as road traffic accidents, habitat loss, habitat fragmentation, inbreeding, intensified agricultural practices, a reduction in biodiversity (and thereby natural food items), lack of suitable nest sites in residential gardens, accidents caused by garden tools, netting and other anthropogenic sources in residential gardens, infections with pathogens and endoparasites, badger predation, and finally, molluscicide and rodenticide poisoning [2,12–32]. These factors combined reduce the mean age of hedgehogs to two years (see Rasmussen et al. [33] Table 1 for an overview), despite their potential to reach up to 16 years of age [33]. To optimise the conservation initiatives directed at this species, there is a need for further investigation of the drivers behind this worrying decline in hedgehog populations.

1.1. Hedgehogs and Robotic Lawn Mowers

Research indicates that hedgehogs are nowadays increasingly associated with human occupation [7,17,18,34]. Unfortunately, sharing habitats with humans comes at a cost, as residential gardens provide many anthropogenic sources of danger to hedgehog survival. One of these potentially harmful features is certain models of robotic lawn mowers [26,35,36]. With robotic lawn mowers becoming increasingly popular throughout the distribution range of hedgehogs in Europe, there is a high likelihood for a hedgehog to encounter numerous robotic lawn mowers throughout its lifespan. The risk is heightened because some garden owners let the machines run after sunset, which is convenient for the human residents but coincides with the activity period of the nocturnal hedgehogs. Market insight reports predict that the global robotic lawn mower market will expand from USD 0.8–1.5 billion in 2020–2022 to USD 2.7–4 billion in 2032 with an anticipated compound annual growth rate (CAGR) of 11.5–15.5% during the forecast period [37,38]. This calls for an effort to eliminate any models of robotic lawn mowers which can potentially harm hedgehogs, to mitigate the negative effect these products may pose on hedgehog conservation. However, this endeavour requires research to inform the manufacturers in their development of more hedgehog-friendly robotic lawn mowers, alongside the design of a standardised safety test to evaluate and approve new models of robotic lawn mowers for the market, in terms of hedgehog safety, as an addition to the current mandatory general safety guidelines provided in the European Committee for Electrotechnical Standardization (CENELEC) protocol [39].

1.2. Study Aims

In response to the background information introduced in the previous sections, the aims of this study were as follows:

- Gain further insight on the effects on hedgehogs through collision tests of a selection of robotic lawn mowers available for purchase on the European market, representing different technical specifications, brands, and price ranges.
- Define any technical features in the robotic lawn mowers which may increase the safety for hedgehogs to guide the manufacturers in the design of more hedgehog-friendly machines.
- Obtain the necessary knowledge through the tests to design the optimal standardised safety test, such as the following:
  - o   The number of test replications needed to provide reliable results;
  - o   The ideal size and composition of a future hedgehog crash test dummy;

o   The optimal combination of test positions to represent the most realistic scenarios of encounters between hedgehogs and robotic lawn mowers.

-    To propose a protocol for a standardised safety test to measure the effect of a specific model of robotic lawn mower on hedgehogs.

2. Materials and Methods

Prior to the tests, we contacted the different manufacturers of robotic lawn mowers, offering to include their products in our experiments, but only two (STIHL and Husqvarna) provided robotic lawn mowers for the research. We tested a total of 19 models of robotic lawn mowers in this study. The selection was influenced by the availability of the products at the test facilities and was furthermore based on the advice of a product specialist in robotic lawn mowers. The mowers chosen for this study are considered to represent a broad spectrum of brands, models, and specifications of the products available on the European market (Table 1). We also prioritised the inclusion of as many as possible of the models tested by Rasmussen et al. (2021) [26] to facilitate comparisons between the different tests. The cutting heights of the machines were set to represent the standard settings recommended for each product and are described in Table 1.

Table 1. Overview of features of the models of robotic lawn mowers tested and their cutting height settings. In the column "Blades", P indicates "low energy pivoting blades" and F indicates "heavy duty fixed blades".

Brand	Model	Blades	Collision Sensor	Wheel Motor Current Collision Detection	Wheels	Front (F)-/Rear (R)-Wheel Drive	Skid Plate	Headlights	Ultrasonic Sensors	Camera Vision	Cutting Height (cm)
AL-KO	1150	F	No	Yes	4	R	No	No	No	No	50
Gardena	Sileno City	P	No	Yes	3	F	No	No	No	No	58
Gardena	Sileno Life	P	No	Yes	4	F	No	No	No	No	35
Honda	HRM 40 Live	P	No	Yes	4	R	No	No	No	No	47
Husqvarna Automower®	105	P	Yes	No	3	F	Yes	No	No	No	45
Husqvarna Automower®	305 (310)	P	No	Yes	4	R	Yes	No	No	No	52
Husqvarna Automower®	450X	P	Yes	No	4	R	Yes	Yes	Yes	No	60
Husqvarna Automower®	310	P	Yes	No	4	R	Yes	No	No	No	65
Husqvarna Automower®	Nera	P	Yes	No	4	R	Yes	Yes	Yes (Radar)	No	43
Husqvarna Automower®	Aspire R4	P	No	Yes	3	F	No	No	No	No	50
Kress	KR111	P	Yes	No	4	R	No	No	Yes	No	45
LandXcape	LX812i	P	No	Yes	3	R	No	No	Yes	No	40
Segway NaviMow	H3000E	P	Yes	No	4	R	No	No	No	No	67
Stiga Stig-A	1500	P	Yes	No	4	R	No	No	No	No	35
Worx	Landroid L (WR153E)	P	No	Yes	4	R	No	No	No	No	60
Worx	Landroid M (WR143E)	P	No	Yes	4	R	No	No	Yes	No	60
STIHL	iMOW 422P	F	Yes	No	4	R	No	No	No	No	43
STIHL	iMOW 5	P	Yes	No	4	R	No	Yes	Yes	No	40
STIHL	iMOW 7	P	Yes	No	4	R	No	Yes	Yes	No	40

Of the 19 robotic lawn mowers tested, 2 had fixed blades and 17 had pivoting blades (please see Figure 1A,B in Rasmussen et al. (2021) [26] for pictures of the types of blades).

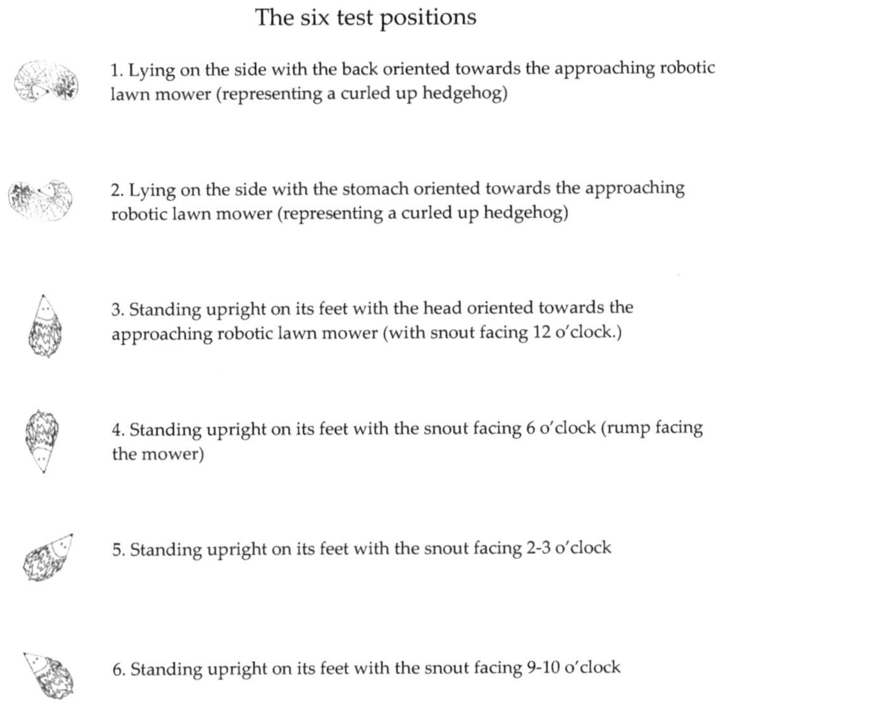

The six test positions

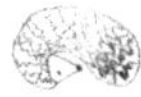

1. Lying on the side with the back oriented towards the approaching robotic lawn mower (representing a curled up hedgehog)

2. Lying on the side with the stomach oriented towards the approaching robotic lawn mower (representing a curled up hedgehog)

3. Standing upright on its feet with the head oriented towards the approaching robotic lawn mower (with snout facing 12 o'clock.)

4. Standing upright on its feet with the snout facing 6 o'clock (rump facing the mower)

5. Standing upright on its feet with the snout facing 2-3 o'clock

6. Standing upright on its feet with the snout facing 9-10 o'clock

Figure 1. An overview of the six different test positions used during the tests. Only hedgehog cadavers were used in these tests.

The robotic lawn mower tests were performed on dead hedgehogs, henceforth referred to as "hedgehogs". These animals had died in care, primarily due to infections, at hedgehog rehabilitation centres in Denmark from May to November 2022. All hedgehogs chosen for this study were intact with no visible injuries prior to the tests. The hedgehogs were stored in freezers at -20 °C and were thawed before the regular tests. The hedgehog cadavers all weighed between 250 and 600 g, representing the age group of recently independent juvenile hedgehogs, equivalent to the weight class 2 described by Rasmussen et al. (2021) [26] Table 2. This weight class was chosen as it yielded the most diverse results, with a larger variation between the different positions compared to individuals of other weight classes, in the tests performed by Rasmussen et al. (2021) [26].

Based on the results reported by Rasmussen et al. (2024) [36] testing the behaviour of live hedgehogs facing a disarmed, robotic lawn mower, each individual was tested in six different positions (Figure 1) in an attempt to mimic the behaviour of a live individual. The most commonly recorded position during the tests on live hedgehogs was "upright position with snout pointing inwards" (43%) [36] which could not be properly mimicked with a dead hedgehog as the head would not bend inwards and stay in place, leaving us to combine this with the second most frequently recorded behaviour (20%), test position 3:

The tests were recorded with two Ring Stick Up Cam® (Ring™, Santa Monica, CA, USA) cameras placed on tripods.

Each model of robotic lawn mower was tested on one hedgehog. If an individual was injured by the mower during a test, the injuries were documented with the cameras. In most cases, the individual would thereafter be discarded to avoid the misinterpretation of previously sustained injuries in the subsequent tests.

The tests of 17 out of 19 machines were carried out in a test hall at Husqvarna head quarters in Huskvarna, Jönköping, Sweden, from 23 to 25 March 2023. The remaining two models (STIHL iMOW 7 and STIHL iMOW 5 (STIHL, Stuttgart, Germany)) were tested in a private garden in Lejre, Denmark, on 10 October 2023, as they could not be made available for the tests taking place in March 2023. All tests were performed during daylight hours.

The tests were performed on a firm base of either concrete flooring, garden tiles (STIHL iMOW 7 and STIHL iMOW 5) or asphalt (Segway NaviMow H3000E (Seqway Inc., Beijing, China) and Stiga Stig-A 1500 (Stiga, Castelfranco Veneto, Italy)), on a coconut mat with a rubber-backed base (dimensions 2 m in width and 5 m in length and 20 mm in height [40]). The coconut mat is the recommended base for the robotic lawn mower safety tests described in the European Committee for Electrotechnical Standardization (CENELEC) protocol [39]. The hedgehog was placed on the coconut mat lying 1 m from the edge of the mat and at a 3 m distance from the robotic lawn mower (Figure 2). The cameras were placed next to the hedgehog on the left-hand side at a 1 m distance from the hedgehog and behind it at a distance of 1 m. The mower was then turned on and manually directed to move towards the hedgehog. The distance of 3 m was sufficient to ensure the machine was operating at maximum speed and the blades were in action before reaching the hedgehog. If the machine did not move in a straight line towards the hedgehog, it was then relocated back to the initial position and turned on again. This was conducted to standardise the tests and to ensure that the hedgehog was located at the centre of the front of each approaching machine. In order to test certain models, the distance between the robotic lawn mower and the hedgehog deviated from the standard 3 m as a longer distance was required before the knives started rotating (3.4 m: Husqvarna Automower[®] Nera, Husqvarna Automower[®] Aspire R4 (Husqvarna, Huskvarna, Sweden)) or there was a need for a shorter distance to ensure the mower approached the hedgehog at the right angle (2 m: Husqvarna Automower[®] 105, Husqvarna Automower[®] 305 (Husqvarna, Huskvarna, Sweden), Gardena Sileno Life (Gardena GMBH, Ulm, Germany)).

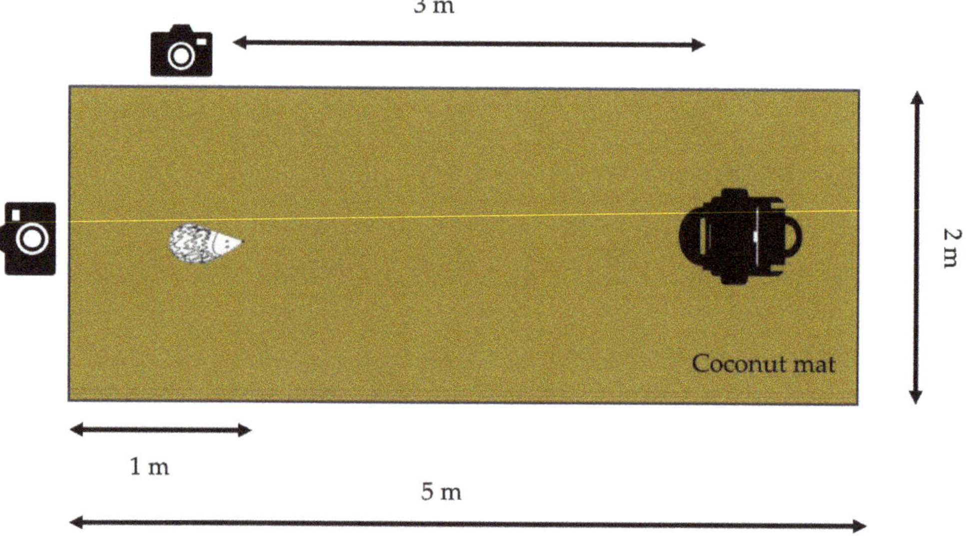

Figure 2. An overview of the setup for the test scenario. Only dead hedgehogs were used in the tests.

In the cases of two models using satellite navigation (tests on Segway NaviMow H3000E and Stiga Stig-A 1500), the tests were performed outdoors with an asphalt concrete base below the coconut mat. In some instances, the machines were switched on at another distance than the standard 3 m from the hedgehog cadaver (Segway NaviMow H3000E all tests at a 5 m distance). This was necessary because this particular model of robotic lawn mower require longer distances to gain momentum for the blades to be rotating at full speed (Segway NaviMow H3000E). For the tests of the Stiga Stig-A 1500 model, the movement algorithm of the machine was unpredictable, forcing the research team to manually place the coconut mat and hedgehog in front of the approaching lawn mower. In all tests, the Stiga Stig-A 1500 was fully up and running when the hedgehog and coconut mat were placed in front of it, at either a 2 m distance (test position 1–5) or a 4 m distance (test position 6).

2.1. Quantifying the Damage

We described the results of the tests, quantifying the severity of damage caused by the robotic lawn mowers, by allocating each outcome to one of five damage categories (Table 2):

Table 2. A description of the five different damage categories used to describe the outcome of the different tests.

Damage Category	Description
0	No physical contact between the machine and the hedgehog. The machine senses the hedgehog from a distance, changes direction, and drives on without touching the hedgehog. No damage is caused to the hedgehog cadaver.
1	The robotic lawn mower approaches the hedgehog, and the front of the machine touches the hedgehog lightly (a "nudge") and thereby detects the corpse. Immediately, the machine changes direction and drives on without touching the hedgehog further. No damage is caused to the hedgehog cadaver.
2	The robotic lawn mower approaches the hedgehog, and the front of the machine touches the hedgehog (a "flip") to detect the hedgehog. The physical interaction causes the hedgehog to be moved into a different body position (flipped from lying on one side of the body to the other side of the body) or be lifted partly from the ground before settling in the same position again. Afterwards, the machine changes direction and drives on without touching the hedgehog further. The damage to the hedgehog is at most minimal and involves no contact with the blades (at worst this might cause a slight bruise).
3	The robotic lawn mower fails to detect the presence of the hedgehog and continues to drive across the hedgehog. The front panel of the machine is lifted as the machine drives over the cadaver, which causes the blades to stop running. In some cases, the machine withdraws and changes direction, so that only part of the dead hedgehog's body was situated underneath the machine. The blades of the robotic lawn mower may have come into contact with the dead hedgehog but have not punctured the skin. The damages observed ranged from undetectable to the cutting of a small number of spines but might have involved minor bruising to a live hedgehog.
4	The robotic lawn mower fails to detect the presence of the hedgehog and continues to drive across it. The blades of the machine have come into contact with the dead hedgehog and have caused injuries to the cadaver. The severity of the injuries ranges from small puncture wounds on the skin (1 cm) to clipping of limbs or complete exposure of the entire abdominal region and decapitation.

2.2. Additional Comparison Tests

It was decided to add two types of comparison tests to the testing procedure. During early tests, it appeared that the size of the hedgehog carcasses used could potentially influence the results, where the smaller ones (<400 g in weight) would more frequently be injured compared to individuals of a larger size (>400 g). Therefore, we decided to perform additional comparison tests on larger hedgehog carcasses for the models of robotic lawn mowers which were previously tested on smaller hedgehogs (<400 g). Due to the limited

number of individuals available for the comparison test, we most often only performed th
test in position 3 to increase the likelihood of having intact carcasses available for tests c
several models of robotic lawn mowers.

To investigate whether the firmness of the carcass would influence the results, we added a test on a frozen hedgehog carcass in test position 3 to the testing procedure for the machines chosen for comparison tests. For the tests performed in Lejre, Denmark, no frozen hedgehogs were available for testing, so this test type was omitted (STIHL iMOW 7 and STIHL iMOW 5).

2.3. Data Analyses

For our analyses, we combined test data from 2020 (published in Rasmussen et al. (2021) [26]) and the results produced in the present experiment in 2023. In contrast to the current experiments, the tests performed in 2020 only used three of the six positions and only tested on thawed hedgehog carcasses. Because of the limited amount of data for each lawn mower model and the categorical nature of the response variable (damage category 0–4), we transformed the response variable y to binary, with y = 1 when a test resulted in damage category 4 (i.e., hedgehog sustained injury in collision with the lawn mower, see definition in Table 2) and y = 0 otherwise (damage category 0–3). We analysed these data with a logistic regression, thus estimating the probability that a test resulted in injury, and the effects of several predictor variables on that probability.

The raw data suggested differences in injury probability among lawn mower models and we used 'lawn mower model' as a random intercept in all models; we refer to the model with the random intercept only as the base model. To evaluate the importance of a predictor for injury probability, we compared models with a predictor to the base model using Akaike's Information Criterion (AIC; [41]). We further gauged the effect strength based on the coefficient estimate and evidence for the effect based on its *p*-value.

All models were fit in R v. 4.2.1 [42] using the lme4 package v. 1.1.30 [43].

2.3.1. Additional Comparison Tests

We compared the outcomes of the tests on frozen hedgehog carcasses to unfrozen hedgehog carcasses. Due to the limited sample size of frozen hedgehogs ($n = 12$), we fit a separate logistic regression comparing injury probability between thawed and frozen hedgehogs, accounting only for lawn mower model, but none of the other predictor variables were found to affect injury probability (see Results). We excluded frozen hedgehogs from our main analysis investigating the effect of predictor variables on injury probability (see Sections 2.3.2–2.3.5). We combined data from the comparison tests using thawed hedgehogs of different sizes (375–419 g) with data from the main tests for our main analysis.

2.3.2. Investigating Potential Differences in Injury Probability Depending on the Position of the Hedgehog

During the tests, the hedgehogs were placed in six positions relative to the direction of approach of the robotic lawn mower (see Figure 1 for a description of the six positions). To test whether there were differences in the probability of injury depending on the position of the hedgehog, we prepared the following model. To reduce the number of levels of this categorical predictor, we grouped positions into 3 categories: 1 + 2 (lying on the side), 3 + 4 (standing, in line with mower), and 5 + 6 (standing, at angle with mower). We included this new position variable as a predictor of injury category.

2.3.3. Comparing the Results of the 2020 and 2023 Tests

Some of the models of robotic lawn mowers were tested in 2020 [26] as well as in the current experiment. The test scenarios did differ slightly between years, as the hedgehog carcasses used in the different tests were not identical, and the tests were performed in different locations with different ground covers (lawns in 2020 and coconut mats in 2023). Therefore, to test whether there were differences in the probability of injury between the 2020 and 2023 tests, we added a categorical year effect to the base model. Because positions 4–6 were not used in tests in 2020 and there was some (albeit weak) evidence that positions

5 + 6 may have a different injury probability (see Results), we also included 'position' in this model. This was to avoid any confounding effect between 'year' and 'position'.

2.3.4. Measuring the Effect of the Size of Hedgehog Carcass Used in the Tests on the Probability of Sustaining Injury

Even though the hedgehog carcasses used for the tests all matched the weight category 250–600 g representing independent juvenile hedgehogs, there was still a large variation in size between them. This led us to test whether the characteristics of the hedgehog carcass affect its probability to sustain injury during the tests. Three measures were collected for the hedgehogs included in the tests: weight (g), height (cm), and circumference (cm). The measurements in cm were recorded as the maximum height and the maximum circumference of the hedgehog. The latter two measures were not collected for any of the hedgehogs tested in 2020. The three measures of size were strongly correlated. To assess via AIC which of the size predictors were the most important, we subset the data to include only those trials where all three measures were taken ($n = 147$ tests). To test whether there were thresholds or optimum relationships between the probability of injury and hedgehog characteristics, we fit models with linear and quadratic effects separately for each predictor (weight, height, and circumference). We compared the six models with linear or quadratic predictors to the base model. We centred and scaled all measures prior to analysis.

2.3.5. Testing Whether the Technical Features of the Robotic Lawn Mowers Affect the Probability of Causing Injury to Hedgehogs

A range of different technical features were registered for the robotic lawn mowers included in the tests (see Table 1). In our analysis to test for effects of these features on the probability of causing injury to hedgehogs, we excluded 'camera vision', as none of the robotic lawn mowers tested had it. We included cutting height as a continuous predictor and all other attributes as categorical (binary) predictors. Because there was very little variation in hedgehog characteristics for each robotic lawn mower (each year, a mower was typically tested only with a single hedgehog), and we wanted to avoid confounding effects of hedgehog characteristics and lawn mower attributes, we included the most important hedgehog characteristic from 2.3.4 (height) in all lawn mower attribute models. To do so, we subset the data to those tests which included records of hedgehog characteristics ($n = 147$); all lawn mower attributes were always recorded for these tests.

2.3.6. Calculating the Optimal Number of Tests to Characterise the Risk of Injury to Hedgehogs Caused by a Specific Robotic Lawn Mower

Measuring categorical data (damage categories), no single damage category alone would characterise a lawn mower model. Rather, each model would have a set of probabilities of how likely each damage category is to occur in a trial. With the current test setup, this would be a set of five probabilities (five damage categories) for each mower, therefore requiring much more data to estimate these probabilities precisely. To reduce this challenge, we again limited the damage categories to injured (category 4) and not injured (category 0–3), focusing on estimating the probability of a lawn mower model causing injury.

One of our goals was to characterise each lawn mower model based on the risk it poses to hedgehogs. Ideally, to do so, we would have included 'lawn mower model' as a fixed effect in the previously described analyses. However, for some models, no tests resulted in injuries (i.e., all $y = 0$), precluding estimating model-specific injury probabilities for these models with fixed effects. More importantly, overall, the data per model were sparse, which leads to uncertain estimates of model-specific injury probabilities. Therefore, for the application in future standardised safety tests, we wanted to apply our test results to determine the optimal number of test repeats (henceforth, sample size) necessary to confidently characterise each robotic lawn mower model's risk of causing injury to hedgehogs.

To define the amount of data (trials per mower) needed to estimate the probability of injury precisely, a simulation-based approach, with the following steps, was used:

(a) We set input injury probabilities for all robotic lawn mower models based on the estimates from a logistic regression with the fixed effect of 'lawn mower model'. We excluded those lawn mower models for which the regression could not estimate a model-specific injury probability.

(b) We created input values for the effects of hedgehog height (the most important characteristic to affect injury probability—see Results) and position on injury probability, using the results from the previously described analyses. Even though the effect of position on injury probability was weak (see Results), we chose to include it in our data simulation to mimic reality, as robotic lawn mowers may encounter hedgehogs in different positions.

(c) We used these input values to simulate new synthetic trial data for different sample sizes per robotic lawn mower model. In the original data, approximately 10 tests were performed per model, depending on whether the model was tested in a comparison test and how many positions were used in that particular test. In the simulations, we explored sample sizes of 25, 50, 75, 100, and 150 per robotic lawn mower model. For each trial, hedgehog height was randomly sampled from all unique heights represented in the dataset from the collision tests (eight different sizes); similarly, position was randomly sampled from the three grouped positions (1 + 2, 3 + 4, and 5 + 6, see Figure 1 for a description of the positions). For each sample size, we created 250 synthetic datasets.

(d) We analysed the synthetic data to estimate the specific injury probability for each model of robotic lawn mower. Specifically, we fit a logistic regression model with a fixed effect of 'lawn mower model', accounting for hedgehog height and height squared. The regression did not account for 'position', as position introduces realistic variability into the synthetic data, and the model estimates the average injury probability across all positions.

(e) We summarised the results across all 250 simulated datasets for each sample size scenario. Specifically, for each dataset, we determined estimated injury probability for each model of robotic lawn mower, which due to scaling of the height variable corresponds to the expected injury probability for an average-sized hedgehog. We calculated 95% confidence intervals (CIs) and the mean CI across all 250 datasets. We plotted average CI width against sample size to visualise how the level of uncertainty declines with increasing sample size.

3. Results

The results from the collision tests between the hedgehog carcasses and the 19 different models of robotic lawn mowers tested can be found in Figure 3. For comparison, the figure includes the results from the tests performed previously on the same weight category size of hedgehog carcasses in 2020 [26]. The results show that some of the robotic lawn mowers did cause injury to the hedgehog carcasses tested (damage category 4), whereas other models of robotic lawn mowers would push the hedgehog prior to detecting it, causing the robotic lawn mower to change direction without harming the hedgehog (damage categories 1–2). There was only one incidence of a damage category 0, where the hedgehog was apparently detected at a distance: the robotic lawn mower changed directions and did not come into contact with the hedgehog. However, the same result could not be replicated; when the test was repeated, it yielded a damage category 3. The full dataset is available in Table S1.

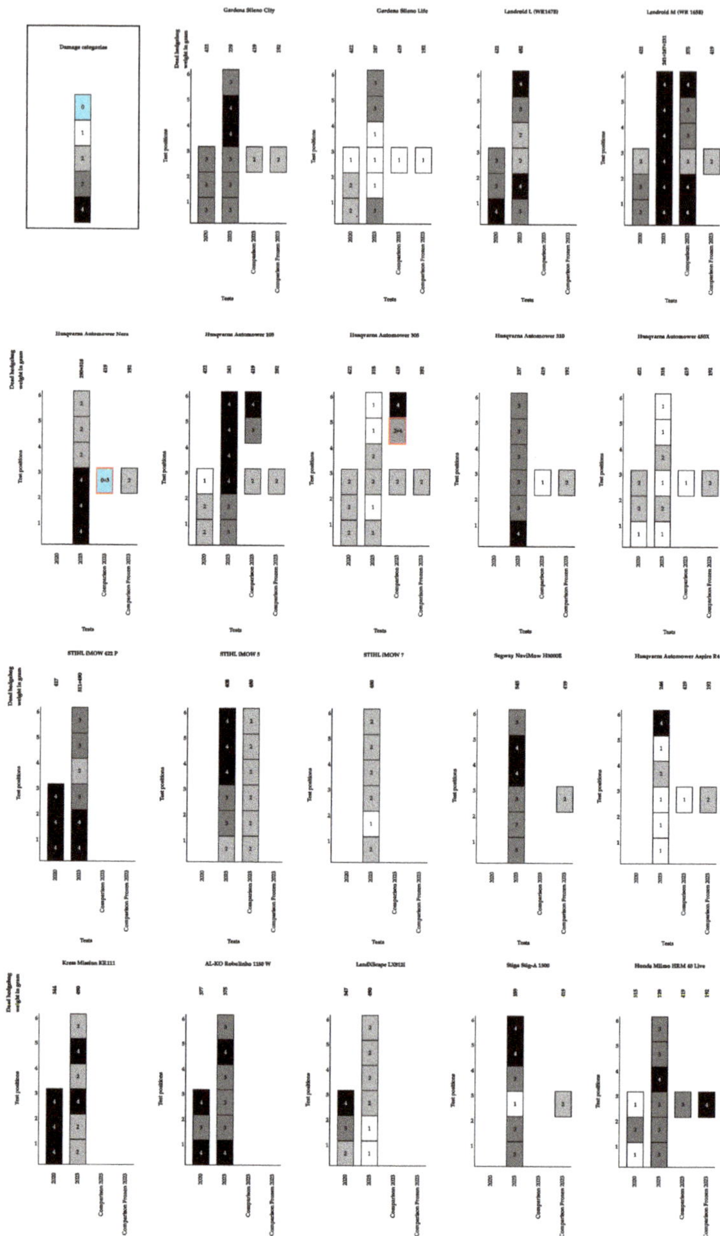

Figure 3. An overview of the test results for each of the 19 tested robotic lawn mowers. The result from the tests performed in 2020 are also visualised for the machines tested by Rasmussen et al. (2021) [26]. The *x*-axis illustrates the test categories (2020, 2023, comparison, and comparison with frozen hedgehog), and the *y*-axis shows the six different test positions. All hedgehog carcasses used in the tests weighed between 250–600 g. The numbers within the fields of the columns denote the damage categories registered for each test position. The numbers above the columns describe the weight in g of the hedgehog carcass used for the specific test. A red highlight marking of the result box (Husqvarna Automower Nera and Husqvarna Automower 305) indicates that this test position and scenario was tested twice and yielded two different results, with colour describing the lowest of the measured damage categories presented in the box.

3.1. Additional Comparison Tests

The model comparing injury probability between frozen and thawed hedgehogs provided little evidence for an effect of the state of the hedgehog. The base model had essentially equal support as the model including the hedgehog state as a predictor (ΔAIC = 0.23); the coefficient estimate had a high uncertainty (1.396, SE = 1.07), and correspondingly, the p-value suggested that evidence in favour of this effect was weak (p = 0.192). We caution, however, that this may be a function of the low sample size of frozen hedgehogs and the resulting inability to account for important sources of variation in injury probability (see following sections) in this comparison.

3.2. Investigating Potential Differences in Injury Probability Depending on the Position of the Hedgehog

Evidence that hedgehog position affected injury probability was weak to nonexistent. Both the 'position' model and the base model had essentially equal support (ΔAIC = 0.5). The effect estimates for positions 3 + 4 had a large standard error, and both coefficients had non-significant p-values (Table 3). However, the effect of positions 5 + 6 appears stronger and more certain than that of positions 3 + 4.

Table 3. Coefficient estimates from the 'position' model.

Position	Coefficient	SE	p
3 + 4	−0.17	0.43	0.69
5 + 6	0.72	0.43	0.1

3.3. Comparing the Results in the 2020 and 2023 Tests

There was no evidence that year affected the injury probability. The model with year had a higher AIC value than the base model (ΔAIC = 1.23), and the coefficient for tests being conducted in 2023 had a large standard error (beta = −0.26, SE = 0.49) and p-value (0.60).

3.4. Measuring the Effect of the Size of Hedgehog Carcass Used in the Tests on the Probability of Sustaining Injury

The AIC values showed that out of the three characteristics, height was the most important predictor of injury probability; the model including a quadratic effect of height was considerably better than the one with the linear effect (Table 4). All other models (of weight and circumference) were similar or worse in AIC than the base model. Coefficient estimates from the quadratic height model showed that injury probability initially increased with height but then declined after about 7 cm of height (Figure 4).

Table 4. AIC-based model selection for testing the effects of size of the hedgehog carcass on the probability of sustaining injury in collision with a robotic lawn mower.

Model	AIC	dAIC
Height sq.	162.9	0
Height	165.26	2.36
Weight	171.23	8.33
Base	171.28	8.38
Circ.	171.42	8.52
Weight sq.	172.47	9.57
Circ. sq.	173.41	10.51

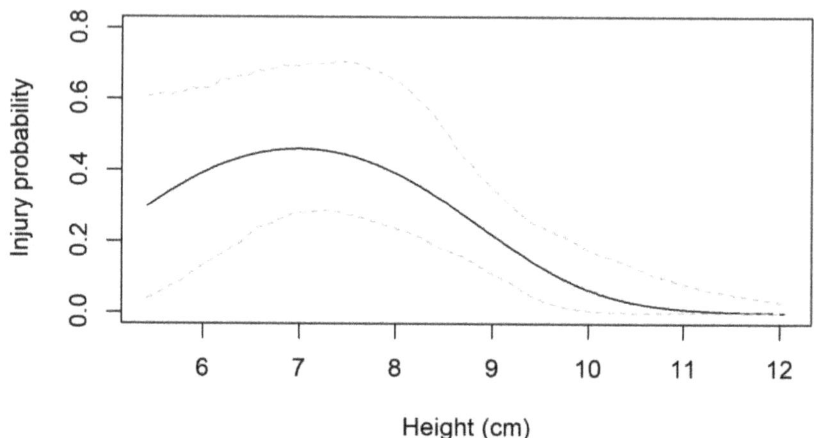

Figure 4. Relationship between hedgehog height and probability of being injured by a lawn mower. Dashed lines show confidence intervals.

3.5. Testing Whether the Technical Features of the Robotic Lawn Mowers Affect the Probability of Causing Injury to Hedgehogs

The only model whose AIC was lower than that of the base model was the one containing an effect of front- or rear-wheel drive, but the ΔAIC was 0.6, thus suggesting that this attribute did not improve the model (Table 5). The model indicated that rear-wheel drive caused higher injury probability, but the standard error was large (1.27, SE = 0.83) and the effect was non-significant (p = 0.13).

Table 5. AIC-based model selection for testing the influence of technical features on the injury probability of the robotic lawn mowers on hedgehogs.

Model	AIC	dAIC
Drive (front vs. rear wheel)	162.28	0
Base	162.9	0.62
# wheels (3 or 4)	163.6	1.32
Wheel motor current collision detection (Y/N)	163.79	1.51
Ultrasonic sensors (Y/N)	164.18	1.9
Cutting height (mm)	164.23	1.95
Collision sensor (Y/N)	164.44	2.16
Skid plate (Y/N)	164.61	2.33
Blades (pivoting vs. fixed)	164.71	2.43
Headlights (Y/N)	164.87	2.59

3.6. Determining the Optimal Number of Tests to Characterise the Risk of Injury to Hedgehogs Caused by a Specific Robotic Lawn Mower

Out of the 19 lawn mower models, it was possible to estimate a mower-specific injury probability (as input value for the simulation) for 15. For the remaining four models (Gardena Sileno Life, Husqvarna Automower 450X, LandXScape LX812i, and STIHL iMOW 7), no (or extremely few) trials resulted in injury, rendering the model unable to estimate injury probabilities for these mowers, causing an exclusion of these models from the simulation. As expected, for the remaining models of robotic lawn mowers, the level of uncertainty around their injury probability declined as the sample size (trials per model) was increased (Figure 5). However, that decline depended on the injury probability: the decline was less pronounced in models with very high or very low injury probability. Generally, the gains in certainty declined considerably after sample sizes of 50, causing us to suggest a test number of 50 to confidently characterise the risk of injury to hedgehogs caused by a specific lawn mower.

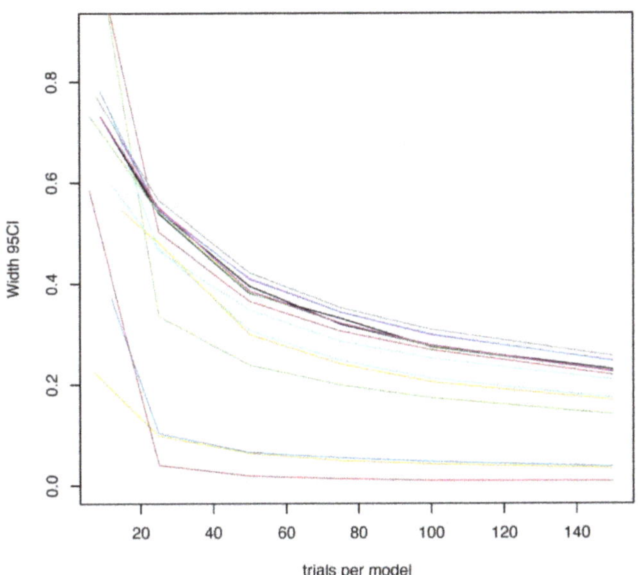

Figure 5. A visualisation of the reduction in confidence interval width (indicating a better representation) as a function of an increased sample size (number of tests) for all 15 robotic lawn mower models included in the analysis. Each mower is represented by a specific colour.

3.7. Using the Results to Design a Standardised Safety Test
3.7.1. Size of the Hedgehog Crash Test Dummies

Based on the test results, indicating that the size of the hedgehog affects the outcome of a collision test, we suggest that a future standardised safety test measuring the effect of a specific model of robotic lawn mower on hedgehogs would include two sizes of hedgehog crash test dummies: one representing an independent juvenile hedgehog <400 g and 7 cm in height and another representing an adult hedgehog >600 g and $\geq$10 cm in height.

3.7.2. Positions Used in the Tests

The findings of Rasmussen et al. (2024) [36] showed that live hedgehogs tend to either run away or position themselves in positions 3, 5, and 6 (Figure 1) when approached by a robotic lawn mower. As our tests showed that the position did not significantly influence the outcome of the collision tests, and as position 1 and 2 (curled up hedgehogs) would be challenging to mimic with a non-flexible hedgehog crash test dummy, we suggest excluding these two positions from the standardised safety test. As the hedgehog crash test dummy currently being prepared is designed without the features of a head, having completely similar front and back design, the position 4 could be considered redundant if the features of the front and back of the model are identical, as position 3 would therefore already have represented that position. As our results indicated a tendency for a higher probability of injury in positions 5 and 6, we recommend including both of these positions in a standardised safety test. Accordingly, we suggest using three positions for future standardised safety tests, namely positions 3, 5, and 6 (described in Figure 1).

3.7.3. The Test Setup

Even though our results showed consistency between the 2020 and 2023 tests, performed on different surfaces (grass compared to a coconut mat placed on top of either solid base of concrete, asphalt, or garden tiles), we recommend the test setup described in Figure 2. This furthermore serves to standardise the design, as lawns may differ in softness and grass height and plant composition. The proposed coconut mat is already recommended standard base for the tests described in the CENELEC protocol [39].

3.7.4. Number of Tests

As described previously, the optimal number of tests to characterise the risk of injury to hedgehogs caused by a specific robotic lawn mower is 50 or above. Therefore, we suggest that the standardised safety test should consist of 60 trials per size hedgehog crash test dummy to accommodate the three test positions chosen, testing each position 20 times.

3.7.5. The Proposed Standardised Safety Test

Our suggestion is that the framework of a standardised safety test to measure the effect of a specific model of robotic lawn mower on hedgehogs is as follows (Figure 6):

Figure 6. The suggested framework for a standardised safety test to measure the effect of a specific model of robotic lawn mower on hedgehogs.

- The tests shall be performed on concrete flooring on a coconut mat with a rubber-backed base (dimensions 2 m in width, 5 m in length, and 20 mm in height).
- The hedgehog crash test dummy shall be placed on the coconut mat lying 1 m from the edge of the mat and at a 3 m distance from the robotic lawn mower.
- Two cameras shall be positioned next to the hedgehog crash test dummy on the left-hand side at a 1 m distance from the dummy and behind the dummy at a distance of 1 m.
- Two sizes of hedgehog crash test dummies shall be used: <400 g and 7 cm in height and >600 g and ≥10 cm in height.
- Each hedgehog crash test dummy shall be tested in 60 trials:
  - 20 trials: Standing upright on its feet with the head oriented towards the approaching robotic lawn mower with the snout facing 12 o'clock;
  - 20 trials: Standing upright on its feet with the snout facing 2–3 o'clock;
  - 20 trials: Standing upright on its feet with the snout facing 9–10 o'clock.
- The interpretation of the results should be conducted as follows:
  - Robotic lawn mowers yielding only damage categories 0–2 in the tests (see Table 2 for a description of damage categories) should be labelled as safe for hedgehogs;
  - Models of robotic lawn mowers showing any results belonging to damage categories 3 and 4 cannot be labelled as safe for hedgehogs;
  - A robotic lawn mower fails the safety test if any of the results are classified as damage category 4.

4. Discussion

Our results showed that some of the robotic lawn mowers tested may injure hedgehogs, whereas others gave no evidence of being harmful to hedgehogs. Apart from one incidence, where we recorded damage category 0, all robotic lawn mowers had to physically interact with the hedgehog carcasses to detect them, whereafter some robotic lawn mowers changed direction and did not cause injury to the hedgehogs. Larger-sized hedgehogs were less likely to be injured, with height being the measure of size most useful in predicting injury. The firmness of the hedgehog cadavers (thawed or frozen) did not affect the outcome of the collision tests. There were no differences in test outcomes between the years 2020 and 2023, showing consistency in the results produced. There was little evidence that hedgehog position influenced injury probability and even less evidence that any of the selected technical features of the lawn mowers tested affected the probability of injury.

4.1. Hedgehog Crash Test Dummies as Alternatives to Hedgehog Carcasses in Future Tests

We found no difference between the outcomes of tests on frozen compared to thawed hedgehog carcasses, although keeping in mind that this comparison was based on a limited sample size. We decided to include the comparison test of frozen carcasses, as we considered a potential bias in the results which could arise if the hedgehog carcasses were softer than live hedgehogs due to the latter curling up and thereby tightening the muscles during a confrontation with a robotic lawn mower.

Work is currently underway to design an optimal hedgehog crash test dummy, mimicking a real, live hedgehog, to be used in the test setup. The dummy will be offered as a standard model for testing the safety of a robotic lawn mower on hedgehogs in the future. The ultimate goal is to provide an open-access recipe allowing relevant stakeholders, such as manufacturers of robotic lawn mowers and test institutes, to 3D print the crash test dummy and use this in the development of hedgehog-friendly robotic lawn mowers and for the standardised safety test we propose.

It is important to ensure that the hedgehog crash test dummy offered is realistic. The results of the present tests should inform the process of designing the dummy. In the worst-case scenario, a faulty dummy could cause misguided adjustments to the robotic lawn mower designs to the detriment of the hedgehogs. The degree of firmness of the hedgehog cadaver did not significantly affect the outcomes of the collision tests (frozen or thawed), encouraging us to conclude that a crash test dummy could be representative even though its composition is not identical to a real hedgehog. However, our results showed that the size of the hedgehog carcass does influence the risk of injury, which leads us to suggest that the hedgehog crash test dummy should be produced in two size categories: (1) <400 g and 7 cm in height (mimicking an independent juvenile hedgehog) and (2) >600 g and ≥10 cm in height (mimicking an adult hedgehog). Further work on the crash test dummy currently being developed is needed to ensure that the injury prediction model from the tests on dead hedgehogs applies directly to the dummy.

4.2. A Standardised Safety Test to Measure the Effect of a Specific Model of Robotic Lawn Mower on Hedgehogs

The goal is to have the standardised hedgehog safety test implemented in the CEN-ELEC protocol [39] as a test offered for all robotic lawn mowers being approved for sale on the European market. The intention is to use this standardised safety test to establish an official labelling system for hedgehog-safe robotic lawn mowers, guiding the consumers to make the hedgehog-friendly choice when purchasing robotic lawn mowers.

One of the purposes of the present study was to compile sufficient data to provide a solid suggestion for a protocol for a standardised safety test to measure the effect of a specific model of robotic lawn mower on hedgehogs. Based on the information gathered in the present study, we have described our suggestion for a standardised safety test. This protocol should now be tested and validated, before being implemented in the CENELEC protocol (International Electrotechnical Commission (IEC), Technical Committee (TC) 116 Working Group (WG) 10, IEC 62841-4-X: Particular requirements for robotic lawnmowers.

4.3. The Safety of Robotic Lawn Mowers for Hedgehogs

Three years after the first tests on the effect of robotic lawn mowers on hedgehogs [26] and an increased focus on improving the safety of the robotic lawn mowers for hedgehogs in new models designed, we were pleased to see how the new models, tested for the first time, in general involved less harmful encounters with the hedgehog carcasses in our experiments. The damage category 0, where the robotic lawn mower detects the hedgehog at a distance and changes direction without coming into physical contact with it, should be regarded as the desired outcome. We did observe the damage category 0 in a single test. However, this result could not be replicated by repeating the test, which causes us to suggest that the robotic lawn mower may have detected (and avoided) a larger obstacle by chance, such as the camera recording the event, in the background. Regardless, the general reduction in harmful outcomes for the hedgehogs in the collision tests with new models and designs of robotic lawn mowers gives cause for optimism about the future. For now, our advice remains to restrict the running of robotic lawn mowers to daylight hours and check the lawn for any wildlife species which may be vulnerable in the encounter with a robotic lawn mower before turning on the machine.

5. Conclusions

Based on the experiments presented in this study, we conclude that some models of robotic lawn mowers may injure hedgehogs, whereas others are not harmful to them. Apart from one single incidence, all robotic lawn mowers had to physically touch the hedgehog carcasses to detect them. Height affected the risk of injury, with larger hedgehog cadavers being less likely to be damaged. The firmness of the tested hedgehog cadavers (frozen or thawed) did not influence the outcome of the collision tests. Neither the position of the hedgehog cadavers nor the selected technical features of the lawn mowers affected the probability of injury. The level of uncertainty regarding injury probability declined as the number of trials per model of robotic lawn mower was increased to a level of 50 tests, causing us to suggest a test number of 50 to characterise confidently the risk of injury to hedgehogs caused by a specific lawn mower.

The insights provided by our results have enabled the design of a protocol for a standardised hedgehog safety test to quantify the effect of a given robotic lawn mower on hedgehogs. Used in combination with specially designed hedgehog crash test dummies, this protocol will hopefully lead to the development of more hedgehog-friendly robotic lawn mowers, thereby reducing the negative impact on hedgehogs.

Supplementary Materials: The following supporting information can be downloaded at: https://www.mdpi.com/article/10.3390/ani14010122/s1, Table S1: The full dataset from the tests of collisions between hedgehog carcasses and robotic lawn mowers.

Author Contributions: Conceptualisation, S.L.R., A.B., C.P., D.W.M. and A.K.O.A.; methodology, S.L.R., A.K.O.A. and A.B.; validation, S.L.R., A.K.O.A., B.T.S. and A.B.; formal analysis, S.L.R., R.S. and C.P., investigation, S.L.R., A.K.O.A., B.T.S. and A.B.; resources, S.L.R.; data curation, S.L.R., A.K.O.A., B.T.S. and A.B.; writing—original draft preparation, S.L.R.; writing—review and editing, S.L.R., A.K.O.A., B.T.S., A.B., R.S., C.P. and D.W.M.; visualisation, S.L.R. and R.S.; supervision, C.P., D.W.M. and A.K.O.A.; project administration, S.L.R.; funding acquisition, S.L.R. All authors have read and agreed to the published version of the manuscript.

Funding: This research was funded by Husqvarna; STIHL; British Hedgehog Preservation Society; Carlsberg Foundation, grant number CF20_0443.

Institutional Review Board Statement: Ethical review and approval were waived for this study as the experiments only included dead animals, dying from natural causes at wildlife rehabilitation centres.

Informed Consent Statement: Not applicable.

Data Availability Statement: All relevant data from the research have been made available in the publication.

Acknowledgments: The authors are grateful to hedgehog rehabilitator Dorthe Madsen from Animal Protection Denmark (Dyrenes Beskyttelse) for assisting during the test procedures. We would furthermore like to thank hedgehog rehabilitators representing the organisations Pindsvineplejerne, Pindsvinevennerne i Danmark, and Animal Protection Denmark (Dyrenes Beskyttelse) for collecting deceased hedgehogs, dying in care at their facilities, for the tests.

Conflicts of Interest: The authors declare no conflicts of interest. The funders had no role in the design of this study; in the analyses or interpretation of data; in the writing of this manuscript; or in the decision to publish the results.

References

Müller, F. Langzeit-Monitoring der Strassenverkehrsopfer beim Igel (*Erinaceus europaeus* L.) zur Indikation von Populationsdichteveränderungen entlang zweier Teststrecken im Landkreis Fulda. *Beiträge Naturkunde Osthess.* **2018**, *54*, 21–26.

British Trust for Ornithology (BTO); Commissioned by British Hedgehog Preservation Society and People's Trust for Endangered Species. *The State of Britain's Hedgehogs 2011*; British Hedgehog Preservation Society and People's Trust for Endangered Species: London, UK, 2011.

British Hedgehog Preservation Society; People's Trust for Endangered Species. *The State of Britain's Hedgehogs 2015*; British Hedgehog Preservation Society and People's Trust for Endangered Species: London, UK, 2015.

British Hedgehog Preservation Society; People's Trust for Endangered Species. *The State of Britain's Hedgehogs 2018*; British Hedgehog Preservation Society and People's Trust for Endangered Species: London, UK, 2018.

Hof, A.R.; Bright, P.W. Quantifying the long-term decline of the West European hedgehog in England by subsampling citizen-science datasets. *Eur. J. Wildl. Res.* **2016**, *62*, 407–413. [CrossRef]

Krange, M. Change in the Occurrence of the West European Hedgehog (*Erinaceus europaeus*) in Western Sweden during 1950–2010. Master's Thesis, Karlstad University, Karlstad, Sweden, 2015.

van de Poel, J.L.; Dekker, J.; van Langevelde, F. Dutch hedgehogs *Erinaceus europaeus* are nowadays mainly found in urban areas, possibly due to the negative effects of badgers Meles meles. *Wildlife Biol.* **2015**, *21*, 51–55. [CrossRef]

Williams, B.M.; Baker, P.J.; Thomas, E.; Wilson, G.; Judge, J.; Yarnell, R.W. Reduced occupancy of hedgehogs (*Erinaceus europaeus*) in rural England and Wales: The influence of habitat and an asymmetric intra-guild predator. *Sci. Rep.* **2018**, *8*, 12156. [CrossRef] [PubMed]

Taucher, A.L.; Gloor, S.; Dietrich, A.; Geiger, M.; Hegglin, D.; Bontadina, F. Decline in Distribution and Abundance: Urban Hedgehogs under Pressure. *Animals* **2020**, *10*, 1606. [CrossRef] [PubMed]

10. Wembridge, D.; Johnson, G.; Al-Fulaij, N.; Langton, S. *The State of Britain's Hedgehogs 2022*; British Hedgehog Preservation Socie and People's Trust for Endangered Species: London, UK, 2022.

11. Mathews, F.; Harrower, C. *IUCN—Compliant Red List for Britain's Terrestrial Mammals. Assessment by the Mammal Society und contract to Natural England, Natural Resources Wales and Scottish Natural Heritage*; Natural England: Peterborough, UK, 2020.

12. Brakes, C.R.; Smith, R.H. Exposure of non-target small mammals to rodenticides: Short-term effects, recovery and implicatior for secondary poisoning. *J. Appl. Ecol.* **2005**, *42*, 118–128. [CrossRef]

13. Haigh, A.; O'Riordan, R.M.; Butler, F. Nesting behaviour and seasonal body mass changes in a rural Irish population of th Western hedgehog (*Erinaceus europaeus*). *Acta Theriol.* **2012**, *57*, 321–331. [CrossRef]

14. Hof, A.R.; Bright, P.W. The value of agri-environment schemes for macro-invertebrate feeders: Hedgehogs on arable farms Britain. *Anim. Conserv.* **2010**, *13*, 467–473. [CrossRef]

15. Huijser, M.P.; Bergers, P.J.M. The effect of roads and traffic on hedgehog (*Erinaceus europaeus*) populations. *Biol. Conserv.* **2000**, *9* 111–116. [CrossRef]

16. Young, R.P.; Davison, J.; Trewby, I.D.; Wilson, G.J.; Delahay, R.J.; Doncaster, C.P. Abundance of hedgehogs (*Erinaceus europaeus*) relation to the density and distribution of badgers (*Meles meles*). *J. Zool.* **2006**, *269*, 349–356. [CrossRef]

17. Hubert, P.; Julliard, R.; Biagianti, S.; Poulle, M.-L. Ecological factors driving the higher hedgehog (*Erinaceus europeaus*) density an urban area compared to the adjacent rural area. *Landsc. Urban Plan.* **2011**, *103*, 34–43. [CrossRef]

18. Pettett, C.E.; Moorhouse, T.P.; Johnson, P.J.; Macdonald, D.W. Factors affecting hedgehog (*Erinaceus europaeus*) attraction to rur villages in arable landscapes. *Eur. J. Wildl. Res.* **2017**, *63*, 54. [CrossRef]

19. Dowding, C.V.; Harris, S.; Poulton, S.; Baker, P.J. Nocturnal ranging behaviour of urban hedgehogs, *Erinaceus europaeus*, in relatic to risk and reward. *Anim. Behav.* **2010**, *80*, 13–21. [CrossRef]

20. Dowding, C.V.; Shore, R.F.; Worgan, A.; Baker, P.J.; Harris, S. Accumulation of anticoagulant rodenticides in a non-targe insectivore, the European hedgehog (*Erinaceus europaeus*). *Environ. Pollut.* **2010**, *158*, 161–166. [CrossRef] [PubMed]

21. Rasmussen, S.L.; Nielsen, J.L.; Jones, O.R.; Berg, T.B.; Pertoldi, C. Genetic structure of the European hedgehog (*Erinaceus europaeu* in Denmark. *PLoS ONE* **2020**, *15*, e0227205. [CrossRef] [PubMed]

22. Rasmussen, S.L.; Yashiro, E.; Sverrisdóttir, E.; Nielsen, K.L.; Lukassen, M.B.; Nielsen, J.L.; Asp, T.; Pertoldi, C. Applying th GBS Technique for the Genomic Characterization of a Danish Population of European Hedgehogs (*Erinaceus europaeus*). *Gen. Biodivers. J.* **2019**, *3*, 78–86. [CrossRef]

23. Rasmussen, S.L.; Berg, T.B.; Dabelsteen, T.; Jones, O.R. The ecology of suburban juvenile European hedgehogs (*Erinaceus europaeu* in Denmark. *Ecol. Evol.* **2019**, *9*, 13174–13187. [CrossRef] [PubMed]

24. Rasmussen, S.L.; Larsen, J.; van Wijk, R.E.; Jones, O.R.; Berg, T.B.; Angen, O.; Larsen, A.R. European hedgehogs (*Erinace. europaeus*) as a natural reservoir of methicillin-resistant *Staphylococcus aureus* carrying *mecC* in Denmark. *PLoS ONE* **201** 14, e0222031. [CrossRef]

25. Rasmussen, S.L.; Hallig, J.; van Wijk, R.E.; Petersen, H.H. An investigation of endoparasites and the determinants of parasi infection in European hedgehogs (*Erinaceus europaeus*) from Denmark. *Int. J. Parasitol. Parasites Wildl.* **2021**, *16*, 217–227. [CrossRe [PubMed]

26. Rasmussen, S.L.; Schrøder, A.E.; Mathiesen, R.; Nielsen, J.L.; Pertoldi, C.; Macdonald, D.W. Wildlife Conservation at a Garde Level: The Effect of Robotic Lawn Mowers on European Hedgehogs (*Erinaceus europaeus*). *Animals* **2021**, *11*, 1191. [CrossRe [PubMed]

27. Moore, L.J.; Petrovan, S.O.; Baker, P.J.; Bates, A.J.; Hicks, H.L.; Perkins, S.E.; Yarnell, R.W. Impacts and potential mitigation of roa mortality for hedgehogs in Europe. *Animals* **2020**, *10*, 1523. [CrossRef]

28. Jota Baptista, C.; Oliveira, P.A.; Gonzalo-Orden, J.M.; Seixas, F. Do Urban Hedgehogs (*Erinaceus europaeus*) Represent a Relevar Source of Zoonotic Diseases? *Pathogens* **2023**, *12*, 268. [CrossRef]

29. Burroughes, N.D.; Dowler, J.; Burroughes, G. Admission and Survival Trends in Hedgehogs Admitted to RSPCA Wildli Rehabilitation Centres. *Proc. Zool. Soc.* **2021**, *74*, 198–204. [CrossRef]

30. Lukešová, G.; Voslarova, E.; Vecerek, V.; Vucinic, M. Trends in intake and outcomes for European hedgehog (*Erinaceus europaeu* in the Czech rescue centers. *PLoS ONE* **2021**, *16*, e0248422. [CrossRef]

31. Garces, A.; Soeiro, V.; Loio, S.; Sargo, R.; Sousa, L.; Silva, F.; Pires, I. Outcomes, Mortality Causes, and Pathological Findings European Hedgehogs (*Erinaceus europaeus*, Linnaeus 1758): A Seventeen Year Retrospective Analysis in the North of Portuga *Animals* **2020**, *10*, 1305. [CrossRef]

32. Martínez, J.C.; Rosique, A.I.; Royo, M.S. Causes of admission and final dispositions of hedgehogs admitted to three Wildli Rehabilitation Centers in eastern Spain. *Hystrix* **2014**, *25*, 107.

33. Rasmussen, S.L.; Berg, T.B.; Martens, H.J.; Jones, O.R. Anyone Can Get Old—All You Have to Do Is Live Long Enoug Understanding Mortality and Life Expectancy in European Hedgehogs (*Erinaceus europaeus*). *Animals* **2023**, *13*, 626. [CrossRef

34. Doncaster, C.P.; Rondinini, C.; Johnson, P.C.D. Field test for environmental correlates of dispersal in hedgehogs *Erinaceus europaeu J. Anim. Ecol.* **2001**, *70*, 33–46.

35. Berger, A. Occurrence and Characteristics of Cut Injuries in Hedgehogs in Germany: A Collection of Individual Cases. *Anima* **2024**, *14*, 57. [CrossRef]

6. Rasmussen, S.L.; Schrøder, B.T.; Berger, A.; Macdonald, D.W.; Pertoldi, C.; Briefer, E.F.; Alstrup, A.K.O. Facing Danger: Exploring Personality and Reactions of European Hedgehogs (*Erinaceus europaeus*) towards Robotic Lawn Mowers. *Animals* **2024**, *14*, 2. [CrossRef]

7. Future Market Insights 2022. Robotic Lawn Mower Market Outlook (2022–2032). Future Market Insights. p. 342. Available online: https://www.futuremarketinsights.com/reports/robotic-lawn-mower-market (accessed on 1 October 2023).

8. Fortune Business Insights 2022. Robotic Lawn Mower Market Size, Share & COVID-19 Impact Analysis. p. 150. Available online: https://www.fortunebusinessinsights.com/robotic-lawn-mower-market-106531 (accessed on 1 October 2023).

9. *EN 50636-2-107*; Safety of Household and Similar Appliances—Part 2-107: Particular Requirements for Robotic Battery Powered Electrical Lawnmowers. CENELEC European Committee for Electrotechnical Standardization: Brussels, Belgium, 2015.

10. Kilands Mattor. Available online: https://www.kilandsmattor.se/kokosborst-natur-entrematta-pa-metervara?gclid=CjwKCAjw3ueiBhBmEiwA4BhspGR0HEu3vqzjS1pmCB0Z3EL9wgETD3pBcjh3gHiHlhQsyD_rrkCgMRoCW90QAvD_BwE&gclsrc=aw.ds (accessed on 1 November 2023).

11. Burnham, K.P.; Anderson, D.R. *Model Selection and Multimodel Inference, A Practical Information-Theoretic Approach*, 2nd ed.; Springer: New York, NY, USA, 2002.

12. R Core Team. *R: A Language and Environment for Statistical Computing*; R Foundation for Statistical Computing: Vienna, Austria, 2022.

13. Bates, D.; Mächler, M.; Bolker, B.; Walker, S. Fitting Linear Mixed-Effects Models Using lme4. Available online: https://doi.org/10.18637/jss.v067.i01 (accessed on 1 October 2023).

Article

Facing Danger: Exploring Personality and Reactions of European Hedgehogs (*Erinaceus europaeus*) towards Robotic Lawn Mowers

Sophie Lund Rasmussen [1,2,3,*], Bettina Thuland Schrøder [4], Anne Berger [5], David W. Macdonald [1], Cino Pertoldi [2,6], Elodie Floriane Briefer [4] and Aage Kristian Olsen Alstrup [7,8]

1 Wildlife Conservation Research Unit, The Recanati-Kaplan Centre, Department of Biology, University of Oxford, Tubney House, Abingdon Road, Tubney, Abingdon OX13 5QL, UK; david.macdonald@biology.ox.ac.uk
2 Department of Chemistry and Bioscience, Aalborg University, Fredrik Bajers Vej 7H, 9220 Aalborg, Denmark; cp@bio.aau.dk
3 Linacre College, University of Oxford, St. Cross Road, Oxford OX1 3JA, UK
4 Behavioral Ecology Group, Section for Ecology and Evolution, Department of Biology, University of Copenhagen, Universitetsparken 15, 2100 Copenhagen, Denmark; catsandconservation@gmail.com (B.T.S.); elodie.briefer@bio.ku.dk (E.F.B.)
5 Leibniz Institute for Zoo and Wildlife Research, Alfred-Kowalke-Straße 17, 10315 Berlin, Germany; berger@izw-berlin.de
6 Aalborg Zoo, Mølleparkvej 63, 9000 Aalborg, Denmark
7 Department of Nuclear Medicine and PET, Aarhus University Hospital, Palle Juul-Jensens Boulevard 165, 8200 Aarhus, Denmark; aagealst@rm.dk
8 Department of Clinical Medicine, Aarhus University, Palle Juul-Jensens Boulevard 165, 8200 Aarhus, Denmark
* Correspondence: sophie.rasmussen@biology.ox.ac.uk or sophielundrasmussen@gmail.com

Citation: Rasmussen, S.L.; Schrøder, B.T.; Berger, A.; Macdonald, D.W.; Pertoldi, C.; Briefer, E.F.; Alstrup, A.K.O. Facing Danger: Exploring Personality and Reactions of European Hedgehogs (*Erinaceus europaeus*) towards Robotic Lawn Mowers. *Animals* 2024, 14, 2. https://doi.org/10.3390/ani14010002

Academic Editor: Brian L. Cypher

Received: 23 November 2023
Revised: 11 December 2023
Accepted: 14 December 2023
Published: 19 December 2023

Correction Statement: This article has been republished with a minor change. The change does not affect the scientific content of the article and further details are available within the backmatter of the website version of this article.

Simple Summary: The European hedgehog is a generally welcomed but nowadays less common guest in residential gardens, as the species is in decline. Sharing habitats with humans comes at a cost: a residential garden holds many potential dangers for hedgehogs. Previous research has shown that certain models of robotic lawn mowers may harm hedgehogs. This study sought to investigate the personality and reactions of live hedgehogs towards a disarmed, approaching robotic lawn mower. Personality tests revealed that the hedgehogs could be divided into categories of "shy" and "bold" individuals, independently of age and sex. The encounter tests with a disarmed robotic lawn mower showed that they behaved and positioned themselves in seven different ways, and the individuals with a bold personality reacted in a more unpredictable way. Adult hedgehogs tended to react in a shyer manner, and the tested hedgehogs, generally, acted less boldly the second time they encountered a robotic lawn mower. This knowledge will be used in the process of designing a standardised hedgehog safety test to eventually produce and approve hedgehog-friendly robotic lawn mowers that pose no hazards to hedgehogs, ultimately, serving to eliminate their influence on hedgehog survival and, thereby, improve hedgehog conservation.

Abstract: The populations of European hedgehog (*Erinaceus europaeus*) are in decline, and it is essential that research identifies and mitigates the factors causing this. Hedgehogs are increasingly sharing habitats with humans, being exposed to a range of dangers in our backyards. Previous research has documented that some models of robotic lawn mowers can cause harm to hedgehogs. This study explored the personality and behaviour of 50 live hedgehogs when facing an approaching disarmed robotic lawn mower. By combining a novel arena and novel object test, we found that 27 hedgehogs could be categorised as "shy" and 23 as "bold", independently of sex and age. The encounter tests with a robotic lawn mower showed that the hedgehogs positioned themselves in seven different ways. Personality did not affect their reactions. Adult hedgehogs tended to react in a shyer manner, and the hedgehogs, generally, acted less boldly during their second encounter with the robotic lawn mower. Additionally, our results show that bold individuals reacted in a more unpredictable way, being more behaviourally unstable compared to the shy individuals. This knowledge will be applied in the design of a standardised hedgehog safety test, eventually serving to produce and approve hedgehog-friendly robotic lawn mowers.

Keywords: *Erinaceus europaeus*; animal personality; applied animal behaviour research; shyness–boldness; wildlife conservation; anthropogenic disturbance; robotic lawn mowers; garden technology; lawn care; behavioural instability

1. Introduction

The European hedgehog (*Erinaceus europaeus*), hereafter referred to as "hedgehog", is a widely distributed species that can survive across a range of diverse habitat types [1,2]. Despite the species' ability to adapt to many different settings [3], recent research has either documented or suggested declines in the populations of European hedgehogs in several Western European countries [4–14]. For the time being, the investigated factors contributing to this decline include habitat loss; habitat fragmentation; inbreeding; intensified agricultural practices; road traffic accidents; a reduction in biodiversity and, hence, food items; lack of suitable nest sites in residential gardens; accidents caused by garden tools; netting and other anthropogenic sources in residential gardens; infections with pathogens and endoparasites; molluscicide and rodenticide poisoning; and, in some areas, badger predation [3,5,15–31]. With the potential to reach 16 years of age [32], it is of concern that the majority of research into the lifespan of European hedgehogs has found a mean age of only two years (see Rasmussen, Berg, Martens and Jones [32], Table 1 for an overview). It is, therefore, essential to investigate the possible causes for this early mortality and the overall population decline in order to optimise the conservation initiatives directed at this species.

1.1. Hedgehogs and Robotic Lawn Mowers

Robotic lawn mowers are becoming increasingly popular in Europe and the US. According to market insight reports, it is expected that the global robotic lawn mower market will expand from US 0.8–1.5 billion in 2020–2022 to US 2.7–4 billion in 2032, and the market is anticipated to develop at a compound annual growth rate (CAGR) of 11.5–15.5% during the forecasted period [33,34]. With the increasing number of robotic lawn mowers operating in residential gardens throughout the distribution range of hedgehogs in Europe, combined with the fact that research indicates that hedgehogs are, nowadays, increasingly associated with human-occupied areas [10,20,31,35], it seems likely that many individual hedgehogs will encounter several robotic lawn mowers during their lifetime.

Human activities negatively impact the welfare of countless wild vertebrates [36], and hedgehogs are no exception. Injured hedgehogs are frequently found by members of the public and are admitted to hedgehog rehabilitation centres with different types of cuts and injuries, most often caused by garden tools and netting, or predators such as dogs, foxes, or badgers [37–40]. In some cases, the injuries are fatal, whilst others necessitate euthanasia. However, because of a growing concern reported by hedgehog carers and members of the public that an increasing number of these incidents could have been caused by robotic lawn mowers, Rasmussen et al. (2021) [29] investigated the effects of robotic lawn mowers on hedgehogs. Tests were carried out on deceased hedgehogs and showed that some models of robotic lawn mowers did, indeed, injure hedgehog carcasses, whereas other models only had to touch a hedgehog carcass lightly to detect it before changing direction accordingly to avoid it. Consequently, it was observed that all models of robotic lawn mowers included in the tests had to physically interact with the hedgehog carcasses to detect them [29]. This led to the suggestion that research should be initiated to develop more hedgehog-friendly robotic lawn mowers and that a standardised safety test should be designed for the evaluation and approval of new models of robotic lawn mowers for the market in terms of hedgehog safety as an addition to the current mandatory general safety guidelines [29,41]. This test should, ideally, be performed on a specially designed hedgehog crash test dummy placed in realistic fixed positions.

Having previously tested the effects of robotic lawn mowers on dead hedgehogs positioning the carcasses in the tests based on the knowledge of hedgehog behaviour, we decided to test the reactions of live hedgehogs to an approaching robotic lawn mower to optimise the design of a future standardised safety test by ensuring that it is realistic. Furthermore, testing and quantifying the behaviour of live animals should also consider the effects of personality on the outcome.

1.2. Measuring Personality in Animals

When facing danger, hedgehogs tend to stand completely still, as if they are in a frozen state, often in an upright position with the snout pointing inwards (i.e., head bent inwards in a partially curled-up position, later referred to as behavioural category 3) whilst deciding whether the next step should be to curl up or run away [42]. The strategy of curling up in front of an approaching car or robotic lawn mower appears to be less successful than running away. Could personality determine the reaction of hedgehogs towards an approaching robotic lawn mower?

Personality, defined as individual differences that are stable over time and across situations, affects how individuals react to challenging situations [43] and may influence their survival [44]. Several studies have shown that it is possible to estimate the shyness/boldness of individuals, including hedgehogs [45], by analysing how they explore a novel environment or arena, or by measuring their latency to approach a novel object in a familiar environment [44,46–48]. Previous research has explored and documented the occurrence of a shyness–boldness gradient in natural populations [44,49–51]. In addition, it has been demonstrated that personality potentially influences fitness through reproductive success and survival [44,52,53]. Natural selection affects factors such as boldness at population level [54], which is why individuals with inappropriate levels of boldness may suffer reduced fitness in the wild due to extensive risk-taking behaviour [44]. Accordingly, Bremner-Harrison, Prodohl, and Elwood (2004) [44] found that bolder juvenile captive-bred swift foxes had a lower postrelease survival compared to their shyer conspecifics.

Another important aspect that characterises the personality of an individual is the degree of behavioural instability, which quantifies the degree of unpredictability of behavioural response. The notion of behavioural instability, as suggested by Pertoldi et al. (2016) and others [55–57], goes beyond being solely described by the variance and/or interquartile range (IQR). It also encompasses the kurtosis and skewness (i.e., asymmetry) of the distributions. These parameters collectively impact the median absolute deviation, a measure of the variability in a dataset, which is estimated by the median distance of the data values from the median (MAD). The higher levels of behavioural instability exhibited by certain individuals could have relevant implications, as an increased variability in the behavioural repertoire could enhance the probability of survival in a dangerous situation such as an encounter between a hedgehog and a robotic lawn mower [55].

1.3. Aim of the Research

In this study, we tested the reactions of live hedgehogs towards a disarmed, approaching robotic lawn mower to optimise the test design of a future standardised realistic safety test. Additionally, we investigated the effects of personality, measured as shyness/boldness, on behavioural responses and associated predictability and, thereby, the risk-taking behaviour of these animals when facing the approaching robotic lawn mower to ensure that this test would account for differences in the hedgehog reactions linked to their personality. We predicted that shy individuals would have a higher tendency to run away and that bolder individuals would be more inquisitive towards the approaching robotic lawn mower.

Measuring the responses of live hedgehogs towards an approaching robotic lawn mower will facilitate the optimal design of a standardised safety test for robotic lawn mowers, which will become an important tool for enhancing the safety of hedgehogs entering gardens. This will, ultimately, result in the improved conservation of the declining populations of European hedgehogs.

2. Materials and Methods

The research included both personality tests and encounter tests with a robotic lawn mower (Figure 1), performed on 50 live rehabilitated Danish hedgehogs that had been assessed as ready for release back into the wild. All individuals were released at suitable sites within a few days after the tests. Some individuals were adults originally admitted into care because of disease or injury (N = 15), and some were orphaned juveniles (N = 35) that were hand-raised by a hedgehog rehabilitator. They had reached the age and capabilities similar to wild juveniles at the age of independence, having been raised under natural conditions, and had been deemed ready for release back into the wild.

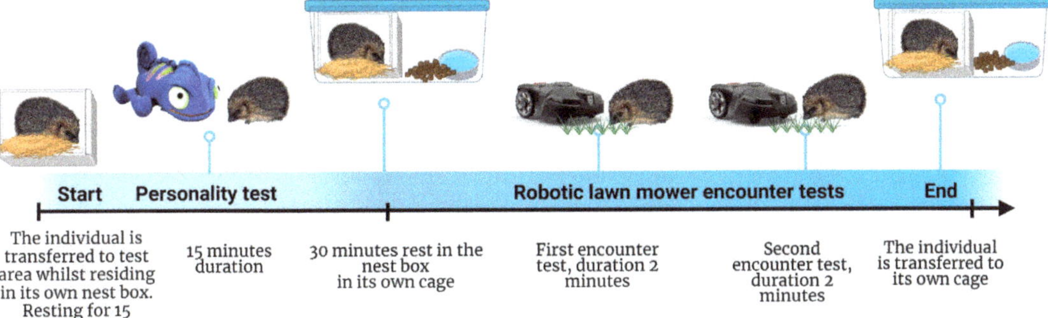

Figure 1. Experimental design and timeline. Design adapted from the "Mouse Experimental Timeline", template by BioRender.com, agreement number: RS262OWRC1. Photographs and illustrations by Cloud b; Encyclopaedia Britannica, and Husqvarna.

The experimental design was created with the welfare of the hedgehogs in mind, taking precautions to reduce the levels of stress caused by transportation and handling. Therefore, all tests took place in the garden surrounding the wildlife rehabilitation centre, where the hedgehogs had been in care, and the hedgehogs were exclusively handled by the wildlife rehabilitator, who had been nursing them back to health.

The tests were divided into a pilot study on ten individuals, which took place on two consecutive nights from the 3rd–5th of June 2022, and further tests on 40 individuals taking place from the 6th–8th of September 2022 (N = 20) and 15th–17th of September 2022 (N = 20). As hedgehogs are nocturnal, the experiments were performed during their natural activity period, from sunset to sunrise, varying based on the time of year and the number of individuals tested on the specific nights of testing.

2.1. Personality Tests

The personality tests were performed as a combination of a novel object test [48,58,59] and a novel arena test [59,60] for the purpose of assessing each individual's tendency to respond with either curiosity or fearfulness and, hence, exploration or avoidance of the novel object in a novel environment. The novel object was a 45 × 25 × 20 cm blue plush toy (Cloud b "Charley the Chameleon"). The toy stimulated both visual and auditory senses through a sound emission, a soothing melody at a maximum of 5 dB, and blinking RGB LED lights in the colours red, blue, and green.

The novel arena consisted of a wooden pallet frame with a solid, wooden pallet base (dimensions inside the arena: length 104 cm, width 76 cm, and height 38 cm). The base was covered by a large piece of cardstock thick paper in a light grey colour, covering the whole base of the arena. The novel object was placed in the end of the test arena, 60 cm from the nest box containing the hedgehog subjected to the personality test, which was placed in the opposite end of the novel arena. The nest box was added to the novel arena just before the test started. The novel arena was placed under a cover by a dark-blue garden gazebo for the purpose of standardising the weather conditions and shielding the arena against rain.

The procedure for each personality test lasted for 30 min (including rest) in total (Figure 1). Before the test started, the hedgehog was transported to the novel arena in its individual, familiar nest box, from its own enclosure situated in the housing ward of the hedgehog rehabilitation centre. The closed nest box was placed in the novel arena, allowing the hedgehog to rest and acclimatise for 15 min after transportation. During this resting period, the novel object was not activated and was not visible to the hedgehog. After 15 min of rest, the nest box entrance was opened, and the novel object's light and sound effects were switched on (Figure 2). Over the following 15 min, two cameras recorded any activity in the novel arena. A Ring Stick Up Cam video camera (Ring™, Santa Monica, CA, USA), which was suspended from the ceiling of the garden gazebo at around the centre of the novel arena, provided a complete overview of the novel arena. The Ring camera was controlled through the Ring app and was set to record with a live view continuously for 12 min at a time, as this was the maximum possible duration per recording. The Ring camera was manually re-activated after 12 min using the Ring app. A Nedis HD Wildlife CameraWCAM130GN (Nedis, 's-Hertogenbosch, The Netherlands) was also set to record for 30 s after the detection of any locomotion and to record back to back with no delay between recordings. The Nedis camera was placed on a tripod halfway along the long side of the novel arena allowing for all activity during a test to be detected and recorded.

Figure 2. A photo from the Ring surveillance camera, which was used during the personality tests, illustrating the experimental design.

An experimenter (the same person for all personality tests) silently monitored all tests in complete darkness at a distance of 4 m from the test arena.

2.2. Encounter Tests with a Robotic Lawn Mower

The robotic lawn mower encounter tests were conducted on the lawn of the residential garden surrounding the wildlife rehabilitation centre, where the hedgehogs were housed. The experimental area was fenced off with a green, 45 cm tall wire mesh fence to ensure the hedgehogs tested could not run away (Figure 3). The area was 1.5 m in width and 6.75 m in length. Green tentor poles were placed alongside the fence inside the test area and marked each 1 m distance from the hedgehog. A black folding fence, which was 1.2 m in height, consisting of three panels that were 0.8 m in width, was placed outside the fenced area to surround the zone in which the hedgehog was placed. This allowed for the experimenters recording the behaviour and stepping into the test zone to manually stop the robotic lawn mower and to remain visually hidden from the hedgehog during the tests.

Figure 3. A photo from the surveillance camera used during the encounter tests, illustrating the experimental design. A range of safety precautions for the tests can be seen: (1) a person standing on the right side of the test area stepping in to manually turn off the robotic lawn mower by pushing the (visual) button on the shield of the mower; (2) the rope attached to the back of the mower (with a person pulling it in the other end); and (3) the hedgehog rehabilitator standing inside the test area next to the hedgehog ready to intervene, if necessary. The two persons standing in the vicinity of the hedgehog tested (the hedgehog rehabilitator and the person recording with the handheld FLIR E54 Thermal Imaging camera) remained the same throughout all tests and were cautious to behave in the same manner during all tests.

The robotic lawn mower (Husqvarna Automower® 415X, Husqvarna, Huskvarna, Sweden) was positioned on a spot marked with duct tape at a distance of 5.75 m from the hedgehog, allowing it to move in a straight line through the test area in the direction of the hedgehog. At the beginning of each test, the hedgehog was placed inside the fenced area, on a spot on the lawn marked with duct tape, at a distance of 1 m from the surrounding fence. The hedgehog was positioned in an upright standing position with the front facing the mower and the head folding inwards (later referred to as behavioural category 3). When the hedgehog was in place, the mower was turned on, causing it to move in the direction of the hedgehog and gradually approaching the individual at 380 mm/s, until a distance of 0.5 m, at which the mower was stopped. During the pilot test on the first ten individuals, the robotic lawn mower was stopped at a 1 m distance from the hedgehogs, but this was changed to 0.5 m for the tests with the last 40 individuals due to a concern that the 1 m distance was too far away to properly trigger a reaction from the hedgehogs.

Each hedgehog was tested with the robotic lawn mower twice: one test in which the headlights of the mower were switched off and another with the headlights switched on. The second test was performed immediately after the first (Figure 1). The order of the tests was counterbalanced among the individuals (Figure 1). In the case that the test was interrupted because of a technical failure, or if the hedgehog ran away before the robotic lawn mower was turned on, we restarted the test once, resulting in a maximum of three tests and a duration of a maximum of 15 min in total, during which the hedgehog was situated in the test area. In the majority of cases, we only tested the hedgehogs twice.

The tests were recorded with a Ring Stick Up Cam video camera placed on a tripod overlooking the experimental area, as well as a handheld FLIR E54 Thermal Imaging camera (Teledyne FLIR, Wilsonville, OR, USA) recording from behind the folding fence surrounding the hedgehog being tested. With the use of these cameras, the reactions (i.e., behavioural categories) of each individual towards the approaching robotic lawn mower during the two tests, were recorded.

After the tests, the hedgehog was returned to its enclosure in the housing ward at the hedgehog rehabilitation centre by the hedgehog rehabilitator.

Safety Precautions during the Encounter Tests

Multiple initiatives were implemented to ensure that the safety measures were met during the tests. The pivoting blades (i.e., knives) were removed from the robotic lawn mower, and the mower was controlled via remote control (Husqvarna Automower® Connect App, Husqvarna, Huskvarna, Sweden). Additionally, a rope was fastened to the robotic lawn mower, and a staff member held this rope during the tests, generating manual control to ensure the mower never got closer to the hedgehog than the 0.5 m (1 m in the pilot study) distance allowed in the test. Furthermore, a person standing alongside the test area, hidden behind the folding fence, stepped in to stop the robotic lawn mower manually when it reached the 0.5 m (1 m in pilot study) limit by pressing the red stop button on the dorsal shield of the robotic lawn mower. The hedgehog carer remained standing next to the hedgehog during the tests and was ready to intervene and pick up the hedgehog in case it tried to run away or the other safety precautions failed (which never occurred during the tests).

2.3. Data Analysis

The data analyses and graphs for publication were prepared in R v 4.2.3. [61] using the packages lme4, DHARMa, pbkrtest, multcomp, ordinal, car, and tidyverse.

2.3.1. Personality Tests

On the basis of their behaviour during the combined novel arena and novel object test, the hedgehogs were categorised as either shy or bold as follows: individuals that remained in their nest box during the full 15 min duration of the test were categorised as "shy", whilst those that left the nest box and entered the arena to explore the novel arena and novel object during the test were categorised as "bold". This categorisation was established during the pilot study after observing a large number (6/10) of hedgehogs remaining in their nest box for the entire test.

We used a generalised linear model (GLM) to analyse the results of the personality tests due to a binary outcome (shy versus bold, analysed as 0 vs. 1) and having no repeated measures of the same individuals. The GLM with binomial family included personality (0 versus 1) as a response variable and age (adult versus juvenile), sex (female versus male), and weight (in kg, z-score transformed) as fixed effects.

On the basis of the variance inflation factor (VIF), using the VIF function in R, testing for multicollinearity, we found that the fixed effects of age and weight were correlated, as VIF > 5. Therefore, we removed the response variable "weight" from the initial model (VIF = 5.42). Accordingly, the final model was: glm (personality_category ~ sex + age, family = binomial, data = personality).

2.3.2. Encounter Tests with a Robotic Lawn Mower

During the encounter tests with a robotic lawn mower, the behaviour of each individual was scored in five ordinal behavioural categories, with 1 being the more exploratory behavioural (bold) reaction and 5 being the more cautious (shy) reaction. Behaviour was recorded at the point in time at which the robotic lawn mower was situated just in front of the hedgehog and before the mower was manually stopped.

The ordinal categories were:

(1)   Upright position with snout facing 9–10 o'clock, 2–3 o'clock, or 6 o'clock (rump towards the mower).
(2)   Upright position with snout facing the mower.
(3)   Upright position with snout pointing inwards (head bent inwards in a partially curled-up position). This was the position the hedgehog was originally placed in when the test was initiated.
(4)   Hedgehog running away from the robotic lawn mower.
(5)   Hedgehog curling up.

To analyse the results of the encounter tests, a cumulative link mixed model (CLMM) was chosen, as the response is ordinal (ordered categorical data), and no assumption was made concerning the spacing between behavioural categories 1 and 5.

The CLMM included the ordinal behavioural category (1–5) as a response variable; lights (on vs. off), personality (shy vs. bold, coded as 0 vs. 1), age (adult versus juvenile), sex (female vs. male), and test number (1 vs. 2 as a control) as fixed effects; and individual ID crossed with the date of the test as random effects (to control for repeated measures, as each hedgehog was tested twice, and the differences between the days). The variance inflation factor was below 5 for all fixed effects, indicating the absence of multicollinearity, and allowing us to include all factors in the same model. The final model was: clmm (responseF ~ lights + personality_cat + weight + sex + test_number + (1 | individual) + (1 | date), data = behaviour).

Furthermore, by comparing a model with to a model without the random effect "individual" using a likelihood ratio test, we tested whether individuals differed in their responses in general. Finally, the degree of behavioural instability, estimated as the median absolute deviation (MAD) of the ordinal behavioural category was tested with the Mann–Whitney U test and Levene's test from the medians for differences among the individuals belonging to the two categories of personality (shy vs. bold). In the two categories of personality, individuals of different sex, weight, and age were pooled, as these factors did not seem to affect personality (see Section 3, results below).

2.4. Ethical Approval

Ethical approval for this study was provided by the Animal Experiments Inspectorate under the Danish National Committee for the Protection of Animals used for Scientific Purposes (license number: 2021-15-0201-00865) in accordance with 2010/63/EU [62]. We followed the 3R concept for use of animals in research: we used a paired design to reduce the number of experimental animals needed (i.e., reduction), and we ensured that no animal suffered harm during the study through specified safety measures (i.e., refinements). It was not possible to replace the studies with alternative methods (i.e., replacement). Permission was also obtained from the Danish Nature Agency to work with this protected species. All animals completed the tests without injury and were released back into the wild within a few days.

3. Results

The 50 individuals tested comprised 15 adults and 35 independent juveniles, 30 females and 20 males.

3.1. Personality Tests

On the basis of their behaviour during the combined novel arena and novel object test, 27 (54%) individuals were categorised as shy and 23 (46%) as bold.

Our data analysis showed that neither sex (GLM: $Z = -0.93$, $p = 0.35$) nor age (and thereby weight) (GLM: $Z = 1.31$, $p = 0.19$) had a significant effect on the personality of the hedgehogs (see Appendix A for the raw data).

3.2. Encounter Tests with a Robotic Lawn Mower

The distribution of the behaviours shown by the 50 individuals tested during the total 100 encounter tests with a robotic lawn mower is illustrated in Figure 4 (see Appendix A for a full overview of the results). The most frequent behavioural response (43%) was the upright position with the snout pointing inwards, which was also the position the hedgehogs were placed in when the tests started. In 15% of the cases, the strategy of the hedgehogs was to run away. Curling up was only observed once in the 100 tests that were conducted.

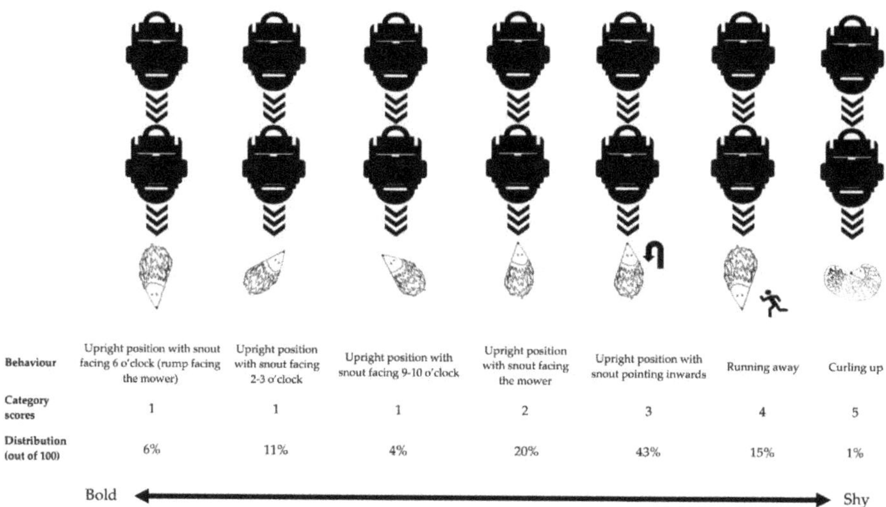

Figure 4. An overview of the results from the encounter tests with a robotic lawn mower. "Behaviour" describes the behaviour exhibited by the hedgehogs. "Category scores" indicate the degree of shyness or boldness of the behaviour, with 1 being the boldest and 5 being the shyest. "Distribution" denotes the frequency of the specific behaviours, in percentage, out of 100 tests conducted with a total of 50 different hedgehogs. The line with arrows indicates the gradient of shyness or boldness of the behaviour exhibited by the hedgehogs tested. At the beginning of each test, the hedgehog was placed in an "upright position with snout pointing inwards".

We found that age had an effect on the behavioural responses of the hedgehogs (CLMM: $Z = -3.71$, $p = 0.0002$), with younger and, thereby, lighter individuals showing a bolder behavioural response to the approaching robotic lawn mower compared to the older and heavier individuals (Figure 5). In addition, the behavioural responses were also affected by the test number (CLMM: $Z = 2.32$, $p = 0.021$), with hedgehogs reacting in a shyer manner during the second of the two encounter tests with the robotic lawn mower regardless of whether the lights were turned on or off during the first test (Figure 6). The other fixed factors (lights on/off, personality, and sex) did not reach a low p-value (CLMM: $p \geq 0.54$ for all).

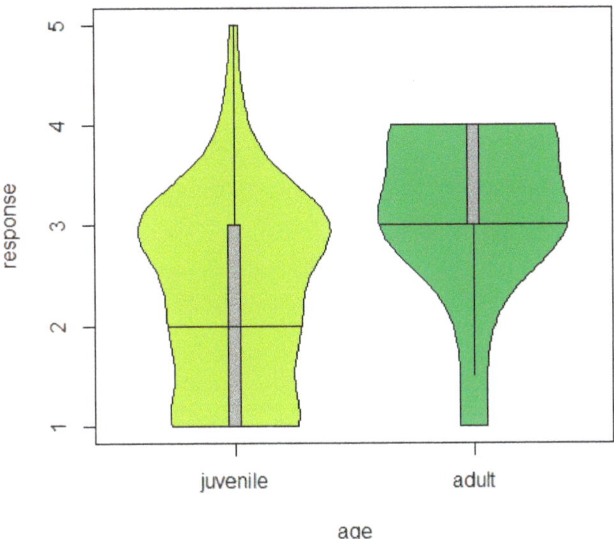

Figure 5. Behavioural responses as a function of the age category of the individuals during the encounter test with a robotic lawn mower (CLMM: $p = 0.0002$). "Response" shows behavioural categories 1–5, with 1 being the most inquisitive (i.e., bold) and 5 being the most timid (i.e., shy) reactions. Violin and box plots: the horizontal line shows the median, with the box extending from the lower to the upper quartiles and the whisker to the data extremes.

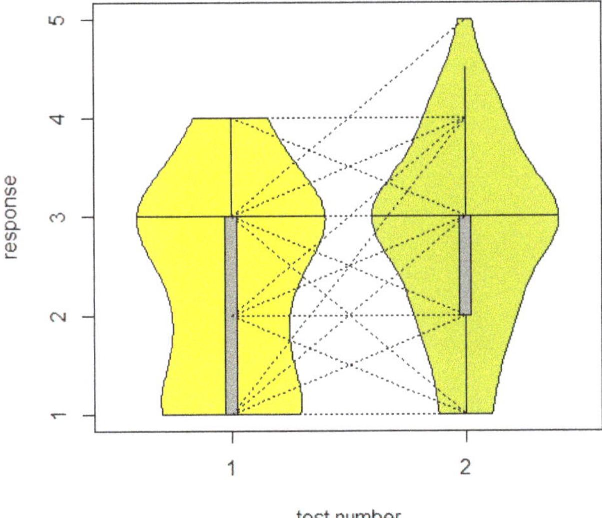

Figure 6. Behavioural responses of the individuals during the first and the second encounter tests with a robotic lawn mower (CLMM: $p = 0.021$). "Response" indicates the behavioural categories 1–5, with 1 being the boldest and 5 being the shyest. Violin and box plots: the horizontal line shows the median, with the box extending from the lower to the upper quartiles and the whiskers to the data extremes. The dashed lines indicate repeated measures (i.e., connecting responses by the same individual between tests 1 and 2).

The random effect "individual" did not significantly affect the behavioural response, indicating an absence of clear differences among subjects (CLMM: $\chi^2 = 0.50$, df = 1, $p = 0.4$; Figure 6).

The behavioural instability of the ordinal behavioural category estimated with the MAD showed a significantly higher variability in the individuals belonging to the bold category compared to the individuals belonging to the shy category (MAD bold > MAD shy; Mann–Whitney U-test; H = 5.26; $p = 0.022$ Levene's test: $p < 0.0029$).

4. Discussion

Our research showed that the 50 hedgehogs tested could be categorised as shy and bold, regardless of sex and age. Personality was not found to influence their reaction towards an approaching robotic lawn mower. However, we observed a higher level of behavioural instability exhibited by the bold individuals. During the encounter tests, several different behavioural reactions were observed in the hedgehogs. There was a tendency for adults and, therefore, more experienced individuals to behave in a more eluding manner when facing an approaching robotic lawn mower. The hedgehogs, in general, acted less boldly during their second encounter test. This could be important from a conservation and welfare perspective if the experience gained from an encounter with a robotic lawn mower could reduce the risk of this individual coming into contact with such a potential risk in the future.

4.1. Distribution of Personality and Behavioural Instability

We found that 54% (N = 27/50) of the individuals could be categorised as shy, as they did not exit their nest box during the whole test, whilst the rest (46%; N = 23/50) could be categorised as bold. Neither sex nor weight (and, thereby, age) had an effect on personality. This is in line with the suggestion that individuals in a given population can be divided into a shy–bold continuum across age classes [48]. The findings of a balanced distribution of shy and bold individuals in the test group are also supported by the current lack of evidence for the pace-of-life syndrome (POLS) in populations of hedgehogs. The POLS predicts that behavioural traits such as high boldness, exploration, aggressiveness, or activity increase the acquisition of resources at the expense of life span, causing individuals expressing these traits to exhibit a faster life history, i.e., a higher growth rate compared to other conspecifics [45,63–65].

However, the higher levels of behavioural instability in the bold individuals could have some relevant implications, as a higher variability in the behavioural repertoire could enhance the probability of survival when an individual encounters a dangerous situation, as proposed by Pertoldi et al. (2016) [55], which suggests that heightened behavioural instability could possess adaptive value in an unpredictable environment. If higher variability in the behavioural repertoire can increase the probability of surviving an encounter with a robotic lawn mower, then bold individuals could be favoured.

4.2. Insight Gained from the Encounter Tests with a Robotic Lawn Mower

After the 100 encounter tests performed with 50 live hedgehogs exposed to an approaching robotic lawn mower, it became evident that the reactions of the hedgehogs could be divided into seven different behavioural categories, resulting in six different test positions (excluding the behaviour of running away). These positions adopted by the hedgehogs in reaction to the robotic lawn mower will be applied in a test design with the purpose of describing the safety and effects of particular models of robotic lawn mower on hedgehogs, using both dead hedgehogs and hedgehog crash test dummies to provide realistic test scenarios for the standardised safety tests.

Interestingly, we found that the tested individuals tended to behave in a more timid manner during the second trial. This could indicate that the hedgehogs adjusted their behaviour towards the robotic lawn mower based on their experience, which might imply that they learned to avoid future encounters with the robotic lawn mower after a single undramatic episode. If this is, indeed, the case in real life, this could potentially prove very beneficial to the survival of hedgehogs, provided that they are not critically harmed in an encounter with a robotic lawn mower. In testing for differences in the behavioural responses among the individuals, in general, no significant effects were found. This suggests that there were no marked differences in the responses among the individuals and/or that each individual might have reacted differently in the two encounter tests. This is in line with the findings that the hedgehogs reacted in a manner that was more timid during the second of the two encounter tests with the robotic lawn mower. Similar results were found, for example, in rainbow trout, whereby boldness was reduced as an adaptive response to negative experiences in previous tests [47].

Older individuals exhibited shyer behaviour in the encounter tests with the robotic lawn mower. We suggest that this could be due to experience and that the independent juveniles we tested were, in general, more naïve because of their young age and lack of experience. The independent juveniles we tested (N = 35/50 individuals in total) had been hand raised in captivity. They, hence, likely had never encountered a robotic lawn mower before. Unfortunately, we have no knowledge of whether the adults (N = 15/50) included in the tests had had any previous confrontations with robotic lawn mowers in the wild before they were admitted into care. This finding could suggest that there should be a focus on the size/age category of independent juveniles in the experimental design of future hedgehog safety tests in case they are more likely to come into physical contact with robotic lawn mowers.

4.3. Potential Biases in the Test Designs

When working with live animals, especially wildlife, it is often challenging to obtain a balanced experimental test design, as the distribution of, for example, the sex and age of the individuals being tested depend on factors such as the season, as well as chance. To prioritise animal welfare and minimise the stress caused to the hedgehogs included in the tests, by eliminating factors such as capture, transport, and handling by strangers, all of the tests took place at a single wildlife rehabilitation centre, where the hedgehogs had been in care and were considered ready for release back into the wild. We conducted the experiments on three occasions, limiting the availability and selection of individuals for the study, ultimately, resulting in a somewhat unbalanced composition of the 50 hedgehogs tested (15 adults and 35 independent juveniles; 30 females and 20 males). This should be taken into consideration when interpreting the results of our tests.

As the novel object was covered in fabric and had permanent electronic parts distributed throughout, the plush toy was nonwashable. It was, therefore, decided that it could not be cleaned between the tests. This might have caused the odours of hedgehogs previously exposed to and exploring the novel object to accumulate, potentially rendering the novel object more stimulating for the last individuals tested, as a few cases of self-anointing were observed in the vicinity of the novel object [42]. The same applied to the cardstock paper covering the floor of the novel arena, as it remained the same throughout the tests. However, no hedgehogs defecated on the floor, and any nesting material or other foreign objects were removed in between the tests. As we decided to exclusively base our categorisation of shyness–boldness on the distribution of the individuals entering the arena or staying in the nest box for the duration of the 15 min test, we assumed that these factors (i.e., the novel object and the floor cover being more olfactory stimulating for the last individuals tested) likely did not influence the outcome of the tests.

During the encounter tests with the robotic lawn mower, it was also not possible to clean the designated spot where each individual was placed, which potentially caused the area to become more olfactorily stimulating for the last individuals tested. This could, in principle, have influenced the behaviour of the hedgehogs tested. However, we did not notice any excessive sniffing activity from the hedgehogs directed at the ground where the individuals were placed. The hedgehogs, when deciding to expose their noses, away from the original position with their heads folded inwards, investigated their surroundings by smelling the air with the noses pointing upwards and not pointed in the direction of the ground.

The wildlife rehabilitation centre and its surrounding garden, where the tests took place, was situated in a residential area near Aarhus in Denmark. During the tests we observed variability in the sensory stimuli coming from the surroundings. Some nights had heavy rainfall, whilst other nights had a range of different anthropogenic noise disturbances. These factors were beyond our control but, nevertheless, constituted a realistic scenario for hedgehogs inhabiting suburban residential areas. To the best of our abilities, we tried to adjust the timing of the tests to reduce the influence of these disturbances on our results.

It has been suggested that confidently determining personality in individuals requires repeated measures to determine whether these personality traits remain consistent during different test scenarios [66]. As we prioritised an experimental design to reduce the duration of the tests and handling of the hedgehogs to reduce stress [45] and allowing for the hedgehogs to be released into the wild as fast as possible, it was not possible to accommodate more repeated measures in our test design. It would be relevant to consider a method to increase the repeated measures of personality extending beyond the combined novel object and novel arena test and the encounter tests with a robotic lawn mower.

Previous traumatic experiences may have influenced the reactions of the individual hedgehogs during the tests. However, the lack of knowledge about the life history of the hedgehogs prior to admission to the wildlife rehabilitation centre prevents any meaningful analysis of the influence of different traumas on the outcomes of the personality and encounter tests.

4.4. Next Steps

Knowing that certain models of robotic lawn mowers may cause injuries or even kill hedgehogs, understanding the extent of the problem is critical for hedgehog conservation and welfare. Therefore, we encourage the establishment of an open access international hedgehog database that can function as a daily record-keeping system for hedgehog carers gathering and storing vital information on the hedgehogs coming into care. Additionally, collecting photographic evidence of hedgehogs injured or killed by electronic garden tools such as trimmers and robotic lawn mowers, in order to quantify, document, and describe the types of damage caused by these machines is important (Berger et al. in prep.).

On the basis of the present behavioural study, through the knowledge gained on how live hedgehogs position themselves when confronted by an approaching robotic lawn mower, it is now possible to prepare a realistic framework for a standardised safety test to measure the impacts of robotic lawn mowers on hedgehogs. Furthermore, complete standardisation requires the use of hedgehog crash test dummies, which can be 3D printed and applied by manufacturers of robotic lawn mowers in the process of designing and testing prototypes of new and more hedgehog-friendly machines. Ultimately, hedgehog crash test dummies should be used as proxies for dead hedgehogs in the standardised safety test and be designed to mimic reality and yield the same results compared to tests on dead hedgehogs.

The final step is to have the standardised hedgehog safety test implemented in the CENELEC protocol [41], testing and approving robotic lawn mowers for sale on the European market. This test would, furthermore, allow for a labelling system for hedgehog-safe robotic lawn mowers to be established, guiding the consumers to make the hedgehog friendly choice when purchasing these tools.

Work is currently well underway to achieve these described goals. But, for now, our advice remains to restrict the running of robotic lawn mowers to daylight hours and to check lawns for any wildlife species that may be vulnerable to an encounter with a robotic lawn mower, before turning on the machine.

5. Conclusions

The robotic lawn mower market is growing rapidly, and, consequently, it is essential to help inform manufacturers on how to design more hedgehog-friendly machines in the future if we wish to eliminate this potentially negative influence on hedgehog survival in our backyards. This study sought to explore the personality and reactions of live hedgehogs when facing a disarmed, approaching robotic lawn mower to inform future standardised hedgehog safety tests allowing manufacturers to evaluate the performance of the new models being designed and developed at their facilities.

In testing the reactions of hedgehogs towards an approaching (disarmed) robotic lawn mower, we conclude that that personality did not appear to affect the outcome; that adult and, thereby, more experienced hedgehogs tended to react in a more timid manner; and that the hedgehogs generally acted less boldly during their second encounter with the robotic lawn mower.

The important insights gained from this study will be applied in the process of testing and refining the design of a hedgehog crash test dummy to be used in a future, standardised hedgehog safety test, which will be informed by the present results. This test will eventually serve to produce and approve hedgehog-friendly robotic lawn mowers and, ultimately, improve hedgehog conservation.

Author Contributions: Conceptualisation, S.L.R., C.P., D.W.M. and A.K.O.A.; methodology, S.L.R., C.P., E.F.B. and A.K.O.A.; validation, S.L.R., A.B., C.P., D.W.M. and A.K.O.A.; formal analysis, B.T.S., S.L.R. and E.F.B.; investigation, S.L.R., B.T.S., A.B., C.P., D.W.M. and A.K.O.A.; resources, S.L.R.; data curation, B.T.S., S.L.R. and E.F.B.; writing—original draft preparation, S.L.R.; writing—review and editing, S.L.R., D.W.M., B.T.S., A.B., C.P., E.F.B. and A.K.O.A.; visualisation, B.T.S., S.L.R. and E.F.B.; supervision, C.P., E.F.B., D.W.M. and A.K.O.A.; project administration, S.L.R.; funding acquisition, S.L.R. and D.W.M. All authors have read and agreed to the published version of the manuscript.

Funding: This research was funded by Husqvarna; STIHL; British Hedgehog Preservation Society; Carlsberg Foundation, grant number: CF20_0443.

Institutional Review Board Statement: The animal study protocol was approved by the Animal Experiments Inspectorate under the Danish National Committee for the Protection of Animals used for Scientific Purposes (2021-15-0201-00865, dates of approval: 20th of September 2021 and 29th of June 2022).

Informed Consent Statement: Not applicable.

Data Availability Statement: All relevant data from this research have been made available in the publication.

Acknowledgments: The authors are grateful to hedgehog rehabilitator Dorthe Madsen from Animal Protection Denmark (Dyrenes Beskyttelse) for hosting the tests and caring for the hedgehogs that participated in the research project.

Conflicts of Interest: The authors declare no conflict of interest. The funders had no role in the design of the study; in the analyses, or interpretation of data; in the writing of the manuscript; or in the decision to publish the results.

Appendix A

Table A1. Results from the encounter tests with a disarmed robotic lawn mower. "Personality" refers to the categories shy = 0 and bold = 1. Test number refers to the order of the two tests on each individual. Light describes whether the headlights of the robotic lawn mower were turned on or off during the specific test. Response denotes the behavioural response category of the individual, as defined in Section 2, "Materials and Methods", with category 1 being the boldest and category 5 being the shyest.

Individual	Personality	Age	Sex	Weight (Gram)	Date	Time	Test Number	Lights	Response
Hedgehog 1	1	Adult	Female	1154	03.06.2022	23:17	Two	Off	3
Hedgehog 1	1	Adult	Female	1154	03.06.2022	23:15	One	On	4

Table A1. *Cont.*

Individual	Personality	Age	Sex	Weight (Gram)	Date	Time	Test Number	Lights	Response
Hedgehog 2	1	Adult	Female	1071	04.06.2022	00:00	One	Off	3
Hedgehog 2	1	Adult	Female	1071	04.06.2022	00:02	Two	On	3
Hedgehog 3	0	Adult	Male	774	04.06.2022	01:07	One	On	4
Hedgehog 3	0	Adult	Male	774	04.06.2022	01:10	Two	Off	4
Hedgehog 4	0	Juvenile	Female	497	04.06.2022	01:33	One	Off	3
Hedgehog 4	0	Juvenile	Female	497	04.06.2022	01:35	Two	On	3
Hedgehog 5	0	Adult	Female	996	04.06.2022	02:09	One	On	4
Hedgehog 5	0	Adult	Female	996	04.06.2022	02:13	Two	Off	4
Hedgehog 6	0	Adult	Female	924	04.06.2022	23:05	One	Off	4
Hedgehog 6	0	Adult	Female	924	04.06.2022	23:07	Two	On	4
Hedgehog 7	1	Adult	Female	1137	04.06.2022	23:40	One	On	3
Hedgehog 7	1	Adult	Female	1137	04.06.2022	23:42	Two	Off	4
Hedgehog 10	0	Adult	Female	772	05.06.2022	01:32	Two	On	1
Hedgehog 10	0	Adult	Female	772	05.06.2022	01:30	One	Off	3
Hedgehog 8	0	Adult	Female	645	05.06.2022	00:17	One	Off	3
Hedgehog 8	0	Adult	Female	645	05.06.2022	00:19	Two	On	3
Hedgehog 9	1	Adult	Female	699	05.06.2022	00:55	One	On	3
Hedgehog 9	1	Adult	Female	699	05.06.2022	00:56	Two	Off	3
Hedgehog 11	0	Adult	Male	1225	06.09.2022	21:03	Two	Off	3
Hedgehog 11	0	Adult	Male	1225	06.09.2022	21:01	One	On	4
Hedgehog 12	0	Adult	Male	1361	06.09.2022	21:37	One	Off	3
Hedgehog 12	0	Adult	Male	1361	06.09.2022	21:38	Two	On	3
Hedgehog 13	1	Juvenile	Female	361	06.09.2022	22:02	One	On	1
Hedgehog 13	1	Juvenile	Female	361	06.09.2022	22:06	Two	Off	3
Hedgehog 14	1	Juvenile	Male	381	06.09.2022	22:57	Two	On	2
Hedgehog 14	1	Juvenile	Male	381	06.09.2022	22:56	One	Off	3
Hedgehog 15	1	Juvenile	Male	328	06.09.2022	23:23	One	On	1
Hedgehog 15	1	Juvenile	Male	328	06.09.2022	23:27	Two	Off	2
Hedgehog 16	1	Juvenile	Female	361	06.09.2022	23:57	One	Off	1
Hedgehog 16	1	Juvenile	Female	361	06.09.2022	23:58	Two	On	1
Hedgehog 17	1	Juvenile	Female	328	07.09.2022	00:27	One	On	3
Hedgehog 17	1	Juvenile	Female	328	07.09.2022	00:30	Two	Off	3
Hedgehog 18	1	Juvenile	Male	216	07.09.2022	00:58	One	Off	3
Hedgehog 18	1	Juvenile	Male	216	07.09.2022	01:02	Two	On	3
Hedgehog 19	1	Juvenile	Female	321	07.09.2022	01:35	One	On	2
Hedgehog 19	1	Juvenile	Female	321	07.09.2022	01:35	Two	Off	2
Hedgehog 20	1	Juvenile	Female	264	07.09.2022	02:05	One	Off	1
Hedgehog 20	1	Juvenile	Female	264	07.09.2022	02:06	Two	On	4
Hedgehog 21	0	Juvenile	Female	318	07.09.2022	02:34	One	On	2
Hedgehog 21	0	Juvenile	Female	318	07.09.2022	02:36	Two	Off	3
Hedgehog 22	1	Juvenile	Female	343	07.09.2022	21:07	One	On	2
Hedgehog 22	1	Juvenile	Female	343	07.09.2022	21:10	Two	Off	2
Hedgehog 23	0	Juvenile	Male	383	07.09.2022	21:39	One	Off	1
Hedgehog 23	0	Juvenile	Male	383	07.09.2022	21:41	Two	On	3

Table A1. *Cont.*

Individual	Personality	Age	Sex	Weight (Gram)	Date	Time	Test Number	Lights	Response
Hedgehog 24	0	Juvenile	Male	398	07.09.2022	22:10	One	On	1
Hedgehog 24	0	Juvenile	Male	398	07.09.2022	22:11	Two	Off	1
Hedgehog 25	1	Juvenile	Female	308	07.09.2022	22:45	One	Off	1
Hedgehog 25	1	Juvenile	Female	308	07.09.2022	22:45	Two	On	4
Hedgehog 26	1	Juvenile	Male	387	07.09.2022	23:11	One	On	3
Hedgehog 26	1	Juvenile	Male	387	07.09.2022	23:13	Two	Off	3
Hedgehog 27	1	Juvenile	Female	383	07.09.2022	23:43	One	Off	1
Hedgehog 27	1	Juvenile	Female	383	07.09.2022	23:43	Two	On	4
Hedgehog 28	1	Juvenile	Female	231	08.09.2022	00:11	One	On	1
Hedgehog 28	1	Juvenile	Female	231	08.09.2022	00:13	Two	Off	2
Hedgehog 29	1	Juvenile	Male	221	08.09.2022	00:41	One	Off	1
Hedgehog 29	1	Juvenile	Male	221	08.09.2022	00:42	Two	On	1
Hedgehog 30	1	Juvenile	Male	196	08.09.2022	01:11	One	On	3
Hedgehog 30	1	Juvenile	Male	196	08.09.2022	01:12	Two	Off	3
Hedgehog 31	0	Adult	Male	1301	15.09.2022	20:34	One	On	2
Hedgehog 31	0	Adult	Male	1301	15.09.2022	20:35	Two	Off	4
Hedgehog 32	0	Juvenile	Female	230	15.09.2022	21:06	Two	On	1
Hedgehog 32	0	Juvenile	Female	230	15.09.2022	21:05	One	Off	2
Hedgehog 33	0	Adult	Female	718	15.09.2022	21:34	One	On	1
Hedgehog 33	0	Adult	Female	718	15.09.2022	21:36	Two	Off	2
Hedgehog 34	1	Juvenile	Female	322	15.09.2022	22:05	One	Off	2
Hedgehog 34	1	Juvenile	Female	322	15.09.2022	22:07	Two	On	2
Hedgehog 35	1	Juvenile	Male	300	15.09.2022	22:35	One	On	1
Hedgehog 35	1	Juvenile	Male	300	15.09.2022	22:36	Two	Off	2
Hedgehog 36	1	Juvenile	Male	316	15.09.2022	23:05	One	Off	1
Hedgehog 36	1	Juvenile	Male	316	15.09.2022	23:07	Two	On	2
Hedgehog 37	0	Juvenile	Female	298	15.09.2022	23:37	Two	Off	1
Hedgehog 37	0	Juvenile	Female	298	15.09.2022	23:35	One	On	2
Hedgehog 38	0	Adult	Female	668	16.09.2022	00:05	One	Off	3
Hedgehog 38	0	Adult	Female	668	16.09.2022	00:07	Two	On	3
Hedgehog 39	1	Adult	Female	1529	16.09.2022	00:35	One	On	4
Hedgehog 39	1	Adult	Female	1529	16.09.2022	00:36	Two	Off	4
Hedgehog 40	0	Juvenile	Female	318	16.09.2022	01:06	One	Off	2
Hedgehog 40	0	Juvenile	Female	318	16.09.2022	01:07	Two	On	3
Hedgehog 41	0	Juvenile	Male	236	16.09.2022	01:35	One	On	2
Hedgehog 41	0	Juvenile	Male	236	16.09.2022	01:36	Two	Off	2
Hedgehog 42	0	Juvenile	Male	213	16.09.2022	20:36	One	Off	3
Hedgehog 42	0	Juvenile	Male	213	16.09.2022	20:37	Two	On	3
Hedgehog 43	0	Juvenile	Male	478	16.09.2022	21:01	One	Off	1
Hedgehog 43	0	Juvenile	Male	478	16.09.2022	21:02	Two	On	2
Hedgehog 44	0	Juvenile	Male	226	16.09.2022	21:32	One	Off	3
Hedgehog 44	0	Juvenile	Male	226	16.09.2022	21:33	Two	On	5
Hedgehog 45	0	Juvenile	Female	326	16.09.2022	22:02	One	On	3

Table A1. *Cont.*

Individual	Personality	Age	Sex	Weight (Gram)	Date	Time	Test Number	Lights	Response
Hedgehog 45	0	Juvenile	Female	326	16.09.2022	22:04	Two	Off	3
Hedgehog 46	0	Juvenile	Male	413	16.09.2022	22:32	One	Off	3
Hedgehog 46	0	Juvenile	Male	413	16.09.2022	22:33	Two	On	3
Hedgehog 47	0	Juvenile	Female	302	16.09.2022	23:02	One	On	3
Hedgehog 47	0	Juvenile	Female	302	16.09.2022	23:03	Two	Off	3
Hedgehog 48	0	Juvenile	Female	295	16.09.2022	23:33	One	Off	3
Hedgehog 48	0	Juvenile	Female	295	16.09.2022	23:35	Two	On	3
Hedgehog 49	0	Juvenile	Female	300	17.09.2022	00:03	One	On	1
Hedgehog 49	0	Juvenile	Female	300	17.09.2022	00:04	Two	Off	3
Hedgehog 50	0	Juvenile	Male	168	17.09.2022	00:33	One	Off	3
Hedgehog 50	0	Juvenile	Male	168	17.09.2022	00:34	Two	On	3

References

1. Reeve, N. *Hedgehogs*; Poyser: London, UK, 1994.
2. Morris, P. *Hedgehogs*; Whittet Books Ltd.: Stansted, UK, 2014.
3. Rasmussen, S.L.; Berg, T.B.; Dabelsteen, T.; Jones, O.R. The ecology of suburban juvenile European hedgehogs (*Erinaceus europaeus*) in Denmark. *Ecol. Evol.* **2019**, *9*, 13174–13187. [CrossRef] [PubMed]
4. Müller, F. Langzeit-Monitoring der Strassenverkehrsopfer beim Igel (*Erinaceus europaeus* L.) zur Indikation von Populations dichteveränderungen entlang zweier Teststrecken im Landkreis Fulda. *Beiträge Zur Naturkunde Osthess.* **2018**, *54*, 21–26.
5. British Trust for Ornithology (BTO). *The state of Britain's Hedgehogs 2011*; Commissioned by British Hedgehog Preservation Society and People's Trust for Endangered Species; British Trust for Ornithology (BTO): London, UK, 2011.
6. British Hedgehog Preservation Society; People's Trust for Endangered Species. *The State of Britain's Hedgehogs 2015*; British Hedgehog Preservation Society and People's Trust for Endangered Species: London, UK, 2015.
7. British Hedgehog Preservation Society; People's Trust for Endangered Species. *The State of Britain's Hedgehogs 2018*; British Hedgehog Preservation Society and People's Trust for Endangered Species: London, UK, 2018.
8. Hof, A.R.; Bright, P.W. Quantifying the long-term decline of the West European hedgehog in England by subsampling citizen science datasets. *Eur. J. Wildl. Res.* **2016**, *62*, 407–413. [CrossRef]
9. Krange, M. Change in the Occurrence of the West European Hedgehog (*Erinaceus europaeus*) in Western Sweden during 1950–2010. Master's Thesis, Karlstad University, Karlstad, Sweden, 2015.
10. van de Poel, J.L.; Dekker, J.; van Langevelde, F. Dutch hedgehogs *Erinaceus europaeus* are nowadays mainly found in urban areas possibly due to the negative effects of badgers Meles meles. *Wildl. Biol.* **2015**, *21*, 51–55. [CrossRef]
11. Williams, B.M.; Baker, P.J.; Thomas, E.; Wilson, G.; Judge, J.; Yarnell, R.W. Reduced occupancy of hedgehogs (*Erinaceus europaeus*) in rural England and Wales: The influence of habitat and an asymmetric intra-guild predator. *Sci. Rep.* **2018**, *8*, 12156. [CrossRef] [PubMed]
12. Taucher, A.L.; Gloor, S.; Dietrich, A.; Geiger, M.; Hegglin, D.; Bontadina, F. Decline in Distribution and Abundance: Urban Hedgehogs under Pressure. *Animals* **2020**, *10*, 1606. [CrossRef] [PubMed]
13. Wembridge, D.; Johnson, G.; Al-Fulaij, N.; Langton, S. *The State of Britain's Hedgehogs 2022*; British Hedgehog Preservation Society and People's Trust for Endangered Species: London, UK, 2022.
14. Mathews, F.; Harrower, C. *IUCN—Compliant Red List for Britain's Terrestrial Mammals. Assessment by the Mammal Society under contract to Natural England, Natural Resources Wales and Scottish Natural Heritage*; Natural England: Peterborough, UK, 2020.
15. Brakes, C.R.; Smith, R.H. Exposure of non-target small mammals to rodenticides: Short-term effects, recovery and implications for secondary poisoning. *J. Appl. Ecol.* **2005**, *42*, 118–128. [CrossRef]
16. Haigh, A.; O'Riordan, R.M.; Butler, F. Nesting behaviour and seasonal body mass changes in a rural Irish population of the Western hedgehog (*Erinaceus europaeus*). *Acta Theriol.* **2012**, *57*, 321–331. [CrossRef]
17. Hof, A.R.; Bright, P.W. The value of agri-environment schemes for macro-invertebrate feeders: Hedgehogs on arable farms in Britain. *Anim. Conserv.* **2010**, *13*, 467–473. [CrossRef]
18. Huijser, M.P.; Bergers, P.J.M. The effect of roads and traffic on hedgehog (*Erinaceus europaeus*) populations. *Biol. Conserv.* **2000**, *95*, 111–116. [CrossRef]
19. Young, R.P.; Davison, J.; Trewby, I.D.; Wilson, G.J.; Delahay, R.J.; Doncaster, C.P. Abundance of hedgehogs (*Erinaceus europaeus*) in relation to the density and distribution of badgers (*Meles meles*). *J. Zool.* **2006**, *269*, 349–356. [CrossRef]
20. Hubert, P.; Julliard, R.; Biagianti, S.; Poulle, M.-L. Ecological factors driving the higher hedgehog (*Erinaceus europeaus*) density in an urban area compared to the adjacent rural area. *Landsc. Urban Plan.* **2011**, *103*, 34–43. [CrossRef]

Dowding, C.V.; Harris, S.; Poulton, S.; Baker, P.J. Nocturnal ranging behaviour of urban hedgehogs, *Erinaceus europaeus*, in relation to risk and reward. *Anim. Behav.* **2010**, *80*, 13–21. [CrossRef]

Dowding, C.V.; Shore, R.F.; Worgan, A.; Baker, P.J.; Harris, S. Accumulation of anticoagulant rodenticides in a non-target insectivore, the European hedgehog (*Erinaceus europaeus*). *Environ. Pollut.* **2010**, *158*, 161–166. [CrossRef] [PubMed]

Rasmussen, S.L.; Nielsen, J.L.; Jones, O.R.; Berg, T.B.; Pertoldi, C. Genetic structure of the European hedgehog (*Erinaceus europaeus*) in Denmark. *PLoS ONE* **2020**, *15*, e0227205. [CrossRef] [PubMed]

Rasmussen, S.L.; Yashiro, E.; Sverrisdóttir, E.; Nielsen, K.L.; Lukassen, M.B.; Nielsen, J.L.; Asp, T.; Pertoldi, C. Applying the GBS Technique for the Genomic Characterization of a Danish Population of European Hedgehogs (*Erinaceus europaeus*). *Genet. Biodivers. (GABJ)* **2019**, *3*, 78–86. [CrossRef]

Rasmussen, S.L.; Vorkamp, K.; Zhu, L.; Roslev, P.; Pertoldi, C.; Nielsen, J.L. Pesticides and Biocides in the Danish Population of European Hedgehogs (*Erinaceus europaeus*). In Proceedings of the Abstract Book SETAC Europe 33rd Annual Meeting, Dublin, Ireland, 30 April–4 May 2023; p. 213.

Larsen, J.; Raisen, C.L.; Ba, X.; Sadgrove, N.J.; Padilla-González, G.F.; Simmonds, M.S.J.; Loncaric, I.; Kerschner, H.; Apfalter, P.; Hartl, R.; et al. Emergence of methicillin resistance predates the clinical use of antibiotics. *Nature* **2022**, *602*, 135–141. [CrossRef]

Rasmussen, S.L.; Larsen, J.; van Wijk, R.E.; Jones, O.R.; Berg, T.B.; Angen, O.; Larsen, A.R. European hedgehogs (*Erinaceus europaeus*) as a natural reservoir of methicillin-resistant *Staphylococcus aureus* carrying *mecC* in Denmark. *PLoS ONE* **2019**, *14*, e0222031. [CrossRef]

Rasmussen, S.L.; Hallig, J.; van Wijk, R.E.; Petersen, H.H. An investigation of endoparasites and the determinants of parasite infection in European hedgehogs (*Erinaceus europaeus*) from Denmark. *Int. J. Parasitol. Parasites Wildl.* **2021**, *16*, 217–227. [CrossRef]

Rasmussen, S.L.; Schrøder, A.E.; Mathiesen, R.; Nielsen, J.L.; Pertoldi, C.; Macdonald, D.W. Wildlife Conservation at a Garden Level: The Effect of Robotic Lawn Mowers on European Hedgehogs (*Erinaceus europaeus*). *Animals* **2021**, *11*, 1191. [CrossRef]

Moore, L.J.; Petrovan, S.O.; Baker, P.J.; Bates, A.J.; Hicks, H.L.; Perkins, S.E.; Yarnell, R.W. Impacts and potential mitigation of road mortality for hedgehogs in Europe. *Animals* **2020**, *10*, 1523. [CrossRef]

Pettett, C.E.; Moorhouse, T.P.; Johnson, P.J.; Macdonald, D.W. Factors affecting hedgehog (*Erinaceus europaeus*) attraction to rural villages in arable landscapes. *Eur. J. Wildl. Res.* **2017**, *63*, 12. [CrossRef]

Rasmussen, S.L.; Berg, T.B.; Martens, H.J.; Jones, O.R. Anyone Can Get Old—All You Have to Do Is Live Long Enough: Understanding Mortality and Life Expectancy in European Hedgehogs (*Erinaceus europaeus*). *Animals* **2023**, *13*, 626. [CrossRef] [PubMed]

Future Market Insights. Robotic Lawn Mower Market Outlook (2022–2032). *Future Market Insights.* 2022, p. 342. Available online: https://www.futuremarketinsights.com/reports/robotic-lawn-mower-market (accessed on 1 October 2023).

Future Business Insights. Robotic Lawn Mower Market Size, Share & COVID-19 Impact Analysis. 2022, p. 150. Available online: https://www.fortunebusinessinsights.com/robotic-lawn-mower-market-106531 (accessed on 1 October 2023).

Doncaster, C.P.; Rondinini, C.; Johnson, P.C.D. Field test for environmental correlates of dispersal in hedgehogs *Erinaceus europaeus*. *J. Anim. Ecol.* **2001**, *70*, 33–46.

Macdonald, D.W. Mitigating human impacts on wild animal welfare. *Animals* **2023**, *13*, 2906. [CrossRef] [PubMed]

Martínez, J.C.; Rosique, A.I.; Royo, M.S. Causes of admission and final dispositions of hedgehogs admitted to three Wildlife Rehabilitation Centers in eastern Spain. *Hystrix* **2014**, *25*, 107.

Lukešová, G.; Voslarova, E.; Vecerek, V.; Vucinic, M. Trends in intake and outcomes for European hedgehog (*Erinaceus europaeus*) in the Czech rescue centers. *PLoS ONE* **2021**, *16*, e0248422. [CrossRef] [PubMed]

Garcês, A.; Soeiro, V.; Lóio, S.; Sargo, R.; Sousa, L.; Silva, F.; Pires, I. Outcomes, mortality causes, and pathological findings in European hedgehogs (*Erinaceus europeus*, Linnaeus 1758): A seventeen year retrospective analysis in the north of Portugal. *Animals* **2020**, *10*, 1305. [CrossRef]

Burroughes, N.D.; Dowler, J.; Burroughes, G. Admission and Survival Trends in Hedgehogs Admitted to RSPCA Wildlife Rehabilitation Centres. *Proc. Zool. Soc.* **2021**, *74*, 198–204. [CrossRef]

EN 50636-2-107; Safety of Household and Similar Appliances-Part 2-107: Particular Requirements for Robotic Battery Powered Electrical Lawnmowers. CENELEC European Committee for Electrotechnical Standardization: Brussels, Belgium, 2015.

Morris, P. *Hedgehog*; Collins: London, UK, 2018; pp. 1–404.

Reale, D.; Martin, J.; Coltman, D.W.; Poissant, J.; Festa-Bianchet, M. Male personality, life-history strategies and reproductive success in a promiscuous mammal. *J. Evol. Biol.* **2009**, *22*, 1599–1607. [CrossRef]

Bremner-Harrison, S.; Prodohl, P.A.; Elwood, R.W. Behavioural trait assessment as a release criterion: Boldness predicts early death in a reintroduction programme of captive-bred swift fox (*Vulpes velox*). *Anim. Conserv.* **2004**, *7*, 313–320. [CrossRef]

Rasmussen, S.L.; Kalliokoski, O.; Dabelsteen, T.; Abelson, K. An exploratory investigation of glucocorticoids, personality and survival rates in wild and rehabilitated hedgehogs (*Erinaceus europaeus*) in Denmark. *BMC Ecol. Evol.* **2021**, *21*, 1–16. [CrossRef] [PubMed]

Verbeek, M.E.M.; Drent, P.J.; Wiepkema, P.R. Consistent individual differences in early exploratory behavior of male great tits. *Anim. Behav.* **1994**, *48*, 1113–1121. [CrossRef]

Frost, A.J.; Winrow-Giffen, A.; Ashley, P.J.; Sneddon, L.U. Plasticity in animal personality traits: Does prior experience alter the degree of boldness? *Proc. R. Soc. B-Biol. Sci.* **2007**, *274*, 333–339. [CrossRef] [PubMed]

48. Wilson, D.S.; Coleman, K.; Clark, A.B.; Biederman, L. Shy bold continuum in pumpkinseed sunfish *Lepomis gibbosus—A* ecological study of a psychological trait. *J. Comp. Psychol.* **1993**, *107*, 250–260. [CrossRef]
49. Smith, B.R.; Blumstein, D.T. Fitness consequences of personality: A meta-analysis. *Behav. Ecol.* **2008**, *19*, 448–455. [CrossRef]
50. Boon, A.K.; Reale, D.; Boutin, S. Personality, habitat use, and their consequences for survival in North American red squirrel Tamiasciurus hudsonicus. *Oikos* **2008**, *117*, 1321–1328. [CrossRef]
51. Stuart-Smith, A.K.; Boutin, S. Behavioural differences between surviving and depredated juvenile red squirrels. *Ecoscience* **1995**, 34–40. [CrossRef]
52. Reale, D.; Festa-Bianchet, M. Predator-induced natural selection on temperament in bighorn ewes. *Anim. Behav.* **2003**, *65*, 463–47 [CrossRef]
53. Dingemanse, N.J.; Both, C.; Drent, P.J.; Tinbergen, J.M. Fitness consequences of avian personalities in a fluctuating environmen *Proc. R. Soc. B-Biol. Sci.* **2004**, *271*, 847–852. [CrossRef]
54. Dingemanse, N.J.; Réale, D. Natural selection and animal personality. *Behaviour* **2005**, *142*, 1159–1184. [CrossRef]
55. Pertoldi, C.; Bahrndorff, S.; Kurbalija Novicic, Z.; Duun Rohde, P. The novel concept of "behavioural instability" and its potenti applications. *Symmetry* **2016**, *8*, 135. [CrossRef]
56. Linder, A.C.; Lyhne, H.; Laubek, B.; Bruhn, D.; Pertoldi, C. Modeling Species-Specific Collision Risk of Birds with Wind Turbine A Behavioral Approach. *Symmetry* **2022**, *14*, 2493. [CrossRef]
57. Linder, A.C.; Lyhne, H.; Laubek, B.; Bruhn, D.; Pertoldi, C. Quantifying raptors' flight behavior to assess collision risk an avoidance behavior to wind turbines. *Symmetry* **2021**, *4*, 2245. [CrossRef]
58. Takola, E.; Krause, E.T.; Müller, C.; Schielzeth, H. Novelty at second glance: A critical appraisal of the novel object paradig based on meta-analysis. *Anim. Behav.* **2021**, *180*, 123–142. [CrossRef]
59. Carter, A.J.; Feeney, W.E.; Marshall, H.H.; Cowlishaw, G.; Heinsohn, R. Animal personality: What are behavioural ecologis measuring? *Biol. Rev.* **2013**, *88*, 465–475. [CrossRef]
60. Walsh, R.N.; Cummins, R.A. The open-field test: A critical review. *Psychol. Bull.* **1976**, *83*, 482. [CrossRef]
61. R Core Team. *R: A Language and Environment for Statistical Computing*; R Core Team: Vienna, Austria, 2022.
62. European Parliament. Directive 2010/63/EU of the European Parliament and of the Council of 22 September 2010. Availab online: https://eur-lex.europa.eu/LexUriServ/LexUriServ.do?uri=OJ:L:2010:276:0033:0079:en:PDF (accessed on 1 October 202
63. Wolf, M.; Van Doorn, G.S.; Leimar, O.; Weissing, F.J. Life-history trade-offs favour the evolution of animal personalities. *Natu* **2007**, *447*, 581–584. [CrossRef]
64. Réale, D.; Garant, D.; Humphries, M.M.; Bergeron, P.; Careau, V.; Montiglio, P.-O. Personality and the emergence of the pace-of-li syndrome concept at the population level. *Philos. Trans. R. Soc. B Biol. Sci.* **2010**, *365*, 4051–4063. [CrossRef]
65. Stamps, J.A. Growth-mortality tradeoffs and 'personality traits' in animals. *Ecol. Lett.* **2007**, *10*, 355–363. [CrossRef]
66. Roche, D.G.; Careau, V.; Binning, S.A. Demystifying animal 'personality' (or not): Why individual variation matters to experime tal biologists. *J. Exp. Biol.* **2016**, *219*, 3832–3843. [CrossRef] [PubMed]

 *animals*

Article

Effects of Artificial Light at Night (ALAN) on European Hedgehog Activity at Supplementary Feeding Stations

Domhnall Finch [1], Bethany R. Smith [2], Charlotte Marshall [2], Frazer G. Coomber [1,2], Laura M. Kubasiewicz [2], Max Anderson [1], Patrick G. R. Wright [1,2] and Fiona Mathews [1,2,*]

[1]   School of Life Sciences, University of Sussex, Falmer BN1 9QG, UK; d.finch@sussex.ac.uk (D.F.); science@themammalsociety.org (F.G.C.); Max.Anderson@sussex.ac.uk (M.A.); P.G.R.Wright@sussex.ac.uk (P.G.R.W.)
[2]   Mammal Society, London E9 6EJ, UK; bethanys935@gmail.com (B.R.S.); charlotte.marshall401@gmail.com (C.M.); laurakuba@googlemail.com (L.M.K.)
*   Correspondence: F.Mathews@sussex.ac.uk

Received: 23 March 2020; Accepted: 14 April 2020; Published: 28 April 2020

Simple Summary: Owing to the rapid expansion of urbanisation, light pollution has increased dramatically in the natural environment causing significant negative effects on species fitness, abundance, foraging and roosting behaviours. However, very little research has examined the impacts of artificial light at night (ALAN) on mammal species other than bats. Using a large-scale citizen science project, we examined the potential impact of ALAN on European hedgehogs (*Erinaceus europaeus*) at supplementary feeding stations. Our results show that there were no significant effects of ALAN on the presence, feeding activity or activity patterns of hedgehogs throughout the experiment, although some variations in individual hedgehogs were observed. This suggests that while there was no significant impact of ALAN found at supplementary feeding stations, there could be other costs associated with lighting, e.g., reproductive success, territory maintenance and natural prey availability, which need to be considered.

Abstract: Artificial light at night (ALAN) can have negative consequences for a wide range of taxa. However, the effects on nocturnal mammals other than bats are poorly understood. A citizen science camera trapping experiment was therefore used to assess the effect of ALAN on the activity of European hedgehogs (*Erinaceus europaeus*) at supplementary feeding stations in UK gardens. A crossover design was implemented at 33 gardens with two treatments—artificial light and darkness—each of which lasted for one week. The order of treatment depended on the existing lighting regime at the feeding station: dark treatments were applied first at dark feeding stations, whereas light treatments were used first where the station was already illuminated. Although temporal changes in activity patterns in response to the treatments were noted in some individuals, the direction of the effects was not consistent. Similarly, there was no overall impact of ALAN on the presence or feeding activities of hedgehogs in gardens where supplementary feeding stations were present. These findings are somewhat reassuring insofar as they demonstrate no net negative effect on a species thought to be in decline, in scenarios where the animals are already habituated to supplementary feeding. However, further research is needed to examine long-term effects and the effects of lighting on hedgehog prey, reproductive success and predation risk.

Keywords: activity pattern; ALAN; camera trap; citizen science; fragmentation; hedgehogs; *Erinaceus europaeus*; light pollution; lightscape; urbanisation

---

1. Introduction

Urbanisation is increasing rapidly across the globe [1], with important impacts for biodiversity and ecosystem functioning [2,3]. One of the major environmental changes that accompanies urbanisation is an increase in the amount and intensity of artificial light at night (ALAN), and this has become a significant conservation concern [4]. Reported impacts of ALAN include the disruption of predator–prey interactions [5], seed dispersal [6], foraging [7], and migratory behaviour [8].

Lunar cycles have been shown to elicit a variety of behavioural and physiological responses in nocturnal mammals [9–11]. Given that the illuminance from streetlights is usually more than an order of magnitude greater than that from a full moon, it is likely that nocturnal mammals are susceptible to the effects of ALAN. Most research to date has focused on bats, partly because of their high legislative protection in some parts of the world (e.g., Europe) [12,13]. However, there has been limited research on other nocturnal mammals.

The European hedgehog (*Erinaceus europaeus*) is a relatively small nocturnal insectivore that has a wide distribution across Europe. However, populations are thought to be in decline. In Britain, for example, populations have reduced from approximately 1.5 million individuals in 1995 to 522,000 in 2016 [14], and concerns have been raised in other countries [15–18]. The mechanisms underlying these declines remain unclear, though a wide variety of pressures—including predation by and competition with badgers [19–21], road collisions [22,23], and agricultural intensification [24]—have been suggested.

Hedgehogs make extensive use of parks and gardens, particularly in peri-urban and suburban areas, where they can obtain shelter and food (including from supplementary feeding by householders, which is common in Britain) [15,25]. However, while these areas are widely viewed as strongholds for the species, close interactions with humans also present challenges. For example, roads act as an important barrier to movement, resulting in isolated populations [26] and direct hedgehog mortalities [27]. The amount and intensity of ALAN associated with built environments is increasing [28,29], and the implications of this change for hedgehogs and their prey are relatively unknown. This study uses a citizen science approach to assess whether the presence of ALAN in gardens is linked to (1) the amount of time hedgehogs are present at supplementary feeding stations; (2) the amount of time hedgehogs spend feeding at supplementary feeding stations and (3) hedgehog activity patterns in those gardens.

2. Materials and Methods

2.1. Survey

Initial questionnaires were sent out to any member of the public interested in taking part in this experiment, asking for details about their gardens and the environmental features surrounding them, e.g., streetlights. Any participant who had features likely to interfere with the experiment was excluded. Camera traps (Bushnell Trophy CAM HD Max/Bushnell NatureView CAM HD Max; Bushnell Corporation, Overland Park, Kansas, United States) were then sent to 33 volunteer citizen scientists throughout England and Wales who had indicated that hedgehogs visited feeding stations in their gardens (Figure S1). Each volunteer received one camera trap that they deployed in their garden (hereafter referred to as site) facing pre-existing feeding stations. To prevent disturbance to the hedgehogs beyond the experimental treatment, volunteers were asked to continue to supply the same fish-free food they usually provided. The camera traps were placed 2–4 m away from feeding stations and were elevated approximately 0.5 m from the ground to avoid interference with the artificial light source used during the experiment. The cameras were set to record 60 s of video, with an interval of at least 60 s between videos.

At each site, two lighting treatments were used for one week each—'dark' (no artificial lighting illuminating the feeding station) and 'light' (constant artificial light source illuminating the feeding station during the night)—and each treatment lasted for seven nights. The two treatments were deployed sequentially, beginning with the treatment that was already present at the site (i.e., sites that already had illuminated feeding stations began with the 'light' treatment, whereas the opposite was true for 'dark' feeding stations). The two treatments were applied in consecutive weeks using a paired design, and the experiments took place between 15 July and 21 August 2017. For sites that were initially 'dark', an LED floodlight of approximately 1000 lumens (Powerline rechargeable flood lights, 10 W 115 Lm/W) was supplied if the participants did not have a suitable bright exterior light available (e.g., a patio light). The lights were placed so that they constantly illuminated the feeding station throughout the entirety of the night.

The times at which hedgehogs were present and feeding from the supplementary feeding stations were then recorded from the camera trap videos. Supporting data for this study have been deposited on Figshare digital repository (10.6084/m9.figshare.11872113). To reduce the impact of pseudo-replication caused by taking multiple measurements of the same individual (for example, if one animal was stationary in front of the camera for 5 min), a presence index was created by classifying 10 min recording blocks as either being positive or negative for hedgehogs, and a similar index for feeding was created depending on whether they fed during this interval. To ensure ease of recording for volunteers, recording blocks were split by clock time (e.g., 21:00–21:10, and then 21:10–21:20). So if a hedgehog was present over both blocks, this counted as two positive recording blocks, even if the hedgehog was present for less than 10 min. It is highly likely that multiple individuals would have visited the same feeder within each night, but it was not possible to recognise individuals in this project. Therefore, for the purposes of analysis, each 10 min recording block was considered to be a replicate, and the nightly count of hedgehog-positive (the 'presence index') and feeding-positive (the 'feeding index') recording blocks were used as outcome variables.

The percentage of urban cover (combined urban and suburban) within a circular buffer centred on each site was calculated from the Land Cover Map 2007 [30] using ArcMAP 10.5 [31]. The area of the buffer was set at 9.7 ha, which has been reported as the mean home range of hedgehogs in England in regions where badgers are present [32].

2.2. Statistical Analysis

Statistical analyses were undertaken using R v.3.5.3 [33]. To assess the relationships between light exposure and the indices of hedgehog presence/feeding activity, Generalised Linear Mixed Models with a negative binomial distribution were fitted to the count data using the 'lme4' package [34]. Using a paired design, the full model had the following predictor variables: treatment, treatment order, the interaction between treatment and treatment order, percentage of urbanisation, the type of light used (supplied by study participant or the researchers) and an interaction between percentage of urbanisation and treatment type. Site was included as a random effect (accounting for some variation between sites), and an offset for night-length was also included to account for the increase in the time available for nocturnal activity that occurred across the course of the study. The percentage of urban cover was included to account for potential differences in hedgehog behaviour between more urban and more rural areas. Treatment, the random effect (site), and the offset were included in all models, and the most parsimonious model was identified using stepwise deletion of the fixed effects and inspection of the AIC values.

Activity patterns for 31 sites were calculated following a nonparametric kernel density approach [35] using the package 'activity' [36,37]. Sites 1 and 14 were removed from this analysis because they had sample sizes of less than 10 [38]. The times of camera detections were converted to radians and were used to build circular kernel Probability Density Functions (PDF), which approximate the underlying activity pattern of the animals [35]. These PDFs were generated for the 'dark' and 'light' treatment at each site in a pairwise manner.

To investigate whether fitted activity patterns differed according to the lighting treatment, the coefficient of overlap (Δ)—a continuous variable that ranges from 0 (no overlap) to 1 (complete overlap) [35]—was calculated for each site. Then, a randomisation test with 1000 bootstrap iterations was run to generate a null distribution of randomised overlap values, followed by a Wald test to estimate the probability that the observed overlap arose by chance [38].

3. Results

Throughout the study period, night length ranged from 459 to 586 minutes (mean: 509 minutes), equating to an average of 51 possible 10 min recording blocks per site, per night. Of the total 22,615 recording blocks surveyed, 3470 (15%) contained hedgehogs and 19,145 did not. Of the 3470 hedgehog-positive recording blocks, 1724 occurred in 'dark' and 1746 in 'light' treatments, respectively. The mean was 7.8 (SD: 5.9) hedgehog-positive blocks per night across all sites, with a maximum number of hedgehog-positive recording blocks per night under dark treatments of 27 (mean: 7.7; SD: 5.5) and 36 blocks (mean: 7.9; SD: 6.2) for light treatments. The maximum number of animals seen within a single recording block was five. There was no evidence of a link between any of the fixed factors and their interactions (treatment order, percentage of urbanisation, type of light used) and hedgehog presence at the supplementary feeding station ($p > 0.1$ in each case of stepwise removal). The odds ratio for just treatment in the final model was 1.00 (95% Confidence Interval: 0.92–1.09, $p = 0.945$; Figure 1).

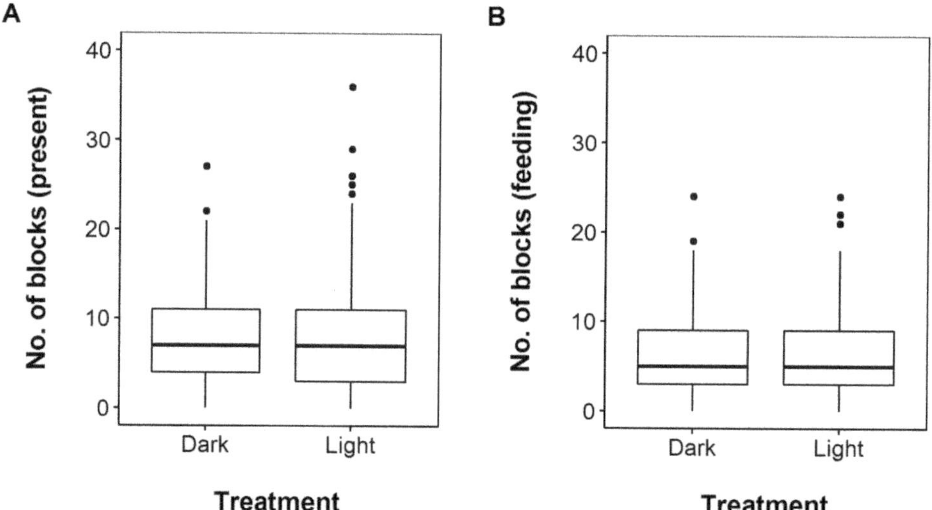

Figure 1. Box plot illustrating the effects of artificial light treatments at supplementary feeding stations on hedgehog (**A**) presence and (**B**) feeding records. 'No. of blocks' represents the number of 10 min periods per site, per night, in which hedgehogs were present and/or feeding. The total possible number of blocks, based on night lengths, ranged from 46 to 59, with an average of 51. Plots are based on the raw data.

Feeding activity was recorded in 2673 of the 3470 10 min recording blocks where hedgehogs were present, with a mean of 6.0 blocks per night across all sites (SD: 4.8). Hedgehogs were recorded feeding in 78% and 76% of the hedgehog-positive recording blocks in the 'dark' and 'light' treatments, respectively. The nightly number of feeding-positive blocks was similar under the different treatments (mean: 6.1; SD: 4.8 for 'dark' and mean: 6.0; SD: 4.8 for 'light' treatments). There was no evidence of a link between any of the fixed factors and the hedgehog feeding activity index ($p > 0.2$ in each case in stepwise removal). In the final model, the odds ratio for treatment was 0.99 (95% CI: 0.91–1.10, $p = 0.992$).

The circadian patterns of hedgehog activity did not vary between treatments at 18 of 31 sites (58%), with high levels of overlap being recorded (mean Δ: 0.93; SD: 0.06 (see Figure 2 for example)). However, at the other 13 sites, activity patterns differed significantly between the 'light' and 'dark' treatments (mean Δ: 0.72; SD: 0.09); see supplementary material for individual test statistics (Table S1) and activity pattern plots (Figure S2). Furthermore, the observed differences in the responses of hedgehog activity at these sites to light treatments were variable, with no consistent directional changes in peak activity times or duration of activity. The peak activity occurred at a similar time at five sites (38.5%), was later in dark compared with light at four sites and was earlier in dark compared with light at four sites. Similarly, the duration of activity in light compared with dark was similar at nine sites (69.2%), was longer at two sites, and was shorter at two other sites. A bimodal pattern of activity was observed during 32% sites during dark treatment and 45% of sites during light treatments (Figure S2).

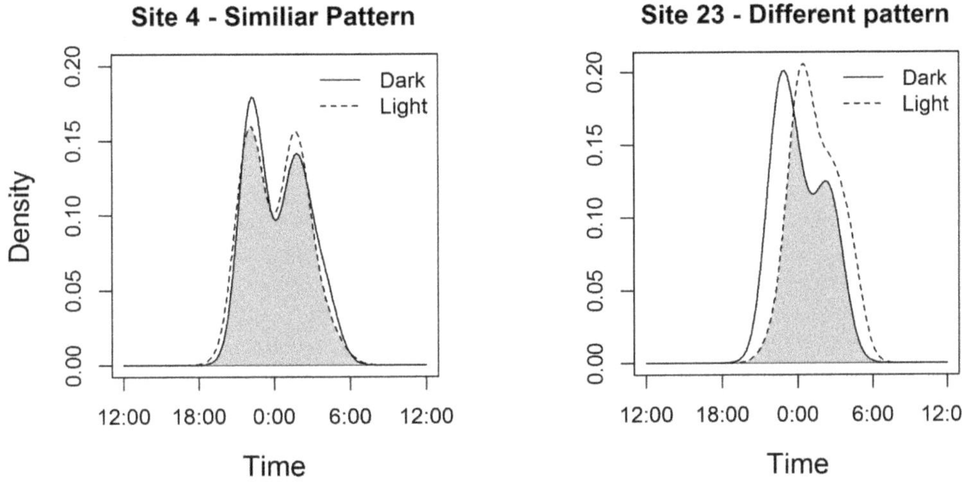

Figure 2. Examples of activity patterns at two different sites: hedgehog activity patterns did not change between treatments at site 4 (Δ = 0.93, *p* = 0.78), but did at site 23 (Δ = 0.70, *p* = 0.02), whereby hedgehogs shifted their activity to become active later at night when the light was on. The area shaded in grey is where the two patterns overlap (Δ).

4. Discussion

The disruptive effects of ALAN on the behaviour and biological processes of a wide range of taxa are well known [39–43]. However, our results indicate that there is no consistent overall effect of lighting on indices of hedgehog activity and feeding at supplementary feeding stations, or on the timing of these behaviours. These findings support those of De Molenaar et al. [44], who found no significant effect of ALAN on hedgehog crossing behaviour at roads. Although it is not possible to rule out the possibility of type-II errors (false negatives) categorically, the use of a controlled crossover trial should have minimised the effect of confounding factors, such as the structure of the garden and local food availability. Our results do, however, offer some support to previous findings that hedgehogs display bimodal feeding activity [45,46].

Although most studies report negative behavioural consequences of ALAN in mammals, some species are able to exploit the foraging opportunities created by lighting, such as the accumulation of insects around streetlights [12,47–52]. In contrast, an experimental study on the beech mouse (*Peromyscus polionotus leucocephalus*) showed that supplementary feeding stations were less likely to be visited, and had less food removed, when they were exposed to ALAN [53]. These results were attributed to a greater predation risk at lit feeding stations, although the degree of this risk will depend on the degree of light-tolerance in the predators. The most common prey items of hedgehogs are ground beetles (*Carabidae*) [54,55], which have been shown to be more abundant in artificially lit areas compared to dark [56]; though other taxa that feature in hedgehog diet, such as woodlice (*Onisicidea*), are light averse. The main predator of hedgehogs in the UK is the European badger (*Meles meles*) [57].

The activity and distribution of badgers has been shown to influence that of hedgehogs in urban and suburban environments [58], with hedgehogs tending to be more active in smaller gardens, which are less likely to be visited by badgers [59]. However, supplementary feeding stations provide very energy-dense resources compared with natural foods, and the drive to obtain these resources may outweigh the risk of predation or light-avoidance behaviours.

No consistent effects of ALAN were reported in this study, and this may in part be because of individual differences, i.e., sex and age. For example, male hedgehogs are bolder [59] and have larger home range sizes than females [60]. Changes in individual responses may explain the marked variability between sites (as illustrated in Figure S2). Differences between the sexes in response to ALAN are reported in great tits (*Parus major*), with females spending more time awake under ALAN conditions than males [61]. In this project, it was not possible to identify individuals or classify their age and sex. Therefore, it would be useful to conduct a similar study, with individuals of known age and sex.

5. Conclusions

In conclusion, this study revealed that the use of artificial light at night had no overall effect on the feeding and general activity of hedgehogs at supplementary feeding stations. There was also no evidence for any overall impact on the periodicity of activity: whilst some individuals delayed their activity when exposed to light, the reverse was true for other individuals. Despite the lack of any difference in hedgehog activity between lit and unlit treatments, there may be costs for reproductive success, territory maintenance, predation rate, and natural prey availability. Future research should focus on these areas.

Supplementary Materials: The following are available online at http://www.mdpi.com/2076-2615/10/5/768/s1, Figure S1: The spatial distribution of the 33 project volunteers. The 31 points in white were included in the activity pattern analyses, whereas the two in black were not due to insufficient sample sizes for activity analyses. Figure S2: Activity patterns of hedgehogs, expressed as a kernel density, at each of the 31 sites in the week where the light was on (dashed line) and when the light was off (solid line). Sites 1 and 14 were removed due to insufficient data. Table S1: Site-level results of activity pattern analyses giving the sample size (hedgehog records) in both the dark and light treatments, the coefficient of overlap between the two activity patterns (Δ), and the p-value test statistic of the Wald test performed. If p < 0.05 (denoted with *) then the activity patterns are significantly different between treatments. Activity analyses were not conducted for sites 1 and 14 due to sample sizes of less than 10.

Author Contributions: Conceptualization D.F., C.M., L.M.K. and F.M.; Methodology, D.F., C.M., L.M.K. and F.M.; Software, D.F. and B.R.S.; Validation, D.F., B.R.S. and F.M.; Formal Analysis, D.F. and B.R.S.; Investigation, D.F., C.M., L.M.K. and F.M.; Resources D.F., C.M., L.M.K. and F.M.; Data Curation, C.M. and L.M.K.; Writing—Original Draft Preparation, D.F., B.R.S., F.G.C., M.A., P.G.R.W. and F.M.; Writing—Review and Editing, D.F., B.R.S., C.M., F.G.C., M.A., L.M.K., P.G.R.W. and F.M.; Visualization, B.R.S. and F.G.C.; Supervision, F.M.; Project Administration, D.F. and F.M.; Funding Acquisition, F.M. All authors have read and agreed to the published version of the manuscript.

Funding: This research was supported by NERC, grant number NE/S006486/1.

Acknowledgments: We thank all of the volunteers who made this experiment possible, and Dan Brown for the loan of camera traps.

Conflicts of Interest: The authors declare no conflicts of interest.

References

1. UN DESA. *World Urbanization Prospects 2018: Highlights*; UN DESA: New York, NY, USA, 2019. [CrossRef]
2. Jung, K.; Kalko, E.K. Where forest meets urbanization: Foraging plasticity of aerial insectivorous bats in an anthropogenically altered environment. *J. Mammal.* **2010**, *91*, 144–153. [CrossRef]
3. Penone, C.; Kerbiriou, C.; Julien, J.F.; Julliard, R.; Machon, N.; Le Viol, I. Urbanisation effect on Orthoptera: Which scale matters? *Insect Conserv. Divers.* **2013**, *6*, 319–327. [CrossRef]
4. Gaston, K.J.; Duffy, J.P.; Gaston, S.; Bennie, J.; Davies, T.W. Human alteration of natural light cycles: Causes and ecological consequences. *Oecologia* **2014**, *176*, 917–931. [CrossRef] [PubMed]

5. Bennie, J.; Davies, T.W.; Cruse, D.; Inger, R.; Gaston, K.J. Cascading effects of artificial light at night: Resource-mediated control of herbivores in a grassland ecosystem. *Philos. Trans. R. Soc. B Biol. Sci.* **2015**, *370*, 20140131. [CrossRef]

6. Lewanzik, D.; Voigt, C.C. Artificial light puts ecosystem services of frugivorous bats at risk. *J. Appl. Ecol.* **2014**, *51*, 388–394. [CrossRef]

7. Becker, A.; Whitfield, A.K.; Cowley, P.D.; Järnegren, J.; Næsje, T.F. Potential effects of artificial light associated with anthropogenic infrastructure on the abundance and foraging behaviour of estuary-associated fishes. *J. Appl. Ecol.* **2013**, *50*, 43–50. [CrossRef]

8. Evans, W.R.; Akashi, Y.; Altman, N.S.; Manville, A.M. Response of night-migrating songbirds in cloud to colored and flashing light. *N. Am. Birds* **2007**, *60*, 476–488.

9. Cresswell, W.; Harris, S. The effects of weather conditions on the movements and activity of badgers (*Meles meles*) in a suburban environment. *J. Zool.* **1988**, *216*, 187–194. [CrossRef]

10. Elangovan, V.; Marimuthu, G. Effect of moonlight on the foraging behaviour of a megachiropteran bat *Cynopterus sphinx*. *J. Zool.* **2001**, *253*, 347–350. [CrossRef]

11. Preston, E.F.; Johnson, P.J.; Macdonald, D.W.; Loveridge, A.J. Hunting success of lions affected by the moon's phase in a wooded habitat. *Afr. J. Ecol.* **2019**, *57*, 586–594. [CrossRef]

12. Mathews, F.; Roche, N.; Aughney, T.; Jones, N.; Day, J.; Baker, J.; Langton, S. Barriers and benefits: Implications of artificial night-lighting for the distribution of common bats in Britain and Ireland. *Philos. Trans. R. Soc. B Biol. Sci.* **2015**, *370*, 20140124. [CrossRef] [PubMed]

13. Day, J.; Baker, J.; Schofield, H.; Mathews, F.; Gaston, K. Part-night lighting: Implications for bat conservation. *Anim. Conserv.* **2015**, *18*, 512–516. [CrossRef]

14. Mathews, F.; Kubasiewicz, L.; Gurnell, J.; Harrower, C.; McDonald, R.; Shore, R. *A Review of the Population and Conservation Status of British Mammals. A Report by the Mammal Society under Contract to Natural England, Natural Resources Wales and Scottish Natural Heritage*; Natural England: Peterborough, UK, 2018.

15. Huijser, M.P.; Bergers, P.J. The effect of roads and traffic on hedgehog (*Erinaceus europaeus*) populations. *Biol. Conserv.* **2000**, *95*, 111–116. [CrossRef]

16. van de Poel, J.L.; Dekker, J.; van Langevelde, F. Dutch hedgehogs *Erinaceus europaeus* are nowadays mainly found in urban areas, possibly due to the negative effects of badgers *Meles meles*. *Wildl. Biol.* **2015**, *21*, 51–55. [CrossRef]

17. Holsbeek, L.; Rodts, J.; Muyldermans, S. Hedgehog and other animal traffic victims in Belgium: Results of a countrywide survey. *Lutra* **1999**, *42*, 111–119.

18. Krange, M. *Change in the Occurrence of the West European Hedgehog* (Erinaceus europaeus) *in Western Sweden during 1950–2010*; Karlstad University: Karlstad, Sweden, 2015.

19. Pettett, C.E.; Johnson, P.J.; Moorhouse, T.P.; Macdonald, D.W. National predictors of hedgehog Erinaceus europaeus distribution and decline in Britain. *Mammal Rev.* **2018**, *48*, 1–6. [CrossRef]

20. Williams, B.M.; Baker, P.J.; Thomas, E.; Wilson, G.; Judge, J.; Yarnell, R.W. Reduced occupancy of hedgehogs (*Erinaceus europaeus*) in rural England and Wales: The influence of habitat and an asymmetric intra-guild predator. *Sci. Rep.* **2018**, *8*, 12156. [CrossRef]

21. Hof, A.R.; Allen, A.M.; Bright, P.W. Investigating the role of the eurasian badger (*Meles meles*) in the nationwide distribution of the western european hedgehog (*Erinaceus europaeus*) in England. *Animals* **2019**, *9*, 759. [CrossRef]

22. Wembridge, D.E.; Newman, M.R.; Bright, P.W.; Morris, P.A. An estimate of the annual number of hedgehog (*Erinaceus europaeus*) road casualties in Great Britain. *Mammal Commun.* **2016**, *2*, 8–14.

23. Wright, P.G.; Coomber, F.G.; Bellamy, C.C.; Perkins, S.E.; Mathews, F. Predicting hedgehog mortality risks on British roads using habitat suitability modelling. *PeerJ* **2020**, *7*, e8154. [CrossRef]

24. Hof, A.R.; Bright, P.W. The value of green-spaces in built-up areas for western hedgehogs. *Lutra* **2009**, *52*, 69–82.

25. Doncaster, C.P. Factors regulating local variations in abundance: Field tests on hedgehogs, *Erinaceus europaeus*. *Oikos* **1994**, 182–192. [CrossRef]

26. Rondinini, C.; Doncaster, C. Roads as barriers to movement for hedgehogs. *Funct. Ecol.* **2002**, *16*, 504–509. [CrossRef]

27. Orlowski, G.; Nowak, L. Road mortality of hedgehogs *Erinaceus* spp. in farmland in Lower Silesia (south-western Poland). *Pol. J. Ecol.* **2004**, *52*, 377–382.

28. Bennie, J.; Davies, T.W.; Duffy, J.P.; Inger, R.; Gaston, K.J. Contrasting trends in light pollution across Europe based on satellite observed night time lights. *Sci. Rep.* **2014**, *4*, 3789. [CrossRef]

29. Kyba, C.C.; Tong, K.P.; Bennie, J.; Birriel, I.; Birriel, J.J.; Cool, A.; Danielsen, A.; Davies, T.W.; Peter, N.; Edwards, W. Worldwide variations in artificial skyglow. *Sci. Rep.* **2015**, *5*, 8409. [CrossRef]

30. Morton, D.; Rowland, C.; Wood, C.; Meek, L.; Marston, C.; Smith, G.; Wadsworth, R.; Simpson, I. *Final Report for LCM2007-the New UK Land Cover Map*; Countryside Survey Technical Report No 11/07; NERC/Centre for Ecology & Hydrology: Bailrigg, UK, 2011.

31. ESRI. *ArcGIS Release 10.5*; Environmental Systems Research Institute: Redlands, CA, USA, 2016.

32. Pettett, C.E.; Moorhouse, T.P.; Johnson, P.J.; Macdonald, D.W. Factors affecting hedgehog (Erinaceus europaeus) attraction to rural villages in arable landscapes. *Eur. J. Wildl. Res.* **2017**, *63*, 54. [CrossRef]

33. R Core Team. *R: A Language and Environment for Statistical Computing*; R Foundation for Statistical Computing: Vienna, Austria; Available online: https://www.R-project.org/ (accessed on 6 December 2018).

34. Bates, D.; Mächler, M.; Bolker, B.; Walker, S. Fitting linear mixed-effects models using lme4. *arXiv* **2014**, arXiv:1406.5823.

35. Ridout, M.S.; Linkie, M. Estimating overlap of daily activity patterns from camera trap data. *J. Agric. Biol. Environ. Stat.* **2009**, *14*, 322–337. [CrossRef]

36. Rowcliffe, J.M.; Kays, R.; Kranstauber, B.; Carbone, C.; Jansen, P.A. Quantifying levels of animal activity using camera trap data. *Methods Ecol. Evol.* **2014**, *5*, 1170–1179. [CrossRef]

37. Rowcliffe, M. *Activity: Animal Activity Statistics*, R package version 1.1; 2016. Available online: https://rdrr.io/cran/activity/ (accessed on 11 March 2019).

38. Lashley, M.A.; Cove, M.V.; Chitwood, M.C.; Penido, G.; Gardner, B.; DePerno, C.S.; Moorman, C.E. Estimating wildlife activity curves: Comparison of methods and sample size. *Sci. Rep.* **2018**, *8*, 4173. [CrossRef]

39. Gaston, K.J.; Bennie, J. Demographic effects of artificial nighttime lighting on animal populations. *Environ. Rev.* **2014**, *22*, 323–330. [CrossRef]

40. Bennie, J.; Davies, T.W.; Cruse, D.; Bell, F.; Gaston, K.J. Artificial light at night alters grassland vegetation species composition and phenology. *J. Appl. Ecol.* **2018**, *55*, 442–450. [CrossRef]

41. Bennie, J.; Davies, T.W.; Cruse, D.; Inger, R.; Gaston, K.J. Artificial light at night causes top-down and bottom-up trophic effects on invertebrate populations. *J. Appl. Ecol.* **2018**, *55*, 2698–2706. [CrossRef]

42. Hoffmann, J.; Palme, R.; Eccard, J.A. Long-term dim light during nighttime changes activity patterns and space use in experimental small mammal populations. *Environ. Pollut.* **2018**, *238*, 844–851. [CrossRef] [PubMed]

43. Yurk, H.; Trites, A. Experimental attempts to reduce predation by harbor seals on out-migrating juvenile salmonids. *Trans. Am. Fish. Soc.* **2000**, *129*, 1360–1366. [CrossRef]

44. De Molenaar, J.C.; Henkens, R.H.C.; ter Braak, C.; van Duyne, C.; Hoefsloot, C.; Jonkers, D.A. *Road Illumination and Nature IV*; Ministry of Transport Public Works and Water Management of The Netherlands: The Hague, The Netherlands, 2013; p. 72.

45. Reeve, N. *Hedgehogs*; T & AD Poyser: Calton, UK, 1994.

46. Wroot, A.J. *Feeding Ecology of the European Hedgehog Erinaceus europaeus*; Royal Holloway, University of London: London, UK, 1984.

47. Schoeman, M. Light pollution at stadiums favors urban exploiter bats. *Anim. Conserv.* **2016**, *19*, 120–130. [CrossRef]

48. Rydell, J. Seasonal use of illuminated areas by foraging northern bats Eptesicus nilssoni. *Ecography* **1991**, *14*, 203–207. [CrossRef]

49. Rydell, J. Exploitation of insects around streetlamps by bats in Sweden. *Funct. Ecol.* **1992**, 744–750. [CrossRef]

50. Azam, C.; Le Viol, I.; Bas, Y.; Zissis, G.; Vernet, A.; Julien, J.-F.; Kerbiriou, C. Evidence for distance and illuminance thresholds in the effects of artificial lighting on bat activity. *Landsc. Urban Plan.* **2018**, *175*, 123–135. [CrossRef]

51. Straka, T.M.; Wolf, M.; Gras, P.; Buchholz, S.; Voigt, C.C. Tree Cover Mediates the Effect of Artificial Light on Urban Bats. *Front. Ecol. Evol.* **2019**, *7*. [CrossRef]

52. Voigt, C.; Rehnig, K.; Lindecke, O.; Pētersons, G. Migratory bats are attracted by red light but not by warm-white light: Implications for the protection of nocturnal migrants. *Ecol. Evol.* **2018**, *8*. [CrossRef] [PubMed]

53. Bird, B.L.; Branch, L.; Miller, D.L. Effects of Coastal Lighting on Foraging Behavior of Beach Mice. *Conserv. Biol.* **2004**, *18*, 1435–1439. [CrossRef]

54. Yalden, D. The food of the hedgehog in England. *Acta Theriol.* **1976**, *21*, 401–424. [CrossRef]

55. Dickman, C. Age-related dietary change in the European hedgehog, *Erinaceus europaeus*. *J. Zool.* **1988**, *215*, 1–14. [CrossRef]

56. Davies, T.W.; Bennie, J.; Gaston, K.J. Street lighting changes the composition of invertebrate communities. *Biol. Lett.* **2012**, *8*, 764–767. [CrossRef]

57. Doncaster, C.P. Testing the role of intraguild predation in regulating hedgehog populations. *Proc. R. Soc. Lond. Ser. B Biol. Sci.* **1992**, *249*, 113–117. [CrossRef]

58. Young, R.P.; Davison, J.; Trewby, I.D.; Wilson, G.J.; Delahay, R.J.; Doncaster, C.P. Abundance of hedgehogs (*Erinaceus europaeus*) in relation to the density and distribution of badgers (*Meles meles*). *J. Zool.* **2006**, *269*, 349–356. [CrossRef]

59. Dowding, C.V.; Harris, S.; Poulton, S.; Baker, P.J. Nocturnal ranging behaviour of urban hedgehogs, Erinaceus europaeus, in relation to risk and reward. *Anim. Behav.* **2010**, *80*, 13–21. [CrossRef]

60. Rautio, A.; Valtonen, A.; Kunnasranta, M. *The Effects of Sex and Season on Home Range in European Hedgehogs at the Northern Edge of the Species Range*; Finnish Zoological and Botanical Publishing Board: Helsink, Finland, 2013; pp. 107–123.

61. Raap, T.; Pinxten, R.; Eens, M. Artificial light at night disrupts sleep in female great tits (*Parus major*) during the nestling period, and is followed by a sleep rebound. *Environ. Pollut.* **2016**, *215*, 125–134. [CrossRef]

 *animals* 

Article

Moving in the Dark—Evidence for an Influence of Artificial Light at Night on the Movement Behaviour of European Hedgehogs (*Erinaceus europaeus*)

Anne Berger [1], **Briseida Lozano** [1,2], **Leon M. F. Barthel** [1] and **Nadine Schubert** [1,3,]*

[1] Department of Evolutionary Ecology, Leibniz-Institute for Zoo and Wildlife Research, Alfred-Kowalke-Straße 17, 10315 Berlin, Germany; berger@izw-berlin.de (A.B.); briseida.lozano@gmx.de (B.L.); Barthel.leon@gmx.de (L.M.F.B.)
[2] Institut für Ökologie, Technische Universität Berlin, Rothenburgstraße 12, 12165 Berlin, Germany
[3] Department of Animal Behaviour, Bielefeld University, Konsequenz 45, 33615 Bielefeld, Germany
* Correspondence: nadine.schubert@uni-bielefeld.de

Received: 3 July 2020; Accepted: 27 July 2020; Published: 30 July 2020

Simple Summary: The European hedgehog is one of the most popular and well-known wild animals, but its numbers are declining throughout Europe, especially in rural areas. Effective hedgehog conservation requires an understanding of the hedgehog's ability to adapt to a changing environment. Due to globally increasing urbanisation, the use of artificial light sources to illuminate the night, called light pollution, has spread dramatically. Light pollution significantly affects the behaviour and ecology of wildlife, but the hedgehog's behaviour towards light pollution remains unknown. We therefore investigated the effects of light pollution on the natural movement behaviour of hedgehogs living in an urban environment. Although hedgehogs can react very variably to environmental influences, the majority of hedgehogs studied here preferred to move in less illuminated rather than in strongly illuminated areas. This apparently rigid behaviour could be used in applied hedgehog conservation to connect isolated hedgehog populations or to safely guide the animals around places dangerous for them via dark corridors that are attractive for hedgehogs.

Abstract: With urban areas growing worldwide comes an increase in artificial light at night (ALAN), causing a significant impact on wildlife behaviour and its ecological relationships. The effects of ALAN on nocturnal and protected European hedgehogs (*Erinaceus europaeus*) are unknown but their identification is important for sustainable species conservation and management. In a pilot study, we investigated the influence of ALAN on the natural movement behaviour of 22 hedgehogs (nine females, 13 males) in urban environments. Over the course of four years, we equipped hedgehogs at three different study locations in Berlin with biologgers to record their behaviour for several weeks. We used Global Positioning System (GPS) tags to monitor their spatial behaviour, very high-frequency (VHF) loggers to locate their nests during daytime, and accelerometers to distinguish between active and passive behaviours. We compared the mean light intensity of the locations recorded when the hedgehogs were active with the mean light intensity of simulated locations randomly distributed in the individual's home range. We were able to show that the ALAN intensity of the hedgehogs' habitations was significantly lower compared to the simulated values, regardless of the animal's sex. This ALAN-related avoidance in the movement behaviour can be used for applied hedgehog conservation.

Keywords: hedgehogs; *Erinaceus europaeus*; light pollution; ALAN; GPS; acceleration; activity; movement behaviour; urbanisation; conservation

---

1. Introduction

The European hedgehog (*Erinaceus europaeus*) is a solitary, hibernating, nocturnal insectivore that is one of the most popular and well-known wild species. Recent studies show that population densities of hedgehogs in cities and suburbs are higher than those in the countryside [1–3]. Moreover, long-term monitoring studies found that the overall hedgehog population in various countries is declining, in some places dramatically [4]. Consequently, in some cases the hedgehog's protection status has been upgraded. In Great Britain, for example, the population has declined from about 1.5 million hedgehogs in 1995 to 522,000 in 2016 [5]. In Germany, only two long-term monitoring studies on a very local scale exist, showing similar decreases to those in the UK [6,7], and concerns about the decline in hedgehogs have also been expressed in other European countries [1,8,9]. The underlying mechanisms causing these declines are certainly complex and multifactorial, including habitat loss, interspecific competition, road collisions, and the intensification of agriculture [4,10–14]. Due to increasing fragmentation and decreasing density of hedgehog populations, the danger of the formation of island populations and inbreeding is already being discussed [15–18].

In order to develop effective protection concepts, threats to hedgehogs as well as their limits of adaptation must be determined precisely. Although the hedgehog's biology and care husbandry are well known, there is—due to its cryptic lifestyle—hardly any knowledge about its behaviour in the wild, the influence of environmental factors on its behaviour, and its adaptation limits.

With the steadily growing human population and the increasing urbanisation worldwide, the amount and intensity of artificial light at night (ALAN) is increasing as well [19]. ALAN has dramatic ecological effects and causes considerable impacts on the migratory behaviour [20], reproduction [21,22], fitness [23], predator-prey interaction [24], activity [25] or phenology [26] of certain species. However, the effects of ALAN on mammalian species other than bats have been studied very little [27,28]. Thus far, one study on the effect of ALAN on hedgehogs' activity at supplementary feeding stations has been carried out [29] and did not find a significant effect. As mentioned by the authors, the drive to obtain high-energy resources at the illuminated feeding stations may outweigh the hedgehogs' light-avoidance behaviour. Thus, this study could not clarify the natural reaction of hedgehogs to ALAN [29].

In our study we investigate to what extent ALAN influences the movement behaviour of wild hedgehogs inhabiting urban spaces. With the rising numbers of hedgehogs living in urban environments, as well as increasing light pollution, it is important to determine the influence of ALAN on this protected species. If hedgehogs show any preference or avoidance, guiding systems for hedgehogs could be developed to safely lead them around risky places (e.g., busy roads) or to link fragmented populations. We have concentrated on the movement behaviour of hedgehogs, as we suspect the greatest opportunity for behavioural adaptability here: hedgehogs that are not moving are either in their nests, which are usually situated in (light-)protected vegetation or constructs, or they are hunting natural food in places where it is available [30,31]. However, the most common natural prey items of hedgehogs [32–34] differ in their behaviour towards light: whether earthworms (Lumbricidae) can perceive light is unknown, but ground beetles (Carabidae) are more abundant in artificially lit compared to dark areas [35], and woodlice (Oniscidea) are light averse. This availability and distribution of prey will in turn affect the movement behaviour of hedgehogs to a certain extent.

In this study we aimed to test the effect of ALAN on the natural movement behaviour of 22 hedgehogs (nine females and 13 males) living in urban environments by tagging their movement using Global Positioning System (GPS) loggers and acceleration data.

2. Materials and Methods

2.1. Study Area

The study was conducted in three different areas in Berlin, the capital of Germany, which has about 4 million inhabitants: in Tierpark Berlin–Friedrichsfelde in the east of Berlin (52°30′11.7″ N,

13°31′47.1′′ E), in Treptower Park in the southeast of the city (52°29′18.5′′ N, 13°28′11.1′′ E) and around the S-Bahn station Tegel in the North (52°35′18.0′′ N, 13°17′23.0′′ E) (Figure 1). All areas contain green spaces (meadows, bushes, hedges, and large trees), playgrounds, larger sealed areas, and footpaths.

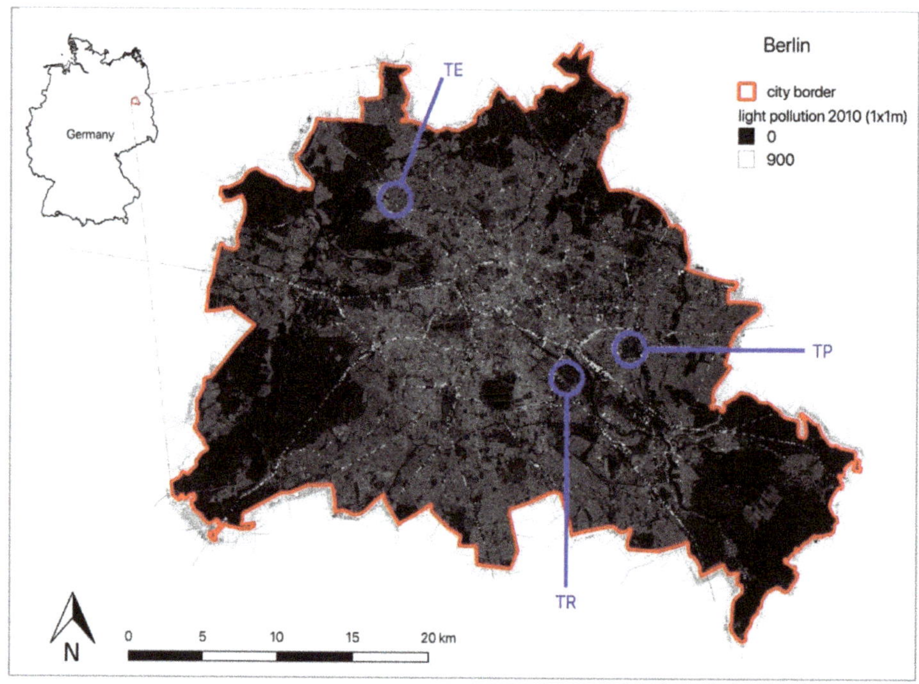

Figure 1. Map of Berlin showing the occurrence of artificial light at night. The three study areas, Tegel (TE), Treptower Park (TR) and Tierpark Berlin–Friedrichsfelde (TP), are marked. Light intensities are indicated by grey shades with increasing brightness in the map corresponding to increasing light intensities. Light intensity is indicated on a relative scale without a measurement unit.

The Tierpark (TP) is a zoological garden of about 160 ha in size. It is open to the public daily from 09:00 to 18:30 CET and dogs are allowed on a leash. Apart from the security service, the TP is free of people and traffic at night. The TP contains numerous animal enclosures, creating a mosaic-like fragmented habitat with many areas inaccessible to hedgehogs.

The Treptower Park (TR) is a public city park of 88.2 ha in size. It is accessible from all sides, open 24 h a day, and is used a lot by citizens for recreation, especially for picnics, sports games or for walking dogs.

The area of the study location in Tegel (TE) is about 52 ha in size and consists mainly of a residential area with public roads and some industrial locations (such as railway facilities), containing multi-floor blocks of houses as well as detached houses surrounded by gardens. Between these houses there are streets and paths that are always open to the public.

The TP and the TR do not contain residential buildings or lights as a source of ALAN. However, both parks are at some sides bordered by big streets that are strongly illuminated at night. In TE, all roads and railway facilities are continuously illuminated at night.

2.2. Field Work

The fieldwork took place from 10th of August to 19th of September 2016 (TR), from 14th of August to 4th of September 2017 (TP), from 22nd of May to 30th of June 2018 (TP) and from 20th of July to 8th of October 2019 (TE).

At the beginning of each field season, several nightly searches (starting around one hour after sunset) were carried out in the corresponding study area to find active hedgehogs using flashlights (P14.2, LED Lenser, Solingen, Germany). Each hedgehog found was weighed, sexed and marked with five shrink tubes glued on the spines [36]. The markings were of different colour in each study year and were given an uprising number starting with 1 to clearly identify each hedgehog in case of subsequent recaptures [37]. We equipped healthy hedgehogs with a body mass of more than 600 g with GPS-acceleration (ACC) loggers (E-obs GmbH, Munich, Germany) and very high-frequency (VHF) transmitters (Dessau Telemetry, Dessau, Germany), which we mounted on a backplate system glued to the hedgehogs' spines [38].

In order to disturb the hedgehogs as little as possible, the tagged individuals were located every day during their stay in the daytime nest by means of the mounted VHF transmitter and a receiver (TRX-1000S, Wildlife Materials Inc., Murphysboro, IL, USA, or Wide Range Receiver AR 8200, AOR Ltd., Tokyo, Japan), and the exact GPS position of the nest was recorded using a Garmin GPSmap 60CSx device (Garmin Deutschland GmbH, Garching, Germany). At night, we only checked, by VHF distance tracking, whether the tagged individuals were moving (away from their nests). Every 4–5 nights, the tagged hedgehogs were located, caught, weighed and checked for potential problems with their backplate or their general health; these opportunities were also used for the necessary recharging of the GPS/ACC data loggers and the GPS/ACC data download.

At the end of the studies, the animals were caught again, weighed, the loggers were removed, and the backplate system was cut off the spines.

All procedures performed involving handling of animals were in accordance with the ethical standards of the institution (IZW permit 2016-02-01) and German federal law (permission number Reg0115/15 and G0104/14).

2.3. Data Logger Setup

In order to be able to find the tagged individuals at any time, the VHF transmitters were programmed to send signals continuously. In addition, GPS positions were recorded during the expected activity times of the nocturnal hedgehogs from 07 p.m. to 07 a.m. in 5-min intervals (TE_2019, TP_2018) or in 10-min intervals (TR_2016, TP_2017). This measurement was done in five-point shifts of one second difference. In addition to the GPS data, three-dimensional acceleration data were recorded. Here, a short burst of high-resolution data was recorded on all 3 axes at the same time. For the present study, the sampling frequency was 100 Hz per axis, a burst lasted 2.5 s and the bursts were recorded every minute of the day.

2.4. Data Analysis and Statistics

The data analysis and presentation were performed using R Studio version 3.5.1 software [39,40] and QGIS version 3.4.4-Madeira [41].

GPS locations that were recorded at times during which the data loggers were not attached to the animals, for example during transport of the loggers for attachment and detachment or during handling of the animals, as well as GPS locations identified as outliers (movement speed > 2 m/s were removed from the data set. Mean values of the GPS data collected within 10 s were calculated and then used for further analyses.

Individual GPS error (GPS_err) was determined by calculating the mean deviation of the GPS positions at times during which the hedgehog was immobile in its daily nest (07 a.m. to 07 p.m.) from the exact position of that nest, which we measured in the field.

Before calculating the home range (HR) with the cleaned GPS dataset (GPS_total), a bootstrap method and a site fidelity test were performed. The bootstrap method shows whether the number of locations is sufficient for calculating the home range, which is indicated by the curve reaching a plateau. HRs were calculated for those individuals that showed site fidelity by calculating the Minimum Convex Polygon (MCP100) for each hedgehog, using the standard bandwidth href for smoothing [42].

The principle of site fidelity means that only an animal whose occupied area is smaller than the estimated area occupied when the animal moves randomly shows site fidelity [43]. Spencer et al. [44] evaluated the site fidelity as a prerequisite for the presence of a home range in an animal.

Afterwards, the GPS_total (measured between 07 p.m. and 07 a.m.) data set was divided into two groups: locations at times the hedgehog was considered to be moving (GPS_act) and at times the hedgehog was considered to be immobile (GPS_pas). As hedgehogs are strongly nocturnal, we expected them to have left the nest 1 h after sunset and to have returned to the nest 1 h before sunrise. Thus, we sorted all GPS locations recorded from 07 p.m. to 1 h after sunset and from 1 h before sunrise until 07 a.m. into GPS_pas as well to ensure that we only considered GPS locations for further analysis that had been recorded outside of the nest. For data collected between 1 h after sunset and 1 h before sunrise, we detected passive phases based on an acceleration data threshold (ACC_thres). We estimated this threshold individually for each hedgehog using the summed standard deviations (of the x, y and z axis of each burst, measured every minute) of the acceleration data. This acceleration threshold is expected to separate active and passive behaviours [45]. Locations recorded during this potential active phase of the hedgehogs (between 1 h after sunset and 1 h before sunrise) for which acceleration values dropped below ACC_thres were assigned to the GPS_pas dataset as well, as the hedgehog was considered immobile and we aimed to investigate movement behaviour.

For the following calculations, we used the publicly available light intensity map of Berlin (resolution 1 sqm) [46] and the GPS locations of active (moving) hedgehogs (GPS_act). In the light map, light intensities measured via flyover are mapped using a relative scale from 0 (absolute darkness) to 48,429 (highest light intensity mapped). Thus, no measurement unit is given and increasing brightness of grey shades corresponds to increasing light intensities in the area (please see [46] for further information on the establishment and properties of the light map). A buffer (radius = mean GPS error) was placed around each of the GPS locations, light intensities were extracted and the mean light intensity of the area of these locations was calculated. Random GPS points with the same buffer size and the same number as the corresponding real measured locations were simulated within the individual's MCP100 and light intensities were calculated in the same way as done for the actual GPS locations.

The mean light intensity of the real hedgehog's locations and that of the random points were then used for comparison via the randomisation method. Utilising Equation (1) below, p-values for each hedgehog were established.

$$p = (\text{sum}(\text{mean H0} \leq \text{meanobs}) + 1)\, k + 1 \tag{1}$$

where mean H0: mean light intensity for simulated random points of individual study animal; meanobs: mean light intensity of individual study animal; k: number of repeats (=1000).

The obtained p-values were then analysed for all hedgehogs combined using Fisher's method to investigate the behaviour towards ALAN for all hedgehogs together. The calculation of the combined p-values for all male hedgehogs was carried out in R Studio using the metap package [47].

3. Results

Data were collected successfully for 22 hedgehogs, including nine females and 13 males (Table 1). Over a period of 14–72 days, interspersed by regular breaks for the recharging of the loggers or due to technical issues with the backplate, we tracked the hedgehogs' spatio-temporal behaviour using GPS tags and three-dimensional accelerometers and recorded between 487 and 2469 GPS locations per animal. Within this timespan, movements were tracked for most animals at least 21 days per individual, except for three animals (TR_08_2016, TR_09_2016 and TP_30_2018). For these hedgehogs, measurements were aborted ahead of schedule due to technical problems with the loggers.

Table 1. Overview of the study animals and data collected. Animal_ID: identification number including the study area (Tegel (TE), Treptower Park (TR) and Tierpark Berlin–Friedrichsfelde (TP)) and study year; sex: m = male, f = female; start date = tagging; end date = detagging; number of Global Positioning System (GPS)_total (total number of locations after cleaning), GPS_act (locations recorded during active behaviour), GPS_pas (locations recorded during passive behaviour); ACC_thres: activity threshold between active/passive behaviour; GPS interval: GPS measurement interval in minutes.

Animal_ID	Sex	Start Date [YYYY-MM-DD]	End Date [YYYY-MM-DD]	Number GPS_Total	Number GPS_Act	Number GPS_Pas	ACC_Thres	GPS Interval [Min]
TR_01_2016	m	2016-08-10	2016-09-19	1313	1140	173	1.87	10
TR_02_2016	f	2016-08-10	2016-09-19	1141	936	205	2.01	10
TR_08_2016	f	2016-08-10	2016-08-27	576	514	62	0.97	10
TR_09_2016	m	2016-08-10	2016-08-24	528	459	69	0.60	10
TR_13_2016	f	2016-08-10	2016-09-19	1357	1237	120	0.82	10
TR_17_2016	f	2016-08-10	2016-09-18	1227	1098	129	0.74	10
TR_19_2016	m	2016-08-10	2016-09-19	1275	1090	185	0.89	10
TR_21_2016	m	2016-08-10	2016-09-19	2566	2450	116	0.84	10
TP_13_2018	m	2018-05-22	2018-06-30	1127	1071	56	0.86	5
TP_18_2018	m	2018-05-22	2018-06-30	1110	994	116	0.99	5
TP_20_2017	f	2017-08-14	2017-09-04	489	449	40	0.84	10
TP_21_2018	m	2018-05-22	2018-06-30	1226	1151	75	0.84	5
TP_26_2017	f	2017-08-16	2017-09-12	487	400	87	0.88	10
TP_28_2017	f	2017-08-14	2017-09-04	684	606	78	0.92	10
TP_29_2018	m	2018-05-28	2018-06-23	941	828	113	0.94	5
TP_30_2018	m	2018-05-23	2018-06-09	991	838	153	0.94	5
TE_02_2019	f	2019-07-20	2019-09-30	973	446	527	1.15	5
TE_03_2019	m	2019-07-10	2019-08-01	870	633	237	1.00	5
TE_04_2019	m	2019-08-30	2019-10-08	1769	1100	669	0.86	5
TE_05_2019	m	2019-07-24	2019-09-16	1318	745	573	1.03	5
TE_07_2019	m	2019-07-24	2019-09-27	1906	1320	586	1.06	5
TE_10_2019	f	2019-08-24	2019-10-08	2469	1595	874	0.69	5

The bootstrap method showed that we clearly exceeded the minimum number of GPS locations per animal necessary to calculate a reliable HR for all tagged hedgehogs (Figures S1–S4). Site fidelity was detected for all 22 hedgehogs; however, eight animals, two females and six males (TR_09_2016, TP_13_2018, TP_21_2018, TP_26_2017, TE_02_2019, TE_03_2019, TE_05_2019, TE_07_2019), exhibited site fidelity not in both criteria, but only with regard to the mean square distance from the activity centre, the main criterion for site fidelity (Figures S5–S8).

According to the individual activity thresholds (ACC_thres), the GPS locations were divided into locations recorded during inactive (GPS_pas) and active behaviour (GPS_act) (Table 1). With the exception of one female (TE_02_2019), who did not leave the nest for a few nights during the study period, all hedgehogs showed predominantly active behaviour during their nocturnal activity time from one hour after sunset until one hour before sunrise. Exemplarily, all buffered GPS locations (GPS_pas = yellow, GPS_act = red) are pictured on a map for one female and one male of each study area (Figures 2–4). Blue spots within the individual HR represent the buffered random positions (its number corresponds to the number of the individual's GPS_act locations). As an example, only a single simulation of the random points is indicated, but this simulation was repeated 1000 times for further analysis using the randomisation test. In the maps, the occurrence and intensity of ALAN is marked by grey shades, with increasing brightness corresponding to areas with higher ALAN intensity in the maps (Figures 2–4).

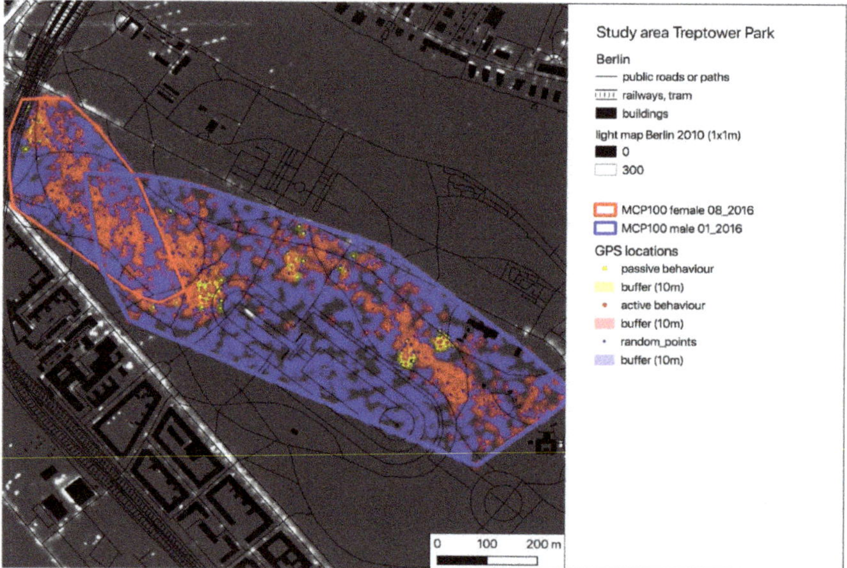

Figure 2. Map of the GPS locations of a female (TR_08_2016) and a male (TR_01_2016) hedgehog in the study area Treptower Park (yellow dots—locations with inactive behaviour = GPS_pas, red dots—locations with active behaviour = GPS_act). Blue dots represent randomly distributed locations within the individual Minimum Convex Polygon (MCP100) (GPS_random) (the number of GPS_act equals the number of GPS_random). All locations are buffered (radius = GPS error = 10 m). The overlayed light map of Berlin displays the occurrence and intensity of artificial light at night (ALAN).

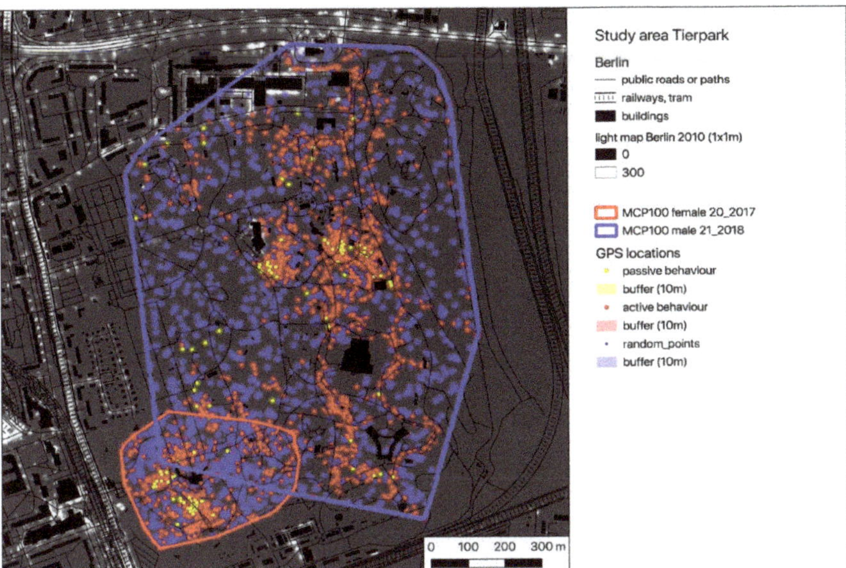

Figure 3. Map of the GPS locations of a female (TP_20_2017) and a male (TP_21_2018) hedgehog in the study area Tierpark (yellow dots—locations with inactive behaviour = GPS_pas, red dots—locations with active behaviour = GPS_act). Blue dots represent randomly distributed locations within the individual MCP100 (GPS_random) (the number of GPS_act equals the number of GPS_random). All locations are buffered (radius = GPS error = 10 m). The overlayed light map of Berlin shows the occurrence and intensity of ALAN.

Figure 4. Map of the GPS locations of a female (TE_02_2019) and a male (TE_05_2019) hedgehog in the study area Tegel (yellow dots—locations with inactive behaviour = GPS_pas, red dots—locations with active behaviour = GPS_act). Blue dots represent randomly distributed locations within the individual MCP100 (GPS_random) (the number of GPS_act equals the number of GPS_random). All locations are buffered (radius = GPS error = 10 m). The overlayed light map of Berlin shows the occurrence and intensity of ALAN.

Table 2 shows the light intensities of the buffered GSP locations measured for each individual (Animal_) and the corresponding value for the buffered random sites (Random_). The simulation of these random locations was repeated 1000 times per animal. The total of all random mean values of light intensity differed slightly between the study areas and ranges from 120.1 ± 77.9 (TP) over 122.2 ± 82.8 (TE) up to 124.8 ± 143.9 (TR) (Table 2). Five out of 22 hedgehogs did not show a mean light intensity at their locations that differed significantly from the randomly distributed locations in their MCP100. Out of these five individuals showing no statistically significant difference in the light intensity comparison, three animals (TE_03_2019, TE_05_2019 and TE_07_2019) exhibited mean light intensity values for their measured GPS locations (animal_mean) that were above the mean value determined for the randomly distributed locations (random_mean). These three individuals were also observed to have their nests or their movement paths situated along small bushes directly bordering the roadside or the track bed. The remaining 17 hedgehogs exhibited a significantly lower mean light intensity compared to that obtained via simulation of random locations (Table 2).

The difference between the random and the animals' mean light intensities was compared between the sexes and study locations. The difference in the light intensities did not vary significantly between the sexes (Mann–Whitney U-Test, difference female = 15.2 ± 6.3, difference male = 5.9 ± 21.4, $p = 0.5$). Comparison of the differences between simulated and experienced light intensities of the hedgehogs in the different study locations (TR, TP and TE) yielded statistically significant differences (Kruskal Wallis rank sum test, $p = 0.0005$). Pairwise comparison revealed that the light intensity difference varied significantly between hedgehogs of TE and TR (Wilcoxon pairwise rank sum exact test, difference TE = −11.1 ± 24.1, difference TR = 20.2 ± 4.6, $p = 0.007$) and the hedgehogs of TE and TP (Wilcoxon pairwise rank sum exact test, difference TE = −11.1 ± 24.1, difference TP = 14.8 ± 4.5, $p = 0.004$).

Table 2. Mean light intensity values. Presented are the results (mean, SD = standard deviation, min = minimum, max = maximum) of the light intensity estimation for the buffered hedgehog's GPS locations (animal_) and for the corresponding random points (random_). from the results of the statistical test for differences between animal_mean and random_mean, *p*-values and the significance level (0.05 = *, 0.01 = **, 0.001 = ***) are listed. Statistical test used: randomisation method.

Animal_ID	Animal_Mean	Animal_SD	Animal_Min	Animal_Max	Random_Mean	Random_SD	Random_Min	Random_Max	*p*-Values	Significance Level
TR_01_2016	102.0	4.2	99.3	140.9	118.9	25.1	101.4	226.8	0.001	***
TR_02_2016	102.8	6.2	99.6	177.0	125.2	150.5	98.6	10,339.3	0.001	***
TR_08_2016	106.5	20.2	100.6	290.2	125.6	159.2	99.0	8220.8	0.03	*
TR_09_2016	102.3	5.4	100.2	205.6	126.4	165.2	99.0	9845.4	0.001	***
TR_13_2016	115.2	39.6	100.2	578.1	125.8	160.3	98.7	11,076.4	0.993	n.s.
TR_17_2016	103.4	5.1	100.1	179.0	125.8	155.6	98.7	10,103.6	0.001	***
TR_19_2016	102.0	2.4	100.0	123.6	125.9	170.3	98.8	11,093.1	0.001	***
TR_21_2016	102.9	7.4	99.8	212.8	125.2	165.3	98.6	11,295.8	0.001	***
mean_TR	104.6	11.3	100.0	238.4	124.8	143.9	99.1	9025.1	-	-
TP_13_2018	105.2	26.9	100.5	544.2	120.6	83.9	43.6	3386.4	0.74	n.s.
TP_18_2018	102.7	1.1	100.7	116.3	120.5	81.8	47.8	3404.3	0.001	***
TP_20_2017	102.7	0.8	101.1	106.2	119.8	70.8	50.0	3269.8	0.001	***
TP_21_2018	103.9	9.9	100.6	235.7	120.2	80.4	44.8	3371.5	0.004	**
TP_26_2017	103.3	6.1	100.7	166.2	118.3	57.0	49.2	2835.1	0.001	***
TP_28_2017	103.0	1.2	100.1	106.6	120.5	86.4	43.9	3360.7	0.001	***
TP_29_2018	116.8	25.0	102.1	366.0	120.7	84.4	47.0	3317.2	0.987	n.s.
TP_30_2018	105.0	7.6	101.3	197.5	120.3	78.7	45.9	3337.0	0.005	**
mean_TP	105.3	9.8	100.9	229.8	120.1	77.9	46.5	3285.2	-	-
TE_02_2019	118.9	25.1	101.4	226.8	122.3	83.7	10.4	3348.0	0.001	***
TE_03_2019	159.8	351.7	101.5	3274.0	122.3	81.1	7.3	3332.2	1.0	n.s.
TE_04_2019	120.8	47.1	100.8	445.0	122.2	84.8	6.0	3346.3	0.02	*
TE_05_2019	160.9	86.3	101.9	443.3	122.2	82.6	5.1	3294.6	1.0	n.s.
TE_07_2019	127.0	44.4	101.2	623.6	122.3	83.2	6.1	3402.3	0.02	*
TE_10_2019	112.7	25.3	101.4	399.7	122.1	81.5	5.4	3396.6	0.001	***
mean_TE	133.4	96.6	101.4	902.1	122.2	82.8	6.7	3353.3	-	-

4. Discussion

A comparison of the light intensity of each animal's GPS locations extracted from a light map of Berlin with a corresponding number of simulated random locations within the animal's HR indicates a preference for movement in locations with lower levels of ALAN compared to the simulated values. Our study is the first to observe such a preference in the movement behaviour of wild European hedgehogs in an unchanged, urban setting.

As ALAN is mostly emitted by streetlamps, it usually occurs locally (spotlike or linelike) with high intensity and especially in the vicinity of streets and crossroads (Figures 2–4). The hedgehogs in our study seemed to avoid these spots of high light intensity when moving in their HRs, as all 22 hedgehogs exhibited lower animal_max values for their GPS locations compared to the random_max values of the simulated points. Additionally, the variance of the experienced light intensities (animal_SD) is, with the exception of three animals (TE_03_2019, TE_05_2019 and TE_07_2019), lower than the variance of the light intensities established through simulation of random locations within their HR (random_SD) (Table 2). These individuals were observed to have nests or movement paths situated along covering vegetation alongside illuminated areas, which might have enabled them to move relatively protected from ALAN sheltered by low bushes despite the presence of artificial light sources (corresponding to high ALAN values in the map).

Overall, a comparison of the intensity of ALAN experienced by the animals with the simulated random values indicated a preference for movement in areas with lower values of ALAN (Table 2). This behavioural pattern was observed regardless of the animal's sex. This finding indicates a certain behavioural stability of ALAN avoidance in hedgehogs, which is interesting given that significant interindividual differences in movement behaviour of hedgehogs were observed in response to a music festival in a park in Berlin [48]. Despite the overall preference for movement in less strongly illuminated areas, comparison of movement behaviour in relation to ALAN between the three study locations indicated that hedgehogs living in TE differed in their behaviour compared to the hedgehogs living in TP and TR. Hedgehogs in TE showed a negative mean difference between simulated and experienced light intensities, meaning that they appear to prefer more strongly illuminated areas compared to the values obtained from a random distribution of locations in their home range. In TE, three hedgehogs exhibited a preference for more strongly illuminated areas compared to the simulated locations, which can be explained by the aforementioned locations of their nests and movement paths alongside roads and track beds, which display high ALAN intensities due to the streetlamps associated with these areas. However, since the light map used for the analysis maps light that was measured via overfly, these three hedgehogs might have actually experienced lower light intensities than those estimated in our analysis when moving through cover such as shrubs and bushes. This might explain the high light intensities estimated for those animals, which appear contradictory to the results of the remaining hedgehogs. However, it is also possible that these hedgehogs prefer more strongly illuminated areas due to other reasons, such as the avoidance of intraspecific competition for food resources or mating opportunities as well as other ecological factors correlating with ALAN intensity.

The preference for movement in areas with lower intensities of ALAN, which has been observed in wild hedgehogs in this study for the first time, is an important finding, since the response of nocturnal insectivores such as the hedgehog to ALAN might vary. First, they might intentionally seek out artificially illuminated areas to increase their nutritional intake. ALAN has the potential to attract invertebrates or affect their community composition [35,49], potentially leading to an altered or increased presence of predators feeding on invertebrates at artificially illuminated sites. This increased availability of prey has been suggested as a potential factor drawing hedgehogs into artificially illuminated areas such as roads [50]. Second, hedgehogs might avoid artificially lit areas to reduce the risk of encountering humans or predators, which can be detrimental to survival [51]. Concordant with this hypothesis, the nocturnal beach mouse (*Peromyscus polionotus*) prefers food patches situated further away from artificial light sources [52]. Similarly, brown rats (*Rattus norvegicus*) avoid artificially lit areas. In contrast, predatory species such as the fox (*Vulpes vulpes*), the stoat (*Mustela erminea*), the polecat

(*Mustela putorius*), and the weasel (*Mustela nivalis*) prefer illuminated areas [53]. Interestingly, the badger (*Meles meles*), intraguild predator to the hedgehog [54,55], avoids urban and recreational areas as well as roads [1]. Roads as well as urban and recreational areas are closely linked to human presence and are thus expected to feature high ALAN intensities and these areas have been shown to have a positive effect on hedgehog presence in a previous study [1]. Together with the observed negative effect of badger presence on hedgehog population density, it has been suggested that hedgehogs might thus prefer illuminated areas due to a decreased risk of predation by badgers [1]. However, our findings did not support attraction to illuminated areas. Nonetheless, since we neither measured prey abundance nor badger distribution in our study locations, we can only hypothesise what caused the light-averse movement behaviour of the hedgehogs in our study. Based on local citizen science projects conducted in Berlin using camera trap data, we are aware that during the time of the study there had been no eagle owls and only few badgers detected in our study areas [56]. Furthermore, we encountered foxes, martens and raccoons during our nightly fieldwork but have not encountered badgers in our study locations. Hence it is likely, that badger densities in Great Britain are higher compared to the ones in Germany and especially compared to our study locations in Berlin. The lower intraguild predator density might hence affect the behaviour of the hedgehogs observed in our study, leading to the absence of a preference for intensely illuminated areas [12]. Taking the missing data on badger distribution into account, we conclude that our results can be best explained by a risk-avoidance strategy which causes hedgehogs to prefer less intensely lit areas.

Previous studies on hedgehog spatial behaviour in relation to ALAN neither found evidence for a preference nor an avoidance of illuminated areas [29,53]. First, such an indiscriminate response can be caused by habituation to ALAN. With ALAN becoming more and more abundant [57], animals thriving in artificially illuminated areas might even be selected for decreasing light sensitivity [19]. Molenaar et al. [53] used experimentally installed streetlights at drainage ditches connecting upland habitat and incorporated a habituation phase to ensure animals got accustomed to the changes. However, it is unknown which additional ecological factors correlate with ALAN intensity and might even have a stronger effect on the movement behaviour of hedgehogs. In concordance with this, Molenaar et al. [53] stated that, in contrast to ALAN, vegetation height indeed had a significant impact on the movement of hedgehogs. However, other ecological factors, such as vegetation height, prey abundance, impervious surface, or traffic, that might correlate with ALAN and impact hedgehog movement were not measured in our study. Thus, we can only hypothesise that a preference for less strongly illuminated areas might have been observed in our study due to other factors correlating with ALAN, such as vegetation height or cover. This is supported by data collected for the two hedgehogs in TE that exhibited exceptionally high light intensity values for their locations but seemed to nest and move in vegetation covering them from ALAN. The potential correlation with and impact of other ecological factors apart from ALAN might also explain why we do not see habituation to ALAN in our study, even though the majority of the light sources in our study are expected to be permanently installed. Without knowing the distribution of these additional ecological factors and their link to ALAN, it is difficult to disentangle the interconnectedness of these factors and compare the results of our study with those of Molenaar et al. [53].

Another study [29] investigated the behaviour of hedgehogs in relation to ALAN at supplementary feeding stations. However, because of the supplementary feeding, the study's experimental set-up does not reflect a natural distribution of food sources and might mask the natural movement behaviour of wild hedgehogs. As decisions related to foraging are based on a trade-off between the benefits of energy intake and the risk of decreasing fitness by jeopardising survival, the risk posed by ALAN might be outweighed by the high amount of energetically valuable nutrition provided at the feeding station. In contrast, our study investigated the movement behaviour of hedgehogs in a natural unaltered setting without the confounding factor of supplementary feeding.

Although our results provide evidence for avoidance of artificially illuminated areas in hedgehogs, there are limitations to their interpretation. As we investigated the movement behaviour of wild hedgehogs in an unaltered urban setting, there are several factors that could not be controlled or are unknown. First, the correlation between the intensity of ALAN and other environmental factors affecting hedgehog behaviour are unknown but likely to exist. Since artificial light sources serve the purpose of illuminating areas for human use and are thus frequented by humans more intensely than dark areas, factors such as human disturbance, traffic, amount of impervious surface and vegetation structure might be linked to ALAN intensity. Furthermore, the distribution of food sources, both natural and anthropogenic, might be linked to ALAN intensity and levels of human presence. Hence, these factors might even have a stronger impact on movement behaviour than ALAN itself. Assessing or controlling these environmental parameters in upcoming studies will help unravel the contribution of these factors to shaping hedgehog behaviour and will aid in evaluating the influence of ALAN on hedgehog movement. Another limitation of this study is the estimation of ALAN intensities. The light map used was obtained via flyover and thus maps the emitted light detectable from above. This means that the light intensities correspond to the amount of light emitted upwards, including direct upward-facing light beams and scattered light. As the beams of streetlights serve to illuminate pathways, they often face downwards. Hedgehogs might thus be exposed to light intensities differing from those indicated in this light map. Furthermore, the light intensities mapped are displayed on a relative scale, which impedes comparisons with light intensities emitted by streetlights as well as other studies.

The measurement acuity of the GPS loggers undoubtedly impacts the acuity of the results as well. The loggers were programmed to record the GPS position every 5 or 10 min. Hedgehogs, however, have been reported to move with speeds of 1–2 m/s during brief sprints [58]. Together with the measurement error of the logger itself, which is approximately 10 m, and the resolution of 1×1 m of the light map, this might cause the measured GPS locations and thus the light intensity mapped to not reflect the exact actual position and ALAN intensity for an animal. However, due to the maximum weight of the loggers, which is limited by the hedgehog's body weight, we were not able to use larger batteries allowing for locational fixes obtained at higher frequencies in order to account for the measurement error with repeated measurements. Bootstrap-analysis showed that for most animals, the number of GPS locations measured clearly exceeded the numbers necessary to calculate a reliable HR. Thus, HR estimation was accurate, but the high number of GPS positions also caused the corresponding number of simulated random points to fill up most of the HR. Thus, the simulated values reflected a mean value of the HR's ALAN intensity rather than a simulated random path. Repetitions of these experiments should thus aim at achieving more accurate GPS positions with smaller inter-measurement intervals and a shorter overall measurement period to limit the total number of GPS positions.

Our study is the first to provide evidence that the movement of hedgehogs is related to ALAN intensity. As the results of this study should reflect natural behaviour, the obtained knowledge could be used to cushion population declines by increasing survival rates through conservation. Hedgehogs could therefore be led around dangerous areas such as roads by building dark corridors using vegetation and reduced illumination through streetlights. Apart from the threat of being killed in traffic, hedgehog populations in urban areas might face genetic isolation [16–18]. Bridging parks with dark corridors could help to safely connect isolated populations from different parks and thus increase genetic diversity. In this regard, empirical studies examining the sensory capabilities of hedgehogs using streetlights of different wavelengths and intensities would be helpful. The results of these studies can help in establishing guidelines for intensity thresholds and properties of streetlights by policymakers. Nonetheless, further studies are needed to confirm the influence of ALAN on the movement of wild European hedgehogs. Additionally, the influence of other environmental factors on the movement behaviour needs to be disentangled from the effect of ALAN. Shedding a light on the cryptic lifestyle of the European hedgehog, an urban adaptor species in serious decline, can help establish effective conservation measures for the protection of this nocturnal insectivore.

5. Conclusions

The results of our study provide unique evidence for an influence of ALAN on the movement behaviour of hedgehogs inhabiting urban spaces. European hedgehogs preferred less intensely lit areas compared to the ALAN intensity obtained via simulation. Moreover, this behaviour was observed regardless of sex and for 17 out of 22 individuals. Further studies are needed to confirm the role of ALAN and to disentangle it from the potential effects of other environmental factors.

Supplementary Materials: The following are available online at http://www.mdpi.com/2076-2615/10/8/1306/s1, Figures S1–S4: Bootstrap results. Figures S5–S8: Site fidelity results.

Author Contributions: Conceptualisation, A.B..; methodology, A.B.; Software, A.B., N.S., B.L.; formal analysis, N.S., B.L., A.B.; investigation, L.M.F.B., N.S., B.L.; data curation, N.S., B.L., L.M.F.B.; writing—original draft preparation, A.B., N.S.; writing—review and editing, A.B., N.S.; visualisation, A.B., B.L., N.S.; supervision, A.B.; project administration, A.B.; funding acquisition, A.B. All authors have read and agreed to the published version of the manuscript.

Funding: This work was partly funded by the German Federal Ministry of Education and Research (BMBF) within the Collaborative Project "Bridging in Biodiversity Science-BIBS" (grant number 01LC1501A-H) and by the Stiftung Naturschutz Berlin (Förderfonds Trenntstadt Berlin). We acknowledge support for the publication costs by the Deutsche Forschungsgemeinschaft and the Open Access Publication Fund of Bielefeld University.

Acknowledgments: We would like to thank Wanja Rast, Doreen Müller, Feona Oltmann, Annika Krüger, Judith Niedersen, Juline Cibis, Catrin Zordick, Dominik Henoch, Hanna-Lea Schmid, Phillip Gutowsky, Martin Stankalla, Sara Gottschalk, Lu von Lengerke, Lenz Bartelt; Maria Kochs, Jennifer Till, Laurenz Gewiss, Linda Pollack, Lynn Pollee, Henrik Eritsland, Shanice Schneider, Sharumathi Natarajan, Saskia Arndt, Judith Wachinger, Alexander Doege, Sandra Iurino, Lena Husemann, and Amy Lee Wincer for their assistance during the fieldwork. We would also like to thank Alexandre Courtiol for his advice on statistical analysis.

Conflicts of Interest: The authors declare no conflict of interest.

References

1. van de Poel, J.L.; Dekker, J.; van Langevelde, F. Dutch hedgehogs *Erinaceus europaeus* are nowadays mainly found in urban areas, possibly due to the negative effects of badgers *Meles meles*. *Wildl. Biol.* **2015**, *21*, 51–55. [CrossRef]

2. Pettett, C.E.; Moorhouse, T.P.; Johnson, P.J.; Macdonald, D.W. Factors affecting hedgehog (*Erinaceus europaeus*) attraction to rural villages in arable landscapes. *Eur. J. Wildl. Res.* **2017**, *63*, 54. [CrossRef]

3. Hubert, P.; Julliard, R.; Biagianti, S.; Poulle, M.L. Ecological factors driving the higher hedgehog (*Erinaceus europaeus*) density in an urban area compared to the adjacent rural area. *Landsc. Urban Plan.* **2011**, *103*, 34–43. [CrossRef]

4. Pettett, C.E.; Johnson, P.J.; Moorhouse, T.P.; Macdonald, D.W. National predictors of hedgehog *Erinaceus europaeus* distribution and decline in Britain. *Mammal Rev.* **2018**, *48*, 1–6. [CrossRef]

5. Mathews, F.; Kubasiewicz, L.; Gurnell, J.; Harrower, C.; McDonald, R.; Shore, R. *A Review of the Population and Conservation Status of British Mammals*; A Report by the Mammal Society under Contract to Natural England, Natural Resources Wales and Scottish Natural Heritage; Natural England: Peterborough, UK, 2018.

6. Müller, F. Langzeit-Monitoring der Straßenverkehrsopfer beim Igel (*Erinaceus europaeus* L.) zur Indikation von Populationsdichteveränderungen entlang zweier Teststrecken in Landkreis Fulda. *Beiträge Nat. Osthess.* **2018**, *54*, 21–26, ISSN 0342-5452-54.

7. Reichholf, J.H. Starker Rückgang der Häufigkeit überfahrener Igel *Erinaceus europaeus* in Südostbayern und seine Ursachen. *Mittl. Zool. Ges. Braunau* **2015**, *11*, 309–314.

8. Krange, M. *Change in the Occurrence of the West European Hedgehog* (Erinaceus europaeus*) in Western Sweden during 1950–2010*; Karlstad University: Karlstad, Sweden, 2015.

9. Holsbeek, L.; Rodts, J.; Muyldermans, S. Hedgehog and other animal traffic victims in Belgium: Results of a countrywide survey. *Lutra* **1999**, *42*, 111–119.

10. Hof, A.R.; Bright, P.W. The value of green-spaces in built-up areas for western hedgehogs. *Lutra* **2009**, *52*, 69–82.

11. Williams, B.M.; Baker, P.J.; Thomas, E.; Wilson, G.; Judge, J.; Yarnell, R.W. Reduced occupancy of hedgehogs (*Erinaceus europaeus*) in rural England and Wales: The influence of habitat and an asymmetric intra-guild predator. *Sci. Rep.* **2018**, *8*, 12156. [CrossRef]

12. Hof, A.R.; Allen, A.M.; Bright, P.W. Investigating the role of the eurasian badger (*Meles meles*) in the nationwide distribution of the western european hedgehog (*Erinaceus europaeus*) in England. *Animals* **2019**, *9*, 759. [CrossRef]

13. Wright, P.G.; Coomber, F.G.; Bellamy, C.C.; Perkins, S.E.; Mathews, F. Predicting hedgehog mortality risks on British roads using habitat suitability modelling. *PeerJ* **2020**, *7*, e8154. [CrossRef] [PubMed]

14. Hof, A.R.; Bright, P.W. The value of agri-environment schemes for macro-invertebrate feeders: Hedgehogs on arable farms in Britain. *Anim. Conserv.* **2010**, *13*, 467–473. [CrossRef]

15. Becher, S.A.; Griffiths, R. Genetic differentiation among local populations of the European hedgehog (*Erinaceus europaeus*) in mosaic habitats. *Mol. Ecol.* **1998**, *7*, 1599–1604. [CrossRef] [PubMed]

16. Rondinini, C.; Doncaster, C. Roads as barriers to movement for hedgehogs. *Funct. Ecol.* **2002**, *16*, 504–509. [CrossRef]

17. Braaker, S.; Kormann, U.; Bontadina, F.; Obrist, M.K. Prediction of genetic connectivity in urban ecosystems by combining detailed movement data, genetic data and multi-path modelling. *Landsc. Urban Plan.* **2017**, *160*, 107–114. [CrossRef]

18. Rasmussen, S.L.; Yashiro, E.; Sverrisdóttir, E.; Nielsen, K.L.; Lukassen, M.B.; Nielsen, J.L.; Asp, T.; Pertoldi, C. Applying the GBS technique for the genomic characterization of a Danish population of European hedgehogs (*Erinaceus europaeus*). *Gen. Biodiv. J.* **2019**, *3*, 78–86.

19. Hölker, F.; Wolter, C.; Perkin, E.K.; Tockner, K. Light pollution as a biodiversity threat. *Trends Ecol. Evol.* **2010**, *25*, 681–682. [CrossRef]

20. Evans, W.R.; Akashi, Y.; Altman, N.S.; Manville, A.M. Response of night-migrating songbirds in cloud to colored and flashing light. *N. Am. Birds* **2007**, *60*, 476–488.

21. Robert, K.A.; Lesku, J.A.; Partecke, J.; Chambers, B. Artificial light at night desynchronizes strictly seasonal reproduction in a wild mammal. *Proc. R. Soc. B Biol. Sci.* **2015**, *282*, 20151745. [CrossRef]

22. Dominoni, D.; Quetting, M.; Partecke, J. Artificial light at night advances avian reproductive physiology. *Proc. R. Soc. B* **2013**, *280*, 20123017. [CrossRef]

23. de Jong, M.; Ouyang, J.Q.; Da Silva, A.; van Grunsven, R.H.A.; Kempenaers, B.; Visser, M.E.; Spoelstra, K. Effects of nocturnal illumination on life-history decisions and fitness in two wild songbird species. *Phil. Trans. R. Soc. B* **2015**, *370*, 20140128. [CrossRef] [PubMed]

24. Bennie, J.; Davies, T.W.; Cruse, D.; Inger, R.; Gaston, K.J. Cascading effects of artificial light at night: Resource-mediated control of herbivores in a grassland ecosystem. *Phil. Trans. R. Soc. B* **2015**, *370*, 20140131. [CrossRef] [PubMed]

25. Titulaer, M.; Spoelstra, K.; Lange, C.Y.M.J.G.; Visser, M.E. Activity patterns during food provisioning are affected by artificial light in free living great tits (*Parus major*). *PLoS ONE* **2012**, *7*, e37377. [CrossRef] [PubMed]

26. De Silva, A.; Valcu, M.; Kempenaers, B. Light pollution alters the phenology of dawn and dusk singing in common European songbirds. *Phil. Trans. R. Soc. B* **2015**, *370*, 20140126. [CrossRef]

27. Stone, E.L.; Wakefield, A.; Harris, S.; Jones, G. The impacts of new street light technologies: Experimentally testing the effects on bats of changing from low-pressure sodium to white metal halide. *Phil. Trans. R. Soc. B* **2015**, *370*, 20140127. [CrossRef]

28. Straka, T.M.; Wolf, M.; Gras, P.; Buchholz, S.; Voigt, C.C. Tree Cover Mediates the Effect of Artificial Light on Urban Bats. *Front. Ecol. Evol.* **2019**, *7*, 201900091. [CrossRef]

29. Finch, D.; Smith, B.R.; Marshall, C.Â.; Coomber, F.G.; Kubasiewicz, L.M.; Anderson, M.; Wright, P.G.R.; Mathews, F. Effects of Artificial Light at Night (ALAN) on European Hedgehog Activity at Supplementary Feeding Stations. *Animals* **2020**, *10*, 768. [CrossRef]

30. Reeve, N.J. A Field Study of the Hedgehog (*Erinaceus europaeus*) with Particular Reference to Movements and Behaviour. Ph.D. Thesis, University of London, London, UK, 1981.

31. Morris, P. *Hedgehog*; William Collins: Sydney, Australia, 2018; ISBN 978-0-00-823573-4.

32. Yalden, D. The food of the hedgehog in England. *Acta Theriol.* **1976**, *21*, 401–424. [CrossRef]

33. Wroot, A.J. *Feeding Ecology of the European Hedgehog* Erinaceus europaeus; Royal Holloway; University of London: London, UK, 1984.

34. Dickman, C. Age-related dietary change in the European hedgehog, *Erinaceus europaeus*. *J. Zool.* **1988**, *215*, 1–14. [CrossRef]

35. Davies, T.W.; Bennie, J.; Gaston, K.J. Street lighting changes the composition of invertebrate communities. *Biol. Lett.* **2012**, *8*, 764–767. [CrossRef]

36. Mori, E.; Menchetti, M.; Bertolino, S.; Mazza, G.; Ancillotto, L. Reappraisal of an old cheap method for marking the European hedgehog. *Mam. Res.* **2015**, *60*, 189–193. [CrossRef]

37. Reeve, N.J.; Bowen, C.; Gurnell, J. *An Improved Identification Marking Method for Hedgehogs*; The Mammal Society: Southampton, UK, 2019.

38. Barthel, L.M.F.; Hofer, H.; Berger, A. An easy, flexible solution to attach devices to hedgehogs (*Erinaceus europaeus*) enables long-term high-resolution studies. *Ecol. Evol.* **2019**, *9*, 672–679. [CrossRef] [PubMed]

39. R Development Core Team. *R: A Language and Environment for Statistical Computing*; R Foundation for Statistical Computing: Vienna, Austria, 2018.

40. RStudio Team. *RStudio: Integrated Development Environment for R*; RStudio, Inc.: Boston, MA, USA, 2016.

41. QGIS. Free Software Foundation, Inc. Available online: http://www.gnu.org (accessed on 22 August 2019).

42. Calenge, C. The package "adehabitat" for the R software: A tool for the analysis of space and habitat use by animals. *Ecol. Model.* **2006**, *197*, 516–519. [CrossRef]

43. Munger, J.C. Home ranges of horned lizards (Phrynosoma): Circumscribed and exclusive? *Oecologia* **1984**, *62*, 351–360. [CrossRef]

44. Spencer, S.R.; Cameron, G.N.; Swihart, R.K. Operationally defining home range: Temporal dependence exhibited by hispid cotton rats. *Ecology* **1990**, *71*, 1817–1822. [CrossRef]

45. Ruby, N.F.; Joshi, N.; Heller, H.C. Constant darkness restores entrainment to phase-delayed Siberian hamsters. *Am. J. Physiol. Regul. Integr. Comp. Physiol.* **2002**, *283*, R1314–R1320. [CrossRef]

46. Kuechly, H.U.; Kyba, C.C.M.; Ruhtz, T.; Lindemann, C.; Wolter, C.; Fischer, J.; Hölker, F. Aerial survey and spatial analysis of sources of light pollution in Berlin, Germany. *Remote Sens. Environ.* **2012**, *126*, 39–50. [CrossRef]

47. Dewey, M.; Metap: Meta-Analysis of Significance Values. R package version 1.0. 2018. Available online: http://www.dewey.myzen.co.uk/meta/meta.html (accessed on 30 July 2020).

48. Rast, W.; Barthel, L.M.F.; Berger, A. Music festival makes Hedgehogs move: How individuals cope behaviorally in response to human-induced stressors. *Animals* **2019**, *9*, 455. [CrossRef]

49. Frank, K.D. Effects of artificial night lighting on moths. In *Ecological Consequences of Artificial Night Lighting*; Rich, C., Longcore, T., Eds.; Island Press: Washington, DC, USA, 2006; pp. 305–344.

50. Molenaar, J.D.; Jonkers, D.A.; Henkens, R.J.H. *Wegverlichting en Natuur*; DWW Ontsnipperingsreeks: Wageningen, The Netherlands, 1997; pp. 1–98.

51. Dowding, C.V.; Harris, S.; Poulton, S.; Baker, P.J. Nocturnal ranging behaviour of urban hedgehogs, *Erinaceus europaeus*, in relation to risk and reward. *Anim. Behav.* **2010**, *80*, 13–21. [CrossRef]

52. Bird, B.L.; Branch, L.C.; Miller, D.L. Effects of coastal lighting on foraging behavior of beach mice. *Conserv. Biol.* **2004**, *18*, 1435–1439. [CrossRef]

53. Molenaar, J.D.; Henkens, R.J.H.; ter Braak, C.; van Duyne, C.; Hoefsloot, G.; Jonkers, D.A. *Effects of Road Lights on the Spatial Behaviour of Mammals*; Alterra, Green World Research: Wageningen, The Netherlands, 2003; pp. 1–72.

54. Reeve, N.J. *Hedgehogs*; T & AD Poyser: London, UK, 1994.

55. Doncaster, C.P.; Krebs, J.R. The wider countryside-principles underlying the responses of mammals to heterogenous environments. *Mamm. Rev.* **1993**, *23*, 113–120. [CrossRef]

56. WTimpact–Citizen Science as a Tool for Knowledge Transfer. Available online: http://www.izw-berlin.de/en/wtimpact-citizen-science-as-a-tool-for-knowledge-transfer.html (accessed on 22 July 2020).

57. Hölker, F.; Moss, T.; Griefahn, B.; Kloas, W.; Voigt, C.C.; Henckel, D.; Hänel, A.; Kappeler, P.M.; Völker, S.; Schwope, A.; et al. The dark side of light: A transdisciplinary research agenda for light pollution policy. *Ecol. Soc.* **2010**, *15*. Available online: https://www.jstor.org/stable/26268230 (accessed on 30 July 2020). [CrossRef]

58. Bontadina, F. Strassenüberquerungen von Igeln (*Erinaceus europaeus*). Ph.D. Thesis, University of Zurich, Zurich, Swiss, 1991.

Article

Urban Hedgehog Behavioural Responses to Temporary Habitat Disturbance versus Permanent Fragmentation

Anne Berger [1,2,*], **Leon M. F. Barthel** [1,2], **Wanja Rast** [1], **Heribert Hofer** [2,3,4] **and Pierre Gras** [2,5]

1   Department of Evolutionary Ecology, Leibniz Institute for Zoo and Wildlife Research, Alfred Kowalke Straße 17, 10315 Berlin, Germany; Barthel.leon@gmx.de (L.M.F.B.); rast@izw-berlin.de (W.R.)
2   Berlin Brandenburg Institute of Advanced Biodiversity Research (BBIB), 14195 Berlin, Germany; hofer@izw-berlin.de (H.H.); pierregras@gmx.de (P.G.)
3   Department of Biology, Chemistry, Pharmacy, Freie Universität Berlin, Takustrasse 3, 14195 Berlin, Germany
4   Department of Veterinary Medicine, Freie Universität Berlin, Oertzenweg 19b, 14195 Berlin, Germany
5   Department of Ecological Dynamics, Leibniz Institute for Zoo and Wildlife Research, Alfred Kowalke Straße 17, 10315 Berlin, Germany
*   Correspondence: berger@izw-berlin.de

Received: 14 October 2020; Accepted: 10 November 2020; Published: 13 November 2020

Simple Summary: Wildlife is exposed to environmental disturbances. Some are limited to a short period and pass by, others are of a permanent nature. Often these two kinds of disturbances occur simultaneously. This makes it difficult to disentangle the specific behavioural response to each disturbance. As species may respond to different disturbances in different ways, it is important to know the species-specific and disturbance-specific responses to develop effective species conservation action. We investigated the behavioural responses of European hedgehogs (*Erinaceus europaeus*) in Berlin to temporary disturbance (in the form of an open-air music festival) and permanent disturbance (in the form of habitat fragmentation). We show that a music festival is a major stressor that strongly influences all investigated behaviours. Urban hedgehogs in a highly fragmented area showed subtle behavioural changes compared to those in low-fragmented areas, suggesting that fragmentation was a moderate challenge which they could cope with. Thus, the temporary disturbance by a music festival had a more serious impact on hedgehog behaviour than permanent disturbance caused by fragmentation. Moreover, we show that males responded stronger to the transient disturbance and females responded stronger to habitat fragmentation.

Abstract: Anthropogenic activities can result in both transient and permanent changes in the environment. We studied spatial and temporal behavioural responses of European hedgehogs (*Erinaceus europaeus*) to a transient (open-air music festival) and a permanent (highly fragmented area) disturbance in the city of Berlin, Germany. Activity, foraging and movement patterns were observed in two distinct areas in 2016 and 2017 using a "Before & After" and "Control & Impact" study design. Confronted with a music festival, hedgehogs substantially changed their movement behaviour and nesting patterns and decreased the rhythmic synchronization (DFC) of their activity patterns with the environment. These findings suggest that a music festival is a substantial stressor influencing the trade-off between foraging and risk avoidance. Hedgehogs in a highly fragmented area used larger home ranges and moved faster than in low-fragmented and low-disturbed areas. They also showed behaviours and high DFCs similar to individuals in low-fragmented, low disturbed environment, suggesting that fragmentation posed a moderate challenge which they could accommodate. The acute but transient disturbance of a music festival, therefore, had more substantial and severe behavioural effects than the permanent disturbance through fragmentation. Our results are relevant for the welfare and conservation measure of urban wildlife and highlight the importance of allowing wildlife to avoid urban music festivals by facilitating avoidance behaviours.

Animals **2020**, *10*, 2109

Keywords: disturbance; fragmentation; anthropogenic habitat change; urban ecology; behavioural plasticity; GPS telemetry; hedgehogs

---

1. Introduction

A disturbance describes a change in an environment that poses a change in the ecosystem [1]. Nowadays, all ecosystems are increasingly subjected to anthropogenic disturbances which include a host of features, from artificial light, noise and air pollution to habitat fragmentation [2–4]. When habitats are disturbed, animals must reconsider trade-offs between foraging or mating success and risk aversion similar to the landscape of fear [5,6]. Humans may change these environments slowly, e.g., through changes over decades, or fast within a few days when building a road, removing a forest patch or harvesting hay from a meadow [4,7,8]. Fast changes require individuals to show appropriate behavioural flexibility or plasticity within their own lifetime to cope. If facing frequent or regular disturbances, such flexibility could contribute to population viability. On the individual scale, there are several options to respond to disturbance: disperse and seek another place to live or stay or adjust through behavioural plasticity [4,6]. On the population scale, genetic changes may be possible to adapt to the new environmental conditions if they are permanent and exert sufficient selection pressure [7].

It can be challenging to disentangle the effects of transient and permanent stressors as different forms of disturbance often operate simultaneously. In this study, we investigate the effects of habitat fragmentation and transient disturbance on individual spatiotemporal behavioural responses in urban European hedgehogs (*Erinaceus europaeus*) by monitoring two populations in Berlin, Germany. One population lives in a habitat with permanent fragmentation, the other faced several severe transient disturbances associated with a music festival.

Festivals often take place on open green areas. Within days, these green areas will be modified and disturbed through heavy machinery and the presence of many people as the festival's infrastructure is being set up. During the festival, a huge crowd of people wanders around the festival site, loud music plays throughout most of day and night and for enjoyment and safety, the whole area is lit up and fenced. Thus, during the festival, wildlife living on the site is confronted with extraordinary amounts of habitat changes, noise, light pollution and people activities.

Fragmentation is created by transforming habitats into smaller patches, thereby creating a permanent mosaic-like landscape [9–11]. Thus, fragmentation can impede movements in the matrix between resource patches [12] and limit access to mating partners [13,14], forcing animals to adjust their behaviour to these changes [15]. Some examples of behavioural plasticity include increasing home ranges and adjusted activity rhythms to cope with fragmentation [16–18].

The European hedgehog (*Erinaceus europaeus*) is a solitary, hibernating, nocturnal insectivore which lives in a wide range of habitat types [19,20]. Long-term monitoring studies found that the overall hedgehog population in several countries is declining, sometimes dramatically so [21–24]. The underlying mechanisms responsible for these declines are likely to be complex and multifactorial, including habitat loss, interspecific competition, traffic accidents and intensification of agriculture [21,25–28]. Increasing fragmentation and decreasing hedgehog densities may cause negative genetic effects associated with isolated populations [29–32]. Usually, urban green spaces such as public parks are favourable environments for urban hedgehogs, providing easily accessible food and resting (nest) sites. If the urban matrix provides a very patchy environment with only small, isolated patches of suitable habitat dispersal out of unfavourable habitats can be problematic and limited [7]. It is at present unclear how hedgehogs cope with habitat fragmentation as an example of environmental disturbance and how this might influence their population structure. In order to develop effective conservation concepts, threats to hedgehogs as well as their limits of adaptability should be known. Due to their cryptic lifestyle, little is known about their behaviour in the wild, the effect of different disturbances on their behaviour, and the limits to hedgehog adaptability [28,33].

In the case of habitat fragmentation, hedgehogs need to trade-off between foraging profitability and safety, by changing the distance to anthropogenic sources of disturbance, and reconsider options for easily accessible food. To investigate the behavioural plasticity of hedgehogs in response to transient changes and habitat fragmentation, we analysed the behavioural response at different levels. First, we used GPS data to investigate the daily home range size and movement behaviour of hedgehogs, both closely related to foraging [34,35]. Second, we assessed the circadian behaviour patterns of hedgehogs, known to be usually strictly nocturnal but adjustable as a response to stress [19,35]. Third, we monitored nesting behaviour. Nests are important resting sites of both sexes and typically used for several days [36,37], females also keep their offspring in nests.

We investigated two scenarios. First, we studied effects of a transient and intense disturbance in the Treptower Park in Berlin, Germany, which has little habitat fragmentation and low levels of night-time disturbance outside the festival period. Second, we investigated the effects of habitat fragmentation in the Tierpark Berlin, which has many daytime visitors but very low nightly disturbance. We hypothesise that both transient disturbance and habitat fragmentation influence space use, activity patterns and nesting behaviour to different degrees.

We predict the following behavioural adjustments in response to (A) transient disturbance:

(1) Regarding space use, we predict that hedgehogs avoid transient disturbed areas in their habitat, by either leaving the area or adjusting movement patterns: (1a) We predict avoidance of the disturbed (festival) area by shifting the centre of the nightly used area. (1b) Additionally, we predict a decreased size of nightly home ranges in or close to the festival area due to avoidance of the disturbed (festival) area.

(2) We predict that hedgehogs adjust their movement behaviour. The animals now have to look for the same amount of food in a potentially less favourable and/or smaller area and thus foraging effort may have to be increased. We, therefore, predict an increase in search intensity, greater turning angles and slower speed under disturbance.

(3) The general levels of activity will be reduced due to increased vigilance behaviour which in hedgehogs is characterised by immobility (little activity).

(4) We predict that high levels of disturbance during the festival induces females and males to switch their nests more often and the number of days spent in the same nest decreases.

We predict the following behavioural adjustments in response to (B) fragmentation:

(1) In urban areas, or areas with predation risk, hedgehog movements are strongly associated with linear structures [35], fragmentation will increase the area of space that is of no interest to hedgehogs and thus increase the distances they have to cover. Thus, in a highly fragmented park, the home range area would be bigger than in the low-fragmented park.

(2) As fragmentation is likely to increase distances for commuting between favourable food patches, movement characteristics should change. We expect a faster speed, a larger number of smaller turning angels and a lower search intensity than in a low-disturbed, low-fragmented habitat.

(3) Fragmentation could influence nesting behaviour in two ways: animals either have to change their nests more frequently to be closer to favourable food patches, or extend their stay in nests if they are close to favourable food patches. During disturbances, we predict a more frequent nest change than in undisturbed areas or at undisturbed time.

2. Materials and Methods

2.1. Study Areas and Study Design

Fieldwork was conducted between 10th of August and 20th of September 2016 in the Treptower Park, in southeast Berlin, Germany (52.48846° N, 13.46974° E) and between 14th of August and 4th of September in 2017 in the Tierpark, a big park containing a zoological garden in East Berlin, Germany (52.50326° N, 13.52976° E). Both parks include big trees and several green spaces of short grass swards with shrubs of various sizes and hedges. Additionally, there are playgrounds, larger sealed areas and footpaths. Treptower Park is a public city park of 88.2 ha in size; it is open to the general public 24 h and 7 days per week whereas the Tierpark is closed to the general public from dusk to the morning. The Tierpark is a zoological garden of about 160 ha in size; it contains numerous animal enclosures, small buildings, water ditches and many concrete footpaths, creating a mosaic-like fragmented habitat with many areas that are non-accessible to hedgehogs. The maintenance of the parks is similar, with leaf litter being removed from some areas, particularly the footpaths, but left in bushes and scrub throughout the year, offering a similar habitat in both parks suitable for hedgehogs and other wildlife.

In 2016, the Lollapalooza Festival with over 140,000 visitors took place in the Treptower Park for which a substantial portion of the park (excluding a war memorial and the southeastern segment) was temporally substantially changed. Music stages, amusement facilities and enclosures were constructed and built between 29th of August and 9th of September, the festival took place on 10th and 11th of September, and deconstruction of all facilities took place from 12th of September onwards. We collected data on hedgehog movements and behaviour before the festival (pre-festival) until construction work for the festival started and during the festival-phase, including the time periods of construction and deconstruction. The festival phase lasted from 29th of August until 16th of September 2016. The pre-festival period is defined from 10th of August until 28th of August and represents the control for both the transient disturbance caused by the festival and, as an example of hedgehogs living in a low-fragmented and low disturbed urban habitat, for the hedgehogs studied in the Tierpark.

To investigate the effect of the transient disturbance, all data from hedgehogs in the Treptower Park were used by comparing those collected before and during the festival. To investigate the effect of fragmentation change, data from the highly fragmented park (Tierpark) were compared with the pre-festival data from the low-fragmented park (Treptower Park).

2.2. Hedgehog Capture and Logger Attachment

At the beginning of each study period, surveys were carried out during two to three nights at least one hour after sunset to find active hedgehogs by spotlighting (P14.2, LED Lenser, Solingen, Germany). Every hedgehog located was marked with five shrink tubes glued to the spines [38]. The tubes were labelled with a number starting with 1 to identify them during recapture [39]. From all previously captured hedgehogs, we selected eight hedgehogs (four of each sex) and equipped them with Global Positioning System (GPS)/Accelerometer (ACC) loggers (e-obs GmbH, München, Germany) and Very High Frequency (VHF) (Dessau Telemetrie, Dessau, Germany) transmitters using a backplate system [40]. We only used hedgehogs with a body mass exceeding 600 g to ensure that the attached logger equipment was below the 5% body mass rule [41,42].

During the study, all hedgehogs were weighted and inspected for any problems once a week; these occasions were also used for the necessary recharging of the GPS/ACC data loggers. At the end of the studies, the animals were caught again, weighed, the loggers removed, and the backplate system was cut off the spines.

All procedures performed involving the handling of animals were in accordance with the ethical standards of the institution (IZW permit 2016-02-01) and German federal law (permission number Reg0115/15 and G0104/14).

2.3. Logger Sampling Setup

In order to find the animals with data loggers at any time, VHF transmitters continuously broadcast signals throughout the whole study period. GPS positions were taken during expected activity times of hedgehogs from 7:00 p.m. to 7:00 a.m. in 10 min intervals, in bursts of five points, which means that the GPS record 5 singular GPS positions of one-second difference. Acceleration data were recorded alongside GPS data every minute. These three-dimensional accelerometers were programmed to record a short burst of simultaneous high-resolution data. A sampling frequency of 100 Hz per axis was chosen for the present study. A burst took 2.64 s for two individuals (01_2016 and 19_2016); the other individuals were recorded with bursts of 2.5 s. This difference in burst length is not ideal, although the burst length is only important for three out of 25 predictors used for the model (see "Behaviour prediction and budget" in Section 2.5.2).

2.4. Nesting Behaviour Monitoring

Nesting behaviour was recorded every day by locating the VHF signals of each hedgehog carrying a logger (TRX-1000S, Wildlife Materials Inc., Murphysboro, IL, USA, or Wide Range Receiver AR 8200, AOR Ltd., Tokyo, Japan). The nest position was recorded using a Garmin GPSmap 60CSx device (Garmin Deutschland GmbH, Garching, Germany) every day for the nest the hedgehogs stayed in during the daytime. If a hedgehog was found in the vicinity (2 m) of a nest without a new nest, the existing nest was noted as the day nest of the hedgehog. Some animals had to be removed from the dataset of 2017 in the Tierpark because they occupied fewer than 5 nests. If a hedgehog lost a transmitter and could not be found to re-attach the transmitter, or night work had been discontinued, the nest surveys for the individual were stopped.

2.5. Analyses

To investigate the effect of the transient disturbance of the festival, all data (GPS, ACC and nest behaviour) from the hedgehogs in the Treptower Park were used by comparing before and during the festival as a temporal control. To investigate the effect of permanent habitat change (fragmentation), data from the highly fragmented park (Tierpark) were compared with data from the low-fragmented park (Treptower Park) from the pre-festival period in a spatial control.

2.5.1. GPS Data

Because of high fluctuations in some GPS points, we excluded all points that were more than 1000 m away from the study site. We then calculated the average of all remaining GPS points measured within a burst and we calculated the distance between two consecutive GPS positions to derive the speed the hedgehogs had to overcome for that distance. Outliers of more than 2 m/s speed from one location to the next were excluded. Overall, the mean error of the GPS position was between 10 and 40 m.

To conserve the natural variability, we decided to use every night as a single event and calculated the following values accordingly using R and Rstudio [43,44]. To assess the nightly used area, we calculated the 95% Minimum Convex Polygon (mcp95) and the Kernel density estimates (kde50) as a core area of use by smoothing with the ad hoc method (href) [45]. In both cases, we used a linear model [46] to perform a linear mixed effects analysis on the relationship between used area (mcp95 or kde50) and our three treatments. As fixed effects, we entered treatment and sex with interaction into the model. As random effects, we had intercepts for individuals. We did not include age as a covariate because age determination in hedgehogs is generally difficult. Moreover, we can guarantee that all hedgehogs were adults, meaning that they were born at least a year before the study To fit these assumptions, we included a power function using varPower() as weights. Visual inspection of residual plots did not reveal any obvious deviations from homoscedasticity or normality. To obtain *p*-values of the mixed-effect model, an ANOVA (Type II) was performed on the fitted model using the 'Anova' function from R-package 'car' and applying a Wald chi-square tests [47]. Subsequently, differences between groups of the fixed-effects of the fitted model were tested using the R package 'multcomp' [48]. We compared the groups: pre vs. pre comparing both sexes, pre vs. festival within sex, pre vs. fragmented within sex, fragmented vs. fragmented comparing both sexes.

We proceeded in a similar manner when analysing the movement speed of hedgehogs as a travelled distance for a time interval between two consecutive GPS positions (m/s). We used the lme4 package to perform a linear mixed effects analysis on the relationship between speed and treatment [49]. As fixed effects, we entered treatment and sex (with interaction) and as random effects, we had intercepts for individuals. Visual inspection of residual plots did not reveal any obvious deviations from homoscedasticity or normality. *p*-values were obtained by applying an ANOVA (Type II) and group comparison as described above.

To evaluate how animals use the available habitat, we calculated a ratio of area used (mcp95 in (m^2)) and distance travel (m) per night (calculated with st_length [50]), resulting in a measure of search intensity with units (m/(m^2 × d)) or moved distance per square metre and day. To evaluate whether treatment or sex had an effect on this parameter, we used the SpaMM package [51] by first finding the right fit and then comparing the null model with different models. The SpaMM package was necessary to counter auto-correlation.

To detect wherever the hedgehogs shifted their utilised area during the festival, the longitude and latitude values of centroids per night were used separately using function centroid() [52]. We normalised the values by subtracting the mean value from the pre-festival phase and worked with the absolute values. For the latitude and longitude values, linear mixed effect models were fitted with treatment as the fixed effect and sex as the random effect. Both time values had to be square-root transformed before fitting the model to meet the assumptions of homoscedasticity or normality. P-values were obtained from ANOVA (Type II) as described above.

Movement of hedgehogs was further characterized by calculating turning angles [53] and plotted as absolute values because we were interested in the general movement. Results were then randomly sampled and compared in a permutation approach 1000 times using a two-sample two-sided Kolmogorov–Smirnov test [43]. This comparison was only done on the treatment level (pre-festival vs. festival and pre-festival vs. fragmentation).

2.5.2. ACC/Acceleration Data

To account for missing data because of recharging or logger malfunctioning, all data with fewer than 1430 measurements between 00:00 and 23:59 were removed from the data set (n = 1440 for complete 24 h). This removal of data ensured that only days with a comparable length and the same number of records during days and nights were considered for the analysis and, therefore, did not favour behaviours that only occurred during a specific time of the day.

Behaviour Prediction and Budget

Acceleration raw data were tested for missing measurements within the bursts. All bursts where fewer data were recorded than intended by the settings (n = 264) were removed. We sued the remaining data for behaviour detection by applying a supervised machine learning algorithm. The train and test dataset for the behaviour recognition were taken from a previous study. The whole procedure is described in [54]. By joining multiple Support Vector Machines (SVM), the selected behaviours were classified. Here, we considered three behaviour classes: resting, balling up and locomotion (referred to as walking). To account for behaviours that are not included in the model but might occur in hedgehogs, a threshold for the probability belonging to a class of 0.7 was set for the SVM. Otherwise, the behaviour was classified as "other" behaviour.

The SVM model was then used to assign a predicted behaviour to every burst and its corresponding timestamp. The behaviour of every individual was treated for the following tests separately. To test for effects on behaviour classes, a general linear model was performed [49] taking a quotient of the behaviour in relation to all behaviours. As a fixed effect, the treatment, as well as the sex and the interaction, were put into the model. Individuals were included as random effect. This was followed by an analysis of variance [47] and general linear hypotheses and multiple comparisons [48] using the same matrix as before.

Stress Detection via the Degree of Functional Coupling (DFC)

The Degree of Functional Coupling (DFC) is a measure for the synchrony of (internal) cyclic behaviour and the (external) environmental 24 h rhythm [55,56]. To calculate DFCs, the standard deviations of raw acceleration data of all three axes were calculated and summed up per measurement interval (burst). Following the protocol of Berger et al. [56,57], this time series was autocorrelated in order to filter out the noise and enhance rhythmic components. Afterwards, a Fourier transformation was used to break it down into its rhythmic components, as described by the percentage of each component in the original time series. The longest Fourier period tested covered the entire length of the autocorrelation function (here three days); the shortest Fourier period tested was twice the sampling interval (here 2 min). The DFC is then calculated by dividing the portion of Fourier transformation components that harmonize with the 24 h rhythm by the entirety of the Fourier spectrum. To gain an adequate statistical power of the 24 h period, DFC were calculated for time series of three days equivalent to the procedure of a moving average (first data set covers day 1 to 3, second data set covers day 2 to 4 and so on). The resulting DFCs were assigned to the day of the three days that entered the calculation for the first time. These data were then analysed using a linear mixed effect model with treatment and sex and their interaction. Values had to be arcsine transformed in order to meet the assumptions of homoscedasticity and normality. Afterwards, an analysis of variance [47] and general linear hypotheses and multiple comparisons [48] were performed.

2.5.3. Nesting Behaviour

For each nest, the duration of occupation was scored as exact if both starting and stopping dates of nest use were recognised, or as right-censored (a minimum estimate), if either the starting date or the stopping date at the beginning and end of study periods were not known. Then, the survivorship function was calculated using package survival [58] separately for both parks and treatment conditions. If significance was found, a post hoc Mantel test was performed to detect the source of the difference.

3. Results

Data were collected for 16 hedgehogs (8 of each sex) over a period of 3 to 41 nights per animal; between 95 and 2566 GPS locations per animal were recorded (Table S1). In total, 39 days from the Tierpark data set and 79 days from the Treptower Park data set were removed from further analyses due to high number of outliers or missing data. In total, we tracked 426 nights including 156 nights for the control, 152 for the festival and 118 for the highly fragmented site. Sexes are represented with 236 females and 190 male data points. In Figures S1 and S2, we mapped GPS locations of some hedgehogs in the two parks to indicate movement characteristics (paths) and spatial obstacles for the hedgehogs which they have to run around.

The daily maximum temperatures were between 18 and 34.5 °C (mean = 25.2 °C) during the 2016 study period (41 days) and between 18 and 30 °C (mean = 22.1 °C) during the 2017 study period (21 days). The lowest nighttime temperatures ranged between 7 and 18.4 °C (mean = 13.8 °C) during the 2016 study period and between 6.3 and 21 °C (mean = 13 °C) during the 2017 study period. The mean rainfall was 0.8 mm/day (11 days with rain) during the 2016 study period and 1.1 mm/day (10 days with rain) during the 2017 study period (https://www.wetteronline.de/, further weather information is listed in Table S2).

3.1. Home Range Size

The nightly used area (Figure 1a) measured by the mcp95 was significantly smaller during the festival ($\chi^2 = 54.82$, df = 2, $p < 0.001$) and in females ($\chi^2 = 6.48$, df = 1, $p = 0.011$, details on all models Table S3). Although treatment was similar between sexes ($\chi^2 = 1.74$, df = 2, $p = 0.42$), females in the control group used smaller areas by 1.9 times than males in the control group (2.55 ha to 4.71 ha, respectively). While both sexes decreased their home ranges during the festival (females acute change estimate E = -1.029 ± 0.22 (standard error), z = -4.648, $p < 0.001$; males -1.467 ± 0.38, z = -3.890, $p < 0.001$), only females increased the area in the highly fragmented habitat (E = 1.942 ± 0.69, z = 2.796, $p = 0.028$). Male hedgehogs occupied a bigger area than females in the control group and showed only a slight increase in the highly fragmented area from the control group (E = 0.730 ± 0.8663, z = 0.843, $p = 0.92$). Thus, males and females in the highly fragmented habitats had similar-sized home ranges

(E = 0.6849 ± 0.8706, z = 0.787, *p* = 0.9390). These were replicated by the used core area measured by the kde50, except for the comparison between pre-festival and fragmented within the females (Figure S1).

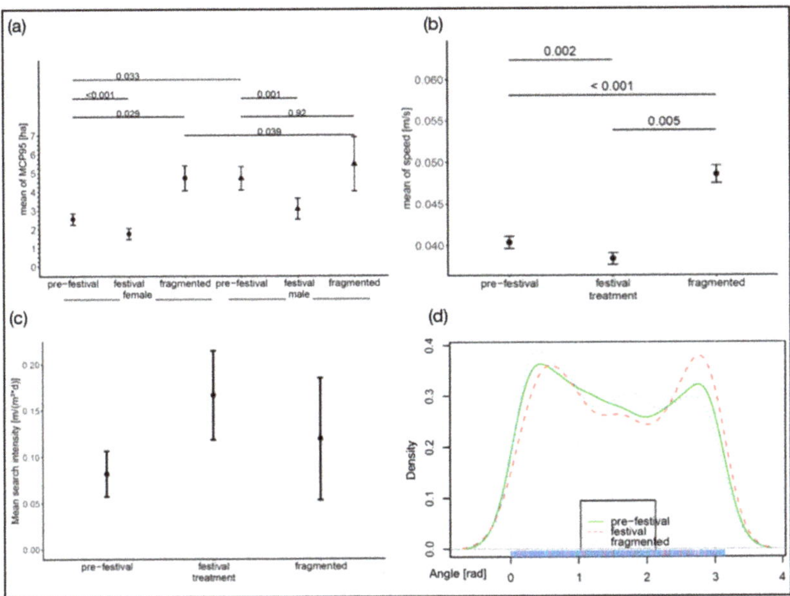

Figure 1. Results from the GPS data analyses. (**a**) MCP 95 measured per treatment and sex per day (dots/triangles indicate mean values, whiskers are confidence interval); (**b**) mean speed (m/s) for each treatment over both sexes (dots indicate mean values, whiskers are confidence interval); (**c**) mean search intensity (m/(m^2 × d)) for each treatment over both sexes (dots indicate mean values, whiskers are confidence interval); (**d**) Distribution of absolute turning angles of one subsample that were tested.

3.2. Movement Speed and Turning Angle

Hedgehogs moved significantly faster in the highly fragmented area (0.049 m/s ± 0.001) than in the low-fragmented area during the control period (0.040 m/s ± 0.008, Figure 1b; χ = 33.3, df = 2, *p* < 0.001). During the festival, hedgehogs moved even more slowly (0.038 m/s ± 0.008, Figure 1b). Mean search intensity (m/m^2 × d) was lowest for the control group and the highest during the festival (χ^2 = 7.42, df = 2, *p* = 0.024); the highly fragmented area has the largest confidence interval in search intensity (Figure 1c). There was no difference between the sexes (χ^2 = 0.11, df = 1, *p* = 0.74) nor was the interaction significant χ^2 = 3.22, df = 2, *p* = 0.20). Although the curves of the subsample of the turning angles appear to look different, there was no significant difference in the characteristics of the turning angles (Figure 1d).

3.3. The Centre of the Nightly Home Range

The centroid values of the daily used area present the mean point of the used area and should have shifted if the hedgehogs used other areas, and the distribution should change if they avoid certain areas. During the control period, hedgehogs focused on a central big open meadow, both sexes had a near to normal distribution in their longitudinal values (Figure 2). During the festival period, hedgehogs moved significantly away from their mean centroids of the control period by on average ~35/30 m (longitudinal/latitudinal) in females and more than 65/105 m in males (effect of treatment latitude χ^2 = 80.59, *p* < 0.001/longitude χ^2 = 80.86, *p* < 0.001, with the interaction of treatment and sex χ^2 = 21.44, *p* < 0.001/χ^2 = 11.79, *p* < 0.001).

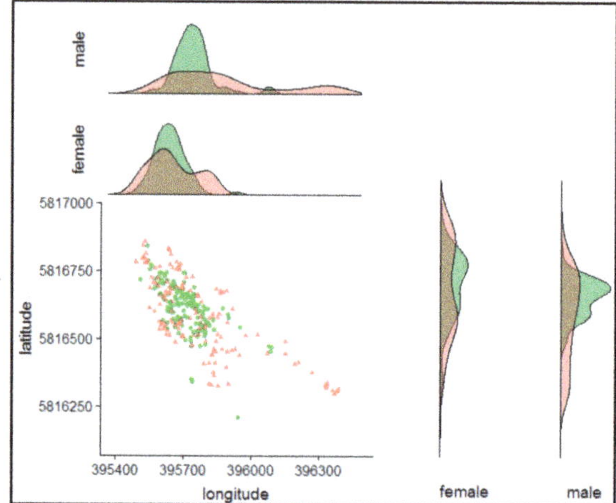

Figure 2. Centroid distribution. Circle/green = pre-festival period, triangle/red = festival period; density plots show the total of kernel density estimates (kde50) in the same range.

3.4. Behaviour Parameters

There were significant differences in the behaviours between the control (pre-festival) and festival period (Figure 3). In females and males, forming a ball (balling behaviour) was detected more frequently during the festival than the control period, females showed an increase of 0.153 and males of 0.2 (Figure 3a). For walking behaviour, only males showed a significant increase of 0.03 (Figure 3b). Resting behaviour was identified less in both sexes during the festival (−0.21 in females, −0.014 in males, Figure 3c). In the DFC values as an indicator of the degree of synchrony of the overall behaviours with diurnal rhythm, both sexes showed the same patterns. Both sexes showed similar values in the control group, whereas the values during the festival were reduced. In the highly fragmented area, we found high DFC values for females and slightly lower DFC values with high variation for males (Figure 3d).

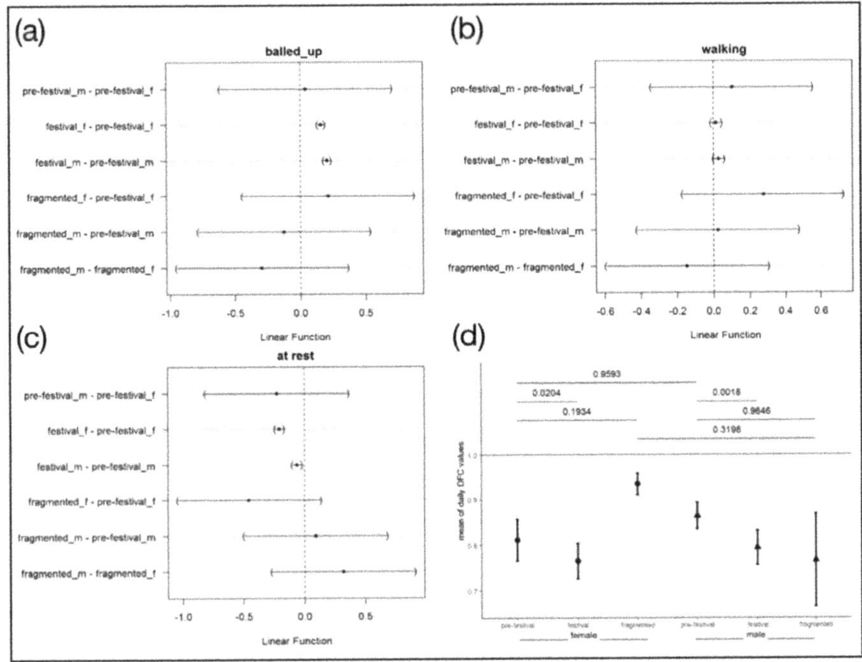

Figure 3. Behaviour parameters in the context of transient and fragmentation habitat changes. Mean values with 95% confidence interval for differences in behaviour parameter counts of balling up (**a**), locomotion (**b**) and immobile behaviour (**c**) of female (f) and male (m) hedgehogs for the different study phases. The difference represents changes in behaviour event counts. Negative values represent a decrease in the behaviour counts while positive values represent an increase. We considered all differences to be significant where the confidence interval does not include 0 (dashed line); (**d**) Mean degrees of functional coupling for each study period and sex (1 means maximal synchrony animal's between behaviour and the environmental 24 h period, 0 means no synchrony), whiskers = 95% family-wise confident intervals.

3.5. Nesting Behaviour

In the control group, female hedgehogs used their nests again on the next day in 66.1% of cases. Nests of male hedgehogs were used with a probability of 57.8% on the next day. The comparison between the control and the festival period showed significant changes in nesting behaviour and differed between the sexes: During the festival, nests of male hedgehogs were used over significantly shorter time periods (Mantel logrank test, N = 156, p = 0.02) and the probability of using a new nest the next day was reduced from 57.8% to 45.5%. After eight days at the latest, males had left the nest and moved to another one. In contrast, for females, values were in general similar to or higher than during the control period (Mantel logrank test, N = 88, p = 0.83, Figure 4).

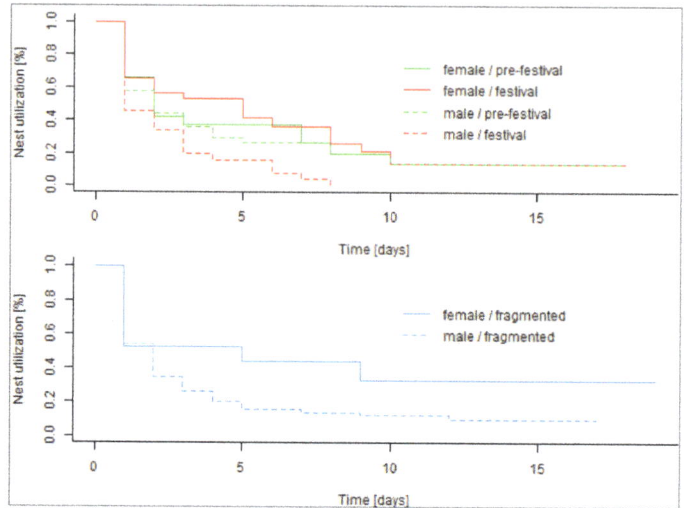

Figure 4. Nest utilization probability for pre-festival (green) and festival (red) phase of nine males (dashed line) and eight females (solid line) and for highly fragmented area (blue) with shortened *x*-axis for comparison.

4. Discussion

Consistent with our predictions, there was an influence of fragmentation as well as of transient changes in habitat on hedgehog spatiotemporal behaviours. However, hedgehogs demonstrated sex-specific responses to different types of disturbances.

By using the same measuring method, the same number of animals, the same gender ratio, the same season, similar park sizes (in parts of the city with comparable urbanity index) and hedgehog population densities, we tried to achieve the greatest possible comparability between the two study areas and years. The two study areas, Treptower Park and Tierpark, are comparable considering their sealing index [59], which closely corresponds to other urbanization indices, such as human population density, disturbance by humans and pets, noise and light pollution [60,61]. In hedgehogs with main natural prey being earthworms and ground beetles, the sealing index will also be linked to food availability.

Compared to Treptower Park, the Tierpark is clearly more fragmented for hedgehogs, which is also the most obvious difference between the two parks: in Treptower Park almost all fences can be slipped through by hedgehogs. Only a long semicircular wall around the Soviet memorial in the middle of the park is impassable for hedgehogs. The Tierpark, on the other hand, consists of a large number of enclosures, most with borders insurmountable for hedgehogs (moats with wall edges, dense fences as protection against rats). Even if the hedgehogs manage to get into the enclosures, it can be more dangerous for them there than outside—for example, keepers once found a pregnant hedgehog kicked to death by takins (*Budorcas bedfordi*) in an enclosure. Hedgehogs therefore usually move around the enclosures, often on visitor paths (see Figure S3, male 2017_31).

Moreover, the weather data of the two study years do not differ significantly from each other: both summers were relatively warm; at least warmer than the average temperatures of 4 reference years (Table S2). Summer 2016 was dryer than average, in 2017 rainfall was above average. The average temperatures of September 2017 were lower than those of September 2016, and the amount of rainfall in August 2017 was higher than in August 2016. However, the variance of weather data during both study periods was greater than the differences between the two. As a warmer drier summer will make earthworms harder to find and, therefore, hedgehogs need to roam further to meet their energy requirements, then hedgehogs in our study should have had smaller home ranges in 2017 than in 2016. However, our measurements on hedgehogs in the Tierpark in 2017 and in 2018 (unpublished data) demonstrated much larger distances and home ranges than those of the hedgehogs in Treptower Park.

Studies on urban foxes also showed that seasonal differences (and thus differences between years) are attenuated due to the permanent availability of food [62]. In this respect, we have come to the conclusion that the slightly different weather conditions in the two study years, although they are boundary conditions to be considered, are not causally relevant for our study results.

Due to transient changes of the festival, hedgehogs decreased home range size, movement speed and rested less but increased search intensity, performed balling up behaviour more frequently and moved further away from their previous home range center. Thus, hedgehogs avoided the core festival area, changed their behavioural budget and their activity patterns indicated stress.

In fragmented areas, hedgehogs increased home range size and movement speed without any significant effects on their behaviour budget or synchrony to environmental 24 h rhythm.

These results suggest that hedgehogs can adjust to permanent disturbances such as fragmentation. It was also interesting to see that disturbance affected females differently to males; males seemed to be more active in avoiding or coping with the environmental changes. These findings show that urban hedgehog populations can be resilient to transient as well as permanent habitat changes. However, care should be taken when extrapolating our results to other urban environments.

4.1. Home Range Sizes

Consistent with our predictions, nightly home ranges decreased during the festival (A1b) and home range size was enlarged in highly fragmented areas (B1). Fragmentation increased the neglected area and thus increased the distances hedgehogs moved between resource patches. If hedgehogs in the highly fragmented area accessed one spacious food patch the size of home ranges did not vary between highly and low-fragmented areas [63].

Our home range sizes are consistent with earlier findings for urban areas [35] and in other hedgehog species in suburban habitats [64] but are smaller than in many studies [65–69]. However, these studies calculated 100% MCP from longer tracking periods, which commonly leads to bigger home range than 95% MCP of high-resolution daily data. Furthermore, home range size may not be a constant over time, as previous studies showed big differences in individuals which were compared between two consecutive years [35]. Our results are consistent with the general conclusion that males used larger areas than females [70]. In a study on hedgehogs in rural habitats, hedgehogs further away from settlements had higher energy expenditure, presumably because they had longer distances to cover [71]. The same study showed that hedgehogs may restrict their movements in the presence of predators such as badgers. It is, therefore, important to know whether the trade-off between the use of spatially fragmented areas (and concomitant high energy expenditure) and the risk of encountering predators is biased in favour of hedgehogs in closer proximity to settlements [71].

4.2. Movement Speed and Turning Angle

A comparative study on movement behaviour of mammals demonstrated a negative effect of anthropogenic disturbance on long distance displacements [72]. In key indicators of movement behaviour, our results were consistent with predictions that during transient disturbance, hedgehogs should increase search intensity, enlarge their turning angles and reduce speed (A2). Also, in the highly fragmented areas, hedgehogs moved at a higher speed and increased the number of (smaller) turning angles but we did not find the predicted lower search intensity than in low-fragmented habitat (B2).

In our study, we calculated mean speed based on GPS positions measured during both periods of active and inactive behaviours. We report similar values to the mean (average over sex and season) speed of Ethiopian hedgehogs of 0.039 m/s [70]. We identified differences between highly and low fragmented areas, showing that hedgehogs in highly fragmented areas moved on average by 20% faster, a substantial increase and relevant for total energy turnover [71]. In former studies, it was shown that hedgehogs move faster when passing bigger roads, which could mean that hedgehogs on tarmacked or concrete paths in the fragmented area in the Tierpark also increase their speed if they use them [73]. Higher speed in the highly fragmented habitat could also be a consequence of commuting between foraging patches, whereas slow movements in hedgehogs are often indicative of foraging [34,74]. These factors will increase the variance in search intensity in the highly fragmented habitat.

4.3. Centre of Nightly Home Range

Our results were consistent with the predictions regarding a shift of the centre of nightly home ranges during the transient disturbance (introduction (A1a)).

The detected spatial shift of the nightly used core area is a clear indicator that hedgehogs avoided the core festival area. As the open park areas, where the local hedgehogs usually foraged, was blocked by festival visitors or, during the construction work, by workers, hedgehogs stayed for longer periods, and sometimes the entire night, in areas on the edge of the park. Or they did not leave the bushes where their nest where located. The observed sudden shift of the home range centroids which is unusual for the season and linked to the timing of the festival is a clear sign of avoidance. Such a response was also shown by hedgehogs living in cultivated land as a response to dramatic changes in resource quality [75]. In a similar semi-experimental approach, koalas (*Phascolarctos cinereus*) were tracked before and during a music festival and changed their home ranges in a manner similar to the hedgehogs in Berlin during the festival [76,77]. Hence, it is important to facilitate avoidance behaviour by creating escape routes for wildlife during events such as music festivals.

4.4. Behaviour Parameter

Where and when an animal conducts certain behaviours are crucial for understanding habitat use as well as elucidating the response of wildlife to disturbances as changes in behavioural patterns can be the consequence of disturbance [57].

Our results were consistent with predictions that during transient disturbance, the general level of hedgehog activity will be reduced as vigilance behaviour should increase (A3). During the festival, we observed an increase in the frequency of curling-up or roll-up behaviour. Disturbance first induces an erection of spikes, and then is followed by a complete roll up if the animal is further stressed [19]. When analysing our data, we have to be careful about the possible misidentification between resting and rolled up behaviour. The huge 95% confidence interval in the behavioural parameters suggests a high variability in the behavioural response. Thus, visitors and noises of the festival as well as natural predators such as red foxes and badgers that roam in both highly and low-fragmented areas could cause curling-up behaviour. An alternative response to disturbance is that hedgehogs can run away from a disturbance or a predator, which in turn would lead to increased activity levels. Hedgehogs are known to move out of unfavourable habitats and thus could avoid disturbances if necessary [78].

Although the size of the area used in the highly fragmented habitat was increased, it was not clear whether female hedgehogs may have a higher energetic investment there [71] as they covered similar distances. In terms of behaviour parameters, there was no obvious general pattern as to how hedgehogs responded to different disturbances. Such high inter-individual behavioural variance in response to anthropogenic disturbance is an expression of behavioural flexibility and may contribute to successfully persist in challenging environments such as big cities [54]. However, behavioural flexibility also has the evolutionary cost that a chosen behavioural response may be inappropriate in that situation [79], and these are more likely if the anthropogenic disturbance is unequal to natural disturbances to which there may be an evolved response [41].

Rhythms in behaviours have evolved as adaptations to the environment and enable organisms to behave at the times most suited to their physiology or ecology. DFCs are calculated as a measure of harmony between behavioural rhythms and the most important environmental rhythm, the 24 h period. High DFCs are often found in healthy animals or those that are strongly diurnal or nocturnal [80]. Low DFCs indicate that the animal is weakly synchronized with the environmental rhythm, which can be an indicator of stressors or disease, but also parturition [57,81]. In our study, DFC clearly decreased during the transient disturbance in both sexes, while there was no change detected with regard to fragmentation. Interestingly, we found the highest DFC values in females in a highly fragmented habitat, suggesting that they had adjusted their biological rhythm to the environmental circumstances. DFC values of male hedgehogs in the highly fragmented area had a higher variation, i.e., displayed higher individual differences during this time of the year. This could be expected as male hedgehogs may vary in their behaviour during and shortly after the mating season. Some of the females in our study gave birth to hoglets during the study period, which could have influenced the DFC values in addition to the disturbances. To disentangle the effect of reproductive status (such as parturition, lactation) from transient disturbances, studies in semi-natural enclosures could give baseline data for an energetic and hormonal assessment and for evaluation of DFC values as a non-invasive tool to assess the stress response in wildlife [57].

4.5. Nesting Behaviour

Consistent with our predictions, male hedgehogs switched their nests more often and spent fewer days in the same nest during the transient disturbance of the festival (A4). In contrast, females used their nests longer, which might be related to the fact that some of them gave birth during the study period and thus were restricted to stay in their nests.

Nocturnal hedgehogs change their day nests regularly. Changing nests more frequently rather than reusing a previously built nest entails a higher expenditure of energy. In our study, we recorded much higher numbers of nest changes within a period of up to 40 days than in Irish populations of rural hedgehogs (our study mean values 7.5 in males, 4.9 in females vs. 2.5 in both sexes [37]) but similar ranges to hedgehogs on a golf course in the suburbs of London (males 5–10 (our study) vs. 2–15, females 2–6 (our study) vs. 2–6 [36]). Our females used fewer nesting sites than males, similar to the London study [36]. There are currently no other survival analyses of nesting studies available. With access to data with day to day records it should be possible to find a baseline for European hedgehogs.

4.6. Limitations and Outlook

Overall, we could show that males responded stronger to the transient disturbance, although caring for hoglets may have diminished the response of some females to the transient disturbance of the festival. In contrast, female hedgehogs responded stronger to habitat fragmentation, possibly a consequence of their higher energy requirements for lactation. Details on individual differences in coping strategies as a function of reproductive status or other aspects of individual life histories could help predict responses to specific disturbances. Individual behavioural flexibility is, therefore, an important issue for further studies [17,82].

In an anthropogenically changed environment, some species maybe have already adjusted their behaviour to suboptimal habitats to avoid direct conflicts with people [83,84]. Thus, hedgehogs living in urban areas may have already adjusted their behaviour to urban environments. The responses to the transient disturbance of the festival clearly indicate that this constitutes a habitat change which would worsen the situation.

Female hedgehogs in the highly fragmented areas increased their speed and the size of their home ranges, which might either indicate higher energy requirements satisfied by abundant resources, or energetically more efficient locomotion [7]. The right combination of state-of-the-art technologies will enable us in future to study such subtle coping strategies. Nowadays, advanced technologies offer a level of detail in behavioural studies of free-ranging wild animals that has previously been impossible. This will improve our understanding of the role of behavioural mechanisms in ecological and evolutionary processes [85], provided care is taken to combine on the ground close population monitoring with advanced remote technologies [86,87].

Supplementary Materials: The following are available online at http://www.mdpi.com/2076-2615/10/11/2109/s1. Figure S1: Mean kde95 and confidence intervals measured per treatment and sex. Figure S2: Map of GPS locations of a female (2016_17, yellow dots = locations with inactive behaviour, red dots = locations with active behaviour) and a male (2016_01, light blue dots = locations with inactive behaviour, dark blue dots = locations with active behaviour) hedgehog in the study area Treptower Park. Figure S3: Map of GPS locations of a female (2017_20), a male measured in 2017 (2017_30), and a male measured in 2018 (2018_30) in the study area Tierpark. Table S1: Overview of the study animals and data collected per animal. Table S2: Overview of mean yearly weather data (source: https://www.wetteronline.de/). Table S3: Overview of model results. All GPS and ACC data are available through the Movebank platform (www.movebank.org) by contacting the authors.

Author Contributions: Conceptualization, H.H.; methodology, L.M.F.B., P.G., W.R. and A.B.; software, L.M.F.B., P.G., W.R. and A.B.; formal analysis, L.M.F.B., P.G., W.R., A.B. and H.H.; investigation, L.M.F.B.; data curation, L.M.F.B.; writing—original draft preparation, L.M.F.B.; writing—review and editing, A.B., P.G., W.R. and H.H.; visualization, L.M.F.B. and A.B.; supervision, A.B. and H.H.; project administration, A.B. and L.M.F.B.; funding acquisition, H.H. and A.B. All authors have read and agreed to the published version of the manuscript.

Funding: This work was partly funded by the German Federal Ministry of Education and Research BMBF within the Collaborative Project "Bridging in Biodiversity Science—BIBS" (grant number 01LC1501). All responsibility for the content of this publication is assumed by the author.

Acknowledgments: We would like to thank: Feona, Annika, Judith, Juline, Catrin, Dominik, Hanna Lea, Phillip, Selina, Martin, Sara, Lu, Lenz, Andres, Maria, Jennifer, Laurenz, Linda, Dana, Nadine, Lynn, Sharumathi, Saskia and Henrik for assistance during fieldwork. A special thanks go to Feona and Sara for preparing the nesting data for the analyses. We thank the Lollapalooza organising committee for allowing us to access the festival ground during the festival phase. We also thank, and for their improvement of the manuscript. For statistical analysis, we thank Alexandre Courtiol who discussed with us the results. For proofreading and help, we thank Indra Möllenkotte and Shannon Currie.

Conflicts of Interest: The authors declare no conflict of interest.

References

1. Rykiel, E.J. Towards a definition of disturbance. *Aust. J. Ecol.* **1985**, *10*, 361–365. [CrossRef]
2. Acevedo-Whitehouse, K.; Duffus, A.L.J. Effects of environmental change on wildlife health. *Phil. Trans. R. Soc. B* **2009**, *364*, 3429–3438. [CrossRef]
3. Walker, L.R. *The Biology of Disturbed Habitats*; Oxford University Press: Oxford, UK, 2012.
4. McDonnell, M.J.; Hahs, A.K. Adaptation and adaptedness of organisms to urban environments. *Ann. Rev. Ecol. Evol. Syst.* **2015**, *46*, 261–280. [CrossRef]
5. Bleicher, S.S. The landscape of fear conceptual framework: Definition and review of current applications and misuses. *PeerJ* **2017**, *5*, e3772. [CrossRef] [PubMed]
6. Stillfried, M.; Gras, P.; Börner, K.; Göritz, F.; Painer, J.; Röllig, K.; Wenzler, M.; Hofer, H.; Ortmann, S.; Kramer-Schadt, S. Secrets of success in a landscape of fear: Urban wild boar adjust risk perception and tolerate disturbance. *Front. Ecol. Evol.* **2017**, *5*, 157. [CrossRef]
7. Wong, B.B.M.; Candolin, U. Behavioral responses to changing environments. *Behav. Ecol.* **2015**, *26*, 665–673. [CrossRef]
8. Hastings, A.; Abbott, K.C.; Cuddington, K.; Francis, T.; Gellner, G.; Lai, Y.C.; Morozov, A.; Petrovskii, S.; Scranton, K.; Zeeman, M.L. Transient phenomena in ecology. *Science* **2018**, *361*. [CrossRef]
9. Doncaster, C.P.; Dickman, C.R. The ecology of small mammals in urban habitats. I. Population in a patchy environment. *J. Anim. Ecol.* **1987**, *56*, 629–640.
10. Wilcove, D.S.; McLellan, C.H.; Dobson, A.P. Habitat fragmentation in the temperate zone. In *Conservation Biology—The Science of Scarcity and Diversity*; Soulé, M.E., Ed.; Sinauer Associates, Inc.: Sunderland, MA, USA, 1986; pp. 237–256.
11. Fahrig, L. Effects of habitat fragmentation on biodiversity. *Ann. Rev. Ecol. Evol. Syst.* **2003**, *34*, 487–515. [CrossRef]
12. Baker, P.J.; Harris, S. Urban mammals: What does the future hold? An analysis of the factors affecting patterns of use of residential gardens in Great Britain. *Mammal Rev.* **2007**, *37*, 297–315. [CrossRef]
13. Banks, S.C.; Piggott, M.P.; Stow, A.J.; Taylor, A.C. Sex and sociality in a disconnected world: A review of the impacts of habitat fragmentation on animal social interactions. *Can. J. Zool.* **2007**, *85*, 1065–1079. [CrossRef]
14. Shepard, D.B.; Kuhns, A.R.; Dreslik, M.J.; Phillips, C.A. Roads as barriers to animal movement in fragmented landscapes. *Anim. Conserv.* **2008**, *11*, 288–296. [CrossRef]

15. Lowry, H.; Lill, A.; Wong, B.B.M.M. Behavioural responses of wildlife to urban environments. *Biol. Rev.* **2013**, *88*, 537–549. [CrossRef]

16. Hertel, A.G.; Steyaert, S.M.J.G.; Zedrosser, A.; Mysterud, A.; Lodberg-Holm, H.K.; Gelink, H.W.; Kindberg, J.; Swenson, J.E. Bears and berries: Species-specific selective foraging on a patchily distributed food resource in a human-altered landscape. *Behav. Ecol. Sociobiol.* **2016**, *70*, 831–842. [CrossRef] [PubMed]

17. Hertel, A.G.; Swenson, J.E.; Bischof, R. A case for considering individual variation in diel activity patterns. *Behav. Ecol.* **2017**, *28*, 1524–1531. [CrossRef]

18. Soanes, K.; Taylor, A.C.; Sunnucks, P.; Vesk, P.A.; Cesarini, S.; van der Ree, R. Evaluating the success of wildlife crossing structures using genetic approaches and an experimental design: Lessons from a gliding mammal. *J. Appl. Ecol.* **2018**, *55*, 129–138. [CrossRef]

19. Reeve, N. *Hedgehogs*; T. & A.D. Poyser: London, UK, 1994.

20. Morris, P. *Hedgehogs*; Whittet Books Ltd.: Stansted, UK, 2014.

21. Pettett, C.E.; Johnson, P.J.; Moorhouse, T.P.; Macdonald, D.W. National predictors of hedgehog *Erinaceus europaeus* distribution and decline in Britain. *Mammal Rev.* **2018**, *48*, 1–6. [CrossRef]

22. Krange, M. *Change in the Occurrence of the West European Hedgehog (Erinaceus europaeus) in Western Sweden During 1950–2010*; Karlstad University: Karlstad, Sweden, 2015.

23. Holsbeek, L.; Rodts, J.; Muyldermans, S. Hedgehog and other animal traffic victims in Belgium: Results of a countrywide survey. *Lutra* **1999**, *42*, 111–119.

24. Reichholf, J.H. Starker Rückgang der Häufigkeit überfahrener Igel *Erinaceus europaeus* in Südostbayern und seine Ursachen. *Mitt. Zool. Ges. Braunau* **2015**, *11*, 309–314.

25. Hof, A.R.; Bright, P.W. The value of green-spaces in built-up areas for western hedgehogs. *Lutra* **2009**, *52*, 69–82.

26. Williams, B.M.; Baker, P.J.; Thomas, E.; Wilson, G.; Judge, J.; Yarnell, R.W. Reduced occupancy of hedgehogs (*Erinaceus europaeus*) in rural England and Wales: The influence of habitat and an asymmetric intra-guild predator. *Sci. Rep.* **2018**, *8*, 12156. [CrossRef]

27. Wright, P.G.; Coomber, F.G.; Bellamy, C.C.; Perkins, S.E.; Mathews, F. Predicting hedgehog mortality risks on British roads using habitat suitability modelling. *PeerJ* **2020**, *7*, e8154. [CrossRef]

28. Hof, A.R.; Bright, P.W. The value of agri-environment schemes for macro-invertebrate feeders: Hedgehogs on arable farms in Britain. *Anim. Conserv.* **2010**, *13*, 467–473. [CrossRef]

29. Becher, S.A.; Griffiths, R. Genetic differentiation among local populations of the European hedgehog (*Erinaceus europaeus*) in mosaic habitats. *Mol. Ecol.* **1998**, *7*, 1599–1604. [CrossRef]

30. Rondinini, C.; Doncaster, C. Roads as barriers to movement for hedgehogs. *Funct. Ecol.* **2002**, *16*, 504–509. [CrossRef]

31. Braaker, S.; Kormann, U.; Bontadina, F.; Obrist, M.K. Prediction of genetic connectivity in urban ecosystems by combining detailed movement data, genetic data and multi-path modelling. *Landsc. Urban Plan.* **2017**, *160*, 107–114. [CrossRef]

32. Rasmussen, S.L.; Yashiro, E.; Sverrisdóttir, E.; Nielsen, K.L.; Lukassen, M.B.; Nielsen, J.L.; Asp, T.; Pertoldi, C. Applying the GBS technique for the genomic characterization of a Danish population of European hedgehogs (*Erinaceus europaeus*). *Gen. Biodiv. J.* **2019**, *3*, 78–86.

33. Hof, A.R.; Allen, A.M.; Bright, P.W. Investigating the role of the Eurasian Badger (Meles meles) in the nationwide distribution of the Western European hedgehog (*Erinaceus europaeus*) in England. *Animals* **2019**, *9*, 759. [CrossRef]

34. Braaker, S.; Moretti, M.; Boesch, R.; Ghazoul, J.; Obrist, M.K.; Bontadina, F. Assessing habitat connectivity for ground-dwelling animals in an urban environment. *Ecol. Appl.* **2014**, *24*, 1583–1595. [CrossRef]

35. Dowding, C.V.; Harris, S.; Poulton, S.; Baker, P.J. Nocturnal ranging behaviour of urban hedgehogs, *Erinaceus europaeus*, in relation to risk and reward. *Anim. Behav.* **2010**, *80*, 13–21. [CrossRef]

36. Reeve, N.J.; Morris, P.A. Construction and use of summer nests by the hedgehog *Erinaceus europaeus*. *Mammalia* **1985**, *49*, 187–194. [CrossRef]

37. Haigh, A.; O'Riordan, R.M.; Butler, F. Nesting behaviour and seasonal body mass changes in a rural Irish population of the Western hedgehog (*Erinaceus europaeus*). *Acta Theriol.* **2012**, *57*, 321–331. [CrossRef]

38. Mori, E.; Menchetti, M.; Bertolino, S.; Mazza, G.; Ancillotto, L. Reappraisal of an old cheap method for marking the European hedgehog. *Mammal Res.* **2015**, *60*, 189–193. [CrossRef]

39. Reeve, N.J.; Bowen, C.; Gurnell, J. *An Improved Identification Marking Method for Hedgehogs*; The Mammal Society: Southampton, UK, 2019.

40. Barthel, L.M.F.; Hofer, H.; Berger, A. An easy, flexible solution to attach devices to hedgehogs (*Erinaceus europaeus*) enables long-term high-resolution studies. *Ecol. Evol.* **2019**, *9*, 672–679. [CrossRef]

41. Hofer, H.; East, M.L. Biological conservation and stress. *Adv. Study Behav.* **1998**, *21*, 405–525. [CrossRef]

42. Sikes, R.S.; Gannon, W.L.; The Animal Care and Use Committee of the American Society of Mammalogists. Guidelines of the American Society of Mammalogists for the use of wild mammals in research. *J. Mammal.* **2011**, *92*, 235–253. [CrossRef]

43. R Core Team. *R: A Language and Environment for Statistical Computing*; R Foundation for Statistical Computing: Vienna, Austria, 2018.

44. RStudio Team. *RStudio: Integrated Development Environment for R*; RStudio, Inc.: Boston, MA, USA, 2016.

45. Calenge, C. The package adehabitat for the R software: Tool for the analysis of space and habitat use by animals. *Ecol. Model.* **2006**, *197*, 1035. [CrossRef]

46. Pinheiro, J.; Bates, D.; DebRoy, S.; Sarkar, D.; R Core Team. nlme: Linear and Nonlinear Mixed Effects Models. 2018. Available online: https://cran.r-project.org/package=nlme (accessed on 25 March 2019).

47. Fox, J.; Weisberg, S. *An R Companion to Applied Regression*, 2nd ed.; Sage: Thousand Oaks, CA, USA, 2011; Available online: http://socserv.socsci.mcmaster.ca/jfox/Books/Companion (accessed on 25 March 2019).

48. Hothorn, T.; Bretz, F.; Westfall, P. Simultaneous inference in general parametric models. *Biom. J.* **2008**, *50*, 346–363. [CrossRef]

49. Bates, D.; Mächler, M.; Bolker, B.; Walker, S. Fitting linear mixed-effects models using lme4. *J. Stat. Softw.* **2015**, *67*, 1–48. [CrossRef]

50. Pebesma, E. Simple features for R: Standardized support for spatial vector data. *R J.* **2018**, *10*, 439–446. [CrossRef]

51. Rousset, F.; Ferdy, J.-B. Testing environmental and genetic effects in the presence of spatial autocorrelation. *Ecography* **2014**, *37*, 781–790. [CrossRef]

52. Hijmans, R.J. Geosphere: Spherical Trigonometry. 2017. Available online: https://cran.r-project.org/package=geosphere (accessed on 25 March 2019).

53. Michelot, T.; Langrock, R.; Patterson, T.A. moveHMM: An R package for the statistical modelling of animal movement data using hidden Markov models. *Methods Ecol. Evol.* **2016**, *7*, 1308–1315. [CrossRef]

54. Rast, W.; Barthel, L.M.F.; Berger, A. Music festival makes Hedgehogs move: How individuals cope behaviorally in response to human-induced stressors. *Animals* **2019**, *9*, 455. [CrossRef] [PubMed]

55. Scheibe, K.M.; Berger, A.; Langbein, J.; Streich, W.J.; Eichhorn, K. Comparative analysis of ultradian and circadian behavioural rhythms for diagnosis of biorhythmic state of animals. *Biol. Rhyth. Res.* **1999**, *30*, 216–233. [CrossRef]

56. Berger, A.; Scheibe, K.-M.; Michaelis, S.; Streich, W.J. Evaluation of living conditions of free-ranging animals by automated chronobiological analysis of behavior. *BRMIC* **2003**, *35*, 458–466. [CrossRef]

57. Berger, A. Activity patterns, chronobiology and the assessment of stress and welfare in zoo and wild animals. *Int. Zoo Yearb.* **2011**, *45*, 80–90. [CrossRef]

58. Therneau, T.M. A Package for Survival Analysis in R. 2015. Available online: https://cran.r-project.org/package=survival (accessed on 1 March 2019).

59. Von der Lippe, M.; Buchholz, S.; Hiller, A.; Seitz, B.; Kowarik, I. CityScapeLab Berlin: A research platform for untangling urbanization effects on biodiversity. *Sustainability* **2020**, *12*, 2565. [CrossRef]

60. Buchholz, S.; Hannig, K.; Möller, M.; Schirmel, J. Reducing management intensity and isolation as promising tools to enhance ground-dwelling arthropod diversity in urban grasslands. *Urban Ecosyst.* **2018**, *21*, 1139–1149. [CrossRef]

61. Seress, G.; Lipovits, Á.; Bókony, V.; Czúni, L. Quantifying the urban gradient: A practical method for broad measurements. *Landsc. Urban Plan.* **2014**, *131*, 42–50. [CrossRef]

62. Harris, S. The food of suburban foxes (*Vulpes vulpes*), with special reference to London. *Mammal Rev.* **2008**, *11*, 151–168. [CrossRef]

63. Luniak, M. Synurbization—Adaptation of animal wildlife to urban development. In Proceedings of the 4th International Symposium on Urban Wildlife Conservation, Tucson, AZ, USA, 1–5 May 1999.

64. Schoenfeld, M.; Yom-Tov, Y. The biology of two species of hedgehogs, *Erinaceus europaeus* concolor and *Hemiechinus auritus aegyptius*, in Israel. *Mammalia* **1985**, *49*, 339–354. [CrossRef]

65. Boitani, L.; Reggiani, G. Movements and activity patterns of Hedgehogs (*Erinaceus europaeus*) in Mediterranean coastal habitats. *Z. Säugetierkunde* **1984**, *49*, 193–206.

66. Kristiansson, H. *Ecology of a Hedgehog Erinaceus europaeus Population in Southern Sweden*; University of Lund: Lund, Sweden, 1984.

67. Morris, P.A. A study of home range and movements in the hedgehog (*Erinaceus europaeus*). *J. Zool.* **1988**, *214*, 433–449. [CrossRef]

68. Reeve, N.J. The survival and welfare of hedgehogs (*Erinaceus europaeus*) after release back into the wild. *Anim. Welf.* **1997**, *7*, 189–202.

69. Rautio, A.; Valtonen, A.; Kunnasranta, M. The effects of sex and season on home range in European hedgehogs at the northern edge of the species range. *Ann. Zool. Fenn.* **2013**, *50*, 107–123. [CrossRef]

70. Baker, M.A.A.; Reeve, N.; Conkey, A.A.T.; Macdonald, D.W.; Yamaguchi, N. Hedgehogs on the move: Testing the effects of land use change on home range size and movement patterns of free-ranging Ethiopian hedgehogs. *PLoS ONE* **2017**, *12*, e0180826. [CrossRef]

71. Pettett, C.E.; Johnson, P.J.; Moorhouse, T.P.; Hambly, C.; Speakman, J.R.; Macdonald, D.W. Daily energy expenditure in the face of predation: Hedgehog energetics in rural landscapes. *J. Exp. Biol.* **2017**, *220*, 460–468. [CrossRef]

72. Tucker, M.A.; Böhning-Gaese, K.; Fagan, W.F.; Fryxell, J.M.; van Moorter, B.; Alberts, S.C.; Ali, A.H.; Allen, A.M.; Attias, N.; Avgar, T.; et al. Moving in the Anthropocene: Global reductions in terrestrial mammalian movements. *Science* **2018**, *359*, 466–469. [CrossRef]

73. Bontadina, F. *Strassenüberquerungen von Igeln Zürich, Institute of Zoology*; University of Zurich: Zurich, Switzerland, 1991.

74. Zingg, R. Aktivitaet Habitat- und Raumnutzung von Igeln (Erinaceus Europaeus) in Einem Ländlichen Siedlungsgebiet. Ph.D. Thesis, University of Zurich, Zurich, Switzerland, 1994.

75. Doncaster, C.P.; Krebs, J.R. The wider countryside-principles underlying the responses of mammals to heterogeneous environments. *Mammal Rev.* **1993**, *23*, 113–120. [CrossRef]

76. Phillips, S. Aversive behaviour by koalas (*Phascolarctos cinereus*) during the course of a music festival in northern New South Wales, Australia. *Aust. Mammal.* **2016**, *38*, 158. [CrossRef]

77. Fitzgibbon, S.I.; Gillett, A.K.; Barth, B.J.; Taylor, B.; Ellis, W.A. Do koalas really get the blues? Critique of aversive behaviour by koalas (*Phascolarctos cinereus*) during the course of a music festival in northern New South Wales, Australia. *Aust. Mammal* **2017**, *39*, 108. [CrossRef]

78. Doncaster, C.P.; Rondinini, C.; Johnson, P.C.D. Field test for environmental correlates of dispersal in hedgehogs *Erinaceus europaeus*. *J. Anim. Ecol.* **2001**, *70*, 33–46. [CrossRef]

79. Lott, D.F. *Intraspecific Variation in the Social Systems of Wild Vertebrates*; Cambridge University Press: Cambridge, UK, 1991.

80. Krone, O.; Berger, A.; Schulte, R. Recording movement and activity pattern of a White-tailed Sea Eagle (*Haliaeetus albicilla*) by a GPS datalogger. *J. Ornithol.* **2009**, *150*, 273–280. [CrossRef]

81. Langbein, J.; Scheibe, K.-M.; Eichhorn, K. Investigations on periparturient behaviour in free-ranging mouflon sheep (*Ovis orientalis musimon*). *J. Zool.* **1998**, *244*, 553–561. [CrossRef]

82. Santini, L.; González-Suárez, M.; Russo, D.; Gonzalez-Voyer, A.; von Hardenberg, A.; Ancillotto, L. One strategy does not fit all: Determinants of urban adaptation in mammals. *Ecol. Lett.* **2018**, *22*, 365–376. [CrossRef]

83. Kerley, G.I.H.; Kowalczyk, R.; Cromsigt, J.P.G.M. Conservation implications of the refugee species concept and the European bison: King of the forest or refugee in a marginal habitat? *Ecography* **2012**, *35*, 519–529. [CrossRef]

84. Kuemmerle, T.; Hickler, T.; Olofsson, J.; Schurgers, G.; Radeloff, V.C. Refugee species: Which historic baseline should inform conservation planning? *Divers. Distrib.* **2012**, *18*, 1258–1261. [CrossRef]

85. Kays, R.; Crofoot, M.C.; Jetz, W.; Wikelski, M. Terrestrial animal tracking as an eye on life and planet. *Science* **2015**, *348*, aaa2478. [CrossRef]

86. Hebblewhite, M.; Haydon, D.T. Distinguishing technology from biology: A critical review of the use of GPS telemetry data in ecology. *Philos. Trans. R. Soc. B* **2010**, *365*, 2303–2312. [CrossRef]

87. Shamoun-Baranes, J.; Bom, R.; van Loon, E.E.; Ens, B.J.; Oosterbeek, K.; Bouten, W. From sensor data to animal behaviour: An oystercatcher example. *PLoS ONE* **2012**, *7*, e37997. [CrossRef]

Publisher's Note: MDPI stays neutral with regard to jurisdictional claims in published maps and institutional affiliations.

Article

Patterns of Feeding by Householders Affect Activity of Hedgehogs (*Erinaceus europaeus*) during the Hibernation Period

Abigail Gazzard * and Philip J. Baker

School of Biological Sciences, University of Reading, Whiteknights, Reading, Berkshire RG6 6AS, UK; p.j.baker@reading.ac.uk
* Correspondence: Abigail.Gazzard@pgr.reading.ac.uk

Received: 30 June 2020; Accepted: 1 August 2020; Published: 4 August 2020

Simple Summary: Urban areas are thought to represent a stronghold habitat for the West European hedgehog population in the UK. However, little is known about hibernation patterns in residential areas and if overwinter activity is influenced by any "urban-associated" factors. We monitored hedgehog activity in gardens during the winter hibernation period of 2017–2018 using weekly presence/absence surveys. Hedgehogs were more likely to be present in gardens where householders had provided food in previous seasons or where food was supplied more regularly in a given season. Such relationships could have positive or negative effects on the survival or condition of hedgehogs across the hibernation period. Consequently, further research is needed to identify the effects of supplementary feeding on hibernation biology to help inform conservation guidelines for householders.

Abstract: West European hedgehogs (*Erinaceus europaeus*) are likely to encounter unusual ecological features in urban habitats, such as anthropogenic food sources and artificial refugia. Quantifying how these affect hedgehog behaviour is vital for informing conservation guidelines for householders. We monitored hedgehog presence/absence in gardens in the town of Reading, UK, over the winter of 2017–2018 using a volunteer-based footprint tunnel survey, and collected data on garden characteristics, supplementary feeding (SF) habits, and local environmental conditions. Over a 20-week survey period, hedgehog presence was lowest between January and March. Occupancy analysis indicated that SF significantly affected hedgehog presence/absence before, during, and after hibernation. The number of nesting opportunities available in gardens, average temperatures, and daylength were also supported as important factors at different stages. In particular, our results suggest that SF could act to increase levels of activity during the winter when hedgehogs should be hibernating. Stimulating increased activity at this sensitive time could push hedgehogs into a net energy deficit or, conversely, help some individuals survive which might not otherwise do so. Therefore, further research is necessary to determine whether patterns of feeding by householders have a positive or negative effect on hedgehog populations during the hibernation period.

Keywords: conservation; urban ecology; hedgehogs; *Erinaceus europaeus*; citizen science; gardens; occupancy

---

1. Introduction

Hibernation is critical for the overwinter survival of a range of vertebrate and invertebrate species [1–4]. A reduced core body temperature and lowered metabolic rate allows individuals to conserve energy during periods of harsh environmental conditions and low food supply at the cost of becoming physically inactive for periods lasting days, weeks, or months [5]. To ensure success, mammalian hibernators must increase food intake prior to entering hibernation to accumulate sufficient fat reserves which will later provide energy for day-to-day body maintenance and inducing arousal [5,6]. If too little fat is accumulated, individuals are in danger of depleting their reserves before the hibernation season is over [7–9]. In addition, survival during hibernation is also likely to be linked to nest quality [10] and local environmental conditions [11].

The West European hedgehog (*Erinaceus europaeus*) is a small (<1.5 kg) winter-hibernating mammal that is thought to be in decline in the UK [12,13]. The specific drivers of this decline are unclear, although a wide range of threats can be recognized, including the following: habitat loss, fragmentation, and degradation [14–18]; road traffic accidents [19–21]; the application of chemical herbicides, pesticides, and molluscicides, as well as the use of anticoagulant rodenticides [6,7,22]; competition with and predation by badgers (*Meles meles*) [23–26], and climate-driven changes in invertebrate prey availability and hibernation success [7].

Although timings differ in relation to climate, sex, body size, and condition, hedgehogs typically hibernate between November and April in the UK [6,7]. It is not unusual for hedgehogs to temporarily rouse during the hibernation period and active individuals may relocate to alternative nests [6,10,27]. These partial arousals can last anywhere from several hours to several days [7,28,29]. Since hedgehog hibernation timings are variable, it is difficult to pinpoint which factors trigger the process of entering and arousing from hibernation, although it is likely to involve environmental and hormonal cues related to lower ambient temperatures, shorter days, and reduced invertebrate prey availability [7].

Evidence suggests that hedgehogs are increasingly associated with areas of human habitation [26,30] with substantially higher densities observed in towns and cities than in rural habitats [31–33]. Despite a relative plethora of studies on the winter activity of captive, rehabilitated or rural-dwelling hedgehogs [9,10,27–29,34–38], our understanding of the behaviour of urban-dwelling hedgehogs during this period is limited [11].

Urban areas are associated with a range of factors that could potentially positively or negatively affect patterns of hibernation. For example, in addition to potential nesting sites in patches of remnant natural or semi-natural vegetation, hedgehogs can access cavities beneath buildings, gardens sheds, or decking within residential gardens; urban residents may also supply artificial refugia in the form of homemade or commercially available "hedgehog houses" [7,31]. However, within each of these habitats/locations, hedgehogs are exposed to different levels of disturbance from humans or companion animals [39,40], road traffic [19], and artificial light [41] and sound. Similarly, temperatures within different microhabitats are likely to vary in relation to, for example, the density and composition of surrounding buildings and associated structures [31,42]. It is possible that such "urban-associated" factors could have direct impacts upon the onset of and patterns of arousal during hibernation. For example, warmer temperatures in urban areas [43] may stimulate early arousal from hibernation which, in turn, could increase fat consumption, thereby posing a risk to overwinter survival [7,11].

It has been suggested that supplementary feeding could, in particular, negatively affect natural patterns of hibernation behaviour in hedgehogs [44]. In the UK, many wildlife organisations actively encourage householders to leave out food for hedgehogs in gardens during the colder months in an effort to aid the accumulation of fat prior to hibernation but also to provide sustenance during periodic arousals when natural food availability is low (e.g., [45–47]). The effects of anthropogenic feeding on some aspects of the ecology of urban wildlife (e.g., density, health, and reproductive output) have been investigated extensively (e.g., [48–50]), but data on the impacts on hibernating species are limited. Key observations are that overwinter supplementary feeding is linked to the increased probability of sighting animals [51], interruptions to denning behaviour [52], and accelerated telomere attrition [53]. Conversely, artificial food sources could provide invaluable additional sustenance for individuals in need [6].

Overall, urban areas act as significant strongholds for the UK hedgehog population and expanding our knowledge of overwinter activity and the parameters affecting it is fundamental to developing robust conservation management strategies. Therefore, studies are needed which investigate the following: (a) the activity patterns of urban hedgehogs throughout the hibernation season and (b) how these are affected by external factors. In this study, we quantified patterns of hedgehog occupancy within residential gardens before, during, and after the winter season (see Methods for our definition of the winter season) in relation to within garden and surrounding habitat characteristics, environmental conditions (e.g., daylength and temperature), and patterns of anthropogenic feeding.

2. Materials and Methods

2.1. Footprint Tunnel Survey

Hedgehog presence/absence surveys were carried out in the back gardens of private households in the town of Reading, UK (51°, 27′ N and 0°, 58′ W; population >230,000; and area >60 km^2) and its outskirts from 18 November 2017–7 April 2018. Badgers and foxes (*Vulpes vulpes*) which are both potential predators of hedgehogs and competitors for hedgehog food, are present in Reading, although records of the former indicate that they are limited to the northern section of the town [54]. Domestic dogs (*Canis familiaris*) were present in some gardens surveyed, but these were typically confined to the owner's garden from approximately 11 p.m., whereas hedgehogs could be active throughout the night; consequently, dogs have been shown to not affect patterns of hedgehog occupancy [54]. Similarly, there has been no evidence to suggest that domestic cats (*Felis catus*) are likely to affect hedgehog occupancy either; cats pose little direct threat to hedgehogs and there is an abundance of anecdotal evidence of both species using the same garden at the same time.

Volunteers (citizen scientists) were recruited through an advert on social media in October 2017. Interested participants were asked to provide information on their garden location, current hedgehog-feeding habits, and, to the best of their knowledge, the frequency with which hedgehogs used their garden (ranging from "never" to "every night"). This information was used to categorise volunteers as those who were feeding hedgehogs prior to the start of the study itself (and, by default, who had hedgehogs in their garden), those who were not feeding hedgehogs at this time but who had them visiting their garden, and those who were not feeding hedgehogs at this time and who did not think they visited their garden. As we were interested in investigating the patterns of behaviour of hedgehogs in relation to the existing pattern of feeding by householders (i.e., this was an observational study), and because we were reliant on members of the public agreeing to participate, the distribution of households relative to one another and garden size were dependent on the volunteers themselves; these issues are considered further below.

Prior to the start of the study, householders that had been feeding hedgehogs were asked to either continue feeding them for the duration of the study (November–April) or to stop feeding completely; asking them to maintain a consistent pattern of feeding throughout the study simplified the analyses, especially as we had to assign the start and end of the hibernation period retrospectively. Consequently, the sample of householders consisted of the following four groups: (i) people that had been feeding hedgehogs previously and who continued to feed throughout the study, (ii) people that had been feeding hedgehogs previously but who stopped feeding for the duration of the study, (iii) householders that did not feed hedgehogs before and during the study but who did think they had hedgehogs in their garden, and (iv) householders that did not feed hedgehogs before and during the study and did not think they had hedgehogs in their garden. For those people that elected to continue feeding hedgehogs throughout the project, we asked that they carried on feeding at the same frequency, give the same volume of food each time, and not alter the type of food. This approach was adopted to avoid unduly affecting patterns of hedgehog behaviour in relation to changes in the amount of food available.

Gardens were surveyed using footprint tunnels, which have been used previously to survey hedgehogs in both rural and urban environments (e.g., [19,26,54–56]). Each householder was given one footprint tunnel and instructed to place the tunnel in their rear garden in a position where they thought hedgehogs would be likely to encounter it (e.g., parallel to fences at points where animals could enter the garden). Tunnels consisted of folded corrugated plastic in the form of a triangular tunnel (1200 × 210 × 180 mm) [55]. Ink (carbon powder mixed with vegetable oil) was applied to two strips of masking tape on either side of a food bait (~30 g of commercially available dry hedgehog food) in the centre of a removable plastic insert inside the tunnel; two sheets of A4 paper were fastened at either end of the insert to "capture" footprints of any hedgehogs that traversed through. In order to attract animals without significantly influencing their behaviour, the pot containing the food was sealed but pierced with small holes to allow the scent of the bait to escape; this would prevent hedgehogs (and foxes or domestic cats) from depleting the food bait within a given survey period. Volunteers were given sufficient supplies for the footprint tunnel (e.g., food bait, ink, and paper) to last the duration of the study, as well as an instruction booklet and animal tracks identification guide.

Volunteers checked their tunnels every Saturday and submitted weekly presence/absence results of all tracks recorded through an online survey form (SurveyMonkey.com). Any suspected hedgehog footprints were photographed and sent digitally to one of the authors (AG) for verification. The study was terminated after 20 weeks when volunteer interest had started to decline (weekly reminders to prompt the submission of results needed to be increased markedly in the latter stages).

2.2. Dividing the Data into Seasons

Whilst it is understood that hibernation timings vary between individuals, we opted to subdivide the data into "seasons" that broadly reflected stages before, during, and after the principal hibernation period (henceforth denoted as autumn, winter, and spring). The purpose of this approach was to allow us to analyse the influence of different factors across the contrasting phases of the hibernation season when hedgehogs would be expected to place different emphasis on those factors. For example, the availability of anthropogenic food sources could be more important in the autumn season than the winter season, whereas the reverse could be true when considering access to a secure long-term nest site. Additionally, one assumption of occupancy analysis which we used to analyse these data was that sites remain closed to changes in occupancy between sampling visits [57]. This assumption would have been violated had the data been analysed as one continuous season as, for example, hedgehogs could have consistently used gardens during autumn and spring but not during winter.

The cut-off dates encompassing each season were informed by the pattern of occupancy observed during the 20-week survey. When ≤15% sites were occupied each week, the majority of hedgehogs were considered to be inactive and any data collected during that time were allocated to the winter category. Thus, the three time periods were identified as Weeks 1–7 (18 November 2017–05 January 2018), Weeks 8–16 (06 January 2018–09 March 2018), and Weeks 17–20 (10 March 2018–06 April 2018). Although we concede that this is an a posteriori approach to defining the hibernation period, the timing of low occupancy is in line with that reported elsewhere for hibernation in Britain at this latitude [7,10,21]. Analyses were, however, also conducted with an alternative cut-off threshold (≤20% sites occupied, Weeks 6–16) to investigate the consistency of the occupancy models; no marked differences in the results were evident (see Supplementary Table S1).

2.3. Data Analyses

Pearson Chi-squared tests were used initially to assess whether hedgehogs tended to be consistently present or absent in the same gardens between seasons. The effects of the variables listed in Table 1 on hedgehog presence/absence within each season were investigated using occupancy analysis, a technique which has been used successfully in previous studies of hedgehogs [26,54,55]. In occupancy modelling, an optimisation process is used to find the maximum likelihood of an event occurring. Data from each season were initially analysed independently of any covariates to identify whether the best-fitting baseline models were ones where weekly detection rates (p) were considered to be constant (detection probability did not vary between weeks within that season) or survey specific (detection probability did vary between weeks within that season). These initial analyses were also used to compare naïve occupancy (the proportion of sites where hedgehogs were detected) and true occupancy (Ψ, an estimate of the proportion of sites where hedgehogs were present, accounting for false absences). Analyses were conducted using Presence 12.24.

Variables were quantified using an online questionnaire at the end of the study, from the data itself or from external sources (Table 1). The questionnaire survey requested information about features within the participant's back garden, the proportion of neighbouring gardens that were accessible to hedgehogs from their own garden, patterns of feeding during the study, and the number of potential nesting sites.

Three variables were used to investigate the potential effects of garden size and proximity to other survey gardens on patterns of detection and occupancy (Table 1). For example, garden size (mean $\pm$ SD = 238.5 $\pm$ 244.8 m^2) could have potentially affected detection rates as we only used one footprint tunnel in each garden, although the majority of gardens (93.7%) covered <550 m^2, three gardens (4.8%) covered 786–870 m^2, and one garden (1.6%) was 1520 m^2 in area. Within each season, the straight-line distance to the nearest other house and the straight-line distance to the nearest other house where hedgehogs were detected were incorporated to determine whether hedgehogs were more likely to be detected in houses close to one another, which would potentially indicate that patterns of detection were not independent.

Habitat characteristics in the area around each house were quantified using the straight-line distances to the nearest arable, grassland, and woodland habitats, and the total area of habitats within 250 and 500 m radii of each garden. These measures were quantified from Natural Environment Research Council land class datasets [58] with QGIS 3.4.4 (Table 1). Radii of 250 and 500 m were selected based upon existing data of hedgehog nightly ranges outside the hibernation season [15,59]. Minimum grass level and air temperatures, and weekly rainfall volume, were taken from a weather station on the University of Reading's Whiteknights campus [60]. Mean weekly daylength was quantified from sunset and sunrise measurements from Benson weather station, approximately 18 km north of Reading [61]. As these data were taken from sites in proximity to the survey gardens, but not in the gardens themselves, they reflected general environmental conditions and not the specific micro-habitat characteristics of each garden.

Table 1. Summary of the variables used in analysis, collected from questionnaire surveys and external data sources. Q indicates that the data were derived from a questionnaire survey of the householder; D indicates that the variable was extracted from the occupancy data itself; E indicates data from an external source (see text).

Covariate	Source	Description
FEEDHOG	Q	An ordinal measure of whether food was left out for hedgehogs during the study: 1 = never, 2 = less frequently (monthly or less), 3 = more frequently (nightly or weekly)
FEDBEFORE	Q	A binary measure of whether the participant usually left out food for hedgehogs prior to the commencement of the study
FEEDOTHERS	Q	A binary measure of whether food was left out by the participant for birds or other animals at some point during the study
NESTSITES	Q	The number of potential types of nest sites available in the participant's garden as assessed by the participant. Tick-box options of possible nesting sites were listed on the questionnaire as "hedgehog house", "under a shed or decking", "under bushes or shrubs", "under a compost heap" or "other (please provide more information)". The total number of potential nest sites were converted to z-scores
CONNECTIVITY	Q	The proportion of front and back gardens neighbouring the participant's household that is accessible for hedgehogs from the participant's own gardens

Table 1. *Cont.*

Covariate	Source	Description
FRONT2BACK	Q	A binary measure of whether a hedgehog could access the participant's back garden from their front garden
GOODHABITAT	Q	The proportion of habitat in the participant's back garden only that is considered "good" for wildlife, including lawn, shrubs, flowerbeds and ponds
HOUSETYPE	Q	A binary measure of whether houses were: (i) semi-detached, link-detached or detached; or (ii) other (e.g., terraced)
GARDENSIZE	E	The area of each garden (m^2) converted to z-scores
NEARESTOTHER	D	Distance from each site to the next nearest site (m) converted to z-scores
NEAREST + VE	D	Distance from each site to the next nearest hedgehog-positive site (m) per season (autumn, winter or spring) converted to z-scores
ARABLEDIST	E	Distance from each site to the nearest area of arable land (m) converted to z-scores
ARABLE 500 m	E	The area of arable land (m^2) within a 500 m radius of each site converted to z-scores (Note: As only 4 sites fell within 250 m of arable land, the potential variable ARABLE250 m was not considered for analyses)
WOODDIST	E	Distance from each site to the nearest area of woodland (m) converted to z-scores
WOOD 250 m and WOOD 500 m	E	The area of woodland (m^2) within 250 and 500 m radii of each site converted to z-scores
GRASSDIST	E	Distance from each site to the nearest area of grassland (m) converted to z-scores
GRASS 250 m and GRASS 500 m	E	The area of grassland (m^2) within 250 and 500 m radii of each site converted to z-scores
URBAN 250 m and URBAN 500 m	E	The area of urban and suburban habitat (m^2) within 250 and 500 m radii of each site converted to z-scores (Note: As all sites fell within the urban habitat classification, the straight-line distance from each site to urban habitat was not considered for analysis)
DAYTIME	E	Mean daylength (time between sunrise and sunset) per week, converted to z-scores
AIRTEMP	E	Minimum air temperature (°C) averaged per survey week based on hourly recordings taken between 21:00 and 09:00, converted to z-scores
GRASSTEMP	E	Minimum grass temperature (°C) averaged per survey week based on daily recordings taken at 09:00, converted to z-scores

Following checks for multicollinearity, single-species, single-season models were fitted; all variables were first considered in single-covariate models. Then, multi-covariate models were constructed based upon the known ecology of hedgehogs, as well as the hypothesised importance of different variables on occupancy during each season. Supplementary feeding before and during the study, as well as feeding intended for other species, were considered to be important in all seasons, and for autumn, models included the availability of and proximity to potential winter nesting sites; for winter and spring, models included environmental conditions that were likely to affect the timing of hibernation, i.e., daylength, ground temperature, and air temperature. A maximum of three covariates was considered in each model because of the relatively small sample sizes. This approach was favoured to produce a realistic set of candidate models, avoiding the shortcomings of algorithm-based model selection [62,63].

The goodness-of-fit of the most global model for each season was tested using the bootstrap method with 1000 replicates. Bootstrapping simulates detection histories for each site and produces a test statistic (Pearson Chi-squared) for each of the 1000 runs [64]. A measure of "lack of fit", defined as a variance inflation factor $\hat{c}$, is calculated by dividing the observed test statistic by the average bootstrap statistic [65]. When $\hat{c} > 1$, there is evidence of poor fit and it is recommended that (a) Akaike's information criterion (AIC) values should be converted into quasi-likelihood adjusted AIC (QAIC) and (b) standard errors of beta estimates should be inflated by a factor of $\sqrt{\hat{c}}$ [62,64,65]. Models that did not converge were excluded. Those with ΔQAIC values <2 were considered to be top-ranking models [62], and covariates were regarded as significant when their associated 95% confidence intervals did not cross 0 [66].

3. Results

3.1. General Trends

Overall, 63 householders completed the study (Figure 1). During Week 1, results were obtained for 26 (41.2%) sites as compared with 100% in subsequent weeks; this was associated with the challenges of getting volunteers started but is not likely to have affected the results since occupancy analysis is robust to missing data [67]. In autumn, or "pre-hibernation", hedgehogs were and were not being fed in 25 (39.7%) and 38 (60.3%) gardens, respectively (Figure 1).

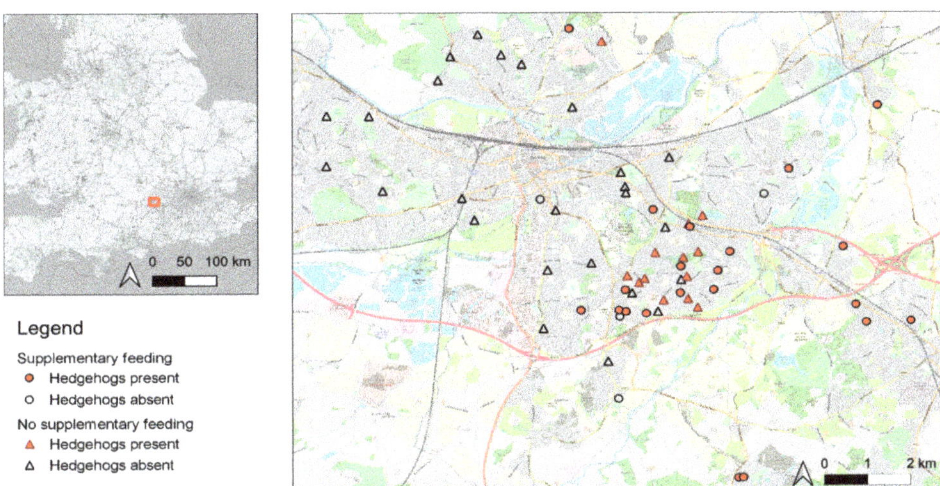

Figure 1. The locations of gardens (n = 63) in Reading and surrounding areas surveyed for hedgehogs between November 2017 and April 2018 inclusive. Circles denote gardens where hedgehogs were fed by householders prior to the study; diamonds denote gardens where hedgehogs were not fed prior to the study; filled and open symbols denote gardens where hedgehogs were and were not detected at any point during the current study, respectively.

Hedgehogs were active throughout all survey periods (Figure 2) and were recorded on 247 occasions (19.6% of the 1260 surveyor weeks). In autumn, hedgehogs were detected in 34 (54.0%) gardens, i.e., 21 of 25 (84.0%) gardens where they had been fed previously and 13 of 38 (34.2%) gardens where they had not been fed previously (Figure 1). Cumulatively, 97.1% of hedgehog-positive sites were detected by the third week of surveying. Occupancy (Ψ) and detection probability (p) were lowest between January and March (autumn true Ψ = 0.54, winter true Ψ = 0.32, and spring true Ψ = 0.39); full occupancy estimates from the baseline models are given in Table 2. False-absence error rates were very low (autumn 0.1%, winter 2.1%, and spring 0.6%).

Of the 34 hedgehog-positive gardens, 18 gardens (52.9%) were used every season, 9 (26.5%) were used during the autumn period only, and none were used exclusively during winter or spring. Consequently, there was a strong association in the pattern of presence/absence of hedgehogs in individual gardens between successive seasons, i.e., autumn-winter (Chi-squared test $\chi^2_1 = 23.204$, $p < 0.001$) and winter-spring ($\chi^2_1 = 37.010$, $p < 0.001$).

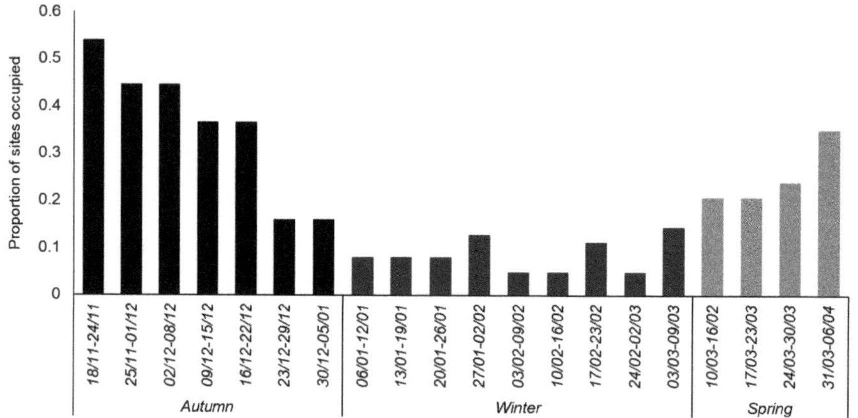

Figure 2. The proportion of all gardens surveyed ($n = 26$ in Week 1 and $n = 63$ for Weeks 2–20) where hedgehogs were recorded each week. Weekly survey dates are given in the format *dd/mm*, running from November 2017 to April 2018 inclusive.

Table 2. Summary of baseline hedgehog occupancy models where detection rate was modelled as constant (did not vary between weeks within each season) versus survey specific (did vary between weeks within each season). Seasons are illustrated in Figure 2.

Season	Model	QAIC	ΔQAIC	AIC Weight	Model Likelihood	K	Detection Rate	Naïve Ψ	True Ψ
Autumn	Ψ(.), *p*(survey-specific)	270.59	0.00	1	1.0000	8	0.8234 0.8225 0.8225 0.6756 0.6756 0.2938 0.2938	0.5397	0.5403
	Ψ(.), *p*(.)	294.26	23.67	0.0000	0.0000	2	0.6138	0.5397	0.5411
Winter	Ψ(.), *p*(.)	213.63	0.00	0.9852	1.0000	2	0.2626	0.3016	0.3224
	Ψ(.), *p*(survey-specific)	222.02	8.39	0.0148	0.0151	10	0.2481 0.2481 0.2481 0.397 0.1489 0.1489 0.3474 0.1489 0.4467	0.3016	0.3198
Spring	Ψ(.), *p*(.)	64.13	0.00	0.8006	1.0000	2	0.6459	0.3810	0.3870
	Ψ(.), *p*(survey-specific)	66.91	2.78	0.1994	0.2491	5	0.5377 0.5377 0.6204 0.9100	0.3810	0.3838

Ψ = occupancy, *p* = detection probability, *K* = number of parameters. ΔQAIC is the change in quasi-likelihood adjusted Akaike's information criterion. For each season, the variance inflation factor ĉ was adjusted based on goodness-of-fit tests of the most parameterised models (1.3226, 1.3385, and 3.5534 for autumn, winter, and spring, respectively). Naïve occupancy is the number of gardens where hedgehogs were detected and true occupancy is the number of gardens estimated to be occupied by hedgehogs after accounting for the false-absence error rate.

3.2. Factors Affecting Hedgehog Occupancy

For analyses incorporating covariates (Table 1), all top-ranking models included a feeding variable (Table 3). Occupancy in autumn and winter was associated with supplementary feeding prior to the hibernation period (FEDBEFORE), whereas in spring it was most associated with feeding in that season

(FEEDHOG). There was also some support for detection probability being positively influenced by DAYTIME and FEEDOTHER during spring, but the effect was not significant. All other covariates reported in the best-fitting models, in each season, had statistically significant positive effects on occupancy or detection probability. Full model results can be found in Supplementary Tables S2–S4. Garden size, proximity to other gardens per se, and proximity to the nearest other garden where hedgehogs were detected in that season were not included in the top-ranked models in any season.

Table 3. A summary of the top-ranking models (ΔQAIC < 2) produced in single-season occupancy analyses. Seasons are illustrated in Figure 2.

Season	Model	QAIC	ΔQAIC	AIC Weight	Model Likelihood	K
Autumn	Ψ(FEDBEFORE + WOOD500 m), p(survey + NESTSITES)	283.85	0.00	0.8966	1.0000	11
Winter	Ψ(FEDBEFORE), p(FEEDHOG + FEEDOTHERS)	214.42	0.00	0.4561	1.0000	5
	Ψ(FEDBEFORE), p(FEEDHOG + GRASSTEMP)	215.41	0.99	0.2780	0.6096	5
	Ψ(FEDBEFORE), p(FEEDHOG + AIRTEMP)	215.51	1.09	0.2645	0.5798	5
Spring	Ψ(FEEDHOG), p(DAYTIME + FEEDOTHERS)	71.83	0.00	0.2413	1.0000	5
	Ψ(FEDBEFORE), p(.)	71.97	0.14	0.2250	0.9324	3
	Ψ(FEEDHOG), p(.)	72.82	0.99	0.1471	0.6096	3

Ψ = occupancy, p = detection probability, K = number of parameters. ΔQAIC is the change in quasi-likelihood adjusted Akaike's information criterion. For each season, the variance inflation factor ĉ was adjusted based on goodness-of-fit tests of the most parameterised models (1.1309, 1.1531, and 2.8138 for autumn, winter, and spring, respectively).

In winter, hedgehogs were recorded in 16 of 25 (64.0%) gardens where the householder had been feeding them in autumn as compared with 3 of 38 (7.9%) gardens where they had not been fed. Overall, of the hedgehog-positive sites within each season, gardens where householders had previously put out food were visited, on average, for 4.4 weeks in autumn (n = 21 gardens, 62.9% of weeks in the 7-week season), 2.6 weeks in winter (n = 16 gardens, 28.9% of the nine-week season), and 2.7 weeks in spring (n = 18 gardens, 67.5% of the four-week season). Comparable figures for gardens where they were not fed were 3.3 weeks (n = 13 gardens, 47.1%), 2.0 weeks (n = 3 gardens, 22.2%), and 2.5 weeks (n = 6 gardens, 62.5%), respectively. Consequently, hedgehogs were much more likely to be present in gardens where food was supplied by householders (Figure 3).

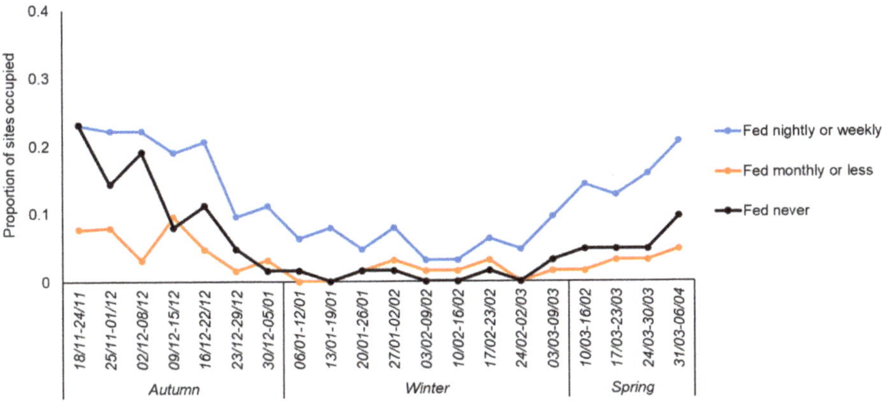

Figure 3. The proportion of gardens (n = 26 in Week 1 and n = 63 for Weeks 2–20) where hedgehogs were detected in relation to the frequency with which householders provided food at the outset of the study. Weekly survey dates are given in the format *dd/mm*, running from November 2017 to April 2018 inclusive.

4. Discussion

Hibernation is an adaptive physiological response to reduce energetic requirements during periods of low food availability. Hedgehogs, therefore, need to accumulate sufficient fat reserves prior to hibernation, and then minimize expenditure of energy during this period. In behavioural terms, this essentially means that hedgehogs need to avoid rousing unnecessarily from hibernation. However, they do need to retain the ability to be able to respond if environmental conditions become unfavourable or, for example, if they are detected by predators or disturbed. Consequently, individuals need to find locations that afford them protection, but which are also in proximity to alternative locations, with appropriate building materials, if they need to move.

In this study, hedgehog occupancy and detection in autumn were significantly linked to the area of woodland habitat within 500 m (WOOD 500 m) of focal gardens and the number of potential nest sites available within gardens (NESTSITES), respectively. Previous studies have reported that a significant proportion of winter nests are constructed in wooded areas [9,10] and the nearby woodland measured in this study area may have provided valuable pockets of semi-natural nesting habitat within an otherwise built-up area. However, the relative qualities of woodland and within-garden nesting sites are unknown. For example, wooded areas can be associated with a higher abundance of favoured building materials (the leaves of broadleaved trees: [7]) but urban woodlands are often open to the public and are likely to be associated with high levels of disturbance by walkers and especially their dogs. Alternatively, gardens offer potentially advantageous nesting sites such as beneath sheds and decking, but where natural nesting materials could be scarce. Future studies of urban hedgehog populations, therefore, need to focus on quantifying where hibernacula are located and if this is linked to over-winter survival rates.

Urban areas also pose one additional challenge. Research to date has indicated that hedgehogs tend to enter hibernation in response to the combination of a reduction in temperatures and a decline in food availability [7]. This was also evident in this study, with hedgehog detection during winter reduced as grass and air temperatures declined. In urban areas, however, food supplied by householders was not directly linked to prevailing temperatures. As a result, hedgehogs could be getting "mixed messages", that is, food availability is still high even though temperatures are low. Ultimately, this could result in maladaptive responses leading to reduced over-winter survival rates and longevity.

In autumn, hedgehog occupancy was correlated with whether they had been fed in the previous season. Hedgehogs were detected in 54.0% of gardens overall, with a marked difference between those houses where they had (84.0%) and had not (34.2%) been fed. Similarly, occupancy in winter (30.2% of gardens overall) was also correlated with the pattern of feeding at the outset of the study, with an increase in the disparity between gardens where they had (64.0%) and had not been fed (7.9%). This was consistent with the radio-tracking data reported by Rasmussen et al. [11] which indicated that urban-dwelling hedgehogs tended to stay in the vicinity of local feeding stations during both active and inactive seasons, but also potentially suggested that patterns of feeding prior to hibernation could increase the likelihood that hedgehogs visited gardens during the hibernation period. In contrast, in spring, hedgehog occupancy tended to be associated with the frequency with which animals were being fed in that season, with occupancy higher where they were being fed more frequently.

Winter activity is, however, not unusual, and hedgehogs typically relocate nests at least once during the hibernation period [6,10,27]. As we used footprint tunnels to record hedgehog activity on a weekly basis, it was not possible to determine if detections during the winter season reflected individual animals in the normal process of relocating nests, or if they reflected the behaviour of several animals in the same garden. For example, the continued use of a single garden by individual hedgehogs overwinter has been recorded previously [9,11]. That being said, hedgehogs were detected for an average of 2.6 weeks in winter in gardens where they had been fed previously ($n = 16$) as compared with 2.0 weeks in other gardens ($n = 3$). Again, this is suggestive of the fact that householder feeding patterns could be influencing over-winter activity.

However, although anthropogenic feeding could negatively affect hedgehogs during hibernation [52,68], it is possible that it could be beneficial [9,28]. For example, it could enable animals that have not accumulated sufficient body fat to delay the point at which they enter hibernation [6], especially juveniles born in late summer [7]. Similarly, it could also help animals that have roused from hibernation to replenish some of their reserves. This could be important for animals that experience an increasing number of arousal events in relation to changing climatic conditions and anthropogenic influences.

Conversely, as hedgehogs are capable of surviving losses of up to 44% of their pre-hibernation weight [9,11,27,69], it is not clear if access to food during winter is beneficial. For example, it has been suggested that hedgehogs might only enter into a "partial hibernation" where food is available [7]. Since even a single rousing can consume the same amount of energy required to survive 3–4 days of hibernation [36], animals can experience proportionately larger losses in mass overwinter if they cannot access sufficient food [7]. Furthermore, animals that are active during the winter also face the additional risks associated with, for example, road traffic, companion animals, and domestic gardens [11,21,39]. "Shortenings" of the hibernation period, caused by increased arousals or a delay in hibernation commencement, have also been linked to accelerated cellular aging in mammals [53,70–72], thereby potentially having implications for longevity [73–75]. In addition, there is a need to consider the nutritional value of foods being provided by householders. For example, should animals come to rely on non-natural foods as a principal source of energy, it is possible that these do not fulfill their nutritional requirements and could compromise their condition [50]. Therefore, additional information is required on the types of food used by householders and its nutritional content relative to the needs of hedgehogs at this time.

Despite the absence of any definitive data that feeding hedgehogs overwinter is beneficial, it is encouraged by several wildlife organisations in Britain [45–47]. Given that arguments can be made that overwinter feeding could negatively impact hedgehogs, there is an urgent need to study its effects in more detail so that accurate advice can be given to householders. Such investigations will require the study of the activity and movement patterns, body mass changes, reproductive success, and longevity of individual hedgehogs before, during, and after the hibernation period in an experimental framework (i.e., controlling the frequency and volume of food supplied by randomly selected householders). These studies are, however, likely to be associated with significant challenges since they would require the cooperation of large numbers of householders for extended periods of time.

5. Conclusions

This study has indicated that residential gardens may be used frequently by hedgehogs throughout all stages of hibernation. Supplementary feeding in preceding seasons was found to be a key factor associated with hedgehog presence/absence during the hibernation period. This potentially indicates that supplementary feeding could affect key components of the hibernation behaviour of urban-dwelling hedgehogs, which could be detrimental or beneficial to overwinter survival and reproduction. Therefore, further intensive studies of known individuals before, during, and after the hibernation period are required.

Supplementary Materials: The following are available online at http://www.mdpi.com/2076-2615/10/8/1344/s1, Table S1: Model results of Weeks 6–16, Table S2–S4: Full occupancy model results.

Author Contributions: Conceptualization, A.G. and P.J.B.; Data curation, A.G.; Formal analysis, A.G.; Supervision, P.J.B.; Writing—original draft, A.G.; Writing—review and editing, A.G. and P.J.B. All authors have read and agreed to the published version of the manuscript.

Funding: This study was supported by the People's Trust for Endangered Species, British Hedgehog Preservation Society and the University of Reading.

Acknowledgments: We thank the volunteers who took part in this study as well as Cathy Holwill and Samuel Nutt who assisted significantly with equipment distribution and data collection.

Conflicts of Interest: The authors declare no conflict of interest.

References

1. Leather, S.R.; Walters, K.F.A.; Bale, J.S. *The Ecology of Insect Overwintering*; Cambridge University Press: Cambridge, UK, 1993.

2. Cáceres, C.E. Dormancy in invertebrates. *Invert. Biol.* **1997**, *116*, 371–383. [CrossRef]

3. Wells, K.D. *The Ecology and Behavior of Amphibians*; University of Chicago Press: Chicago, IL, USA, 2007.

4. Ruf, T.; Geiser, F. Daily torpor and hibernation in birds and mammals. *Biol. Rev.* **2015**, *90*, 891–926. [CrossRef] [PubMed]

5. Martin, S.L.; Yoder, A.D. Theme and variations: Heterothermy in mammals. *Integ. Comp. Biol.* **2014**, *54*, 439–442. [CrossRef] [PubMed]

6. Reeve, N. *Hedgehogs*; Academic Press: London, UK, 1994.

7. Morris, P. *Hedgehog*; William Collins: London, UK, 2018.

8. Kristiansson, H. Population variables and causes of mortality in a hedgehog (*Erinaceus europaeus*) population in southern Sweden. *J. Zool.* **1990**, *220*, 391–404. [CrossRef]

9. Jensen, A.B. Overwintering of European hedgehogs *Erinaceus europaeus* in a Danish rural area. *Acta Theriol.* **2004**, *49*, 145–155. [CrossRef]

10. Morris, P. Winter nests of the hedgehog (*Erinaceus europaeus* L.). *Oecologia* **1973**, *11*, 299–313. [CrossRef]

11. Rasmussen, S.L.; Berg, T.B.; Dabelsteen, T.; Jones, O.R. The ecology of suburban juvenile European hedgehogs (*Erinaceus europaeus*) in Denmark. *Ecol. Evol.* **2019**, *9*, 13174–13187. [CrossRef]

12. Roos, S.; Johnston, A.; Noble, D. *UK Hedgehog Datasets and Their Potential for Long-Term Monitoring. BTO Research Report, 598*; The British Trust for Ornithology: Thetford, UK, 2012.

13. Mathews, F.; Kubasiewicz, L.; Gurnell, J.; Harrower, C.; McDonald, R.; Shore, R. *A Review of the Population and Conservation Status of British Mammals. A Report by the Mammal Society under Contract to Natural England, Natural Resources Wales and Scottish Natural Heritage*; Natural England: Peterborough, UK, 2018.

14. Becher, S.A.; Griffiths, R. Genetic differentiation among local populations of the European hedgehog (*Erinaceus europaeus*) in mosaic habitats. *Mol. Ecol.* **1998**, *7*, 1599–1604. [CrossRef]

15. Rondinini, C.; Doncaster, C. Roads as barriers to movement for hedgehogs. *Funct. Ecol.* **2002**, *16*, 504–509. [CrossRef]

16. Hof, A.R.; Bright, P.W. The value of green-spaces in built-up areas for western hedgehogs. *Lutra* **2009**, *52*, 69–92.

17. Hof, A.R.; Bright, P.W. The value of agri-environment schemes for macro-invertebrate feeders: Hedgehogs on arable farms in Britain. *Anim. Conserv.* **2010**, *13*, 467–473. [CrossRef]

18. Moorhouse, T.P.; Palmer, S.C.F.; Travis, J.M.J.; Macdonald, D.W. Hugging the hedges: Might agri-environment manipulations affect landscape permeability for hedgehogs? *Biol. Conserv.* **2014**, *176*, 109–116. [CrossRef]

19. Huijser, M.P.; Bergers, P.J.M. The effect of roads and traffic on hedgehog (*Erinaceus europaeus*) populations. *Biol. Cons.* **2000**, *95*, 111–116. [CrossRef]

20. Wembridge, D.E.; Newman, M.R.; Bright, P.W.; Morris, P.A. An estimate of the annual number of hedgehog (*Erinaceus europaeus*) road casualties in Great Britain. *Mamm. Comm.* **2016**, *2*, 8–14.

21. Wright, P.G.R.; Coomber, F.G.; Bellamy, C.C.; Perkins, S.E.; Mathews, F. Predicting hedgehog mortality risks on British roads using habitat suitability modelling. *PeerJ* **2020**, *7*, e8154. [CrossRef]

22. Dowding, C.V.; Shore, R.F.; Worgan, A.; Baker, P.J.; Harris, S. Accumulation of anticoagulant rodenticides in a non-target insectivore, the European hedgehog (*Erinaceus europaeus*). *Environ. Pollut.* **2010**, *158*, 161–166. [CrossRef]

23. Young, R.P.; Davison, J.; Trewby, I.D.; Wilson, G.J.; Delahay, R.J.; Doncaster, C.P. Abundance of hedgehogs (*Erinaceus europaeus*) in relation to the density and distribution of badgers (*Meles meles*). *J. Zool.* **2006**, *269*, 349–356. [CrossRef]

24. Trewby, I.D.; Young, R.; McDonald, R.A.; Wilson, G.J.; Davison, J.; Walker, N.; Robertson, A.; Doncaster, C.P.; Delahay, R.J. Impacts of removing badgers on localised counts of hedgehogs. *PLoS ONE* **2014**, *9*, e95477. [CrossRef]

25. Pettett, C.E.; Moorhouse, T.P.; Johnson, P.J.; Macdonald, D.W. Factors affecting hedgehog (*Erinaceus europaeus*) attraction to rural villages in arable landscapes. *Eur. J. Wildl. Res.* **2018**, *63*, 54. [CrossRef]

26. Williams, B.; Baker, P.J.; Thomas, E.; Wilson, G.; Judge, J.; Yarnell, R.W. Reduced occupancy of hedgehogs (*Erinaceus europaeus*) in rural England and Wales: The influence of habitat and an asymmetric intraguild predator. *Sci. Rep.* **2018**, *8*, 1–10. [CrossRef]

27. Yarnell, R.W.; Surgery, J.; Grogan, A.; Thompson, R.; Davies, K.; Kimbrough, C.; Scott, D.M. Should rehabilitated hedgehogs be released in winter? A comparison of survival, nest use and weight change in wild and rescued animals. *Eur. J. Wildl. Res.* **2019**, *65*, 6. [CrossRef]

28. Walhovd, H. Partial arousals from hibernation in hedgehogs in outdoor hibernacula. *Oecologia* **1979**, *40*, 141–153. [CrossRef]

29. Webb, P.I.; Ellison, J. Normothermy, torpor, and arousal in hedgehogs (*Erinaceus europaeus*) from Dunedin, New Zealand. *J. Zool.* **1998**, *25*, 85–90. [CrossRef]

30. Doncaster, C.P. Factors regulating local variations in abundance: Field tests on hedgehogs, *Erinaceus europaeus*. *Oikos* **1994**, *69*, 182–192. [CrossRef]

31. Hubert, P.; Julliard, R.; Biagianti, S.; Poulle, M.L. Ecological factors driving the higher hedgehog (*Erinaceus europeaus*) density in an urban area compared to the adjacent rural area. *Landsc. Urban Plan.* **2011**, *103*, 34–43. [CrossRef]

32. Van de Poel, J.L.; Dekker, J.; van Langevelde, F. Dutch hedgehogs *Erinaceus europaeus* are nowadays mainly found in urban areas, possibly due to the negative effects of badgers *Meles meles*. *Wild. Biol.* **2015**, *21*, 51–55. [CrossRef]

33. Schaus, J.; Uzal, A.; Gentle, L.K.; Baker, P.J.; Bearman-Brown, L.; Bullion, S.; Gazzard, A.; Lockwood, H.; North, A.; Reader, T.; et al. Application of the Random Encounter Model in citizen science projects to monitor animal densities. *Remote Sens. Ecol. Cons.* **2020**. [CrossRef]

34. Soivio, A.; Tähti, H.; Kristoffersson, R. Studies on the periodicity of hibernation in the hedgehog (*Erinaceus europaeus* L.): III. Hibernation in a constant ambient temperature of −5° C. *Ann. Zoo. Fenn.* **1968**, *5*, 224–226.

35. Parkes, J. Some aspects of the biology of the hedgehog (*Erinaceus europaeus* L.) in the Manawatu, New Zealand. *New Zeal. J. Zool.* **1975**, *2*, 463–472. [CrossRef]

36. Tähti, H.; Soivio, A. Respiratory and circulatory differences between induced and spontaneous arousals in hibernating hedgehogs (*Erinaceus europaeus* L.). *Ann. Zool. Fenn.* **1977**, *14*, 198–203.

37. Dmi'el, R.; Schwarz, M. Hibernation patterns and energy expenditure in hedgehogs from semi-arid and temperate habitats. *J. Comp. Physiol. B* **1984**, 117–123. [CrossRef]

38. Fowler, P.A.; Racey, P.A. Daily and seasonal cycles of body temperature and aspects of heterothermy in the hedgehog *Erinaceus europaeus*. *J. Comp. Physiol. B* **1984**, *160*, 299–307. [CrossRef] [PubMed]

39. Stocker, L. *The Complete Hedgehog*; Chatto and Windus: London, UK, 1987.

40. Rast, W.; Barthel, L.M.F.; Berger, A. Music festival makes hedgehogs move: How individuals cope behaviorally in response to human-induced stressors. *Animals* **2019**, *9*, 455. [CrossRef] [PubMed]

41. Finch, D.; Smith, B.R.; Marshall, C.; Coomber, F.G.; Kubasiewicz, L.M.; Anderson, M.; Wright, P.G.R.; Mathews, F. Effects of Artificial Light at Night (ALAN) on European hedgehog activity at supplementary feeding stations. *Animals* **2020**, *10*, 768. [CrossRef] [PubMed]

42. Perini, K.; Magliocco, A. Effects of vegetation, urban density, building height, and atmospheric conditions on local temperatures and thermal comfort. *Urban For. Urban Green.* **2014**, *13*, 495–506. [CrossRef]

43. Chapman, S.; Watson, J.E.; Salazar, A.; Thatcher, M.; McAlpine, C.A. The impact of urbanization and climate change on urban temperatures: A systematic review. *Landsc. Ecol.* **2017**, *32*, 1921–1935. [CrossRef]

44. Keep Feeding Hedgehogs in the Autumn—University of Brighton. Available online: https://www.brighton.ac.uk/about-us/news-and-events/news/2017/09-13-keep-feeding-hedgehogs-in-the-autumn.aspx (accessed on 6 June 2020).

45. Hedgehog Street: Should I Keep Feeding Hedgehogs Overwinter? Available online: https://www.hedgehogstreet.org/should-i-keep-feeding-hedgehogs-over-winter/ (accessed on 16 May 2020).

46. British Hedgehog Preservation Society: Feeding. Available online: https://www.britishhedgehogs.org.uk/feeding/ (accessed on 16 May 2020).

47. Tiggywinkles Wildlife Hospital Hedgehog Fact Sheet. Available online: https://www.sttiggywinkles.org.uk/hedgehog-fact-sheet/ (accessed on 16 May 2020).

48. Robb, G.N.; McDonald, R.A.; Chamberlain, D.E.; Bearhop, S. Food for thought: Supplementary feeding as a driver of ecological change in avian populations. *Front. Ecol. Environ.* **2008**, *6*, 476–484. [CrossRef]

49. Ewen, J.G.; Walker, L.; Canessa, S.; Groombridge, J.J. Improving supplementary feeding in species conservation. *Conserv. Biol.* **2014**, *29*, 341–349. [CrossRef]

50. Murray, M.H.; Becker, D.J.; Hall, R.J.; Hernandez, S.M. Wildlife health and supplemental feeding: A review and management recommendations. *Biol. Conserv.* **2016**, *205*, 163–174. [CrossRef]

51. Bojarskaa, K.; Drobniak, S.; Jakubiec, Z.; Zyśk-Gorczyńsk, E. Winter insomnia: How weather conditions and supplementary feeding affect the brown bear activity in a long-term study. *Glob. Ecol. Conserv.* **2019**, *17*, e00523. [CrossRef]

52. Krofel, M.; Spacapan, M.; Jerina, K. Winter sleep with room service: Denning behaviour of brown bears with access to anthropogenic food. *J. Zool.* **2017**, *302*, 8–14. [CrossRef]

53. Kirby, R.; Johnson, H.E.; Alldredge, M.W.; Pauli, J.N. The cascading effects of human food on hibernation and cellular aging in free-ranging black bears. *Sci. Rep.* **2019**, *9*, 1–7. [CrossRef] [PubMed]

54. Williams, B.; Mann, N.; Neumann, J.L.; Yarnell, R.W.; Baker, P.J. A prickly problem: Developing a volunteer-friendly tool for monitoring populations of a terrestrial urban mammal, the West European hedgehog (*Erinaceus europaeus*). *Urban Ecosyst.* **2018**, *21*, 1075–1086. [CrossRef]

55. Yarnell, R.W.; Pacheco, M.; Williams, B.; Neumann, J.L.; Rymer, D.J.; Baker, P.J. Using occupancy analysis to validate the use of footprint tunnels as a method for monitoring the hedgehog *Erinaceus europaeus*. *Mamm. Rev.* **2014**, *44*, 234–238. [CrossRef]

56. Williams, R.L.; Stafford, R.; Goodenough, A. Biodiversity in urban gardens: Assessing the accuracy of citizen science data on garden hedgehogs. *Urban Ecosyst.* **2014**, *18*, 1–15. [CrossRef]

57. MacKenzie, D.I.; Royle, J.A. Designing occupancy studies: General advice and allocating survey effort. *J. Appl. Ecol.* **2005**, *42*, 1105–1114. [CrossRef]

58. Land Cover Map 2015 (Vector, GB). Available online: https://catalogue.ceh.ac.uk/documents/6c6c9203-7333-4d96-88ab-78925e7a4e73 (accessed on 18 April 2020).

59. Dowding, C.V.; Harris, S.; Poulton, S.; Baker, P.J. Nocturnal ranging behaviour of urban hedgehogs, *Erinaceus europaeus*, in relation to risk and reward. *Anim. Behav.* **2010**, *80*, 13–21. [CrossRef]

60. Met Office Integrated Data Archive System (MIDAS) Land and Marine Surface Stations Data (1853-Current). Available online: https://catalogue.ceda.ac.uk/uuid/220a65615218d5c9cc9e4785a3234bd0 (accessed on 10 February 2020).

61. Reading, England, United Kingdom—Sunrise, Sunset, and Daylength. Available online: https://www.timeanddate.com/sun/uk/reading (accessed on 10 February 2020).

62. Burnham, K.P.; Anderson, D.R. *Model Selection and Inference*, 2nd ed.; Springer: New York, NY, USA, 1998.

63. Whittingham, M.J.; Stephens, P.A.; Bradbury, R.B.; Freckleton, R.P. Why do we still use stepwise modelling in ecology and behaviour? *J. Anim. Ecol.* **2006**, *75*, 1182–1189. [CrossRef]

64. Mackenzie, D.I.; Bailey, L.L. Assessing the fit of site-occupancy models. *JABES* **2004**, *9*, 300–318. [CrossRef]

65. Program MARK: A Gentle Introduction. Available online: http://www.phidot.org/software/mark/docs/book/ (accessed on 22 May 2020).

66. Exercises in Occupancy Modelling and Estimation. Exercise 4: Single-Species, Single-Season Model with Site Level Covariates. Available online: http://www.uvm.edu/rsenr/vtcfwru/spreadsheets/occupancy/occupancy.htm (accessed on 22 May 2020).

67. Mackenzie, D.I.; Nichols, J.D.; Royle, J.A.; Pollock, K.H.; Bailey, L.L.; Hines, J.E. *Occupancy Estimation and Modeling Inferring Patterns and Dynamics of Species Occurrence*; Academic Press: Cambridge, MA, USA, 2006.

68. Baldwin, R.A.; Bender, L.C. Denning chronology of black bears in eastern rocky mountain national park, Colorado. *West. North Am. Nat.* **2010**, *70*, 48–54. [CrossRef]

69. Haigh, A.; O'Riordan, R.M.; Butler, F. Nesting behaviour and seasonal body mass changes in a rural Irish population of the Western hedgehog (*Erinaceus europaeus*). *Acta Theriol.* **2012**, *57*, 321–331. [CrossRef]

70. Turbill, C.; Smith, S.; Deimel, C.; Ruf, T. Daily torpor is associated with telomere length change over winter in Djungarian hamsters. *Biol. Lett.* **2012**, *8*, 304–307. [CrossRef] [PubMed]

71. Turbill, C.; Ruf, T.; Smith, S.; Bieber, C. Seasonal variation in telomere length of a hibernating rodent. *Biol. Lett.* **2013**, *9*, 20121095. [CrossRef] [PubMed]

72. Hoelzl, F.; Cornils, J.S.; Smith, S.; Moodley, Y.; Ruf, T. Telomere dynamics in free-living edible dormice (*Glis glis*): The impact of hibernation and food supply. *J. Exp. Biol.* **2016**, *219*, 2469–2474. [CrossRef] [PubMed]

73. Lyman, C.P.; Brien, R.C.O.; Greene, G.C.; Papafrangos, E.D. Hibernation and longevity in the Turkish hamster *Mesocricetus brandi*. *Science* **1981**, *212*, 668–670. [CrossRef]

74. Blanco, M.B.; Zehr, S.M. Striking longevity in a hibernating lemur. *J. Zool.* **2015**, *296*, 177–188. [CrossRef]

75. Wu, C.; Storey, K.B. Life in the cold: Links between mammalian hibernation and longevity. *Biomol. Concepts* **2016**, *7*, 41–52. [CrossRef]

Article

Garden Scraps: Agonistic Interactions between Hedgehogs and Sympatric Mammals in Urban Gardens

Dawn Millicent Scott [1,*], **Robert Fowler** [2], **Ariadna Sanglas** [3] and **Bryony Anne Tolhurst** [4]

[1] School of Animal Rural and Environment Sciences, Brackenhurst Campus, Nottingham Trent University, Southwell NG25 0QF, UK
[2] School of Life Sciences, John Maynard Smith Building, University of Sussex, Brighton BN1 9QG, UK
[3] Department of Conservation Biology, Estación Biológica Doñana, CSIC, Américo Vespucio, 26, 41092 Sevilla, Spain
[4] School of Applied Sciences, University of Brighton, Brighton BN2 4GJ, UK
* Correspondence: dawn.scott@ntu.ac.uk

Simple Summary: Hedgehogs are one of several mammals that occur in urban areas in the United Kingdom and are fed by people. Food provided by people may help wild animals but may also attract animals together that could compete, injure, or predate each other. To understand the impact of food on urban animals we need to investigate how they interact when food is available. In this study, we assessed the type of interaction between hedgehogs, foxes, badgers, and cats using videos submitted by the public. We analyzed interactions between pairs of species to determine interaction type, hierarchical relationships, and the effect of food. We found that agonistic interactions (aggression and/or submission between animals) were more common than neutral interactions and that between-species interactions showed greater 'agonism' than those within the same species. Of interactions within a species, those between hedgehogs were the most agonistic (54.9%) and between badgers the least (6.7%). The species interacting affected the level of agonism, with cats and foxes showing the highest level when together (76.7%). Badgers also outcompeted cats where there were contests over food, but cats were equally as successful as foxes, which were more successful than hedgehogs. However, hedgehogs dominated access to food over cats. We discuss the need to understand interactions between urban animals and the effects of providing food, to inform practice and ensure any potential risks are minimized.

Abstract: Hedgehogs occur within an urban mammal guild in the United Kingdom. This guild commonly utilizes anthropogenic food provision, which is potentially beneficial to wild animal populations, but may also bring competitors and predators into proximity, raising the question of how these species interact in urban gardens. In this study, we determined interactions between hedgehogs, foxes, badgers, and domestic cats using videos submitted via citizen science. We analyzed interactions within and between species to determine interaction type, hierarchical relationships, and effect of supplementary food presence/amount. We found that overall agonistic interactions between individuals occurred more frequently (55.4%) than neutral interactions (44.6%) and that interspecific interactions showed greater agonism (55.4%) than intraspecific ones (36%). Within intraspecific interactions, those between hedgehogs were the most agonistic (54.9%) and between badgers the least (6.7%). Species composition of the interaction affected agonism, with interactions between cats and foxes showing the highest level (76.7%). In terms of overall "wins", where access to garden resources was gained, badgers dominated cats, which were dominant or equal to foxes, which dominated hedgehogs. However, hedgehogs exhibited a greater overall proportion of wins (39.3%) relative to cats. Our findings are important in the context of the documented impact of patchy resources on urban wildlife behavior, and we show that provision of anthropogenic food can potentially result in unintended consequences. We recommend actions to reduce proximity of guild competitors in space and time to limit negative effects.

Citation: Scott, D.M.; Fowler, R.; Sanglas, A.; Tolhurst, B.A. Garden Scraps: Agonistic Interactions between Hedgehogs and Sympatric Mammals in Urban Gardens. *Animals* **2023**, *13*, 590. https://doi.org/10.3390/ani13040590

Academic Editors: Anne Berger, Nigel Reeve and Sophie Lund Rasmussen

Received: 21 December 2022
Revised: 4 February 2023
Accepted: 6 February 2023
Published: 8 February 2023

Keywords: hedgehog; *Erinaceus europaeus*; red fox; *Vulpes vulpes*; Eurasian badger; *Meles meles*; supplementary feeding; urban mammals; domestic cat; *Felis catus*; citizen science

1. Introduction

Increasing urbanization, where habitats have been highly modified for intense human use and residence, typically has a negative effect on biodiversity [1], yet a few species appear to benefit, as evidenced by their urban colonization, and high relative densities (e.g., [2]). These species are termed "synurbic" [3]. Although urban development can present challenges for wildlife, it can also offer benefits. For example, thermal climatic conditions are more stable in urban environments, and ground and air temperatures can be higher than in surrounding rural areas [4], while towns and cities provide numerous sheltering opportunities, and abundant supplementary food from anthropogenic sources [5,6]. Among the many reasons to colonize urban landscapes, food availability is perhaps one of the most important [7].

The intentional provision of supplementary food to wild animals by urban residents, either as surplus household food or commercially purchased for this purpose, has increased in the UK [8], where an estimated 87% of urban dwellers have access to a back or front garden/yard of their residence [9]. Supplementary feeding can benefit wild populations by providing food sources when natural food sources are low. However, it can also create clustered, abundant, spatio-temporally predictable food resources in urban areas [10], leading to behavioral modifications of wildlife, such as changes in foraging and spatial behavior [6,11], diet [12], and social/territorial configuration [11,13] with corresponding changes in density. In the United Kingdom (UK), synurbic mammal species include the Western European hedgehog (*Erinaceus europaeus*) (hereafter "hedgehog"), the red fox (*Vulpes vulpes*) (hereafter "fox"), and the Eurasian badger (*Meles meles*) (hereafter "badger"). These three species form a guild within which they compete for similar food resources; however, badgers and foxes can prey on hedgehogs, i.e., intraguild predation (IGP) also occurs [14,15]. The presence of both badgers and foxes negatively predicts UK hedgehog presence [16,17], but it is unclear whether this is due to predation, competition, or a combination of both. Modeling studies suggest that intraguild predation dynamics may additionally be confounded by the presence of supplementary food [18].

Higher densities relative to rural counterparts, have been documented in hedgehogs [5,16], foxes [6], and to some extent badgers [19]. Factors affecting hedgehogs in the UK are of particular interest due to recent widespread population declines [20–23]. Within population studies, sub/urban habitats are considered potential refuges for hedgehogs [16,24,25], typically providing a range of ecologically favorable attributes, including high food availability and low overall populations of badgers, although high badger sett (burrow network) densities have been documented in some sub/urban areas [26]. This raises the question of how these species interact and co-occur in towns and cities.

Exploitation of supplementarily provisioned food in urban gardens has been observed in all three species, and although this often manifests in individuals feeding alone, co-occurrence and aggregation of multiple individuals and species can also result [8,27]. In addition to synurbic wildlife, domestic cats (*Felis catus*) occur at high densities in urban areas [28] and utilize anthropogenic food sources, spatially overlapping with urban wildlife [29]. Therefore, cats can also be considered part of the urban mammalian guild when investigating species interactions.

Most urban mammals, including cats, forage alone, despite some living in social or family groups, except for hedgehogs, which are considered solitary [30]. Hierarchy within social groups, such as those comprised of badgers or foxes, reduces competition and the risk of aggressive interactions [31]. Evidence of aggression between conspecifics is limited in hedgehogs except for in the context of mating [32]. A garden in which supplementary food is provided could be considered a 'high-quality' patch containing predictable resources in abundance. Individuals may compete to defend, or acquire, high-quality patches and engage in intra- and interspecific agonistic behavior (fighting or conflict behavior, such as threatening, aggressive, and/or submissive behaviors) to dominate access for their own benefit [33]. Each dyadic encounter (paired encounter between individuals) represents the balance of the possible fitness costs and benefits from competing with the opponent/s for access to the resource [34] and will also reflect hierarchical positioning within groups and between competing species.

Currently, there is a gap in our knowledge of how these species interact when coming into proximity at focal food resources. Aggregation of multiple individuals of several species in gardens attracted by food could potentially result in higher encounter rates between competitors and predators. Increasing interference competition, aggression, stress and predation risk may result. There have been reported incidents of urban hedgehogs admitted to rehabilitation centers with injuries from suspected encounters with urban predators, including 2.3% of admissions attributed to injuries from dogs and cats [35]. Therefore it is essential we understand the interactions between synurbic mammals, what factors drive interactions and access to food, and how coexistence in these habitats can be supported.

The aims of the study were to determine if the type of interaction between sympatric urban mammal species varied depending on which species were interacting, which of the co occurring species was most likely to 'win' dyadic competitions for access to supplementary food, and finally, if presence and amount of food affected the type of interaction. We hypothesized that there would be lower levels of agonism within species, especially those with established hierarchies, and greater agonism between species. We hypothesized that the largest of the species (badger) would dominate and be the most successful at 'winning' access to food, and that increased food availability would increase levels of agonistic interactions between and within species. Understanding interactions between synurbic mammals and the effect of supplementary feeding on interactions can help inform best practice around food provision to prevent unintended costs to the species concerned.

2. Materials and Methods

2.1. Data Collection and Cleaning

Data on urban mammal interactions were obtained from the public following a national appeal for such video footage on a UK television broadcast (British Broadcasting Commission [BBC] "Springwatch" series) in May 2017. This program is part of a long standing seasonal series on British wildlife that has approximately 2.5 million viewers. The broadcast explained the study aims and provided details of a link to a data capture site where videos and associated metadata could be submitted. The link was also available on the associated program website and the University of Brighton website. We did not provide instructions of an experimental set-up to follow, but instead called for people who had existing footage of multiple animals within their gardens to submit their existing footage. Once on the site, the public (volunteers) could upload their videos and answer a questionnaire on the location, video content, and food provision where the video was taken. The link was available between 29 May and 15 June 2017, during which time 683 files were submitted. Video analysis at feeding sites has previously been used as a method to determine relationships and interactions between sympatric species [36].

2.2. Data Handling

All data collected were downloaded to an Excel sheet with a linking unique ID code assigned to each video. Prior to analysis, data were removed that were not in video format or did not contain a hedgehog, fox, badger, or cat. We included videos where at least two adult individuals were visible, recorded in a UK residential garden between Jan 2010 and May 2017. Urban habitat was verified by using Google Maps for the postcode data submitted with the video. Multiple files from the same address/date were assumed to be consecutive recordings, so only one representative video was used. Of the 683 files submitted, 586 dyads (interactions between two individuals) [30] were extracted during behavioral analysis.

2.3. Behavioral Analysis

Interactions between species were assessed per dyad. When interactions involved three or more individuals, multiple dyads were derived. For each dyadic interaction, the following were recorded from the video analysis: interaction number (for videos with more than two dyadic interactions); duration of video (in seconds); visible food present (yes or no); duration of the interaction (in seconds); species; predominant behavior of animal; interaction type (neutral or agonistic) [36]; and outcome for both individuals in the dyad ("win" or "no win") [31]. The amount of food left out ("high" or "low", where "high" was two handfuls or more and "low" was a single handful or less) was also included from the questionnaire response. An ethogram was compiled containing detailed descriptions of six typical dyadic behaviors [37] during encounters (passive, submissive, avoidance, defensive, aggressive, attack; see Table 1). As the study focused on multiple species, some definitions included reference to a particular species. During the analysis, a predominant behavior was assigned to each member of the dyad and an interaction type then chosen to summarize the encounter. Behaviors that appeared to have no impact on either animal were deemed *neutral*, i.e., animals were passive towards each other, there was no defensive or aggressive behavioral change in the presence of another, and/or there was no observed agonistic behavior. Conversely, behaviors involving submission, threat, aggression, defense, or attack were classified as *agonistic* [36]. Dyadic outcomes for each animal were classified as either "win" (if one animal was seen eating the food or dominating the space close to the food during the video clip), "draw" if the two animals continued feeding or stayed in the garden together, or "unclear" if the video or observation ended before an outcome could be determined. A loss, draw, or unclear outcome was classified as "no win" for the individual in the dyad.

2.4. Data Analysis

To test the effect of dyad composition (i.e., species), food presence, and amount on interaction type and probability of wins, Generalized Linear Models (GLMs) were computed with binomial error structures in R Studio (RStudio, 2012) using R v3.6.1 [38] and packages MASS and lme4. Inter- and intraspecific dyads were modeled separately. For interaction type, both models contained dyad composition as the independent variable and interaction type (Neutral = 0, Agonistic = 1) as the dependent variable. For the intraspecific model, the highest neutral interaction dyad (badger-badger) was used as a reference in the model. For probability of wins, the independent variables comprised interaction type (Neutral = 0, Agonistic = 1), whether supplementary food was provided (No = 0, Yes = 1), and the amount of food (Low = 0, High = 1) and dyad composition (all the interspecific dyads that contain at least one of the target species for that specific model). The hedgehog-cat dyad was used as a reference for the interspecific model, as the highest neutral interaction dyad that was not deemed a potentially predatory interaction. The dependent variable in each case was win (eats the food = 1) or no win (lose, draw/shares food or is unclear = 0) for each test species. Finally, the reference level was switched for each dyad comparison (using the relevel function) so that all pairwise dyad comparisons were tested against each other.

Prior to applying models, proposed explanatory variables were checked for multi-collinearity using Variance Inflation Factors (VIFs). If variables had VIFs greater than 3 or correlation coefficients more than 0.6 with other variables, they were excluded from models [39]. Model residuals were assessed for normality and heteroscedasticity. Omnidirectional stepwise selection was undertaken to build the best model in each case using the step function with the command direction = both. This procedure sequentially keeps or drops variables starting with the null (intercept only) model through testing the significance of each independent variable in a linear regression model.

Table 1. Behavior categories and types of interaction based on analysis of encounters between four species of urban mammal—Western European Hedgehog (*Erinaceus europaeus*), Eurasian badger (*Meles meles*), Red fox (*Vulpes vulpes*), and domestic cat (*Felis catus*). Symbols represent each behavior and how they relate to the two broad interaction types.

Animal Behaviour		Interaction Type	
Passive ○	Animal continues feeding or remains in the area when approached by the other individual. Departs without retaliation if aggression occurs.		
Submissive ☐	An act or posture that does not challenge the incoming animal. Body position or response indicting lower hierarchy or submission, e.g. in foxes the body and head lower. In cats crouching with ears flattened, avoiding, retreating or fleeing.	Neutral ○ ☐ —	Behaviour of both animals has no impact on the other, e.g. ignore each other.
Avoid —	Animal moves or backs away from the area or other animal before close proximity or physical contact. Often with body held low, casting repeated glances at the one which stays.		
Defensive ▌	A defensive posture or positioning e.g. piloerection, frowning or complete rolling in hedgehogs. In cats, hissing or piloerection.	Agonistic ▌ ▶ ● ☐ —	Includes submissive, threat, attack and aggression behaviour. Could result in injury, or death, of at least one individual.
Aggressive ▶	Vocalisation or aggressive posture. Action of initiating physical contact with another animal including lunging, biting, scratching etc. Foxes side profile body position with arched back and head up.		
Attack ●	One animals runs towards the other in an aggressive manner. Chases, lunges, bites etc. In badgers, head is lowered in a threat posture and pursues the challenger with physical contact. Cat striking with paw.		

3. Results

A total of 683 files were received from volunteers. The files collected dated from May 2010 to June 2017. Using the filtering method stated in the methods section 586 separate dyads were analyzed, representing both intra- and interspecific interactions. Two datasets were created with these data; the 'all dyads' dataset, which included all 586 separate dyads, and the 'outcome' dataset, which consisted of 331 interspecific dyads with wins, draws, and unclear outcomes recorded, alongside data on food presence and amount.

3.1. Interactions between Sympatric Species

Badger-badger dyads had the lowest level of agonistic interactions and were recorded to be neutral in 93.3% of interactions (Table 2). In comparison, fox-fox and hedgehog-hedgehog dyads exhibited neutral behavior in 63.6% and 45.1% of interactions, respectively. Hedgehog-hedgehog dyads had the highest number of intraspecific agonistic interactions (54.9%). The most frequent interspecific dyad was fox-hedgehog, with 143 separate instances of this dyad recorded (Table 2). The split between agonistic and neutral interactions for fox-hedgehog dyads was relatively even at 49% and 51%, respectively. Interspecific interactions between cats and foxes had the highest level of agonistic interactions (76.7%). Badger-cat dyads were the least recorded, being observed only eight times, but also showed a high proportion of agonistic interactions (75%). In every dyad containing a cat, the proportion of agonistic interactions was greater than neutral; however, the incidence of these was relatively low compared to other dyads (Table 2). Hedgehogs displayed higher agonistic interactions with foxes than badgers, although these comprised <50% of interactions.

Table 2. The number of events recorded from videos sent in by volunteers for each dyad of four species of urban mammal (Western European Hedgehog (*Erinaceus europaeus*), Eurasian badger (*Meles meles*), Red fox (*Vulpes vulpes*), and domestic cat (*Felis catus*)) and the number and relative percentage of agonistic or neutral interactions.

	Dyad	No. of Events	No. of Agonistic Interactions N (%)	No. of Neutral Interactions N (%)
Intraspecific Dyads	Hedgehog-Hedgehog	142	78 (54.9)	64 (45.1)
	Badger-Badger	95	7 (6.7)	88 (93.3)
	Fox-Fox	33	12 (36.4)	21 (63.6)
	Total	270	97 (36.0)	173 (64.0)
Interspecific Dyads	Fox-Hedgehog	143	70 (49.0)	73 (51.0)
	Badger-Hedgehog	16	5 (31.3)	11 (68.7)
	Cat-Hedgehog	28	16 (57.1)	12 (42.9)
	Badger-Fox	78	45 (57.7)	33 (42.3)
	Fox-Cat	43	33 (76.7)	10 (23.3)
	Badger-Cat	8	6 (75.0)	2 (25.0)
	Total	316	175 (55.4)	141 (44.6)

3.2. Comparisons between Species Dyads

Statistical models revealed a greater level of agonistic behavior in all other intra- and interspecific dyads compared to badger-only dyads (Table 3). Statistically, only cat-fox dyads had higher agonistic interactions than hedgehog-hedgehog dyads. Cat-fox and badger-fox showed significantly higher agonistic interactions than fox-fox. Both badger-hedgehog and badger-fox agonistic interactions were greater than badger-badger but less than cat-fox. Badger-fox agonism was less than cat-fox but greater than fox-fox.

Table 3. Generalized Linear Models (GLM) of the test results investigating the effects of intraspecific and interspecific dyadic composition on interaction type (Neutral = 0; Agonistic = 1), using the relevel command in R to switch the levels to investigate all dyad comparisons against each other. Displayed are parameter estimates $\pm$ standard error, z-value and p-value for each pairwise comparison. '<' agonistic interactions are greater than the reference dyad and '>' are less than the reference dyad.

Reference Dyad	Explanatory Variables	Parameter Estimate $\pm$ SE	z-Value	p-Value
Badger-Badger	<Cat-Hedgehog	2.819 $\pm$ 0.548	5.147	<0.001
	<Cat-Fox	3.725 $\pm$ 0.533	6.984	<0.001
	<Hedgehog-Hedgehog	2.729 $\pm$ 0.427	6.386	<0.001
	<Badger-Cat	3.63 $\pm$ 0.906	4.007	<0.001
	<Badger-Fox	2.845 $\pm$ 0.455	6.249	<0.001
	<Badger-Hedgehog	1.743 $\pm$ 0.668	2.612	<0.01
	<Fox-Fox	1.972 $\pm$ 0.534	3.692	<0.001
	<Fox-Hedgehog	2.490 $\pm$ 0.427	5.832	<0.001
Hedgehog-Hedgehog	<Cat-Fox	0.996 $\pm$ 0.398	2.5	<0.05
Fox-Fox	<Cat-Fox	1.754 $\pm$ 0.511	3.431	<0.001
	<Badger-Fox	0.870 $\pm$ 0.428	2.031	<0.05
Badger-Hedgehog	<Cat-Fox	1.982 $\pm$ 0.649	3.054	<0.01
Fox-Hedgehog	<Cat-Fox	1.236 $\pm$ 0.398	3.106	<0.01
	>Badger-Badger	−2.490 $\pm$ 0.427	−5.832	<0.001
Badger-Fox	<Cat-Fox	0.884 $\pm$ 0.428	2.067	<0.05
	>Fox-Fox	−0.870 $\pm$ 0.428	−2.031	<0.05
Cat-Fox	>Hedgehog-Hedgehog	−0.996 $\pm$ 0.398	−2.5	<0.05
	>Badger-Badger	−3.725 $\pm$ 0.533	−6.984	<0.001

Table 3. *Cont.*

Reference Dyad	Explanatory Variables	Parameter Estimate ± SE	z-Value	p-Value
	>Badger-Fox	−0.884 ± 0.428	−2.067	<0.05
	>Badger-Hedgehog	−1.982 ± 0.649	−3.054	<0.01
	>Fox-Fox	−1.754 ± 0.511	−3.431	<0.001
	>Fox-Hedgehog	−1.236 ± 0.398	−3.106	<0.01

3.3. Species Dominance and Hierarchy

In terms of overall percentage of wins (access to garden resources), the order of dominance was firstly badgers, then cats, and then foxes. Foxes dominated hedgehogs but hedgehogs dominated cats; hence, the hierarchy was non-linear (Table 4; Figure 1 for diagrammatical representation). However, there were a large proportion of draws and unclears in these dyads, so it is not always apparent which species won in each combination. Badgers tended to be more successful at 'winning' interactions than the other species, winning 45.9%, 42.1%, and 66.6% of interactions against foxes, hedgehogs, and cats, respectively. Foxes had the largest proportion of wins of any species when the dyad was with hedgehogs (46.2%). While cats won 44.4% of their dyads with foxes, foxes won 31.1% of these dyads (Table 4). Unexpectedly, hedgehogs won 39.3% of the time in dyads with cats, compared to cats winning only 10.7% of these dyads.

Table 4. The number of events recorded from videos sent in by volunteers between each dyad of four species of urban mammal (Western European Hedgehog (*Erinaceus europaeus*), Eurasian badger (*Meles meles*), Red fox (*Vulpes vulpes*), and domestic cat (*Felis catus*)), and the number and relative percentage (%) of 'wins' for each species within the dyad.

Dyad	No. of Events	Win Hedgehog	Win Badger	Win Fox	Win Cat	Draw	Unclear
Fox-Hedgehog	145	14 (9.6)		67 (46.2)		41 (28.3)	23 (15.9)
Badger-Hedgehog	19	2 (10.5)	8 (42.1)			7 (36.8)	2 (10.5)
Hedgehog-Cat	28	11 (39.3)			3 (10.7)	9 (32.1)	5 (17.9)
Badger-Fox	85		39 (45.9)	8 (9.4)		30 (35.3)	8 (9.4)
Cat-Fox	45			14 (31.1)	20 (44.4)	7 (15.6)	4 (8.9)
Badger-Cat	9		6 (66.6)		1 (11.1)	1 (11.1)	1 (11.1)
Total	331	27 (14.1)	53 (46.9)	79 (28.7)	24 (29.3)	101 (30.5)	43 (12.9)

3.4. Impact of the Presence of Food

When data were applied to models to test the effect of interspecific dyad, interaction type and presence/amount of food on probability of wins, badgers won more interactions when agonistic behavior was involved in the dyad (parameter est. ± S.E. = 1.395 ± 0.413; z = 3.375; p < 0.001). Zero-inflated datasets prevented the inclusion of food presence and amount in some models. However, foxes exhibited a greater chance of winning when the interaction was agonistic (parameter est. ± S.E. = 2.734 ± 0.397; z = 6.878; p < 0.001) and food was present (parameter est. ± S.E. = 2.143 ± 0.830; z = 2.581; p < 0.01).

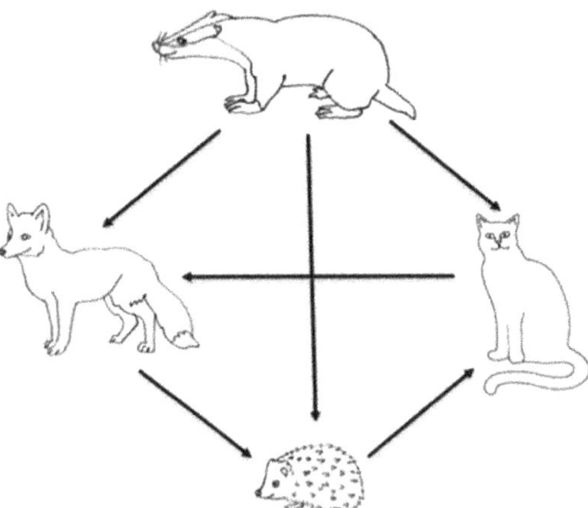

Figure 1. A diagrammatical representation of the hierarchical relationship in 'winning' access to food during paired dyads. The direction of the arrow shows the direction of the dominance is the species with the higher % of wins within dyads.

4. Discussion

Our study is the first to quantify behavioral interactions between hedgehogs and three other intraguild mammal species in sub/urban gardens in the UK. We have shown that food provided in gardens by human residents is utilized by a range of intraguild species with associated evidence of interactions within and between species. Based on the assessment of 'wins' during interactions over food access, we have derived a hierarchical relationship between a community of sympatric urban mammals. The study findings supported our hypothesis that the largest of the species (badgers) in the guild would be most successful at 'winning' access to food. We also found some support for our hypothesis that food availability affected interaction type, with higher levels of agonistic interactions when food was present than when it was absent, although analysis of this element was limited due to an unbalanced dataset. Spatially predictable and/or abundant food patches can create focal activity hotspots of resource exploitation, causing aggregation of sympatric species, which likely lead to higher occurrences of interactions [40]. The consequences of interactions between garden mammals are potentially numerous. Direct interactions between competing species can be aggressive, leading to injury or death, with increased competition or competitive exclusion reducing access to resources for subordinate species or individuals, with knock-on welfare effects. Furthermore, the dynamics of space use overlap and interaction rates within wild populations and between domestic and wild species has implications for pathogen transmission, particularly in urban areas at the interface between humans and wildlife, where zoonoses can emerge [41].

4.1. Intraspecific Interactions

Our findings support the hypothesis that agonistic behavior would be higher between species compared to within species. As expected from previous studies [31], we also showed there were few agonistic interactions between badgers at feeding sites (7%). It appears that this observation is consistent between urban and rural badger populations, despite differences in urban badger group size, density, and territory size compared to rural [19]. As expected, foxes showed higher intraspecific agonism than badgers. Although foxes typically forage alone, there is a hierarchy within fox social groups [42]. Subordinate foxes have been shown to use supplementary food patches in urban gardens in a different way to dominant foxes, using fewer patches, spending less time in predictable patches, and feeding later [43]. This is explained as an evolved behavioral strategy to reduce competition/antagonism within social groups.

Hedgehogs are considered solitary and non-territorial [32]. Field studies of behavior have shown that hedgehogs tend to avoid each other, and adults are usually only found together during courtship, or when attracted to a localized food source. Other than during courtship, when females are typically aggressive towards males, and competing male suitors may fight each other, overt aggression is rarely seen [44]. Contrastingly, we observed high levels of agonism between wild hedgehogs. This included a characteristic behavior where one hedgehog attacked another by running at it, causing the victim to roll up, after which the attacker pushed it away. Typically, the function of this behavior appeared to involve moving a competitor away from the food source, such as to the edge of the garden. In one case, an individual was pushed down a flight of concrete, and another into water. We termed this behavior 'barge and roll' and deemed it to be competitive.

Many of these observations of agonism between hedgehogs occurred outside of the breeding period. Anecdotal evidence from rescue centers suggests aggression can occur when food is provisioned to a group; thus, the most likely explanation for agonism in our study is defense of food patches. This behavioral disparity may be a consequence of patch distribution and abundance of urban food resources relative to rural or wilderness areas. Previous studies on red deer (*Cervus elaphus*) have shown increased aggression at food patches when provided with spatially and temporally predictable supplementary food during winter [45]. Access to food that will allow for an increase in body weight may be more critical for hibernating species such as hedgehogs than species that are active all year as very low body weight could affect overwinter survival [46,47]. As the focus of our study was to investigate intraguild interactions of urban wildlife, we did not request videos of cat-cat interactions in this study, and thus, could not determine such interactions at garden feeding sites.

4.2. Interspecific Interactions and Hierarchy

Between-species interactions were more agonistic overall than those within species, and badgers tended to be more successful at securing food than all other species. In our study, badgers were dominant over foxes where contests occurred, but if initial interactions did not escalate to aggression, each species was unaffected by the presence, proximity, or orientation of the other (consistent with [34]). Nonetheless, our study is a temporal snapshot, and previous experience may affect what was observed [36]. Interspecific interactions involving hedgehogs showed the highest levels of agonism with cats, although cats were predominantly observed to be submissive during these interactions, allowing hedgehogs access to food without contest. Hedgehogs co-occurred in urban gardens with potential predators (badgers and foxes) and approximately half of encounters between hedgehogs and foxes were agonistic. Hedgehogs have previously been reported to avoid predators [48], but in urban gardens where food is provided, they may not always do so, with the benefits of access to food provision potentially outweighing predation risk.

As cats are typically fed by owners, access to additional food is less critical to their survival or fitness compared to wild counterparts, although they are highly territorial [49] and likely to defend the gardens within their territories. We observed high levels of agonism between cats and foxes, with the former appearing dominant over the latter overall. The relatively high proportion of hedgehog wins over cats was unexpected but may relate to hedgehog spines, which domestic cats are not physically or behaviorally adapted to defend themselves against, as compared to wild predators. Although competition could impact access to food, previous studies have shown that there is more food available in urban areas than some species metabolically require (e.g., foxes [12,42,43]), which suggests that if an animal unsuccessfully forages at one patch, there are likely to be multiple alternative patches where competition could be lower due to temporal activity partitioning [43]. These may be readily accessible to more mobile species, such as foxes [50], whereas hedgehog movement can be affected by barriers in urban areas [51]. This may account for the higher than-expected agonism observed between hedgehogs, i.e., if the value of a feeding patch is perceived to be high and movement between patches is restricted, defense may be adaptive

Hierarchical positioning of the garden mammal guild is dominated by badgers. Foxes and cats then rank approximately equally, with cats dominating access to food over foxes, and hedgehogs dominating food over cats. However, just under half of contest outcomes overall were either unclear or a draw, and the same proportion involved no contest at all, i.e., neutral co-occurrence. Thus, "noise" in the dataset was substantial.

4.3. Disease Transmission Risk

Increases in space use overlap can increase both intra- and interspecific pathogen transmission risk [41]. We showed evidence of animals coming into proximity and utilizing similar feeding sites within urban gardens and interactions between and within species, as well as between wild and domestic animals. Common pathogens spread by the species in this study include the sarcoptic mange mite (*Sarcoptes scabiei*) [52] and the dog roundworm (*Toxocara canis*) [53]. Rabies lyssavirus [54] and *Echinococcus multilocularis* [53,55] are transmitted between foxes, cats, and badgers in Europe, and have the potential to emerge in the UK, although *E. multilocularis* has not yet been detected in European hedgehogs [56]. Food provision practice that reduces spatio-temporal species overlap could, thus, also reduce pathogen transmission.

4.4. Application of Findings to Hedgehog Conservation

This study has important implications for hedgehog conservation, in the context of both intraguild predation, and competition. Previous studies report hedgehog abundance to be negatively associated with badger abundance [57], with experimental evidence showing a doubling of hedgehog populations when badgers were absent or present at very low density, i.e., meso-predator release [58]. In our study, badgers and hedgehogs co-occurred at feeding sites, although most interactions between them were classed as neutral, including occasions where both species ate simultaneously close together. Where agonistic interactions did occur, approximately 2/3rd of these were seen as competitive, with 1/3rd considered potentially predatory, representing a small proportion of overall interactions involving these two species (10%). Although we did not record any direct hedgehog predation events by foxes in our study, 5% of agonistic interactions between the two species were classed as potential predatory behavior, with the remaining agonistic interactions considered competitive. Levels of fox-hedgehog agonism were, thus, greater than badger-hedgehog agonism, but in the form of competition rather than predation. Research conducted in Regents Park in London, UK, found foxes to be a major cause of hedgehog mortality due to injury, although this may also arise from competitive interactions rather than direct predation [59]. Rehabilitation centers report occasional admission of hedgehogs with injuries from wild predators, but these may also be caused by domestic pets [35]. Any agonistic interaction could potentially result in injury or death, and mechanisms to prevent co-occurrence at feeding sites are likely to reduce both outcomes.

However, feeding sources are not always utilized by multiple animals simultaneously, with solitary feeding regularly observed. In our study, feeding did not always lead to interactions and many co-occurrences were neutral. Supplementary feeding can benefit hedgehogs by providing food at times when natural food resources are low, potentially contributing to increased abundance [5]. Nonetheless, it is pertinent to adopt strategies to minimize negative effects of agonism towards hedgehogs according to the precautionary principle. Variable food availability in space or time, for example, reduces its predictability and the corresponding spatio-temporal overlap between or among species. Strategies to do so could include altering timing of food provision and providing food in locations with restricted (species-specific) access, such as hedgehog feeding boxes (although this would not reduce intraspecific hedgehog agonism) or at multiple sites within the garden.

As urban areas are considered refuges for hedgehogs [23], understanding how to minimize negative impacts of interactions in urban habitats will aid conservation management. In general, hedgehogs do not commonly inflict injury on conspecifics [35], although injury/other consequences of the frequently observed 'barge and roll' behavior are currently unknown. High levels of agonism and the associated risk of lethal/sublethal effects can be disproportionately high where hedgehogs aggregate around feeding sites in urban gardens. Sublethal effects such as exacerbated health conditions and reduced fecundity and breeding success [60] may follow competitive exclusion from food resources and associated stress. It is, therefore, recommended that feeding practices also aim to reduce hedgehog intraspecific agonism.

4.5. Use of Citizen Science and Data Limitations

Citizen science data collected from volunteers can elucidate species interactions in inaccessible areas, such as urban gardens [61]. Collecting data in private gardens can be intrusive to owners, involve a lot of time, and be logistically challenging [62]. To achieve a sufficiently large sample size of independent videos throughout UK sub/urban areas would be challenging in a short time frame; hence, national citizen science approaches were deemed preferable in our study. One limitation of this approach is lack of standardization in video collection, although standardized video processing can partially compensate. Participant responses to questions on presence and abundance of food provision were also difficult to standardize; thus, the trustworthiness of questionnaires is a limitation in their use [63]. Overall representativeness of datasets can also be compounded by the lower likelihood of citizen scientists submitting less 'interesting' data, which in our study could mean a bias toward dramatic interactions, potentially overestimating levels of agonism. In addition, in our study an unbalanced dataset arose from cameras typically being deployed at locations where animals were already encouraged such as feeding sites, hence, footage from sites with food absent is limited. The robustness of the model including food presence/absence and amount was, therefore, inevitably reduced relative to other analyses. Finally, grouping of the "no-win" category in video processing to include 'draws' and 'unclear' interactions as well as losses to allow computation of the binomial regression analysis, reduced our capacity to model nuance in species interactions.

5. Conclusions

Citizen science is a useful method of investigating urban wildlife, where access to private spaces is limited, with the main challenge being access to participants across a broad geographic range, although media recruiting can partially compensate for this. Animal interaction studies have benefited from the increasing availability, affordability, and ease of use of remote monitoring cameras to the public in recent years. Urban wildlife behavior is a suitable focus to engage the public, as species are familiar and interesting interactions can be easily observed at close quarters within private spaces.

Our study is the first to quantify interactions within a sympatric urban mammal community and document hierarchical relationships between wild and domestic mammals in urban gardens. We show that badgers tend to dominate this hierarchy and that high levels of agonism can occur between hedgehogs within sub/urban populations. We also report relatively high levels of agonism between hedgehogs and their potential predators, although the majority of these were competitive interactions, and actual predation events were rare. Clearly, clustering of supplementary food sources in urban environments through anthropogenic feeding can lead to multi-species co-occurrence (including both domestic animals and wildlife) and high species abundance at feeding sites. Where species co-occur at food patches, agonism is typically higher, with potential for increases in transmission rates of some pathogens. Distribution of food sources in urban areas affects the spatial ecology of synurbic animals, in addition to interaction dynamics within and between co-occurring species. Further research is necessary on the impact of supplementary food in the urban environment on animal health, ecology, and disease risk via changes in species interactions. As urban areas are considered hedgehog refuges, understanding interactions within the guild they occupy is critical for informing conservation and welfare management, including feeding practices, for this declining species.

Author Contributions: Conceptualization D.M.S. and B.A.T.; methodology, D.M.S. and B.A.T.; supervision, D.M.S.; validation, D.M.S. and A.S.; formal analysis, R.F.; investigation, D.M.S.; resources, D.M.S.; data curation, D.M.S. and A.S.; visualization, D.M.S. and B.A.T.; writing—original draft preparation, D.M.S.; writing—review and editing, D.M.S., R.F., A.S. and B.A.T.; project administration, D.M.S.; funding acquisition, D.M.S. All authors have read and agreed to the published version of the manuscript.

Funding: This research was funded internally by the School of Applied Sciences, University of Brighton, UK. A. Sanglas was funded by ERASMUS exchange from the Universitat de Barcelona, Spain.

Institutional Review Board Statement: The questionnaire and data collection protocol was approved by the School of Pharmacy and Biomolecular Sciences (Applied Sciences) Ethics Committee of the University of Brighton and given a 'favorable' outcome on 1 March 2017.

Informed Consent Statement: Informed consent was obtained from all subjects involved in the study, via an online process. Prior to uploading videos and accompanying metadata into the submission form on the Springtails website, participants were required to tick an "approve consent" box. It was, thus, not possible for participants to proceed to the next stage of submitting their data without giving this consent for the authors to use and publish the data.

Data Availability Statement: Restrictions apply to the availability of these data. The data are not publicly available due to data use restrictions stated when obtained data from the general public.

Acknowledgments: We are immensely grateful to all the students at the University of Brighton who helped with the initial data processes and behavioral analysis, in particular, M. Williams, C. Wilkinson, J. Fincham, C. Panter, D. Davies, and E. Atkinson. Further checking of the dataset was undertaken by I. Mattioli. Our thanks to the University of Brighton IT department, especially K. Piatt, M. Cundy, and J. Bailey for helping set up the website and data capture processes, and to the BBC Springwatch Team that allowed us to mention the study during broadcast to encourage data submission. We are indebted to all the people who submitted videos and data for the study, without which the study would not be possible.

Conflicts of Interest: The authors declare no conflict of interest.

References

McKinney, M.L. Effects of urbanization on species richness: A review of plants and animals. *Urban Ecosyst.* **2008**, *11*, 161–176.

Prange, S.; Gehrt, S.D.; Wiggers, E.P. Demographic factors contributing to high raccoon densities in urban landscapes. *J. Wild. Manag.* **2003**, *67*, 324–333.

Parker, T.S.; Nilon, C.H. Gray squirrel density, habitat suitability, and behaviour in urban parks. *Urban Ecosyst.* **2008**, *11*, 243–255.

Grimm, N.B.; Faeth, S.H.; Golubiewski, N.E.; Redman, C.L.; Jianguo, W.; Xuemei, B.; Briggs, J.M. Global change and the ecology of cities. *Science* **2008**, *319*, 756–760. [CrossRef]

Hubert, P.; Julliard, R.; Biagianti, S.; Poulle, M.L. Ecological factors driving the higher hedgehog (*Erinaceus europaeus*) density in an urban area compared to the adjacent rural area. *Landsc. Urban Plan.* **2011**, *103*, 34–43. [CrossRef]

Šálek, M.; Drahnikova, L.; Tkadlec, E. Changes in home range sizes and population densities of carnivore species along the natural to urban habitat gradient. *Mamm. Rev.* **2015**, *45*, 1–14.

Newsome, S.D.; Garbe, H.M.; Wilson, E.C.; Gehrt, S.D. Individual variation in anthropogenic resources use in an urban carnivore. *Oecologia* **2014**, *178*, 115–128. [CrossRef]

Baker, P.J.; Harris, S. Urban mammals: What does the future hold? An analysis of the factors affecting patterns of use of residential gardens in Great Britain. *Mamm. Rev.* **2007**, *37*, 297–315. [CrossRef]

Davies, Z.G.; Fuller, R.A.; Loram, A.; Irvine, K.N.; Sims, V.; Gaston, K.J. A national scale inventory of resource provision for biodiversity within domestic gardens. *Biol. Conserv.* **2009**, *142*, 761–771. [CrossRef]

10. Oro, D.; Genovart, M.; Tavecchia, G.; Fowler, M.S.; Martínez-Abraín, A. Ecological and evolutionary implications of food subsidies from humans. *Ecol. Lett.* **2013**, *16*, 1501–1514. [CrossRef]

11. Ditchkoff, S.S.; Saalfeld, S.T.; Gibson, C.J. Animal behaviour in urban ecosystems: Modications due to human-induced stress. *Urban Ecosyst.* **2006**, *9*, 5–12. [CrossRef]

12. Contesse, P.; Hegglin, D.; Gloor, S.; Bontadina, F.; Deplazes, P. The diet of urban foxes (*Vulpes vulpes*) and the availability of anthropogenic food in the city of Zurich, Switzerland. *Mamm. Biol.* **2004**, *69*, 81–95.

13. Prange, S.; Gehrt, S.D.; Wiggers, E.P. Influences of anthropogenic resources on raccoon (*Procyon lotor*) movements and spatial distribution. *J. Mammal.* **2004**, *85*, 483–490. [CrossRef]

14. Holt, R.D.; Huxel, G.R. Alternative prey and the dynamics of intraguild predation: Theoretical perspectives. *Ecology* **2007**, *88*, 2706–2712. [CrossRef] [PubMed]

15. Doncaster, C.P. Testing the role of intraguild predation in regulating hedgehog populations. *Proc. R. Soc. B* **1992**, *249*, 113–117. [PubMed]

16. Pettet, C.E.; Johnson, P.J.; Moorhouse, T.P.; Macdonald, D.W. National predictors of hedgehog *Erinaceus europaeus* distribution and decline in Britain. *Mamm. Rev.* **2018**, *48*, 1–6.

17. Hof, A.R.; Allen, A.M.; Bright, P.W. Investigating the role of the Eurasian badger (*Meles meles*) in the nationwide distribution of the Western European Hedgehog (*Erinaceus europaeus*) in England. *Animals* **2019**, *9*, 759.

18. Briggs, C.J.; Borer, E.T. Why short-term experiments may not allow long term predictions about intraguild predation. *Ecol. Appl.* **2005**, *15*, 1111–1117. [CrossRef]

19. Davison, J.; Huck, M.; Delahay, R.J.; Roper, T.J. Restricted range behaviour in a high-density population of urban badgers. *J. Zool.* **2009**, *277*, 45–53. [CrossRef]

20. Roos, S.; Johnston, A.; Noble, D. *UK Hedgehog Datasets and Their Potential for Long-Term Monitoring*; BTO Research Report 598; British Trust for Ornithology: Thetford, UK, 2012; pp. 1–63.

21. Wilson, E.; People's Trust for Endangered Species (PTES). *Conservation Strategy for West-European Hedgehog (Erinaceus europaeus) in the United Kingdom (2015–2025)*; People's Trust for Endangered Species: London, UK, 2018.

22. Williams, B.; Baker, P.J.; Thomas, E.; Wilson, G.; Judge, J.; Yarnell, R.W. Reduced occupancy of hedgehogs (*Erinaceus europaeus*) in rural England and Wales: The influence of habitat and an asymmetric intra-guild predator. *Sci. Rep.* **2018**, *8*, 12156. [CrossRef]

23. Wembridge, D.; Johnson, G.; Al-Fulaij, N.; Langton, S. *The State of Britain's Hedgehogs*; Peoples Trust for Endangered Species: London, UK, 2022.

24. Van De Poel, J.L.; Dekker, J.; Van Langevelde, F. Dutch hedgehogs *Erinaceus europaeus* are nowadays mainly found in urban areas, possibly due to the negative effects of badgers *Meles meles*. *Wildl. Biol.* **2015**, *21*, 51–55. [CrossRef]

25. Turner, J.; Freeman, R.; Carbone, C. Using citizen science to understand and map habitat suitability for a synurbic mammal in an urban landscape: The hedgehog *Erinaceus europaeus*. *Mamm. Rev.* **2022**, *52*, 291–303.

26. Davison, J.; Huck, M.; Delahay, R.J.; Roper, T.J. Urban badger setts: Characteristics, patterns of use and management. *J. Zool.* **2008**, *275*, 90–200.

27. Gazzard, A.; Baker, P.J. Patterns of feeding by householders affect activity of hedgehogs (*Erinaceus europaeus*) during the hibernation period. *Animals* **2020**, *10*, 1344. [PubMed]

28. van Heezik, Y.; Smyth, A.; Adams, A.; Gordon, J. Do domestic cats impose an unsustainable harvest on urban bird populations? *Biol. Conserv.* **2010**, *143*, 121–130. [CrossRef]

29. van Heezik, Y. Pussyfooting around the issue of cat predation in urban areas. *Oryx* **2010**, *44*, 153–154. [CrossRef]

30. Riber, A.B. Habitat use and behaviour of European hedgehog *Erinaceus europaeus* in a Danish rural area. *Acta Therio.* **2006**, *51*, 363–371. [CrossRef]

31. MacDonald, D.W.; Stewart, P.D.; Johnson, P.J.; Porkert, J.; Buesching, C. No evidence of social hierarchy amongst feeding badgers *Meles meles*. *Ethology* **2002**, *108*, 613–628. [CrossRef]

32. Haigh, A.J.; Butler, F.; O'Riordan, R.M. Courtship behaviour of western hedgehogs (*Erinaceus europaeus*) in a rural landscape in Ireland and the first appearance of offspring. *Lutra* **2012**, *55*, 41–54.

33. Huntingford, F.; Turner, A.K. *Animal Conflict*; Chapman and Hall: New York, NY, USA, 1987.

34. MacDonald, D.; Buesching, C.; Stopka, P.; Henderson, J.; Ellwood, S.A.; Baker, S. Encounters between two sympatric carnivores: Red foxes (*Vulpes vulpes*) and European badgers (*Meles meles*). *J. Zool.* **2004**, *263*, 385–392.

35. Reeve, N.J.; Huijser, M.P. Mortality factors affecting wild hedgehogs: A study of records from wildlife rescue centres. *Lutra* **1999**, *42*, 7–24.

36. McGlone, J.J. Agonistic behaviour in food animals: Review of research and techniques. *J Anim. Sci.* **1986**, *62*, 1130–1139. [CrossRef] [PubMed]

37. Stanton, L.A.; Sullivan, M.S.; Fazio, J.M. A standardized ethogram for the felidae: A tool for behavioural researchers. *App. Anim. Behav. Sci.* **2015**, *173*, 3–16.

38. Zuur, A.E.; Ieno, E.N.; Walker, N.J.; Saveliev, A.A.; Smith, G.M. *Mixed Effects Model and Extensions in Ecology with R*; Springer: New York, NY, USA, 2009.

39. R Core Team. *R: A Language and Environment for Statistical Computing*; R Foundation for Statistical Computing: Vienna, Austria, 2021. Available online: https://www.R-project.org/ (accessed on 1 March 2022).

40. Goldberg, J.L.; Grant, J.W.; Lefebvre, L. Effects of the temporal predictability and spatial clumping of food on the intensity of competitive aggression in the Zenaida dove. *Behav. Ecol.* **2001**, *12*, 490–495.

41. Hassell, J.M.; Begon, M.; Ward, M.J.; Fèvre, E.M. Urbanization and Disease Emergence: Dynamics at the Wildlife-Livestock-Human Interface. *Trends Ecol. Evol.* **2017**, *32*, 55–67. [CrossRef]

42. Harris, S.; Baker, P. *Urban Foxes*; Whittet Books Ltd.: Cotton, Suffolk, UK, 2001; 152p.

43. Dorning, J.; Harris, S. Dominance, gender, and season influence food patch use in a group-living, solitary foraging canid. *Behav. Ecol.* **2017**, *28*, 1302–1313.

44. Reeve, N. *Hedgehogs*; T. & A.D. Poysner: London, UK, 1994; 313p.

45. Schmidt, K.T.; Seivwright, L.J.; Hoi, H.; Staines, B.W. The effect of depletion and predictability of distinct food patches on the timing of aggression in red deer stags. *Ecography* **1998**, *21*, 415–422.

46. Kristiansson, H. Population variables and causes of mortality in a hedgehog (*Erinaceus europaeus*) population in southern Sweden. *J. Zool.* **1990**, *220*, 391–404.
47. Morris, P. An estimate of the minimum body weight necessary for hedgehogs (*Erinaceus europaeus*) to survive hibernation. *J. Zool.* **1984**, *203*, 291–294.
48. Ward, J.F.; MacDonald, D.W.; Doncaster, C.P. Responses of foraging hedgehogs to badger odour. *Anim. Behav.* **1997**, *53*, 709–720. [CrossRef]
49. Liberg, O. Spacing patterns in a population of rural free roaming domestic cats. *Oikos* **1980**, *35*, 336. [CrossRef]
50. CIEH. *Urban Foxes: Guidelines on Their Management*; Chartered Institute for Environmental Health: London, UK, 2014; 12p.
51. Rondinini, A.C.; Doncaster, C.P. Roads as barriers to movement for hedgehogs. *Funct. Ecol.* **2002**, *16*, 504–509.
52. Scott, D.M.; Baker, R.; Tomlinson, A.; Berg, M.J.; Charman, N.; Tolhurst, B.A. Spatial distribution of sarcoptic mange (*Sarcoptes scabiei*) in urban foxes (*Vulpes vulpes*) in Great Britain as determined by citizen science. *Urban Ecosyst.* **2020**, *23*, 1127–1140.
53. Brochier, B.; De Blander, H.; Hanosset, R.; Berkvens, D.; Losson, B.; Saegerman, C. *Echinococcus multilocularis* and *Toxocara canis* in urban red foxes (*Vulpes vulpes*) in Brussels, Belgium. *Prev. Vet. Med.* **2007**, *80*, 65–67. [CrossRef] [PubMed]
54. Singer, A.; Smith, G.C. Emergency rabies control in a community of two high-density hosts. *BMC Vet. Res.* **2012**, *8*, 79. [CrossRef]
55. Learmount, J.; Zimmer, I.A.; Conyers, C.; Boughtflower, V.D.; Morgan, C.P.; Smith, G.C. A diagnostic study of *Echinococcus multilocularis* in red foxes (*Vulpes vulpes*) from Great Britain. *Vet. Parasitol.* **2012**, *190*, 447–453.
56. Rasmussen, S.L.; Hallig, J.; van Wijk, R.E.; Petersen, H.H. An investigation of endoparasites and the determinants of parasite infection in European hedgehogs (*Erinaceus europaeus*) from Denmark. *Int. J. Parasit. Parasites Wildl.* **2021**, *16*, 217–227. [CrossRef]
57. Young, R.P.; Davison, J.; Trewby, I.D.; Wilson, G.J.; Delahay, R.J.; Doncaster, C.P. Abundance of hedgehogs (*Erinaceus europaeus*) in relation to the density and distribution of badgers (*Meles meles*). *J. Zool.* **2006**, *269*, 349–356.
58. Trewby, I.D.; Young, R.; McDonald, R.A.; Wilson, G.J.; Davison, J.; Walker, N.; Robertson, A.; Patrick Doncaster, C.; Delahay, R.J. Impacts of removing badgers on localized counts of hedgehogs. *PLoS ONE* **2014**, *9*, 2–5. [CrossRef]
59. Gurnell, J.; Reeve, N.; Bowen, C.; Pettinger, T.; Cross, B. *Surveys of Hedgehogs in The Regent's Park, London 2014–2020*; The Royal Parks and Zoological Society of London (ZSL): London, UK, 2021. Available online: https://www.royalparks.org.uk/__data/assets/pdf_file/0005/129479/The-Regents-Park-Hedgehog-Research-Report-2014-2020.pdf (accessed on 12 December 2022).
60. Uphouse, L. Chapter 7—Stress and Reproduction in Mammals. In *Hormones and Reproduction of Vertebrates*; Norris, D.O., Lopez, K.H., Eds.; Academic Press: Cambridge, MA, USA, 2011; pp. 117–138.
61. Schaus, J.; Uzal, A.; Gentle, L.K.; Baker, P.J.; Bearman-Brown, L.; Bullion, S.; Gazzard, A.; Lockwood, H.; North, A.; Reader, T.; et al. Application of the Random Encounter Model in citizen science projects to monitor animal densities. *Rem. Sense. Ecol. Conser.* **2020**, *6*, 514–528. [CrossRef]
62. Williams, B.; Mann, N.; Neumann, J.; Yarnell, R.; Baker, P. A prickly problem: Developing a volunteer-friendly tool for monitoring populations of a terrestrial urban mammal, the west European hedgehog (*Erinaceus europaeus*). *Urban Ecosys.* **2018**, *21*, 1075–1086.
63. Hunter, J.; Alabri, A.; van Ingen, C. Assessing the quality and trustworthiness of citizen science data. *Concurr. Comput.* **2013**, *25*, 454–466.

Communication

Consumption of Rodenticide Baits by Invertebrates as a Potential Route into the Diet of Insectivores

Emily J. Williams [1,2,*], Sheena C. Cotter [1] and Carl D. Soulsbury [1]

[1] School of Life and Environmental Sciences, University of Lincoln, Lincoln LN6 7TS, UK; scotter@lincoln.ac.uk (S.C.C.); csoulsbury@lincoln.ac.uk (C.D.S.)
[2] UK Centre for Ecology & Hydrology, Wallingford, Oxon OX10 8BB, UK
* Correspondence: emiwil@ceh.ac.uk

Simple Summary: Anticoagulant rodenticides are commonly used as a method of rodent population control. Unfortunately, many non-target species are exposed to rodenticides. The route by which non-target animals are poisoned is not always clear, which can hinder conservation efforts. It has been suggested that insectivorous species may be exposed to rodenticides via the consumption of contaminated insect prey. This study examined whether rodenticide baits mixed with the biomarker rhodamine B can be used to track invertebrate consumption of rodenticide baits in a natural environment, and, in doing so, we assessed whether insects could be a source of rodenticide poisoning in insectivores. The rhodamine B baits created an observable response; molluscs were the most frequent consumers of bait. Maximum temperature, distance from baits, the addition of copper tape to boxes, and proximity to buildings were all found to affect their rate of uptake. Other invertebrates rarely showed signs of uptake. This has provided valuable insights into the mechanisms by which insectivores experience rodenticide poisoning, which is necessary in developing effective mitigation measures to aid conservation efforts. We suggest that further investigation into using mollusc repellents around bait boxes should be considered.

Abstract: Non-target species are commonly exposed to anticoagulant rodenticides worldwide, which may pose a key threat to declining species. However, the main pathway of exposure is usually unknown, potentially hindering conservation efforts. This study aimed to examine whether baits mixed with the biomarker rhodamine B can be used to track invertebrate consumption of rodenticides in a field environment, using this to observe whether invertebrate prey are a potential vector for anticoagulant rodenticides in the diet of insectivores such as the European hedgehog (*Erinaceus europaeus*). Rhodamine B baits were found to create an observable response. Uptake was negligible in captured insects; however, 20.7% of slugs and 18.4% of snails captured showed uptake of bait. Maximum temperature, distance from bait, proximity to buildings, and the addition of copper tape to bait boxes all influenced the rate of bait uptake in molluscs. Based on these data, it seems likely that molluscs could be a source of rodenticide poisoning in insectivores. This research demonstrates which prey may pose exposure risks to insectivores and likely environmental factors, knowledge of which can guide effective mitigation measures. We suggest that further investigation into using mollusc repellents around bait boxes should be considered.

Keywords: insectivores; rodenticides; route of exposure; wildlife conservation; non target; invertebrates; secondary exposure; molluscs

Citation: Williams, E.J.; Cotter, S.C.; Soulsbury, C.D. Consumption of Rodenticide Baits by Invertebrates as a Potential Route into the Diet of Insectivores. *Animals* **2023**, *13*, 3873. https://doi.org/10.3390/ani13243873

Academic Editors: Anne Berger, Nigel Reeve and Sophie Lund Rasmussen

Received: 30 November 2023
Revised: 11 December 2023
Accepted: 14 December 2023
Published: 16 December 2023

1. Introduction

Rodents are estimated to cost the UK economy £60–200 million every year, mainly through disease transmission and food spoilage [1]. Anticoagulant rodenticides are a preferred control method, working by inhibiting vitamin K1-2,3 epoxide reductase, which in turn inhibits the ability to clot blood. This leads to delayed death via haemorrhaging [2,3]. Second generation anticoagulant rodenticides (SGARs), developed in response to the genetic resistance that emerged after extensive usage of first generation anticoagulant rodenticides (FGARs), are routinely used worldwide with five licensed for use in the UK [1,4,5]. SGARs are more potent than their predecessors, with a greater affinity for binding sites in the liver which results in increased persistence, toxicity, and accumulation [5,6].

Although anticoagulant rodenticides have many advantages, as they work by targeting a biochemical pathway that occurs in all mammals and birds, they pose a risk to many non-target species [7]. Small mammals [6], predatory birds [8], and even passerines [9] suffer from exposure to anticoagulant rodenticides worldwide. Typically, non-target species are exposed to rodenticides either by directly consuming baits (primary exposure) or by consuming contaminated prey (secondary exposure), and can even be exposed at further levels [10].

Exposure can occur at multiple levels in a single organism, accumulating from different sources [11]. The high persistence of SGARs allows for both bioaccumulation and biomagnification of rodenticides in non-target predatory species, such as red foxes (*Vulpes vulpes*) and polecats (*Mustela putorius*)—rodenticide residues were found in 82% of avian and mammalian scavenger and predatory species sampled in Finland [12–14]. Only recently has the extent to which rodenticides travel through the food chain become clear; baits are consumed not only by non-target mammals and birds but by also reptiles and invertebrates [10,15,16]. Rodenticides could easily accumulate in invertebrates as they possess different blood clotting mechanisms to target species, decreasing the likelihood of death after consumption [5]. As previous research on non-target carnivores suggests a major route of rodenticide exposure is via consumption of contaminated small mammals it has been suggested that contaminated invertebrate prey may similarly expose non-target insectivore predators to rodenticides through food chain transfer [5,10]. However, research is limited in this area—data on invertebrate uptake is rarely linked to insectivore exposure and tends to focus on risks individual species pose rather than assessing the dietary spectrum.

Few studies observe whether secondary poisoning occurs in insectivores; however, what research there is indicates that exposure is widespread. European starlings (*Sturnus vulgaris*), dunnocks (*Prunella modularis*), and the common shrew (*Sorex araneus*) have been found to experience rodenticide uptake [6,10,17–19]. Samples taken from dead Stewart Island robin nestlings (*Petroica australis rakiura*) and New Zealand dotterels (*Charadrius obscurus aquilonius*) contained brodifacoum residues, suggested to be a result of ingesting contaminated invertebrate prey [9,20]. Rodenticide residues have been found in high numbers of European hedgehog (*Erinaceus europaeus*) liver samples from Spain, Britain, and New Zealand, often testing positive for several different rodenticides [5,14,21–23].

Information about the routes of rodenticide exposure in threatened non-target species is necessary to develop effective mitigation measures and aid conservation efforts [11]. Several insectivorous species are experiencing significant population declines across the UK, including European hedgehogs, spotted flycatchers (*Muscicapa striata*), and common swifts (*Apus apus*) [24–26]. A long term study of British woodland birds found that 27% of foliage insectivores and 57% of ground insectivores could be classed as a 'declining species' [27]. Pesticides such as rodenticides may be contributing towards these declines; however, the lack of in-depth information available on the subject may be obscuring this threat and hindering conservation efforts.

This study aims to help address the knowledge gap by testing whether multiple invertebrate species consume rodenticide baits in field environments across UK locations. The handful of studies that have focused on the detection of rodenticides in invertebrates have usually used HPLC or LC-MSMS, though as these techniques utilise rodenticides, they risk environmental contamination [10,16]. We tested whether a rodenticide-free non-toxic indicator paste bait mixed with rhodamine B (rhdB), a xanthene dye with fluorescent properties often used in bait uptake studies and shown to create an observable fluorescent response in invertebrates [28,29], can be used to track invertebrate bait consumption in the field. Using this paste, we assessed the effects of environmental variables (temperature rainfall) and UK habitats (close to and far from buildings) on bait uptake across invertebrat groups. Assessing rodenticide uptake in a broad range of invertebrate species in a natura environment will provide a more comprehensive idea of which insectivore prey item pose an exposure risk. This methodology will also provide new information on this under researched topic, including whether rhdB can be used as a low-cost and non-toxic metho of tracking bait consumption in invertebrates, and whether invertebrate uptake of an activity around rodenticide baits is influenced by environmental variables. When take together, these data will have the potential to provide valuable insights into the mechanism by which insectivores experience rodenticide poisoning.

2. Materials and Methods

2.1. Laboratory Pilot Study

Previous work has not shown whether rhodamine B can be detected in molluscs. T test this, we carried out a brief pilot study using garden snails (*Cornu aspersum*) collecte from various locations around the city of Lincoln (UK) by hand during February 2021. Th snails were housed in tanks in the University of Lincoln insectary between February an April 2021, kept at ~20 degrees Celsius and fed with lettuce, cabbage, and ProRep Bu Gel, a mixture of water and Polyacrylamide gel which provides a source of hydration t invertebrates without the risk of drowning [30].

We mixed Deadline's Non-tox Indicator Paste (Deadline®, Professional Pest Contro Crawley, UK) with rhdB powder at 0.5% and 1% concentration; both provided an observabl response in the foot and body under light from a handheld UV torch after 48 h exposur when tested on a group of five snails.

Following this, 70 snails were divided into two groups (N = 35 each). Each snail wa then weighed and placed inside a plastic salad box containing 8–10 g of rhdB paste bai or no paste. Within each group, 14 snails were exposed to 0.5% rhdB paste, 14 snails t 1% rhdB paste, and 7 controls to no paste. Group one and control snails were left for 24 before the baits were removed. Group two and control snails were left for 48 h before bait were removed. Following bait removal, the presence or absence (Y/N) of snail fluorescenc under UV light was recorded. The snails were then returned to their boxes and provide with lettuce and bug gel. Every day for the seven following days the snails that had showe uptake were observed for fluorescence under UV light to measure the persistence of rhd in their system. Following this all snails no longer showing fluorescence were moved bac to the tank, while the remainder (N = 5) underwent continued observation for five days.

Following analysis, rhdB mixed with non-toxic rodenticide paste at 0.5% concentratio was determined to be suitable for tracking consumption of bait in invertebrates whe exposed for 24 h or longer.

2.2. Field Exposure Study

The field study took place across two farms—Riseholme Farm in Lincoln (53.26819 −0.52743664) from April to August 2021 and Malthouse Livery and Stables in Oxfordshir (51.728834, −1.4448872) during October and November 2021 (Figure 1). Rat rodenticid boxes from Rentokil were set out, each containing two dishes with ~8 g of 0.5% rhdB past bait, in various habitats (farm buildings, urban, hedgerow, forest (deciduous trees) an pine forest (coniferous trees)). Bait boxes are a common and recommended method use in rodenticide baiting [31] and so were used here to match the method design as much a possible to real life baiting. The boxes were left in place for 72 h, after which they wer opened and checked for invertebrates. Any invertebrates found were viewed under UV torch to assess whether the consumption of the bait had occurred (indicated by bod fluorescence). The number of captures and the UV responses were noted. Invertebrate were released at the site of capture. The bait boxes were then removed and moved t the next location, and four to six pitfall traps were placed around where the bait boxe had been. Maximum distance was measured by using a tape measure to mark 0–1 m an

1–2 m away from the bait stations, and half the pitfall traps were placed within each of these ranges. Yeast paste was used as bait in half of the traps as mixtures using yeast are known to attract molluscs [32]. In the remaining traps, a non-toxic indicator paste that did not contain rhdB was used, as it had already been established that invertebrates will consume paste bait, and it was determined from this that it may act as an attractant. The traps were left for 24 h, after which, they were checked for invertebrate activity. Again, any invertebrates found were observed under UV light to check for bait uptake, which along with the number of captures was noted before they were released. Maximum temperature and rainfall values for the postcodes of the sites on data collection days were taken from World Weather Online [33].

Figure 1. Sampling locations: Riseholme Farm in Lincoln, Lincolnshire (April to August 2021), and Malthouse Livery and Stables in Standlake, Oxfordshire (October to November 2021). Urban and Farm Building sites are classified as 'near' buildings (within 30 metres) and Hedgerow and Forest sites as 'far' from buildings (over 85 metres away).

Early on, it was found that molluscs were frequently present in bait boxes and frequently consuming bait. If a mollusc repellent could be added to bait boxes, this could significantly reduce the number of molluscs entering and exiting the boxes and so reduce their bait consumption. Copper is a known mollusc repellent [34] and so copper tape (Evergreen Goods copper tape, width 20 mm) was adhered to the boxes, covering the full outline of each entrance, on certain baiting occasions, to test its effectiveness as a deterrent. Around 30 cm of tape was used to cover each entrance.

2.3. Statistical Analysis

For the pilot results, effects of snail weight prior to exposure (g), exposure length (hours), and bait concentration (%) on the uptake of the baits were analysed by running a generalised linear model. Fluorescence of the body and foot or faeces that were pink in colour to the naked eye, indicating consumption of bait, were counted as a 'Y' uptake response. For those snails with a 'Y' response, we then tested persistence of rhdB using a binomial generalised linear mixed effects model (GLMM). We ran two models due to convergence issues: in the first, we included time (hours) as a continuous covariate with the interactions of time x weight and time x exposure length. In the second model, we included time (hours) as a continuous covariate with the interactions of time $\times$ weight and time $\times$ concentration.

The field data from slugs and snails were combined and run through a binomial GLMM to test how distance from bait (m), maximum temperature (°C), and precipitation (mm) influenced the percentage of molluscs showing uptake. In each model, sampling event nested within site was included as a random effect. Data from other invertebrates were not analysed, as their bait uptake was too low. We then then used a Poisson GLMM to test how distance from bait, maximum temperature, and precipitation influenced the total numbers of slugs and snails that ingested the bait. Again, we included sampling event nested within site as a random effect.

A binomial GLMM was run to test how proximity to buildings affected the percentage of molluscs showing uptake in the field study, while a Poisson GLMM was run to test how it affected the total number of molluscs showing uptake of bait. Sampling locations were divided into 'near' to and 'far' from buildings; 'near' locations were within 30 metres of buildings, while the 'far' locations were all found > 85 metres from buildings.

Finally, using a subset of sites where boxes with and without copper tape attached were set out, we tested whether using copper tape affected the number of molluscs showing uptake or the total number of molluscs present in the bait boxes using paired-Wilcoxon tests

GLMMs were run using the package 'lme4' [35], in R version 4.3.1 [36]. We calculated the Wald stats using the *car* package [37]. Figures were plotted using the package ggplot2 [38].

3. Results

3.1. Laboratory Pilot Study

Most snails exposed to baits ingested rhdB within 48 h (41/56). Snails exposed to 0.5% rhdB bait were significantly more likely to show bait uptake (89.3%, 25/28 snails) compared to those exposed to 1% (42.8%, 12/28 snails; X^2 (1, N = 70) = 18.80, p = 0.004), but there was no difference in uptake between those exposed for 24 and 48 h (X^2 (1, N = 70) = 0.34, p = 0.557). Heavier snails were significantly more likely to show uptake of the baits than lighter snails (X^2 (1, N = 70) = 14.56, p < 0.001; Figure 2a).

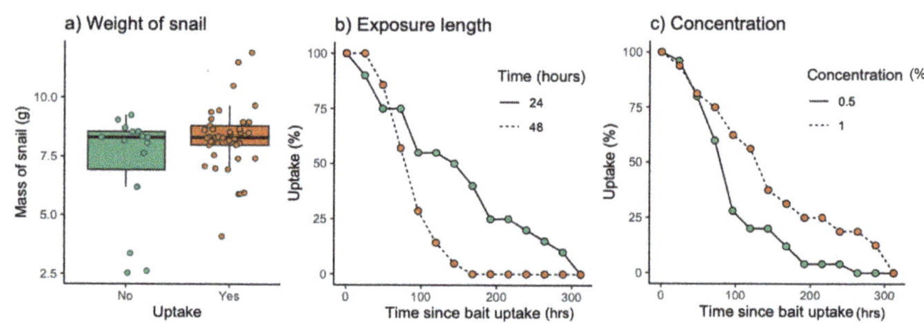

Figure 2. (**a**) The median +/− IQR starting mass (g) of the snails that did (yes, orange) and did not (no, green) consume bait. Plots (**b**,**c**) show the percentage of snails that showed visible signs of rhdB consumption following (**b**) exposure to rhdB for different time periods (24 h, green, or 48 h, orange) and (**c**) exposure to different concentrations of rhdB (0.5%, green, or 1%, orange), up to 300 h after exposure.

Following uptake, rhdB persisted in the system of those exposed to baits for 24 h significantly longer than those exposed for 48 h (exposure length $\times$ time: $X^2 = 5.47$, $p = 0.019$; Figure 2b). There was no effect of weight on persistence (weight $\times$ time: $X^2 = 0.76$, $p = 0.380$). Persistence of rhdB was significantly longer for snails consuming 1% rhdB bait than those exposed to 0.5% rhdB bait ($X^2 = 12.92$, $p < 0.001$; Figure 2c), though the retention was similar up to about 96 h for both concentrations (Figure 2c).

3.2. Field Exposure Study

Across 70 baiting events, 1588 invertebrates were captured (Table 1). Of these, 20.7% of slugs and 18.4% of snails showed evidence of bait uptake. In contrast, only a handful of other invertebrates (two springtails, four earwigs, and one carabid beetle) showed uptake of bait (Table 1).

Table 1. The total number of each invertebrate species caught over the course of the field study, and the percentage of those that showed uptake of bait.

Category	Total Caught	% Bait Uptake
Ant	32	0.0
Aphid	2	0.0
Beetle	589	0.2
Caterpillar	8	0.0
Earthworm	6	0.0
Earwig	63	6.8
Lepidoptera	1	0.0
Millipede	23	0.0
Slug	239	20.7
Snail	251	18.4
Springtail	178	1.1
Woodlouse	196	1.0

The percentage of slugs and snails showing uptake of bait was significantly higher closer to the bait boxes (Table 2; Figure 3a) and at cooler temperatures (Table 2; Figure 3b). Precipitation had no significant effect. However, while the total number of molluscs showing uptake of rhodamine B baits was also higher closer to the bait boxes (Table 2; Figure 3c), temperature or precipitation had no effect (Table 2).

Table 2. The X^2, df, and p values for the effects of tested factors on (**a**) the percentage of molluscs caught showing visible signs of bait uptake and (**b**) the total number of molluscs caught showing visible signs of bait uptake. Significant explanatory variables are in bold.

Model	Parameter	X^2	df	p
(a) Percentage uptake	Distance	37.51	1	**<0.001**
	Max. temperature	4.50	1	**0.034**
	Precipitation	0.55	1	0.460
(b) Total number showing uptake	Distance	4.98	1	**0.026**
	Max. temperature	1.63	1	0.202
	Precipitation	3.73	1	0.053

Proximity to buildings had a significant positive effect on mollusc bait uptake, with both the percentage ($X^2 = 4.336$, $p = 0.037$; Figure 4a) and the total number ($X^2 = 4.910$, $p = 0.027$; Figure 4b) of molluscs showing uptake of bait being higher closer to buildings. The total number of molluscs captured was also significantly higher closer to buildings ($X^2 = 8.167$, $p = 0.004$).

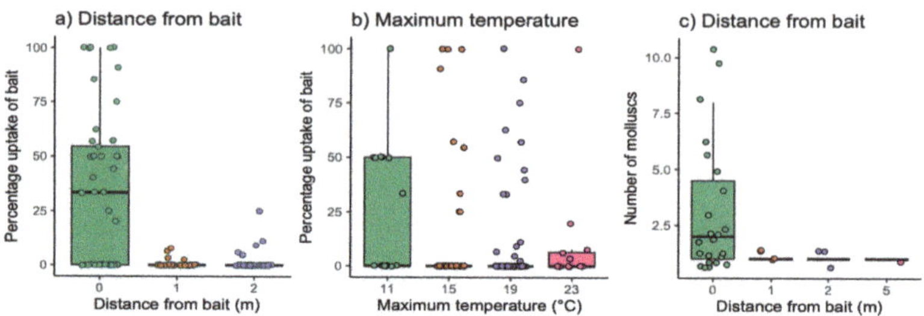

Figure 3. The median $+/-$ IQR percentage of molluscs found to consume bait in relation to (**a**) maximum distance from the bait (m) and (**b**) maximum temperature (°C). Plot (**c**) shows the median $+/-$ IQR total number of molluscs captured that showed uptake of bait in relation to maximum distance (m).

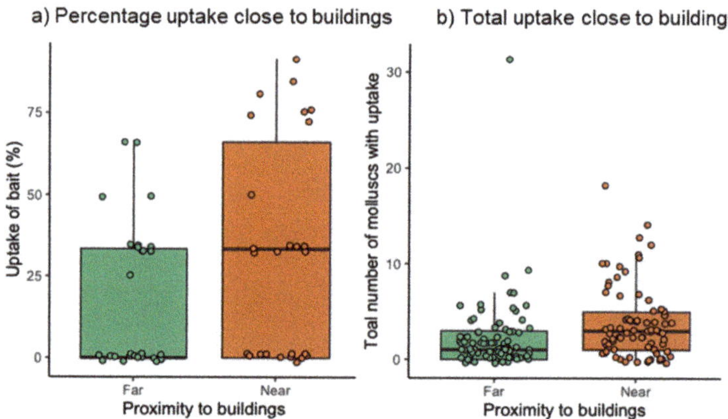

Figure 4. (**a**) The median $+/-$ IQR percentage of molluscs found to consume bait in relation to proximity to buildings, and (**b**) the median $+/-$ IQR total number of molluscs found to show uptake in relation to proximity to buildings.

There was no significant difference in the percentage of molluscs showing uptake of rhdB when copper tape was present (V = 23.5, p = 0.472); however, the total number of molluscs found in boxes was significantly lower when copper tape was added (V = 128.5 p = 0.014; Figure 5).

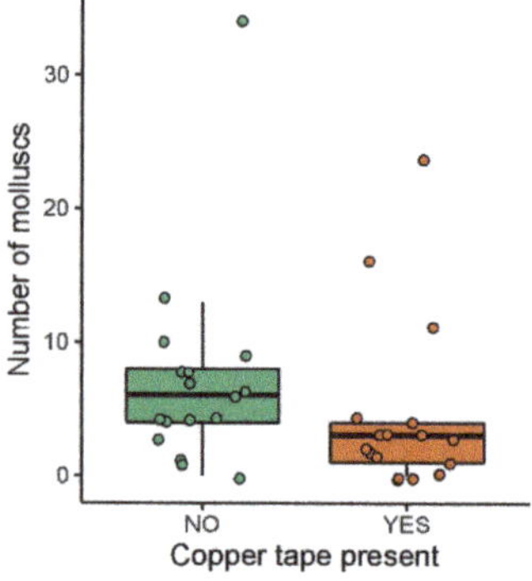

Figure 5. The median $+/-$ IQR total number of molluscs caught when copper tape was not (NO green) and was (YES, orange) present.

4. Discussion

The most common consumers of rodenticide bait in this study were snails and slugs, indicating that consumption of contaminated molluscs may be a key source of rodenticide contamination in UK insectivores (Table 1). These findings are backed up by previous data—multiple studies have found gastropods to be present on or to consume rodenticide baits [10,16,39,40]. Mollusc bait consumption may be even higher in areas where boxes are not used, and bait is therefore easier to access, putting insectivorous species at higher risk of exposure. This is concerning, as mollusc predators range beyond insectivores from birds to mammals, including species vulnerable to decline in the UK such as Scottish wildcats (*Felis s. silvestris*) [41].

Very few other invertebrate species were found to consume bait in this study (Table 1). However, rodenticide residues have been detected in earthworms following exposure to baits [16], and Orthoptera, Arthropods, Collembola, and Dermaptera have all been observed to feed on baits and bait mimics in previous studies [6,40,42,43]. Furthermore, multiple beetle species have been found in bait trays or directly feeding on rodenticides [42,43]. Families such as Carabidae which consume vertebrate carrion may consume rodenticides via their prey and so pose a risk to insectivores through tertiary poisoning, another pathway which should be considered [44]. Although using non-toxic baits mixed with rhdB could prove a convenient, cost-effective, and non-lethal way of investigating the movement of rodenticides through food chains, using UV detection rather than HPLC may miss uptake if it is in particularly low amounts, such as in tertiary feeders. Further investigation is needed to confirm whether rhdB is sufficiently detectable in other invertebrates, and further research using more sensitive equipment and analysis would be valuable in gathering a comprehensive picture of the poisoning risks all invertebrate species pose to insectivores.

Bait Uptake

Higher maximum temperatures significantly decreased the percentage of molluscs ingesting bait (Table 2; Figure 3b), but had no significant effect on the total number of molluscs showing uptake of bait (Table 2). The decrease in the percentage uptake at warmer temperatures is therefore likely a result of there being a greater number of molluscs present overall at higher temperatures. Warm temperatures and high moisture levels have been found to correlate with increased slug activity in multiple species, and snail climbing behaviours have been found to increase at higher temperatures [45,46]. Furthermore, it has been reported that when temperatures are high, terrestrial gastropods tend to restrict activity to favourable times such as nights and mornings, so high levels of activity may be maintained [47]. In addition, data were collected between April and November. Multiple species of land snails and slugs enter a state of dormancy in colder months [48]; as a result, it is likely that mollusc activity was much lower in the latter months when maximum temperatures were at their lowest. The influence of maximum temperature on mollusc activity may be important when considering the risk they pose to insectivores via secondary poisoning, as greater prey availability at warmer temperatures may reduce the risk to non-target insectivorous species via a dilution effect.

Both the percentage of molluscs and the total number of molluscs showing uptake of bait decreased with distance from the baited boxes (Table 2; Figure 3a,c). This reflects previous findings in multiple other species; however, spatial data on non-target rodenticide poisoning is limited [3,18]. It is likely those further away from baits had not encountered the bait. Supporting this, the rate of secondary poisoning in predators appears to be lower the further away from bait stations they are, although again, data are limited [49]. Alternatively, the shelter provided by the bait boxes may have attracted molluscs; in turn, molluscs may attract insectivores to areas around bait boxes.

Proximity to buildings significantly increased the percentage and total number of molluscs ingesting bait (Figure 4). In this study, the sample sites closest to buildings were the urban and farm building sites. What is particularly concerning is that although in this study the same number of bait boxes were used in each location, in reality, urban habitats and locations near rural buildings likely provide particularly high accessibility of rodenticide baits to molluscs, further increasing the likelihood that contaminated molluscs are present in these areas. Multiple studies have found that exposure in predators is positively correlated with human population densities, thought to reflect the high utilisation of rodenticides against commensal urban rodents [22,50]. One study found that the majority of commercial and industrial buildings associated with urban areas within the study area had permanent SGAR bait stations on the perimeter [51]. Additionally, although the use of rodenticides on farms is not particularly well documented or monitored [49], permanent baiting is common, especially in and around farm buildings; one study found almost 40% of farms they investigated permanently baited with SGARs [31]. Urban habitats or areas near rural buildings therefore have the potential to become hotspots for insectivore consumption of contaminated molluscs.

The presence of copper tape around the entrances to the bait boxes did not significantly affect the number of molluscs ingesting bait but significantly reduced the number of molluscs found in boxes (Figure 5). These results suggest that copper tape may be useful in limiting the movement of slugs and snails around baits; however, it does not appear to be powerful enough to limit bait uptake. Further research using copper tape around bait boxes is needed across a wider variety of seasons and locations to conclusively investigate its efficacy in preventing mollusc bait uptake. Testing the effectiveness of different widths of tape may also be useful. Additionally, other forms of copper such as copper hydroxide fungicides should be tested, along with other formulations found to act as mollusc repellents such as cinnamamide crystals and sodium silicate [52].

Molluscs appear to be key consumers of rodenticide baits; insectivorous species in which molluscs make up a large part of the diet may, therefore, be at especially high risk of secondary poisoning. Molluscs make up a notable portion of European hedgehog diet across their geographic range (25–51% of total diet in mainland Europe [53,54] and up 59% in the UK [55–58]), comprise 6.4% of the diet of European starlings (Shrey 1981, cited in South, 1992 [59]), and account for between 14.2 and 18.4% of the diet of common shrews [60], all insectivorous species previously found to suffer from exposure to anticoagulant rodenticides [10]. Although multiple insectivorous species have been reported to be exposed to rodenticides, the route of exposure and the potential threat rodenticides pose to insectivore populations are areas of limited research. This has resulted in gaps in our knowledge of the ecology of insectivores, which in turn could be limiting conservation efforts; a concern that needs to be addressed if we are to halt continuing population declines of species such as European hedgehogs. Exposure to rodenticides could lead to reduced mobility, impaired hazard awareness and reaction speeds, and clotting disorders [3,6]. Findings from this study provide new information on the route by which insectivores are exposed to rodenticides, and the factors that influence the threat contaminated prey pose. This information is necessary to develop effective mitigation measures and aid conservation efforts. Limiting the access of molluscs to baits should be considered as a measure to decrease sources of contamination and so insectivore exposure. Other mitigation measures, including reducing permanent baiting and adding bittering agents to baits, may also help decrease non-target exposure. Furthermore, efforts should be made to find a less toxic alternative to second generation rodenticides regardless of the route of exposure. Third generation anticoagulants using less persistent diastereomers have been proposed, as these would clear from the system of non-target animals more quickly [61].

Data on the percentage of diet that individual invertebrate species make up would be useful to further assess the risks that rodenticide contaminated invertebrates pose to insectivores. If reliable data on the concentrations of rodenticide residues in invertebrate prey were also available, toxicity exposure ratios could be calculated to determine whether contaminated prey pose a significant threat to individual insectivore species [10].

5. Conclusions

Despite a strong case to suggest that insectivores are exposed to rodenticides via consumption of contaminated invertebrate prey, this theory had not been subject to extensive testing. Our work assessing rodenticide uptake in multiple invertebrate species within a natural environment has provided a more comprehensive idea of which prey items pose a potential exposure risk to insectivores, and which environmental factors influence invertebrate bait uptake. Analysis showed that molluscs consumed rodenticide paste bait at a far higher rate than any other invertebrate group, making up nearly a third of invertebrates captured. Uptake decreased with maximum temperature and increased with proximity to buildings and bait boxes. As such, it appears that molluscs could pose a particular risk to insectivores via secondary poisoning. Using rhdB baits proved successful in the tracking of invertebrate uptake of rodenticides and, following further research, may prove a useful method of investigating the movement of rodenticides through the food chain. Supplementary research using more sensitive analytic techniques such as HPLC would also be valuable, as would further investigation into using mollusc repellents around bait boxes including copper tape which, although ineffective in limiting mollusc bait uptake here, decreased the number of molluscs entering the bait boxes. These data have the potential to provide useful insights into the mechanisms by which insectivores experience rodenticide poisoning, generating an enhanced understanding of their ecology and the threats they face. With many insectivorous species in worrying decline, such research may prove invaluable.

Author Contributions: Conceptualization, C.D.S. and E.J.W.; methodology, C.D.S., S.C.C. and E.J.W.; formal analysis, C.D.S. and E.J.W.; writing—original draft preparation, C.D.S. and E.J.W.; writing—review and editing, C.D.S., S.C.C. and E.J.W. All authors have read and agreed to the published version of the manuscript.

Funding: This research received no external funding.

Institutional Review Board Statement: The animal study protocol was approved by the Non-Human Ethics Committee of the University of Lincoln (ref: 6550 on the 28/4/2021).

Informed Consent Statement: Not applicable.

Data Availability Statement: All data are contained within the article and has been uploaded to FigShare (https://figshare.com/articles/dataset/Williams_et_al_-_Consumption_of_rodenticide_baits_by_invertebrates_as_a_potential_route_into_the_diet_of_insectivores_-_field_data_and_pilot_data/24842055, accessed on 10 October 2023; DOI: 10.6084/m9.figshare.24842055).

Acknowledgments: Thank you to the teams at Riseholme Farm and Malthouse Livery and Stables for facilitating the setup of the field studies at their sites.

Conflicts of Interest: The authors declare no conflict of interest.

References

Sainsbury, K.A.; Shore, R.F.; Schofield, H.; Croose, E.; Pereira, M.G.; Sleep, D.; Kitchener, A.C.; Hantke, G.; McDonald, R.A. Long-term increase in secondary exposure to anticoagulant rodenticides in European polecats Mustela putorius in Great Britain. *Environ. Pollut.* **2018**, *236*, 689–698. [CrossRef] [PubMed]

Witmer, G.W. The Changing Role of Rodenticides and Their Alternatives in the Management of Commensal Rodents. *Hum. Wildl. Interact.* **2019**, *13*, 186–199.

Geduhn, A.; Esther, A.; Schenke, D.; Mattes, H.; Jacob, J. Spatial and temporal exposure patterns in non-target small mammals during brodifacoum rat control. *Sci. Total Environ.* **2014**, *496*, 328–338. [CrossRef] [PubMed]

Shore, R.F.; Walker, L.A.; Potter, E.D.; Chaplow, J.S.; Pereira, M.G.; Sleep, D.; Hunt, A.G. *Second Generation Anticoagulant Rodenticide Residues in Barn Owls 2018*; NERC/Centre for Ecology & Hydrolog: Lancaster, UK, 2019.

Dowding, C.V.; Shore, R.F.; Worgan, A.; Baker, P.J.; Harris, S. Accumulation of anticoagulant rodenticides in a non-target insectivore, the European hedgehog (*Erinaceus europaeus*). *Environ. Pollut.* **2010**, *158*, 161–166. [CrossRef] [PubMed]

Brakes, C.R.; Smith, R.H. Exposure of non-target small mammals to rodenticides: Short-term effects, recovery and implications for secondary poisoning. *J. Appl. Ecol.* **2005**, *42*, 118–128. [CrossRef]

Smith, R.; Shore, R. *Environmental Impacts of Rodenticides*; CABI: Wallingford, UK, 2015; pp. 330–345.

Walker, L.A.; Turk, A.; Long, S.M.; Wienburg, C.L.; Best, J.; Shore, R.F. Second generation anticoagulant rodenticides in tawny owls (*Strix aluco*) from Great Britain. *Sci. Total Environ.* **2008**, *392*, 93–98. [CrossRef]

9. Masuda, B.M.; Fisher, P.; Jamieson, I.G. Anticoagulant rodenticide brodifacoum detected in dead nestlings of an insectivorou passerine. *N. Z. J. Ecol.* **2014**, *38*, 110–115.
10. Alomar, H.; Chabert, A.; Coeurdassier, M.; Vey, D.; Berny, P. Accumulation of anticoagulant rodenticides (chlorophacinone bromadiolone and brodifacoum) in a non-target invertebrate, the slug, Deroceras reticulatum. *Sci. Total Environ.* **2018**, *610–61* 576–582. [CrossRef]
11. Quinn, N. Assessing Individual and Population-Level Effects of Anticoagulant Rodenticides on Wildlife. *Hum. Wildl. Interac* **2019**, *13*, 7.
12. Lohr, M.T.; Davis, R.A. Anticoagulant rodenticide use, non-target impacts and regulation: A case study from Australia. *Sci. Tota Environ.* **2018**, *634*, 1372–1384. [CrossRef]
13. Tosh, D.G.; McDonald, R.A.; Bearhop, S.; Lllewellyn, N.R.; Fee, S.; Sharp, E.A.; Barnett, E.A.; Shore, R.F. Does small mammal pre* guild affect the exposure of predators to anticoagulant rodenticides? *Environ. Pollut.* **2011**, *159*, 3106–3112. [CrossRef] [PubMed]
14. Koivisto, E.; Santangeli, A.; Koivisto, P.; Korkolainen, T.; Vuorisalo, T.; Hanski, I.K.; Loivamaa, I.; Koivisto, S. The prevalence an correlates of anticoagulant rodenticide exposure in non-target predators and scavengers in Finland. *Sci. Total Environ.* **2018**, *64.* 701–707. [CrossRef] [PubMed]
15. Lettoof, D.C.; Lohr, M.T.; Busetti, F.; Bateman, P.W.; Davis, R.A. Toxic time bombs: Frequent detection of anticoagulant rodenticide in urban reptiles at multiple trophic levels. *Sci Total Env.* **2020**, *724*, 138218. [CrossRef] [PubMed]
16. Booth, L.H.; Fisher, P.; Heppelthwaite, V.; Eason, C.T. *Toxicity and Residues of Brodifacoum in Snails and Earthworms*; Department o Conservation: Wellington, New Zealand, 2003; p. 14.
17. Barker, G.M. *Natural Enemies of Terrestrial Molluscs*; CABI: Wallingford, UK, 2004.
18. Elmeros, M.; Bossi, R.; Christensen, T.K.; Kjær, L.J.; Lassen, P.; Topping, C.J. Exposure of non-target small mammals to ant: coagulant rodenticide during chemical rodent control operations. *Environ. Sci. Pollut. Res.* **2019**, *26*, 6133–6140. [CrossRef] [PubMed]
19. Walther, B.; Geduhn, A.; Schenke, D.; Jacob, J. Exposure of passerine birds to brodifacoum during management of Norway rats o farms. *Sci. Total Environ.* **2021**, *762*, 144160. [CrossRef] [PubMed]
20. Dowding, J.E.; Lovegrove, T.i.G.; Ritchie, J.; Kast, S.N.; Puckett, M. Mortality of northern New Zealand dotterels (Charadriu obscurus aquilonius) following an aerial poisoning operation. *Notornis* **2006**, *53*, 235.
21. Spurr, E.; Maitland, M.; Taylor, G.; Wright, G.; Radford, C.; Brown, L. Residues of brodifacoum and other anticoagulant pesticide in target and non-target species, Nelson Lakes National Park, New Zealand. *N. Z. J. Zool.* **2005**, *32*, 237–249. [CrossRef]
22. López-Perea, J.J.; Camarero, P.R.; Molina-López, R.A.; Parpal, L.; Obón, E.; Solá, J.; Mateo, R. Interspecific and geographica differences in anticoagulant rodenticide residues of predatory wildlife from the Mediterranean region of Spain. *Sci. Total Enviror* **2015**, *511*, 259–267. [CrossRef]
23. The Royal Parks. *Appendix 8 Post-Mortem Analysis of Hedgehog Liver Tissue Samples for Second-Generation Rodenticides*; The Roya Parks Foundation: London, UK, 2016.
24. Wilson, E.; Wembridge, D. *The State of Britain's Hedgehogs 2018*; People's Trust for Endangered Species: London, UK, 2018.
25. BTO. Spotted Flycatcher. Available online: https://www.bto.org/understanding-birds/birdfacts/spotted-flycatcher (accessed o 24 November 2023).
26. Finch, T.; Bell, J.R.; Robinson, R.A.; Peach, W.J. Demography of Common Swifts (*Apus apus*) breeding in the UK associated wit local weather but not aphid biomass. *Ibis* **2023**, *165*, 420–435. [CrossRef]
27. Fuller, R.J.; Noble, D.G.; Smith, K.W.; Vanhinsbergh, D. Recent declines in populations of woodland birds in Britain. *Br. Bird* **2005**, *98*, 116–143.
28. Blanco, C.A.; Perera, O.; Ray, J.D.; Taliercio, E.; Williams, L., III. Incorporation of rhodamine B into male tobacco budworm moth Heliothis virescens to use as a marker for mating studies. *J. Insect Sci.* **2006**, *6*, 5. [CrossRef] [PubMed]
29. Mascari, T.M.; Foil, L.D. Evaluation of Rhodamine B as an Orally Delivered Biomarker for Rodents and a Feed-Throug Transtadial Biomarker for Phlebotomine Sand Flies (Diptera: Psychodidae). *J. Med. Entomol.* **2009**, *46*, 1131–1137. [CrossRef] [PubMed]
30. ProRep. Bug Gel. Available online: https://www.pro-rep.co.uk/portfolio/bug-gel/ (accessed on 25 November 2023).
31. Tosh, D.G.; Shore, R.F.; Jess, S.; Withers, A.; Bearhop, S.; Ian Montgomery, W.; McDonald, R.A. User behaviour, best practice an the risks of non-target exposure associated with anticoagulant rodenticide use. *J. Environ. Manag.* **2011**, *92*, 1503–1508. [CrossRef] [PubMed]
32. Cranshaw, W. *Attractiveness of Beer and Fermentation Products to the Gray Garden Slug, Agriolimax Reticulatum (Muller) (Mollusca Limacidae)*; Colorado State University: Colorado State University Agricultural Experiment Station: Fort Collins, CO, USA, 1997
33. World Weather Online. World Weather API and Weather Forecast. Available online: https://www.worldweatheronline.com (accessed on 1 September 2021).
34. Schüder, I.; Port, G.; Bennison, J. Barriers, repellents and antifeedants for slug and snail control. *Crop Prot.* **2003**, *22*, 1033–103 [CrossRef]
35. Bates, D.; Mächler, M.; Bolker, B.; Walker, S. Fitting Linear Mixed-Effects Models Using {lme4}. *J. Stat. Softw.* **2015**, *67*, 1–4 [CrossRef]
36. RStudioTeam. *RStudio: Integrated Development Environment for R, RStudio*; PBC: Boston, MA, USA, 2020.
37. Fox, J.; Weisberg, S. *An {R} Companion to Applied Regression*, 3rd ed.; Sage: Thousand Oaks, CA, USA, 2019.

38. Wickham, H. *ggplot2: Elegant Graphics for Data Analysis*; Springer: Berlin/Heidelberg, Germany, 2006. [CrossRef]
39. Elliott, J.E.; Hindmarch, S.; Albert, C.A.; Emery, J.; Mineau, P.; Maisonneuve, F. Exposure pathways of anticoagulant rodenticides to nontarget wildlife. *Environ. Monit. Assess.* **2014**, *186*, 895–906. [CrossRef] [PubMed]
40. Dunlevy, P.A.; Campbell, E.W.; Lindsey, G.D. Broadcast application of a placebo rodenticide bait in a native Hawaiian forest. *Int. Biodeterior. Biodegrad.* **2000**, *45*, 199–208. [CrossRef]
41. Széles, G.L.; Purger, J.J.; Molnár, T.; Lanszki, J. Comparative analysis of the diet of feral and house cats and wildcat in Europe. *Mammal Res.* **2018**, *63*, 43–53. [CrossRef]
42. Spurr, E.B.; Drew, K.W. Invertebrates feeding on baits used for vertebrate pest control in new zealand. *N. Z. J. Ecol.* **1999**, *23*, 167–173.
43. Morgan, D.R.; Wright, G.; Ogilvie, S.C.; Pierce, R.; Thomson, P. Assessment of the Environmental Impact of Brodifacoum during Rodent Eradication Operations in New Zealand. 1996. Available online: https://digitalcommons.unl.edu/vpc17/38/ (accessed on 17 October 2021).
44. Evans, M.J.; Wallman, J.F.; Barton, P.S. Traits reveal ecological strategies driving carrion insect community assembly. *Ecol. Entomol.* **2020**, *45*, 966–977. [CrossRef]
45. Morii, Y.; Ohkubo, Y.; Watanabe, S. Activity of invasive slug Limax maximus in relation to climate conditions based on citizen's observations and novel regularization based statistical approaches. *Sci. Total Environ.* **2018**, *637–638*, 1061–1068. [CrossRef] [PubMed]
46. Rosin, Z.M.; Kwieciński, Z.; Lesicki, A.; Skórka, P.; Kobak, J.; Szymańska, A.; Osiejuk, T.S.; Kałuski, T.; Jaskulska, M.; Tryjanowski, P. Shell colour, temperature, (micro)habitat structure and predator pressure affect the behaviour of Cepaea nemoralis. *Sci. Nat.* **2018**, *105*, 35. [CrossRef] [PubMed]
47. Schweizer, M.; Triebskorn, R.; Köhler, H.-R. Snails in the sun: Strategies of terrestrial gastropods to cope with hot and dry conditions. *Ecol. Evol.* **2019**, *9*, 12940–12960. [CrossRef] [PubMed]
48. Nicolai, A.; Ansart, A. Conservation at a slow pace: Terrestrial gastropods facing fast-changing climate. *Conserv. Physiol.* **2017**, *5*, cox007. [CrossRef] [PubMed]
49. Rial-Berriel, C.; Acosta-Dacal, A.; Cabrera Pérez, M.Á.; Suárez-Pérez, A.; Melián Melián, A.; Zumbado, M.; Henríquez Hernández, L.A.; Ruiz-Suárez, N.; Rodriguez Hernández, Á.; Boada, L.D.; et al. Intensive livestock farming as a major determinant of the exposure to anticoagulant rodenticides in raptors of the Canary Islands (Spain). *Sci. Total Environ.* **2021**, *768*, 144386. [CrossRef]
50. van den Brink, N.W.; Elliott, J.E.; Shore, R.F.; Rattner, B.A. *Anticoagulant Rodenticides and Wildlife*; Springer International Publishing AG: Cham, Switzerland, 2017.
51. Hindmarch, S.; Elliott, J.E.; McCann, S.; Levesque, P. Habitat use by barn owls across a rural to urban gradient and an assessment of stressors including, habitat loss, rodenticide exposure and road mortality. *Landsc. Urban Plan.* **2017**, *164*, 132–143. [CrossRef]
52. Schüder, I.; Port, G.; Bennison, J. The behavioural response of slugs and snails to novel molluscicides, irritants and repellents. *Pest Manag. Sci.* **2004**, *60*, 1171–1177. [CrossRef]
53. Rautio, A.; Isomursu, M.; Valtonen, A.; Hirvelä-Koski, V.; Kunnasranta, M. Mortality, diseases and diet of European hedgehogs (*Erinaceus europaeus*) in an urban environment in Finland. *Mammal Res.* **2016**, *61*, 161–169. [CrossRef]
54. Grosshans, W. The food of the hedgehog erinaceus europaeus studies of stomach contents of hedgehogs from schleswig holstein west germany. *Zool. Anz.* **1983**, *211*, 364–384.
55. Yalden, D. The food of the hedgehog in England. *Acta Theriol.* **1976**, *21*, 401–424. [CrossRef]
56. Wroot, A.J. Feeding ecology of the European hedgehog *Erinaceus europaeus*. Ph.D. Thesis, Royal Holloway, University of London, London, UK, 1984.
57. Howes, C. Field Notes: Notes on Suburban Hedgehogs. *Naturalis* **1976**, *101*, 147–148.
58. Kruuk, H. Predators and anti-predator behaviour of the black-headed gull (*Larus ridibundus* L.). *Behaviour. Suppl.* **1964**, *11*, III-129.
59. South, A. Predators, parasites and disease. In *Terrestrial Slugs: Biology, Ecology and Control*; Springer: Dordrecht, The Netherlands, 1992; pp. 220–241.
60. Churchfield, S.; Rychlik, L.; Taylor, J.R.E. Food resources and foraging habits of the common shrew, Sorex araneus: Does winter food shortage explain Dehnel's phenomenon? *Oikos* **2012**, *121*, 1593–1602. [CrossRef]
61. Damin-Pernik, M.; Espana, B.; Lefebvre, S.; Fourel, I.; Caruel, H.; Benoit, E.; Lattard, V. Management of Rodent Populations by Anticoagulant Rodenticides: Toward Third-Generation Anticoagulant Rodenticides. *Drug Metab. Dispos.* **2017**, *45*, 160–165. [CrossRef]

Commentary

Beneficial Land Management for Hedgehogs (*Erinaceus europaeus*) in the United Kingdom

Richard W. Yarnell [1,*] and Carly E. Pettett [2]

[1] School of Animal, Rural and Environmental Science, Brackenhurst Campus, Nottingham Trent University, Southwell NG25 0QF, UK
[2] Wildlife Conservation Research Unit, Department of Zoology, Recanati-Kaplan Centre, University of Oxford, Tubney House, Abingdon Road, Tubney, Oxfordshire OX13 5QL, UK; carly.pettett@gmail.com
* Correspondence: richard.yarnell@ntu.ac.uk; Tel.: +44-(0)115-8485333

Received: 30 April 2020; Accepted: 27 August 2020; Published: 3 September 2020

Simple Summary: Hedgehogs are declining in the United Kingdom and are now absent from large areas of agriculture land. This commentary discusses the requirements of hedgehogs and links these to land management options that are currently used to benefit wildlife in agricultural areas. Using our knowledge of hedgehog requirements for population persistence, we suggest which land management practices are likely to be of benefit to hedgehogs in the hope that land owners will adopt some of the suggestions to help maintain and expand existing hedgehog populations across agricultural landscapes of the United Kingdom.

Abstract: Hedgehogs (*Erinaceus europaeus*) are traditionally thought of as being a rural dwelling species, associated with rural and agricultural landscapes across Europe. However, recent studies have highlighted that hedgehogs are more likely to be found in urban than rural habitats in the United Kingdom. Here, we review the status of rural hedgehog populations across the UK and evaluate the potential benefits of agri-environment schemes for hedgehog persistence, while highlighting a lack of empirical evidence that agri-environment options will benefit hedgehog populations. Our synthesis has implications for future conservation strategies for hedgehogs and insectivorous mammals living in agricultural landscapes, and calls for more empirical studies on agri-environment options and their potential benefits to hedgehogs.

Keywords: insectivore; agri-environment schemes; habitat preference; farmland biodiversity; small mammal; conservation

---

1. Introduction

The western European hedgehog (*E. europaeus*) (hereafter referred to as 'hedgehog') is a species of conservation concern in the United Kingdom [1–3], due to reported population declines [4–6]. Possible reasons for decline may include agricultural intensification leading to the loss of habitat complexity due to the removal of hedgerows and increased field sizes [7–9]; limited connectivity and fragmentation in rural landscapes [10]; reductions in food availability due to wide scale pesticide use and cultivation, and possible climate mediated effects [11,12]. The Eurasian badger (*Meles meles*) is also implicated in the decline of hedgehogs due to their increasing abundance and through the mechanism of intra-guild predation [13,14]. Mortality risk associated with road traffic collisions may also have increased due to an increase in the density of road networks and associated traffic [15–19]. Conservation actions are therefore required to identify key habitats for hedgehogs and habitat management to ensure hedgehog persistence in the wider countryside.

Hedgehogs are traditionally thought of as being a rural dwelling species, associated with Europe's countryside and agricultural landscape. However, recent studies have highlighted that hedgehogs are more likely to be found in towns and cities [20–23], and have higher densities in urban areas [24,25] compared to agricultural landscapes.

Other studies have also struggled to find hedgehogs on agricultural land, and more often closely associated with villages surrounded by an agricultural matrix [9,20,21,23]. Indeed, hedgehogs were only found in 22% of 262 survey sites across the UK [26]. This is particularly worrying since 70% of the land across the United Kingdom is made up of rural agricultural habitats [27], suggesting that hedgehogs are no longer ubiquitous across these landscapes, but rather have a much more patchy discontinuous distribution.

Where hedgehogs do exist in the rural landscape, they display clear habitat preferences for pasture fields, and avoidance of arable and woodland habitats [26–28]. Arable areas may be used if bordered by field margins and extensive hedgerows as prescribed by agri-environmental schemes [8,29,30]. Furthermore, hedgehogs in southern Ireland actively selected arable fields for foraging after the autumn harvest [31]. Therefore, hedgehogs can and do persist in agricultural landscapes. However, no studies have empirically tested whether changes in agricultural land management can lead to changes in the density and distribution of hedgehogs.

Agri-environment schemes are designed to help land managers manage their land in an environmentally friendly way and many options have been suggested as being beneficial to hedgehogs. The uptake of agri-environment schemes by land managers in the UK has increased since their inception in the late 1980s [32]. In 2018, 3.2 million hectares were managed under some form of agri-environment scheme, which is 18% of the total agricultural land area of the UK [33]. Despite the increase in the uptake of agri-environment schemes, hedgehogs have continued to decline [6], which suggests that either agri-environment schemes are not being implemented across sufficient areas of land or that they are not beneficial to hedgehogs. Here, we will review literature on hedgehog ecology to provide an informed insight as to which agri-environment options are likely to benefit hedgehogs and evaluate the evidence pertaining to these.

2. The Importance of Food Availability

Hedgehogs require an abundant and varied macro-invertebrate prey base and any habitats that provide this spatially and temporally are likely to benefit hedgehogs. Hedgehogs are principally insectivores, but they also take a wide range of more unusual food items such as bird eggs, smal mammals, amphibians, and occasionally fruit [23,34]. What a hedgehog eats on any given night will depend on local food availability which is determined by a range of factors such as the local habitat, daily weather, and temperature [34]. Therefore, a hedgehog's diet will vary from night to night and throughout the year [35,36], suggesting that a diversity of food will be beneficial.

Food availability is often cited as having a positive influence on hedgehog distribution [37] and has been shown to influence hedgehog habitat selection [8,11,30,31]. The ability of hedgehogs to associate with food rich patches has also been demonstrated experimentally [37]. Therefore, where prey abundance is high or becomes more readily available at a site, they are likely to be exploited.

Sites of high food availability have also been used to explain variation in hedgehog density [23,24,37]. For example, supplementary feeding in villages has been suggested as the cause of observed higher hedgehog densities in villages compared to the surrounding agricultural landscape [24]. Natural prey availability may also be higher in villages due to lower pesticide application relative to agricultural landscapes [24]. Lower food availability in agricultural landscapes has been suggested as causing larger hedgehog home ranges, where individual movements need to be larger to meet daily energy requirements [38]. Therefore, it appears the daily resources required by hedgehogs are more widely distributed on agricultural land compared to villages that have higher food availability. The impacts of pesticides on hedgehog prey abundance and the concurrent role of supplementary feeding on hedgehog movement and abundance requires further investigation.

Competition for food resources may also impact hedgehogs, although no empirical studies have investigated this [34]. The degree to which competition influences population size is unknown [39], but hedgehogs have been shown to co-exist with larger competitors such as badgers and foxes (*Vulpes vulpes*) in urban areas where food resources are thought to be abundant. Abundant food resources therefore are likely to allow co-existence across a range of mammals and may allow greater niche differentiation. Individual hedgehogs also face intra-specific competition with population size regulated via density dependence [29], further supporting the idea that increases in food availability will increase population size.

If food availability is limiting hedgehog density and distribution in rural agricultural habitats [9,23], it follows that any land management that increases hedgehog prey abundance is likely to be beneficial to hedgehogs locally.

3. The Need for Habitat Connectivity

Habitat fragmentation is one of the major drivers of global biodiversity loss. Small areas of fragmented habitat result in small populations which face higher probabilities of extinction. Recent studies on hedgehogs suggest that their populations are highly fragmented in the rural agricultural landscape [26]. Furthermore, rural local populations are genetically differentiated [15], suggesting hedgehogs are either poor dispersers or that barriers to dispersal are fragmenting populations [40]. Therefore, identifying and creating suitable connecting habitat such as hedgerows or field margins can reduce the probability of local extinction [41].

Some of the insights into hedgehog movements come from studies that have released rehabilitated hedgehogs into the rural environment. These studies have shown that hedgehogs move quickly through agricultural landscapes, presumably in search of areas occupied by other hedgehogs [28,29,42,43]. Hedgehogs released into novel surroundings undergo highly variable exploratory movements, showing a significant preference for urban areas while avoiding agriculturally dominated areas [42,44]. Whether naturally dispersing hedgehogs follow these patterns is uncertain and further research into how and over what distances hedgehogs disperse is needed to fully inform the functionality of habitat corridors. However, based on the available evidence, it is plausible that hedgehog movement through the landscape is likely to be facilitated by greater habitat complexity, with increases in the number of linear features such as hedgerows, and smaller field sizes that would facilitate movement and improve population connectivity [41].

4. Shelter Resources

One of the most important resources for hedgehogs is the nest site. During the day, hedgehogs will rest in a nest which provides security and protection from the elements. The nest is also important during winter hibernation [45,46], and for giving birth to young [34]. Nests vary in construction materials and location, but typically comprise broad leaves and or grass constructed in a supporting structure of brambles or hedge [34]. In the UK, the majority of nests are sited within thorny or stinging vegetation, under bramble, holly, hawthorn, or nettles [46–48]. Hedgehogs will rest up in any dense vegetation and in summer, may forego the construction of nests and simply sleep in dense vegetation [34].

Winter nests for hibernation are also particularly important to hedgehogs. In the UK, hibernation usually starts around mid-November and continues through to Mid-March, but the exact timings vary due to local weather conditions, with a mean hibernation period of 149 nights in southern Ireland [12]. Most individuals will become active for short periods over winter, typically in response to warmer temperatures, and on average hedgehogs, will use four to five nests per winter [45,46]. In warmer climates such as Italy, hibernation lasts for two months between January and February [49]. Therefore, nest availability or structures that could support nests are an important resource for hedgehogs and any enhancements of such features are likely to be beneficial.

5. The Role of Habitat Mediating Badger Predation

Predation is a natural process that can regulate prey densities. In natural systems where prey and predator co-exist, the predation of prey does not result in the extinction of the prey due to a large number of complex interactions such as: the density of predators and prey; predators being able to predate a wide range of prey; and prey having evolved strategies to reduce individual chances of predation, such as camouflage and avoidance of areas with predators. These complex interactions work together to make it difficult to predict how the predation by one species will influence the prey populations.

Hedgehogs could be prey to a small number of predators in the UK. The main predator is the badger, but foxes and pet dogs (*Canis familiaris*) can also cause some mortality, especially in juveniles [34]. The degree to which predation could limit hedgehog populations is unknown, as are the badger predation rates experienced by hedgehogs across their distribution. However, there is evidence that hedgehogs spatially avoid badgers [20,50], and that in areas where badger numbers have been reduced via culling, local hedgehog numbers have responded positively [39].

Despite these trends and relationships, no empirical study has been able to demonstrate the mechanism by which badgers would exert a negative population response on hedgehogs, i.e., whether this is through direct predation, competition for food resources, or via a landscape of fear. Indeed, there are many areas in the UK where badgers and hedgehogs are known to co-exist [21,26], and hedgehogs have also been shown to decline in areas with low badger sett density [5]. Therefore, it seems that badgers may exert a negative influence of hedgehogs, but factors other than badgers are also having an impact on hedgehog distributions.

Whether particular features in the landscape can help hedgehogs avoid or reduce predation is unknown. It may be that at a fine scale, badgers and hedgehogs prefer different habitats across the landscape. In the Netherlands for example, the distribution of hedgehogs was positively influenced by recreational areas (parks), urban areas, and roads, whilst these factors negatively influenced the distribution of badgers [22], suggesting that either hedgehogs are spatially avoiding badgers or that both species have differing habitat preferences. Further research on each species habitat preference is needed to untangle these affects.

Where both badgers and hedgehogs co-exist, one may speculate that habitat features such as dense vegetation, intact hedgerows, and areas of scrub may help provide suitable refuges that make it harder for badgers to find and predate hedgehogs [30]. Hedgehogs do stay closer to edge habitats in areas frequented by badgers than in areas without them [30]. Under such circumstances, any increase in edge habitat such as smaller fields, increased hedgerow extent, and possible extensive field margins could facilitate hedgehog persistence in areas with badger presence.

6. Beneficial Management Actions

For hedgehog populations to persist in the UK's rural countryside, management action that helps provide suitable sites for shelter, connectivity and abundant food are essential. However, as this review indicates, very few studies have empirically tested whether increases in food availability, or improvements in habitats result in hedgehog population improvements. This knowledge gap needs addressing before resources are devoted to habitat improvements specifically for hedgehogs. In the meantime, many of the options included in current agri-environment schemes designed to improve habitat for farmland birds and invertebrates could also benefit hedgehogs (Table 1). In many cases, food, connectivity, and shelter can be increased under the same management actions. For example, maintaining and increasing hedgerow density should provide nest sites, shelter, and refuge from predators while increasing invertebrate prey biomass and provide corridors for dispersal between neighboring populations [30]. Using the information provided in the review above, we contend that the following management actions are likely to be beneficial to hedgehogs, as well as wider biodiversity targeted beneficiaries such as farmland birds, and re-assure landowners that the implementation of these agri-environmental schemes is likely to have genuine biodiversity benefits.

Table 1. Summary of agri-environment management options that benefit biodiversity and are likely to be of benefit to hedgehogs (*Erinaceus europaeus*) in agriculturally dominated landscapes in the United Kingdom.

Management Action	Key Requirements	Benefits to Other Species	Potential Benefits for Hedgehogs
Hedgerow management	1. Maintain a hedge at least 2 m tall and 1.5 m wide. 2. Cut hedgerows: either no more than 1 year in 3 or no more than 1 year in 2. 3. In-fill any length of hedge with more than 10% gaps.	Increases blossom for invertebrates, food for overwintering birds and nesting habitat.	Shelter, corridors for movement connecting populations and invertebrate food.
Field margin availability and management	1. Establish or maintain a 4 to 6m wide grass buffer strip. 2. Cut between 1 and 3 m of the strip next to the crop edge every year after 15 July. 3. Only cut the remaining width to control woody growth.	Provides habitat and movement corridors for wildlife.	Corridors for movement connecting populations and invertebrate food.
Beetle banks	1. Create or maintain an earth ridge, measuring between 3 m to 5 m wide and at least 0.4 m high. 2. Establish or maintain a tussocky grass mixture.	Provides nesting and foraging habitat, benefiting invertebrate biodiversity, small mammals and barn owls	Corridors for movement connecting populations and invertebrate food.
Areas of scrub and decaying vegetation	1. Only cut to maintain the scrub and grass mosaic and to control the spread of noxious weeds and invasive non-native species. 2. Protect growing trees from livestock and wild animals.	Provides enhanced habitat for wildlife such as birds and invertebrates.	Shelter, corridors for movement connecting populations and invertebrate food.

6.1. Hedgerow Availability and Density

Increasing the density, width, height, and length of hedgerows on agricultural land will benefit many species [51] including hedgehogs. To be of benefit, species rich hedgerows with stands of trees will improve hedge structure increasing invertebrate abundance and diversity [52] and improve food availability. Maintaining and re-establishing well connected hedgerows with bramble understorey and good ground cover within arable habitat is recommended to enhance the suitability of fields for hedgehogs both during summer foraging and winter hibernation [12]. Species such as brambles and rose will also improve the structure of the hedge for nest sites [34]. Mature trees provide additional nesting material and are also beneficial to invertebrate prey [53]. If combined with appropriate field margins, the hedgerow matrix across the landscape will allow dispersal and movement linking hedgehog populations [41], reducing fragmentation and increasing population viability.

The size and management of hedgerows may also improve the wider landscape characteristics that are beneficial for hedgehogs. Larger hedges will provide more shelter and foraging opportunities for hedgehogs than small ones. Hedges that are over 3 m in height and flayed in winter (January onwards) on a 3-year rotation will provide robust and healthy hedges that have high flower and fruit yields [54], which has wider biodiversity benefits for many invertebrates [55] and their predators. Cutting on rotation will also ensure that two thirds of hedges will be uncut in any year, reducing the levels of mulch at the base of the hedge which can hamper vegetative growth needed for nest construction in the understorey. Ideally, the base of the hedge should be greater than 2 m wide, with dense vegetation at the base of the hedge with no gaps. When cutting hedges, care must be taken not to cut the vegetation at the base of the hedge to protect nesting and hibernating hedgehogs.

Hedges are often fenced to prevent damage from livestock, and farmers can receive payments for fencing which can ensure a good hedge for nesting, foraging, and dispersal. However, the size of fence mesh needs consideration to ensure they are large enough to allow hedgehogs to pass through and access the hedge.

6.2. Field Margin Availability and Management

Agri-environment schemes that (re)create hedgerows and establish field margins are recommended for hedgehog conservation, particularly in intensively farmed arable landscapes where food availability and nesting sites are sparse [11]. Of great importance is the presence of a hedge buffer or headland that will provide additional cover, nest material, and invertebrate prey. The margin of grass should be allowed to extend by at least 2 m from the base of the hedge into the crop. Where possible, wide grassy/bushy margins around agricultural fields should be established to help hedgehogs disperse and access to suitable resources [41]. Unmanaged grassy margins in pasture fields should also be encouraged to provide summer nesting/resting up areas [41]. Establishing and/or maintaining 4–6 m grassy field margins and incorporating conservation headlands in arable dominated landscapes is recommended and supported by agri-environment schemes to provide refuge and foraging habitat [30].

Tussocky grass can be used by hedgehogs for cover and daytime nests during summer, and where this is particularly thick, it may be used for hibernacula in winter. Such dense undergrowth also provides good habitat for beetles, spiders, and caterpillars [56] which are important food.

Beetle banks that cross large arable fields will improve food availability for hedgehogs and potentially act as corridors for movement. The benefits to the landowner for maintaining a beetle bank is that it will support predatory spiders and beetles that will migrate into the crop and feed on aphids and other pest species that feed on crops [57]. This will reduce pest damage and reduce the need for insecticides, providing economic benefits for the farmer [58].

6.3. Field Size in Relation to General Habitat Availability in the Landscape

Large field sizes are likely to hinder hedgehog movement due to their propensity for using field boundaries. Therefore, landscapes with high density of linear features and small land parcels will be advantageous for movement. Where field sizes are large, the addition of field margins, robust hedges, and features such as beetle banks will make these habitats more penetrable, while also providing additional food resources [41,43,44].

Hedgehogs are capable of long-distance dispersal (>4 km) and thus populations located within this range are unlikely to be isolated [42]. Habitat edges, particularly roads and hedgerows, are utilised as dispersal corridors by hedgehogs, thus management activities should be sensitive to the potential presence of hedgehogs and should focus on increasing suitability by maintaining connectivity between linear features [30].

Management actions that increase the diversity of invertebrate species are recommended to allow for niche partitioning amongst different age classes of hedgehogs and include but are not limited to agri-environment options that include buffer strips, establishing and managing hedgerows, organic farming, and beetle banks [36]. Having diverse land use will also increase heterogeneity in the landscape, creating habitat for a greater diversity of wildlife. As such, mixed farms with areas of pasture, arable crops, and set-aside fields have the potential to be beneficial to hedgehogs and are more likely to support a viable hedgehog population than single use arable farms [59]. Increasing heterogeneity in the landscape by increasing edge habitat, copses, different land uses with amenity, and garden habitats surrounding buildings will all help to provide a diverse array of shelter and foraging resources.

6.4. Cropping and Ploughing Regimes in Relation to Prey Availability

Traditional farming methods that include mosaics of pasture and arable, well connected hedgerows, over-winter stubble, and fodder crops are recommended to increase the suitability of arable land for hedgehogs [30]. The greater the diversity of land types will provide a greater array of habitats types from which to support invertebrate prey that is needed throughout the year [36]. Where possible, areas that are dominated by arable crops could enhance habitat in less economically important habitats in the wider arable landscape. For example, habitat around farm buildings could be enhanced to support local populations, with enhancement of gardens, amenity grassland, and small pasture fields that could provide refuge habitats.

Organic farming will increase prey availability and has great potential as rural hedgehog habitat due to the increased invertebrate abundance associated with organic farms [60]. Many taxa have been shown to benefit from organic farming practices through increases in abundance and species richness which are principally driven by reductions in the use of pesticides and herbicides, sympathetic management of non-cropped areas, and utilisation of mixed farming. Such practices are likely to be beneficial to hedgehogs as well and may be equally as beneficial if targeted at specific areas on non-organic farms [59].

Reduced tillage will increase earthworm abundance [61,62] and may increase arable field use by hedgehogs [31]. There is a direct negative relationship between earthworm abundance and the depth of tillage, with no-till and conservation agriculture providing highest earthworm abundance. Switching from conventional tillage will also improve levels of soil organic matter, reduce the depth of the soil organic layer, and reduce soil compaction [62]. However, it is acknowledged that switching from conventional tillage to no-till or conservation agriculture will take up to 10 years for improvements in soil structure and health and associated earthworm biomass [62].

6.5. Areas of Scrub and Decaying Vegetation

Unkept areas providing cover and leaf litter are often utilised by hedgehogs principally for nest building [47] but also as a habitat for invertebrate prey [11]. Leaf litter is an important nesting resource for hedgehogs throughout the winter and should be either left or collected into piles near potential hedgehog nesting sites, such as tree lines, copses, or hedgerows [48]. Sheltered areas with bramble should be established and/or maintained as these provide important hibernacula sites that have increased longevity and lower daytime temperatures which may reduce arousal and thus risk of mortality of hedgehogs over winter [48]. Management of scrub during winter should be sensitive to hibernating hedgehogs, whereby areas of scrub and piles of leaf litter should be left intact [47,48].

7. Conclusions

Rural hedgehog populations across the UK tend to be located in villages more than the surrounding agricultural matrix of habitats [44] and it is important for their future persistence that the wider rural landscape can be utilised by hedgehogs. Based on our understanding of hedgehog ecology, we contend that many agricultural schemes designed to combat biodiversity loss would also benefit hedgehogs including enhancements to the extent and size of hedgerows and field margins, less intensive agriculture, and more diverse farming types. These changes will likely improve food availability, shelter, connectivity, and possibly reduce predation by badgers. Unfortunately, there is a lack of conservation evidence that such changes would result in tangible benefits to hedgehogs and more research is required to test these suggestions so that more specific targeted actions for hedgehogs can be proposed and implemented. Without such action, it is likely that hedgehogs will only be found in urban habitats in the future.

Author Contributions: Conceptualization, R.W.Y.; investigation, R.W.Y. and C.E.P.; writing—original draft preparation, R.W.Y.; writing—review and editing, R.W.Y. and C.E.P.; funding acquisition, R.W.Y. All authors have read and agreed to the published version of the manuscript.

Funding: This research was funded by the People's Trust for Endangered Species, The British Hedgehog Preservation Society, and Nottingham Trent University.

Acknowledgments: The authors would like to thank the comments from four reviewers who helped to improve the manuscript.

Conflicts of Interest: The authors declare no conflict of interest. The funders had no role in the design of the study; in the collection, analyses, or interpretation of data; in the writing of the manuscript, or in the decision to publish the results.

References

1. Morris, P.A.; Reeve, N.J. Hedgehog *Erinaceus europaeus*. In *Mammals of the British Isles: Handbook*, 4th ed.; Harris, S., Yalden, D.W., Eds.; The Mammal Society: Southampton, UK, 2008; pp. 241–248.
2. JNCC. UK Priority Species Pages: *Erinaceus Europaeus* Version 2 Updated on 15/12/2010. Available online: http://jncc.defra.gov.uk/_speciespages/2253.pdf (accessed on 4 December 2012).
3. Mathews, F.; Kubasiewicz, L.M.; Gurnell, J.; Harrower, C.A.; McDonald, R.A.; Shore, R.F. *A Review of the Population and Conservation Status of British Mammals: Technical Summary. A Report by the Mammal Society under Contract to Natural England, Natural Resources Wales and Scottish Natural Heritage*; Natural England: Peterborough, UK, 2018.
4. Battersby, J.E.; Greenwood, J.J.D. Monitoring terrestrial mammals in the UK: Past, present and future, using lessons from the bird world. *Mamm. Rev.* **2004**, *34*, 3–29. [CrossRef]
5. Wembridge, D. The State of Britain's Hedgehogs. People's Trust for Endangered Species, 2011. Available online: https://ptes.org/wp-content/uploads/2014/06/SOBH2011lowres.pdf (accessed on 28 April 2020).
6. Roos, S.; Johnston, A.; Noble, D. *UK Hedgehog Datasets and Their Potential for Long-Term Monitoring*; BTO Research Report No. 598; British Trust for Ornithology: Thetford, UK, 2012.
7. Krebs, J.R.; Wilson, J.D.; Bradbury, R.B.; Siriwardena, G.M. The second Silent Spring? *Nature* **1999**, *400*, 611–612. [CrossRef]
8. Hof, A.R.; Bright, P.W. The value of agri-environment schemes for macro-invertebrate feeders: Hedgehogs on arable farms in Britain. *Anim. Conserv.* **2010**, *13*, 467–473. [CrossRef]
9. Hof, A.R. A Study of the Current Status of the Hedgehog (*Erinaceus Europaeus*), and Its Decline in Great Britain Since 1960. Ph.D. Thesis, University of London, London, UK, 2010.
10. Hof, A.R.; Bright, P.W. The value of green-spaces in built-up areas for western hedgehogs. *Lutra* **2009**, *52*, 69–82.
11. Hof, A.R.; Bright, P.W. The impact of grassy field margins on macro-invertebrate abundance in adjacent arable fields. *Agric. Ecosyst. Environ.* **2010**, *139*, 280–283. [CrossRef]
12. Haigh, A.; O'Riordan, R.M.; Butler, F. Nesting behaviour and seasonal body mass changes in a rural Irish population of the Western hedgehog (*Erinaceus europaeus*). *Acta Thériol.* **2012**, *57*, 321–331. [CrossRef]
13. Judge, J.; Wilson, G.J.; MacArthur, R.; Delahay, R.J.; McDonald, R.A. Density and abundance of badger social groups in England and Wales in 2011–2013. *Sci. Rep.* **2015**, *4*, 3809. [CrossRef]
14. Judge, J.; Wilson, G.J.; MacArthur, R.; McDonald, R.A.; Delahay, R.J. Abundance of badgers (Meles meles) in England and Wales. *Sci. Rep.* **2017**, *7*, 276. [CrossRef]
15. Becher, S.A.; Griffiths, R. Genetic differentiation among local populations of the European hedgehog (*Erinaceus europaeus*) in mosaic habitats. *Mol. Ecol.* **1998**, *7*, 1599–1604. [CrossRef]
16. Rondinini, C.; Doncaster, C.P. Roads as barriers to movement for hedgehogs. *Funct. Ecol.* **2002**, *16*, 504–509. [CrossRef]
17. Wembridge, D.E.; Newman, M.R.; Bright, P.W.; Morris, P.A. An estimate of the annual number of hedgehog (*Erinaceus europaeus*) road casualties in Great Britain. *Mamm. Commun.* **2016**, *2*, 8–14.
18. Wright, P.G.R.; Coomber, F.G.; Bellamy, C.C.; Perkins, S.E.; Mathews, F. Predicting hedgehog mortality risks on British roads using habitat suitability modelling. *PeerJ* **2020**, *7*, e8154. [CrossRef] [PubMed]
19. Moore, L.J.; Petrovan, S.O.; Baker, P.J.; Bates, A.J.; Hicks, H.L.; Perkins, S.E.; Yarnell, R.W. Impacts and potential mitigation of road mortality for hedgehogs in Europe. *Animals* **2020**, *10*, 1523. [CrossRef]
20. Young, R.P.; Davison, J.; Trewby, I.D.; Wilson, G.J.; Delahay, R.J.; Doncaster, C.P. Abundance of hedgehogs (*Erinaceus europaeus*) in relation to the density and distribution of badgers (*Meles meles*). *J. Zool.* **2005**, *269*, 349–356. [CrossRef]
21. Yarnell, R.W.; Pacheco, M.; Williams, B.; Neumann, J.L.; Rymer, D.J.; Baker, P.J. Using occupancy analysis to validate the use of footprint tunnels as a method for monitoring the hedgehog *Erinaceus europaeus*. *Mamm. Rev.* **2014**, *44*, 234–238. [CrossRef]

22. Van de Poel, J.L.; Dekker, J.J.A.; Van Langevelde, F. Dutch hedgehogs *Erinaceus europaeus* are nowadays mainly found in urban areas, possibly due to the negative Effects of badgers Meles meles. *Wildl. Biol.* **2015**, *21*, 51–55. [CrossRef]

23. Pettett, C. Factors Affecting Hedgehog Distribution and Habitat Selection in Rural Landscapes. Ph.D. Thesis, Oxford University, Oxford, UK, 2016.

24. Hubert, P.; Julliard, R.; Biagianti, S.; Poulle, M.-L. Ecological factors driving the higher hedgehog (*Erinaceus europaeus*) density in an urban area compared to the adjacent rural area. *Landsc. Urban Plan.* **2011**, *103*, 34–43. [CrossRef]

25. Schaus, J.; Uzal, A.; Gentle, L.K.; Baker, P.J.; Bearman-Brown, L.; Bullion, S.; Gazzard, A.; Lockwood, H.; North, A.; Reader, T.; et al. Application of the Random Encounter Model in citizen science projects to monitor animal densities. *Remote Sens. Ecol. Conserv.* **2020**. [CrossRef]

26. Williams, B.M.; Baker, P.J.; Thomas, E.; Wilson, G.J.; Judge, J.; Yarnell, R.W. Reduced occupancy of hedgehogs (*Erinaceus europaeus*) in rural England and Wales: The influence of habitat and an asymmetric intra-guild predator. *Sci. Rep.* **2018**, *8*, 12156. [CrossRef] [PubMed]

27. National Statistics. Agriculture in the United Kingdom. Available online: https://www.gov.uk/government/statistics/agriculture-in-the-united-kingdom-2014 (accessed on 2 September 2020).

28. Doncaster, C.P. Testing the role of intraguild predation in regulating hedgehog populations. *Proc. R. Soc. B* **1992**, *249*, 113–117. [CrossRef]

29. Doncaster, C.P. Factors Regulating Local Variations in Abundance: Field Tests on Hedgehogs, *Erinaceus europaeus*. *Oikos* **1994**, *69*, 182. [CrossRef]

30. Hof, A.R.; Snellenberg, J.; Bright, P.W. Food or fear? Predation risk mediates edge refuging in an insectivorous mammal. *Anim. Behav.* **2012**, *83*, 1099–1106. [CrossRef]

31. Haigh, A.; O'Riordon, R.M.; Butler, F. Habitat selection, philopatry and spatial segregation in rural Irish hedgehogs (*Erinaceous europaeus*). *Mammalia* **2013**, *77*, 163–172. [CrossRef]

32. Agricultural and Forest Area in Environmental Management Schemes. Available online: https://assets.publishing.service.gov.uk/government/uploads/system/uploads/attachment_data/file/829197/22__agri_environment_and_forestry_2019_rev.pdf (accessed on 2 September 2020).

33. Farming Statistics. Provisional Crop Areas, Yields and Livestock Populations. Available online: https://assets.publishing.service.gov.uk/government/uploads/system/uploads/attachment_data/file/747210/structure-jun2018prov-UK-11oct18.pdf (accessed on 2 September 2020).

34. Morris, P.A. *Hedgehogs*; Whittet Books: Stansted, UK, 2015.

35. Wroot, A.J. Foraging in the European Hedgehog, *Erinaceus Europaeus*. Ph.D. Thesis, University of London, London, UK, 1984.

36. Dickman, C.R. Age-Related dietary change in the European hedgehog, *Erinaceus europaeus*. *J. Zool.* **1988**, *215*, 1–14. [CrossRef]

37. Cassini, M.H.; Krebs, J.R. Behavioural responses to food addition by hedgehogs. *Ecography* **1994**, *17*, 289–296. [CrossRef]

38. Pettett, C.E.; Johnson, P.J.; Moorhouse, T.P.; Hambly, C.; Speakman, J.R.; Macdonald, D.W. Daily energy expenditure in the face of predation: Hedgehog energetics in rural landscapes. *J. Exp. Biol.* **2017**, *220*, 460–468. [CrossRef]

39. Trewby, I.D.; Young, R.P.; MacDonald, R.A.; Wilson, G.J.; Davison, J.; Walker, N.; Robertson, A.; Doncaster, C.P.; Delahay, R.J. Impacts of Removing Badgers on Localised Counts of Hedgehogs. *PLoS ONE* **2014**, *9*, e95477. [CrossRef]

40. Braaker, S.; Kormann, U.; Bontadina, F.; Obrist, M.K. Prediction of genetic connectivity in urban ecosystems by combining detailed movement data, genetic data and multi-path modelling. *Landsc. Urban Plan.* **2017**, *160*, 107–114. [CrossRef]

41. Moorhouse, T.P.; Palmer, S.C.F.; Travis, J.M.J.; Macdonald, D.W. Hugging the hedges: Might agri-environment manipulations affect landscape permeability for hedgehogs? *Biol. Conserv.* **2014**, *176*, 109–116. [CrossRef]

42. Doncaster, C.P.; Rondinini, C.; Johnson, P.C.D. Field test for environmental correlates of dispersal in hedgehogs *Erinaceus europaeus*. *J. Anim. Ecol.* **2001**, *70*, 30–46. [CrossRef]

43. Driezen, K.; Adriaensen, F.; Rondinini, C.; Doncaster, C.P.; Matthysen, E. Evaluating least-cost model predictions with empirical dispersal data: A case-study using radiotracking data of hedgehogs (*Erinaceus europaeus*). *Ecol. Model.* **2007**, *209*, 314–322. [CrossRef]

44. Pettett, C.E.; Moorhouse, T.P.; Johnson, P.J.; Macdonald, D.W. Factors affecting hedgehog (*Erinaceus europaeus*) attraction to rural villages in arable landscapes. *Eur. J. Wildl. Res.* **2017**, *63*, 54. [CrossRef]

45. Yarnell, R.W.; Surgey, J.; Grogan, A.; Thompson, R.; Davies, K.; Kimbrough, C.; Scott, D.M. Should rehabilitated hedgehogs be released in winter? A comparison of survival, nest use and weight change in wild and rescued animals. *Eur. J. Wildl. Res.* **2019**, *65*, 6. [CrossRef]

46. Bearman-Brown, L.E.; Baker, P.J.; Scott, D.M.; Uzal, A.; Evans, L.; Yarnell, R.W. Over-winter survival and best site selection of the West-European hedgehog (Erinaceus europaeus) in arable dominated landscapes. *Animals* **2020**, *10*, 1449. [CrossRef] [PubMed]

47. Reeve, N.J.; Morris, P.A. Construction and use of summer nests by the hedgehog (*Erinaceus europaeus*). *Mammalia* **1985**, *49*, 187–194. [CrossRef]

48. Morris, P.A. Winter nests of the hedgehog (*Erinaceus europaeus* L.). *Oecologia* **1973**, *11*, 299–313. [CrossRef]

49. Boitani, L.; Reggiani, G. Movement and activity patterns of Hedgehogs. *Z. Säugetierkd.* **1984**, *49*, 193–206.

50. Dowding, C.V.; Harris, S.; Poulton, S.; Baker, P.J. Nocturnal ranging behaviour of urban hedgehogs, *Erinaceus europaeus*, in relation to risk and reward. *Anim. Behav.* **2010**, *80*, 13–21. [CrossRef]

51. Garratt, M.P.D.; Senapathi, D.; Coston, D.J.; Mortimer, S.R.; Potts, S.G. The benefits of hedgerows for pollinators and natural enemies depends on hedge quality and landscape context. *Agric. Ecosyst. Environ.* **2017**, *247*, 363–370. [CrossRef]

52. Amy, S.R.; Heard, M.S.; Hartley, S.E.; George, C.T.; Pywell, R.F.; Staley, J.T. Hedgerow rejuvenation management affects invertebrate communites through changes to habitat structure. *Basic Appl. Ecol.* **2015**, *16*, 443–451. [CrossRef]

53. Merckx, T.; Marini, L.; Feber, R.E.; Macdonald, D.W. Hedgerow trees and extended-width field margins enhance macro-moth diversity: Implications for management. *J. Appl. Ecol.* **2012**, *49*, 1396–1404. [CrossRef]

54. Staley, J.T.; Sparks, T.H.; Croxton, P.J.; Baldock, K.C.R.; Heard, M.S.; Hulmes, S.; Hulmes, L.; Peyton, J.; Amy, S.R.; Pywell, R.F. Long-Term effects of hedgerow management policies on resource provision for wildlife. *Biol. Conserv.* **2012**, *145*, 24–29. [CrossRef]

55. Maudsley, M.J. A review of the ecology and conservation of hedgerow invertebrates in Britain. *J. Environ. Manag.* **2000**, *60*, 65–76. [CrossRef]

56. Meek, B.; Loxton, D.; Sparks, T.H.; Pywell, R.F.; Pickett, H.; Nowakowski, M. The effect of arable field margin composition on invertebrate biodiversity. *Biol. Conserv.* **2002**, *106*, 259–271. [CrossRef]

57. Collins, K.L.; Boatman, N.D.; Wilcox, A.; Holland, J.M.; Chaney, K. Influence of beetle banks on cereal aphid predation in winter wheat. *Agric. Ecosyst. Environ.* **2002**, *93*, 337–350. [CrossRef]

58. Wratten, S.D. The role of field boundaries as reservoirs of beneficial insects. In *Environmental Management in Agriculture: European Perspectives*; Burn, A.J., Coaker, T.H., Jepson, P.C., Eds.; EEC/Pinter Publishers: London, UK, 1988; pp. 144–150.

59. Macdonald, D.W.; Tattersall, F.H.; Service, K.M.; Firbank, L.G.; Feber, R.E. Mammals, agri-environment schemes and set-aside—What are the putative benefits? *Mamm. Rev.* **2007**, *37*, 259–277. [CrossRef]

60. Hole, D.G.; Perkins, A.J.; Wilson, J.D.; Alexander, I.H.; Grice, P.V.; Evans, A.D. Does organic farming benefit biodiversity? *Biol. Conserv.* **2005**, *122*, 113–130. [CrossRef]

61. Peigné, J.; Cannavaciuolo, M.; Gautronneau, Y.; Aveline, A.; Giteau, J.; Cluzeau, D. Earthworm populations under different tillage systems in organic farming. *Soil Tillage Res.* **2009**, *104*, 207–214. [CrossRef]

62. Briones, M.J.I.; Schmidt, O. Conventional tillage decreases the abundance and biomass of earthworms and alters their community structure in a global meta-analysis. *Glob. Chang. Biol.* **2017**, *23*, 4396–4419. [CrossRef]

MDPI
St. Alban-Anlage 66
4052 Basel
Switzerland
www.mdpi.com

Animals Editorial Office
E-mail: animals@mdpi.com
www.mdpi.com/journal/animals